Compendium of
Chemical Terminology
IUPAC RECOMMENDATIONS

International Union of Pure and Applied Chemistry

Compendium of Chemical Terminology

IUPAC RECOMMENDATIONS

COMPILED BY ALAN D. McNAUGHT
AND ANDREW WILKINSON
The Royal Society of Chemistry, Cambridge, UK

SECOND EDITION

Blackwell
Science

© 1987, 1997 International Union
of Pure and Applied Chemistry
published by Blackwell Science Ltd
Editorial Offices:
Osney Mead, Oxford OX2 0EL
25 John Street, London WC1N 2BL
23 Ainslie Place, Edinburgh EH3 6AJ
350 Main Street, Malden
　MA 02148 5018, USA
54 University Street, Carlton
　Victoria 3053, Australia

Other Editorial Offices:
Blackwell Wissenschafts-Verlag GmbH
Kurfürstendamm 57
10707 Berlin, Germany

Blackwell Science KK
MG Kodenmacho Building
7–10 Kodenmacho Nihombashi
Chuo-ku, Tokyo 104, Japan

All rights reserved. No part of
this publication may be reproduced,
stored in a retrieval system, or
transmitted, in any form or by any
means, electronic, mechanical,
photocopying, recording or otherwise,
except as permitted by the UK
Copyright, Designs and Patents Act
1988, without the prior permission
of the copyright owner.

First published 1987
Second edition 1997

Set by A. Wilkinson, Cambridge, UK
using Corel Ventura Publisher V. 4.2
and facilities kindly provided by
the Royal Society of Chemistry.
Printed and bound in Great Britain
by Hartnolls Ltd, Bodmin, Cornwall.

The Blackwell Science logo is a
trade mark of Blackwell Science Ltd,
registered at the United Kingdom
Trade Marks Registry

DISTRIBUTORS

Marston Book Services Ltd
PO Box 269
Abingdon, Oxon OX14 4YN
(*Orders*: Tel: 01235 465500
　　　　　Fax: 01235 465555)

USA
Blackwell Science, Inc.
Commerce Place
350 Main Street
Malden, MA 02148 5018
(*Orders*: Tel: 800 759 6102
　　　　　 617 388 8250
　　　　Fax: 617 388 8255)

Canada
Copp Clark Professional
200 Adelaide St West, 3rd Floor
Toronto, Ontario M5H 1W7
(*Orders*: Tel: 416 597-1616
　　　　　 800 815-9417
　　　　Fax: 416 597-1617)

Australia
Blackwell Science Pty Ltd
54 University Street
Carlton, Victoria 3053
(*Orders*: Tel: 3 9347 0300
　　　　Fax: 3 9347 5001)

A catalogue record for this title
is available from the British Library

ISBN 0-86542-684-8

Contents

Preface, vii
Acknowledgements, vii
Alphabetical Entries, 1
Source Documents, 447

Preface

The IUPAC *Compendium of Chemical Terminology* is popularly referred to as the 'Gold Book', in recognition of the contribution of the late Victor Gold, who initiated work on the first edition. It is one of the series of IUPAC 'Colour Books' on chemical nomenclature, terminology, symbols and units (see the list of source documents at the end of the present volume), and collects together terminology definitions from IUPAC recommendations already published in *Pure and Applied Chemistry* and in the other Colour Books. This second edition has been compiled in a similar way to the first, by the undersigned, at the offices of the Royal Society of Chemistry in Cambridge, UK.

Terminology definitions published by IUPAC are drafted by international committees of experts in the appropriate chemistry sub-disciplines, and ratified by IUPAC's Interdivisional Committee on Nomenclature and Symbols. In this second edition of the Compendium these IUPAC-approved definitions are supplemented with some definitions from ISO and from the International Vocabulary of Basic and General Terms in Metrology; both these sources are recognised by IUPAC as authoritative. Since the first edition was published eleven years ago, the majority of the material in it has been revised and updated by IUPAC; thus many of the first-edition definitions have been modified or replaced. Furthermore, since 1986 IUPAC has published new terminology documents covering many areas of chemistry not previously treated, and definitions from these documents are now included. Among new areas covered are polymer science, photochemistry, stereochemistry, clinical chemistry, biotechnology, toxicology, class names for organic chemicals and atmospheric chemistry. The result is a collection of nearly 7000 terms, with authoritative definitions, spanning the whole range of chemistry.

All IUPAC recommendations published up to the end of 1995 were considered for inclusion, together with some particularly significant material published in 1996 (class names, kinetics, clinical chemistry quantities and units, stereochemistry, photochemistry and basic polymer terms). A selection was made on the basis of general utility: some terms were omitted as being of interest only to a highly specialist audience.

Some minor editorial changes were made to the originally published definitions, to harmonise the presentation and to clarify their applicability, if this is limited to a particular sub-discipline. For example, -ization rather than -isation (as well as the corresponding verb endings) has been used throughout. Verbal definitions of terms from the Green Book (in which definitions are generally given as mathematical expressions) were developed specially for this Compendium by the Physical Chemistry Division of IUPAC. Definitions of a few physicochemical terms not mentioned in the Green Book were added at the same time (referred to here as Physical Chemistry Division, unpublished).

The first reference given at the end of each definition is to the page of *Pure Appl. Chem.* or other source where the original definition appears; other references given designate other places where compatible definitions of the same term or additional information may be found, in other IUPAC documents. Italicized terms within individual definitions refer to other entries in this Compendium where additional information is available.

Extensive cross-referencing has been included.

A cross (+) against an entry implies that use of the term is discouraged.

Acknowledgements

We thank the Royal Society of Chemistry for the use of office facilities, and in particular David Stout (Information Technology Consultant, RSC Information Services) for advice and assistance with processing and manipulation of the electronic data files involved in the final stages of production. We are also grateful to W. Val Metanomski (Chemical Abstracts Service) for proof reading the entire volume.

Alan D. McNaught
Andrew Wilkinson

Alphabetical Entries

α-addition (alpha-addition)

A *chemical reaction* resulting in a single reaction product from two or three reacting chemical species, with formation of two new chemical *bonds* to the same atom in one of the reactant *molecular entities*. The synonymous term 1/1/addition is also used. For example:

$$Cl_2C: + CH_3OH \longrightarrow \underset{Cl}{\overset{Cl}{\diagup}}C\underset{OCH_3}{\overset{H}{\diagdown}}$$

(This particular example can also be viewed as an *insertion reaction*). In inorganic chemistry such α-addition reactions, generally to a metallic central atom, are known as 'oxidative additions'. α-Addition is the reverse of α-elimination or 1/1/elimination.

See also *addition, elimination*.
1994, *66*, 1087

α- (β-, γ-)ray spectrometer

A measuring assembly incorporating a *radiation detector* and a *pulse amplitude analyser*, used for determining the energy *spectrum* of α (β, γ) radiation.
1982, *54*, 1552

α (alpha), β (beta)

Stereodescriptors, used in a number of different ways.

1. Relative stereodescriptors used in carbohydrate nomenclature to describe the configuration at the anomeric carbon by relating it to the anomeric reference atom. For simple cases the anomeric reference atom is the same as the configurational reference atom. Thus in α-D-glucopyranose the reference atom is C-5 and the OH at C-1 is on the same side as the OH at C-5 in the Fischer projection.

β-D-glucose   β-L-rhamnose

2. Relative stereodescriptors used by Chemical Abstracts Service to describe the configuration of a cyclic molecule (including suitable polycyclic systems) with several stereogenic centres whereby the α side of the reference plane is the side on which the substituent with CIP priority lies at the lowest numbered stereogenic centre. The other side is β.

tricyclo[3.2.1.0^{2,4}]octan-2-ol, 5-chloro, (1α,2α,4α,5β)-

3. Absolute stereodescriptors originally devised for steroid nomenclature. However in this sense it is only meaningful if there is an agreed absolute configuration and orientation of the structure so as to define the plane and which way up the molecule is represented. Substituents above the plane of the steroid are described as β and are shown as a solid line (◂ or ▬); those below the plane are described as α and are shown by a broken line (""" or ---). The extension of this system to tetrapyrroles has been documented and it has been widely used elsewhere.

5α-androstan-3β-ol

1996, *68*, 2197; see also 1989, *61*, 1783; 1987, *59*, 779

α-cleavage (alpha-cleavage)

1. (in mass spectrometry)

The fission of a bond originating at an atom which is adjacent to one assumed to bear the charge; the definition of β-, γ-, cleavage then follows automatically. The process:

$$R^1-\overset{\overset{+\cdot}{O}}{\underset{\|}{C}}-R^2 \longrightarrow \overset{+}{O}\equiv C-R^2 + R^{1\cdot}$$

would thus be described as α-fission of a ketone with expulsion of a radical $R^{1\cdot}$. The carbon atoms of the radical $R^{1\cdot}$ are called the α-, β-, γ-, carbons, starting with the atom nearest the functional group.
1991, *63*, 1558

2. (in photochemistry)

Homolytic cleavage of a bond connecting an atom or group to a specified group. Often applied to a bond connected to a carbonyl group, in which case it is called a *Norrish type I photoreaction*. This reaction should be distinguished from an *alpha- (α-)expulsion*.
1996, *68*, 2228

α-decay (alpha-decay)

Radioactive decay in which an *alpha particle* is emitted.
1982, *54*, 1535

α-effect

A positive deviation of an α-nucleophile (a nucleophile bearing an unshared pair of electrons on an atom adjacent to the nucleophilic site) from a Brønsted-type plot of log k_{nuc} vs. pK_a constructed for a series of related normal nucleophiles. More generally, it is the influence of the atom bearing a lone pair of electrons on the reactivity at the adjacent site.

See also *Brønsted relation*.

The use of the term has been extended to include the effect of any substituent on an adjacent reactive centre, for example in the case of the 'α-silicon effect'.
1994, *66*, 1088

α-elimination

A *transformation* of the general type:

$$\begin{array}{c} R \\ R' \end{array}\!\!Z\!\!\begin{array}{c} X \\ Y \end{array} \longrightarrow \begin{array}{c} R \\ R' \end{array}\!\!Z\ + XY\ (or\ X + Y,\ or\ X^+ + Y^-)$$

where the central atom Z is commonly carbon. The reverse reaction is called α-*addition*.
  1994, *66*, 1088

α-expulsion (alpha-expulsion)(in photochemistry)

A general reaction by which a group attached to the alpha carbon of an excited chromophore is expelled either as an odd electron species or as an anionic species. This reaction should be distinguished from an α- (alpha-) *cleavage*.
  1988, *60*, 1059

α-oxo carbenes

Synonymous with *acyl carbenes*.
  See also *keto carbenes*.
  1995, *67*, 1354

α-particle (alpha-particle)

Nucleus of the ^{4}He atom.
  G.B. 93

A-factor

See *pre-exponential factor*.
  1994, *66*, 1082

'A' value

The conformational preference of an equatorial compared to an axial substituent in a monosubstituted cyclohexane. This steric substituent parameter equals $\Delta_r G°$ in kcal mol^{-1} for the equatorial to axial equilibration on cyclohexane. The values are also known as 'Winstein–Holness' A values.
  1994, *66*, 1087

abatement (in atmospheric chemistry)

Action taken to reduce *air pollution* which involves the use of control equipment or some new process. This refers to a reduction or lessening as opposed to elimination of a type of discharge or pollutant.
  1990, *62*, 2169

abeo-

Bond migration, i.e. the formal transfer of one end of a skeletal bond to another position, with a compensating transfer of a hydrogen atom, can be indicated by a prefix of the form '$x(y \rightarrow z)$abeo-'. The prefix is compiled as follows: a numeral denoting the stationary (unchanged) end of the migrating bond (x) is followed by parentheses enclosing the locant denoting the original position (y) from which the other end of this bond has migrated, an arrow, and the locant (z) denoting the new position to which the bond has moved. The closing parenthesis is followed by the italicized prefix *abeo-* (Latin: I go away) to indicate bond migration (see diagram **I** for an example). The original numbering is retained for the new compound and is used for the numbers x, y and z. It is always necessary to specify the resulting stereochemistry.
  B.B.(G) 31

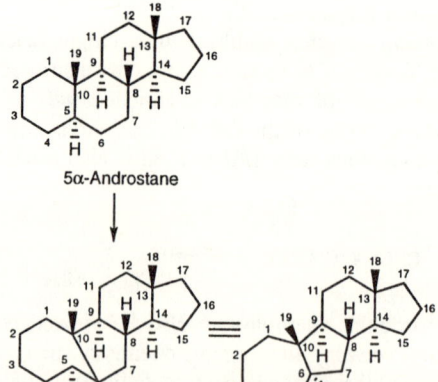

I (See *abeo*)

abiological

See *abiotic*.
  1993, *65*, 2007

abiotic

Not associated with living organisms. Synonymous with abiological.
  1993, *65*, 2007

abiotic transformation

Process in which a substance in the environment is modified by non-biological mechanisms.
  1993, *65*, 2007

absolute activation analysis

A kind of *activation analysis* in which the elemental concentrations in the material are calculated from known nuclear constants, irradiation and measurement parameters, rather than by comparing with known standards.
  1994, *66*, 2515

absolute activity, λ

The exponential of the ratio of the *chemical potential*, μ, to RT where R is the gas constant and T the thermodynamic temperature, $\lambda = \exp(\mu/RT)$.
  G.B. 40; see also 1994, *56*, 569; 1996, *68*, 959

absolute configuration

The spatial arrangement of the atoms of a chiral molecular entity (or group) and its stereochemical description e.g. *R* or *S*.
  See also *relative configuration* and α (alpha), β (beta).
  1996, *68*, 2197; see also B.B. 474

absolute counting (in radioanalytical chemistry)

A measurement under such well-defined conditions that the *activity* of a sample can be derived directly from the observed *counting rate*.
  1994, *66*, 2517

absolute electrode potential

The electrode potential of a metal measured with respect to a universal reference system (not including any additional metal/solution *interface*).
  1986, *58*, 957

absolute full energy peak efficiency
Of a *radiation* spectrometer, the *counting efficiency* when considering only the events recorded in the *full energy peak*.
1982, *54*, 1543

absolute lethal concentration (LC_{100})
Lowest concentration of a substance in an environmental medium which kills 100% of test organisms or species under defined conditions. This value is dependent on the number of organisms used in its assessment.
1993, *65*, 2007

absolute lethal dose (LD_{100})
Lowest amount of a substance which kills 100% of test animals under defined conditions. This value is dependent on the number of organisms used in its assessment.
1993, *65*, 2007

absolute photopeak efficiency
Of a *γ-ray spectrometer*, the *counting efficiency* when only the events recorded in the *photopeak* are considered.
1982, *54*, 1548

absolute preconcentration (in trace analysis)
An operation (process) as the result of which microcomponents are transferred from the sample of larger mass into the sample of smaller mass, so that the concentration of the microcomponents is increased. Examples include the decrease in solvent volume during distillation or evaporation, and the transfer of microcomponents from an aqueous solution into a smaller volume of organic solvent by extraction.
1979, *51*, 1197

absorbance, A
Logarithm of the ratio of incident to transmitted *radiant power* through a sample (excluding the effects on cell walls). Depending on the base of the logarithm a decadic and Napierian absorbance are used. Symbols: A, A_{10}, A_e. This quantity is sometimes called extinction, although the term extinction, better called *attenuance*, is reserved for the quantity which takes into account the effects of luminescence and scattering as well.
G.B. 32; see also 1996, *68*, 2226; 1990, *62*, 2169

absorbance matching (in spectrochemical analysis)
A procedure where the concentration of a known analyte may be determined by diluting the sample with solvent until the absorbance matches the absorbance of the analyte in the reference cell. This method is particularly useful if the *Beer–Lambert law* does not hold.
1988, *60*, 1456

absorbed dose (of a substance)
Amount of a substance absorbed into an organism or into organs and tissues of interest.
1993, *65*, 2007

absorbed dose (of radiation), D
Energy imparted to matter by ionizing radiation in a suitable small element of volume divided by the mass of that element of volume.
1996, *68*, 959; 1993, *65*, 2007; O.B. 220

absorbed electron coefficient (in *in situ* microanalysis)
Number of *absorbed electrons* per *primary electron*.
1983, *55*, 2026

absorbed electrons (in *in situ* microanalysis)
The (excess) electrons present in a specimen under electron bombardment which are led to ground and measured as specimen current. The number of absorbed electrons per unit time (or the specimen current), equals the number of *primary electrons* minus the number of back scattered, secondary and transmitted electrons per unit time. Therefore the fraction of electrons being absorbed depends on many parameters, including the composition and thickness of the specimen, the primary electron energy, the electron incidence angle and local electrostatic fields when present.
1983, *55*, 2026

absorber
1. A device used commonly for sampling by absorption in which a gaseous or liquid material is removed from another gas or liquid by selective absorption; these include: scrubber, impinger, packed column, spray chamber, etc.
1990, *62*, 2169
2. A substance used to absorb energy from any type of *radiation*.
1982, *54*, 1534

absorptance, α
Ratio of the absorbed to the incident *radiant power*. Also called absorption factor. When $\alpha \leq 1$, $\alpha \approx A_e$, where A_e is the Napierian absorbance.
G.B. 32; see also 1985, *57*, 114; 1996, *68*, 959; 1996, *68*, 2226

absorption
1. The process of one material (absorbent) being retained by another (absorbate); this may be the physical solution of a gas, liquid, or solid in a liquid, attachment of molecules of a gas, vapour, liquid, or dissolved substance to a solid surface by physical forces, etc. In spectrophotometry, absorption of light at characteristic wavelengths or bands of wavelengths is used to identify the chemical nature of molecules, atoms or ions and to measure the concentrations of these species.
1990, *62*, 2169
2. A phenomenon in which *radiation* transfers to matter which it traverses some of or all its energy.
1982, *54*, 1534; 1996, *68*, 2226

absorption coefficient
Linear decadic (a, K) and Napierian absorption coefficients (α) are equal to the corresponding *absorbances* divided by the optical path length through the

sample. The molar absorption coefficients (decadic ε, Napierian κ) are the linear absorption coefficients divided by the *amount concentration*.
   G.B. 32; see also 1990, *62*, 2169; 1996, *68*, 2226; 1996, *68*, 959

absorption cross-section, σ
Molar Napierian *absorption coefficient* divided by the Avogadro constant. When integrated against the logarithm of wavenumber (or frequency) it is called the integrated absorption cross-section.
   G.B. 32, 33; see also 1990, *62*, 2169; 1996, *68*, 2226

absorption factor
   See *absorptance*.
   G.B. 32; 1996, *68*, 960

absorption intensity
Differently defined physical quantities describing the absorption of radiation by a sample.
   G.B. 32, 33

absorption line
A narrow range of wavelengths in which a substance absorbs light; a series of discrete absorption lines can be used as an unambiguous identification for many relatively simple chemical species.
   1990, *62*, 2169

absorption pathlength (of a sample cell)
The length of the radiation path through the absorbing medium; it is equal to the cell path length, l, in the case of single-pass cells at normal incidence of radiation.
   1988, *60*, 1453

absorption spectrum
The wavelength dependence of the absorption cross-section (or absorption coefficient); usually represented as a plot of absorption cross-section versus wavelength λ (or $1/\lambda$) of the light.
   1990, *62*, 2169

+ absorptivity
Absorptance divided by the optical path length. For very low attenuance it approximates the *absorption coefficient* [within the approximation $(1 - e^{-A}) \sim A$]. The use of this term is not recommended.
   1996, *68*, 2227

abstraction
A *chemical reaction* or *transformation*, the main feature of which is the bimolecular removal of an atom (neutral or charged) from a molecular entity. For example:

$$CH_3COCH_3 + (i\text{-}C_3H_7)_2N^- \rightarrow (CH_3COCH_2)^- + (i\text{-}C_3H_7)_2NH$$

(proton abstraction from acetone)

$$CH_4 + Cl\cdot \rightarrow H_3C\cdot + HCl$$

(hydrogen abstraction from methane)
   See *detachment*.
   1994, *66*, 1081

abstraction process (in catalysis)
In abstraction and extraction processes, an *adsorptive* or *adsorbate* species extracts an adsorbed atom or a lattice atom respectively.
Abstraction process:

$$*H + H(g) \rightarrow * + H_2(g)$$

Extraction process:

$$O_s^{2-} + CO(g) \rightarrow 2e + CO_2(g)$$

   1976, *46*, 85

abundance, isotopic
   See *isotopic abundance*.
   1982, *54*, 1535

abundance sensitivity (in mass spectrometry)
The ratio of the maximum ion current recorded at a mass m to the ion current arising from the same species recorded at an adjacent mass ($m \pm 1$).
   1991, *63*, 1554

ac
   See *torsion angle*.
   1996, *68*, 2197

accelerating voltage (high voltage, V) scan (in mass spectrometry)
This is an alternative method of producing a momentum (mass) spectrum in magnetic deflection instruments by varying the accelerating voltage. This scan can also be used, in conjunction with a fixed radial electric field, to produce an *ion kinetic energy spectrum*.
   1991, *63*, 1550

acceleration, a
Vector quantity equal to the derivative of *velocity* with respect to time. For the acceleration of free fall the symbol g is used.
   G.B. 11; 1996, *68*, 960

acceleration energy (in *in situ* microanalysis)
   See *excitation energy (in* in situ *microanalysis)*.
   1983, *55*, 2027

acceleration of free fall, g
Acceleration of free fall in vacuum due to gravity. Also called acceleration due to gravity.
   1996, *68*, 960

accelerator (in solvent extraction)
Synonymous with catalyst.
   See *catalyst, kinetic synergist, modifier*.
   1993, *65*, 2380

acceptable daily intake (ADI)
Estimate of the amount of a substance in food or drinking water, expressed on a body mass basis (usually mg kg^{-1} body weight), which can be ingested daily over a lifetime by humans without appreciable health risk. For calculation of the daily intake per person, a standard body mass of 60 kg is used. ADI is normally used for food additives (tolerable daily intake is used for contaminants).
   1993, *65*, 2008

acceptor number (AN)
A quantitative measure of *Lewis acidity*.
1994, *66*, 1081

accommodation coefficient
A measure of the efficiency of capture of molecules or atoms which collide with aerosol particles, cloud droplets, etc. The accommodation coefficient is the fraction of the collisions which result in the capture of the molecules (atoms, radicals, etc.) by the particle, cloud droplet, etc.; fraction of colliding molecules which are not reflected but which enter the surface of an aqueous aerosol.
Synonymous with *sticking coefficient*.
1990, *62*, 2169

accretion (in atmospheric chemistry)
The process by which *aerosols* grow in size by external addition of various chemical species; a form of agglomeration.
1990, *62*, 2169

accuracy (of a measuring instrument)
Ability of a measuring instrument to give responses close to a true value.
VIM; 1990, *62*, 2174

accuracy (of measurement)
Closeness of the agreement between the result of a measurement and a true value of the measurand.
Notes:
1. Accuracy is a qualitative concept.
2. The term *precision* should not be used for accuracy.
VIM; O.B. 6; 1989, *61*, 1663; 1990, *62*, 2170; 1994, *66*, 598

acenes
Polycyclic aromatic *hydrocarbons* consisting of fused benzene rings in a rectilinear arrangement.

1995, *67*, 1310

acetals
Compounds having the structure $R_2C(OR')_2$ ($R' \neq H$) and thus diethers of geminal *diols*. Originally, the term was confined to derivatives of *aldehydes* (one R = H), but it now applies equally to derivatives of *ketones* (neither R = H). Mixed acetals have different R′ groups.
See also *acetonides, ketals, acylals, hemiacetals*.
1995, *67*, 1310

acetogenins
See *polyketides*.
1995, *67*, 1311

acetonides
Cyclic *acetals* derived from acetone and *diols*, usually vicinal diols, or polyhydroxy compounds. E.g.

1995, *67*, 1311

acetylene black
A special type of *carbon black* formed by an exothermic decomposition of acetylene. It is characterized by the highest degree of aggregation and crystalline orientation when compared with all types of carbon black.
Note:
Acetylene black must not be confused with the carbon black produced as a by-product during the production of acetylene in the electric arc process.
1995, *67*, 476

acetylenes
Acyclic (branched or unbranched) and cyclic (with or without side chain) *hydrocarbons* having one or more carbon–carbon triple bonds.
See also *alkynes*.
1995, *67*, 1311

acetylides
Compounds arising by replacement of one or both hydrogen atoms of acetylene (ethyne) by a metal or other cationic group. E.g. $NaC{\equiv}CH$ monosodium acetylide. By extension, analogous compounds derived from terminal *acetylenes*, $RC{\equiv}CH$.
1995, *67*, 1311

Acheson graphite
A *synthetic graphite* made by the Acheson process.
Note:
Reference to Acheson in combination with synthetic graphite honours the inventor of the first technical *graphitization*. Today the term Acheson graphite, however, is of historical interest only because it no longer covers the plurality of *synthetic graphite*.
1995, *67*, 476

achiral
See *chirality*.
1996, *68*, 2197; B.B. 479

achirotopic
See *chirotopic*.
1996, *68*, 2197

acid
A *molecular entity* or *chemical species* capable of donating a *hydron* (proton) (see *Brønsted acid*) or capable of forming a covalent bond with an electron pair (see *Lewis acid*).
See also *hard acid, carboxylic acids, oxoacids, sulfonic acids*.
1994, *66*, 1081; see also 1990, *62*, 2170

acid anhydrides
Compounds consisting of two *acyl groups* bonded to the same oxygen atom acyl–O–acyl. Symmetric and mixed anhydrides have identical and different acyl groups, respectively. E.g. $CH_3C(=O)OC(=O)CH_3$ acetic anhydride, $CH_3C(=O)OS(=O)_2Ph$ acetic benzenesulfonic anhydride, $PhC(=S)OC(=S)Ph$ (thiobenzoic) anhydride.
1995, *67*, 1311

acid–base catalysis, general
See *general acid–base catalysis*.
1996, *68*, 150

acid–base catalysis, specific
See *specific acid–base catalysis*.
1996, *68*, 150

acid–base indicator
An acid or base which exhibits a colour change on neutralization by the basic or acidic titrant at or near the equivalence point of a titration.
O.B. 48

acid–base titration
See *titration*.
O.B. 47

acid deposition (in atmospheric chemistry)
The acids deposited by transfer from the atmosphere either by precipitation (rain, fog, snow, etc.), called wet deposition, or by direct transfer to the surface of the earth induced by the flow of acid-containing air masses over the earth's surface, called dry deposition.
1990, *62*, 2170

acidimetric titration
See *titration*.
O.B. 47

acidity
1. Of a compound:
For *Brønsted acids* the tendency of a compound to act as a *hydron* donor. It can be quantitatively expressed by the acid dissociation constant of the compound in water or some other specified medium. For *Lewis acids* it relates to the association constants of *Lewis adducts* and π *adducts*.
2. Of a medium:
The use of the term is mainly restricted to a medium containing *Brønsted acids*, where it means the tendency of the medium to hydronate a specific reference base. It is quantitatively expressed by the appropriate *acidity function*.
1994, *66*, 1081

acidity constant
The *equilibrium constant* K_{an} for splitting off the nth proton from a charged or uncharged acid, to be defined. One may write K_a for K_{a1}.
O.B. 13

acidity function
Any function that measures the thermodynamic *hydron*-donating or -accepting ability of a solvent system, or a closely related thermodynamic property, such as the tendency of the *lyate ion* of the solvent system to form *Lewis adducts*. (The term 'basicity function' is not in common use in connection with basic solutions.) Acidity functions are not unique properties of the solvent system alone, but depend on the solute (or family of closely related solutes) with respect to which the thermodynamic tendency is measured.
Commonly used acidity functions refer to concentrated acidic or basic solutions. Acidity functions are usually established over a range of composition of such a system by UV/VIS spectrophotometric or NMR measurements of the degree of hydronation (protonation or Lewis adduct formation) for the members of a series of structurally similar indicator bases (or acids) of different strength: the best known of these functions is the Hammett acidity function H_0 (for uncharged indicator bases that are primary aromatic amines).
1994, *66*, 1081

acidosis
Pathological condition in which the hydrogen ion substance concentration of body fluids is above normal and hence the pH of blood falls below the reference interval.
See also *alkalosis*.
1993, *65*, 2009

acid rain (in atmospheric chemistry)
Rain with pH values < about 5; commonly results from acids formed from pollutants. 'Pure' rain water equilibrated with atmospheric CO_2 and naturally occurring acids in relatively clean air usually has a pH > 5.
1990, *62*, 2170

acid thioanhydrides
See *thioanhydrides*.
1995, *67*, 1311

***aci*-nitro compounds**
A class name for hydrocarbylideneazinic acids, $R_2C=N^+(-O^-)OH$. The use of *aci*-nitro as a prefix in systematic nomenclature to name specific compounds is abandoned.
See also *azinic acids*.
1995, *67*, 1351

actinic flux, S_λ
The quantity of light available to molecules at a particular point in the atmosphere and which, on absorption, drives photochemical processes in the atmosphere. It is calculated by integrating the spectral radiance $L(\lambda, \theta, \varphi)$ over all directions of incidence of the light, $\iint\limits_{\theta\,\varphi} L(\lambda, \theta, \varphi)/(hc/\lambda) \sin\theta \, d\theta \, d\varphi$. If the radiance is expressed in J m^{-2} s^{-1} st^{-1} nm^{-1} and hc/λ is the energy per quantum of light of wavelength λ, the actinic flux has units of quanta cm^{-2} s^{-1} nm^{-1}. This important quantity is one of the terms required in the calculation of j-values, the first order rate coefficients for photochemical processes in the sunlight-absorbing, trace gases in the atmosphere. The actinic flux is determined by the solar radiation entering the atmosphere and by any changes in this due to atmospheric gases and particles (e.g. Rayleigh scattering absorption by stratospheric ozone, scattering and absorption by aerosols and clouds), and reflections from the ground. It is therefore dependent on the wavelength of the light, on the altitude and on specific local environmental conditions. The actinic flux has borne many names (e.g. flux, flux density, beam irradiance actinic irradiance, integrated intensity) which has caused some confusion. It is important to distinguish

the actinic flux from the spectral irradiance, which refers to energy arrival on a flat surface having fixed spatial orientation (J m^{-2} nm^{-1}) given by:

$$E(\lambda) = \iint_{\theta\,\varphi} L(\lambda, \theta, \varphi)\cos\theta \sin\theta\, d\theta\, d\varphi.$$

The actinic flux does not refer to any specific orientation because molecules are oriented randomly in the atmosphere. This distinction is of practical relevance: the actinic flux (and therefore a *j*-value) near a brightly reflecting surface (e.g. over snow or above a thick cloud) can be a factor of three higher than that near a non-reflecting surface. The more descriptive name of spectral spheradiance is suggested for the quantity herein called actinic flux.

See also *flux density, photon*.
1990, *62*, 2170

actinometer
A chemical system or physical device which determines the number of *photons* in a beam integrally or per unit time. This name is commonly applied to devices used in the ultraviolet and visible wavelength ranges. For example, solutions of iron(III) oxalate can be used as a chemical actinometer, while bolometers, thermopiles, and photodiodes are physical devices giving a reading that can be correlated to the number of photons detected.

See also *chemical actinometer*.
1996, *68*, 2227

action spectrum
A plot of a relative biological or chemical photoresponse (= Δy) per number of incident photons against wavelength or energy of radiation under the same *radiant power* of light. This form of presentation is frequently used in the studies of biological or solid state systems, where the nature of the absorbing species is unknown. This type of action spectrum is sometimes called *spectral responsivity* or sensitivity spectrum. The precise action spectrum is a plot of the *spectral* (*photon* or quantum) *effectiveness*. By contrast, a plot of the biological or chemical change or response per absorbed photon (*quantum efficiency*) versus wavelength is the *efficiency spectrum*.

See also *excitation spectrum, efficiency spectrum*.
1996, *68*, 2227

activated adsorption process
If the temperature coefficient of the rate of *adsorption* is substantial, an adsorption process is said to be activated (i.e. to have a significant *activation energy*). In this case, the *sticking coefficient* is small. In general, the activation energy of activated adsorption is a function of coverage and it usually increases with increasing coverage.
1976, *46*, 79

activated carbon
A porous *carbon material*, a *char* which has been subjected to reaction with gases, sometimes with the addition of chemicals, e.g. $ZnCl_2$, before, during or after *carbonization* in order to increase its adsorptive properties.

Notes:
Activated carbons have a large adsorption capacity, preferably for small molecules, and are used for purification of liquids and gases. By controlling the process of carbonization and activation, a variety of active carbons having different porosity can be obtained. Activated carbons are used mainly in granular and powdered forms, but can also be produced in textile form by controlled carbonization and activation of textile fibres. Other terms used in the literature: active carbons, active charcoals.
1995, *67*, 476

activated charcoal
A traditional term for *activated carbon*.
1995, *67*, 476

activated complex
An activated complex, often characterized by the superscript ‡, is defined as that assembly of atoms which corresponds to an arbitrary infinitesimally small region at or near the col (saddle point) of a potential energy surface.

See also *transition state*.
1994, *66*, 1081; see also 1993, *65*, 2292; 1996, *68*, 150

activation
This word is used in different senses:
1. Input of external energy into a chemical system is said to bring about activation of the system.
2. An added substance that increases the rate of a catalysed reaction is known as an activator, and the effect is called activation.
If v_0 is the rate of the catalysed reaction in the absence of the activator, and v is the rate in its presence, the degree of activation ε_a is defined by:

$$\varepsilon_a = (v - v_0)/v_0 = (v/v_0) - 1$$

3. When some of the energy required for a reaction to occur is provided by a previous exothermic chemical reaction there is said to be chemical activation.

See *chemical activation, catalysis*.
See also *activator, inhibition*.
1996, *68*, 151

activation (in electrochemical corrosion)
The process of transition from the passive to the active state by removal of the passivating film.
A necessary condition for activation is an electrode potential negative to the equilibrium potential of formation of the passivating film. Activation is achieved by cathodic currents, by a reduced substance in the adjacent solution, or by contact with an electronic conductor having a suitably negative corrosion potential.

See also *passive state, active state*.
1989, *61*, 21

activation (in radiochemistry)
The process of inducing *radioactivity* by *irradiation*. In general, a specification is added of the type of incident *radiation* (e.g nuclear, neutron, photon) or its energy (e.g. thermal, fast).
   1994, *66*, 2515

activation analysis (nuclear)
A kind of elemental or isotopic analysis based on the measurement of characteristic *radiation* from *nuclides* formed directly or indirectly by *activation* of the test portion.
In general, a specification is added of the type of the incident radiation (e.g. neutron, photon) and its *energy* (e.g. thermal, fast).
   1994, *66*, 2515; O.B. 210

activation analysis, absolute
   See *absolute activation analysis*.
   1994, *66*, 2515

activation analysis, instrumental
   See *instrumental activation analysis*.
   1994, *66*, 2515

activation analysis, non-destructive
   See *non-destructive activation analysis*.
   1982, *54*, 1547

activation analysis, radiochemical
   See *radiochemical activation analysis*.
   1994, *66*, 2515

activation cross-section
The *cross-section* for the formation of a *radionuclide* by a specified reaction.
   1994, *66*, 2517; O.B. 217

activation energy (Arrhenius activation energy)
An empirical parameter characterizing the exponential temperature dependence of the *rate coefficient, k*, $E_a = RT^2 (\mathrm{d ln}k/\mathrm{d}T)$, where R is the gas constant and T the thermodynamic temperature. The term is also used for *threshold energies* in electronic potential surfaces, in which case the term requires careful definition.
   G.B. 55; see also 1996, *68*, 151; 1993, *65*, 2294; 1994, *66*, 1112

activation reaction
Process leading from the reactants to the formation of an activated complex or transition state.
   Physical Chemistry Division, unpublished

activator
A substance, other than the catalyst or one of the substrates, that increases the rate of a catalysed reaction without itself being consumed; the process is called *activation*. An activator of an enzyme-catalysed reaction may be called *enzyme activator*, if it acts by binding to the enzyme.
   See also *effector*.
   1993, *65*, 2292

active centre
   See *active site*.
   1976, *46*, 77; see also 1992, *64*, 145

active medium
   See *laser*.
   1995, *67*, 1915

active metal (in electrochemical corrosion)
A metal corroding in the *active state*.
   1989, *61*, 21

active site (in heterogeneous catalysis)
The term is often applied to those sites for *adsorption* which are the effective sites for a particular heterogeneous *catalytic reaction*. The terms active site and active centre are often used as synonyms, but active centre may also be used to describe an ensemble of sites at which a catalytic reaction takes place.
   1976, *46*, 77; see also 1992, *64*, 145

active solid
A porous solid with *adsorptive* properties by means of which chromatographic separations may be achieved. The separations resulting from this action follow laws different from those deriving from the partitioning action of the liquid phase.
   O.B. 98

active species
   See *laser*.
   1995, *67*, 1915

active state (in electrochemical corrosion)
Corrosion in the active state occurs by direct transfer (which may involve one or several steps) of metal ions from the metallic phase to the adjacent electrolyte.
The corrosion current of an active metal usually increases when the electrode potential is made more positive, other conditions remaining constant.
   1989, *61*, 21

active transport (in biology)
The carriage of a solute across a biological membrane from low to high concentration which requires the expenditure of (metabolic) energy.
   1992, *64*, 145

activity
   See *activity (relative activity)*.
   G.B. 49; 1996, *68*, 989

activity (of a radioactive material), A
The number of *nuclear decays* occurring in a given quantity of material in a small time interval, divided by that time interval. Often, this term is referred to as absolute activity. Symbol $A = -\mathrm{d}N/\mathrm{d}t$.
Synonymous with *disintegration rate* or *decay rate*.
   1994, *66*, 2515; G.B. 22; 1996, *68*, 960

activity (relative activity), a
Defined by the equation $a = \exp[(\mu - \mu^o)/RT]$ where R is the gas constant, T the thermodynamic temperature, μ the *chemical potential* and μ^o the standard chemical potential the definition of which depends on the choice of the *standard state*.
   See also *absolute activity*.
   G.B 49; 1996, *68*, 989

activity coefficient, f, γ

The activity coefficient f_B of a substance B in a liquid or solid mixture containing mole fractions x_B, x_C, ... of the substances B, C, ... is a dimensionless quantity defined in terms of the chemical potential μ_B of B in the mixture by:

$$RT \ln (x_B f_B) = \mu_B (\text{cd}, T, P, x) - \mu_B^* (\text{cd}, T, p)$$

where x denotes the set of mole fractions x_B, x_C, ... The activity coefficient γ_B of a solute B in a solution (especially a dilute liquid solution) containing molalities m_B, m_C, ... of solutes B, C, ... in a solvent A is a dimensionless quantity defined in terms of the chemical potential μ_B by:

$$RT \ln (m_B \gamma_B / m^{\ominus}) =$$
$$\mu_B - \{\mu_B - RT \ln (m_B / m^{\ominus})\}^{\infty}$$

1994, 66, 543; 1994, 66, 546; 1990, 62, 2171; 1996, 68, 960

activity, molar
See *molar activity*.
1994, 66, 2515

activity, specific
See *specific activity*.
1994, 66, 2515

acylals
Diesters of geminal *diols*; i.e. $R_2C[OC(=O)R]_2$. E.g.

Ph–CH(OC(=O)CH$_3$)(OC(=O)CH$_3$)

benzylidene diacetate

1995, 67, 1311

acyl anions
See *acyl species*.
1995, 67, 1311

acyl carbenes
Any compound acyl–C̈–R. In organic chemistry, an unspecified acyl carbene is commonly a carboxylic acyl carbene, $RC(=O)\ddot{C}R$.
1995, 67, 1311

acyl cations
See *acyl species*.
1995, 67, 1312

acyl groups
Groups formed by removing one or more hydroxy groups from *oxoacids* that have the general structure $R_kE(=O)_l(OH)_m$ ($l \neq 0$), and replacement analogues of such acyl groups. In organic chemistry an unspecified acyl group is commonly a carboxylic acyl group. E.g. $CH_3C(=O)-$, $CH_3C(=NR)-$, $CH_3C(=S)-$, $PhS(=O)_2-$, $HP(\equiv N)-$, $>P(R)=O$.
1995, 67, 1311

acyl halides
Compounds consisting of an acyl group bonded to halogen, e.g. $CH_3S(=O)_2Cl$, methanesulfonyl chloride; $CH_3C(=O)Cl$, acetyl chloride; cyclohexanecarboximidoyl chloride,

1995, 67, 1311

acylium ions
See *acyl species*.
1995, 67, 1312

acyloins
α-Hydroxy ketones, $RCH(OH)C(=O)R$. So named from the fact that they are formally derived from reductive coupling of carboxylic acyl groups.
See also *ketones*.
1995, 67, 1312

acyloxyl radicals
Oxygen-centered radicals consisting of an acyl radical bonded to an oxygen atom. E.g. $RC(=O)O\cdot$, $RC(=NR)O\cdot$, $RS(=O)O\cdot$.
1995, 67, 1312

acyl radicals
See *acyl species*.
1995, 67, 1311

acyl species
Acyl *intermediates* include acyl anions, acyl radicals and acyl cations (synonym *acylium ions*) which are formally derived from *oxoacids* $R_kE(=O)_l(OH)_m$ ($l \neq 0$) by removal of a hydroxyl cation HO^+, a hydroxyl radical $HO\cdot$ or a hydroxyl anion HO^-, respectively, and replacement analogues of such intermediates. Acyl anions, radicals and cations can formally be represented by canonical forms having a negative charge, an unpaired electron or a positive charge on the acid-generating element of the oxoacid.
Acyl anions. E.g. $RC^-(=O)$, $RS^-(=O)_2$, $RC^-(=S)$, $RC^-(=NH)$.
Acyl radicals. E.g. $RC\cdot(=O)$, $RS\cdot(=O)_2$.
Acyl cations. E.g. $RC^+(=O) \leftrightarrow RC\equiv O^+$.
See also *acyl groups*.
1995, 67, 1312; see also 1993, 65, 1357

added hydrogen
See *indicated hydrogen*.
B.B.(G) 34

addend
See *addition reaction*.
1994, 66, 1081

addition
1. Refers to *addition reaction* or addition *transformation*.
2. Loosely, the formation of an adduct. (For an example, see *Lewis acid*.)
3. Loosely, any *association* or *attachment*.
1994, 66, 1081

addition reaction
A *chemical reaction* of two or more reacting *molecular* entities, resulting in a single reaction product containing all atoms of all components, with formation of two chemical bonds and a net reduction in bond

multiplicity in at least one of the reactants. The reverse process is called an *elimination* reaction. The addition may occur at only one site (α-*addition*, 1/1/addition), at two adjacent sites (1/2/addition) or at two non-adjacent sites (1/3/- or 1/4/addition, etc.). For example:

(a) $H^+ + Br^- + (CH_3)_2C=CH_2 \rightarrow (CH_3)_2CBr–CH_3$
(1/2/addition)
(b) $Br_2 + CH_2=CH–CH=CH_2$
$\rightarrow BrCH_2–CH=CH–CH_2Br$ (1/4/addition)
and $BrCH_2–CH(Br)–CH=CH_2$ (1/2/addition)

If the reagent or the source of the addends of an addition are not specified, then it is called an addition transformation.

See also *addition*, α-*addition*, *cheletropic reaction*, *cycloaddition*.
1994, *66*, 1081

additive
A substance added to a sample for any of a variety of purposes.
O.B. 159

additive name
A name that describes:
i. The formal assembly of names for the components of a compound without loss of atoms or groups of atoms from any component.
ii. The addition or attachment of atoms or groups of atoms.
B.B.(G) 16

additivity of mass spectra
The process by which each chemical species present in the ion source at a certain partial pressure makes a contribution to the total mass spectrum which is the same as that which it would give if that chemical species alone were present in the ion source at a pressure equal to this certain partial pressure.
O.B. 206

additivity principle
The hypothesis that each of several structural features of a *molecular entity* makes a separate and additive contribution to a property of the substance concerned. More specifically, it is the hypothesis that each of the several *substituent* groups in a parent molecule makes a separate and additive contribution to the standard Gibbs energy change (or *Gibbs energy of activation*) corresponding to a particular equilibrium (or *rate of reaction*).

See also *transferability*.
1994, *66*, 1081

adduct
A new *chemical species* AB, each *molecular entity* of which is formed by direct combination of two separate molecular entities A and B in such a way that there is change in *connectivity*, but no loss, of atoms within the *moieties* A and B. Stoichiometries other than 1:1 are also possible, e.g. a bis-adduct (2:1). An '*intramolecular adduct*' can be formed when A and B are groups contained within the same molecular entity.

This is a general term which, whenever appropriate, should be used in preference to the less explicit term *complex*. It is also used specifically for products of an *addition reaction*. For examples, see *Lewis adduct*, *Meisenheimer adduct*, π-*adduct*.
1994, *66*, 1082

adduct ion (in mass spectrometry)
An ion formed by interaction of two species, usually an ion and a molecule, and often within the ion source, to form an ion containing all the constituent atoms of one species as well as an additional atom or atoms.
1991, *63*, 1549

adhesional wetting
A process in which an adhesional joint is formed between two phases.
1972, *31*, 597

adiabatic
This word is used with various different meanings, and when it is used it should be defined.

In thermodynamics 'adiabatic' is used in a macroscopic sense to refer to a process occurring in a thermally insulated system, so that there is no flow of heat to or from the surroundings.

In reaction dynamics, the word has been used in a microscopic sense, with a range of meanings which have only a tenuous relationship to the thermodynamic meaning or the etymology. Whereas the thermodynamic meaning relates to conditions imposed on a process by an observer, the microscopic meaning relates to conditions under which the process occurs naturally.

The microscopic meanings, as used in reaction dynamics, all have in common the feature that quantum states remain unchanged during the course of reaction. Different quantum states may be referred to:
1. A reaction in which there is no change of electronic state or multiplicity has been called adiabatic, or more specifically electronically adiabatic.
2. A reaction in which there is no change of vibrational state during the course of reaction has been said to be vibrationally adiabatic. More loosely, the expression has been applied to a process in which excess vibrational energy in the *reactants* appears as vibrational energy in the *products*, or in which ground-state vibration in the reactants leads to ground-state vibration in the products.
3. A reaction in which excess rotational energy in the reactants appears as rotational energy in the products, or in which ground-state rotation in the reactants leads to ground-state rotation in the products, has been referred to as rotationally adiabatic.
4. In the *Rice–Ramsperger–Kassel–Marcus (RRKM)* theory of unimolecular reactions, a degree of freedom whose quantum number is more or less preserved

during energization and subsequent reaction has been called 'adiabatic'; the word 'inactive' has also been applied to it.

A reaction that is not adiabatic is referred to as nonadiabatic or diabatic, and some workers make a distinction between the two words.

See also *diabatic coupling*.
1996, *68*, 152

adiabatic electron transfer
Electron transfer process in which the reacting system remains on a single electronic surface in passing from reactants to products. For adiabatic electron transfer the electronic transmission factor is close to unity (see *Marcus equation*.)

See also *diabatic electron transfer*.
1996, *68*, 2227

adiabatic ionization (in mass spectrometry)
A process whereby an electron is either removed from or added to an atom or molecule producing an ion in its ground state.
1991, *63*, 1547

adiabatic lapse rate (in atmospheric chemistry)
The rate of decrease in temperature with increase in altitude of an air parcel which is expanding slowly to a lower atmospheric pressure without exchange of heat; for a descending parcel it is the rate of increase in temperature with decrease in altitude. Theory predicts that for dry air it is equal to the acceleration of gravity divided by the specific heat of dry air at constant pressure (approximately 9.8 °C km^{-1}). The moist adiabatic lapse rate is less than the dry adiabatic lapse rate and depends on the moisture content of the air mass.
1990, *62*, 2171

adiabatic photoreaction
Within the 'Born–Oppenheimer' approximation, a reaction of an excited state species that occurs on a single 'potential-energy surface.'

See also *diabatic photoreaction*.
1996, *68*, 2227

adiabatic transition-state theory
A form of *transition-state theory* in which the system is assumed to preserve its internal quantum states as it moves over the *potential-energy surface*. A detailed state-to-state scattering theory version of adiabatic transition-state theory has been referred to as the adiabatic channel model.
1996, *68*, 152

adiabatic treatments of reaction rates
These are treatments of reaction rates in which the system is assumed to remain on a single *potential-energy surface* during the entire course of reaction, or to remain in a conserved internal state.
1996, *68*, 152

adjacent re-entry model (in polymer crystals)
A model of crystallinity in which chain folds regularly connect adjacent stems.
P.B. 84

adjusted retention volume (in chromatography)
See *retention volumes (in chromatography)*.
O.B. 103

adjuvant
1. In pharmacology, a substance added to a drug to speed or increase the action of the main component.
2. In immunology, a substance (such as aluminium hydroxide) or an organism (such as bovine tuberculosis bacillus) which increases the response to an antigen.
1993, *65*, 2010

ADMR
See *ODMR*.
1996, *68*, 2227

adsorbate
A molecular species of gas, dissolved substance or liquid which adheres to or is adsorbed in an extremely thin surface layer of a solid substance.
1990, *62*, 2171; O.B. 85

adsorbent
A condensed phase at the surface of which adsorption may occur.
O.B. 85; 1990, *62*, 2171

adsorber
Solid material used to capture either a gas or liquid; often activated carbon is employed as the solid because of its high surface area per unit mass.
1990, *62*, 2171

adsorption
An increase in the concentration of a dissolved substance at the interface of a condensed and a liquid phase due to the operation of surface forces. Adsorption can also occur at the interface of a condensed and a gaseous phase.
O.B. 85; 1990, *62*, 2171

adsorption capacity
For strongly adsorbed solutes of limited solubility, the value of the amount of adsorbed substance reached in a saturated solution is called the adsorption capacity of the *adsorbent* for a specific solute; its value depends also, in general, on the nature and, in the case of more than two components, on the relative composition of the bulk liquid.
1972, *31*, 594

adsorption chromatography
Chromatography in which separation is based mainly on differences between the adsorption affinities of the sample components for the surface of an active solid.
1993, *65*, 826; O.B. 94

adsorption complex
The entity constituted by the *adsorbate* and the part of the *adsorbent* to which it is bound.
1976, *46*, 74; see also 1972, *31*, 584

adsorption current
A *faradaic current* whose magnitude depends on the *applied potential*, and, at any particular applied potential, on the rate or extent of the adsorption of an electroactive substance (or the product obtained from

the reduction or oxidation of an electroactive substance) onto the surface of the *indicator* or *working electrode*.
See also *limiting adsorption current*.
1985, *57*, 1494

adsorption current, limiting
See *limiting adsorption current*.
1985, *57*, 1494

adsorption hysteresis
Adsorption hysteresis is said to occur when *adsorption* and desorption values deviate from one another.
1972, *31*, 585

adsorption indicator
A substance which is adsorbed or desorbed with concomitant colour change at or near the equivalence point of a titration.
O.B. 48

adsorption isobar
The function relating the amount, mass, or volume, or corresponding excess of substance adsorbed by a given amount of solid, to the temperature at constant pressure.
1972, *31*, 596

adsorption isostere
The function relating the equilibrium pressure to the temperature at a constant value of the amount, or excess amount, of substance adsorbed by a given amount of solid.
1972, *31*, 596

adsorption isotherm
An adsorption isotherm for a single gaseous *adsorptive* on a solid is the function which relates at constant temperature the amount of substance adsorbed at equilibrium to the pressure (or concentration) of the adsorptive in the gas phase. The surface excess amount rather than the amount adsorbed is the quantity accessible to experimental measurement, but, at lower pressures, the difference between the two quantities becomes negligible.
1976, *46*, 77; see also 1972, *31*, 585

adsorptive
The material that is present in one or other (or both) of the bulk phases and capable of being adsorbed.
1972, *31*, 584

advancement
See *extent of reaction*.
G.B. 43

advection (in atmospheric chemistry)
The transport of air, its properties (such as heat), trace materials, fog, cold air, etc., solely by mass motion of the atmosphere, generally in a horizontal direction.
1990, *62*, 2171

aeration (in atmospheric chemistry)
The process by which a volume filled with a liquid becomes permeated with air or another gas; aeration is often accomplished by spraying the liquid into the air, bubbling air through a liquid or agitating the liquid to promote surface absorption of air.
1990, *62*, 2172

aerobe
An organism that needs oxygen for respiration and hence for growth.
1992, *64*, 145

aerobic
Requiring molecular oxygen.
See also *anaerobic*.
1993, *65*, 2011

aerobic conditions
Conditions for growth or *metabolism* in which the organism is sufficiently supplied with oxygen.
1992, *64*, 145

aerogel
A *gel* is a *colloidal* system with a finite, rather small, yield stress. The term aerogel is used when the openness of the structure is largely maintained.
1972, *31*, 606

aeromete (in atmospheric chemistry)
Instrument used to measure the density of gases.
1990, *62*, 2172

aerometric measurement (in atmospheric chemistry)
A measurement of the temperature, pressure, air motion (velocity vectors), or other physical measurement to characterize an air mass.
1990, *62*, 2172

aerosol
Mixtures of small particles (solid, liquid or a mixed variety) and the carrier gas (usually air); owing to their size, these particles (usually less than 100 μm and greater than 0.01 μm in diameter) have a comparatively small settling velocity and hence exhibit some degree of stability in the earth's gravitational field. An aerosol may be characterized by its chemical composition, its radioactivity, the particle size distribution, the electrical charge and the optical properties.
1990, *62*, 2172

affinity chromatography
An expression characterizing the particular variant of chromatography in which the unique biological specificity of the analyte and ligand interaction is utilized for the separation.
1993, *65*, 826; see also 1992, *64*, 145

affinity of reaction, *A*
Negative partial derivative of Gibbs energy with respect to *extent of reaction* at constant pressure and temperature. It is positive for spontaneous reactions.
See also *driving force*.
G.B. 49

after mass analysis (in mass spectrometry)
The sum of all the separate ion currents carried by the different ions contributing to the spectrum.
O.B. 206

ageing (of precipitate)
The time-dependent change of those properties of a precipitate, e.g. loss of water, growth of crystals, recrystallization, decrease of the specific surface, loss of coprecipitated substances, which generally improve the filtering properties. The process of ageing is very often promoted by maintaining the precipitate and precipitation medium together at elevated temperatures for a period of time. The terms chemical, physical and thermal ageing may be used in cases in which some of the (usually combined) effects named above are to be emphasized specifically.
O.B. 86

agglomerate (in catalysis)
See *aggregate (in catalysis)*.
1991, *63*, 1231

agglomeration
The formation and growth of aggregates ultimately leading to phase separation by the formation of precipitates of larger than colloidal size.
O.B. 85; see also 1990, *62*, 2172

agglutination
An immunochemical reaction leading to the aggregation of particulate matter such as *bacteria*, erythrocytes or other *cells*, or synthetic particles such as plastic beads coated with *antigens* or *antibodies*.
1992, *64*, 145

aggregate (in catalysis)
Certain materials used as catalysts or supports consist of spheroids smaller than 10 nm in diameter, cemented into larger entities. A *primary particle* should be defined as the smallest discrete identifiable entity and the method of identification should be mentioned (e.g. transmission electron microscopy, scanning electron microscopy). An assemblage of such primary particles exhibiting an identifiable collective behaviour (e.g. chemical nature of the aggregated primary particles, texture of the aggregate, resistance to mechanical separation upon grinding) constitutes an aggregate. When describing the aggregates the criterion of identification should be mentioned. Strongly bonded aggregates are called *agglomerates*.
1991, *63*, 1231

aglycon (aglycone)
The non-sugar compound remaining after replacement of the *glycosyl* group from a *glycoside* by a hydrogen atom.
1995, *67*, 1312

agonist
Substance which binds to cell receptors normally responding to naturally occurring substances and which produces a response of its own.
See also *antagonist*.
1993, *65*, 2011

agostic
The term designates structures in which a hydrogen atom is bonded to both a carbon atom and a metal atom. The term is also used to characterize the interaction between a CH bond and an unsaturated metal centre, and to describe similar bonding of a transition metal with Si-H compounds. The expression 'μ-hydrido-bridged' is also used to describe the bridging hydrogen.
1994, *66*, 1082

agranular carbon
A monogranular or monolithic *carbon material* with homogeneous microstructure which does not exhibit any structural components distinguishable by optical microscopy.
Notes:
The above definition of a homogeneous microstructure does not pertain to pores and structural components which may be visible by contrast differences in optical microscopy with polarized light. As a consequence, *glass-like carbon* with visible pores is still an agranular carbon. The same is true, for instance, for *pyrolytic carbon* with preferred orientation, such as conical or lamellar structures, visible in optical microscopy with polarized light. Use of the term agranular carbon is not restricted to bulk materials of a minimum size. Only *particulate carbon* should be excluded even if the isolated particles exhibit a homogeneous microstructure.
1995, *67*, 476

air, composition of pure
See *composition of pure air*.
1990, *62*, 2172

air contaminant (in atmospheric chemistry)
A substance, gaseous material or aerosol, which is present in an air mass at levels greater than in clean air. An air contaminant has been added commonly by anthropogenic activity.
See also *air pollutant*.
1990, *62*, 2182

air-lift bioreactor
A bioreactor in which the reaction medium is kept mixed and gassed by introduction of air or another gas (mixture) at the base of a column-like reactor equipped either with a draught tube or another device (e.g. external tube) by which the reactor volume is separated into a gassed and an ungassed region thus generating a vertically circulating flow.
1992, *64*, 146

air mass (in atmospheric chemistry)
A qualitative term to describe a widespread body of air with approximately uniform characteristics which had been identified at a given time over a particular region of the earth's surface. Sometimes an air mass is marked by inert tracers such as SF_6 which may be added to it. The composition of a given air mass undergoes alteration as it migrates, chemical changes occur, compounds are removed by dry and wet deposition and new impurities are added to the mass.
1990, *62*, 2172

air monitoring station (in atmospheric chemistry)
A site at which monitoring of the concentration of one or more pollutants is carried out (e.g. the BAPMoN stations, Background Air Pollution Monitoring Network of the WMO, World Meteorological Organization).
1990, *62*, 2172

air pollutant
A substance, gaseous material or aerosol which has been introduced into the air (either by human activity or by natural processes) in sufficient concentration to produce a measurable effect on humans, animals, vegetation or materials (monuments, etc.): SO_2, NO_2, H_2S, CO, hydrocarbons, etc.
1990, *62*, 2172

air pollution
Usually the presence of substances in the atmosphere, resulting either from human activity or natural processes, present in sufficient concentration, for a sufficient time and under circumstances such as to interfere with comfort, health or welfare of persons or the environment.
1990, *62*, 2172

air pollution index (air quality index)
A mathematical combination of the concentrations of air pollutants (weighted in some fashion to reflect the estimated health impact of the specific pollutant) which gives an approximate numerical measure of the quality of the air at a given time. These indices have little scientific basis but have been used to inform the public (in a qualitative fashion) of the degree of pollution present at a given time. It is recommended that the actual measured pollutant concentrations be used by all information services when possible with simultaneous reference given to the corresponding concentrations which are considered by health authorities to be hazardous to human health.
1990, *62*, 2172

air pollution survey (in atmospheric chemistry)
A study of the concentrations and geographical distribution of specified air pollutants in a given area and an assessment of the damage, if any, which the pollutants have caused.
1990, *62*, 2173

air quality characteristic (in atmospheric chemistry)
One of the quantifiable properties relating to an air sample: concentration of a constituent, wind speed, temperature, etc. The quantity of air quality characteristic is the true value of the characteristic being investigated; it is recognized that in practice, this value can only be approximated by existing methods.
1990, *62*, 2173

air resource management (in atmospheric chemistry)
The detailed planning and the implementation of air pollution control programs designed to preserve the health and welfare of the people in the region, the plant and animal life, physical property, good visibility and other factors which determine the air quality and the maintenance of an aesthetically acceptable environment.
1990, *62*, 2173

air sampling network (in atmospheric chemistry)
A number of air sampling stations which are established in a given geographical region at which measurements of both pollutant concentrations and meteorological quantities (wind speed, direction, rain fall, humidity, etc.) are made to determine the extent and the nature of the air pollution and to establish trends in the concentrations of the air pollutants with time.
1990, *62*, 2173

Aitken particles
Aerosol particles below 0.1 μm in diameter. These generally are the most numerous among all particles in the air. Their concentrations can be determined with the Aitken counter which measures total particle number density. Owing to their small size, Aitken particles contribute little to the total mass concentration of all aerosol particles; this is determined primarily by particles of diameter > 0.1 μm.
1990, *62*, 2173

albedo
The fraction of the energy of electromagnetic radiation reflected from a body (or surface) relative to the energy incident upon it. The reflection of light from a surface is, of course, dependent on the wavelength of the light, the nature of the surface and its angle of incidence with the surface. The term albedo usually connotes a broad wavelength band (visible, ultraviolet or infrared), whereas the terms reflectivity and spectral albedo are used to describe the reflection of monochromatic (single wavelength or small band of wavelengths) radiation.
1990, *62*, 2173

alcoholates
Synonymous with *alkoxides*. Alcoholate should not be used for solvates derived from an *alcohol*, such as $CaCl_2 \cdot nROH$, for the ending -ate often occurs in names for anions.
1995, *67*, 1312

alcohols
Compounds in which a hydroxy group, –OH, is attached to a saturated carbon atom R_3COH. The term 'hydroxyl' refers to the radical species, HO·.
See also *enols, phenols*.
1995, *67*, 1312; see also 1993, *65*, 1357

alcoholysis
See *solvolysis*.
1994, *66*, 1082

aldaric acids
Polyhydroxy dicarboxylic acids having the general formula $HOC(=O)[CH(OH)]_nC(=O)OH$, formally

derived from an aldose by oxidation of both terminal carbon atoms to carboxyl groups.
1995, *67*, 1312; W.B. 142

aldazines
Azines of *aldehydes* RCH=NN=CHR.
1995, *67*, 1312

aldehydes
Compounds RC(=O)H, in which a carbonyl group is bonded to one hydrogen atom and to one R group.
1995, *67*, 1312; see also 1994, *62*, 2173

aldimines
Imines derived from *aldehydes* RCH=NR. E.g. EtCH=NH, PhCH=NMe.
1995, *67*, 1312

alditols
Acyclic polyols having the general formula HOCH$_2$[CH(OH)]$_n$CH$_2$OH (formally derivable from an aldose by reduction of the carbonyl group).
1995, *67*, 1312

aldoketoses
A now less preferred synonym for *ketoaldoses*.
1995, *67*, 1313

aldonic acids
Polyhydroxy acids having the general formula HOCH$_2$[CH(OH)]$_n$C(=O)OH and therefore derived from an aldose by oxidation of the aldehyde function, e.g. D-gluconic acid.

1995, *67*, 1313

aldoses
Aldehydic parent sugars (polyhydroxyaldehydes H[CH(OH)]$_n$C(=O)H, $n \geq 2$) and their intramolecular hemiacetals.

See also *monosaccharides*.
1995, *67*, 1313

aldoximes
Oximes of *aldehydes* RCH=NOH.
1995, *67*, 1313

alert levels (in atmospheric chemistry)
Designated concentrations of certain key *pollutants* at which some degree of danger to public health is expected. In many areas in which a relatively high level of pollution is often encountered, several levels of alert are often established. For example, a first alert may signify a potential problem exists; a second alert becomes a signal for the curtailment of certain significant emission sources; the third alert signifies the need for some specified emergency action which must be taken to ensure the public safety.
1990, *62*, 2173

alicyclic compounds
Aliphatic compounds having a carbocyclic ring structure which may be saturated or unsaturated, but may not be a benzenoid or other aromatic system.
1995, *67*, 1313

aliphatic compounds
Acyclic or cyclic, saturated or unsaturated carbon compounds, excluding *aromatic compounds*.
1995, *67*, 1313

aliquot (in analytical chemistry)
A known amount of a homogeneous material, assumed to be taken with negligible sampling error. The term is usually applied to fluids.
The term 'aliquot' is usually used when the fractional part is an exact divisor of the whole; the term 'aliquant' has been used when the fractional part is not an exact divisor of the whole (e.g. a 15 mL portion is an aliquant of 100 mL). When a laboratory sample or test sample is 'aliquoted' or otherwise subdivided, the portions have been called split samples.
1990, *62*, 1206; 1990, *62*, 2173

alkalimetric titration
See *titration*.
O.B. 47

alkaloids
Basic nitrogen compounds (mostly heterocyclic) occurring mostly in the plant kingdom (but not excluding those of animal origin). Amino acids, *peptides*, *proteins, nucleotides, nucleic acids, amino sugars* and antibiotics are not normally regarded as alkaloids. By extension, certain neutral compounds biogenetically related to basic alkaloids are included.
1995, *67*, 1313

alkalosis
Pathological condition in which the hydrogen ion substance concentration of body fluids is below normal and hence the pH of blood rises above the reference interval.
See also *acidosis*.
1993, *65*, 2012

alkanes
Acyclic branched or unbranched hydrocarbons having the general formula C$_n$H$_{2n+2}$, and therefore con-

sisting entirely of hydrogen atoms and saturated carbon atoms.
> See also *cycloalkanes*.
> 1995, *67*, 1313

alkanium ions
Carbocations derived from *alkanes* by *C*-hydronation containing at least one pentacoordinate carbon atom, E.g. $^+CH_5$ methanium, $[C_2H_7]^+$ ethanium.
> See also *carbonium ions*.
> 1995, *67*, 1313

alkenes
Acyclic branched or unbranched *hydrocarbons* having one carbon–carbon double bond and the general formula C_nH_{2n}. Acyclic branched or unbranched hydrocarbons having more than one double bond are alkadienes, alkatrienes, etc.
> See also *olefins*.
> 1995, *67*, 1313

alkoxides
Compounds, ROM, derivatives of alcohols, ROH, in which R is saturated at the site of its attachment to oxygen and M is a metal or other cationic species.
> See *alcoholates*.
> 1995, *67*, 1314

alkoxyamines
O-Alkyl hydroxylamines (with or without substitution on N) R'ONR$_2$ (R' ≠ H).
> 1995, *67*, 1314

alkylenes
1. An old term, which is not recommended, for *alkenes*, especially those of low molecular weight.
2. An old term for alkanediyl groups commonly but not necessarily having the free valencies on adjacent carbon atoms, e.g. –CH(CH$_3$)CH$_2$– propylene (systematically called propane-1,2-diyl).
> 1995, *67*, 1314

alkyl groups
Univalent groups derived from *alkanes* by removal of a hydrogen atom from any carbon atom –C$_n$H$_{2n+1}$. The groups derived by removal of a hydrogen atom from a terminal carbon atom of unbranched alkanes form a subclass of normal alkyl (*n*-alkyl) groups H[CH$_2$]$_n$. The groups RCH$_2$, R$_2$CH (R ≠ H), and R$_3$C (R ≠ H) are primary, secondary and tertiary alkyl groups, respectively.
> See also *cycloalkyl groups, hydrocarbyl groups*.
> 1995, *67*, 1314

alkylideneamino carbenes
> See *nitrile ylides*.
> 1995, *67*, 1314

alkylideneaminoxyl radicals
Radicals having the structure R$_2$C=N–O·. Synonymous with *iminoxyl radicals*.
> 1995, *67*, 1314

alkylideneaminyl radicals
Radicals having the structure R$_2$C=N·. Synonymous with *iminyl radicals*.
> 1995, *67*, 1314

alkylidene groups
The divalent groups formed from *alkanes* by removal of two hydrogen atoms from the same carbon atom, the free valencies of which are part of a double bond, e.g. (CH$_3$)$_2$C= propan-2-ylidene.
> 1995, *67*, 1314

alkylidenes
Carbenes R$_2$C: formed by mono or dialkyl substitution of methylene, H$_2$C:, e.g. CH$_3$CH$_2$CH: propylidene.
> 1995, *67*, 1314

alkylidynes
Carbenes RC: containing a univalent carbon atom, e.g. CH$_3$CH$_2$C: propylidyne.
> 1995, *67*, 1314

alkyl radicals
Carbon-centered *radicals* derived formally by removal of one hydrogen atom from an *alkane*, e.g. CH$_3$CH$_2$ĊH$_2$ propyl.
> 1995, *67*, 1314

alkynes
Acyclic branched or unbranched *hydrocarbons* having a carbon–carbon triple bond and the general formula C_nH_{2n-2}, RC≡CR. Acyclic branched or unbranched hydrocarbons having more than one triple bond are known as alkadiynes, alkatriynes, etc.
> See also *acetylenes*.
> 1995, *67*, 1314

allele
One of several alternate forms of a *gene* which occur at the same locus on homologous *chromosomes* and which become separated during *meiosis* and can be recombined following *fusion* of gametes.
> 1992, *64*, 146

allenes
Hydrocarbons (and by extension, derivatives formed by substitution) having two double bonds from one carbon atom to two others R$_2$C=C=CR$_2$. (The simplest member, propadiene, is known as allene).
> See also *cumulenes, dienes*.
> 1995, *67*, 1314

allo- (in amino-acid nomenclature)
Amino acids with two chiral centres were named in the past by allotting a name to the first diastereoisomer to be discovered. The second diastereoisomer, when found or synthesized, was then assigned the same name but with the prefix allo-. This method can be used only with trivial names but not with semisystematic or systematic names. It is now recommended that allo should be used only for alloisoleucine and allothreonine.
> W.B. 46

allosteric enzymes
Enzymes which contain regions to which small, regulatory molecules (cf. *effector*) may bind in addition to and separate from *substrate* binding sites. On binding the effector, the catalytic activity of the enzyme towards the substrate may be enhanced, in which case

the effector is an activator, or reduced, in which case it is an *inhibitor*.
1992, *64*, 146

allostery
A phenomenon whereby the conformation of an *enzyme* or other protein is altered by combination, at a site other than the substrate-binding site, with a small molecule, referred to as an *effector*, which results in either increased or decreased activity by the enzyme.
1994, *66*, 2593

allotriomorphic transition
See *allotropic transition*.
1994, *66*, 579

allotropes
Different structural modifications of an element.
R.B. 35

allotropic transition
A transition of a pure element, at a defined temperature and pressure, from one crystal structure to another which contains the same atoms but which has different properties.
Examples: The transition of graphite to diamond, that of body-centred-cubic iron to face-centred-cubic iron, and the transition of orthorhombic sulfur to monoclinic sulfur.
Synonymous with allotriomorphic transition.
1994, *66*, 579

allylic groups
The group $CH_2=CHCH_2$ (allyl) and derivatives formed by substitution. The term 'allylic position' or 'allylic site' refers to the saturated carbon atom. A group, such as $-OH$, attached at an allylic site is sometimes described as 'allylic'.
1995, *67*, 1315

allylic intermediates
Carbanions, carbenium ions or *radicals*, formally derived by detachment of one hydron, hydride or hydrogen from the CH_3 group of propene or derivatives thereof. E.g. $H_2C=CHCH_2^+$ allyl cation.
1995, *67*, 1315

allylic substitution reaction
A *substitution reaction* occurring at position 1/ of an allylic system, the double bond being between positions 2/ and 3/. The incoming group may be attached to the same atom 1/ as the *leaving group*, or the incoming group becomes attached at the relative position 3/, with movement of the double bond from 2/3 to 1/2. For example:

$$CH_2=CHCH(Me)Br + OH^-$$
$$\downarrow$$
$$HOCH_2CH=CHMe + CH_2=CHCH(Me)OH + Br^-$$

(written as a *transformation*).
1994, *66*, 1082

alpha
For entries see α (beginning of 'a').

alternant
A *conjugated system* of π-electrons is termed alternant if its atoms can be divided into two sets so that no atom of one set is directly linked to any other atom of the same set.

Example of alternant π system:

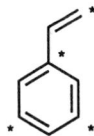

Example of non-alternant π system:

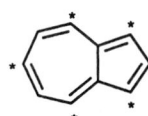

(two atoms of unstarred set are directly linked)
1994, *66*, 1082

alternating copolymer
A *copolymer* consisting of *macromolecules* comprising two species of *monomeric units* in alternating sequence.
Note:
An alternating copolymer may be considered as a homopolymer derived from an implicit or hypothetical monomer.
See also *homopolymer (1)*.
1996, *68*, 2301

alternating copolymerization
A *copolymerization* in which an *alternating copolymer* is formed.
1996, *68*, 2307

alternating current
Current with sinusoidal wave forms; all other wave forms should be termed 'periodic'.
See also *amplitude of alternating current*.
1985, *57*, 1494

alternating current, amplitude of
See *amplitude of alternating current*.
1985, *57*, 1494

alternating voltage
This term should be applied only to sinusoidal phenomena; the term *periodic voltage* should be used for other wave forms.
See also *amplitude of alternating voltage*.
1985, *57*, 1505

alternating voltage, amplitude of
See *amplitude of alternating voltage*.
1985, *57*, 1505

altocumulus cloud (in atmospheric chemistry)
A dappled layer of patch or cloud composed of flattened globules that may be arranged in groups, lines

or waves collectively known as billows; 2 000–6 000 m; vertical velocities of 0.05–0.1 m s^{-1}.
  1990, *62*, 2180

altostratus cloud (in atmospheric chemistry)
A grey, uniform, striated or fibrous sheet but without halo phenomena, and through which the sun is seen only as a diffuse, bright patch or not at all; usually at elevations 2 000–6 000 m; vertical velocities of 0.05–0.1 m s^{-1}.
  1990, *62*, 2180

ambident
A description applied to a *chemical species* whose *molecular entities* each possess two alternative and strongly interacting distinguishable reactive centres, to either of which a *bond* may be made in a reaction: the centres must be connected in such a way that reaction at either site stops or greatly retards subsequent attack at the second site. The term is most commonly applied to *conjugated nucleophiles*, for example the enolate ion:

$$\diagdown_{\diagup}C=C\diagup^{\diagdown}_{O^-} \longleftrightarrow \diagdown_{\diagup}\underset{\beta}{C}-C\diagup^{\diagdown}_{O}$$

(which may react with *electrophiles* either at the β-carbon atom or at oxygen) or γ-pyridones, and also to the vicinally ambident cyanide ion, cyanate ion, thiocyanate ion, sulfinate ion, nitrite ion and unsymmetrical hydrazines. Ambident electrophiles are exemplified by carboxylic esters RC(O)OCR$_3$ which react with nucleophiles either at the carbonyl carbon or the alkoxy carbon.
Molecular entities, such as dianions of dicarboxylic acids, containing two non-interacting (or feebly interacting) reactive centres, are not generally considered to be ambident and are better described as 'bifunctional'.
The Latin root of the word implies two reactive centres, but the term has in the past also incorrectly been applied to chemical species with more than two reactive centres. For such species the existing term 'polydent' (or, better, 'multident') is more appropriate.
  See also *chelation*.
  1994, *66*, 1082

ambient air (in atmospheric chemistry)
The outdoor air in the particular location.
  1990, *62*, 2173

ambient air quality (in atmospheric chemistry)
A general term used to describe the quality of the outside air. Usually adjectives such as good, fair, bad, etc. are used by the media to describe this; often some form of air pollution or air quality index is employed to determine the specific descriptive term to be used. These are very qualitative terms of little or no scientific value.
  See also *air pollution index*.
  1990, *62*, 2173

ambo
A prefix used to indicate that a molecule with two (or more) chiral elements is present as a mixture of the two racemic diastereoisomers in unspecified proportions. For example, the dipeptide formed from L-alanine and DL-leucine is L-alanyl-*ambo*-leucine.
  1996, *68*, 2198; see also 1984, *56*, 595; 1982, *54*, 1507

Ames/salmonella test
A screening test employed in predicting the mutagenic and the potential carcinogenic activities of chemicals in the environment. It employs Ames test strains of salmonella bacteria (his$^-$) which lack the ability to produce histidine. The compound to be tested, the bacteria and a small amount of histidine (insufficient to permit colony growth but enough to allow sufficient growth for expression of mutations) are added to agar. The bacteria are allowed to incubate for about 63 h at 37 °C. If a significant increase in colonies above background is observed in the sample containing the test compound, then it is concluded that the chemical tested is a direct mutagen for the particular Ames strain of bacteria. This is taken as a qualitative indication of the possible carcinogenic activity of this chemical in other biological systems. This procedure seems to be of qualitative value in a preliminary screening of potential carcinogens.
  1990, *62*, 2173

amic acids
Carbamoyl carboxylic acids, i.e. compounds containing a carboxy and a carboxamide group, e.g. 5-carbamoylnicotinic acid.
Note:
In systematic nomenclature replacement of the '-ic' suffix of a dicarboxylic acid by '-amic' is limited to dicarboxylic acids that have a trivial name. e.g. HOC(=O)CH$_2$C(=O)NH$_2$ malonamic acid (2-carbamoylacetic acid).
  1995, *67*, 1315

amide hydrazones
  See *amidrazones*.
  1995, *67*, 1315

amide oximes
Compounds having the structure RC(NH$_2$)=NOH and derivatives formed by substitution; formally the *oximes* of *carboxamides*.
  1995, *67*, 1315

amides
1. Derivatives of *oxoacids* R$_k$E(=O)$_l$(OH)$_m$ ($l \neq 0$) in which an acidic hydroxy group has been replaced by an amino or substituted amino group. Chalcogen replacement analogues are called thio-, seleno- and telluro-amides. Compounds having one, two or three acyl groups on a given nitrogen are generically included and may be designated as primary, secondary and tertiary amides, respectively, e.g. PhC(=O)NH$_2$ benzamide, CH$_3$S(=O)$_2$NMe$_2$ *N,N*-dimethylmethanesulfonamide, [RC(=O)]$_2$NH secondary am-

ides (see *imides*), [RC(=O)]$_3$N tertiary amides, PhP(=O)(OH)NH$_2$ phenylphosphonamidic acid.
Notes:
i. Amides with NH$_2$, NHR and NR$_2$ groups should not be distinguished by means of the terms primary, secondary and tertiary.
ii. Derivatives of certain acidic compounds R$_n$E(OH)$_m$, where E is not carbon (e.g. *sulfenic acids*, RSOH, *phosphinous acids*, R$_2$POH) having the structure R$_n$E(NR$_2$)$_m$ may be named as amides but do not belong to the class amides proper, e.g. CH$_3$CH$_2$SNH$_2$ ethanesulfenamide or ethylsulfanylamine.
2. The term applies also to metal derivatives of ammonia and amines, in which a cation replaces a hydrogen atom on nitrogen. Such compounds are also called azanides, e.g. LiN(Pr^i)$_2$ lithium diisopropylamide, synonym lithium diisopropylazanide.
See also *carboxamides, lactams, peptides, phosphoramides, sulfonamides*.
1995, 67, 1315; see also 1993, 65, 1357

amidines
Derivatives of oxoacids R$_n$E(=O)OH in which the hydroxy group is replaced by an amino group and the oxo group is replaced by =NR. Amidines include carboxamidines, sulfinamidines and phosphinamidines, R$_2$P(=NR)NR$_2$. In organic chemistry an unspecified amidine is commonly a carboxamidine.
See also *carboxamidines, sulfinamidines*.
1995, 67, 1315

amidium ions
Cations formally derived by the addition of one hydron to the N or O atom of an *amide* and N-*hydrocarbyl* derivatives thereof. In organic chemistry an unspecified amidium ion is commonly a carboxamidium ion RC(OH)=N$^+$R$_2$ ↔ RC(=O$^+$H)NR$_2$ or RC(=O)N$^+$R$_3$. The term does not imply knowledge concerning the position of the cationic centre, e.g. PhC(=O)N$^+$Me$_3$ *N,N,N*-trimethylbenzamidium.
1995, 67, 1315; see also 1993, 65, 1357

amidrazones
Compounds having the structure RC(=NH)NHNH$_2$ or RC(NH$_2$)=NNH$_2$, formally derived from *carboxylic acids*. These tautomers are named hydrazide imides and amide hydrazones, respectively. Also included are *N*-hydrocarbyl derivatives.
1995, 67, 1316

aminals
Compounds having two amino groups bonded to the same carbon, R$_2$C(NR$_2$)$_2$. Also called geminal diamines. [The term aminal has also been used, with consequent ambiguity, for α-amino ethers (hemiaminal ethers); such use is discouraged.]
See also *hemiaminals*.
1995, 67, 1316

amine imides
Compounds formally derived from the attachment of an amine R$_3$N to a nitrene RN:. The structure R$_3$N$^+$–N$^-$–R expresses the 1,2-dipolar character of amine imides. They may be named systematically as substituted diazan-2-ium-1-ides, e.g. Me$_3$N$^+$–N$^-$Me, 1,2,2,2-tetramethyldiazan-2-ium-1-ide or trimethylamine *N*-methylimide.
See also *ylides*.
1995, 67, 1316; see also 1993, 65, 1357

+ amine imines
An undesirable synonym for *amine imides*.
See *imides* (2), *ylides*.
1995, 67, 1316

amine oxides
Compounds derived from tertiary amines by the attachment of one oxygen atom to the nitrogen atom R$_3$N$^+$–O$^-$. By extension the term includes the analogous derivatives of primary and secondary amines.
1995, 67, 1316

amines
Compounds formally derived from ammonia by replacing one, two or three hydrogen atoms by hydrocarbyl groups, and having the general structures RNH$_2$ (primary amines), R$_2$NH (secondary amines), R$_3$N (tertiary amines).
1995, 67, 1316

amine ylides
Synonymous with *ammonium ylides*.
1995, 67, 1316

+ aminimides
An undesirable synonym for *amine imides*.
See *ylides*.
1995, 67, 1316

aminium ions
Cations HR$_3$N$^+$ formed by hydronation of an amine R$_3$N. 'Non-quaternary ammonium ions' is a synonymous term, e.g. prolinium, PhN$^+$HMe$_2$, *N,N*-dimethylanilinium.
Note:
If a class X can be hydronated to Xium ions the class Xium ions commonly includes the derivatives formed by the replacement of the added hydron with a hydrocarbyl group. Aminium ions form an exception, made possible by the availability of the class name ammonium ions.
See also *onium compounds*.
1995, 67, 1316

aminiumyl radical ions
Radicals cations, R$_3$N$^{\cdot+}$, derivable from *aminium ions*, R$_3$NH$^+$, by removal of a hydrogen atom. Aminiumyl radical ions are, except for H$_3$N$^{\cdot+}$, synonymous with the ammoniumyl radical ions. As the term ammonium is well known, ammoniumyl radical ions is the more desirable class name.
See *ammoniumyl radical ions*.
1995, 67, 1316

amino-acid residue (in a polypeptide)
When two or more amino acids combine to form a *peptide*, the elements of water are removed, and what remains of each amino acid is called an amino-acid residue. α-Amino-acid residues are therefore structures that lack a hydrogen atom of the amino group

(–NH–CHR–COOH), or the hydroxyl moiety of the carboxyl group (NH$_2$–CHR–CO–), or both (–NH–CHR–CO–); all units of a peptide chain are therefore amino-acid residues. (Residues of amino acids that contain two amino groups or two carboxyl groups may be joined by *isopeptide bonds*, and so may not have the formulas shown.)

The residue in a peptide that has an amino group that is free, or at least not acylated by another amino-acid residue (it may, for example, be acylated or formylated), is called N-terminal; it is at the N-terminus. The residue that has a free carboxyl group, or at least does not acylate another amino-acid residue, (it may, for example, acylate ammonia to give –NH–CHR–CO–NH$_2$), is called C-terminal.

W.B. 48

+ aminonitrenes
An incorrect name for *isodiazenes*.
See *carbene analogues*.
1995, *67*, 1317

+ aminooxyl radicals
See *aminoxyl radicals*.
1995, *67*, 1317; see also 1993, *65*, 1357

+ amino radicals
A non-IUPAC term for *aminyl radicals*.
1995, *67*, 1316

amino sugars
Monosaccharides having one alcoholic hydroxy group (commonly but not necessarily in position 2) replaced by an amino group; systematically known as *x*-amino-*x*-deoxymonosaccharides. (Glycosylamines are excluded.)

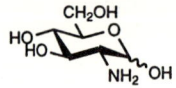

D-glucosamine or 2-amino-2-deoxy-D-glucopyranose

1995, *67*, 1317

aminoxides
The anion H$_2$N–O$^-$, aminoxide and its N-*hydrocarbyl* derivatives R$_2$N–O$^-$; formally derived from hydroxylamines, R$_2$NOH, by removing a hydron from the hydroxy group, e.g. (CH$_3$)$_2$N–O$^-$ dimethylaminoxide.
1995, *67*, 1317; see also 1993, *65*, 1357

aminoxyl radicals
Compounds having the structure R$_2$NO$\cdot$ ↔ R$_2$N$^{\cdot +}$–O$^-$; they are *radicals* derived from hydroxylamines by removal of the hydrogen atom from the hydroxy group, and are in many cases isolable. The synonymous terms 'nitroxyl radicals' and 'nitroxides' erroneously suggest the presence of a nitro group; their use is not desirable. E.g. (ClCH$_2$)N–O$\cdot$ bis(chloromethyl)aminoxyl.
1995, *67*, 1317; see also 1993, *65*, 1357

aminylenes
See *nitrenes*.
1995, *67*, 1317

aminylium ions
See *nitrenium ions*.
1995, *67*, 1317

+ aminyl oxides
Obsolete term for aminoxyl radicals.
1995, *67*, 1317

aminyl radicals
The nitrogen-centered radical H$_2$N$\cdot$, formally derived by the removal of a hydrogen atom from ammonia, and its *hydrocarbyl* derivatives R$_2$N$\cdot$.
1995, *67*, 1317

ammonium compounds
See *onium compounds*.
1995, *67*, 1317

+ ammonium imines
An undesirable synonym for *amine imides*.
See *ylides*.
1995, *67*, 1317

ammonium ylides
1,2-Dipolar compounds of general structure R$_3$N$^+$–C$^-$R$_2$.
See also *ylides*.
1995, *67*, 1317

ammoniumyl radical ions
H$_3$N$^{\cdot +}$ and its *hydrocarbyl* derivatives, e.g. (CH$_3$)$_3$N$^{\cdot +}$ trimethylammoniumyl, PhN$^{\cdot +}$H$_2$ phenylammoniumyl or benzenaminiumyl.
1995, *67*, 1317; see also 1993, *65*, 1357

amorphous carbon
A *carbon material* without long-range crystalline order. Short-range order exists, but with deviations of the interatomic distances and/or interbonding angles with respect to the graphite lattice as well as to the diamond lattice.
Notes:
The term amorphous carbon is restricted to the description of carbon materials with localized π-electrons as described by P.W.Anderson (*Phys. Rev.*, 1958, *109*, 1492). Deviations in the C–C distances greater than 5% (i.e. $\Delta x/x_0 > 0.05$, where x_0 is the inter-atomic distance in the crystal lattice for the sp^2 as well as for the sp^3 configuration) occur in such materials, as well as deviations in the bond angles because of the presence of 'dangling bonds'. The above description of amorphous carbon is not applicable to carbon materials with two-dimensional structural elements present in all pyrolysis residues of carbon compounds as polyaromatic layers with a nearly ideal interatomic distance of a = 142 pm and an extension greater than 1000 pm.
See also *diamond-like carbon films*.
1995, *67*, 477

amount concentration, *c*
Amount of a constituent divided by the volume of the mixture. Also called amount-of-substance concentration, substance concentration (in clinical chemistry) and in older literature molarity. For entities B it is often denoted by [B]. The common unit is mole per

cubic decimetre (mol dm^{-3}) or mole per litre (mol L^{-1}) sometimes denoted by M.
See also *concentration*.
G.B. 42

amount fraction, x (y for gaseous mixtures)
Amount of a constituent divided by the total amount of all constituents in the mixture. It is also called mole fraction. Amount fraction is equal to the number fraction: the number of entities of one constituent divided by the total number of *entities* in the mixture.
See also *fraction*.
G.B. 41

amount of substance, n
Base quantity in the system of quantities upon which SI is based. It is the number of elementary entities divided by the Avogadro constant. Since it is proportional to the number of entities, the proportionality constant being the reciprocal Avogadro constant and the same for all substances, it has to be treated almost identically with the number of entities. Thus the counted elementary entities must always be specified. The words 'of substance' may be replaced by the specification of the entity, for example: amount of chlorine atoms, n_{Cl}, amount of chlorine molecules, $n(Cl_2)$. No specification of the entity might lead to ambiguities [amount of sulfur could stand for $n(S)$, $n(S_8)$, etc.], but in many cases the implied entity is assumed to be known: for molecular compounds it is usually the molecule [e.g. amount of benzene usually means $n(C_6H_6)$], for ionic compounds the simplest formula unit [e.g. amount of sodium chloride usually means $n(NaCl)$] and for metals the atom [e.g. amount of silver usually stands for $n(Ag)$]. In some derived quantities the words 'of substance' are also omitted, e.g. *amount concentration, amount fraction*. Thus in many cases the name of the base quantity is shortened to amount and to avoid possible confusion with the general meaning of the word the attribute chemical is added. The chemical amount is hence the alternative name for amount of substance. In the field of clinical chemistry the words 'of substance' should not be omitted and abbreviations such as substance concentration (for amount of substance concentration) and substance fraction are in use. The quantity had no name prior to 1969 and was simply referred to as the number of moles.
G.B. 46; 1996, *68*, 961

amount-of-substance concentration
Synonymous with *amount concentration*.
G.B. 42

amount-of-substance fraction, x
Synonymous with *amount fraction*.
G.B. 41

ampere
SI base unit for the electric current (symbol: A). The ampere is that constant current which, if maintained in two straight parallel conductors of infinite length, of negligible circular cross-section, and placed 1 metre apart in vacuum, would produce between these conductors a force equal to 2×10^{-7} newton per metre of length (9th CGPM, 1948).
G.B. 70; 1996, *68*, 961

amperometric detection method (in electrochemical analysis)
A detection method in which the current is proportional to the concentration of the species generating the current.
1990, *62*, 2186

amphipathic
The property of surface activity is usually due to the fact that the molecules of the substance are amphipathic or amphiphilic, meaning that each contains both a hydrophilic and a hydrophobic (lipophilic) group. This assumes that one of the two phases is aqueous, and the other non-aqueous. If both are non-aqueous (e.g. oil/air), molecules containing organophilic and organophobic groups may be amphipathic and surface active.
1972, *31*, 612

amphiphilic
A term used to describe a compound containing a large organic cation or anion which possesses a long unbranched hydrocarbon chain, e.g.

$H_3C(CH_2)_nCO_2^-M^+$ $H_3C(CH_2)_nN(CH_3)_3^+X^-$
$H_3C(CH_2)_nSO_3^-M^+$ ($n > 7$).

The existence of distinct polar (hydrophilic) and non polar (hydrophobic) regions in the molecule promotes the formation of *micelles* in dilute aqueous solution.
See also *amphipathic*.
1994, *66*, 1083

amphiprotic (solvent)
Self-ionizing solvent possessing both characteristics of *Brønsted acids* and *bases*, for example H_2O and CH_3OH, in contrast to *aprotic solvent*.
1994, *66*, 1083: O.B. 30

ampholytes
See *zwitterionic compounds*.
1995, *67*, 1317

amphoteric
A *chemical species* that behaves both as an acid and as a base is called amphoteric. This property depends upon the medium in which the species is investigated: H_2SO_4 is an acid when studied in water, but becomes amphoteric in *superacids*.
1994, *66*, 1083

amplification reaction
A reaction which replaces the conventional reaction used in a particular determination so that a more favourable measurement can be made. The sequence can be repeated to provide a further favourable increase in measurement.
1982, *54*, 2554

amplifier, biased linear pulse
See *biased linear pulse amplifier*.
1982, *54*, 1535

amplifier, linear pulse
See *linear pulse amplifier*.
1982, *54*, 1535

amplitude of alternating current
Half the peak-to-peak amplitude of a sinusoidal *alternating current*.
O.B. 53; 1985, *57*, 1494

amplitude of alternating voltage
This term should denote half of the peak-to-peak amplitude. Peak-to-peak and r.m.s. amplitudes should be so specified.
1985, *57*, 1505

AM (0) sunlight
The solar *irradiance* in space just above the earth's atmosphere (air mass, AM, zero). Also called extraterrestrial global irradiance.
See also *AM(1) sunlight*.
1996, *68*, 2228

AM (1) sunlight
The solar *irradiance* traversing the atmosphere when the sun is in a position perpendicular to the earth's surface. Also called terrestrial global irradiance.
See also *AM (0) sunlight*.
1996, *68*, 2228

anabolism
The processes of *metabolism* that result in the synthesis of cellular components from precursors of low molecular weight.
See also *catabolism*.
1992, *64*, 146; 1993, *65*, 2013

anaerobe
An organism that does not need free-form oxygen for growth. Many anaerobes are even sensitive to free oxygen. Obligate (strict) anaerobes grow only in the absence of oxygen. Facultative anaerobes can grow either in the presence or in the absence of molecular oxygen.
1992, *64*, 146

anaerobic
Not requiring molecular oxygen.
See also *aerobic*.
1993, *65*, 2013

anaesthetic
Substance which produces loss of feeling or sensation: general anaesthetic produces loss of consciousness; local or regional anaesthetic renders a specific area insensible to pain.
1993, *65*, 2013

analgesic
Substance which relieves pain, without causing loss of consciousness.
1993, *65*, 2013

analogue metabolism
Process by which a normally non-biodegradable compound is biodegraded in the presence of a structurally similar compound which can induce the necessary enzymes.
1993, *65*, 2013

analogue to digital converter (pulse)
A *pulse amplitude analyser* which, for each pulse processed, produces an integer proportional to the height of that pulse.
1982, *54*, 1535

analyser, dead time
See *dead time (of an analyser)*.
1990, *62*, 2174

analyser, discontinuous
See *discontinuous analyser*.
1990, *62*, 2174

analyser, discontinuous indication
See *discontinuous indication (of an analyser)*.
1990, *62*, 2174

analyser, fall time of
See *response time*.
1990, *62*, 2174; 1995, *67*, 1751

analyser, measurement threshold
See *measurement threshold (of an analyser)*.
1990, *62*, 2174

analyser, range of measurement
See *range of measurement (of an analyser)*.
1990, *62*, 2174

analyser, response time
See *response time (of an analyser)*.
1990, *62*, 2174

analyser, rise time of
See *rise time (of an analyser)*.
1990, *62*, 2174

analyser, sequential
See *sequential analyser*.
1990, *62*, 2174

analyser, sequential indication
See *sequential indication (of an analyser)*.
1990, *62*, 2174

analyte
The component of a system to be analysed.
1989, *61*, 1660

analytical function
A function which relates the measured value $\hat{C}_a$ to the instrument reading, X, with the value of all interferants, C_i, remaining constant. This function is expressed by the following regression of the calibration results.

$$\hat{C}_a = f(X)$$

The analytical function is taken as equal to the inverse of the calibration function.
1990, *62*, 2174

analytical instrument
A device or a combination of devices used to carry out an analytical process. The analytical process is all or part of the analytical procedure that encompasses all steps from the introduction of the sample or the test portion to the production of the result. An analytical instrument may carry out single or multiple analytical procedures. In the latter case it may be selective, i.e. designed to carry out any requested

combination of procedures within the set, on each specimen.
1989, *61*, 1659

analytical intercomparison
A procedure which gives insight to the *accuracy* of results of analytical procedures by comparing the results obtained in the analyses of identical samples at different laboratories and preferably with different analytical methods.
1982, *54*, 1544

analytical portion
See *test portion*.
1990, *62*, 1206

analytical pyrolysis
The characterization, in an inert atmosphere, of a material or a chemical process by a chemical degradation reaction(s) induced by thermal energy.
1993, *65*, 2406

analytical quality control
Procedures which give insight into the *precision* and *accuracy* of analysis results.
1982, *54*, 1535

analytical radiochemistry
That part of analytical chemistry in which the application of *radioactivity* is an essential step in the analytical procedures.
Synonymous with radioanalytical chemistry.
1982, *54*, 1535

analytical sample
See *sample (in analytical chemistry), test sample*.
1990, *62*, 1200, 1206

analytical solution
See *test solution*.
1990, *62*, 1206

analytical solution calorimetry
See *enthalpimetric analysis*.
1994, *66*, 2489

analytical unit (analyser)
An assembly of subunits comprising: suitable apparatus permitting the introduction and removal of the gas, liquid or solid to be analysed and/or calibration materials; a measuring cell or other apparatus which, from the physical or chemical properties of the components of the material to be analysed, gives signals allowing their identification and/or measurement; signal processing devices (amplification, recording) or, if need be, data processing devices.
1990, *62*, 2175

anchimeric assistance
See *neighbouring group participation*.
1994, *66*, 1083

aneroid barometer
An instrument for monitoring the atmospheric pressure in which no liquid is employed, but rather changes in pressure between the atmosphere and a closed vessel bend a diaphram which moves a pointer on a scale.
1990, *62*, 2175

Anger camera
A *camera* in which a large diameter scintillator is coupled to an array of photomultiplier tubes by *fibre optics*. X-ray imaging may also be achieved in multicrystal cameras where many small crystals individually scintillate.
1995, *67*, 1756

angle
The angle between two half-lines terminating at the same point is the ratio of the length of the included arc of the circle (with its centre at that point) to the radius of the circle. Symbols: α, β, γ ... This is a quantity of dimension one with the SI unit *radian* (rad), but very often *degrees of arc* are used.
G.B. 11; 1996, *68*, 961

angle of observation
See *scattering angle*.
P.B. 65

angle of optical rotation, α
Angle through which plane polarized light is rotated clockwise, as seen when facing the light source, in passing through an optically active medium.
G.B. 33

angle strain
Strain due to a departure in bond angle from 'normal' values. The term is often used in the context of non-aromatic cyclic compounds in which the internal angles differ from the regular tetrahedral angle of 109° 28′; in this sense angle strain is also known as *Baeyer strain*.
1996, *68*, 2198

ångström
Non-SI unit of length, $Å = 10^{-10}$ m, widely used in molecular physics.
G.B. 75; 1996, *68*, 961

angular distribution
With reference to the center of mass, the *products* of a bimolecular reaction are scattered with respect to the initial velocity vector, and the distribution of scattering angles is known as the angular distribution.
1996, *68*, 153

angular frequency, ω
Frequency multiplied by 2π. Also called *pulsatance*. It was formerly called circular frequency.
G.B. 11

angular momentum, L
Angular momentum, or moment of *momentum* of a particle about a point is a vector quantity equal to the vector product of the position vector of the particle and its momentum, $L = r \times p$. For special angular momenta of particles in atomic and molecular physics different symbols are used.
G.B. 12

anhydrides
See *acid anhydrides*.
1995, *67*, 1317

anhydro bases
Compounds resulting from internal acid–base neutralization (with loss of water) in iminium hydroxides containing an acidic site conjugated with the iminium function.

See *pseudo bases*.
1995, *67*, 1317

anilides
1. Compounds derived from *oxoacids* $R_kE(=O)_l(OH)_m$ ($l \neq 0$) by replacing an OH group by the NHPh group or derivative formed by ring substitution; *N*-phenyl amides, e.g. $CH_3C(=O)NHPh$ acetanilide.
See *amides*.
2. Salts formed by replacement of a nitrogen-bound hydron of aniline by a metal or other cation, e.g. NaNHPh sodium anilide.
1995, *67*, 1317

anils
A term for the subclass of *Schiff bases* $R_2C=NR'$, where $R' = N$-phenyl or substituted phenyl group. Thus, *N*-phenyl imines.
1995, *67*, 1317

anion
A monoatomic or polyatomic species having one or more elementary charges of the electron.
R.B. 107

anion exchange
The process of exchanging anions between a solution and an *anion exchanger*.
1993, *65*, 854

anion exchanger
Ion exchanger with anions as counter-ions. The term anion-exchange resin may be used in the case of solid organic polymers. The base form of an anion exchanger is the ionic form in which the counter-ions are hydroxide groups (OH-form) or the ionogenic groups form an uncharged base, e.g. –NH.
1993, *65*, 856

anionic polymerization
An *ionic polymerization* in which the kinetic-chain carriers are anions.
1996, *68*, 2308

anionotropic rearrangement (or anionotropy)
A rearrangement in which the migrating group moves with its electron pair from one atom to another.
1994, *66*, 1083

anion radical
See *radical ion*.
1994, *66*, 1083

anisometric
See *isometric*.
1996, *68*, 2198

annelation
Alternative, but less desirable term for *annulation*. The term is widely used in the German and French languages.
1994, *66*, 1083

annihilation
1. (in radiochemistry) An interaction between a particle and its antiparticle in which they both disappear.
O.B. 211
2. (in photochemistry) Two atoms or molecular entities both in an excited electronic state interact often (usually upon collision) to produce one atom or molecular entity in an excited electronic state and another in its ground electronic state. This phenomenon (annihilation) is sometimes referred to as energy pooling.
See *singlet-singlet annihilation, spin-conservation rule, triplet-triplet annihilation*.
1996, *68*, 2228

annulation
A *transformation* involving fusion of a new ring to a molecule via two new bonds.
Some authors use the term 'annelation' for the fusion of an additional ring to an already existing one, and 'annulation' for the formation of a ring from one or several acyclic precursors, but this distinction is not made generally.
See also *cyclization*.
1994, *66*, 1083

annulenes
Mancude monocyclic *hydrocarbons* without side chains of the general formula C_nH_n (n is an even number) or C_nH_{n+1} (n is an odd number). In systematic nomenclature an annulene with seven or more carbon atoms may be named [*n*]annulene, where *n* is the number of carbon atoms, e.g. [9]annulene for cyclonona-1,3,5,7-tetraene.
1995, *67*, 1318; 1994, *66*, 1083

annulenylidenes
Carbenes, derived by formal insertion of a divalent carbon atom into an even-membered annulene, e.g. cycloheptatrienylidene:

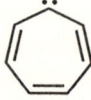

1995, *67*, 1318

anode
Electrode at which oxidation takes place.
G.B. 59

anodic transfer coefficient
Defined by analogy with *cathodic transfer coefficient*.
1974, *37*, 515

anomeric effect

Originally the thermodynamic preference for polar groups bonded to C-1 (the anomeric carbon of a glycopyranosyl derivative) to take up an *axial* position.

This effect is now considered to be a special case of a general preference (the generalized anomeric effect) for *synclinal* (gauche) conformations about the bond C–Y in the system X–C–Y–C where X and Y are heteroatoms having nonbonding electron pairs, commonly at least one of which is nitrogen, oxygen or fluorine. For example in chloro(methoxy)methane the anomeric effect stabilizes the synclinal conformation.

In alkyl glycopyranosides the anomeric effect operates at two sites (i) along the endocyclic C-1 oxygen bond (endo-anomeric effect) and (ii) along the exocyclic C-1 oxygen bond (exo-anomeric effect). The opposite preference is claimed for some systems e.g. glycopyranosyltrialkylammonium salts, and has been referred to as the reverse anomeric effect.

1996, *68*, 2198

anomers

Diastereoisomers of glycosides, hemiacetals or related cyclic forms of sugars, or related molecules differing in *configuration* only at C-1 of an aldose, C-2 of a 2-ketose, etc.

1996, *68*, 2199

ansa compounds

Benzene derivatives having *para* positions (or *meta*) bridged by a chain (commonly 10 to 12 atoms long) (Latin *ansa*, handle). By extension, any arene bridged by a chain constrained to lie over one of the two faces of the arene.

See *cyclophanes*.
1995, *67*, 1318

antagonism

Combined effect of two or more factors which is smaller than the solitary effect of any one of those factors. In bioassays, the term may be used when a specified response is produced by exposure to either of two factors but not by exposure to both together.

See also *synergism*.
1993, *65*, 2014

antagonist

1. Substance that reverses or reduces the effect induced by an agonist.
2. Substance that attaches to and blocks cell receptors that normally bind naturally occurring substances.

See also *agonist*.
1993, *65*, 2014

antarafacial

When a part of a molecule ('molecular fragment') undergoes two changes in bonding (*bond*-making or bond-breaking), either to a common centre or to two related centres, external to itself, these bonding changes may be related in one of two spatially different ways. These are designated as 'antarafacial' if opposite faces of the molecular fragment are involved, and 'suprafacial' if both changes occur at the same face. The concept of 'face' is clear from the diagrams in the cases of planar (or approximately planar) frameworks with isolated or interacting p-orbitals [diagrams (a) and (b)].

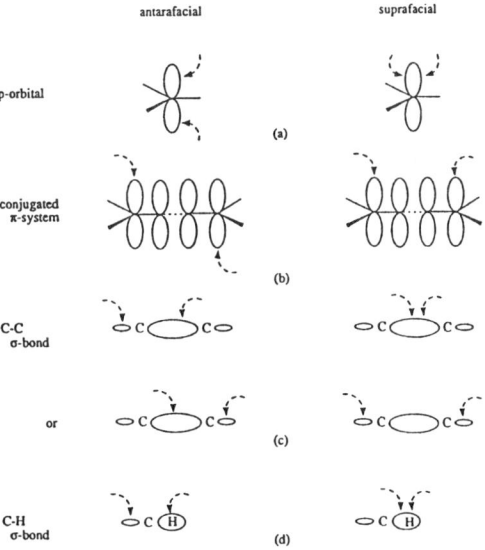

The terms antarafacial and suprafacial are, however, also employed in cases in which the essential part of the molecular fragment undergoing changes in bonding comprises two atoms linked only by a σ-bond. In these cases it is customary to refer to the phases of the local σ-bonding orbital: occurrence of the two bonding changes at sites of like orbital phase is regarded as suprafacial, whereas that at two sites of opposite phase is antarafacial. The possibilities are shown for

C–C and C–H σ-bonds in diagrams (c) and (d). There may be two distinct and alternative stereochemical outcomes of a suprafacial process involving a σ-bond between saturated carbon atoms, i.e. either retention or inversion at both centres. The antarafacial process results in inversion at one centre and retention at the second.

For examples of the use of these terms, see *cycloaddition, sigmatropic rearrangement*.

See also *anti, sigma, pi*.
1994, *66*, 1084

anthelmint(h)ic
Substance intended to kill parasitic intestinal worms, such as helminths.
Synonymous with antihelminth.
1993, *65*, 2014

anthocyanidins
Aglycons of *anthocyanins*; they are oxygenated derivatives of flavylium (2-phenylchromenylium) salts.

See *flavonoids*.
1995, *67*, 1318

anthocyanins
Plant pigments of the *flavonoid* class; they are *glycosides* that on hydrolysis yield coloured *aglycons* called *anthocyanidins*.
1995, *67*, 1318

anti
In the representation of *stereochemical* relationships 'anti' means 'on opposite sides' of a reference plane, in contrast to 'syn' which means 'on the same side', as in the following examples.
(A) Two substituents attached to atoms joined by a single *bond* are anti if the *torsion angle* (dihedral angle) between the bonds to the substituents is greater than 90°, or syn if it is less than 90°. (A further distinction is made between antiperiplanar, synperiplanar, anticlinal and synclinal.)
(B) In the older literature the terms anti and syn were used to designate stereoisomers of oximes and related compounds. That usage was superseded by the terms 'trans' and 'cis' or *E* and *Z*, respectively.
(C) When the terms are used in the context of *chemical reactions* or *transformations*, they designate the relative orientation of substituents in the substrate or product:
(1) Addition to a carbon-carbon double bond:

(2) Alkene-forming elimination:

In the examples described under (1) and (2) anti processes are always *antarafacial*, and syn processes are *suprafacial*.

See also *endo, exo, syn, anti*.
1994, *66*, 1084; 1996, *68*, 2199

antiaromatic compounds
Compounds that contain $4n$ ($n \neq 0$) π-electrons in a cyclic planar, or nearly planar, system of alternating single and double bonds, e.g. cyclobuta-1,3-diene.

See also *aromatic compounds*.
1995, *67*, 1318

antibiotic
Substance produced by, and obtained from, certain living cells (especially bacteria, yeasts and moulds), or an equivalent synthetic substance, which is biostatic or biocidal at low concentrations to some other form of life, especially pathogenic or noxious organisms.
1993, *65*, 2014

antibody
A protein (*immunoglobulin*) produced by the immune system of an organism in response to exposure to a foreign molecule (*antigen*) and characterized by its specific binding to a site of that molecule (antigenic determinant or *epitope*).
1992, *64*, 146; 1993, *65*, 2014

anticholinergic
1. (adjective). Preventing transmission of parasympathetic nerve impulses.
2. (noun). Substance which prevents transmission of parasympathetic nerve impulses.
1993, *65*, 2014

anticircular elution (anticircular development) (in planar chromatography)
The opposite of *circular development*. Here the sample as well as the mobile phase is applied at the periphery of a circle and both move towards the centre.
1993, *65*, 830

anticlinal
See *torsion angle*.
1996, *68*, 2199

anticlined structures (in polymers)
See *isomorphous structures (in polymers)*.
P.B. 43

anticodon
A sequence of three *nucleotides* in the anticodon-loop of a *tRNA*, which recognizes and binds the complementary triplet sequence (*codon*) of the *mRNA*.
1992, *64*, 146

anti-Compton γ-ray spectrometer
A *gamma-ray spectrometer* in which the effect of the *Compton scattering* is at least partly compensated.
1982, *54*, 1543

anticyclone (in atmospheric chemistry)
A large system of winds that rotate about a centre of high atmospheric pressure, clockwise (viewed from above) in the northern hemisphere and counterclockwise in the southern hemisphere.
1990, *62*, 2175

antiferroelectric transition
See *ferroelectric (antiferroelectric) transition*.
1994, *66*, 582

antiferromagnetic transition
See *ferroic transition, magnetic transition*.
1994, *66*, 579

antigen
A substance that stimulates the immune system to produce a set of specific *antibodies* and that combines with the antibody through a specific binding site or *epitope*.
1992, *64*, 146

anti-Hammond effect
See *More O'Ferrall–Jencks diagram*.
1994, *66*, 1085

antihelminth
See *anthelmint(h)ic*.
1993, *65*, 2014

anti-Markownikoff addition
See *Markownikoff rule*.
1994, *66*, 1085

antimetabolite
Substance, structurally similar to a metabolite, which competes with it or replaces it, and so prevents or reduces its normal utilization.
1993, *65*, 2015

antimony–xenon lamp (arc)
An intense source of ultraviolet, visible, and near infra-red radiation produced by an electrical discharge in a mixture of antimony vapour and xenon under high pressure. Its output in the ultraviolet region is higher than that of the *mercury–xenon arc*.
See *lamp*.
1996, *68*, 2228

antimycotic
Substance used to kill a fungus or to inhibit its growth. Synonymous with fungicide.
1993, *65*, 2015

antiparticle
Of a given particle, a particle with the same rest mass, where the two can *annihilate*.
1982, *54*, 1535

antiperiplanar
See *torsion angle*.
1996, *68*, 2199

+ antipodes
Obsolete synonym for *enantiomers*. (Usage strongly discouraged).
1996, *68*, 2199

antiprismo-
An affix used in names to denote eight atoms bound into a rectangular antiprism.
R.B. 245; B.B. 464

antiresistant
Substance used as an additive to a pesticide formulation in order to reduce the resistance of insects to the pesticide.
1993, *65*, 2015

antiserum
Serum containing antibodies to a particular antigen either because of immunization or after an infectious disease.
1993, *65*, 2015

anti-Stokes shift
See *Stokes shift*.
1996, *68*, 2228

anti-Stokes type radiation (fluorescence)
Fluorescence radiation occurring at shorter wavelengths than *absorption*.
1984, *56*, 236

anti-thixotropy
Opposite of *thixotropy*.
See also *work hardening*.
1979, *51*, 1217

ap
See *torsion angle*.
1996, *68*, 2199

apex
See *apex current*.
1985, *57*, 1493

apex current
In measurement of non-faradaic admittance (or tensammetry), a plot of *alternating current* against *applied potential* shows a minimum or maximum when a non-electroactive substance undergoes *adsorption* or desorption at the surface of the *indicator electrode*. Such a maximum or minimum may be called an apex to emphasize its non-faradaic origin and distinguish it from a 'summit', which would result from a charge-transfer process. The highest value of the current on such an apex may be called an apex current, and the corresponding applied potential may be called an apex potential.
1985, *57*, 1494

apex potential
See *apex current*.
1985, *57*, 1494

aphicide
Substance intended to kill aphids.
1993, *65*, 2015

apical (basal, equatorial)
In trigonal bipyramidal structures (e.g. a five-coordinate trigonal bipyramid with phosphorus as central atom) the term apical refers to the two positions that are collinear with the central atom or to the bonds linking these positions to the central atom. The three equivalent bonds (or positions) in a plane passing through the central atom and perpendicular to the direction of the apical bonds are described as equatorial. (See *axial, equatorial* for alternative use). The term apical is also used for the bond pointing from the atom at or near the centre of the base to the apex of a pyramidal structure. The positions at or near the base of the pyramid, or the bonds linking those positions to the central atom of the base are described as basal. The apical bonds have also been called axial.

a = apical
b = basal
e = equatorial

1996, *68*, 2199

apo- (in carotenoid nomenclature)
An unitalicized prefix, preceded by a locant, used to indicate that all of the molecule beyond the carbon atom corresponding to that locant has been replaced by hydrogen atoms.
W.B. 230

apoenzyme
The protein part of an enzyme without the cofactor necessary for catalysis. The cofactor can be a metal ion, an organic molecule (*coenzyme*), or a combination of both.
1994, *66*, 2593

apparent (quantity)
A term, indicated by a superscript ', which means that a process is not well known or that its value carries uncertainties which are not known, e.g. $\Delta G^{o'}$ is the apparent standard Gibbs energy change.
In the context of *partial molar quantities* the word is used in a different sense; as a symbol for 'apparent' in this connection the use of subscript φ, as in Y_φ, is recommended.
Other notations employed for this property include $^\varphi X$ and φ_X.
1986, *58*, 1408

apparent lifetime
See *lifetime*.
1996, *68*, 2228

apparent viscosity (of a liquid)
The ratio of stress to rate of strain, calculated from measurements of forces and velocities as though the liquid were Newtonian. If the liquid is actually non-Newtonian, the apparent viscosity depends on the type and dimensions of the apparatus used.
1979, *51*, 1217

appearance energy (appearance potential)
Refers to ionization of a molecule or atom by electron collision or photon absorption. In mass spectrometry it has often been reported as the voltage which corresponds to the minimum energy of the electrons in the ionizing beam necessary for the production of a given fragment ion. In photoionization it is the minimum energy of the quantum of light which produces ionization of the absorbing molecule.
1990, *62*, 2175; 1991, *63*, 1546

+ appearance potential
See *appearance energy*.
1991, *63*, 1546

appearance temperature, (in electrothermal atomization), T_{app}
The temperature of the atomization surface at which the analyte signal/noise (S/N) ratio reaches a value of 3 when the quantity of analyte in the atomizer is one hundred times the characteristic mass for peak high absorption.
1992, *64*, 257

applied potential
The difference of potential measured between identical metallic leads to two electrodes of a cell. The applied potential is divided into two *electrode potentials*, each of which is the difference of potential existing between the bulk of the solution and the interior of the conducting material of the electrode, an iR or ohmic potential drop through the solution, and another ohmic potential drop through each electrode.
In the electroanalytical literature this quantity has often been denoted by the term *voltage*, whose continued use is not recommended.
1985, *57*, 1502

aprotic (solvent)
Non-*protogenic* (in a given situation). (With extremely strong *Brønsted acids* or *bases*, solvents that are normally aprotic may accept or lose a proton. For example, acetonitrile is in most instances an aprotic solvent, but it is *protophilic* in the presence of concentrated sulfuric acid and *protogenic* in the presence of potassium *tert*-butoxide. Similar considerations apply to benzene, trichloromethane, etc.)
See also *dipolar aprotic solvent*.
1994, *66*, 1085

aquation
The incorporation of one or more integral molecules of water into another species with or without displacement of one or more other atoms or groups. For example, the incorporation of water into the inner *ligand* sphere of an inorganic *complex* is an aquation reaction.
See also *hydration*.
1994, *66*, 1086

arachno-
An affix used in names to designate a boron structure intermediate between *nido-* and *hypho-* in degree of openness.
R.B. 245

Archibald's method
A sedimentation method based on the fact that at the meniscus and at the bottom of the centrifuge cell there is never a flux of the solute across a plane perpendicular to the radial direction and the equations characterizing the sedimentation equilibrium always apply there, even though the system as a whole may be far from equilibrium.
The use of the term 'approach to sedimentation equilibrium' for Archibald's method is discouraged, since it has a more general meaning.
P.B. 62

area (of an electrode-solution interface)
This is understood to be the geometrical or projected area, and to ignore surface roughness.
1985, *57*, 1493

area of interface
In all measurements it is desirable that quantities like the charge and the capacity be related to unit, true, surface area of the interface. While this is relatively simple for a liquid-liquid interface, there are great difficulties when one phase is solid. In any report of these quantities it is essential to give a clear statement as to whether they refer to the true or the apparent (geometric) area and, especially if the former is used, precisely how it was measured.
1974, *37*, 509

area viscosity
See *surface shear viscosity*.
1979, *51*, 1218

areic
Attribute to a physical quantity obtained by division by area. Areic charge is the charge on a surface divided by the surface area.
ISO 31-0: 1992; 1996, *68*, 962

arene epoxides
Epoxides derived from arenes by 1,2-addition of an oxygen atom to a formal double bond, e.g. 5,6-epoxy-cyclohexa-1,3-diene.

(Common usage has extended the term to include examples with the epoxy group bridging nonadjacent atoms.)
1995, *67*, 1318

arene oxides
See *arene epoxides*.
1995, *67*, 1318

arenes
Monoyclic and polycyclic aromatic *hydrocarbons*.
See *aromatic compounds*.
1995, *67*, 1318

arenium ions
Cations derived formally by the addition of a hydron or other cationic species to any position of an arene, e.g. $C_6H_7^+$, benzenium.
The term includes:

1. Arenium σ-adducts (*Wheland intermediates*) (which are considered to be intermediates in electrophilic aromatic substitution reactions) and other cyclohexadienyl cations.

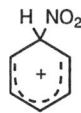

2. Arenium π-adducts, such as:

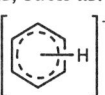

[σ-adduct (sigma-adduct), π-adduct (pi-adduct)].
See *aryl cations*.
1995, *67*, 1318; see also 1994, *66*, 1077

arenols
Synonymous with *phenols* (but rarely used).
1995, *67*, 1319

+ arenonium ions
An obsolescent name for *arenium ions*.
1995, *67*, 1319

argon ion laser
A CW or pulsed laser emitting lines from 334 to 529 nm from singly ionized argon. Principal emissions are at 488.0 and 514.5 nm.
See *gas lasers*.
1996, *68*, 2228

arithmetic mean (average)
The sum of a series of observations divided by the number of observations. Symbol: $\bar{x}$. It can be calculated by the formula:

$$\bar{x} = \frac{\sum x_i}{n}$$

Note:
All summations are taken from 1 to *n*. Note that the arithmetic mean is an unbiased estimate of the population mean, i.e. μ is the limiting value for $\bar{x}$, as $n \to \infty$.
1994, *66*, 599

aromatic
1. In the traditional sense, 'having a chemistry typified by benzene'.
2. A cyclically *conjugated molecular entity* with a stability (due to *delocalization*) significantly greater than that of a hypothetical localized structure (e.g. *Kekulé structure*) is said to possess aromatic character. If the structure is of higher energy (less stable) than such a hypothetical classical structure, the molecular entity is 'antiaromatic'.
The most widely used method for determining aromaticity is the observation of diatropicity in the ^{1}H NMR spectrum.
See also *Hückel (4n + 2) rule, Möbius aromaticity*.
3. The terms aromatic and antiaromatic have been extended to describe the stabilization or destabilization of *transition states* of *pericyclic reactions*. The

hypothetical reference structure is here less clearly defined, and use of the term is based on application of the *Hückel (4n + 2) rule* and on consideration of the topology of orbital overlap in the transition state. Reactions of molecules in the *ground state* involving antiaromatic transition states proceed, if at all, much less easily than those involving aromatic transition states.
1994, 66, 1086; see also 1995, 67, 1319

aromaticity
See *aromatic*
1994, 66, 1086

arrester (in atmospheric chemistry)
Equipment designed to remove particles from a gaseous medium.
1990, 62, 2175

Arrhenius A factor
See *pre-exponential factor*.
1996, 68, 153

Arrhenius energy of activation
See *activation energy*.
1994, 66, 1112

Arrhenius equation
An equation that represents the dependence of the *rate constant k* of a reaction on the absolute temperature T:

$$k = A \exp(-E_a/RT)$$

In its original form the *pre-exponential factor A* and the *activation energy* E_a are considered to be temperature-independent.
See also *modified Arrhenius equation*.
1996, 68, 153; 1994, 66, 1086; 1990, 62, 2175

arsanes
The saturated hydrides of tervalent arsenic, having the general formula As_nH_{n+2}. Individual members having an unbranched arsenic chain are named arsane, diarsane, triarsane, etc. The name of a saturated hydride of arsenic where one or more arsenic atoms have bonding number 5 is formed by prefixing locants and λ^5 symbols to the name of the corresponding arsane, e.g. $H_2AsAsHAsH_2$ triarsane, $H_4AsAsH_3AsH_4$ $1\lambda^5,2\lambda^5,3\lambda^5$-triarsane. Hydrocarbyl derivatives of AsH_3 belong to the class arsines.
1995, 67, 1319

arsanylidenes
Recommended name for *carbene analogues* having the structure R–As: (former IUPAC name is arsinediyls). A common non-IUPAC synonym is *arsinidenes*.
1995, 67, 1319

arsanylium ions
The arsanyl cation, H_2As^+, and derivatives by substitution. The name arsinylium (systematically derived from arsine) is not applied as it already designates $H_2As(=O)^+$, the acylium ion derived from arsinic acid.
1995, 67, 1319

arsenides
Compounds obtained from *arsines* AsR_3 by replacing one or more hydrogen atoms by a metal, e.g. CaAsPh calcium phenylarsenide.
1995, 67, 1319

arsine oxides
$H_3As=O$ and its hydrocarbyl derivatives. (analogously arsine imides and arsine sulfides), e.g. $(CH_3)_3As=O$ trimethylarsine oxide or trimethylarsane oxide.
See *imides (2)*.
1995, 67, 1319

arsines
AsH_3 and compounds derived from it by substituting one, two or three hydrogen atoms by hydrocarbyl groups: R_3As. $RAsH_2$, R_2AsH, R_3As ($R \neq H$) are called primary, secondary and tertiary arsines, respectively. A specific arsine is preferably named as a substituted arsane, e.g. $CH_3CH_2AsH_2$ ethylarsane.
See *arsanes*.
1995, 67, 1319

arsinic acids
$H_2As(=O)OH$ and its As-*hydrocarbyl* derivatives, e.g. $Me_2As(=O)OH$, dimethylarsinic acid.
1995, 67, 1320

arsinidenes
See *arsanylidenes*.
1995, 67, 1319

arsinous acids
H_2AsOH and its As-*hydrocarbyl* derivatives.
1995, 67, 1320

arsonic acids
$HAs(=O)(OH)_2$ and its As-*hydrocarbyl* derivatives.
1995, 67, 1320

arsonium compounds
Salts (including hydroxides) $[R_4As]^+X^-$ containing tetracoordinate arsonium ion and the associated anion.
See also *onium compounds*.
1995, 67, 1320

arsonous acids
$HAs(OH)_2$ and its As-*hydrocarbyl* derivatives.
1995, 67, 1320

arsoranes
The mononuclear hydride AsH_5, systematically named λ^5-arsane, and its hydrocarbyl derivatives. By extension the term also applies to arsonium ylides.
See *ylides, arsanes*.
1995, 67, 1320

+ artificial graphite
A term often used in place of *synthetic graphite*.
1995, 67, 477

artificial radioactivity
Synonymous with *induced radioactivity*.
1982, 54, 1535

aryl cations
Carbocations formally derived by removal of a hydride ion from a ring carbon atom of an *arene*.

See *arenium ions*, e.g. phenyl cation or phenylium:

1995, *67*, 1320

arylene groups
Bivalent groups derived from *arenes* by removal of a hydrogen atom from two ring carbon atoms. A synonym is arenediyl groups.
E.g. *o*-phenylene or benzene-1,2-diyl:

1995, *67*, 1320

aryl groups
Groups derived from *arenes* by removal of a hydrogen atom from a ring carbon atom. Groups similarly derived from *heteroarenes* are sometimes subsumed in this definition. E.g. *o*-tolyl:

See *heteroaryl groups*.
1995, *67*, 1320

aryne
A *hydrocarbon* derived from an *arene* by abstraction of two hydrogen atoms from adjacent carbon atoms; thus 1,2-didehydroarene. Arynes are commonly represented with a formal triple bond. The analogous heterocyclic compounds are called heteroarynes or hetarynes.
For example, benzyne:

Arynes are usually transient species.
See also *benzynes, dehydroarenes, heteroarynes*.
1994, *66*, 1086; 1995, *67*, 1320

ascending elution (ascending development) (in planar chromatography)
A mode of operation in which the paper or plate is in a vertical or slanted position and the mobile phase is supplied to its lower edge; the upward movement depends on capillary action.
1993, *65*, 829

ash (in atmospheric chemistry)
The solid residue which remains after the combustion of a fuel such as coal. Ash consists largely of heat treated mineral matter, but it may contain some products of the incomplete combustion of the fuel as well.
1990, *62*, 2175

ashing (in analysis)
Dry or wet mineralization as a method of *preconcentration* of trace substances.
See also *charring*.
1979, *51*, 1200

aspirator
Any apparatus that produces a movement of a fluid by suction (e.g. a squeeze bulb, pump, Venturi, etc.)
1990, *62*, 2175

assay
A set of operations having the object of determining the value of a quantity. In analytical chemistry, this term is synonymous with measurement.
1994, *66*, 2515; 1993, *65*, 2016

assay kit
A set of components (reagents and other necessary materials) and procedural instructions packaged together and designed for the estimation in vitro of a value of a specified quantity, when used according to the instructions.
1994, *66*, 2515

association
The assembling of separate *molecular entities* into any aggregate, especially of oppositely charged free ions into *ion pairs* or larger and not necessarily well-defined clusters of ions held together by electrostatic attraction. The term signifies the reverse of *dissociation*, but is not commonly used for the formation of definite *adducts* by *colligation* or *coordination*.
1994, *66*, 1086

association reaction (associative combination)(in mass spectrometry)
The reaction of a (slow moving) ion with a neutral species wherein the reactants combine to form a single ionized species.
1991, *63*, 1555

associative desorption
See *dissociative adsorption*.
1976, *46*, 84

associative ionization (in mass spectrometry)
This occurs when two excited gaseous atoms or molecular moieties interact and the sum of their internal energies is sufficient to produce a single, additive ionic product.
See also *chemi-ionization*.
1991, *63*, 1547

associative surface reaction
See *dissociative surface reaction*.
1976, *46*, 84

asym-
An affix used in names to denote asymmetrical.
R.B. 245; B.B. 464

asymmetric
Lacking all *symmetry elements* (other than the trivial one of a one-fold axis of symmetry), i.e. belonging to the symmetry point group C_1. The term has been used loosely (and incorrectly) to describe the absence of a rotation–reflection axis (alternating axis) in a molecule, i.e. as meaning chiral, and this usage persists in the traditional terms *asymmetric carbon atom, asymmetric synthesis, asymmetric induction*, etc.
1996, *68*, 2199; B.B. 480

asymmetric carbon atom
The traditional name (van't Hoff) for a carbon atom that is attached to four different entities (atoms or groups), e.g. Cabcd.
See also *chirality centre*.
1996, *68*, 2200; B.B. 480

asymmetric centre
See *chirality centre*.
1996, *68*, 2200

asymmetric destruction
See *kinetic resolution*.
1996, *68*, 2200

asymmetric film
A *film* bounded by two different bulk phases. When the bulk phases are identical the film is described as symmetric.
1994, *66*, 1671

asymmetric induction
The traditional term describing the preferential formation in a chemical reaction of one *enantiomer* or *diastereoisomer* over the other as a result of the influence of a chiral feature present in the substrate, reagent, catalyst or environment.
1996, *68*, 2200; 1994, *66*, 1086

asymmetric synthesis
A traditional term used for *stereoselective synthesis* of chiral compounds.
1996, *68*, 2200

asymmetric transformation
The conversion of a racemate into a pure *enantiomer* or into a mixture in which one enantiomer is present in excess, or of a *diastereoisomeric* mixture into a single diastereoisomer or into a mixture in which one diastereoisomer predominates. This is sometimes called deracemization.
If the two enantiomers of a *chiral* substrate A are freely interconvertible and if an equal amount or excess of a non-racemizing second enantiomerically pure chemical species, say (*R*)-B, is added to a solution of racemic A, then the resulting equilibrium mixture of adducts A•B will, in general, contain unequal amounts of the diastereoisomers (*R*)-A•(*R*)-B and (*S*)-A•(*R*)-B. The result of this equilibration is called asymmetric transformation of the first kind.
If, in such a system, the two diastereoisomeric adducts differ considerably in solubility so that only one of them, say (*R*)-A•(*R*)-B, crystallizes from the solution, then the equilibration of diastereoisomers in solution and concurrent crystallization will continue so that all (or most) of the substrate A can be isolated as the crystalline diastereoisomer (*R*)-A•(*R*)-B. Such a 'crystallization-induced asymmetric transformation' is called an asymmetric transformation of the second kind.
See also *stereoconvergence*.
1996, *68*, 2200

asymmetry
Denoting absence of any symmetry.
B.B. 479

asymmetry potential (of a glass electrode)
The measured potential difference of a symmetrical cell with identical solutions and reference electrodes on each side of the glass membrane. There is rarely the need, nor the possibility, of measuring the asymmetry potential of commercial glass electrodes. Drifts in glass electrode potentials with time and variations from day-to-day in the potential measured in a standard buffer may be attributed to changes in asymmetry potential.
O.B. 27

atactic macromolecule
A *regular macromolecule* in which the *configurational (base) units* are not all identical.
1996, *68*, 2292

atactic polymer
A polymer composed of *atactic macromolecules*.
1996, *68*, 2303

atmosphere
Non-SI unit of pressure, 1 atm = 101 325 Pa.
G.B. 112; 1990, *62*, 2175; 1996, *68*, 962

atmosphere (of the earth)
The entire mass of air surrounding the earth which is composed largely of nitrogen, oxygen, water vapour, clouds (liquid or solid water), carbon dioxide, together with trace gases and aerosols.
See *air, composition of pure*.
1990, *62*, 2175

+ atmospheric pressure ionization (in mass spectrometry)
An ambiguous term; in essence, it is used to describe chemical ionization at atmospheric pressure. It is recommended that use of the term be discontinued.
1991, *63*, 1547

atom
Smallest particle still characterizing a chemical element. It consists of a nucleus of a positive charge (Z is the proton number and *e* the elementary charge) carrying almost all its mass (more than 99.9%) and Z electrons determining its size.
Physical Chemistry Division, unpublished; R.B. 35

atomic fluorescence
A combined process of photon absorption by an atom followed by spontaneous photon emission.
O.B. 121

atomic laser
A *gas laser* which is pumped using energy transfer from other atoms or molecules. Examples are the helium-neon (HeNe) the copper vapour laser.
1995, *67*, 1919

atomic mass, m_a
Rest mass of an atom in its ground state. The commonly used unit is the *unified atomic mass unit*.
G.B. 20; 1996, *68*, 962

atomic mass constant
One twelfth of the mass of a carbon-12 atom in its nuclear and electronic ground state, m_u = 1.660 5402

(10) × 10^{-27} kg. It is equal to the *unified atomic mass unit*.
  CODATA Bull., 1986, *63*, 1; 1996, *68*, 962

atomic mass unit
  See *unified atomic mass unit*
  CODATA Bull., 1986, *63*, 1; 1996, *68*, 962

atomic number (proton number), Z
The number of protons in the atomic nucleus.
  G.B. 20; R.B. 35; 1996, *68*, 962

atomic orbital, ψ, φ, χ
One-electron wavefunction obtained as a solution of the Schrödinger equation for an atom.
  G.B. 19; 1994, *66*, 1086

atomic ring-sector
  See *ring-sector*.
  B.B. 499

atomic spectral lines
Atomic and ionic spectral lines originate from specified electronic transitions between energy levels of atoms and ions, respectively. In the past it has been common useage to denote atomic lines as arc lines and ionic lines as spark lines. This usage is now considered to be incorrect. The correct way to indicate that lines are due to atomic or ionic transitions is: Element symbol I wavelength e.g. Cu I 324.7 nm; and Element symbol II wavelength e.g. Cu II 213.6 nm. Similarly for higher states of ionization, the type of line is represented by III, IV, etc.
  See also *ionic spectral lines*.
  O.B. 118

atomic symbol
One, two or three letters used to represent the atom in chemical formulae.
  R.B. 36

atomic units
System of units based on four base quantities: length, mass, charge and action (angular momentum) and the corresponding base units the Bohr radius, a_0, rest mass of the electron, m_e, elementary charge, e, and the Planck constant divided by 2π, $\hbar$.
  G.B. 76, 120

atomic weight
  See *relative atomic mass*.
  G.B. 41; 1996, *68*, 962

atomization (in analytical flame spectroscopy)
The conversion of volatilized analyte into free atoms.
  O.B. 165

atomization surface temperature (in electrothermal atomization), T_s
The temperature of the support from which the sample is atomized.
  1992, *64*, 257

atomize
To subdivide a liquid into very small particles; methods include: impact with a jet of gas, use of a spinning disk generator, vibrating orifice generator, etc.
  1990, *62*, 2175

atomizer (in analytical flame spectroscopy)
Any system which is capable of converting the analyte into atomic vapour.
  O.B. 165

atom–molecule complex mechanism
A mechanism that sometimes applies to the combination of atoms, but rarely of free radicals. In this mechanism the atom A first combines with a third body or *chaperon*,

$$A + M \rightarrow AM,$$

and the complex AM then forms $A_2 + M$ by collision with another atom A.
Contrast the *energy-transfer mechanism*.
  1996, *68*, 153

atropisomers
A subclass of *conformers* which can be isolated as separate *chemical species* and which arise from restricted rotation about a single bond, e.g. *ortho*-substituted biphenyl, 1,1,2,2-tetra-*tert*-butylethane.
  See *rotational barrier*.
  1996, *68*, 2200

attachment
A *transformation* by which one *molecular entity* (the *substrate*) is converted into another by the formation of one (and only one) two-centre *bond* between the substrate and another molecular entity and which involves no other changes in *connectivity* in the substrate. For example, the formation of an acyl cation by attachment of carbon monoxide to a *carbenium ion* (R^+):

$$R^+ + CO \rightarrow (RCO)^+$$

The product of an attachment may also be the *adduct* of the two reactants, but not all adducts can be represented as the products of an attachment. (For example, the Diels–Alder cycloaddition:

$$CH_2=CH-CH=CH_2 + H_2C=CH_2 \longrightarrow \bigcirc$$

results in an adduct of buta-1,3-diene and ethene, but the reaction cannot be described as an attachment since bonds are formed between more than two centres.)
  See also *colligation, electron attachment*.
  1994, *66*, 1086

attenuance, D
Analogous to *absorbance*, but taking into account also the effects due to scattering and luminescence. It was formerly called extinction.
  G.B 32; see also 1996, *68*, 2228; 1996, *68*, 962

attenuance filter
An optical device (filter) which reduces the radiant power of a light beam by a constant factor over all wavelengths within its operating range. Sometimes called attenuator or neutral density filter.
  1996, *68*, 2229

attenuation
The reduction of a *radiation* quantity upon passage of radiation through matter resulting from interactions of the radiation with the matter it traverses.
1982, *54*, 1536

attenuation coefficient
Analogous to *absorption coefficient*, but taking into account also the effects due to scattering and luminescence. It was formerly called extinction coefficient.
G.B. 32; see also O.B. 212; 1996, *68*, 963

atto
SI prefix for 10^{-18} (symbol: a).
G.B. 74

attractive–mixed–repulsive (AMR) classification
A classification of *potential-energy surfaces* in which a highly attractive surface is at one extreme and a highly repulsive surface is at the other. The energy release in intermediate cases is referred to as mixed.
1996, *68*, 153

attractive potential-energy surface
A *potential-energy surface* for a process A + B–C in which the initial descent of the system into the product valley is associated with a substantial decrease in the A–B distance and with little separation between the products A–B and C. In terms of a *potential-energy profile*, the energy barrier occurs in the early stage of the reaction path.
Attractive surfaces are also called early-downhill surfaces, and the barrier in such a surface is called a Type-I barrier.
1996, *68*, 153

Auger effect
The emission of an *electron* from an *atom* accompanying the filling of a vacancy in an inner electron shell.
1982, *54*, 1536

Auger electron
Electron originating in the *Auger effect*.
1982, *54*, 1536; see also 1980, *52*, 2546

Auger electron spectroscopy
Any technique in which a specimen is bombarded with keV-energy electrons or X-rays, and the energy distribution of the electrons produced through radiationless de-excitation of the atoms in the sample (*Auger electrons*) is recorded. The derivative curve may also be recorded.
1983, *55*, 2025

Auger electron yield
The fraction of the atoms having a vacancy in an inner orbital which relax by emission of an *Auger electron*.
1979, *51*, 2247

Auger yield
For a given excited state of a specified *atom*, the probability that the de-excitation occurs by the *Auger effect*.
1982, *54*, 1536

autocatalytic reaction
A *chemical reaction* in which a product (or a reaction *intermediate*) also functions as a *catalyst*. In such a reaction the observed *rate of reaction* is often found to increase with time from its initial value.
See *order of reaction*.
1994, *66*, 1087; 1993, *65*, 2292

auto-ionization (in mass spectrometry)
This occurs when an internally supra-excited atom or molecular moiety loses an electron spontaneously without further interaction with an energy source. (The state of the atom or molecular moiety is known as a pre-ionization state.)
1991, *63*, 1547; O.B. 206

automation (in analysis)
Mechanization with process control, where process means a sequence of manipulations. One or several functions in an analytical instrument may be automated. The corresponding adjective is automated and the verb is automate.
See also *mechanization*.
1989, *61*, 1659

automerization
Synonymous with *degenerate rearrangement*.
1994, *66*, 1087

autophobicity
If adsorption equilibrium and mutual saturation of the phases is not achieved instantly, it is possible to distinguish the initial spreading tension, $\sigma_i^{\alpha\beta\delta}$, from the final spreading tension, $\sigma_f^{\alpha\beta\delta}$, when equilibrium has been reached.
In the case in which $\sigma_i^{\alpha\beta\delta}$ is positive, while $\sigma_f^{\alpha\beta\delta}$ is negative, the system is said to exhibit autophobicity.
1972, *31*, 598

autopoisoning (in catalysis)
See *self-poisoning (in catalysis)*.
1976, *46*, 83

autoprotolysis
A *proton* (hydron) *transfer* reaction between two identical molecules (usually a solvent), one acting as a *Brønsted acid* and the other as a *Brønsted base*. For example:

$$2H_2O \rightarrow H_3O^+ + OH^-$$

1994, *66*, 1087

autoprotolysis constant
The product of the activities (or, more approximately, concentrations) of the species produced as the result of *autoprotolysis*. For solvents in which no other ionization processes are significant the term is synonymous with 'ionic product'. The autoprotolysis constant for water, K_w, is equal to the product of activities:

$$a(H_3O^+)a(OH^-) = 1.0 \times 10^{-14} \text{ at } 25 \text{ °C}.$$

1994, *66*, 1087

autoradiograph
A *radiograph* of an object containing *radioactive* substance, produced by placing the object adjacent to a photographic plate or film or a fluorescent screen.
1994, *66*, 2516

autoradiolysis
Radiolysis of a *radioactive* material resulting directly or indirectly from its radioactive decay.
1994, *66*, 2516

auxiliary electrode
Three-electrode cells comprise (1) an indicator (or test) electrode or a working electrode, at the surface of which processes that are of interest may occur, (2) a reference electrode and (3) a third electrode, the auxiliary or counter electrode, which serves merely to carry the current flowing through the cell, and at the surface of which no processes of interest occur.
If processes of interest occur at both the anode and the cathode of a cell (as in differential amperometry or controlled current potentiometric titration with two indicator electrodes), the cell should be said to comprise two indicator (or test) working electrodes.
O.B. 59

+ auxochrome
An atom or group which, when added to or introduced into a chromophore, causes a bathochromic shift and/or a hyperchromic effect in a given band of the chromophore, usually in that of lowest frequency. This term is obsolete.
1996, *68*, 2229

auxotrophy
The inability of a organism to synthesize a particular organic compound required for its growth.
1992, *64*, 147

avalanche photodiode
See *photodiode*.
1995, *67*, 1755

average current density
The average current density, j, is defined by

$$j = A^{-1} \int_A j_x \, dA$$

where A is the electrode area, j_x is the local current density and dA is an infinitesimal surface element.
1981, *53*, 1836

average degree of polymerization
Any average, $\bar{X}_k$, of the degree of polymerization, where k specifies the type of average.
P.B. 55

average life (in nuclear chemistry)
Defined for an *atom* or nuclear system in a specified state. For an exponentially decaying system, it is the average time for the number of atoms or *nuclei* in a specified state to decrease by a factor e.
Synonymous with *mean life*.
1982, *54*, 1536

average rate of flow (in polarography)
The ratio of the mass of a drop, at the instant when it is detached from the tip of the capillary, to the *drop time* t_1; the average value of the *instantaneous rate of flow* over the entire life of the drop.
1985, *57*, 1503

Avogadro constant
Fundamental physical constant (symbols: L, N_A) representing the molar number of entities: $L = 6.022\ 1367\ (36) \times 10^{23}$ mol^{-1}.
CODATA Bull., 1986, *63*, 1; 1996, *68*, 963

avoided crossing (of potential-energy surfaces)
Frequently, two Born–Oppenheimer electronic states (A, B) change their energy order as molecular geometry (x) is changed continuously along a path. In the process their energies may become equal at some points (the surfaces are said to cross, dotted lines in the figure), or only come relatively close (the crossing of the surfaces is said to be avoided). If the electronic states are of the same symmetry, the *surface crossing* is always avoided in diatomics and usually avoided in polyatomics.

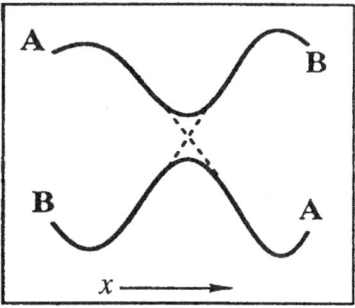

Synonymous with *intended crossing*.
1996, *68*, 2229

Avrami equation
An equation, describing crystallization kinetics, of the form:

$$1 - \varphi_c = \exp(-Kt^n)$$

where φ_c is the crystalline volume fraction developed at time t and constant temperature, and K and n are suitable parameters. K is temperature dependent.
According to the original theory, n should be an integer from 1 to 4, the value of which should depend only on the type of the statistical model; however, it has become customary to regard it as an adjustable parameter that may be non-integral.
P.B. 85

axial (equatorial)
In the chair form of cyclohexane ring bonds to ring atoms (and molecular entities attached to such bonds) are termed axial or equatorial according to whether the bonds make a relatively large or small angle, respectively, with the plane containing or passing closest to a majority of the ring atoms. Thus the axial bonds are approximately parallel to the C_3 axis and the equatorial bonds approximately parallel to two of the ring bonds. These terms are also used for the chair form of other saturated six-membered rings. The cor-

responding bonds occurring at the allylic positions in mono-unsaturated six-membered rings are termed pseudo-axial (or quasi-axial) and pseudo-equatorial (or quasi-equatorial). The terms axial and equatorial have similarly been used in relation to the puckered conformation of cyclobutane, crown conformer of cyclooctane, etc. and the terms pseudo-axial and pseudo-equatorial in the context of the non-planar structures of cyclopentane and cycloheptane.

a = axial
e = equatorial
a' = pseudo-axial
e' = pseudo-equatorial

See *apical, basal, equatorial* for an alternative use of axial and equatorial with bipyramidal structures.
1996, 68, 2200

axial chirality
Term used to refer to stereoisomerism resulting from the non-planar arrangement of four groups in pairs about a *chirality axis*. It is exemplified by allenes abC=C=Ccd (or abC=C=Cab) and by the *atropisomerism* of *ortho*-substituted biphenyls.
The *configuration* in molecular entities possessing axial chirality is specified by the stereodescriptors R_a and S_a (or by P or M).

1996, 68, 2201

axialite (in polymer crystals)
A multilayer aggregate, consisting of lamellar crystals splaying out from a common edge.
P.B. 82

axis of chirality
See *chirality axis*.
1996, 68, 2201

axis of helicity
See *helicity*.
1996, 66, 2201

+ azacarbenes
See *nitrenes*.
1994, 66, 1087

azamines
See *isodiazenes*.
1995, 67, 1320

azanes
Saturated acyclic nitrogen hydrides having the general formula N_nH_{n+2}.
1995, 67, 1320

+ azenes
See *nitrenes*.
1994, 66, 1087

azides
1. Compounds bearing the group N_3, viz. $-N=N^+=N^-$; usually attached to carbon, e.g. PhN_3 phenyl azide or azidobenzene.
2. Salts of hydrazoic acid, HN_3, e.g. NaN_3 sodium azide.
1995, 67, 1320

+ azimines
A commonly used but undesirable term for *azo imides*. (Should not be confused with 'azimino', the name for the bridging group, $-N=NNH-$).
1995, 67, 1321

azines
Condensation products, $R_2C=NN=CR_2$, of two moles of a carbonyl compound with one mole of *hydrazine*. (This term should not be confused with the ending -azine appearing in Hantzsch–Widman names for some heterocycles).
See also *aldazines and ketazines*.
1995, 67, 1321

azinic acids
Derivatives of the parent structure $H_2N^+(O^-)OH$, of which the alkylidene derivatives, $R_2C=N^+(O^-)OH$ (tautomers of nitroalkanes), are the most commonly encountered. The alkylideneazinic acids are known as *nitronic acids* or, synonymously as aci-*nitro compounds*, e.g. $CH_2=N^+(O^-)OH$ methylideneazinic acid.
1995, 67, 1321

azlactones
Oxazol-5(4H)-ones I, compounds derived by cyclization of N-acyl α-amino carboxylic acids, $RC(=O)NHCR_2C(=O)OH$, through formal loss of the elements of water. 4-Hydrocarbylideneazlactones II are often referred to as 'unsaturated azlactones'.

I   II

1995, 67, 1321

azo compounds
Derivatives of diazene (diimide), HN=NH, wherein both hydrogens are substituted by hydrocarbyl groups, e.g. PhN=NPh azobenzene or diphenyldiazene.
1995, 67, 1321

azo imides
N-*Imides* of *azo compounds*, analogous to *azoxy compounds*, having a delocalized structure

RN=N⁺(R)N⁻R ↔ RN⁻N⁺(R)=NR; commonly but undesirably referred to as *azimines*.
   See *dipolar compounds, imides (2), ylides*.
   1995, *67*, 1321

azomethine imides
The 1,3-*dipolar* N-*imides* of *azomethines* having the structure RN⁻–N⁺(R)=CR$_2$ ↔ RN=N⁺(R)C⁻R$_2$. The term azo ylides, derived from the second resonance form, has also been used.
   See *imides (2), ylides*.
   1995, *67*, 1321

azomethine oxides
Synonymous with *nitrones*.
   1995, *67*, 1321

azomethines
Compounds having the structure RN=CR$_2$ (R ≠ H). Many consider the term to include the compounds RN=CRH (R ≠ H), thus making azomethines synonymous with *Schiff bases*.
   See *imines*.
   1995, *67*, 1321

azomethine ylides
1,3-*Dipolar compounds* having the structure R$_2$C⁻–N⁺(R)=CR$_2$ ↔ R$_2$C=N⁺(R)C⁻R$_2$.
   See also *ylides*.
   1995, *67*, 1321

azonic acids
N-Hydrocarbyl derivatives of the parent structure HN⁺(O⁻)(OH)$_2$. Cf. *phosphonic acids*.
   1995, *67*, 1321

azoxy compounds
N-Oxides of *azo compounds*, of structure RN=N⁺(O⁻)R, e.g. PhN=N⁺(O⁻)Ph azoxybenzene or diphenyldiazene oxide.
   See *dipolar compounds*.
   1995, *67*, 1321

azo ylides
   See *azomethine imides*.
   1995, *67*, 1321

+ azylenes
   See *nitrenes*.
   1994, *66*, 1087

β (beta)
See α *(alpha)*, β *(beta)*.
1996, *68*, 2201

β-cleavage (beta-cleavage) (in mass spectrometry)
Fission next but one to a heteroatom or functional group producing a radical and an ion.
See also α-*cleavage*.
O.B. 207

β-decay (beta-decay)
Nuclear decay in which a *beta particle* is emitted or in which orbital *electron capture* occurs.
1982, *54*, 1536

β-particle (beta-particle)
Electron ejected from a radioactive nucleus.
G.B. 93; 1982, *54*, 1536

backbone
See *main chain*.
1996, *68*, 2294

back electron transfer
A term often used to indicate thermal reversal of excited state *electron transfer* restoring the donor and acceptor in their original oxidation level. In using this term one should also specify the resulting electronic state of the donor and acceptor.
1996, *68*, 2229

back extraction
Synonymous with *stripping (by extraction)*.
1993, *65*, 2386

background (of a radiation measuring device)
The term employed to designate the value indicated by a *radiation* measuring device in the absence of the source whose radiation is to be measured, when the device is placed under its normal conditions of operation.
1982, *54*, 1536

background concentration (level) (in atmospheric chemistry)
The concentration of a given species in a pristine air mass in which anthropogenic impurities of a relatively short lifetime are not present. The background concentrations of relatively long-lived molecules, methane, carbon dioxide, halocarbons (CF_3Cl, CF_2Cl_2, etc.) and some other species continue to rise due to anthropogenic input, so the composition of background air is undergoing continual change. Background concentration of a given species is sometimes considered to be the concentration of that impurity in a given air mass when the contribution from anthropogenic sources under study is absent.
Synonymous with baseline concentration.
1990, *62*, 2175

background mass spectrum
The mass spectrum observed when no sample is intentionally introduced into the mass spectrometer or spectrograph.
O.B. 34

background radiation
1. *Radiation* from any source other than the one it is desired to detect or measure.
1994, *66*, 2516; O.B. 213
2. Radiation which originates from the source and reaches the detector when no analyte is present.
1985, *57*, 1462

backscatter
Scattering of radiation in a generally backward direction. In the assay of radioactivity, it applies to the scattering of radiation into the radiation detector from any material except the sample and the detector.
In light scattering, it is said to occur when the scattering angle, θ, is 180°. Conversely, forward scattering occurs when θ is 0°.
O.B. 213; 1983, *55*, 932

back scatter coefficient (in *in situ* microanalysis)
Number of *back scattered electrons* (*BSEs*) generated per *primary electron* for a given specimen and experimental condition. It depends on (mean) atomic number of the excited area of the sample, angle between electron beam and sample surface, primary electron energy and thickness of the sample.
1983, *55*, 2026

back scattered electrons (BSEs) (in *in situ* microanalysis)
All *primary electrons* which are scattered out of the original direction and retransmitted through the surface of the solid. In practice, electrons emitted from the surface of a solid under electron bombardment which have a kinetic energy in the range between 50 eV and *excitation energy* (E_0).
1983, *55*, 2026

backscattering spectroscopy
See *Rutherford backscattering*.
O.B. 250

+ back washing
Often used as a synonym for *stripping*; the term is not recommended.
1993, *65*, 2386

Baeyer strain
See *angle strain*.
1996, *68*, 2201

baffle chamber (in atmospheric chemistry)
A chamber used in incinerator design to promote the settling of fly ash and coarse particulate matter by changing the direction and/or reducing the velocity of the gases produced by the combustion of the refuse.
1990, *62*, 2175

bag filter (in atmospheric chemistry)
A large bag constructed of a suitable fabric which is often tubular in shape, into which a particle-containing air stream flows. Modern bags are constructed of a fabric which is capable of collecting all but very fine particles in the gas stream. The efficiency of the removal of particles of various size ranges changes with the amount of particles captured by the filter and the

filtering time. The bag operates on the same principle as the one on a household vacuum cleaner.
1990, 62, 2175

baghouse (in atmospheric chemistry)
An installation which contains many bag filters in parallel so that the resistance to air flow in a large installation is not seriously increased by the addition of these controls.
1990, 62, 2176

Bainite transition
A term that is sometimes used in metallurgy for the *martensitic transition* that describes the transition between a face-centred-cubic lattice and a body-centred-tetragonal lattice.
Example: The transition between the face-centred-cubic lattice of austenite and the body-centred-tetragonal lattice of a martensite.
1994, 66, 579

baking (in carbon chemistry)
The process in which the carbonaceous *binder*, usually *coal tar pitch* or *petroleum pitch*, as part of a shaped *carbon mix* is converted to *carbon* yielding a rigid carbon body by the slow application of heat. The process can take as little as 14 days in coarse-grained, electrothermic grades (low binder level) and as long as 36 days in ultra-fine-grained, speciality grades (high binder level). The final baking temperature can be in the range of 1100–1500 K, depending on the grade.
1995, 67, 477

Baldwin's rules
A set of empirical rules for certain formations of 3- to 7-membered rings. The predicted pathways are those in which the length and nature of the linking chain enables the terminal atoms to achieve the proper geometries for reaction. The disfavoured cases are subject to severe distortions of bond angles and bond distances.
1994, 66, 1088

bandgap energy, E_g
The energy difference between the bottom of the conduction band and the top of the valence band in a semiconductor or an insulator.
See *conduction band, Fermi level*.
1996, 68, 2229

bandpass filter
An optical device which permits the transmission of radiation within a specified wavelength range and does not permit transmission of radiation at higher or lower wavelengths. It can be an interference filter.
See also *cut-off filter*.
1996, 68, 2229

band spectra
The combination of many different spectral lines resulting from vibrational, rotational and electronic transitions. Unresolved band spectra may appear as a spectral continuum.
O.B. 118

bar
Non-SI unit of pressure, bar = 10^5 Pa.
G.B. 75; 1996, 68, 963

barbiturates
1. Pyrimidine-2,4,6(1H,3H,5H)-trione (trivial name barbituric acid) and derivatives:

(keto tautomeric form)
2. Salts of barbituric acid and its derivatives.
1995, 67, 1322

barn
Non-SI unit of area, b = 10^{-28} m^2, used to express cross sections in nuclear reactions.
G.B. 75

Barton reaction
Photolysis of a nitrite to form a δ-nitroso alcohol. The mechanism is believed to involve a homolytic RO–NO cleavage, followed by δ-hydrogen abstraction and radical coupling.

1996, 68, 2229

basal
See *apical, basal, equatorial*.
1996, 68, 2201

base
A *chemical species* or *molecular entity* having an available pair of electrons capable of forming a *covalent bond* with a *hydron* (proton) (see *Brønsted base*) or with the vacant orbital of some other species (see *Lewis base*).
See also *hard base, superbase*.
1994, 66, 1088; 1990, 62, 2176

base electrolyte
Synonymous with *supporting electrolyte*.
1985, 57, 1501

base kind of quantity
A kind of quantity considered dimensionally independent of other kinds of quantities, i.e. it is not defined by an equation containing other kinds of quantities, e.g. length, mass, amount of substance.
1979, 51, 2459

baseline (in chromatography)
The portion of the chromatogram recording the detector response when only the mobile phase emerges from the column.
1993, *65*, 834; see also 1990, *62*, 2176

baseline concentration (in atmospheric chemistry)
See *background concentration*.
1990, *62*, 2175

baseline error (in spectrochemical analysis)
An error which arises if the two beams in a double-beam spectrometer are not fully equivalent in transmitted power nor are corrected to be so (e.g. due to unmatched cells).
1988, *60*, 1456

base pairing
The specific association between two complementary strands of *nucleic acids* that results from the formation of hydrogen bonds between the base components of the *nucleotides* of each strand: A=T and G≡C in *DNA*, A=U and G≡C (and sometimes G=U) in *RNA* (the lines indicate the number of hydrogen bonds). Single-stranded nucleic acid molecules can adopt a partially double-stranded structure through intrastrand base pairing.
1992, *64*, 147; 1993, *65*, 2017

base peak (in mass spectrometry)
The peak in a mass spectrum corresponding to the separated ion beam which has the greatest intensity. This term may be applied to the spectra of pure substances or mixtures.
1991, *63*, 1554; O.B. 206

base quantity
One of the quantities that, in a system of quantities, are conventionally accepted as functionally independent of one another.
In the SI, these are: *length, mass, time, electric current, temperature, amount of substance*, and *luminous intensity*. All other physical quantities (and units) are regarded as being derived from these base quantities (and *base units*).
1996, *68*, 963; 1986, *58*, 1406

base unit (of measurement)
Unit of measurement of a base quantity in a given system of quantities. By international agreement, a set of seven dimensionally independent units form the SI base units: the *metre, kilogram, second, ampere, kelvin, mole* and *candela*.
See *unit (of measurement)*.
1996, *68*, 963; 1986, *58*, 1406

basicity
For *Brønsted bases*, the tendency of a compound to act as *hydron* (proton) acceptor. The basicity of a *chemical species* is normally expressed by the *acidity* of the conjugate acid (see *conjugate acid–base pair*). For Lewis bases it relates to the association constants of *Lewis adducts* and π-*adducts*.
1994, *66*, 1088

basicity function
See *acidity function*.
1994, *66*, 1088

batch (in analytical chemistry)
A quantity of material which is known or assumed to be produced under uniform conditions.
Some vocabularies assume 'lot' and 'batch' to be synonymous. The distinction made here with respect to knowledge of production history permits a lot to consist of one or more batches and is useful in interpreting the results of analysis.
1990, *62*, 1202

batch injection calorimetry
See *direct injection enthalpimetry*.
1994, *66*, 2490

batch operation (in analysis)
The operation of an analytical instrument in such a way that one or more analytical procedures must be completed for a sequence of samples before the next sequence can be started. This term batch usually implies a sequence of a variably sized group of samples, the size of which is not related to a particular type of instrument.
1989, *61*, 1661

batch reactor
In a batch reactor the reactants and the catalyst are placed in the reactor which is then closed to transport of matter and the reaction is allowed to proceed for a given time whereupon the mixture of unreacted material together with the products is withdrawn. Provision for mixing may be required.
1976, *46*, 80

Bates–Guggenheim convention
See *Debye–Hückel equation*.
1984, *56*, 569

bathochromic shift (effect)
Shift of a spectral band to lower frequencies (longer wavelengths) owing to the influence of substitution or a change in environment. It is informally referred to as a red shift and is opposite to *hypsochromic shift* (blue shift).
1994, *66*, 1088; 1996, *68*, 2230

bead-rod model
A model simulating the hydrodynamic properties of a chain macromolecule consisting of a sequence of beads, each of which offers hydrodynamic resistance to the flow of the surrounding medium and is connected to the next bead by a rigid rod which does not. The mutual orientation of the rods is random.
P.B. 60

bead-spring model
A model simulating the hydrodynamic properties of a chain macromolecule consisting of a sequence of beads, each of which offers hydrodynamic resistance to the flow of the surrounding medium and is connected to the next bead by a spring which does not contribute to the frictional interaction but which is responsible for the elastic and deformational proper-

ties of the chain. The mutual orientation of the springs is random.
P.B. 61

beam current (in *in situ* microanalysis)
The number of *primary electrons* reaching the surface of the specimen per unit time expressed as electrical current. Recommended symbol: i_B; unit: A; typical range: nA. Recommended measurement technique: Faraday cage.
1983, *55*, 2026

beam diameter (in *in situ* microanalysis)
Diameter of the beam between which 68% of the electrons fall. For Gaussian beam shapes this corresponds to the 2σ-value of the intensity distribution of the beam. Measurement can be carried out by scanning a beam across a sufficiently sharp edge and recording the transmitted electrons.
1983, *55*, 2027

beam-gas experiments
See *molecular beams*.
1996, *68*, 153

becquerel
SI derived unit of radioactivity equal to one disintegration per second; admitted for reasons of safeguarding human health, symbol: Bq.
G.B. 72; 1994, *66*, 2516; 1996, *68*, 963

bed volume (in chromatography)
Synonymous with column volume for a packed column.
1993, *65*, 832; O.B. 100

Beer–Lambert law (or Beer–Lambert–Bouguer law)
The absorbance of a beam of collimated monochromatic radiation in a homogeneous isotropic medium is proportional to the absorption pathlength, l, and to the concentration, c, or — in the gas phase — to the pressure of the absorbing species. The law can be expressed as:

$$A = \log_{10}(P_\lambda^0/P_\lambda) = \varepsilon c l$$

or

$$P_\lambda = P_\lambda^0 10^{-\varepsilon c l}$$

where the proportionality constant, ε, is called the molar (decadic) absorption coefficient. For l in cm and c in mol dm^{-3} or M, ε will result in dm^3 mol^{-1} cm^{-1} or M cm^{-1}, which is a commonly used unit. The SI unit of ε is m^2 mol^{-1}. Note that spectral radiant power must be used because the Beer–Lambert law holds only if the spectral bandwidth of the light is narrow compared to spectral linewidths in the spectrum.
See *absorbance, extinction coefficient, Lambert law*.
1996, *68*, 2230; see also 1988, *60*, 1452; 1990, *62*, 2176

before mass analysis (in mass spectrometry)
The sum of all the separate ion currents for ions of the same sign (usually positive) before mass analysis.
O.B. 206

Bell–Evans–Polanyi principle
The linear relation between energy of activation (E_a) and enthalpy of reaction (ΔH_r) sometimes observed within a series of closely related reactions.

$$E_a = A + B\Delta H_r$$

1994, *66*, 1088

bending of energy bands
The distribution of potential in the space charge region of a semiconductor results in a change in the electron energy levels with distance from the interface. This is usually described as 'bending of the energy bands'. Thus the bands are bent, upwards if $\sigma > 0$ and downwards if $\sigma > 0$, where σ is the free charge density. When $\sigma = 0$ the condition of *flat bands* is met, provided no *surface states* are present.
1986, *58*, 443

benzenium ions
Arenium ions, derived from benzene or substituted derivatives.
1995, *67*, 1322

+ benzenonium ions
Obsolescent name for *benzenium ions*.
1995, *67*, 1322

benzylic groups
Arylmethyl groups and derivatives formed by substitution: ArCR$_2$–. Benzyl, C$_6$H$_5$CH$_2$–, is the prototype. The term 'benzylic position' or 'benzylic site' refers to the saturated carbon atom. A group, such as –OH, attached at a benzylic site is sometimes referred to as 'benzylic'.
1995, *67*, 1322

benzylic intermediates
Carbanions, carbenium ions or *radicals*, derived formally by detachment of one hydron, hydride or hydrogen, respectively, from the CH$_3$ group of toluene or substitution derivatives thereof, e.g. C$_6$H$_5$CH$_2$·, benzyl radical.
1995, *67*, 1322

benzynes
1,2-Didehydrobenzene (the *aryne* derived from benzene) and its derivatives formed by substitution; see *arynes, dehydroarenes*. The terms *m*- and *p*-benzyne are occasionally, but erroneously, used for 1,3- and 1,4-didehydrobenzene, respectively.
1995, *67*, 1322; 1994, *66*, 1088

Berry pseudorotation
See *pseudorotation*.
1996, *68*, 2201

beta
For entries see β (beginning of 'b').

betaines
Originally, the compound betaine, (CH$_3$)$_3$N$^+$–CH$_2$C(=O)O$^-$ *N,N,N*-trimethylammonioacetate, and

similar zwitterionic compounds derived from other amino acids. By extension, neutral molecules having charge-separated forms with an *onium* atom which bears no hydrogen atoms and that is not adjacent to the anionic atom. Betaines cannot be represented without formal charges.

E.g. $(CH_3)_3P^+CH_2S(=O)_2O^-$, $(Ph)_3P^+CH_2CH_2O^-$.

See also *dipolar compounds, mesoionic compounds, ylides, zwitterionic compounds.*

1995, *67*, 1322

betweenanenes

Bicyclic *alkenes* having a double bond between the bridgehead atoms and a *trans* attachment of each branch to the double bond. Thus *trans*-bicyclo[$m.n.0$]alk-1(m+2)-enes.

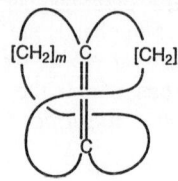

1995, *67*, 1322

bias

See *measurement result.*
1995, *67*, 1705

biased linear pulse amplifier

A *pulse amplifier* which, within the limits of its normal operating characteristics, has a constant gain for that portion of an input pulse that exceeds the threshold value and that produces no output for pulses whose amplitude is below the threshold.

1982, *54*, 1535

bias error

See *measurement result.*
1995, *67*, 1705

bifunctional catalysis

Catalysis (usually for *hydron* transfer) by a bifunctional *chemical species* involving a mechanism in which both *functional groups* are implicated in the *rate-controlling step*, so that the corresponding *catalytic coefficient* is larger than that expected for catalysis by chemical species containing only one of these functional groups.

The term should not be used to describe the *concerted* action of two different catalysts ('concerted catalysis').

1994, *66*, 1089

bilayer

See *multilayer, monolayer.*
1994, *66*, 1672

bimodal distribution

The occurrence of two maxima in a frequency distribution.

1990, *62*, 21

bimolecular

See *molecularity.*
1994, *66*, 1089

binder

Usually a *coal tar pitch* or *petroleum pitch* (but may include thermosetting resins or *mesophase pitch* powders) which, when mixed with a *binder coke* or a *filler*, constitutes a *carbon mix*. This is used in preparation of the formation of shaped green bodies and subsequently *carbon artifacts*.

1995, *67*, 477

binder (in chromatography)

An additive used to hold the solid stationary phase to the inactive plate or sheet in thin-layer chromatography.

1993, *65*, 830

binder coke

A constituent of a *carbon* (or ceramic) artifact resulting from *carbonization* of the *binder* during baking.
Notes:
Pitches are mainly used as binders, i.e. as precursors for binder cokes, but the term binder should include any carbonaceous binder material, for example thermosetting resins such as poly(furfuryl alcohol) or phenolics and similar compounds which may form a *char* during *carbonization*.

1995, *67*, 478

binding site

A specific region (or atom) in a *molecular entity* that is capable of entering into a stabilizing interaction with another molecular entity. An example of such an interaction is that of an active site in an enzyme with its *substrate*. Typical forms of interaction are by *hydrogen bonding*, *coordination* and *ion pair* formation. Two binding sites in different molecular entities are said to be complementary if their interaction is stabilizing.

1994, *66*, 1089

Bingham flow

Many colloidal dispersions show Bingham flow which is characterized by a σ-D diagram as shown. At rates of *shear* greater than that at point A, the following relation applies:

$$\sigma - \sigma_B = \eta_\Delta D$$

where σ_B (or τ_B) is called the Bingham *yield stress*, η_Δ is the *differential viscosity*, D is the *shear rate*, and σ is the average of three normal stress components if the deformation is purely dilatational.

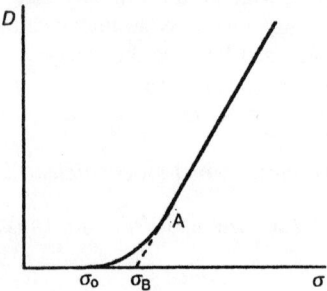

1979, *51*, 1217

bioassay
A procedure for determining the concentration or biological activity of a substance (e.g. vitamin, hormone, plant growth factor, antibiotic) by measuring its effect on an organism or tissue compared to a standard preparation.
1992, *64*, 147

biocatalyst
An *enzyme* or enzyme complex consisting of, or derived from, an organism or *cell culture* (in cell-free or whole-cell forms) that catalyses metabolic reactions in living organisms and/or *substrate* conversions in various chemical reactions.
1992, *64*, 147

biochemical (biological) oxygen demand (BOD)
The amount of oxygen, divided by the volume of the system, taken up through the respiratory activity of microorganisms growing on the organic compounds present in the sample (e.g. water or sludge) when incubated at a specified temperature (usually 20 °C) for a fixed period (usually 5 days, BOD_5). It is a measure of that organic pollution of water which can be degraded biologically. In practice, it is usually expressed in milligrams O_2 per litre.
1992, *64*, 148

biochip
An integrated circuit whose electrical and logical functions are performed by protein molecules appropriately manipulated.
1992, *64*, 148

bioconversion
See *biotransformation*.
1992, *64*, 148

biodegradation
Breakdown of a substance catalysed by enzymes *in vitro* or *in vivo*. This may be characterized for purposes of hazard assessment as:
1. Primary. Alteration of the chemical structure of a substance resulting in loss of a specific property of that substance.
2. Environmentally acceptable. Biodegradation to such an extent as to remove undesirable properties of the compound. This often corresponds to primary biodegradation but it depends on the circumstances under which the products are discharged into the environment.
3. Ultimate. Complete breakdown of a compound to either fully oxidized or reduced simple molecules (such as carbon dioxide/methane, nitrate/ammonium and water. It should be noted that the products of biodegradation can be more harmful than the substance degraded.
1993, *65*, 2020; 1992, *64*, 148

bioelectronics
The application of biomolecular principles to microelectronics such as in *biosensors* and *biochips*.
1992, *64*, 148

biological half life
For a substance the time required for the amount of that substance in a biological system to be reduced to one half of its value by biological processes, when the rate of removal is approximately exponential.
1994, *66*, 2520

bioluminescence
Luminescence produced by living systems.
See *luminescence*.
1996, *68*, 2230

biomass
Material produced by the growth of *microorganisms*, plants or animals.
1992, *64*, 148

biopolymers
Macromolecules (including *proteins, nucleic acids* and *polysaccharides*) formed by living organisms.
1992, *64*, 148

bioreactor
An apparatus used to carry out any kind of bioprocess; examples include *fermenter* or enzyme reactor.
1992, *64*, 148

biosensor
A device that uses specific biochemical reactions mediated by isolated *enzymes,* immunosystems, tissues, *organelles* or whole cells to detect chemical compounds usually by electrical, thermal or optical signals.
1992, *64*, 148

biosphere (in atmospheric chemistry)
That part of the globe that encompasses all forms of life on the earth. It extends from the ocean depths to a few thousand metres of altitude in the atmosphere, and includes life forms at the earth's surface, in soils and constituents which exchange materials with atmospheres, oceans or surfaces.
1990, *62*, 2176

biosynthesis
The production of a chemical compound by a living organism (cf. *biotransformation*).
1992, *64*, 148

biotechnology
The integration of natural sciences and engineering sciences in order to achieve the application of organisms, *cells*, parts thereof and molecular analogues for products and services.
1992, *64*, 148

biotransformation
Any chemical conversion of substances that is mediated by living organisms or *enzyme* preparations derived therefrom.
1992, *64*, 148

biphotonic excitation
Also called two-photon excitation. The simultaneous (coherent) absorption of two photons (either same or different wavelength) the energy of excitation being the sum of the energies of the two photons.
1996, *68*, 2230

biphotonic process
A process resulting from *biphotonic excitation*.
See *multiphoton process*.
1996, 68, 2230

bipolymer
See *copolymer*.
1996, 68, 2300

biradical
An even-electron *molecular entity* with two (possibly delocalized) radical centres which act nearly independently of each other, e.g.

(C$_6$H$_5$)$_2$Ċ—[2,6-Cl$_2$-C$_6$H$_2$]—[2,6-Cl$_2$-C$_6$H$_2$]—Ċ(C$_6$H$_5$)$_2$

Species in which the two radical centres interact significantly are often referred to as 'biradicaloids'. If the two radical centres are located on the same atom, the species are more properly referred to by their generic names: carbenes, nitrenes, etc.

The lowest-energy triplet state of a biradical lies below or at most only a little above its lowest singlet state (usually judged relative to $k_B T$, the product of the Boltzmann constant k_B and the absolute temperature T). The states of those biradicals whose radical centres interact particularly weakly are most easily understood in terms of a pair of local doublets.

Theoretical descriptions of low-energy states of biradicals display the presence of two unsaturated valences (biradicals contain one fewer bond than permitted by the rules of valence): the dominant valence bond structures have two dots, the low energy molecular orbital *configurations* have only two electrons in two approximately nonbonding molecular orbitals, two of the natural orbitals have occupancies close to one, etc.

Although this term has been recommended in the past for *diradicals*, specialists working in the field prefer the latter term.
See also *carbene, nitrene*.
1994, 66, 1089; 1995, 67, 1322

biradicaloid
Biradical-like.
1996, 68, 2231

bisecting conformation (eclipsing conformation)
For a structure containing the grouping R$_3$C–C(Y)=X (with identical or different groups R) the conformation in which the *torsion angle* is such that X is *antiperiplanar* to one of the groups R, and, in a *Newman projection*, the double bond C=X bisects one of the R–C–R angles. In this conformation the bond C–Y *eclipses* one of the C–R bonds. The other conformation, in which X is *synperiplanar* to one of the groups R, is called an eclipsing conformation.
1996, 68, 2201

bisecting conformation   eclipsing conformation

bismuthanes
The saturated hydrides of tervalent bismuth, having the general formula Bi$_n$H$_{n+2}$. *Hydrocarbyl* derivatives of BiH$_3$ belong to the class *bismuthines*.
1995, 67, 1322

bismuthines
BiH$_3$ and compounds derived from it by substituting one, two or three hydrogen atoms by hydrocarbyl groups: R$_3$Bi. RBiH$_2$, R$_2$BiH, R$_3$Bi (R ≠ H) are called primary, secondary and tertiary bismuthines, respectively. A specific bismuthine is preferably named as a substituted bismuthane, e.g. (CH$_3$)$_3$Bi trimethylbismuthane.
See *bismuthanes*.
1995, 67, 1323

bisphenols
By usage, the methylenediphenols, HOC$_6$H$_4$-CH$_2$C$_6$H$_4$OH, commonly *p,p'*-methylenediphenol, and their substitution products (generally derived from condensation of two equivalent amounts of a *phenol* with an *aldehyde* or *ketone*).
1995, 68, 1323

bivane (in atmospheric chemistry)
A wind vane used in turbulence studies to obtain horizontal and vertical components of the wind vector.
1990, 62, 2176

black film
A film thinner than about 1/4 wavelength of visible light. Black films are often *equilibrium films*, but equilibrium films may be considerably thicker under some conditions.
In soap films, two types of equilibrium film are often observed: the one characterized by thickness of the order of 7 nm or more which varies significantly with minor changes in composition such as ionic strength, is called common black film.
See also *Newton black film*.
1972, 31, 614

blank value (in analysis)
A reading or result originating from the matrix, reagents and any residual bias in the measurement device or process, which contributes to the value obtained for the quantity in the analytical procedure.
1989, 61, 1662; 1990, 62, 2176

blaze-angle (in spectrochemical analysis)
The angle (β) between the operating facet of the grooves and the overall plane of a grating.
O.B. 157

bleaching (in photochemistry)
The loss of absorption or emission intensity.
1996, 68, 2231

block
A portion of a *macromolecule*, comprising many *constitutional units* that has at least one feature which is not present in the adjacent portions.
Where appropriate, definitions relating to 'macromolecule' may also be applied to 'block'.
1996, *68*, 2298

block copolymer
A *copolymer* that is a *block polymer*.
In the constituent macromolecules of a block copolymer, adjacent blocks are constitutionally different, i.e. adjacent blocks comprise *constitutional units* derived from different species of *monomer* or from the same species of monomer but with a different composition or sequence distribution of constitutional units.
1996, *68*, 2303

block macromolecule
A *macromolecule* which is composed of *blocks* in linear sequence.
1996, *68*, 2292

block polymer
A polymer composed of *block macromolecules*.
1996, *68*, 2303

blotting (in biotechnology)
A technique used for transferring DNA, RNA or protein from gels to a suitable binding matrix, such as nitrocellulose or Nylon paper, while maintaining the same physical separation.
1992, *64*, 148

blowdown (in atmospheric chemistry)
Hydrocarbons purged during refinery shutdowns and startups which should be piped to storage systems for safe venting, flaring or recovery. This term also applies to the purging of water in boiler operation, and serves in the control of dissolved solids in the boiler water.
1990, *62*, 2176

blue shift
Informal expression for *hypsochromic shift*.
1996, *68*, 2231

boat
See *chair, boat, twist*; and *half-chair, half-boat*.
1996, *68*, 2201

Bodenstein approximation
See *steady state*.
1994, *66*, 1089

bohr
Atomic unit of length, $a_0 \approx 5.291\,772\,49(24) \times 10^{-11}$ m.
G.B. 76

Bohr magneton
Electromagnetic fundamental physical constant: $\mu_B = e\hbar/2m_e = 9.274\,0154\,(31) \times 10^{-24}$ J T^{-1}, where e is the elementary charge, $\hbar$ the Planck constant divided by 2π, and m_e the electron rest mass.
CODATA Bull., 1986, *63*, 1

Bohr radius
Atomic fundamental physical constant used as *atomic unit* of length: $a_0 = 4\pi\varepsilon_0\hbar^2/m_e e^2 = 5.291\,772\,49\,(24) \times 10^{-11}$ m, where e is the elementary charge, $\hbar$ the Planck constant divided by 2π, and m_e the electron rest mass.
CODATA Bull., 1986, *63*, 1

bolometer
A *detector* constructed from a material having a large *temperature coefficient of resistance*. Absorption of radiation gives rise to a change in resistance. A bolometer is named according to its active component, e.g. *thermistor bolometer, semiconductor bolometer, superconductor bolometer*.
1995, *67*, 1752

Boltzmann constant
Fundamental physical constant: $k = R/L = 1.380\,658\,(12) \times 10^{-23}$ J K^{-1}, where R is the gas constant and L the Avogadro constant.
CODATA Bull., 1986, *63*, 1; 1996, *68*, 963

bomb-digestion (in spectrochemical analysis)
Materials which are not fully dissolved by acid-digestion at atmospheric pressure may require a more vigorous treatment in pressure vessels lined with polytetrafluoroethylene (PTFE) glass, silica or vitreous (glassy) carbon or in sealed silica tubes; this treatment is called bomb-digestion. The *test sample* and acids are heated in such a closed vessel, so that the digestion is carried out at higher temperature and pressure.
1988, *60*, 1469

bond
There is a chemical bond between two atoms or groups of atoms in the case that the forces acting between them are such as to lead to the formation of an aggregate with sufficient stability to make it convenient for the chemist to consider it as an independent 'molecular species'.
See also *agostic, coordination, hydrogen bond, multi-centre bond*.
1994, *66*, 1089

bond dissociation
See *heterolysis, homolysis*.
(In ordinary usage the term refers to homolysis only.)
1994, *66*, 1089

bond-dissociation energy, D
The *enthalpy* (per mole) required to break a given bond of some specific molecular entity by homolysis, e.g. for $CH_4 \rightarrow H_3C\cdot + H\cdot$, symbolized as $D(CH_3-H)$ (cf. heterolytic bond dissociation energy).
See also *bond energy*.
1994, *66*, 1089; 1990, *62*, 2176

bonded phase (in chromatography)
A stationary phase which is covalently bonded to the support particles or to the inside wall of the column tubing.
1993, *65*, 823

bond energy (mean bond energy)
The average value of the gas-phase *bond dissociation energies* (usually at a temperature of 298 K) for all *bonds* of the same type within the same *chemical species*. The mean bond energy for methane, for example, is one-fourth the enthalpy of reaction for:

$$CH_4(g) \rightarrow C(g) + 4H(g)$$

Tabulated bond energies are generally values of bond energies averaged over a number of selected typical chemical species containing that type of bond.
1994, *66*, 1090

bond-energy-bond-order method
An empirical procedure for estimating *activation energies*, involving empirical relationships between bond length, bond dissociation energy, and bond order.
1996, *68*, 154

bonding number
The bonding number n of a skeletal atom is the sum of the total number of bonding equivalents (valence bonds) of that skeletal atom to adjacent skeletal atoms in a parent hydride, if any, and the number of attached hydrogen atoms, if any.
Examples:
SH_2: for S, $n = 2$; SH_6: for S, $n = 6$.
B.B.(G) 21; see also 1984, *56*, 774

bond migration
See *migration*.
1994, *66*, 1090

bond number
The number of electron-pair *bonds* between two nuclei in any given *Lewis formula*. For example, in ethene the bond number between the carbon atoms is two, and between the carbon and hydrogen atoms it is one.
1994, *66*, 1090

bond opposition strain
See *eclipsing strain*.
1996, *68*, 2201

bond order, p_{rs}
Theoretical index of the degree of bonding between two atoms relative to that of a single bond, i.e. the bond provided by one localized electron pair. In molecular orbital theory it is the sum of the products of the corresponding atomic orbital coefficients (weights) over all the occupied molecular spin-orbitals.
G.B. 17; see also 1994, *66*, 1090

bond ring-sector
See *ring-sector*.
B.B. 499

boranes
The molecular hydrides of boron, e.g. B_5H_9 pentaborane(9).
1995, *67*, 1323

boranylidenes
The species RB: containing an electrically neutral univalent boron atom with two formally non-bonding electrons.
See *carbene analogues*.
1995, *67*, 1323

borderline mechanism
A mechanism intermediate between two extremes, for example a nucleophilic substitution intermediate between S_N1 and S_N2, or intermediate between electron transfer and S_N2.
1994, *66*, 1090

borenes
A traditional term for *boranylidenes*.
1995, *67*, 1323

borinic acids
Compounds having the structure R_2BOH.
1995, *67*, 1323

boronic acids
Compounds having the structure $RB(OH)_2$.
1995, *67*, 1323

borylenes
A traditional term for *boranylidenes*.
1995, *67*, 1323

boson
Particle of integer spin quantum number following Bose–Einstein statistics.
Physical Chemistry Division, unpublished

boundary layer (in atmospheric chemistry)
That well-mixed region of the lower atmosphere in which the turbulence is maintained largely by convective buoyancy induced by the upward heat flux originating from the solar-heated surface of the earth. During the afternoon this often extends from 1 to 5 km in height. The surface boundary layer is that region of the lower atmosphere where the shearing stress is constant. It is separated by the Ekman layer from the free atmosphere, where the behaviour of the atmosphere approaches that of an ideal fluid in approximate geostrophic equilibrium (horizontal coriolis force balances the horizontal pressure force at all points in the field).
1990, *62*, 2176

bound fraction (in radioanalytical chemistry)
The fraction of the incubation mixture which, after separation, contains the analyte bound to the binding reagent.
1994, *66*, 2519

bowsprit (flagpole)
In the boat form of cyclohexane and related structures there are two ring atoms lying out of the plane of the other four; exocyclic bonds to these two atoms point-

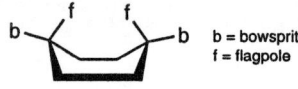

ing in a direction roughly parallel to that plane are called bowsprit, the other two are called flagpole.
1996, *68*, 2201

branch (side chain, pendant chain) (in polymers)
An *oligomeric* or *polymeric* offshoot from a *macromolecular chain*.
Notes:
1. An oligomeric branch may be termed a *short-chain branch*.
2. A polymeric branch may be termed a *long-chain branch*.
1996, *68*, 2297

branched chain (in polymers)
A *chain* with at least one *branch point* intermediate between the boundary units.
1996, *68*, 2293

branched polymer (in polymers)
A *polymer* the molecules of which are *branched chains*.
1996, *68*, 2304

branching chain reaction
A *chain reaction* which includes a *propagation step* or steps in which there is an increase in the number of active intermediates; for example, $H + O_2 \rightarrow OH + O$ is a branching reaction since one active intermediate (H) gives rise to two (OH and O).
1981, *53*, 762

branching decay
Nuclear decay which can proceed in two or more different ways.
1982, *54*, 1536

branching fraction
In *branching decay* the fraction of nuclei which decay in a specified way.
1982, *54*, 1536

branching index (in polymers)
A parameter, g, characterizing the effect of long-chain branches on the size of a branched macromolecule in solution and defined as the ratio of the mean-square radius of gyration of a branched molecule, $<s_b^2>$, to that of an otherwise identical linear molecule, $<s_l^2>$, with the same relative molecular mass in the same solvent and at the same temperature, i.e. $g = <s_b^2>/<s_l^2>$.
P.B. 51

branching ratio
The ratio of alternative *products* formed in a reaction (e.g. the HF/DF ratio in the reaction between F and HD).
1996, *68*, 154; 1982, *54*, 1536

branch point (in polymers)
A point on a *chain* at which a *branch* is attached.
Notes:
1. A branch point from which f linear chains emanate may be termed an f-*functional branch point*, e.g. five-functional branch point. Alternatively, the terms trifunctional, tetrafunctional, pentafunctional, etc. may be used, e.g. pentafunctional branch point.

2. A branch point in a network may be termed a *junction point*.
1996, *68*, 2297

branch unit (in polymers)
A *constitutional unit* containing a *branch point*.
Note:
A branch unit from which f linear chains emanate may be termed an f-*functional branch unit*, e.g. five-functional branch unit. Alternatively, the terms trifunctional, tetrafunctional, pentafunctional, etc. may be used, e.g. pentafunctional branch unit.
1996, *68*, 2297

break (of a foam)
The process involving the *coalescence* of gas bubbles.
1972, *31*, 611

break (of an emulsion)
The *flocculation* of an *emulsion*, viz. the formation of *aggregates* may be followed by *coalescence*. If coalescence is extensive it leads to the formation of a macrophase and the emulsion is said to break.
1972, *31*, 611

Bredt's rule
A double bond cannot be placed with one terminus at the bridgehead of a bridged ring system unless the rings are large enough to accommodate the double bond without excessive *strain*. For example, while bicyclo[2.2.1]hept-1-ene is only capable of existence as a *transient*, its higher homologues having a double bond at the bridgehead position have been isolated: e.g.

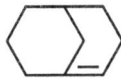 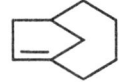

bicyclo[3.3.1]non-1-ene   bicyclo[4.2.1]non-1(8)-ene

1994, *66*, 1077; see also 1996, *68*, 2201

breeching (in atmospheric chemistry)
The passage or conduit through which the exhaust products of combustion are carried to the stack or chimney.
1990, *62*, 2176

breeze (in atmospheric chemistry)
A term sometimes used to describe very fine particles of coke. Also a meteorological term for a gentle wind.
1990, *62*, 2177

Bremsstrahlung
X-radiation emitted as a result of the retardation of high energy particles by matter.
1991, *63*, 737

bridge
A valence bond or an atom or an unbranched chain of atoms connecting two different parts of a molecule. The two tertiary carbon atoms connected through the bridge are termed bridgeheads.
B.B. 31

bridged carbocation
A *carbocation* (real or hypothetical) in which there are two (or more) carbon atoms that could in alternative *Lewis formulae* be designated as *carbenium centres* but which is instead represented by a structure in which a group (a hydrogen atom or a hydrocarbon residue, possibly with substituents in non-involved positions) bridges these potential carbenium centres. One may distinguish 'electron-sufficient bridged carbocations' and 'electron-deficient bridged carbocations'. Examples of the former are phenyl-bridged ions (for which the trivial name 'phenonium ion' has been used), such as (A). These ions are straightforwardly classified as *carbenium ions*. The latter type of ion necessarily involves three-centre bonding. Structures (C) and (D) contain five-coordinate carbon atoms. The 'hydrogen-bridged carbocation' (B) contains a two-coordinate hydrogen atom. Hypercoordination, which includes two-coordination for hydrogen and five- but also higher coordination for carbon is generally observed in bridged carbocations.

See also *carbonium ion, multi-centre bond, neighbouring group participation*.
1994, 66, 1090

bridgehead
See *bridge*.
B.B. 31

bridge index
In a coordination entity, the number of central atoms linked by a particular bridging ligand.
R.B. 149

bridge solution (in pH measurement)
Solution of high concentration of inert salt, preferably comprising cations and anions of equal mobility, optionally interposed between the reference electrode filling and both the test and standard solution, when the test solution and filling solution are chemically incompatible. This procedure introduces into the operational cell a second liquid junction formed usually in a similar way to the first.
(This definition refers to a bridge solution of a double junction reference electrode.)
1985, 57, 541

bridging ligand
A *ligand* attached to two or more, usually metallic, central atoms.
1994, 66, 1091; R.B. 149; B.B. 338

bromohydrins
See *halohydrins*.
1995, 67, 1323

bromonium ions
See *halonium ions*.
1995, 67, 1323

Brønsted acid (Brönsted acid)
A *molecular entity* capable of donating a *hydron* (proton) to a base, (i.e. a 'hydron donor') or the corresponding *chemical species*. For example: H_2O, H_3O^+, CH_3CO_2H, H_2SO_4, HSO_4^-, HCl, CH_3OH, NH_3.
See also *conjugate acid-base pair*.
1994, 66, 1091

Brønsted base (Brönsted base)
A *molecular entity* capable of accepting a *hydron* (proton) from an acid (i.e. a 'hydron acceptor') or the corresponding *chemical species*. For example: OH^-, H_2O, $CH_3CO_2^-$, HSO_4^-, SO_4^{2-}, Cl^-.
See also *conjugate acid–base pair*.
1994, 66, 1091

Brønsted relation (Brönsted relation)
The term applies to either of the equations:
$$k_{HA}/p = G(K_{HA}q/p)^\alpha$$
$$k_A/q = G(K_{HA}q/p)^{-\beta}$$
(or their logarithmic forms) where α, β and G are constants for a given reaction series (α and β are called 'Brønsted exponents'), k_{HA} and k_A are *catalytic coefficients* (or rate coefficients) of reactions whose rates depend on the concentrations of HA and/or of A^-. K_{HA} is the acid dissociation constant of the acid HA, p is the number of equivalent acidic protons in the acid HA, and q is the number of equivalent basic sites in its conjugate base A^-. The chosen values of p and q should always be specified. (The charge designations of HA and A are only illustrative.)
The Brønsted relation is often termed the 'Brønsted catalysis law' (or the 'Catalysis Law'). Although justifiable on historical grounds, this name is not recommended, since Brønsted relations are known to apply to many uncatalysed and pseudo-catalysed reactions (such as simple proton (*hydron*) *transfer reactions*). The term 'pseudo-Brønsted relation' is sometimes used for reactions which involve *nucleophilic catalysis* instead of acid-base catalysis. Various types of Brønsted parameters have been proposed such as β_{lg}, β_{nuc}, β_{eq} for leaving group, nucleophile and equilibrium constants, respectively.
See also *linear free-energy relation*.
1994, 66, 1091; 1996, 68, 154

Brooks and Taylor structure
The structure of the anisotropic spheres which precipitate from isotropic *pitch* during pyrolysis. The

structure of the spheres consists of a lamellar arrangement of aromatic molecules in parallel layers which are perpendicular to the polar axis of the sphere and which are perpendicular to the mesophase-isotropic phase interface.
Note:
The term Brooks and Taylor structure is recommended to describe the particular lamellar morphology of the spherules most commonly precipitated from pyrolysed *pitch*. The term honours the workers who first recognized the significance of *carbonaceous mesophase* to carbon science and technology and who first defined this spherical morphology. The term Brooks and Taylor structure does not cover all structures found in the spherical mesophase, because other lamellar arrangements have been observed.
1995, *67*, 478

Brownian motion
The movement of particles in a colloidal system such as an aerosol caused by collision with the molecules in the fluid in which the particles are imbedded.
1990, *62*, 2177

brush macromolecule
See *comb macromolecule*.
1996, *68*, 2296

bubble column
A *bioreactor*, in the shape of a column, in which the reaction medium is kept mixed and aerated by introduction of air into the bottom (cf. *air lift bioreactor*).
1992, *64*, 149

bubbler (in atmospheric chemistry)
An apparatus used to absorb certain water soluble components in a gas stream for later analysis. Usually it involves the use of a glass fritted tube which forces the air into small bubbles of high surface area during operation.
1990, *62*, 2177

buffer-addition technique (in analytical flame spectroscopy)
A technique in which an additive (called a *spectrochemical buffer*) is added to both the sample and reference solutions for the purpose of making the measure of the analyte less sensitive to variations in *interferent* concentration.
O.B. 172

bulk concentration (in electroanalysis)
In any technique that involves the establishment of a concentration gradient, either within the material from which an electrode is made or in the solution that is in contact with an electrode, the bulk concentration of a substance B is the total or analytical concentration of B at points so remote from the electrode-solution interface that the concentration gradient for B is indistinguishable from zero at the instant under consideration. In common practice the bulk concentration of B is taken to be the total or analytical concentration of B that would be present throughout the electrode or solution if there were no current flowing through the cell and if the electrode and solution did not interact in any way. In the absence of any homogeneous reaction or other process that produces or consumes B, the bulk concentration of B is the total or analytical concentration of B that is present before the excitation signal is applied.
1985, *57*, 1493

bulk mesophase
A continuous anisotropic phase formed by coalescence of mesophase spheres. Bulk mesophase retains fluidity and is deformable in the temperature range up to about 770 K, and transforms into *green coke* by further loss of hydrogen or low-molecular-weight species.
This bulk mesophase can sometimes be formed directly from the isotropic *pitch* without observation of intermediate spheres.
1995, *67*, 478

bulk rheology
Rheology may be conveniently divided into bulk rheology, in which effects due to the surface of the system can be neglected, and surface rheology, in which such effects are predominant. It should be noted that in surface rheology the neglect of bulk behaviour is permissible only in exceptional circumstances, such as for very thin films surrounded by a gas.
1979, *51*, 1215

bulk sample
The sample resulting from the planned aggregation or combination of *sample units*.
1988, *60*, 1463

bulk strain
See *volume strain*.
G.B. 12

Bunnett–Olsen equations
The equations for the relation between $\lg([SH^+]/[S]) + H_0$ and $H_0 + \lg [H^+]$ for base S in aqueous mineral acid solution, where H_0 is Hammett's *acidity function* and $H_0 + \lg [H^+]$ represents the activity function $\lg (\gamma_S \gamma_{H^+}/\gamma_{SH^+})$ for the nitroaniline reference bases to build H_0.

$$\lg ([SH^+]/[S]) - \lg [H^+] = (\Phi - 1)(H_0 + \lg [H^+]) + pK_{SH^+}$$

$$\lg ([SH^+]/[S]) + H_0 = \Phi(H_0 + \lg [H^+]) + pK_{SH^+}$$

See also *Cox–Yates equation*.
1994, *66*, 1091

Bunsen burner
Premix burners are distinguished as Bunsen-, Meker-, or slot-burners according to whether they have one large hole, a number of small holes, or a slot as outlet port(s) for the gas mixture, respectively. When several parallel slots are present, they are identified as multislot burners (e.g. a three-slot burner).
O.B. 166

+ Bunte salts
Salts (usually sodium salts) of S-alkylthiosulfuric acid, of general structure $RSS(=O)_2O^-\ M^+$. Use of this term is discouraged.
   1995, *67*, 1323

burning tension (of an electrical arc)
   See *electrical arc*.
   1985, *57*, 1464

burning velocity (of a flame front) (in flame emission and absorption spectrometry), v_b
The mean velocity of the flame front (in mm s^{-1}) towards the unburnt gas mixture (usually vertically downwards).

The quantity applies to gas mixtures and not to injection burners, and depends on the flame temperature and the solvent nebulized.
   1986, *58*, 1742

burn-up
Induced *nuclear transformation* of *atoms* during reactor operation.
   1982, *54*, 1537

burn-up fraction
The fraction of an initial quantity of a given *nuclide* that has undergone *burn-up*.
   1982, *54*, 1537

burn-up, specific
   See *specific burn-up*.
   1982, *54*, 1537

bypass injector (in gas chromatography)
A *sample injector* by means of which the *eluent* (*carrier gas*) may be temporarily diverted through a sample chamber so that the sample is carried to the *column*.
   O.B. 99

χ-parameter (chi-parameter)
A numerical parameter, χ, employed in the *Flory–Huggins theory*, which accounts in the main for the contribution of the non-combinatorial entropy of mixing and for the enthalpy of mixing.
P.B. 59

cadium–helium laser
See *helium–cadmium laser*.
1996, *68*, 2231

cage
An aggregate of molecules, generally in the condensed phase, that surrounds the fragments formed, for example, by thermal or photochemical dissociation. Because the cage hinders the separation of the fragments by diffusion, they may preferentially react with one another ('*cage effect*') but not necessarily to reform the precursor species. For example:

$$R-N{=}N-R \xrightarrow{\Delta} \left[R^{\cdot} + N{\equiv}N + R^{\cdot} \right]_{cage} \longrightarrow R-R + N_2$$

See *geminate recombination*.
1994, *66*, 1091

cage compound
A polycyclic compound having the shape of a *cage*. The term is also used for *inclusion compounds*.
1994, *66*, 1092

cage effect
When in a condensed phase, or in a dense gas, *reactant* molecules come together, or species are formed in proximity to one another, and are caged in by surrounding molecules, they may undergo a set of collisions known as an *encounter*; the term 'cage effect' is then applied.
The cage effect is also known as the Franck–Rabinowitch effect.
See also *cage, encounter*.
1996, *68*, 155

Cahn–Ingold–Prelog system
See *CIP priority*.
1996, *68*, 2201

calcination
Heating in air or oxygen; the term is most likely to be applied to a step in the preparation of a *catalyst*.
1976, *46*, 80

calcined coke
A *petroleum coke* or *coal-derived pitch coke* obtained by heat treatment of *green coke* to about 1600 K. It will normally have a hydrogen content of less than 0.1 wt.%.
Note:
Calcined coke is the main raw material for the manufacture of *polygranular carbon* and *polygranular graphite* products (e.g. *carbon* and *graphite electrodes*).
1995, *67*, 479

calibration (in analysis)
The set of operations which establish, under specified conditions, the relationship between values indicated by the analytical instrument and the corresponding known values of an analyte.
See *calibration material*.
1989, *61*, 1662

calibration component
A component of a calibration mixture, present in the gaseous or vapour state, quantitatively and qualitatively defined, and used directly for testing and for calibration.
1990, *62*, 2177

calibration curve
See *calibration function*.
1995, *67*, 1703

calibration function (in analysis)
The functional (not statistical) relationship for the *chemical measurement process*, relating the expected value of the observed (gross) signal or response variable $E(y)$ to the analyte amount x. The corresponding graphical display for a single analyte is referred to as the *calibration curve*. When extended to additional variables or analytes which occur in multicomponent analysis, the 'curve' becomes a calibration surface or hypersurface.
1995, *67*, 1703; see also 1990, *62*, 2177

calibration gas mixture (in atmospheric chemistry)
A gas mixture of known composition, generally comprising one or more calibration components and a complementary gas.
1990, *62*, 2177

calibration material (in analysis)
A material of known composition or properties which can be presented to the analytical instrument for calibration purposes.
See also *control material*.
1989, *61*, 1660

calibration mixture (in analysis)
A gaseous or liquid mixture of known composition, generally comprising one or more calibration components and an inert diluent, used directly for testing and calibration of analytical instruments.
1990, *62*, 2178

calibration sample (in analysis)
The test portion or test solution used for calibration of an analytical procedure.
The calibration sample is normally of known weight or volume and is prepared according to specifications.
See also *test portion, test solution, measurement solution*.
1989, *61*, 1660

calixarenes
Originally macrocyclic compounds capable of assuming a basket (or 'calix') shaped conformation. They are formed from p-*hydrocarbyl phenols* and formaldehyde. The term now applies to a variety of derivatives by substitution of the hydrocarbon cyclo-{oligo[(1,3-phenylene)methylene]}, see **I**.
1995, *67*, 1323

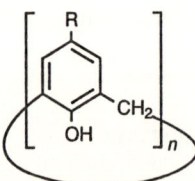

I (See **calixarenes**)

calorie
Non-SI unit of energy. There are three differently defined calories: thermochemical calorie (cal_{th} = 4.184 J), international steam table calorie (cal_{IT} = 4.1868 J), 15 °C calorie (cal_{15} = 4.1855 J), however only the symbol cal is usually used.
  G.B. 112; 1996, *68*, 964

calorimetric titration
A titration performed in a calorimeter which produces a plot of heat change versus volume of titrant. This is the preferred term for experiments in which the main goal is the measurement of thermodynamic parameters. Because such experiments may be performed in any of a variety of types of calorimeters with varying degrees of calorimetric accuracy, it is important that the reporting of such results include an assessment of accuracy and precision. The term stands in contrast to the main emphasis in *enthalpimetric analysis*.
  1994, *66*, 2490

calorimetry
A general term describing any experiment in which heat is measured as some chemical reaction or physical process occurs.
  1994, *66*, 2488

candela
SI base unit for luminous intensity (symbol: cd). The candela is the luminous intensity, in a given direction, of a source that emits monochromatic radiation of frequency 540×10^{12} hertz and that has a radiant intensity in that direction of (1/683) watt per steradian.
  G.B. 70; 1996, *68*, 964

canonical form
  See *contributing structure*.
  1994, *66*, 1092

canonical rate constant
The *rate constant* for a system in which the *reactants* are in thermal equilibrium at a given temperature.
In statistical mechanics the expression 'canonical ensemble' is used to refer to a closed system in thermal equilibrium, the species being present in a statistical distribution. By contrast, a microcanonical ensemble is composed of systems all having the same energy: a canonical ensemble therefore consists of a statistical distribution of microcanonical ensembles. The canonical or thermal rate constant can be obtained from the microcanonical rate constant by summing over the energy, taking into account the statistical distribution.
  1996, *68*, 155

canonical variational transition-state theory (CVTST)
A development of *transition-state theory* in which the position of the *dividing surface* is varied so as to minimize the *rate constant* at a given temperature.
  See also *microcanonical variational transition-state theory*
  1996, *68*, 155

capacitance (of a plate capacitor), *C*
The charge on a plate divided by the potential difference between the plates.
  G.B. 14; 1996, *68*, 964

capacitance hygrometer
Hygrometer using the capacitance variations of a capacitor whose dielectric medium consists of the gas or of a material in contact with this gas.
  1990, *62*, 2195

capillary column (in chromatography)
A general term for columns having a small diameter. A capillary column may contain a packing or have the stationary phase supported on its inside wall. The former case corresponds to packed capillary column while the latter case corresponds to an open-tabular column. Due to the ambiguity of this term its use without an adjective is discouraged.
  1993, *65*, 831

capillary condensation
Capillary condensation is said to occur when, in porous solids, multilayer adsorption from a vapour proceeds to the point at which pore spaces are filled with liquid separated from the gas phase by menisci.
The concept loses its sense when the dimensions of the pores are so small that the term meniscus ceases to have a physical significance. Capillary condensation is often accompanied by *hysteresis*.
  1972, *31*, 588; see also 1976, *46*, 76

captodative effect
Effect on the stability of a carbon-centred *radical* determined by the combined action of a captor (electron withdrawing) and a dative (electron releasing) substituent, both attached to the radical centre. The term is also used for certain unsaturated compounds.
  1994, *66*, 1092

capture
A process in which an atomic or nuclear system acquires an additional particle. In general, a specification is added of the type of the captured particle or its *energy*.
  1994, *66*, 2516

capture (in nuclear chemistry)
A process in which an atomic or nuclear system acquires an additional *particle*. In general a specification is added of the type of the captured particle or its energy.
  1982, *54*, 1537

capture cross-section
The *cross-section* for *capture*.
  1994, *66*, 2516

capture, electron
See *electron capture*.
1982, *54*, 1537

capture γ-radiation
The *γ-radiation* emitted in radiative capture.
1994, *66*, 2516

capture, radiative
See *radiative capture*.
1982, *54*, 1537

carbaboranes
A class of boron compound of general formula $[(CH)_a(BH)_mH_b]^c$ where c can be positive, negative or zero. The CH groups occupy polyhedron vertices, and other hydrogen atoms are either bridging or terminal.
Note:
The contraction 'carboranes' is also well established as the generic name for this class of compound.
R.B. 228; 1995, *67*, 1326

carbamates
Salts or esters of carbamic acid, $H_2NC(=O)OH$, or of *N*-substituted carbamic acids $R_2NC(=O)OR'$, (R' = hydrocarbyl or a cation). The esters are often called *urethanes* or urethans, a usage that is strictly correct only for the ethyl esters.
1995, *67*, 1323

carbanion
Generic name for anions containing an even number of electrons and having an unshared pair of electrons on a tervalent carbon atom (e.g. Cl_3C^- or $HC\equiv C^-$) or — if the ion is mesomeric (see *mesomerism*) — having at least one significant *contributing structure* with an unshared pair of electrons on a tervalent carbon atom, for example:

$$CH_3\overset{O^-}{\underset{|}{C}}=CH-\overset{O}{\underset{\|}{C}}CH_3 \longleftrightarrow CH_3\overset{O}{\underset{\|}{C}}-\overset{-}{C}H-\overset{O}{\underset{\|}{C}}CH_3$$

See also *radical ion*.
1994, *66*, 1092; see also 1995, *67*, 1324

+ carbena
This prefix is now rejected. It has been used in the past, especially in cyclic systems, to name *carbenes*, e.g. 'carbenacycloheptane', now cycloheptylidene.
1995, *67*, 1324

carbenes
The electrically neutral species $H_2C:$ and its derivatives, in which the carbon is covalently bonded to two univalent groups of any kind or a divalent group and bears two nonbonding electrons, which may be spin-paired (singlet state) or spin-non-paired (triplet state). In systematic name formation, carbene is the name of the parent hydride $:CH_2$; hence, the name dichlorocarbene for $:CCl_2$. However, names for acyclic and cyclic hydrocarbons containing one or more divalent carbon atoms are derived from the name of the corresponding all-λ^4-hydrocarbon using the suffix -ylidene.

E.g. prop-2-en-1-ylidene, $CH_2=CHCH:$ ethenylidene, $H_2C=\ddot{C}$; cyclohexylidene,

Subclasses of carbenes include *acyl carbenes* $RC(=O)\ddot{C}R$, *imidoyl carbenes*, $RC(=NR)\ddot{C}R$ and *vinyl carbenes*.
1995, *67*, 1324; see also 1994, *66*, 1092

carbene analogues
The electrically neutral mononuclear hydrides of group 14 having two non-bonding electrons, the electrically neutral mononuclear hydrides of group 15 having four non-bonding electrons, and also the compound $HB:$. The names of a number of these are shown below.

RB: boranylidenes	RAs: arsanylidenes
RN: nitrenes	R_2Sn: stannylidenes
R_2Si: silylenes	RSb: stibanylidenes
RP: phosphanylidenes	R_2Pb: plumbylidenes
R_2Ge: germylidenes	

Note that if R = H, these compounds are parent hydrides; derivatives formed by substitution are named accordingly. However, if the substituent's first atom, bearing the free valence, is of the same element as the atom of the carbene analogue, other parent compounds may be required (see *carbenes*), e.g. $HN:$ nitrene; $CH_3N:$ methylnitrene; $H_2NN:$ *diazanylidene* (synonym *isodiazene*, not *aminonitrene*).
1995, *67*, 1324

carbene metal complexes
See *metal–carbene complexes*.
1995, *67*, 1324

carbene radical anions
Species $R_2\ddot{C}\cdot^-$, having three non-bonding electrons, formally derived by addition of an electron to a *carbene*.
1995, *67*, 1324

carbene radical cations
Species $R_2C\cdot^+$, having one non-bonding electron, formally derived by subtraction of an electron from a *carbene*, e.g. $H_2C\cdot^+$ methyliumyl.
1995, *67*, 1324

carbenium centre
The three-coordinate carbon atom in a *carbenium ion* to which the excess positive charge of the ion (other than that located on heteroatoms) may be formally considered to be largely attributed, i.e. which has one vacant p-orbital. (N.B. It is not always possible to uniquely identify such an atom.) This formal attribution of charge often does not express the real charge distribution.
1994, *66*, 1092

carbenium ion
A generic name for *carbocations*, real or hypothetical, that have at least one important *contributing structure* containing a tervalent carbon atom with a vacant p-

orbital. (The name implies a protonated carbene or a substitution derivative thereof.)

The term was proposed (and rejected) as a replacement for the traditional usage of the name *carbonium ion*.

To avoid ambiguity, the name should not be used as the root for the systematic nomenclature of carbocations. The corresponding difficulty confused carbonium ion nomenclature for many years. For example, the term 'ethylcarbonium ion' has at times been used to refer either to $CH_3CH_2^+$ (ethyl cation) or (correctly) to $CH_3CH_2CH_2^+$ (propyl cation).

1994, *66*, 1092; 1995, *67*, 1324; 1994, *66*, 1092

carbenoids

Complexed carbene-like entities that display the reactivity characteristics of carbenes, either directly or by acting as sources of carbenes.

1995, *67*, 1325; 1994, *66*, 1092

+ carbinolamines

An obsolete term for *hemiaminals*.

1995, *67*, 1325

+ carbinols

An obsolete term for substituted methanols, in which the name carbinol is synonymous with methanol.

1995, *67*, 1325

+ carbinyl cations

An obsolete term, derived from carbinol, once used for *carbenium ions*.

1995, *67*, 1325

carbocation

A cation containing an even number of electrons with a significant portion of the excess positive charge located on one or more carbon atoms. This is a general term embracing *carbenium ions*, all types of *carbonium ions*, vinyl cations, etc. Carbocations may be named by adding the word 'cation' to the name of the corresponding radical. Such names do not imply structure (e.g. whether three-coordinated or five-coordinated carbon atoms are present).

See also *bridged carbocation, radical ion*.

1994, *66*, 1093; 1995, *67*, 1325

carbocyclic compounds

Cyclic compounds in which all of the ring members are carbon atoms, e.g. 1,2-dihydronaphthalene.

See also *homocyclic compounds*

1995, *67*, 1325

carbodiimides

Carbodiimide, HN=C=NH, and its *hydrocarbyl* derivatives.

1995, *67*, 1325

carbohydrates

Originally, compounds such as *aldoses* and *ketoses*, having the stoichiometric formula $C_n(H_2O)_n$, hence 'hydrates of carbon'. The generic term carbohydrate includes *monosaccharides, oligosaccharides* and *polysaccharides* as well as substances derived from monosaccharides by reduction of the carbonyl group (*alditols*), by oxidation of one or more terminal groups to carboxylic acids, or by replacement of one or more hydroxy group(s) by a hydrogen atom, an amino group, thiol group or similar groups. It also includes derivatives of these compounds.

1995, *67*, 1325

carbon

Element number 6 of the periodic table of elements (electronic ground state $1s^2 2s^2 2p^2$).

For a description of the various types of carbon as a solid the term carbon should be used only in combination with an additional noun or a clarifying adjective.

See also *amorphous carbon, carbon fibres, carbon material, glass-like carbon, graphitic carbon, non-graphitic carbon, pyrolytic carbon.*

1995, *67*, 479

carbonaceous mesophase

A liquid-crystalline state of *pitch* which shows the optical birefringence of disc-like (discotic) nematic liquid crystals. It can be formed as an intermediate phase during thermolysis (pyrolysis) of an isotropic molten pitch or by precipitation from pitch fractions prepared by selective extraction. Generally, the spherical mesophase precipitated from a pyrolysing pitch has the *Brooks and Taylor structure*. With continuous heat treatment the *carbonaceous mesophase* coalesces to a state of *bulk mesophase* before solidification to *green coke* with further loss of hydrogen or low-molecular-weight compounds.

Notes:

In the formation of carbonaceous mesophase by thermolysis (pyrolysis) of isotropic molten pitch, the development of a liquid-crystalline phase is accompanied by simultaneous aromatic polymerization reactions. The reactivity of pitch with increasing heat treatment temperature and its thermosetting nature are responsible for the lack of a true reversible thermotropic phase transition for the bulk mesophase in most pitches. Due to its glass-like nature most of the liquid-crystalline characteristics are retained in the super-cooled solid state.

1995, *67*, 483

carbon artifact

An 'artificially produced' solid body which consists mainly of carbonaceous material in a distinct shape.

Notes:

Sometimes this term is also used for artificially (in the sense of technically) produced non-shaped *carbon materials* such as *coke*, blacks, etc. This application of the term carbon artifact is not recommended. Synonyms to the term carbon artifact are: 'artificial carbon article' or 'artificial carbon body'.

1995, *67*, 479

carbon black

An industrially manufactured *colloidal carbon* material in the form of spheres and of their fused aggregates with sizes below 1000 nm.

Notes:
Carbon black is a commercial product manufactured by thermal decomposition, including detonation, or by incomplete combustion of carbon hydrogen compounds and has a well-defined morphology with a minimum content of tars or other extraneous materials.

For historical reasons, however, carbon black is popularly but incorrectly regarded as a form of *soot*. In fact, in many languages, the same word is used to designate both materials. Carbon black is manufactured under controlled conditions, whereas soot is randomly formed. They can be distinguished on the basis of tar, ash content and impurities.

Attempts in the literature to create a general term, 'aciniform carbon', which would cover both carbon black and soot, are not yet generally accepted.

1995, *67*, 479; 1990, *62*, 2178

carbon–carbon composite

A *carbon fibre*-reinforced carbon matrix material. The carbon matrix phase is typically formed by solid, liquid or gaseous pyrolysis of an organic precursor material. The matrix is either a *graphitizable carbon* or *non-graphitizable carbon*, and the carbonaceous reinforcement is fibrous in form. The composite may also contain other components in particulate or fibrous forms.

1995, *67*, 480

carbon cenospheres

Porous or hollow carbonaceous sphere-like particles (frequently in the size range of a few to several hundreds of μm diameter) formed during pyrolysis, also in the course of combustion, of carbonaceous liquid droplets (e.g. heavy fuel) or solid particles (e.g. coal).

1995, *67*, 480

carbon cloth

A textile material consisting of *carbon fibres* oriented at least in two directions. Carbon cloth is not necessarily woven.

1995, *67*, 480

carbon dioxide laser (CO_2 laser)

A continuous or pulsed source of coherent radiation normally tunable through the CO_2 vibration–rotation band centred near 10.6 μm.

See also *gas lasers, laser*.

1996, *68*, 2233

carbon electrode

An electrode for an electrical application. In its green state it comprises *granular carbon* material bound with pitch. The granular carbon material may be either *needle coke*, fine-grained or isotropic coke or reclaimed *graphite* powder. Electrodes for use in steel production can only be manufactured from needle coke, and the green electrodes are fired at temperatures above around 2800 K to produce highly graphitic electrodes (*graphite electrodes*). The other granular carbon materials may be used for aluminium electrodes where the duty is not so severe, and the green electrodes are generally fired to lower temperatures.

Notes:
In both cases it is essential that the granular carbons and the *pitch* binders used in the production of the green electrodes have a low sulfur content as the release of sulfur during the high-temperature firing can lead to the production of significant porosity.

See also *puffing, graphitic carbon, graphitization heat treatment, isotropic carbon*.

1995, *67*, 480

carbon felt

A textile material consisting of, in approximation, randomly oriented and intertwined *carbon fibres*.

Notes:
Carbon felts are usually fabricated by *carbonization* of organic felts but they can also be produced from short carbon fibres.

1995, *67*, 481

carbon fibre

Fibres (filaments, tows, yarns, rovings) consisting of at least 92% (mass fraction) *carbon*, usually in the non-graphitic state.

Notes:
Carbon fibres are fabricated by pyrolysis of organic precursor fibres or by growth from gaseous hydrocarbons. The use of the term *graphite fibres* instead of carbon fibres as often observed in the literature is incorrect and should be avoided. The term graphite fibres is justified only if three-dimensional crystalline order is confirmed, e.g. by X-ray diffraction measurements.

1995, *67*, 480

carbon fibre fabrics

Woven textile materials made of *carbon fibres*.

1995, *67*, 481

carbon fibres type HM

Carbon fibres type HM (high modulus) are *carbon fibres* with a value of Young's modulus (tensile modulus) larger than 300 GPa (nearly 30% of the C_{11} elastic constant of a graphite single crystal).

Notes:
The level of the tensile modulus of carbon fibres is controlled by the degree of preferred orientation of the layer planes in the direction parallel to the fibre axis. C_{11}, the elastic constant of graphite single crystals in the direction of the layer planes, is 1060 ± 20 GPa. In general, the ratio of tensile strength to tensile modulus is smaller than 1×10^{-2} for carbon fibres type HM (but the tensile strength is influenced by flaws in the fibres and may be improved in the future). Carbon fibres type UHM (ultra-high modulus) have moduli of elasticity in excess of 600 GPa, surpassing 50% of the theoretical C_{11} number. Such high values of Young's modulus can be achieved most readily in *mesophase pitch-based carbon fibres* (MPP-based carbon fibres).

1995, *67*, 481

carbon fibres type HT
Carbon fibres type HT are *carbon fibres* with values of Young's modulus between 150 and 275 to 300 GPa. The term HT, referring to high tensile strength, was early applied because fibres of this type display the highest tensile strengths.
Notes:
The disposition of boundaries between the fibre types is somewhat arbitrary. For carbon fibres type HT, the values of the strength-to-stiffness ratio are typically larger than 1.5×10^{-2}. The tensile strength of carbon fibres is flaw-controlled, however, and therefore the measured values increase strongly as the diameter of the filaments is decreased.
1995, *67*, 481

carbon fibres type IM
Carbon fibres type IM (intermediate modulus) are related to *carbon fibres type HT* because of the comparable values of tensile strength, but are characterized by greater stiffness (Young's modulus up to approximately 35% of the theoretical C_{11} value).
Notes:
The tensile modulus (Young's modulus) varies between *ca.* 275 and 350 GPa, but the disposition of the boundaries is somewhat arbitrary. The relatively high ratio of tensile strength to tensile modulus, typcally above 1×10^{-2}, in carbon fibres type IM, in spite of an increase of Young's modulus, requires a further increase of strength, which is achievable by a significant reduction of the monofilament diameter down to about 5 μm. Such small filament diameters are typical of carbon fibres type IM.
1995, *67*, 482

carbon fibres type LM
Carbon fibres type LM (low modulus) are *carbon fibres* with isotropic structure, tensile modulus values as low as 10% of the C_{11} values of the graphite single crystal, and low strength values.
Notes:
The term carbon fibres type LM is sometimes used for various types of isotropic carbon fibres known as *pitch-based* or *rayon-based carbon fibres* that have not been subjected to hot-stretching. Such fibres are not used for reinforcement purposes in high-performance composites.
1995, *67*, 482

carbon fibres type UHM
Carbon fibres type UHM (ultra-high modulus) designates a class of *carbon fibres* having very high values of Young's modulus larger than 600 GPa (i.e. greater than 55% of the theoretical C_{11} value of *graphite*).
1995, *67*, 482

carbonitriles
In systematic nomenclature, the suffix -carbonitrile is used to name compounds RC≡N where the suffix includes the carbon atom of the -CN. However, carbonitrile is not a class name for nitriles.
1995, *67*, 1325

carbonium ion
The term should be used with great care since several incompatible meanings are currently in use. It is not acceptable as the root for systematic nomenclature for *carbocations*.
1. In most of the existing literature the term is used in its traditional sense for what is here defined as *carbenium ion*.
2. A carbocation, real or hypothetical, that contains at least one five-coordinate carbon atom.
3. A carbocation, real or hypothetical, whose structure cannot adequately be described by two-electron two-centre *bonds* only. (The structure may involve carbon atoms with a *coordination number* greater than five.)
1994, *66*, 1093; 1995, *67*, 1325

carbonization
A process by which solid residues with increasing content of the element carbon are formed from organic material usually by *pyrolysis* in an inert atmosphere.
Notes:
As with all pyrolytic reactions, carbonization is a complex process in which many reactions take place concurrently such as dehydrogenation, condensation, hydrogen transfer and isomerization. It differs from *coalification* in that its reaction rate is faster by many orders of magnitude. The final pyrolysis temperature applied controls the degree of carbonization and the residual content of foreign elements, e.g. at $T \sim 1200$ K the carbon content of the residue exceeds a mass fraction of 90 wt.%, whereas at $T \sim 1600$ K more than 99 wt.% carbon is found.
1995, *67*, 484

carbon material
A solid high in content of the element *carbon* and structurally in a non-graphitic state.
Notes:
The use of the term carbon as a short term for a material consisting of *non-graphitic carbon* is incorrect. The use of the term carbon without a second noun or a clarifying adjective should be restricted to the chemical element carbon. The term carbon can be used in combination with other nouns or clarifying adjectives for special types of *carbon materials* (*carbon electrode, carbon fibres, pyrolytic carbon, glass-like carbon* and others).
1995, *67*, 482

carbon mix
A mixture of *filler coke*, e.g. grains and/or powders of solid *carbon materials*, and a carbonaceous *binder* and selected additives, prepared in heated mixers at temperatures in the range of 410–445 K as a preliminary step for the formation of shaped green bodies.
1995, *67*, 483

carbon whiskers
See *graphite whiskers*.
1995, *67*, 483

carbonyl compounds
1. Compounds containing the carbonyl group, C=O. The term is commonly used in the restricted sense of *aldehydes* and *ketones*, although it actually includes *carboxylic acids* and derivatives.
  See also *oxo compounds*.
2. Metal carbonyls, in which carbon monoxide is a formal ligand.
  1995, *67*, 1325

carbonyl imides
1,3-Dipolar compounds having the structure $R_2C=O^+-N^-R \leftrightarrow R_2C^+-O-N^-R$.
  See also *oxonium ylides (2)*.
  1995, *67*, 1325

+ carbonyl imines
An undesirable synonym for *carbonyl imides*.
  See *ylides*.
  1995, *67*, 1325

carbonyl oxides
1,3-*Dipolar compounds* having the structure $R_2C^- - O^+ = O \leftrightarrow R_2C=O^+-O^-$. Also called peroxo compounds.
  See also *oxonium ylides (2)*.
  1995, *67*, 1325

carbonyl ylides
1,3-*Dipolar compounds* having the structure $R_2C=O^+-C^-R_2 \leftrightarrow R_2C^+-OC^-R_2$.
  See also *oxonium ylides (2)*.
  1995, *67*, 1325

carboranes
  See *carbaboranes*.
  R.B. 228; 1995, *67*, 1326

carboxamides
Amides of *carboxylic acids*, having the structure $RC(=O)NR_2$. The term is used as a suffix in systematic name formation to denote the -C(=O)NH$_2$ group including its carbon atom.
  1995, *67*, 1326

carboxamidines
Compounds having the structure $RC(=NR)NR_2$. The term is used as a suffix in systematic nomenclature to denote the -C(=NH)NH$_2$ group including its carbon atom, e.g. acetamidine, $CH_3C(=NH)NH_2$; 2-butyl-4,5-dihydroimidazole,

  1995, *67*, 1326

carboxylic acids
Oxoacids having the structure $RC(=O)OH$. The term is used as a suffix in systematic name formation to denote the -C(=O)OH group including its carbon atom.
  1995, *67*, 1326

+ carbylamines
An obsolete term, which should not be used, for isocyanides.
  1995, *67*, 1326

carbynes
The neutral species $H\dot{C}$: and its derivatives formed by substitution in which a univalent carbon atom is covalently bonded to one group and also bears three nonbonding electrons. (This term carries no implication about spin-pairing).
  1995, *67*, 1326; see also 1994, *66*, 1093

carbyne metal complexes
  See *metal–carbyne complexes*.
  1995, *67*, 1326

carbynium ions
The cationic species $H_2C^{.+}$ or substitution derivatives thereof, formally derived by adding a hydron to a *carbyne* or subtracting an electron from a *carbene*.
  1996, *68*, 1326

carotenes
Hydrocarbon carotenoids (a subclass of tetraterpenes).
  See also *terpenes, carotenoids*.
  1995, *67*, 1326

carotenoids
Tetraterpenoids (C_{40}), formally derived from the acyclic parent ψ,ψ-carotene I by hydrogenation, dehydrogenation, cyclization, oxidation, or combination of these processes. This class includes *carotenes*, *xanthophylls* and certain compounds that arise from rearrangement of the skeleton of I or by loss of part of this structure. *Retinoids* are excluded.

lycopene, ψ,ψ-carotene (I)

zeaxanthin, β,β-carotene-3,3'-diol (a xanthophyll)

8'-apo-β-caroten-8'-ol

  See also *retro*.
  1995, *67*, 1326

carrier
A substance in appreciable amount which, when associated with a trace of a specified substance, will carry the trace with it through a chemical or physical process.
  See also *support (of a catalyst)*.
  1982, *54*, 1537; see also 1976, *46*, 79

carrier (in radioanalytical chemistry)
A substance in appreciable amount which, when associated with a *tracer* of a specified substance, will carry the tracer with it through a chemical or physical

process, or prevent the tracer from undergoing non-specific processes due to its low concentration.
1994, *66*, 2516

carrier atom (in organic reaction mechanisms)
A participant atom which is neither core nor peripheral and which carries other groups to and from the molecule containing the *core atom(s)* but which does not itself become covalently attached to that molecule.
1989, *61*, 27

+ carrier-free
See *no carrier added*, which term should be preferred.
1994, *66*, 2516

carrier gas
A gas introduced in order to transport a sample for analytical purposes. In gas chromatography it is the gas which is passed continuously through the column and whose passage promotes the elution of the components of the sample. The carrier gas together with the portions of the sample present in this phase constitutes the *mobile phase*.
1990, *62*, 2178; O.B. 97 and 136

carrier, hold-back
See *hold-back carrier*.
1994, *66*, 2516

carrier, isotopic
See *isotopic carrier*.
1994, *66*, 2516

carry-over
A process by which materials are carried into a reaction mixture to which they do not belong. These materials can be either parts of a specimen, or reagents including the diluent or wash solution. In such cases, carry-over means the transfer of material (specimen or reagents) from one container, or from one reaction mixture, to another one. It can be either unidirectional or bidirectional in a series of specimens or assays. The term *carry-over effect* is used for carry-over from specimen to specimen.
1991, *63*, 302; 1989, *61*, 1663

carry-over effect
See *carry-over*.
1991, *63*, 302

cascade, γ
See *γ-cascade*.
1982, *54*, 1543

cascade impactor
An instrument used for the classification of aerosols according to size and for possible subsequent chemical analysis. Air is drawn through a series of orifices of decreasing size; the air flow is normal to collecting surfaces on which aerosols are collected by inertial impaction. The particles, separated stepwise by their momentum differences into a number of size ranges, are collected simultaneously.
1990, *62*, 2196

catabolism
1. Reactions involving the oxidation of organic *substrates* to provide chemically available energy (e.g. ATP) and to generate metabolic intermediates.
2. Generally, the process of breakdown of complex molecules into simpler ones, often providing biologically available energy.
See also *anabolism*.
1993, *65*, 2013; 1992, *64*, 149

catabolite repression
A decrease in the activity of certain auxiliary catabolic *enzymes* when a surplus of an (easily metabolizable) substrate is available. Commonly this effect is caused by glucose (glucose *repression*) or by *metabolites* produced from glucose.
1992, *64*, 149

catalymetric titration
A titration process, which involves a catalyst, where the end-point is detected by the sudden increase or decrease of the rate of a reaction.
1993, *65*, 2293

catalysed rate of reaction, $\dot{\xi}_{cat}$
The observed rate of reaction, $\dot{\xi}$, minus the spontaneous rate of reaction, $\dot{\xi}_0$, observed under otherwise similar conditions in the absence of added catalytic material.
1979, *51*, 2464

catalysis
The action of a *catalyst*.
1994, *66*, 1093

catalysis law
See *Brønsted relation*.
1994, *66*, 1093

catalyst
A substance that increases the rate of a reaction without modifying the overall standard Gibbs energy change in the reaction; the process is called catalysis. The catalyst is both a *reactant* and *product* of the reaction. The words catalyst and catalysis should not be used when the added substance reduces the rate of reaction (see *inhibitor*).
Catalysis can be classified as homogeneous catalysis, in which only one phase is involved, and heterogeneous catalysis, in which the reaction occurs at or near an interface between phases. Catalysis brought about by one of the products of a reaction is called autocatalysis. Catalysis brought about by a group on a reactant molecule itself is called intramolecular catalysis.
The term catalysis is also often used when the substance is consumed in the reaction (for example: base-catalysed hydrolysis of esters). Strictly, such a substance should be called an activator.
See also *autocatalytic reaction, bifunctional catalysis, catalytic coefficient, electron-transfer catalysis, general acid catalysis, general base catalysis, intramolecular catalysis, micellar catalysis, Michaelis–Menten kinetics, phase-trans-*

fer catalysis, pseudo-catalysis, rate of reaction, specific catalysis.
1996, *68*, 155; see also 1994, *66*, 1093; 1993, *65*, 2293; 1990, *62*, 2178

catalyst (in solvent extraction)
A substance included in the solvent to increase the rate of transfer without affecting the position of equilibrium. The term accelerator may also be used but kinetic synergist is not recommended.
1993, *65*, 2380

catalyst ageing
Catalyst deactivation resulting from changes in the structure or in the texture of the *catalyst*. Changes of this kind are usually irreversible and the catalyst cannot be regenerated.
1976, *46*, 84

catalyst deactivation
See *catalyst decay*.
1976, *46*, 83

catalyst decay
The conversion in a *catalytic reaction* performed under constant conditions of reaction often decreases with *time of run* or *time on stream*. This phenomenon is called catalyst deactivation or catalyst decay.
1976, *46*, 83

catalytic activity (of an enzyme), z
Increase in the rate of reaction of a specified chemical reaction that an enzyme produces in a specific assay system.
See *rate of reaction*.
1996, *68*, 964; 1994, *66*, 2593

catalytic activity concentration, b
Catalytic activity of the component divided by the volume of the system.
Notes:
1. In clinical chemistry *litre* is recommended as unit for the volume.
2. In clinical chemistry the component is usually an enzyme.
3. The term *catalytic concentration* is accepted for use in clinical chemistry.
4. Use of the term level as a synonym for concentration is deprecated.
5. In describing a quantity, *concentration* must be clearly differentiated from *content*.
1996, *68*, 964

catalytic activity content, z/m
Catalytic activity of the component divided by the mass of the system.
Notes:
1. In clinical chemistry the component is usually an enzyme.
2. The term *catalytic content* is accepted for use in clinical chemistry.
3. Use of the term level as a synonym for concentration is deprecated.
4. In describing a quantity, *content* must be clearly differentiated from concentration.
1996, *68*, 964

catalytic activity fraction (of an isozyme)
Quotient of the catalytic activity of the isozyme and the catalytic activity of all the isozymes of the same enzyme in the system.
Notes:
1. The term *catalytic fraction* is accepted for use in clinical chemistry.
2. The definition is based on the definitions of *catalytic activity* and *fraction*.
3. The definition also applies to other multiple forms of an enzyme that are not isoenzymes.
1996, *68*, 964

catalytic coefficient
If the *rate of reaction*, v, is expressible in the form:
$$v = (k_0 + \sum_i k_i [C_i]^{n_i}) [A]^\alpha [B]^\beta \ldots$$
where A, B, ... are *reactants* and C_i represents one of a set of catalysts, then the proportionality factor k_i is the catalytic coefficient of the particular catalyst C_i. Normally the partial *order of reaction* n_i with respect to a catalyst is unity, so that k_i is an $(\alpha + \beta + \ldots + 1)$th order rate coefficient. The proportionality factor k_0 is the $(\alpha + \beta + \ldots)$th order rate coefficient of the uncatalysed component of the total reaction.
For example, if there is catalysis by hydrogen and hydroxide ions, and the rate constant can be expressed in the form:
$$k = k_0 + k_{H+}[H^+] + k_{OH-}[OH^-],$$
then k_{H+} and k_{OH-} are the catalytic coefficients for H^+ and OH^- respectively. The constant k_0 relates to the uncatalysed reaction.
1996, *68*, 156; 1994, *66*, 1093

catalytic concentration
See *catalytic activity concentration*.
1996, *68*, 965

catalytic content
See *catalytic activity content*.
1996, *68*, 965

catalytic cracking
See *cracking*.
1990, *62*, 2183

catalytic current
The *faradaic current* that is obtained with a solution containing two substances B and A may exceed the sum of the faradaic currents that would be obtained with B and A separately, but at the same concentrations and under the same experimental conditions. In either of the two following situations the increase is termed a catalytic current.
B is reduced or oxidized at the electrode-solution interface to give a product B' that then reduces or oxidizes A chemically. The reaction of B' with A may yield either B or an intermediate in the overall half-reaction by which B' was obtained from B. In this situation the increase of current that results from the addition of A to a solution of B may be termed a regeneration current.

The presence at the electrode-solution interface of one substance, which may be either A or the product A′ of its reduction or oxidation, decreases the over-potential for the reduction or oxidation of B.
In either case the magnitude of the catalytic current depends on the *applied potential*.
If the current observed with a mixture of A and B is smaller than the sum of the separate currents, the term non-additive current should be used.
1985, *57*, 1494

catalytic current, limiting
See *limiting catalytic current*.
1985, *57*, 1495

catalytic dehydrocyclization
A reaction in which an alkane is converted into an aromatic hydrocarbon and hydrogen. For example,

heptane → toluene + 4H$_2$

1976, *46*, 86

catalytic domain (of a polypeptide)
Any part of a *polypeptide* chain that possesses a catalytic function. It may contain more than one structural domain.
W.B. 107

catalytic fraction
See *catalytic activity fraction*.
1996, *68*, 964

catalytic graphitization
The transformation of *non-graphitic carbon* into *graphite* by heat treatment in the presence of certain metals or minerals.
Notes:
Catalytic graphitization gives a fixed degree of *graphitization* at lower temperature and/or for a shorter heat treatment time than in the absence of the catalytic additives (or a higher degree of graphitization at fixed heat treatment conditions). Often it involves dissolution of *carbon* and precipitation of *graphite* at the catalyst particles so that non-graphitizing carbons can be graphitized by this procedure.
1995, *67*, 484

catalytic hydrocracking
A process similar to *catalytic cracking* in its industrial purpose but effected under hydrogen pressure and on a *catalyst* containing an ingredient with a hydrogenating function.
1976, *46*, 86

catalytic hydrodesulfurization
A process in which, in the presence of hydrogen, sulfur is removed as hydrogen sulfide.
1976, *46*, 86

catalytic hydrogenolysis
Ordinarily, a reaction in which **A** + H$_2$ gives **B**.

$$-\overset{|}{\underset{|}{C}}-\overset{|}{\underset{|}{C}}-$$
A

$$-\overset{|}{\underset{|}{CH}} + \overset{|}{\underset{|}{HC}}-$$
B

For example,

propane + H$_2$ → ethane + methane

toluene + H$_2$ → benzene + methane

butane + H$_2$ → 2 × ethane

The term may also be used for cleavage of bonds other than **A**, e.g.

benzyl acetate + H$_2$ → toluene + acetic acid

benzylamine + H$_2$ → toluene + NH$_3$

1976, *46*, 86

catalytic methanation
A process for removing carbon monoxide from gas streams or for producing methane by the reaction

CO + 3H$_2$ → CH$_4$ + H$_2$O

1976, *46*, 86

catalytic reforming
A process for increasing the octane number of naphthas. It involves *isomerization* of alkanes, dehydrogenation of cyclohexanes to aromatic hydrocarbons, isomerization and dehydrogenation of alkylcyclopentanes, and dehydrocyclization of alkanes.
1976, *46*, 86

+ **catalytic thermometric titration**
See *thermometric titration*.
1994, *66*, 2490

+ **cataphoresis (usage discouraged)**
See *electrophoresis*.
1972, *31*, 619

catecholamines
4-(2-Aminoethyl)pyrocatechol [4-(2-aminoethyl)-benzene-1,2-diol] and derivatives formed by substitution.

1995, *67*, 1327

catena-
An affix used in inorganic nomenclature to denote a chain structure; often used to designate linear polymeric substances.
R.B. 245; B.B. 464

catenanes (catena) compounds
Hydrocarbons having two or more rings connected in the manner of links of a chain, without a covalent bond. More generally, the class catena compounds embraces functional derivatives and hetero analogues, e.g:

a [2]catenane

1995, *67*, 1327

cathode
Electrode at which reduction takes place.
G.B. 59

cathodic transfer coefficient, α_c
For a reaction with a single rate-determining step
$$\alpha_c/v = -(RT/nF)(\partial \ln |I_c|/\partial E)_{T, p, c_i \ldots}$$
where α_c is the cathodic transfer coefficient (number), R is the gas constant, T is the thermodynamic temperature, and v is the stoichiometric number giving the number of identical activated complexes formed and destroyed in the completion of the overall reaction as formulated with the transfer of n electrons.
1974, *37*, 515

cation
A monoatomic or polyatomic species having one or more elementary charges of the proton.
R.B. 102

cation exchange
The process of exchanging cations between a solution and a cation exchanger.
1993, *65*, 854

cation exchanger
Ion exchanger with cations as counter-ions. The term cation-exchange resin may be used in the case of solid organic polymers. The acid form of a cation exchanger is the ionic form of a cation exchanger in which counter-ions are hydrogen ions (H-form) or the ionogenic groups have added a proton forming an undissociated acid.
1993, *65*, 855

cationic polymerization
An *ionic polymerization* in which the kinetic-chain carriers are cations.
1996, *68*, 2308

cationotropic rearrangement (cationotropy)
See *tautomerism*.
1994, *66*, 1093

cationotropy
See *tautomerism*.
1994, *66*, 1093

cation radical
See *radical ion*.
1994, *66*, 1093

cavitands
Compounds constrained by structure to having or accommodating a cavity large enough to host other molecules.
See *inclusion compounds*.
1995, *67*, 1327

cavity dumping (in photochemistry)
Periodic removal of *coherent radiation* from a laser cavity.
1996, *68*, 2231

cDNA
See *complementary DNA*.
1992, *64*, 149

ceilometer (in atmospheric chemistry)
An automatic, recording instrument for reading the height of the cloud-base.
1990, *62*, 2178

cell constant (of a conductivity cell)
Defined as:
$$\kappa_{\text{cell}} = \kappa R$$
where R is the measured resistance of the cell and κ is the conductivity (formerly called the specific conductance).
1974, *37*, 511

cell, continuous-flow (in spectrochemical analysis)
See *continuous-flow cell (in spectrochemical analysis)*.
1988, *60*, 1454

cell, continuous measuring
See *continuous measuring cell*.
1990, *62*, 2178

cell, discontinuous measuring
See *discontinuous measuring cell*.
1990, *62*, 2178

cell error (in spectrochemical analysis)
The error which results if the incident beam does not fall perpendicularly to the windows of the cell, or if the cell windows are contaminated or have some other imperfection.
1988, *60*, 1457

cell, multiple-pass (in spectrochemical analysis)
See *multiple-pass cell (in spectrochemical analysis)*.
1988, *60*, 1454

cell, sequential measuring
See *sequential measuring cell*.
1990, *62*, 2178

cells, matched (in spectrochemical analysis)
See *matched cells (in spectrochemical analysis)*.
1988, *60*, 1453

cell, stopped-flow (in spectrochemical analysis)
See *stopped-flow cell (in spectrochemical analysis)*.
1988, *60*, 1454

cell, variable pathlength (in spectrochemical analysis)
See *variable pathlength cell (in spectrochemical analysis)*.
1988, *60*, 1454

Celsius temperature, t, θ
Thermodynamic temperature minus 273.15 kelvin, invariably expressed in the SI unit degree Celsius (°C) which is equal to the *kelvin*. Sometimes the misnomer centigrade temperature is used for Celsius temperature.
G.B. 48; 1996, *68*, 965

centi
SI prefix for 10^{-2} (symbol: c).
G.B. 74

central atom
The atom in a coordination entity which binds other atoms or groups of atoms (ligands) to itself, thereby occupying a central position in the coordination entity.
R.B. 146

centre (of a Mössbauer spectrum)
The centre of a Mössbauer spectrum is defined as the Doppler-velocity at which the resonance maximum is (or would be) observed when all magnetic dipole, electric quadrupole, etc. hyperfine interactions are (or would be) absent. The contribution of the second order Doppler shift (δ_T) should be indicated, if possible
1976, *45*, 215

centre of chirality
See *chirality centre*.
1996, *68*, 2202

centrifugal (centripetal) acceleration, a_{rot}
The acceleration of a component as a result of a uniform rotational motion. Centrifugal acceleration is a vector quantity.
1994, *66*, 900; 1996, *68*, 965

centrifugal barrier
In a reaction without an electronic energy barrier, or its reverse, the rotational energy of the transition state gives rise to a reaction barrier, which is known as the centrifugal barrier.
1996, *68*, 156

centrifugal flow
See *transport*.
1989, *61*, 1661

centrifugal force, F_{rot}
Fictitious force acting on a body as a result of centripetal acceleration.
1996, *68*, 965

centrifugal radius, *r*
The radius at which a component is spinning at the end of the period of centrifugation. For a component sedimented from a dilute suspension, it can be equated with radius of rotation at the bottom of the centrifuge tube.
1994, *66*, 901; 1996, *68*, 965

+ cephalins (kephalins)
Compounds derived from glycerol in which a primary and the secondary hydroxy groups are esterified with long-chain *fatty acids*, and the remaining primary one with the mono(2-aminoethyl) ester of phosphoric acid, or with the monoserine ester of phosphoric acid (see **I**). The term is not recommended. These compounds are preferably designated as (3-phosphatidyl)ethanolamines and (3-phosphatidyl)serines respectively.
See also *phosphatidic acids*.
1995, *67*, 1327

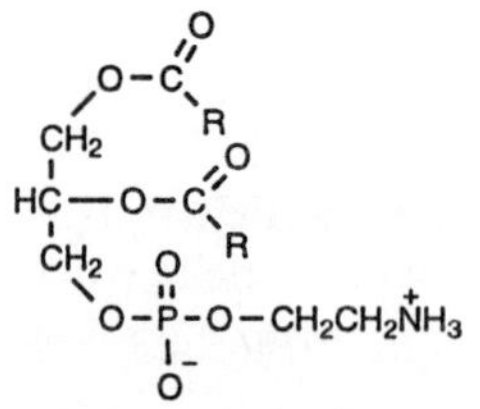

I (See **cephalins**)

cephalosporins
Cephems having the basic structure shown.

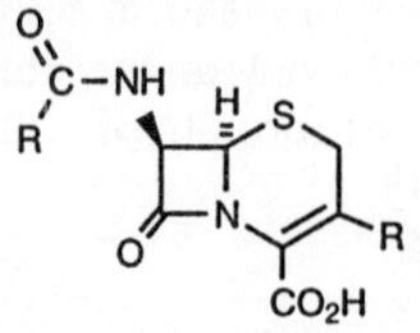

See also *cephams, cephems, penams, penems, penicillins*.
1995, *67*, 1327

cephams
Natural and synthetic antibiotics containing the 5-thia-1-azabicyclo[4.2.0]octan-8-one nucleus; generally assumed to have the 6*R* configuration, unless otherwise specified. The numbering used differs from that of the von Baeyer named bicyclic system. Where they differ, the von Baeyer numbering is shown in parentheses.

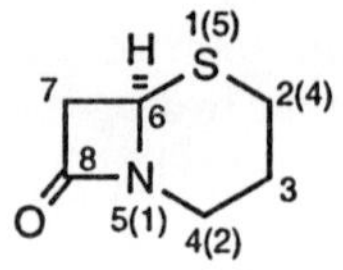

1995, *67*, 1327

cephems
2,3-Didehydrocephams: 5-thia-1-azabicyclo-[4.2.0]oct-2-en-8-ones.
See *cephalosporins, cephams*.
1995, *67*, 1327

ceramic filter
A component of a stack sampling system which is suitable for high temperature use; also known as a ceramic thimble.
1990, *62*, 2178

Cerenkov detector
A charged *particle detector* based on the *Cerenkov effect*.
1982, *54*, 1537

Cerenkov effect
Emission of *radiation* in the visible and ultraviolet spectrum arising when a charged particle crosses a medium with a velocity greater than that of light in the same medium.
1982, *54*, 1537

Cerenkov radiation
Radiation resulting from the *Cerenkov effect*.
1982, *54*, 1537

chain (in polymers)
The whole or part of a *macromolecule*, an *oligomer molecule* or a *block*, comprising a linear or branched

sequence of *constitutional units* between two boundary constitutional units, each of which may be either an *end-group*, a *branch point* or an otherwise-designated characteristic feature of the macromolecule.
Notes:
1. Except in linear single-strand macromolecules, the definition of a chain may be somewhat arbitrary.
2. A cyclic macromolecule has no end groups but may nevertheless be regarded as a chain.
3. Any number of branch points may be present between the boundary units.
4. Where appropriate, definitions relating to 'macromolecule' may also be applied to 'chain'.
1996, *68*, 2293

chain axis (of a polymer)
The straight line parallel to the direction of chain extension, connecting the centres of mass of successive blocks of chain units, each of which is contained within an identity period.
P.B. 77

chain branching
When in a *chain reaction* there is a net increase in the number of *chain carriers* there is said to be chain branching. A simple example of a chain-propagating reaction leading to chain branching is:

$O + H_2 \rightarrow OH + H$

in which there is one *chain carrier* (an oxygen atom) on the left and two chain carriers (a hydrogen atom and a hydroxyl radical) on the right.
See also *degenerate chain branching*.
1996, *68*, 156

chain carrier
A species, such as an atom or free radical, which is involved in *chain-propagating reactions* is known as a chain carrier.
1996, *68*, 156

(chain) conformational repeating unit (of a polymer)
The smallest structural unit of a polymer chain with a given conformation that is repeated along that chain through symmetry operations.
P.B. 77

chain-ending step
See *initiation*.
1981, *53*, 762

chain fission yield
The fraction of *fissions* giving rise to *nuclei* of a particular *mass number*.
1982, *54*, 1537

chain folding (in polymer crystals)
The conformational feature in which a loop connects two parallel stems belonging to the same crystal.
P.B. 83

(chain) identity period (of a polymer)
The shortest distance along the chain axis for translational repetition of the chain structure.

The chain identity period is usually denoted by c. Synonymous with (chain) repeating distance.
P.B. 77

chain initiation
The process in a *chain reaction* that is responsible for the formation of a *chain carrier*.
1996, *68*, 156

chain length, δ
In a *chain reaction*, the average number of times the closed cycle of reactions (involving the *chain-propagating reactions*) is repeated. It is equal to the rate of the overall reaction divided by the rate of the *initiation reaction*.
1996, *68*, 157

chain-orientational disorder (in polymer crystals)
Structural disorder resulting from the statistical coexistence within the crystals of identical chains with opposite orientations.
A typical example is provided by the up-down statistical coexistence of anticlined chains in the same crystal structure.
P.B. 80

chain polymerization
A chain reaction in which the growth of a *polymer chain* proceeds exclusively by reaction(s) between *monomer(s)* and reactive site(s) on the polymer chain with regeneration of the reactive site(s) at the end of each growth step.
Notes:
1. A chain polymerization consists of initiation and propagation reactions, and may also include termination and *chain transfer* reactions.
2. The adjective 'chain' in 'chain polymerization' denotes a 'chain reaction' rather than a 'polymer chain'.
3. Propagation in chain polymerization usually occurs without the formation of small molecules. However, cases exist where a low-molar-mass by-product is formed, as in the polymerization of oxazolidine-2,5-diones derived from amino acids (commonly termed amino-acid *N*-carboxy anhydrides). When a low-molar-mass by-product is formed, the adjective 'condensative' is recommended to give the term *condensative chain polymerization*.
4. The growth steps are expressed by:

$P_x + M \rightarrow P_{x+1} (+L) \{x\} \in \{1, 2, ... \infty\}$

where P_x denotes the growing chain of degree of polymerization x, M a monomer and L a low-molar-mass by-product formed in the case of condensative chain polymerization.
5. The term 'chain polymerization' may be qualified further, if necessary, to specify the type of chemical reactions involved in the growth step, e.g. ring-opening chain polymerization, cationic chain polymerization.
6. There exist, exceptionally, some polymerizations that proceed *via* chain reactions that, according to the

definition, are not chain polymerizations. For example, the polymerization:

$$HS-X-SH + H_2C=CH-Y-CH=CH_2 \rightarrow$$
$$S-X-S-CH_2-CH_2-Y-CH_2-CH_2$$

proceeds *via* a radical chain reaction with intermolecular transfer of the radical centre. The growth step, however, involves reactions between molecules of all degrees of polymerization and, hence, the polymerization is classified as a *polyaddition*. If required, the classification can be made more precise and the polymerization described as a chain-reaction polyaddition.
  1996, *68*, 2306

chain-propagating reaction
A chain-propagating reaction, or more simply a propagating reaction, is an elementary step in a chain reaction in which one *chain carrier* is converted into another. The conversion can be a unimolecular reaction or a bimolecular reaction with a reactant molecule.
  1996, *68*, 157

chain reaction
A reaction in which one or more reactive reaction *intermediates* (frequently radicals) are continuously regenerated, usually through a repetitive cycle of elementary steps (the 'propagation step'). For example, in the chlorination of methane by a radical *mechanism*, Cl· is continuously regenerated in the chain propagation steps:

$$Cl\cdot + CH_4 \rightarrow HCl + H_3C\cdot \quad \text{Propagation steps}$$
$$H_3C\cdot + Cl_2 \rightarrow CH_3Cl + Cl\cdot$$

In chain polymerization reactions, reactive intermediates of the same types, generated in successive steps or cycles of steps, differ in relative molecular mass, as in:

$$RCH_2\dot{C}HPh + H_2C=CHPh$$
$$\rightarrow RCH_2CHPhCH_2\dot{C}HPh$$

See also *chain branching, chain transfer, degenerate chain branching, initiation, termination*.
  1994, *66*, 1094; 1993, *65*, 2293; 1996, *68*, 157; see also 1990, *62*, 2179

chain scission (of a polymer)
A chemical reaction resulting in the breaking of *skeletal bonds*.
  1996, *68*, 2309

chain-termination reaction
  See *termination*.
  1996, *68*, 157; 1994, *66*, 1173

chain transfer
The abstraction, by the *radical* end of a growing chain-polymer, of an atom from another molecule. The growth of the polymer chain is thereby terminated but a new radical, capable of chain propagation and polymerization, is simultaneously created. For the example of alkene polymerization cited for a *chain reaction*, the reaction:

$$RCH_2\dot{C}HPh + CCl_4 \rightarrow RCH_2CHClPh + Cl_3\dot{C}$$

represents a chain transfer, the radical $Cl_3C\cdot$ inducing further polymerization:

$$CH_2=CHPh + Cl_3C\cdot \rightarrow Cl_3CCH_2\dot{C}HPh$$
$$Cl_3CCH_2\dot{C}HPh + CH_2=CHPh$$
$$\rightarrow Cl_3CCH_2CHPhCH_2\dot{C}HPh \quad \text{etc.}$$

The phenomenon occurs also in other chain reactions such as cationic polymerization.
  See also *telomerization*.
  1994, *66*, 1094

chair, boat, twist
If carbon atoms 1, 2, 4 and 5 of cyclohexane occupy coplanar positions and when carbon atoms 3 and 6 are on opposite sides of the plane the *conformation* (of symmetry group D_{3d}) is called a chair form.
The same term is applied to similar conformations of analogous saturated six-membered ring structures containing hetero-atoms and/or bearing substituent groups, but these conformations may be distorted from the exact D_{3d} symmetry. For cyclohexane and most such analogues, the chair form is the most stable conformation. If the cyclohexane conformation has no centre of symmetry but possesses two planes of symmetry, one of them bisecting the bonds between atoms 1 and 2 and between 4 and 5 and the other plane passing through atoms 3 and 6 (which lie out of the plane and on the same side of the plane containing 1, 2, 4 and 5), that conformation (of symmetry group C_{2v}) is called a boat form and it is generally not a stable form. Again, this term is also applied to structural analogues.
The conformation of D_2 symmetry passed through in the interconversion of two boat forms of cyclohexane is called the twist form (also known as skew boat, skew form and stretched form).
  See also *half-chair*.

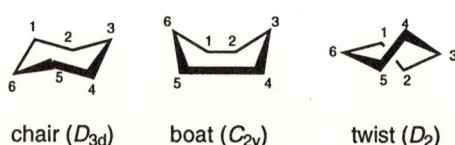

chair (D_{3d})   boat (C_{2v})   twist (D_2)

In a five-membered ring a conformation in which two adjacent atoms are maximally displaced, in opposite directions, relative to the plane containing the other three carbon atoms has been called a half-chair but is better called a twist conformation.
  See also *envelope conformation*.
In carbohydrate chemistry the term twist refers to a five-membered ring and the D_2 symmetry six-membered ring is referred to as skew.
  1996, *68*, 2202

chair–chair interconversion
  See *ring reversal*.
  1996, *68*, 2202

chalcones
1,3-Diphenylpropenone (benzylideneacetophenone), ArCH=CHC(=O)Ar, and its derivatives formed by substitution.
1995, 67, 1327

chamber saturation (in gas chromatography)
Uniform distribution of the *eluent* vapour throughout the chamber prior to *chromatography*.
O.B. 99

change of a quantity
The increment of the value of a quantity Q with time. The change may be expressed either infinitesimally at time t by the differential dQ or $dQ(t)$, or in practice it may be expressed by a finite increment over the time interval $(t_1;t_2)$, i.e. $Q(t_2) - Q(t_1)$, which may be written ΔQ or $\Delta Q\,(t_1; t_2)$

$$\Delta Q = Q(t_1; t_2) = Q(t_2) - Q(t_1)$$

Examples are: mass change, Δm; amount of substance change, Δn; volume change, ΔV; substance concentration change, Δc.
1992, 64, 1571

change ratio of a quantity
A term which may be expressed infinitesimally at time t by a ratio of differentials $dQ_1(t)/dQ_2(t)$ where the kind of quantities are the same but for different components in the same system. In practice, the ratio for a finite interval is:

$$\Delta Q_1(t_1: t_2)/\Delta Q_2(t_1; t_2)$$

Examples are: mass change ratio, $dm_1(t)/dm_2(t)$; amount of substance change ratio, $dn_1(t)/dn_2(t)$.
1992, 64, 1571

channel
1. Regions of *potential-energy surfaces* where there are valleys are sometimes referred to as arrangement channels. The *reactant* channel or entrance channel is that corresponding to configurations similar to those of the reactants; the *product* channel or exit channel relates to configurations similar to those of the products. In collision theory, the specification of a complete set of quantum numbers in a given arrangement channel is called a channel.
   See also *entrance channel, exit channel*.
1996, 68, 157
2. (in analysis)
That part of an analytical instrument that is dedicated to a single analytical procedure, including the transducer.
This term relates to the internal operation of certain types of instruments, particularly in clinical chemistry. The general term channel has a much broader meaning, and the term is not recommended.
1989, 61, 1661

channel black
   See *gas black*.
1990, 62, 2191

channel photomultiplier tube
   See *photomultiplier tube*.
1995, 67, 1753

chaperon
A species that facilitates a combination reaction between atoms or radicals, or the reverse process, is known as a chaperon. It is a special case of a *third body*.
1996, 68, 157

char
A solid decompostion product of a natural or synthetic organic material.
Notes:
If the precursor has not passed through a fluid stage, char will retain the characteristic shape of the precursor (although becoming of smaller size). For such materials the term 'pseudomorphous' has been used. Some simple organic compounds, e.g. sugar, melt at an early stage of decomposition and then polymerize during *carbonization* to produce chars.
1995, 67, 484

characteristic (in analytical chemistry)
A property or attribute of a material that is measured, compared or noted.
Attributes are ordinarily qualitative characteristics, but quantitative characteristics (variables) may be converted into attributes by assigning items to certain categories on the basis of the measured values. The value of the characteristic determined as a result of an observation or test is the observed value; when determined by a specified test method, it is called the test result. The concentration or quantity of an analyte as estimated by use of a sample is usually the characteristic of interest in analytical chemistry.
1990, 62, 1201

characteristic group (in organic nomenclature)
A single heteroatom, for example, –Cl and =O; a heteroatom bearing one or more hydrogen atoms or other heteroatoms, for example, $-NH_2$, $-OH$, $-SO_3H$, $-PO_3H_2$ and IO_2; or a heteroatomic group attached to or containing a single carbon atom, for example, $-CHO$, $-C{\equiv}N$, $-COOH$ and $-NCO$, attached to a parent hydride.
B.B.(G) 13

characteristic length (in thin films)
The term characteristic length or scale refers, in general, to the parameter which characterizes the density profile (of a given physical quantity). The static (equilibrium) or dynamic character of a characteristic length must be specified.
The terms out of plane and in plane refer to characteristic lengths normal or parallel to the interface, respectively. Since interfacial layer 'thickness' and characteristic length correspond to various concepts, the current usage where an out of plane characteristic length is referred to as the interfacial layer thickness, is confusing and should be abandoned.
1994, 66, 1674

characteristic mass for integrated absorbance (in electrothermal atomization), m_o
That mass of analyte which produces an integrated absorbance signal whose net area is equal to 0.0044 s.
 1992, *64*, 258

characteristic mass for peak absorption (in electrothermal atomization), m_p
That mass of analyte which produces a defined peak absorbance of 0.0044 (or 1% absorbance).
 1992, *64*, 258

characteristic mass (in electrothermal atomization), m_c
The mass of analyte which, when atomized electrothermally, produces the uniquely defined absorption signal.
 1992, *64*, 258

characteristic potential
An *applied potential* that is characteristic of a charge-transfer process and the experimental conditions (such as the composition of the solvent and *supporting electrolyte* and the temperature) under which it is investigated, and whose nature depends on the technique that is employed. Some typical characteristic potentials are the half-wave potential in *polarography*, the quarter-transition-time potential in *chronopotentiometry*, the *peak potential* and the *half-peak potential* in linear-sweep *voltammetry*, and the *summit potential* in ac polarography.
 1985, *57*, 1493

characteristic ratio (in polymers)
The ratio of the mean-square end-to-end distance, $<r^2>_0$, of a linear polymer chain in a theta state to $N \cdot L^2$, where N is the number of rigid sections in the main chain, each of length L; if all of the rigid sections are not of equal length, the mean-square value of L is used, i.e.

$$L^2 = \sum_i \overline{L_i^2} / N$$

In simple single-strand chains, the bonds are taken as the rigid sections.
The recommended symbol is: C_N (C_∞ when $N \to \infty$).
 P.B. 49

characteristic scale (in thin films)
 See *characteristic length*.
 1994, *66*, 1674

characteristic X-radiation
 See *characteristic X-ray emission*.
 1991, *63*, 737

characteristic X-ray emission
X-ray emission originates from the radiative decay of electronically highly excited states of matter. Excitation by electrons is called primary excitation and by photons, secondary or fluorescence excitation. Particle induced X-ray emission (PIXE) is produced by the excitation of heavier particles such as protons, deuterons or heavy atoms in varying degrees of ionization. Emission of photons in the X-ray wavelength region also occurs from ionized gases or plasmas at high temperatures, from nuclear processes (low-energy end of the gamma-ray spectrum) and from radiative transitions between muonic states.
Characteristic X-ray emission consists of a series of X-ray spectral lines with discrete frequencies, characteristic of the emitting atom. Other features are emission bands from transitions to valence levels. In a spectrum obtained with electron or photon excitation the most intense lines are called diagram lines or normal X-ray lines. They are dipole allowed transitions between normal X-ray *diagram levels*.
 1991, *63*, 737

charcoal
A traditional term for a *char* obtained from wood, peat, coal or some related natural organic materials.
Notes:
Charcoal has highly reactive inner surfaces and a low sulfur content. It has or has had, therefore, a variety of uses, e.g. in ferrous metallurgy and for gunpowder (minor uses: medical purposes and paint materials).
 1995, *67*, 484

charge
 See *electric charge, elementary charge*.
 1996, *68*, 965

charge density, ρ
Electric charge divided by the volume it occupies.
 See also *electron density*.
 G.B. 14

charge-exchange ionization (in mass spectrometry)
This occurs when an ion/atom or ion/molecule reaction takes place in which the charge on the ion is transferred to the neutral species without any dissociation of either.
Synonymous with *charge exchange ionization*.
 1991, *63*, 1547

charge-exchange reaction
Synonymous with *charge-transfer reaction*.
 1991, *63*, 1555

charge hopping
Electron or *hole transport* between equivalent sites.
 1996, *68*, 2231

charge-inversion mass spectrum
Charge inversion processes of the types:

$$M^+ + X \to M^- + X^{2+}$$
$$\text{or } M^- + X \to M^+ + X + 2e$$

respectively, occuring in a collision cell (containing a gas, X) located in a field-free region preceding a magnetic and electric sector combination placed in either order, may be detected as follows.
If the instrument slits are wide, and if the connections to the two sectors, appropriate to transmission of either positive or negative main-beam ions, are simply reversed, the negative or positive product ions of the two processes, respectively, will be transmitted. If the magnetic field is scanned, a spectrum of such

product ions will be obtained, and this spectrum is called a charge-inversion mass spectrum. These spectra are sometimes referred to as $-E$ and $+E$ spectra, respectively. The terms '$2E, E/2, -E$ or $+E$ mass spectrum' should not be used without prior explanation of the meaning $2E, E, +E$ or $-E$.
1991, *63*, 1551

charge number, *z*
Ratio of the charge of a particle to the elementary charge.
G.B. 44; 1996, *68*, 965; 1991, *63*, 1544

charge number (in inorganic nomenclature)
The magnitude of the charge on an ion, given in arabic numerals followed by the sign of the charge, in parentheses without a space, immediately after the name of an ion, e.g. iron(2+), hexacyanoferrate(4–).
Note:
The term Ewens–Bassett number is no longer recommended.
R.B. 66

charge number (of a cell reaction), n, v_e, z
Number of electrons transferred according to the cell reaction equation.
G.B. 58

charge-permutation reaction
A general term to describe an ion/neutral species reaction where there is a change in the magnitude and/or sign of the charges of the reactant.
1991, *63*, 1555

charge population
The net electric charge on a specified atom in a *molecular entity*, as determined by some prescribed definition.
See also *electron density*.
1994, *66*, 1094

charge recombination
Reverse of *charge separation*. In using this term it is important to specify the resulting electronic state of the donor and acceptor.
1996, *68*, 2231

charge separation
A process in which, under the influence of a suitable *driving force* (e.g. provided by photoexcitation), electronic charge moves in a direction that increases the difference in local charges between donor and acceptor sites. *Electron transfer* between neutral species is one of the most important examples.
1996, *68*, 2231

charge shift
A process in which under the influence of a suitable *driving force* (e.g. provided by photoexcitation) electronic charge moves without changing the difference in local charges between donor and acceptor sites. *Electron transfer* reversing the charges in a system composed of a neutral donor and a cationic acceptor or of a neutral acceptor and an anionic donor provide prominent examples.
1996, *68*, 2231

charge-stripping reaction
An ion/neutral species reaction wherein the charge on the reactant ion is made more positive.
1991, *63*, 1555

charge-transfer adsorption
Oxidative or reductive *chemisorption* where 'reductive' and 'oxidative' refer to electron gain or loss on species in the solid. In simple cases it is non-dissociative, i.e. there is a mere transfer of charge between *adsorptive* and *adsorbent* in forming the *adsorbate*.
1976, *46*, 76

charge-transfer complex
An electron-donor–electron-acceptor complex, characterized by electronic transition(s) to an excited state in which there is a partial transfer of electronic charge from the donor to the acceptor moiety.
1995, *67*, 1328; 1994, *66*, 1094

charge-transfer device (in radiation detection)
A charge-transfer device has a metal oxide semiconductor (MOS) structure that is composed of many independent pixels where charge is stored in such a way that the charge pattern corresponds to the irradiation pattern. These devices can be linear or two-dimensional. According to the method used to detect the charge pattern, two types of charge-transfer devices can be distinguished: charge-coupled devices (CCDs) and charge-injection devices (CIDs).
1995, *67*, 1757

charge-transfer reaction (in mass spectrometry)
An ion/neutral species reaction wherein the *total* charge on the reactant ion is transferred initially to the reactant neutral species so that the reactant ion becomes a neutral entity. Considering some of the possible reactions of ions M^{2+}, M^+ and M^- with a neutral species X, these would be further categorized as follows:

$$M^{2+} + X \rightarrow M^+ + X^+$$
(Partial charge transfer)
$$M^+ + X \rightarrow M^{2+} + X + e$$
(Charge stripping)
$$M^- + X \rightarrow M^+ + X + 2e$$
(Charge stripping and charge inversion)

All are ion/neutral species reactions and also charge permutation reactions.
1991, *63*, 1555

charge-transfer (CT) state
A state related to the ground state by a *charge-transfer transition*.
1996, *68*, 2232

charge-transfer step (of an electrode reaction)
An *elementary step* in which charge is transferred from one phase to the other.
1980, *52*, 235

charge-transfer (CT) transition
An electronic transition in which a large fraction of an electronic charge is transferred from one region

of a molecular entity, called the electron donor, to another, called the electron acceptor (intramolecular CT) or from one molecular entity to another (intermolecular CT). Typical for donor–acceptor complexes or multichromophoric molecular entities. In some cases the charge-transfer absorption band may be obscured by the absorption of the partners.
1996, *68*, 2232

charge transfer transition to solvent (CTTS)
Electronic transition which is adequately described by single *electron transfer* between a solute and the solvent, as opposed to excitation followed by electron transfer to solvent.
See also *charge-transfer transition*.
1996, *68*, 2232

charring
Pyrolysis of samples containing organic matter. In the presence of oxygen this is referred to as *ashing*.
1992, *64*, 256

chelate
See *chelation*.
1994, 66, 1094; R.B. 147

chelation
The formation or presence of *bonds* (or other attractive interactions) between two or more separate *binding sites* within the same ligand and a single central atom. A *molecular* entity in which there is chelation (and the corresponding *chemical species*) is called a 'chelate'. The terms bidentate (or didentate), tridentate, tetradentate, ... multidentate are used to indicate the number of potential binding sites of the ligand, at least two of which must be used by the ligand in forming a 'chelate'. For example, the bidentate ethylenediamine forms a chelate with CuI in which both nitrogen atoms of ethylenediamine are bonded to copper. (The use of the term is often restricted to metallic central atoms.)

The phrase 'separate binding sites' is intended to exclude cases such as $[PtCl_3(CH_2=CH_2)]^-$, ferrocene and (benzene)tricarbonylchromium in which ethene, the cyclopentadienyl group and benzene, respectively, are considered to present single binding sites to the respective metal atom, and which are not normally thought of as chelates.
See also *cryptand*, η *(eta or hapto)*.
1994, *66*, 1094; R.B. 147

chelatometric titration
See *titration*.
O.B. 47

cheletropic reaction
A form of *cycloaddition* across the terminal atoms of a fully *conjugated system* with formation of two new σ-bonds to a single atom of the ('monocentric') reagent. There is formal loss of one π-bond in the substrate and an increase in *coordination number* of the relevant atom of the reagent. An example is the *addition* of sulfur dioxide to butadiene:

$CH_2=CH—CH=CH_2 + SO_2 \longrightarrow$ [cyclopentene-SO$_2$ structure]

The reverse of this type of reaction is designated 'cheletropic elimination'.
1994, *66*, 1094

chelotropic reaction
Alternative (and etymologically more correct) name for cheletropic reaction.
1994, *66*, 1095

chemical actinometer
A detector in which the amount of a chemical formed is proportional to the numbers of photons absorbed.
See also *actinometer*.
O.B. 193

chemical activation
When some of the energy required for a reaction is provided by a preceding exothermic chemical reaction there is said to be chemical activation. For example, in the scheme:

$A + B \rightarrow X^*$
$X^* \rightarrow Y + Z$

some or all of the energy required for X to decompose is provided by the first reaction.
1996, *68*, 158

chemical adsorption
See *chemisorption*.
1972, *31*, 586

chemical amount
See *amount of substance*.
1996, *68*, 965; G.B. 46

chemical decomposition
The breakdown of a single entity (normal molecule, reaction *intermediate*, etc.) into two or more fragments.
1994, *66*, 1104

chemical dosimeter
A dosimeter in which the dose is measured by observing the extent, under specified conditions, of a chemical reaction caused by the ionizing radiation to be measured.
O.B. 215

chemical element
1. A species of atoms; all atoms with the same number of protons in the atomic nucleus.
2. A pure chemical substance composed of atoms with the same number of protons in the atomic nucleus. Sometimes this concept is called the elementary substance as distinct from the chemical element as defined under 1, but mostly the term chemical element is used for both concepts.
Physical Chemistry Division, unpublished; R.B. 35

chemical equilibrium
Reversible processes [processes which may be made to proceed in the forward or reverse direction by the

(infinitesimal) change of one variable], ultimately reach a point where the rates in both directions are identical, so that the system gives the appearance of having a static composition at which the Gibbs energy, G, is a minimum. At equilibrium the sum of the chemical potentials of the reactants equals that of the products, so that:

$$\Delta G_r = \Delta G_r^o + RT \ln K = 0$$

$$\Delta G_r^o = -RT \ln K$$

The equilibrium constant, K, is given by the *mass-law effect*.
1994, *66*, 1114

chemical flux, φ
A concept related to *rate of reaction*, particularly applicable to the progress in one direction only of component reaction steps in a complex system or to the progress in one direction of reactions in a system at dynamic equilibrium (in which there are no observable concentration changes with time). Chemical flux is a derivative with respect to time, and has the dimensions of amount of substance per unit volume transformed per unit time.
The sum of all the chemical fluxes leading to destruction of B is designated the 'total chemical flux out of B' (symbol $\Sigma\varphi_{-B}$); the corresponding formation of B by concurrent elementary reactions is the 'total chemical flux into B or A' (symbol $\Sigma\varphi_B$).
For the mechanism:

$$A + B \underset{-1}{\overset{1}{\rightleftharpoons}} C$$

$$C + D \overset{2}{\rightleftharpoons} E$$

the total chemical flux into C is caused by the single reaction (1):

$$\Sigma\varphi_C = \varphi_1$$

whereas the chemical flux out of C is a sum over all reactions that remove C:

$$\Sigma\varphi_{-C} = \varphi_{-1} + \varphi_2$$

where φ_{-1} is the 'chemical flux out of C into B (and/or A)' and φ_2 is the 'chemical flux out of C into E'. The rate of appearance of C is then given by:

$$d[C]/dt = \Sigma\varphi_C - \Sigma\varphi_{-C}$$

In this system φ_1 (or $\Sigma\varphi_{-A}$) can be regarded as the hypothetical rate of decrease in the concentration of A due to the single (unidirectional) reaction (1) proceeding in the assumed absence of all other reactions.
For a non-reversible *reaction*:

$$A \overset{1}{\rightarrow} P$$

$$-d[A]/dt = \varphi_1$$

If two substances A and P are in *chemical equilibrium*:

$$A \rightleftharpoons P$$

then:

$$\Sigma\varphi_A = \Sigma_{-A} = \Sigma\varphi_P = \Sigma\varphi_{-P}$$

and

$$-d[A]/dt = d[P]/dt = 0$$

See also *order of reaction, rate-limiting step, steady state*.
1994, *66*, 1095

chemical induction (coupling)
When in a chemical system one reaction accelerates another there is said to be chemical induction or coupling. It is due to an intermediate or *product* of the inducing reaction participating in the second reaction. Chemical induction is often observed in oxidation–reduction reactions, *chain reactions*, and biological reactions. Sometimes a reaction having $\Delta G^o > 0$ is induced by a simultaneous process having $\Delta G^o < 0$.
1996, *68*, 158

chemical ionization (in mass spectrometry)
This concerns the process whereby new ionization species are formed when gaseous molecules interact with ions. The process may involve transfer of an electron, proton or other charged species between the reactants. When a positive ion results from chemical ionization, the term may be used without qualification. When a negative ion is involved, the term negative ion chemical ionization should be used; note that negative ion formation by attachment of a free electron does not fall within this definition. Chemical ionization and chemi-ionization are two terms which should not be used interchangeably.
See *electron attachment*; see also *chemi-ionization*.
1991, *63*, 1547; O.B. 204; 1990, *62*, 2179

chemical isotope exchange
Exchange of *isotopes* between different types of molecules or ions in the course of a *chemical reaction*
1982, *54*, 1537

chemical laser
A continuous wave or pulsed laser in which the excitation and population inversion of the emitting species result from a chemical reaction. Typical examples are HF and DF lasers emitting many lines in the IR region.
1996, *68*, 2232

chemically induced dynamic nuclear polarization (CIDNP)
See *CIDNP*.
1994, *66*, 1097

chemical measurement process (CMP)
An analytical method of defined structure that has been brought into a state of statistical control, such that its imprecision and bias are fixed, given the measurement conditions. This is prerequisite for the evaluation of the performing characteristics of the method, or the development of meaningful uncertainty statements concerning analytical results.
1995, *67*, 1702

chemical oxygen demand (COD)
A measure of the amount of oxygen, divided by the volume of the system, required to oxidize the organic (and inorganic) matter in wastewater using a chemically oxidizing agent. In practice, it is usually expressed in milligrams O_2 per litre.
1992, *64*, 149

chemical potential, μ_B
The chemical potential of a substance B in a mixture of substances B, C ... is related to the Gibbs energy G of the mixture by:

$$\mu_B = (\partial G/\partial n_B)_{T, p, n_{C \neq B}}$$

where T is the thermodynamic temperature, p is the pressure and $n_B, n_C, ...$ are the amounts of substance of B, C, ...
For a pure substance B, the chemical potential μ_B^* is given by:

$$\mu_B^* = G^*/n_B = G_m^*$$

where G_m^* is the molar Gibbs energy, and where the superscript * attached to a symbol denotes the property of the pure substance. The superscript ⦵ or ° attached to a symbol may be used to denote a *standard thermodynamic quantity*.

See also *standard thermodynamic quantity, standard chemical potential*.
1994, *66*, 535; G.B. 49; 1996, *68*, 966

chemical reaction
A process that results in the interconversion of *chemical species*. Chemical reactions may be *elementary reactions* or *stepwise reactions*. (It should be noted that this definition includes experimentally observable interconversions of conformers.)
Detectable chemical reactions normally involve sets of *molecular entities*, as indicated by this definition, but it is often conceptually convenient to use the term also for changes involving single molecular entities (i.e. 'microscopic chemical events').

See also *identity reaction*.
1994, *66*, 1096

chemical reaction equation
Symbolic representation of a chemical reaction where the reactant entities are given on the left hand side and the product entities on the right hand side. The coefficients next to the symbols and formulae of entities are the absolute values of the stoichiometric numbers. Different symbols are used to connect the reactants and products with the following meanings: = for a stoichiometric relation; → for a net forward reaction; ⇄ for a reaction in both directions; ⇌ for equilibrium.
G.B. 45

chemical relaxation
If the equilibrium mixture of a *chemical reaction* is disturbed by a sudden change, especially of some external parameter (such as temperature, pressure or electrical field strength), the system will readjust itself to a new position of the chemical equilibrium or return to the original position, if the perturbation is temporary. The readjustment is known as chemical relaxation.
In many cases, and in particular when the displacement from equilibrium is slight, the progress of the system towards equilibrium can be expressed as a first-order law:

$$[C_t - (C_{eq})_2] = [(C_{eq})_1 - (C_{eq})_2]e^{-t/\tau}$$

where $(C_{eq})_1$ and $(C_{eq})_2$ are the equilibrium concentrations of one of the chemical species involved in the reaction before and after the change in the external parameter, and C_t is its concentration at time t.
The time parameter τ, named *relaxation time*, is related to the rate constants of the chemical reaction involved.
Measurements of the relaxation times by relaxation methods [involving a temperature jump (T-jump), pressure jump, electric field jump or a periodic disturbance of an external parameter, as in ultrasonic techniques] are commonly used to follow the kinetics of very fast reactions.
1994, *66*, 1096

chemical shift (in NMR), δ
The variation of the resonance frequency of a nucleus in nuclear magnetic resonance (NMR) spectroscopy in consequence of its magnetic environment. The chemical shift of a nucleus, δ, is expressed in ppm by its frequency, ν_{cpd}, relative to a standard, ν_{ref}, and defined as:

$$\delta = \frac{\nu_{cpd} - \nu_{ref}}{\nu_0} \times 10^6$$

where ν_0 is the operating frequency of the spectrometer. For ^{1}H and ^{13}C NMR the reference signal is usually that of tetramethylsilane ($SiMe_4$). Other references are used in the older literature and in other solvents, such as D_2O.
If a resonance signal occurs at lower frequency or higher applied field than an arbitrarily selected reference signal, it is said to be upfield, and if resonance occurs at higher frequency or lower applied field, the signal is downfield. Resonance lines upfield from $SiMe_4$ have positive, and resonance lines downfield from $SiMe_4$ have negative, δ-values.
1994, *66*, 1096; G.B. 25

chemical shift (in photoelectron and Auger spectra)
The displacement of photoelectron or Auger peak energies originating from changes in electron binding energies as a consequence of differences in the chemical environment of the atoms.
1979, *51*, 2247

chemical species
An ensemble of chemically identical *molecular entities* that can explore the same set of molecular energy levels on the time scale of the experiment. The term is applied equally to a set of chemically identical atomic or molecular structural units in a solid array.

For example, two conformational *isomers* may be interconverted sufficiently slowly to be detectable by separate NMR spectra and hence to be considered to be separate chemical species on a time scale governed by the radiofrequency of the spectrometer used. On the other hand, in a slow chemical reaction the same mixture of conformers may behave as a single chemical species, i.e. there is virtually complete equilibrium population of the total set of molecular energy levels belonging to the two conformers.

Except where the context requires otherwise, the term is taken to refer to a set of molecular entities containing isotopes in their natural abundance.

The wording of the definition given in the first paragraph is intended to embrace both cases such as graphite, sodium chloride or a surface oxide, where the basic structural units may not be capable of isolated existence, as well as those cases where they are. In common chemical usage generic and specific chemical names (such as *radical* or hydroxide ion) or chemical formulae refer either to a chemical species or to a *molecular entity*.

1994, *66*, 1096; 1996, *68*, 2202

chemical substance
Matter of constant composition best characterized by the entities (molecules, formula units, atoms) it is composed of. Physical properties such as density, refractive index, electric conductivity, melting point etc. characterize the chemical substance.

Physical Chemistry Division, unpublished

chemical vapour generation (in spectrochemical analysis)
A system in which the analyte is separated from the sample matrix by the generation of gaseous species as a result of a chemical reaction.

1992, *64*, 262

chemical yield
The fraction of the amount of an element or chemical compound following a specified *chemical reaction* or separation.

1982, *54*, 1537

chemiexcitation
Generation, by a chemical reaction, of electronically excited molecular entities from reactants in their ground electronic states.

1996, *68*, 2232

chemiflux
See *chemical flux*.
1994, *66*, 1095

chemi-ionization (in mass spectrometry)
This refers to a process whereby gaseous molecules are ionized when they interact with other internally excited gaseous molecules or molecular moieties. Chemi-ionization and chemical ionization are two terms which should not be used interchangeably.
See also *associative ionization*, *chemical ionization*.

1991, *63*, 1547; O.B. 204

chemiluminescence
Emission of radiation resulting from a chemical reaction. The emitting species may be a reaction *product* or a species excited by energy transfer from an excited reaction product. The excitation may be electronic, vibrational or rotational; if the luminescence occurs in the infrared the expression infrared chemiluminescence is used.

1996, *68*, 158; 1996, *68*, 2232; 1984, *56*, 234

chemiluminescence analyser
An instrument consisting of a reaction chamber with separate inlets for the sample and reagent gas, an optical filter, a photomultiplier and a signal processing device. The reactive gas is introduced in excess. The quantity of light produced is proportional to the sample flow rate and the concentration of the measured substance in the sample under specified temperature and pressure conditions. The filter limits the wavelength to the region of interest and helps to eliminate interferences.

1990, *62*, 2179

chemiluminescent methods of detection (in analysis)
Methods which utilize gas or liquid phase reactions between two molecules which produce a third molecule in an excited state. The wavelength distribution of the light emission from the excited molecule is characteristic of the species; in some circumstances this may be used to identify and to determine the concentration of one of the reacting species.

1990, *62*, 21

chemisorption (chemical adsorption)
Adsorption which results from chemical bond formation (strong interaction) between the adsorbent and the adsorbate in a monolayer on the surface.

1990, *62*, 2179

chemistry, nuclear
See *nuclear chemistry*.
1994, *66*, 2516

chemistry, radiation
See *radiation chemistry*.
O.B. 236

chemoselectivity (chemoselective)
Chemoselectivity is the preferential reaction of a chemical reagent with one of two or more different functional groups. A reagent has a high chemoselectivity if reaction occurs with only a limited number of different functional groups. For example, sodium tetrahydroborate is a more chemoselective reducing agent than is lithium tetrahydroaluminate. The concept has not been defined in more quantitative terms. The term is also applied to reacting molecules or intermediates which exhibit selectivity towards chemically different reagents.

Some authors use the term chemospecificity for 100% chemoselectivity. However, this usage is discouraged.

See also *regioselectivity, stereoselectivity, stereospecificity*.
1994, *66*, 1097

+ chemospecificity
See *chemoselectivity*.
1994, *66*, 1097

chemostat
A *bioreactor* in which constant growth conditions for *microorganisms* are maintained over prolonged periods of time by supplying the reactor with a continuous input of nutrients and continuous removal of medium.
1992, *64*, 149

chi
See entries under χ (beginning of 'c').

chimney effect (in atmospheric chemistry)
A vertical movement of a localized mass of air or other gases which occurs due to local temperature differences.
1990, *62*, 2179

chimney height (stack height), effective
See *effective chimney height (stack height)*.
1990, *62*, 2179

chiral
Having the property of *chirality*.
As applied to a molecule the term has been used differently by different workers. Some apply it exclusively to the whole molecule, whereas others apply it to parts of a molecule. For example, according to the latter view, a *meso*-compound is considered to be composed of two chiral parts of opposite chirality sense; this usage is to be discouraged.
See *enantiomorph*.
In its application to an assembly of molecules, some restrict the term to an assembly in which all of the molecules have the same *chirality sense*, which is better called *enantiopure*. Others extend it to a *racemic* assembly, which is better just called a *racemate*. Use of the term to describe molecular assemblies should be avoided.
1996, *68*, 2202

chirality
The geometric property of a rigid object (or spatial arrangement of points or atoms) of being non-superposable on its mirror image; such an object has no symmetry elements of the second kind (a mirror plane, $\sigma = S_1$, a centre of inversion, $i = S_2$, a rotation-reflection axis, S_{2n}). If the object is superposable on its mirror image the object is described as being achiral.
See also *handedness, superposability*.
1996, *68*, 2203; B.B. 479

chirality axis
An axis about which a set of ligands is held so that it results in a spatial arrangement which is not superposable on its mirror image. For example with an allene abC=C=Ccd the chiral axis is defined by the C=C=C bonds; and with an *ortho*-substituted biphenyl the atoms C-1, C-1′, C-4 and C-4′ lie on the chiral axis.
1996, *68*, 2203

chirality centre
An atom holding a set of ligands in a spatial arrangement which is not superposable on its mirror image. A chirality centre is thus a generalized extension of the concept of the *asymmetric carbon atom* to central atoms of any element, for example N^+abcd, Pabc as well as Cabcd.
1996, *68*, 2203

chirality element
General name for a *chirality axis, chirality centre* or *chirality plane*. Also referred to as an element of chirality.
1996, *68*, 2203

chirality plane
A planar unit connected to an adjacent part of the structure by a bond which results in restricted torsion so that the plane cannot lie in a symmetry plane. For example with (*E*)-cyclooctene the chiral plane includes the double bond carbon atoms and all four atoms attached to the double bond; with a monosubstituted paracyclophane the chiral plane includes the monosubstituted benzene ring with its three hydrogen atoms and the three other atoms linked to the ring (i.e. from the substituent and the two chains linking the two benzene rings).
1996, *68*, 2203

chirality sense
The property that distinguishes enantiomorphs. The specification of two enantiomorphic forms by reference to an oriented space e.g. of a screw, a right threaded one or a left threaded one. The expression 'opposite chirality' is short for 'opposite chirality sense'.
1996, *68*, 2203

chiroptic/chiroptical
A term referring to the optical techniques (using refraction, absorption or emission of anisotropic radiation) for investigating chiral substances [e.g. measurements of optical rotation at a fixed wavelength, optical rotatory dispersion (ORD), circular dichroism (CD), and circular polarization of luminescence (CPL)].
1996, *68*, 2203

chirotopic
The description of an atom (or point, group, face, etc. in a molecular model) that resides within a *chiral* environment. One that resides within an achiral environment has been called achirotopic.
1996, *68*, 2203

chloramines
Amines substituted at nitrogen with one or two chlorine atoms (a contracted form of *N*-chloroamines).
1995, *67*, 1328

chlorocarbons
Compounds consisting wholly of chlorine and carbon.
1995, *67*, 1328

chlorohydrins
See *halohydrins*.
1995, *67*, 1328

chloronium ions
See *halonium ions*.
1995, *67*, 1328

cholesteric phase
See *liquid-crystal transitions*.
1994, *66*, 579

chromatogram
A graphical or other presentation of detector response, concentration of analyte in the effluent or other quantity used as a measure of effluent concentration versus effluent volume or time. In planar chromatography 'chromatogram' may refer to the paper or layer with the separated zones.
1993, *65*, 823; 1990, *62*, 2179

chromatograph (noun)
The assembly of apparatus for carrying out chromatographic separation.
1993, *65*, 823

chromatograph (verb)
To separate by *chromatography*.
1993, *65*, 823

chromatographic detector
A device that measures the change of composition of the effluent.
O.B. 99; 1993, *65*, 829

chromatography
A physical method of separation in which the components to be separated are distributed between two phases, one of which is stationary (stationary phase) while the other (the mobile phase) moves in a definite direction.

See also *adsorption chromatography, affinity chromatography, column chromatography, displacement chromatography, elution chromatography, exclusion chromatography, frontal chromatography, gas chromatography, ion-exchange chromatography, isothermal chromatography, liquid chromatography, normal-phase chromatography, partition chromatography, planar chromatography, programmed-flow chromatography (flow programming), programmed-pressure chromatography (pressure programming), programmed-temperature chromatography (temperature programming), pyrolysis-gas chromatography, reaction chromatography, reversed-phase chromatography, supercritical fluid chromatography, two-dimensional chromatography.*
1993, *65*, 823; O.B. 92; 1990, *62*, 2179

chromophore
The part (atom or group of atoms) of a *molecular entity* in which the electronic transition responsible for a given spectral band is approximately localized. The term arose in the dyestuff industry, referring originally to the groupings in the molecule that are responsible for the dye's colour.
1994, *66*, 1097; 1996, *68*, 2232

chromosome
A self-replicating structure consisting of *DNA* complexed with various proteins and involved in the storage and transmission of genetic information; the physical structure that contains *genes* (cf. *plasmid*). Eukaryotic *cells* have a characteristic number of *chromosomes* per cell (cf. *ploidy*) and contain DNA as linear duplexes. The chromosomes of bacteria consist of double-standed, circular DNA molecules.
1992, *64*, 150

CIDEP (Chemically Induced Dynamic Electron Polarization)
Non-Boltzmann electron spin state population produced in thermal or photochemical reactions, either from a combination of radical pairs (called radical-pair mechanism), or directly from the triplet state (called triplet mechanism), and detected by ESR spectroscopy.
1996, *68*, 2232

CIDNP (Chemically Induced Dynamic Nuclear Polarization)
Non-Boltzmann nuclear spin state distribution produced in thermal or photochemical reactions, usually from *colligation* and diffusion, or *disproportionation* of *radical pairs*, and detected by NMR spectroscopy by enhanced absorption or emission signals.
1994, *66*, 1097; 1996, *68*, 2232

CIEEL (Chemically Initiated Electron Exchange Luminescence)
A type of luminescence resulting from a thermal electron-transfer reaction. Also called catalysed chemiluminescence.
1996, *68*, 2232

cine-substitution
A *substitution reaction* (generally *aromatic*) in which the entering group takes up a position adjacent to that occupied by the *leaving group*. For example,

See also *tele-substitution*.
1994, *66*, 1097

CIP priority
In the CIP rules the conventional order of ligands established for the purpose of unambiguous designation of *stereoisomers*. It is deduced by application of sequence rules, the authoritative statement of which appears in R.S. Cahn, C.K. Ingold and V. Prelog, *Angew. Chem.* **78**, 413–447 (1966), *Angew. Chem. Internat. Ed. Eng.* **5**, 385–415, 511 (1966); and V. Prelog and G. Helmchen, *Angew. Chem.* **94**, 614–631

(1982), *Angew. Chem. Internat. Ed. Eng. 21*, 567–583 (1982).
1996, *68*, 2203

C. I. P. system
Short for Cahn–Ingold–Prelog system.
See *CIP priority*.
1996, *68*, 2201

circular development
See *radial development*.
1993, *65*, 829

circular elution
See *radial elution*.
1993, *65*, 829

circular frequency
See *angular frequency*.
G.B. 11

cis- (in inorganic nomenclature)
A prefix designating two groups occupying adjacent positions in a coordination sphere, not now generally recommended for precise nomenclature purposes.
R.B. 245

cis conformation (in polymers)
Conformations referring to torsion angles θ (A, B, C, D), where A, B, C, D are main-chain atoms, can be described as: *cis* or *synperiplanar* (C); *gauche* or *synclinal* (G); anticlinal (A); and *trans* or *antiperiplanar* (T), corresponding to torsion angles within ± 30° of, respectively, 0°, ± 120° and ± 180°. The letters shown in parentheses (upper case C, G, A, T) are the recommended abbreviations.

The symbols G^+, G^- (or A^+, A^-, for example) refer to torsion angles of similar type but opposite known sign, i.e. ~ +60°, ~ –60° (or ~ +120°, –120°). The notation G, $\bar{G}$; A, $\bar{A}$ (and T, $\bar{T}$; C, $\bar{C}$ whenever the torsion angles are not exactly equal to 180° and 0°, respectively) is reserved for the designation of enantiomorph conformations, i.e. conformations of opposite but unspecified sign. Where necessary, a deviation from the proper value of the torsion angle can be indicated by the sign (~), as in the following examples: G(~); G(~); G^+(~); G^-(~).
P.B. 41

cis-fused
Steric relations at saturated *bridgeheads* common to two rings are denoted by *cis* or *trans*, followed by a hyphen and placed before the name of the ring system, according to the relative positions of the exocyclic atoms or groups attached to the bridgeheads. Such rings are said to be *cis*-fused or *trans*-fused.
B.B. 479

+ cisoid conformation
(Usage strongly discouraged).
See s-*cis*, s-*trans*. The terms *cisoid* and *transoid* are also used to describe the stereochemistry of fused ring systems.
1996, *68*, 2204; B.B. 478

cistactic polymer
A *tactic polymer* in which the main-chain double bonds of the *configurational base units* are entirely in the *cis* arrangement.
P.B. 33

cis, trans
Descriptors which show the relationship between two ligands attached to separate atoms that are connected by a double bond or are contained in a ring. The two ligands are said to be located *cis* to each other if they lie on the same side of a plane. If they are on opposite sides, their relative position is described as *trans*. The appropriate reference plane of a double bond is perpendicular to that of the relevant σ-bonds and passes through the double bond. For a ring (the ring being in a conformation, real or assumed, without re-entrant angles at the two substituted atoms) it is the mean plane of the ring(s). For alkenes the terms *cis* and *trans* may be ambiguous and have therefore largely been replaced by the *E, Z* convention for the nomenclature of organic compounds. If there are more than two entities attached to the ring the use of *cis* and *trans* requires the definition of a reference substituent.
1996, *68*, 2203; B.B.(G) 149

cis-trans **isomers**
Stereoisomeric olefins or cycloalkanes (or heteroanalogues) which differ in the positions of atoms (or groups) relative to a reference plane: in the *cis*-isomer the atoms are on the same side, in the *trans*-isomer they are on opposite sides.

1996, *68*, 2204

class (a) metal ion
A metal ion that combines preferentially with *ligands* containing ligating atoms that are the lightest of their Periodic Group.
See also *class (b) metal ion, hard acid*.
1994, *66*, 1097

class (b) metal ion
A metal ion that combines preferentially with *ligands* containing ligating atoms other than the lightest of their Periodic Group.
See also *class (a) metal ion, hard acid*.
1994, *66*, 1097

class of helix (in polymers)
The number of skeletal chain atoms contained within the helix residue.
P.B. 79

clathrates
Inclusion compounds in which the guest molecule is in a cage formed by the host molecule or by a lattice of host molecules.
1995, *67*, 1328; 1994, *66*, 1097

clausius
Non-SI unit of entropy, Cl = 4.184 J K^{-1}.
 G.B. 113

clean surface
An experimental surface having no surface contamination observable by means of the used method.
 1979, *51*, 2247

clean up (in spectrochemical analysis)
A process caused by sputtering, trapping (or adsorbing) metallic atoms, gas atoms or molecules on the walls of the vessel or anode. This has the effect of reducing the gas pressure in sealed lamps, and may be useful for cleaning up or gettering unwanted gases.
 O.B. 149

clearance, $\Delta V/\Delta t$
Product of the concentration of a component in an output system and the volume flow rate of that output system divided by the concentration of this component in the input system.
Note:
The term *mean volume rate* is recommended for this quantity.
 1996, *68*, 966

clinal
 See *torsion angle*.
 1996, *68*, 2204

clone
1. A population of genetically identical cells produced from a common ancestor.
2. Sometimes, 'clone' is also used for a number of *recombinant DNA* molecules all carrying the same inserted sequence.
 1992, *64*, 150

closed film
 See *open film*.
 1994, *66*, 1671

closed system (in spectrochemical analysis)
 See *mercury flow system (in spectrochemical analysis)*.
 1992, *64*, 263

closo-
An affix used in names to denote a *cage* or closed structure, especially a boron skeleton that is a polyhedron having all triangular faces.
 R.B. 245; B.B. 464; 1972, *30*, 686

cloud (in atmospheric chemistry)
An aerosol of the atmosphere which is dense enough to be perceptible to the eye. Usually refers to an assembly of water droplet aerosols suspended in the atmosphere, although dust clouds are also recognized.
 1990, *62*, 2180

cloud, altocumulus (in atmospheric chemistry)
 See *altocumulus cloud*.
 1990, *62*, 2180

cloud, altostratus (in atmospheric chemistry)
 See *altostratus cloud*.
 1990, *62*, 2180

cloud, cumulonimbus (in atmospheric chemistry)
 See *cumulonimbus cloud*.
 1990, *62*, 2180

cloud, cumulus (in atmospheric chemistry)
 See *cumulus cloud*.
 1990, *62*, 2180

cloud, nimbostratus (in atmospheric chemistry)
 See *nimbostratus cloud*.
 1990, *62*, 2180

cloud, stratocumulus (in atmospheric chemistry)
 See *stratocumulus cloud*.
 1990, *62*, 2180

cloud, stratus (in atmospheric chemistry)
 See *stratus cloud*.
 1990, *62*, 2180

cluster ion (in mass spectrometry)
An ion formed by the combination of more ions or atoms or molecules of a chemical species often in association with a second species. For example, $[(H_2O)_nH]^+$ is a cluster ion.
 1991, *63*, 1549

coacervation
The separation into two *liquid phases* in *colloidal* systems. The phase more concentrated in *colloid* component is the coacervate, and the other phase is the equilibrium solution.
 1972, *31*, 611

coagulation (flocculation) (in colloid chemistry)
When a *sol* is colloidally unstable (i.e. the rate of *aggregation* is not negligible) the formation of *aggregates* is called coagulation or flocculation.
 1972, *31*, 609; O.B. 85; see also 1990, *62*, 2180

coal-derived pitch coke
The primary industrial solid *carbonization* product obtained from *coal tar pitch*, mainly produced in chamber or *delayed coking processes*.
Notes:
Coal-derived pitch coke, although it exhibits a pregraphitic microstructure, has often a lower graphitizability than *petroleum coke*. Fractions of *coal tar pitches* (obtained by extraction or filtration) may form *cokes* with needle-like structures and have an improved graphitizability. The usually lower graphitizability compared to petroleum coke is due to an inhibition of mesophase growth because of chemical and physical differences of the cokes.
 1995, *67*, 485

coalescence (in colloid chemistry)
The disappearance of the boundary between two particles (usually droplets or bubbles) in contact, or between one of these and a bulk phase followed by changes of shape leading to a reduction of the total surface area. The *flocculation* of an *emulsion*, viz. the formation of *aggregates*, may be followed by coalescence.
 1972, *31*, 610

coalification
A geological process of formation of materials with increasing content of the element carbon from organic materials that occurs in a first, biological stage into peats, followed by a gradual transformation into coal by action of moderate temperature (about 500 K) and high pressure in a geochemical stage.
Notes:
Coalification is a dehydrogenation process with a reaction rate slower by many orders of magnitude than that of *carbonization*.
Some specific reactions approach completion before others have started. The dehydrogenation remains incomplete. The degree of coalification reached by an organic material in the process of coalification increases progressively and can be defined by means of the measured C/H ratio and of the residual contents of oxygen, sulfur and nitrogen.
1995, *67*, 485

coal tar pitch
A residue produced by distillation or heat treatment of coal tar. It is a solid at room temperature, consists of a complex mixture of numerous predominantly aromatic hydrocarbons and heterocyclics, and exhibits a broad softening range instead of a defined melting temperature.
Notes:
The hydrogen aromaticity in coal tar pitch (ratio of aromatic to total content of hydrogen atoms) varies from 0.7 to 0.9.
1995, *67*, 485

cobalamines
See *corrinoids*.
1995, *67*, 1328

codon
The sequence of three consecutive *nucleotides* that occurs in mRNA which directs the incorporation of a specific amino acid into a protein and includes also the starting or termination signals of protein synthesis.
1992, *64*, 150

coefficient
Proportionality constant between two quantities of different dimension
ISO 31-0: 1992; see also 1996, *68*, 966

coefficient of haze (COH)(in atmospheric chemistry)
One technique of measurement of the amount of filterable particulate matter suspended in air which has been used in the past depends upon drawing a measured sample of air (usually 1 000 linear feet) through a paper or membrane filter. A measurement is made of the intensity of light transmitted through the dust spot formed relative to that transmitted through an identical clean filter. The dirtiness of the air is reported in terms of the COH unit. This relates to the quantity of particulate material which produces an optical density, $\lg(I_0/I)$ of 0.01 when measured by light transmission at 400 nm and relative to the transmission of an identical dust-free filter taken as 100%. Thus a filter which transmitted 50% relative to the clean filter has an absorbance of 0.301 or 30.1 COH units. This is not a recommended measure of filterable particulate matter since the size, colour and other properties of the aerosol and the air in which it is suspended affect the results.
1990, *62*, 2180

coenzyme
The dissociable, low-relative-molecular-mass active group of an *enzyme* which transfers chemical groups, hydrogen or electrons. A *coenzyme* binds with its associated protein (apoenzyme) to form the active enzyme (*holoenzyme*).
1994, *66*, 2593; see also 1992, *64*, 150

coextraction
Formation of mixed-species aggregates in a low-polarity organic phase.
1993, *65*, 2377

cofactors
Organic molecules (cf. *coenzymes*) or ions (usually metal ions) that are required by an *enzyme* of its activity. They may be attached either loosely or tightly (*prosthetic group*) to the enzyme. A cofactor binds with its associated protein (apoenzymes), which is functionally inactive, to form the active enzyme (holoenzyme).
1992, *64*, 150

coherence length (in thin films), ζ
The distance over which order is maintained. As an example, there can be long-range atomic or molecular order, i.e. coherence. Coherence lengths are significantly larger than molecular size. Normally, coherence length is estimated from electron, neutron or X-ray scattering and scales the size of ordered domains in material where long range ordering occurs (as in liquid crystals, for example). The term coherence length is also used for the scale characterizing the profile of molecular axis orientation in the distorted, so-called, transition layers formed at a solid/liquid-crystal boundary when an appropriate external field is applied (e.g. when the orientation of anchored surface molecules of a nematic liquid crystal is unaffected by an external field — magnetic or electric — whereas the bulk molecules reorient freely). This scale is referred to as *electric coherence length*, ζ_E or *magnetic coherence length*, ζ_M, depending on the nature of the applied field.
1994, *66*, 1675

coherent radiation
Radiation emitted by a source when all the elementary waves emitted have a phase difference constant in space and time.
1996, *68*, 2233

coherent scattering
Scattering is coherent whenever the phases of the signals arising from different scattering centres are

correlated and incoherent whenever these phases are uncorrelated.
1983, *55*, 932

coherent source (in spectrochemistry)
Those sources where the *radiation* has a constant phase relationship between waves spatially as well as temporally, e.g. lasers.
Non-coherent optical sources emit radiation which is randomly distributed in phase, spatially as well as temporally. Most sources which are used in spectroscopy and for spectrochemical analysis conform to this latter group.
1985, *57*, 1456

coherent structure
If the net of an ordered adsorbed phase is in registry with the lattice of the *adsorbent* the structure is called coherent, if not it is called incoherent.
1976, *46*, 76

coherent system of units (of measurement)
System of units of measurement in which all of the derived units are coherent.
See *coherent (derived) unit (of measurement)*, *derived unit (of measurement)*.
1996, *68*, 966

coherent (derived) unit (of measurement)
Derived unit of measurement that may be expressed as a product of powers of base units with the proportionality factor one.
VIM; 1996, *68*, 966

coincidence circuit
An electronic circuit which produces a usable output pulse only when each of its inputs receives a pulse within a specified time interval.
1982, *54*, 1538

coincidence, delayed
See *delayed coincidence*.
1982, *54*, 1538

coincidence, prompt
See *prompt coicidence*.
1982, *54*, 1537

coincidence resolving time
The greatest time interval that can elapse between the occurrence of two or more consecutive signal pulses, in order that the measuring device processes them as a coincidence.
1982, *54*, 1538

coincidence, true
See *true coincidence*.
O.B.216

co-ions
1. *Ions* of low relative molecular mass with a charge of the same sign as that of a given *colloidal* ion.
1972, 31, 607
2. In an *ion exchanger*, mobile ionic species with a charge of the same sign as the *fixed ions*.
O.B. 88

coke
A solid high in content of the element carbon and structurally in the non-graphitic state. It is produced by pyrolysis of organic material which has passed, at least in part, through a liquid or liquid-crystalline state during the *carbonization* process. Coke can contain mineral matter.
Notes:
As some parts, at least, of the carbonization product have passed through a liquid or liquid-crystalline state, the resulting *non-graphitic carbon* is of the graphitizable variety. From a structural viewpoint, the term coke characterizes the state of graphitizable carbon before the beginning of *graphitization*.
1995, *67*, 485

coke breeze
A by-product of *coke* manufacture; it is the residue from the screening of heat-treated coke. The particle size is less than 10 mm. Generally, coke breeze has a volatile matter content of <3 wt.%.
1995, *67*, 486

coking
Many reactions involving hydrocarbons and particularly those run at higher temperatures lead to the deposition on the *catalyst* of high molecular weight compounds of carbon and hydrogen which deactivate the catalyst. This phenomenon is called coking or fouling. Catalysts so deactivated can often be regenerated.
1976, *46*, 84

col (saddle point)
A mountain-pass in a *potential-energy surface* is known as a col or saddle point. It is a point at which the gradient is zero along all coordinates, and the curvature is positive along all but one coordinate, which is the reaction coordinate, along which the curvature is negative.
1996, *68*, 158

CO_2 laser
See *carbon dioxide laser*.
1996, *68*, 2233

cold neutrons
Neutrons with a *neutron temperature* considerably lower than normal room temperature.
1982, *54*, 1546

collection
The planned removal from solution of a micro- or macro-component by the intentional formation of a contaminated host precipitate, or by the *adsorption* or entrapment of the component by an added solid.
O.B. 84

collection efficiency (in atmospheric chemistry)
A term which characterizes an entire sampling and sample pretreatment procedure, usually represented as a percentage of the original amount of the analyte which is left for measurement (signal formation) after having passed through this procedure. This term also applies to the efficiency of collection of an air pollut-

ant by an arrestment plant. For particulates, the collection efficiency is size related.
1990, *62*, 2181

collection gas flow system (in spectrochemical analysis)
A system in which the analyte is carried to a collecting device called the collector which may be an amalgamator for mercury, a cold trap for some of the more stable hydrides, an absorbing solution in which the analyte is absorbed by some chemical reaction, an electrothermal atomizer or a reservoir.
1992, *64*, 263

collector (in atmospheric chemistry)
A device for removing and retaining contaminants from air or other gases; a term which is often applied to cleaning devices in exhaust systems. Also used to designate a device for removing and retaining samples from media in different environmental compartments to be investigated. A collector is sometimes used to describe a *scavenger*.
1990, *62*, 2181

collector (scavenger)
A solid substance added to or formed within a solution to collect a micro- or macro-component.
O.B. 84

colligation
The formation of a covalent bond by the combination or recombination of two *radicals* (the reverse of *unimolecular homolysis*). For example:

·OH + H$_3$C· → CH$_3$OH
1994, *66*, 1097

collimation
The limiting of a beam of *radiation* to the required dimensions and angular spread.
1982, *54*, 1538

collimator
An arrangement of *absorbers* used for *collimation*.
1982, *54*, 1538

collinear reaction
This is a reaction assumed to occur via an *activated complex* in which all of the atoms directly involved in the process lie along a straight line.
1996, *68*, 158

collisional activation (collisionally activated dissociation)(in mass spectrometry)
An ion/neutral species process wherein excitation of a projectile ion of high translational energy is brought about by the same mechanism as in collision-induced dissociation. (The ion may decompose subsequently.)
1991, *63*, 1554

collisional broadening (of a spectral line)
Collisions of the emitting or absorbing particle with other particles cause collisional broadening as well as collisional shift of the spectral line. When collisions occur between unlike, neutral particles the term foreign-gas broadening is used, or van der Waals broadening when both collision partners are neutral. When the colliding particles are of the same species, the term resonance broadening is employed.
The term Lorentz broadening was previously used for neutral particle collision broadening, and Holtsmark broadening for cases of van der Waals broadening when collisions took place with like particles. Both terms are now discouraged.
Stark broadening refers to collisions with charged particles or particles with a strong permanent electrical dipole moment. Whereas a strong chaotic electrical field causes Stark broadening, an applied static electrical field induces a Stark shift.
1985, *57*, 1463; O.B. 122

collisional excitation (in mass spectrometry)
An ion/neutral species process wherein there is an increase in the reactant ion's internal energy at the expense of the translational energy of either (or both) of the reacting species. The scattering angle may be large.
1991, *63*, 1555

collision complex
An ensemble formed by two reaction partners for which the distance is the sum of their van der Waals radii. As such it constitutes a subclass of the species indicated as *encounter complex*.
1996, *68*, 2233

collision cross-section, σ
In simple hard sphere collision theory the area of the circle with radius equal to the *collision diameter*.
G.B. 56; see also 1996, *68*, 158

collision density, Z_{AA}
Total *collision frequency* for all molecules within a volume divided by that volume.
G.B. 56; see also 1996, *68*, 158

collision diameter, d
In simple hard sphere collision theory the sum of radii of two colliding spherical particles.
See also *collision cross-section*.
G.B. 56

collision efficiency, B_c
The collision efficiency, or de-energization efficiency, is defined by:

$$k_{-1} = B_c k_{-1}^{sc}$$

where k_{-1} is the *rate constant* for a particular substance M when it brings about the de-energization process

$$M + A^* \rightarrow M + A,$$

and k_{-1}^{sc} is the corresponding rate constant for a reference molecule M_r that de-energizes A^* on every collision; that is, the reference molecule M_r undergoes *strong collisions*, and by definition has a collision efficiency B_c of unity. The species A^* is usually in a vibrationally-excited state, and A has energy less than that required for reaction to occur.
1996, *68*, 159

collision, elastic
See *elastic collision*.
1996, *68*, 159

collision frequency, $z_A(A)$
Average number of collisions experienced by a molecule A with other molecules A (or B) in a time interval divided by that time interval.
G.B. 56; see also 1996, *68*, 159

collision-induced dissociation (in mass spectrometry)
An ion/neutral species process wherein the projectile ion is dissociated as a result of interaction with a target neutral species. This is brought about by conversion of part of the translational energy of the ion to internal energy in the ion during the collision.
1991, *63*, 1555

collision, inelastic
See *elastic collision*.
1996, *68*, 159

collision number
Now called *collision density*.
G.B. 56; 1996, *68*, 160; see also 1990, *62*, 2181

collision theory
Various collision theories, dealing with the frequency of collision between *reactant* molecules, have been put forward. In the earliest theories reactant molecules were regarded as hard spheres, and a collision was considered to occur when the distance d between the centres of two molecules was equal to the sum of their radii. For a gas containing only one type of molecule, A, the collision density is given by simple collision theory as:

$$Z_{AA} = \tfrac{1}{2}\sqrt{2}\,\pi\sigma^2 u N_A^2$$

Here N_A is the number density of molecules and u is the mean molecular speed, given by kinetic theory to be $(8k_B T/\pi m)^{1/2}$, where m is the molecular mass, and $\sigma = \pi d_{AA}^2$. Thus:

$$Z_{AA} = 2N_A^2 \sigma^2 (\pi k_B T/m)^{1/2}$$

The corresponding expression for the collision density Z_{AB} for two unlike molecules A and B, of masses m_A and m_B is:

$$Z_{AB} = N_A N_B \sigma^2 (\pi k_B T/\mu)^{1/2}$$

where μ is the reduced mass $m_A m_B/(m_A + m_B)$, and $\sigma = \pi d_{AB}^2$.
For the collision frequency factor these formulations lead to the following expression:

$$z_{AA} \text{ or } z_{AB} = L\sigma^2 (8\pi k_B T/\mu)^{1/2}$$

where L is the Avogadro constant.
More advanced collision theories, not involving the assumption that molecules behave as hard spheres, are known as *generalized kinetic theories*.
1996, *68*, 160

colloid
A short synonym for *colloidal* system.
1972, *31*, 605

colloidal
The term refers to a state of subdivision, implying that the molecules or polymolecular particles dispersed in a medium have at least in one direction a dimension roughly between 1 nm and 1 μm, or that in a system discontinuities are found at distances of that order.
1972, *31*, 605

colloidal carbon
A *particulate carbon* with particle sizes below *ca.* 1000 nm in at least one dimension.
Note:
Colloidal carbon exists in several morphologically distinct forms.
1995, *67*, 486

colloidal dispersion
A system in which particles of *colloidal* size of any nature (e.g. solid, liquid or gas) are dispersed in a *continuous phase* of a different composition (or state). The name *dispersed phase* for the particles should be used only if they have essentially the properties of a bulk phase of the same composition.
1972, *31*, 605

colloidal electrolyte
An electrolyte which gives ions of which at least one is of *colloidal* size. This term therefore includes hydrophobic *sols*, ionic association *colloids*, and polyelectrolytes.
1972, *31*, 607

colloidally stable
Particles which do not *aggregate* at a significant rate: the precise connotation depends on the type of *aggregation* under consideration. For example, a concentrated paint is called stable by some people because oil and pigment do not separate out at a measurable rate, and unstable by others because the pigment particles aggregate into a continuous network.
1972, *31*, 609

colloidal suspension
A *suspension* in which the size of the particles lies in the *colloidal* range.
1972, *31*, 606

colloid osmotic pressure (Donnan pressure)
The pressure difference which has to be established between a *colloidal* system and its equilibrium liquid to prevent material transfer between the two phases when they are separated by a membrane, permeable to all components of the system, except the colloidal ones.
1972, *31*, 615

colorimeter
An instrument used for colour measurement based on optical comparison with standard colours.
1990, *62*, 2181

colour indicator
An indicator which is classified as one- or two-colour, depending on whether it is colourless on one side of the transition interval or possesses a different colour on each side of this range.
O.B. 48

column (in chromatography)
The tube and the stationary phase contained within, through which the mobile phase passes.
1993, *65*, 831; O.B. 97

column chromatography
A separation technique in which the stationary bed is within a tube. The particles of the solid stationary phase or support coated with a liquid stationary phase may fill the whole inside volume of the tube (packed column) or be concentrated on or along the inside tube wall leaving an open, unrestricted path for the mobile phase in the middle part of the tube (open-tubular column).
1993, *65*, 825

column volume (in chromatography), V_c
The geometric volume of the part of the tube that contains the packing:
$$V_c = A_c L$$
where A_c is the internal cross-sectional area of the tube and L is the length of the packed part of the column.
In the case of wall-coated open-tubular columns the column volume corresponds to the geometric volume of the whole tube having a liquid or a solid stationary phase on its wall.
1993, *65*, 832; see also O.B. 100

combination electrode
An *ion-selective electrode* and an external reference electrode combined in a single unit, thereby avoiding the need for a separate external reference electrode.
1994, *66*, 2528

combined sample
A sample obtained by removing specific fractions by separation or selection techniques (e.g. heavy liquid, magnetic, sieving, etc.), analysing the fractions separately, and combining the results mathematically.
When not combined, the sample is a modified sample. This term should not be confused with composite sample.
1990, *62*, 1204

comb macromolecule
A *macromolecule* comprising a *main chain* with multiple trifunctional *branch points* from each of which a linear *side-chain* emanates.
Notes:
1. If the subchains between the branch points of the main chain and the terminal subchains of the main chain are identical with respect to constitution and degree of polymerization, and the side chains are identical with respect to constitution and *degree of polymerization*, the macromolecule is termed a *regular comb macromolecule*.
2. If at least some of the branch points are of functionality greater than three, the macromolecule may be termed a *brush macromolecule*.
1996, *68*, 2296

comb polymer
A polymer composed of *comb macromolecules*.
See *comb macromolecule, regular polymer*.
1996, *68*, 2304

combustion gas
A gas or vapour produced in furnaces, combustion chambers or in open burning.
1990, *62*, 2181

commensurate–incommensurate transition
A transition between two states having magnetic or crystallographic structures with a basic lattice and an imposed super-periodicity. In the commensurate (lock-in) state this super-periodicity is a simple rational multiple of the basic unit cell. In the incommensurate state the ratio of the super-periodicity repeat distance to the basic lattice repeat distance is irrational and may show continuous variation with variation in some degree of freedom (e.g. temperature, pressure, composition) of the system.
Example:
The transition of Rb_2ZnCl_4 from an incommensurate structure to a commensurate structure at the lock-in temperature, T_L, which is dependent on the crystal growth method and varies in the range 128 to 190 K.
Note:
A commensurate–incommensurate transition also occurs in liquid-crystal systems where there is an incommensurability in the packing of dimers in relation to monomers.
1994, *66*, 579

common black film
See *black film*.
1972, *31*, 614

common-ion effect (on rates)
A reduction in the *rate* of certain reactions of a *substrate* RX in solution [by a path that involves a *pre-equilibrium* with formation of R^+ (or R^-) ions as reaction intermediates] caused by the addition to the reaction mixture of an electrolyte solute containing the 'common ion' X^- (or X^+). For example, the rate of solvolysis of diphenylmethyl chloride in acetone–water is reduced by the addition of salts of the common ion Cl^- which causes a decrease in the quasi-equilibrium concentration of the diphenylmethyl cation in the scheme:

$$Ph_2CHCl \underset{-1}{\overset{1}{\rightleftharpoons}} Ph_2CH^+ + Cl^-$$
(free ions, *not* ion pairs)

$$Ph_2CH^+ + OH_2 \overset{2}{\longrightarrow} Ph_2CHOH + H^+ \quad \text{(solvated)}$$

This phenomenon is a direct consequence of the *mass-law effect* on ionization equilibria in electrolytic solution.
More generally, the common-ion effect is the influence of the 'common ion' on the reactivity due to the shift of the dissociation equilibrium. It may also lead to an enhancement of the *rate of reaction*.
1994, *66*, 1098

communities (consortia, syntrophic associations, synergistic associations)
Naturally occuring groups of different (micro)organisms inhabiting a common environment, interacting with each other, especially through food relationships and relatively independent of other groups. Communities may be of varying sizes, and larger communities may contain smaller ones.
1992, *64*, 150

compact layer (in electrochemistry)
See *inner layer (in electrochemistry)*.
1983, *55*, 1261

comparator (in radioanalytical chemistry)
A known amount of element that is simultaneously irradiated with the test portion in the context of activation *analysis*. If one comparator is used (single comparator method), it is essentially identical to a *flux monitor* (expect that this term is not necessarily linked to *activation analysis*).
1994, *66*, 2516

compensation (in catalysis)
For the process:
$$k = A\exp(-E/RT)$$
where k is a rate constant, A the frequency factor, and E the *activation energy*, A and E sometimes exhibit compensation, i.e. they change in the same direction with change in *catalyst* for a given reaction or with change in reaction for a given catalyst.
1976, *46*, 82

compensation (in stereochemistry)
1. For *internal compensation*, see *meso*-compound.
2. See *external compensation*.
1996, *68*, 2204

compensation effect
In a considerable number of cases plots of $T\Delta^{\ddagger}S$ vs. $\Delta^{\ddagger}H$, for a series of reactions, e.g. for a reaction in a range of different solvents, are straight lines of approximately unit slope. Therefore, the terms $\Delta^{\ddagger}H$ and $T\Delta^{\ddagger}S$ in the expression partially compensate, and $\Delta^{\ddagger}G = \Delta^{\ddagger}H - T\Delta^{\ddagger}S$ often is a much simpler function of solvent (or other) variation than $\Delta^{\ddagger}H$ or $T\Delta^{\ddagger}S$ separately.
See also *isokinetic relationship*.
1994, *66*, 1098

competition
See *composite mechanism*.
1996, *68*, 160

competitive binding assay
An *assay* based on the competition between a *labelled* and an unlabelled *ligand* in the reaction with a receptor binding agent (e.g. *antibody*, receptor, transport protein).
1994, *66*, 2516

competitive inhibition (of catalysis)
The reduction in rate due to reduction or loss of activity of a *catalyst*, caused by *adsorption* of poison which is not greatly preferred to adsorption of reactant.
1976, *46*, 83

complementary binding sites
See *binding site*.
1994, *66*, 1098

complementary DNA (cDNA)
A single-stranded DNA molecule with a *nucleotide* sequence that is complementary to an RNA molecule; cDNA is formed by the action of the *enzyme* reverse transcriptase on an RNA *template*. After conversion to the double-stranded form, cDNA is used for molecular cloning or for *hybridization* studies.
1992, *64*, 150

complex
A *molecular entity* formed by loose *association* involving two or more component molecular entities (ionic or uncharged), or the corresponding *chemical species*. The bonding between the components is normally weaker than in a covalent *bond*.
The term has also been used with a variety of shades of meaning in different contexts: it is therefore best avoided when a more explicit alternative is applicable. In inorganic chemistry the term 'coordination entity' is recommended instead of 'complex'.
See also *activated complex, adduct, charge transfer complex, electron-donor-acceptor complex, encounter complex, inclusion complex, σ-adduct, π-adduct, transition state*.
1994, *66*, 1098

complex coacervation
Coacervation caused by the interaction of two oppositely charged *colloids*.
1972, *31*, 611

complex mechanism
This term is sometimes applied to an *elementary reaction* which proceeds via an intermediate species having a lifetime longer than a few rotational periods, in contrast to a *direct reaction*. However, in view of the danger of confusion with a reaction occurring in more than one step (a *composite reaction*) this usage is not to be recommended; it is better to use the terms complex-mode mechanism or *complex-mode reaction*.
1996, *68*, 160

complex-mode reaction
An elementary reaction which proceeds via an intermediate species having a lifetime longer than a few rotational periods is known as a complex-mode reaction, or as an *indirect* reaction.
See also *direct reaction*.
1996, *68*, 160

complexometric titration
See *titration*.
O.B. 47

complex reaction
See *composite mechanism*.
1996, *68*, 161

component
Constituent of a mixture the *amount* or *concentration* of which can be varied independently. The number of components in a given system is the minimum number of independent species necessary to define the composition in all the phases of a system. It may vary with external conditions since additional chemical equilibria reduce the number of components. The term component is also often used in the more general sense as defined here under constituent.
    Physical Chemistry Division, unpublished; see also 1996, *68*, 966

composite mechanism
A reaction that involves more than one elementary reaction is said to occur by a composite mechanism. The terms *complex mechanism indirect mechanism*, and *step-wise mechanism* are also commonly used.
There are two main kinds of evidence for a composite mechanism:
1. The kinetic equation for the reaction does not correspond to its stoichiometry.
2. There is experimental evidence, direct or indirect, for intermediates of such a nature that it is necessary to conclude that more than one elementary reaction is involved.
There are many types of composite mechanisms, for example:
a. Reactions occurring in parallel, such as:

$$A \rightarrow Y$$
$$A \rightarrow Z$$

are called *parallel reactions* or *simultaneous reactions*. When there are simultaneous reactions there is sometimes competition, as in the scheme:

$$A + B \rightarrow Y$$
$$A + C \rightarrow Z$$

where B and C compete with one another for A.
b. Reactions occurring in forward and reverse directions are called opposing reactions:

$$A + B \rightleftarrows Z$$

c. Reactions occurring in sequence, such as

$$A \rightarrow X \rightarrow Y \rightarrow Z$$

are known as consecutive reactions.
d. Reactions are said to exhibit feedback if a substance formed in one step affects the rate of a previous step. For example, in the scheme:

$$A \rightarrow X \rightarrow Y \rightarrow Z$$

The intermediate Y may catalyse the reaction $A \rightarrow X$ (positive feedback) or it may inhibit it (negative feedback).
e. *Chain reactions*.
    1996, *68*, 161

composite reaction
A *chemical reaction* for which the expression for the rate of disappearance of a reactant (or rate of appearance of a product) involves rate constants of more than a single *elementary reaction*. Examples are 'opposing reactions' (where rate constants of two opposed *chemical reactions* are involved), 'parallel reactions' (for which the rate of disappearance of any reactant is governed by the rate constants relating to several simultaneous reactions to form different respective products from a single set of reactants), and *stepwise reactions*.
    1994, *66*, 1098; 1993, *65*, 2293

composite sample
Often prepared as a representative mixture of several different (usually *bulk*) *samples*, and from which the *laboratory sample* is taken.
    1988, *60*, 1465

compositional heterogeneity (of polymers)
The variation in elemental composition from molecule to molecule usually found in copolymers.
    P.B. 52

composition of pure air (in atmospheric chemistry)
The composition of air is variable with respect to several of its components (e.g. CH_4, CO_2, H_2O) so 'pure' air has no precise meaning; it is commonly considered to be air which is free of dust, aerosols and reactive gaseous contaminants of anthropogenic origin. The composition of the major components in dry air is relatively constant (percent by volume given): nitrogen, 78.084; oxygen, 20.946; argon, 0.934; carbon dioxide, 0.033; neon, 0.0018; helium, 0.000524; methane, 0.00016; krypton, 0.000114; hydrogen 0.00005; nitrous oxide, 0.00003; xenon, 0.0000087. The concentrations of carbon dioxide, methane, nitrous oxide, the chlorofluorocarbons and some other species of anthropogenic origin are increasing measurably with time. Relative clean air which is free of most reactive anthropogenic pollution (NO, NO_2, SO_2, non-methane hydrocarbons, etc.), often used as a reference sample in the calibration and operation of instruments, is purchased under the designation of zero air.
    1990, *62*, 2172

compressibility factor
    See *compression factor*.
    G.B. 49

compression factor, Z
Product of pressure and molar volume divided by the gas constant and thermodynamic temperature. For an ideal gas it is equal to 1.
    G.B. 49

comproportionation
The reverse of *disproportionation*. The term 'symproportionation' is also used.
    1994, *66*, 1098

Compton effect
The *elastic scattering* of a *photon* by an *electron* which afterwards occurs as a free electron. Part of the energy and momentum of the incident photon is trans-

ferred to the electron and the remaining part is carried away by the scattered photon.
Synonymous with *Compton scattering*.
1982, *54*, 1538

Compton electron
The energetic *electron* resulting from the *Compton effect*.
1982, *54*, 1538

Compton scattering
See *Compton effect*.
1982, *54*, 1538

concentrated phase (in polymer chemistry)
See *polymer-rich phase*.
P.B. 68

concentration
1. Group of four quantities characterizing the composition of a mixture with respect to the volume of the mixture (*mass, amount, volume* and *number concentration*).
2. Short form for *amount (of substance) concentration* (*substance concentration* in clinical chemistry).
G.B. 42; 1996, *68*, 966; see also 1990, *62*, 2181

concentration, bulk (in electroanalysis)
See *bulk concentration (in electroanalysis)*.
1985, *57*, 1493

concentration-cell corrosion
The local variation of the corrosion rate due to the action of a corrosion cell resulting from inhomogeneous composition of the environment.
1989, *61*, 22

concentration distribution ratio (in chromatography), D_c
The ratio of the analytical concentration of a component in the *stationary phase* to its analytical concentration in the *mobile phase*.

$$D_c = \frac{\text{amount of component}/\text{cm}^3 \text{ of stationary phase}}{\text{amount of component}/\text{cm}^3 \text{ of mobile phase}}$$

O.B. 106

+ concentration factor (in solvent extraction)
Use of this term is not recommended.
See *extraction factor*.
1993, *65*, 2381

concentration gradient, grad C
Differential change in concentration of a component in a given direction in a small distance divided by the distance in that direction.
Notes:
1. It is a vector quantity.
2. May be expressed as amount-of-substance, mass, number, volume, concentration gradient.
1996, *68*, 967

concentration, ground level (in atmospheric chemistry)
See *ground level concentration*.
1990, *62*, 2181

concentration in experimental surface (or surface concentration)
The amount of the material of interest divided by the total amount of substances in the volume of interest. Concentration may be defined in terms of numbers of atoms (particles) (ppma) or in terms of weight (μg/g).
1979, *51*, 2246

concentration overpotential
The concentration overpotential of an *electrode reaction* at a given *electrode current density* (c.d.) is basically the difference in equilibrium potentials across the *diffusion layer*. More precisely, it is the potential of a *reference electrode* (of the same electrode reaction as the *working electrode*) with the interfacial concentrations which establish themselves at c.d., relative to the potential of a similar reference electrode with the concentrations of the bulk solution. From such a measured potential difference, with c.d. flowing, one needs to subtract the ohmic potential drop prevailing between the two electrodes.
1981, *53*, 1838

concentration, particle
See *particle concentration*.
1990, *62*, 2181

concentration-sensitive detector (in chromatography)
A device the response of which is proportional to the concentration of a sample component in the eluent.
1993, *65*, 849

+ concentration thermometric technique (in enthalpimetric analysis)
See *direct injection enthalpimetry*.
1994, *66*, 2490

concerted process
Two or more *primitive changes* are said to be concerted (or to constitute a concerted process) if they occur within the same *elementary reaction*. Such changes will normally (though perhaps not inevitably) be 'energetically coupled'. (In the present context the term 'energetically coupled' means that the simultaneous progress of the primitive changes involves a *transition state* of lower energy than that for their successive occurrence.) In a concerted process the primitive changes may be *synchronous* or asynchronous.
See also *bifunctional catalysis, potential-energy (reaction) surface*.
1994, *66*, 1098

condensation (in atmospheric chemistry)
The physical process of converting a material from a gaseous or vapour phase to a liquid or solid phase; this commonly results when the temperature is lowered and/or the vapour pressure of the material is increased. The tendency exists for condensation to occur when the partial pressure of a given component of a gaseous mixture at a given temperature exceeds the vapour pressure of the liquid or solid form of that component at the given temperature.
1990, *62*, 2182

condensation nuclei (CN) (in atmospheric chemistry)
Particles, either liquid or solid, or ions upon which condensation of water vapour (or other substances) begins in the atmosphere. Condensation nuclei are usually very small hygroscopic aerosols (0.001 to 0.1 µm in diameter), but these are not as abundant as the smaller particles. The number of CN which are active (initiate condensation) in a given air mass may be a function of the relative humidity. Usually CN are counted as the active nuclei at about 300% relative humidity, while cloud condensation nuclei (CCN) are counted as the number of active nuclei at relative humidity less than or equal to 102%.
1990, *62*, 2182

condensation polymerization (polycondensation)
Polymerization by a repeated condensation process (i.e. with elimination of simple molecules).
P.B. 18

condensation reaction
A (usually stepwise) reaction in which two or more reactants (or remote reactive sites within the same *molecular entity*) yield a single main product with accompanying formation of water or of some other small molecule, e.g. ammonia, ethanol, acetic acid, hydrogen sulfide.
The mechanism of many condensation reactions has been shown to comprise consecutive *addition* and *elimination reactions*, as in the base-catalysed formation of (*E*)-but-2-enal (crotonaldehyde) from acetaldehyde, *via* 3-hydroxybutanal (aldol). The overall reaction in this example is known as the aldol condensation.
The term is sometimes also applied to cases where the formation of water or another simple molecule does not occur, as in 'benzoin condensation'.
1994, *66*, 1099

condensative chain polymerization
See *chain polymerization*.
1996, *68*, 2306

conditional (formal) potential
Conditional (formal) potential, $E^{0\prime}$, is related to the potential of a cell reaction, E_{cell}, by an equation analogous to that relating E_{cell} to the standard potential E_0, with the activity, a_i, replaced by any composition variable, c_i (to be indicated by a subscript). For example:

$$E_{cell} = E_c^{0\prime} - (RT/nF) \sum_i v_i \ln c_i$$

1974, *37*, 505

conditional rate constant of an electrode reaction
See *electrode reaction rate constants*.
1974, *37*, 515

conditioning (in solvent extraction)
Synonymous with *pre-equilibration*.
1993, *65*, 2377

conductance, *G*
See *electric conductance*.
G.B. 15; 1996, *68*, 970

conduction band
A vacant or only partially occupied set of many closely spaced electronic levels resulting from an array of a large number of atoms forming a system in which the electrons can move freely or nearly so. This term is usually used to describe the properties of metals and semiconductors.
See *bandgap energy, Fermi level, valence band*.
1996, *68*, 2233

conductivity, γ, σ
Reciprocal of *resistivity*. This quantity is a tensor in an anisotropic medium. It was formerly called specific conductance.
G.B. 15; 1996, *68*, 967

confidence level
The probability of covering the expected value of an estimated parameter with an interval estimated for the parameter (symbol $1 - \alpha$). The confidence level can be expressed as a number between 0 and 1, or in percent. The complementary quantity is known as the significance level.
Comment:
In some cases the confidence level is dictated by the needs of the situation. In all other instances, use of $1 - \alpha = 0.95$ is recommended.
1994, *66*, 599

confidence limits (about the mean)
Symmetric confidence limits ($\pm C$) about the estimated mean, which cover the population mean with probability $1 - \alpha$. The quantity C is calculated by the formula:

$$C = t_{p,v} s/\sqrt{n}$$

Here $t_{p,v}$ is the critical value from the *t*- (or Student) distribution function corresponding to the confidence level $1 - \alpha$ and degrees of freedom v. The symbol p represents the percentile (or percentage point) of the *t*-distribution. For 1-sided intervals, $p = 1 - \alpha$; for 2-sided intervals, $p = 1 - \alpha/2$. In each case, the confidence level is $1 - \alpha$. The confidence interval is given as $\bar{x} \pm C$.
Comment:
If the population standard deviation σ is known, confidence limits about a single result may be calculated with the formula:

$$C = t_{p,\infty} \sigma$$

The coefficient $t_{p,\infty}$ is the limiting value of the *t*-distribution function for $v = \infty$ at confidence level $1 - \alpha$. This is identical to z_p, the *p*th percentage point of the standard normal variate.
1994, *66*, 601

configuration (electronic)
A distribution of the electrons of an atom or a *molecular entity* over a set of one-electron wavefunctions called *orbitals*, according to the Pauli principle.

From one configuration several states with different multiplicities may result. For example, the ground electronic configuration of the oxygen molecule (O_2) is:

$$1\sigma_g^2, 1\sigma_u^2, 2\sigma_g^2, 2\sigma_u^2, 1\pi_u^4, 3\sigma_g^2, 1\pi_g^2$$

resulting in the $^3\Sigma_g^-$, $^1\Delta_g$ and $^1\Sigma_g^+$ multiplets.

1994, *66*, 1099; 1996, *68*, 2233

configuration (stereochemical)
In the context of stereochemistry, the term is restricted to the arrangements of atoms of a molecular entity in space that distinguishes *stereoisomers*, the *isomerism* between which is not due to *conformation* differences.

See also *absolute configuration*; *relative configuration*.
1996, *68*, 2204; 1994, *66*, 1099

configurational base unit (in polymers)
A *constitutional repeating unit* in a *regular macromolecule*, a *regular oligomer molecule*, a *regular block* or a *regular chain*, the configuration of which is defined at least at one site of stereoisomerism in the *main chain*.
1996, *68*, 2291

configurational disorder (in polymers)
Structural disorder resulting from the statistical co-crystallization of different configurational repeating units.
P.B. 80

configurational homosequence (in polymers)
A *constitutional homosequence* in which the relative or *absolute configuration* is defined at one or more sites of stereoisomerism in each *constitutional unit* in the main chain of a *polymer* molecule.
P.B. 36

configurational repeating unit (in polymers)
The smallest set of successive *configurational base units* that prescribes configurational repetition at one or more sites of stereoisomerism in the main chain of a *regular macromolecule*, a *regular oligomer molecule*, a *regular block* or a *regular chain*.
1996, *68*, 2291

configurational sequence (in polymers)
The whole or part of a *chain* comprising one or more species of *configurational unit(s)* in defined sequence.
Note:
Configurational sequences comprising two configurational units are termed diads, those with three such constitutional units triads, and so on. In order of increasing sequence lengths they are called tetrads, pentads, hexads, heptads, octads, nonads, decads, undecads, etc.
1996, *68*, 2299

configurational unit (in polymers)
A *constitutional unit* having at least one site of defined stereoisomerism.
1996, *68*, 2291

configuration interaction (CI)
The mixing of many-electron wavefunctions constructed from different electronic configurations to obtain an improved many-electron state.
1996, *68*, 2233

confined atomizer
See *electrothermal atomizer*.
1992, *64*, 254

conformation
The spatial arrangement of the atoms affording distinction between stereoisomers which can be interconverted by rotations about formally single bonds. Some authorities extend the term to include inversion at trigonal pyramidal centres and other *polytopal rearrangements*.

See also *conformer*; *bisecting conformation*, *eclipsing conformation*; *chair, boat, twist*; *crown conformation*; *envelope conformation*; *half-chair*; *staggered conformation*; *tub conformation*.
1996, *68*, 2204; 1994, *66*, 1099

conformational analysis
The assessment of the relative energies (or thermodynamic stabilities), reactivities, and physical properties of alternative *conformations* of a molecular entity, usually by the application of qualitative or semi-quantitative rules or by semi-empirical calculations.
1996, *68*, 2204

conformational disorder (in polymers)
Structural disorder resulting from the statistical coexistence within the crystals of identical configurational units with different conformations.
P.B. 80

conformational repeating unit (of a polymer)
See *(chain) conformational repeating unit (of a polymer)*.
P.B. 77

conformer
One of a set of *stereoisomers*, each of which is characterized by a *conformation* corresponding to a distinct potential energy minimum.
See also *rotamer*.
1996, *68*, 2204

conglomerate
See *racemic conglomerate*.
1996, *68*, 2204

congruent transition
A transition in which the two-phase equilibrium of *melting*, vaporization or allotropism of a compound involves phases of the same composition.
1994, *66*, 579

coning and quartering (in analytical chemistry)
The reduction in size of a granular or powdered sample by forming a conical heap which is spread out into a circular, flat cake. The cake is divided radially into quarters and two opposite quarters are combined. The other two quarters are discarded. The process is repeated as many times as necessary to obtain the quan-

tity desired for some final use (e.g. as the laboratory sample or as the test sample).

If the process is performed only once, coning and quartering is no more efficient than taking alternate portions and discarding the others.

1990, *62*, 1204

conjugate acid–base pair

The *Brønsted acid* BH^+ formed on protonation of a base B is called the conjugate acid of B, and B is the conjugate base of BH^+. (The conjugate acid always carries one unit of positive charge more than the base, but the absolute charges of the species are immaterial to the definition.) For example: the Brønsted acid HCl and its conjugate base Cl^- constitute a conjugate acid–base pair.

1994, *66*, 1099

conjugated system (conjugation)

In the original meaning a conjugated system is a molecular entity whose structure may be represented as a system of alternating single and multiple bonds: e.g. $CH_2=CH-CH=CH_2$, $CH_2=CH-C\equiv N$.

In such systems, conjugation is the interaction of one p-orbital with another across an intervening σ-bond in such structures. (In appropriate molecular entities d-orbitals may be involved.) The term is also extended to the analogous interaction involving a p-orbital containing an unshared electron pair, e.g. $:Cl-CH=CH_2$.

See also *delocalization, homoconjugation, resonance*.

1994, *66*, 1099

conjugate solutions

Two solutions that co-exist in equilibrium at a given temperature and pressure and, at constant pressure (temperature), change their compositions and relative proportions with a variation of temperature (pressure). The term usually refers to two immiscible liquids, but it is also applicable to two immiscible solid solutions.

1994, *66*, 580

conjugation

See *conjugated system*.
1994, *66*, 1099

conjugation (in gene technology)

The contact-dependent transfer of a part or all of its *genome* from one bacterial *cell* (donor) to another (recipient).

1992, *64*, 150

conjugation labelling (in radioanalytical chemistry)

Labelling of a substance by conjugation with a labelled molecule.

1994, *66*, 2521

conjugative mechanism

See *electronic effect*.
1994, *66*, 1099

conjunctive name

A name for assemblies of functionalized acyclic parent hydrides and cyclic systems implying the loss of an appropriate number of hydrogen atoms from each.

B.B.(G) 15

connectivity

In a chemical context, the information content of a *line formula*, but omitting any indication of *bond* multiplicity.

1994, *66*, 1099

conrotatory

See *electrocyclic reaction*.
1994, *66*, 1100

consecutive reactions

Reactions occurring in sequence, such as:

$$A \xrightarrow{1} X \xrightarrow{2} Y \xrightarrow{3} Z$$

The overall process is said to occur by consecutive steps.

1981, *53*, 760

consecutive steps

See *consecutive reactions*.
1981, *53*, 760

consignment (in analytical chemistry)

A quantity of material transferred on one occasion and covered by a single set of shipping documents. It may consist of one or more lots or portions of lots.

The presence of different lots in a consignment is important from the point of view of the sampling plan and the interpretation of the results of analysis. The term 'population' is used as the general term for the quantity of parent material being sampled when it is immaterial if the parent body is a consignment, lot, batch, entity, etc.

1990, *62*, 1201

consistency

A general term to describe the property of a material by which it resists permanent change of shape.

1979, *51*, 1217

consolute point

See *critical solution point*.
1994, *66*, 580

constituent

Chemical species present in a system; often called a component, although the term *component* has a more restricted meaning in physical chemistry.

Physical Chemistry Division, unpublished

constitution

The description of the identity and connectivity (and corresponding bond multiplicities) of the atoms in a molecular entity (omitting any distinction arising from their spatial arrangement).

1996, *68*, 2204; 1994, *66*, 1100

constitutional heterogeneity (of polymers)

The variation in constitution from molecule to molecule in polymers with molecules uniform with respect to elemental composition.

An example is a polymer composed of linear and branched molecules; another example is a statistical copolymer comprising two isomeric constitutional units.
P.B. 52

constitutional homosequence
A *constitutional sequence* which contains *constitutional units* of only one species and in one sequential arrangement.
P.B. 35

constitutional isomerism
Isomerism between structures differing in *constitution* and described by different line formulae e.g. CH_3OCH_3 and CH_3CH_2OH.
1996, *68*, 2205

constitutional repeating unit (CRU) (in polymers)
The smallest *constitutional unit* the repetition of which constitutes a regular *macromolecule*, a regular *oligomer molecule*, a regular block or a *regular chain*.
1996, *68*, 2291

constitutional sequence (in polymers)
The whole or part of a *chain* comprising one or more species of *constitutional unit(s)* in defined sequence.
Note:
Constitutional sequences comprising two constitutional units are termed diads, those comprising three constitutional units triads, and so on. In order of increasing sequence lengths they are called tetrads, pentads, hexads, heptads, octads, nonads, decads, undecads, etc.
1996, *68*, 2299

constitutional unit
An atom or group of atoms (with pendant atoms or groups, if any) comprising a part of the essential structure of a *macromolecule*, an *oligomer molecule*, a *block* or a *chain*.
1996, *68*, 2291

constitutive enzymes
Enzymes which are produced constitutively by the *cell* under all physiological conditions. Therefore, they are not controlled by *induction* or *repression*.
1992, *64*, 150

contact angle
When a liquid does not spread on a *substrate* (usually a solid), a contact angle (θ) is formed which is defined as the angle between two of the interfaces at the three-phase line of contact. It must always be stated which interfaces are used to define θ.
It is often necessary to distinguish between the 'advancing contact angle' (θ_a), the 'receding contact angle' (θ_r) and the 'equilibrium contact angle' (θ_e). When $\theta_r \neq \theta_a$ the system is said to exhibit contact angle *hysteresis*.
1972, *31*, 598

contact corrosion
The enhancement of the corrosion rate of the material with the more negative corrosion cell resulting from the contact between different electron-conducting phases.
1989, *61*, 22

contact ion pair
See *ion pair*.
1994, *66*, 1100

contact potential difference (Volta potential difference)
The electric potential difference between one point in the vacuum close to the surface of M1 and another point in the vacuum close to the surface of M2, where M1 and M2 are two uncharged metals brought into contact.
1983, *55*, 1257

contaminant, air
See *air contaminant*.
1990, *62*, 2182

contamination
Carry-over from a preceding sample probe into a following specimen cup, which will influence not just one result, but all assays on that specimen, or with that reagent.
1991, *63*, 302

content
See *substance content*.
1996, *68*, 967

continuity inversion (in solvent extraction)
A change in the mutual dispersion of two phases in contact.
See *inversion*.
1993, *65*, 2386

continuous analyser
An analyser in which subassemblies operate continuously.
1990, *62*, 2174

continuous flow
See *transport*.
1989, *61*, 1661

continuous-flow cell (in spectrochemical analysis)
A cell which allows liquid (or gaseous) samples to pass through the cell continuously while absorption measurements are made.
1988, *60*, 1454

continuous flow enthalpimetry
A term used to describe methods wherein a reagent is continuously fed into a flowing analyte stream and the temperature difference is measured before and after a reactor (mixing chamber). Alternatively, the temperature may be measured in a differential manner between a reference and a reactor chamber. The analyte concentration is directly proportional to the measured temperature difference.
Synonymous with continuous flow enthalpimetric analysis.
1994, *66*, 2491

continuous indication analyser
An analysis with a permanent indication related to the sample concentration. To obtain a continuous indication, the sampling and measuring cell need to be continuous. A time lag may exist between sampling and indication of measured concentration.
See *dead time of analyser*.
1990, *62*, 2174

continuously curved chain (in polymers)
See *worm-like chain*.
P.B. 51

continuous measuring cell
A measuring cell which operates continuously.
1990, *62*, 2178

continuous operation (in analysis)
The operation of an analytical instrument in such a way that no analytical procedure needs to be completed on any sample before the next procedure can be started.
1989, *61*, 1661

continuous precipitation
A diffusional reaction in a multi-component system in which atoms are transported to the growing nuclei by diffusion over relatively large distances in the parent phase and during which the mean composition of the parent phase changes continuously towards its equilibrium value.
Synonymous with *nucleation and growth*.
1994, *66*, 580

continuous transition
See *second-order phase transition*.
1994, *66*, 580

+ contour length (in polymers)
The maximum end-to-end distance of a linear polymer chain.
For a single-strand polymer molecule, this usually means the end-to-end distance of the chain extended to the all-*trans* conformation. For chains with complex structure, only an approximate value of the contour length may be accessible. The sum of the lengths of all skeletal bonds of a single-strand polymer molecule is occasionally termed 'contour length'.
This use of this term is discouraged.
P.B. 50

contributing structure
The definition is based on the valence-bond formulation of the quantum mechanical idea of the wavefunction of a molecule as composed of a linear combination of wavefunctions, each representative of a formula containing bonds that are only single, double or triple with a particular pairing of electron spins. Each such formula represents a contributing structure, also called 'resonance structure' to the total wavefunction, and the degree to which each contributes is indicated by the square of its coefficient in the linear combination. The contributing structures, also called 'canonical forms', themselves thus have a purely formal significance: they are the components from which wavefunctions can be built. Structures may be covalent (or non-polar) or ionic (or polar). The representation is frequently kept qualitative so that we speak of important or major contributing structures and minor contributing structures. For example, two major non-equivalent contributing structures for the conjugate base of acetone are:

$$H_2C=\underset{CH_3}{\overset{O^-}{\diagup}} \longleftrightarrow H_2C^-\underset{CH_3}{\overset{O}{\diagup}}$$

See also *delocalization, Kekulé structure, resonance*.
1994, *66*, 1100

controlled atmosphere (in atmospheric chemistry)
A synthetic gaseous sample of pure air which may contain carefully determined amounts of certain contaminants; this may be used as a standard for the calibration of analytical techniques, as a simulated environment for the study of biological responses, or for other purposes.
1990, *62*, 2182

control material (in analysis)
A material to be used for the assessment of the performance of an analytical procedure or part thereof.
See *calibration material*.
1989, *61*, 1660

control sample (in analysis)
The test portion or test solution used for assessment of the performance of an analytical procedure.
1989, *61*, 1660

convection (as applied to air motion)
Vertical motion of the air induced by the expansion of the air heated by the earth's surface and its resulting buoyancy.
1990, *62*, 2182

convenience sample
A sample chosen on the basis of accessibility, expediency, cost, efficiency or other reason not directly concerned with sampling parameters.
1990, *62*, 1203

conventional transition-state theory
See *transition-state theory*.
1996, *68*, 162

conventional true value
Value attributed to a particular quantity and accepted, sometimes by convention, as having an uncertainty appropriate for a given purpose. Examples are the CODATA recommended values of fundamental physical constants.
VIM; 1996, *68*, 162

conversion electron
Electron ejected from the *atom* in the process of *internal conversion*.
1982, *54*, 1538

conversion, internal
See *internal conversion*.
O.B. 101

conversion spectrum
A plot of a quantity related to the absorption (absorbance, cross section, etc.) multiplied by the quantum yield for the considered process against a suitable measure of photon energy, such as frequency, ν, wavenumber, σ, or wavelength, λ, e.g. the conversion cross section, $\sigma\varphi$, has the SI unit m^2.
See also *action spectrum, efficiency spectrum, spectral effectiveness*.
1996, *68*, 2234

converter, wavelength
See *wavelength converter*.
1995, *67*, 1758

co-oligomer
An *oligomer* derived from more than one species of *monomer*.
1996, *68*, 2300

co-oligomerization
Oligomerization in which a *co-oligomer* is formed.
1996, *68*, 2306

cooling, radioactive
See *radioactive cooling*.
1982, *54*, 1538

cooperative transition
A transition that involves a simultaneous, collective displacement or change of state of the atoms and/or electrons in the entire system.
Examples:
An *order-disorder transition* of atoms or electrons, as in an alloy, a ferromagnet or superconductor; a *Jahn-Teller* or *ferroic transition*; a *martensitic transition*.
1994, *66*, 580

cooperativity
Interaction between the *substrate* binding sites of an *allosteric* enzyme. Binding of a substrate molecule to one binding site changes the affinity of the binding sites on the other subunits (cf. *allosteric enzymes*) to the substrate by induction of a conformation change at the other binding sites. Cooperative enzymes typically display a sigmoid (S-shaped) plot of the reaction rate against substrate concentration.
1992, *64*, 151

+ coordinate covalence
See *coordination*.
1994, *66*, 1100

coordinate covalence (coordinate link)
See *coordination*.
1994, *66*, 1100

+ coordinate link
See *coordination*.
1994, *66*, 1100

coordination
The formation of a covalent *bond*, the two shared electrons of which have come from only one of the two parts of the *molecular entity* linked by it, as in the reaction of a Lewis acid and a Lewis base to form a Lewis adduct; alternatively, the bonding formed in this way. In the former sense, it is the reverse of *unimolecular heterolysis*. 'Coordinate covalence' and 'coordinate link' are synonymous (obsolescent) terms. The synonym 'dative bond' is obsolete.
(The origin of the bonding electrons has by itself no bearing on the character of the bond formed. Thus, the formation of methyl chloride from a methyl cation and a chloride ion involves coordination; the resultant bond obviously differs in no way from the C–Cl bond in methyl chloride formed by any other path, e.g. by colligation of a methyl radical and a chlorine atom.) The term is also used to describe the number of ligands around a central atom without necessarily implying two-electron bonds.
See also *dipolar bond, π-adduct*.
1994, *66*, 1100

coordination entity
An assembly consisting of a central atom (usually metallic) to which is attached a surrounding array of other groups of atoms (ligands).
R.B. 145

coordination number
1. The coordination number of a specified atom in a *chemical species* is the number of other atoms directly linked to that specified atom. For example, the coordination number of carbon in methane is four, and it is five in protonated methane, CH_5^+. (The term is used in a different sense in the crystallographic description of ionic crystals.)
1994, *66*, 1100
2. In an inorganic coordination entity, the number of sigma bonds between ligands and the central atom. Pi-bonds are not considered in determining the coordination number.
R.B. 146

coordination polyhedron (polygon)
In a coordination entity, the solid figure defined by the positions of the ligand atoms directly attached to the central atom.
R.B. 146

coordinatively saturated complex
A transition metal complex that has formally 18 outer shell electrons at the central metal atom.
1994, *66*, 1100

coordinatively unsaturated complex
A transition metal complex that possesses fewer ligands than exist in the coordinatively saturated complex. These complexes usually have fewer than 18 outer shell electrons at the central metal atom.
1994, *66*, 1100

copolymer
A *polymer* derived from more than one species of *monomer*.
Note:
Copolymers that are obtained by *copolymerization* of two monomer species are sometimes termed bipo-

lymers, those obtained from three monomers terpolymers, those obtained from four monomers quaterpolymers, etc.
1996, *68*, 2300

copolymerization
Polymerization in which a *copolymer* is formed.
1996, *68*, 2306

copolymer micelle
A micelle formed by one or more block or graft copolymer molecules in a selective solvent.
P.B. 52

copper vapour laser
A pulsed source of coherent radiation emitting at 578.2 and 510.5 nm from excited copper atoms.
See *gas lasers, laser*.
1996, *68*, 2234

coprecipitation
The simultaneous *precipitation* of a normally soluble component with a macro-component from the same solution by the formation of mixed crystals, by *adsorption,* occlusion or mechanical entrapment.
O.B. 86; see also 1979, *51*, 1199

copy number (in biotechnology)
The number of copies of a *plasmid* or a gene within a *cell*.
1992, *64*, 151

core atom (in organic reaction mechanisms)
The primary *reference atom* in a mechanistic change; usually either of the two atoms of a transformed multiple bond or the single atom at which substitution occurs.
1989, *61*, 55

coronands (coronates)
See *crown*.
1995, *67*, 1328; 1994, *66*, 1100

corrected emission spectrum
Obtained after correcting for instrumental and sample effects and usually represented by a graph of Φ_λ against wavelength, where Φ_λ is the spectral (radiant) energy flux.
1984, *56*, 242

corrected excitation spectrum
Obtained if the photon flux incident on the sample is held constant. If the solution is sufficiently dilute that the fraction of the exciting *radiation* absorbed is proportional to the *absorption coefficient* of the analyte, and if the *quantum yield* is independent of the exciting wavelength, the corrected excitation spectrum will be identical in shape to the absorption spectrum.
1984, *56*, 242

corrected retention volume (in gas chromatography)
The corrected retention volume, V_R^0, is given by
$$V_R^0 = jV_R$$
where j is the pressure gradient correction factor for a homogeneously filled column of constant diameter and V_R is the retention volume (corrected).

This quantity is of limited use because it is influenced by the volumes of the sample injector and detector, as well as the interstitial volume of the column.
O.B. 104

correlation analysis
The use of empirical correlations relating one body of experimental data to another, with the objective of finding quantitative estimates of the factors underlying the phenomena involved. Correlation analysis in organic chemistry often uses *linear free-energy relations* for rates or equilibria of reactions, but the term also embraces similar analysis of physical (most commonly spectroscopic) properties and of biological activity.
See also *quantitative structure-activity relationships (QSAR)*.
1994, *66*, 1101

correlation coefficient
A measure of the degree of interrelationship which exists between two measured quantities, x and y; the correlation coefficient (r) is defined by the following relation:

$$r = \frac{\sum_{i=1}^{n}(x_i - \bar{x})(y_i - \bar{y})}{[\sum_{i=1}^{n}(x_i - \bar{x})^2 \sum_{i=1}^{n}(y_i - \bar{y})^2]^{1/2}}$$

where x_i and y_i are the measured values in the ith experiment of n total experiments, $\bar{x}$ and $\bar{y}$ are the arithmetic means of x_i and y_i:

$$\bar{x} = (\sum_{i=1}^{n}/n) \quad \text{(similar expression for } \bar{y}\text{)}$$

The linear correlation coefficient indicates the degree to which two quantities are linearly related. If $x = ay$ is followed then $r = 1$, and departures from this relationship decrease r; if interpretations of data based on the linear correlation coefficient are to be made, one should consult a book on statistics.
1990, *62*, 2182

correlation diagram
A diagram which shows the relative energies of orbitals, configurations, valence bond structures, or states of reactants and products of a reaction, as a function of the molecular geometry, or another suitable parameter. An example involves the interpolation between the energies obtained for the united atoms and the values for the separated atoms limits.
1996, *68*, 2234

correlation energy
The difference between the Hartree–Fock energy calculated for a system and the exact non-relativistic energy of that system. The correlation energy arises

from the approximate representation of the electron-electron repulsions in the Hartree–Fock method.
1996, *68*, 2234

corrinoids (cobalamines, corphyrins, corrins, vitamin B12 compounds)
Derivatives of the corrin nucleus, which contains four reduced or partly reduced pyrrole rings joined in a macrocycle by three =CH– groups and one direct carbon–carbon bond linking α positions.

1995, *67*, 1328; see also W.B. 272

corrosion
An irreversible interfacial reaction of a material (metal, ceramic, polymer) with its enviroment which results in consumption of the material or in dissolution into the material of a component of the enviroment. Often, but not necessarily, corrosion results in effects detrimental to the usage of the material considered. Exclusively physical or mechanical processes such as melting or evaporation, abrasion or mechanical fracture are not included in the term corrosion.
1989, *61*, 20

corrosion cell
A galvanic cell resulting from inhomogeneities in the material or in its environment.
1989, *61*, 22

corrosion, concentration-cell
See *concentration-cell corrosion*
1989, *61*, 22

corrosion current
See *corrosion rate*.
1989, *61*, 20

corrosion, non-uniform
See *non-uniform corrosion*.
1989, *61*, 21

corrosion, pitting
See *pitting corrosion*.
1989, *61*, 22

corrosion potential, E_{cor}
The electrode potential spontaneously acquired by a corroding material in a particular environment.
1989, *61*, 20

corrosion rate
The amount of substance transferred per unit time at a specified surface.
Using Faraday's law, the corrosion rate, v_{cor}, can be formally expressed as an electric current which at the corrosion potential is called the corrosion current, I_{cor}, e.g. for the anodic dissolution of one component of a material with v_{cor} in mol s^{-1} and I_{cor} in A one obtains $I_{cor} = nFv_{cor}$, n being the charge number of the electrode reaction and F the Faraday constant.
1989, *61*, 20

corrosion, selective
See *selective corrosion*.
1989, *61*, 21

corrosion, uniform
See *uniform corrosion*.
1989, *61*, 21

co-solvency (in polymers)
The dissolution of a polymer in a solvent comprising more than one component, each component of which by itself is a non-solvent for the polymer.
P.B. 60

cosolvent
See *diluent*.
1993, *65*, 2380

cosphere
See *cybotactic region*.
1994, *66*, 1101

cotectic
The conditions of pressure, temperature and composition under which two or more solid phases crystallize at the same time from a single liquid over a finite range of decreasing temperature.
1994, *66*, 580

coulomb
SI derived unit of electric charge, C = A·s.
G.B. 72; 1996, *68*, 967

coulomb integral, H_{rr}
Integral over space of the type $\int \psi_r^* \hat{H} \psi_r d\tau$ where $\hat{H}$ is the hamiltonian operator containing electrostatic potential energy terms and ψ is a one-electron wavefunction.
G.B. 17

coulometric detection method (in electrochemical analysis)
A detection method in which the current is directly proportional to the flow rate of the substance involved in the electrochemical reaction, and the amount of charge which flows is proportional to the amount of substances taking part in the reaction.
1990, *62*, 2186

coulometric titration
See *titration*.
O.B. 47

coumarins
2*H*-Chromen-2-one (older name 1,2-benzopyrone), trivially named coumarin, and its derivatives formed by substitution.

See *isocoumarins*.
1995, *67*, 1328

count
1. Information corresponding to a pulse processed for *counting*.
2. Number of pulses recorded during a measurement.
1994, *66*, 2516

counter-ions
1. (in an ion exchanger): the mobile exchangeable ions.
   1993, *65*, 854
2. (in colloid chemistry): ions of low relative molecular mass, with a charge opposite to that of the colloidal ion.
   1972, *31*, 607

counter, radiation
   See *radiation counter*.
   1982, *54*, 1538

counter tube
Radiation detector consisting of a gas-filled tube or valve whose gas amplification is much greater than one, and in which the individual ionizing events give rise to discrete electrical pulses. Often an expression is added indicating the geometry (e.g. end window), composition of the gas (e.g. helium) or the physical process for its operation (e.g. proton recoil, fission).
   1982, *54*, 1538

counter tube, Geiger–Muller
   See *Geiger–Muller counter tube*.
   1982, *54*, 1538

counter tube, proportional
   See *proportional counter tube*.
   1982, *54*, 1538

counting, absolute
   See *absolute counting*
   1994, *66*, 2517

counting efficiency
The ratio between the number of particles or photons counted with a *radiation counter* and the number of particles or photons of the same type and *energy* emitted by the radiation *source*.
   1994, *66*, 2517

counting loss
A reduction of the *counting rate* resulting from phenomena such as the resolving time or the *dead time*.
   1994, *66*, 2517; O.B. 217

counting rate
The number of *counts* occurring in unit time.
   1994, *66*, 2517

coupled (indicator) reaction (in analysis)
A reaction which follows the (slower) reaction of kinetic interest, so as to provide means of monitoring the formation of a reaction product. This reaction is sometimes referred to as the *indicator reaction*.
   1993, *65*, 2294

coupled simultaneous techniques (in analysis)
The application of two or more techniques to the same sample when the two instruments involved are connected through an interface, e.g. simultaneous thermal analysis and mass spectrometry.
   O.B. 39

coupling
   See *chemical induction*.
   1996, *68*, 162

coupling constant (spin-spin), J
A quantitative measure for nuclear spin-spin, nuclear-electron (hyperfine coupling) and electron-electron (fine coupling in EPR) coupling in magnetic resonance spectroscopy. The 'indirect' or scalar NMR coupling constants are in a first approximation independent of the external magnetic field and are expressed in Hz.
   1994, *66*, 1101

covalent bond
A region of relatively high electron density between nuclei which arises at least partly from sharing of electrons and gives rise to an attractive force and characteristic internuclear distance.
   See also *agostic, coordination, hydrogen bond, multi-centre bond*.
   1994, *66*, 1101

covalent network
   See *network*.
   1996, *68*, 2298

Cox–Yates equation
A modification of the *Bunnett–Olsen equation* of the form:

$$\lg ([SH^+]/[S]) - \lg [H^+] = m^*X + pK_{SH^+}$$

where X is the activity function $\lg (\gamma_S \gamma_{H^+}/\gamma_{SH^+})$ for an arbitrary reference base. The function X is called the excess acidity because it gives a measure of the difference between the acidity of a solution and that of an ideal solution of the same concentration. In practice $X = -(H_0 + \lg [H^+])$ and $m^* = 1 - \Phi$.
   See also *Bunnett–Olsen equation*.
   1994, *66*, 1101

cracking
The thermal or catalytic decomposition of a compound such as a hydrocarbon into chemical species of smaller molecular weight.
   1990, *62*, 2183

cream
A highly concentrated *emulsion* formed by *creaming* of a dilute emulsion.
   1972, *31*, 611

creaming
The macroscopic separation of a dilute *emulsion* into a highly concentrated emulsion, in which interglobular contact is important, and a continuous phase under the action of gravity or a centrifugal field. This separation usually occurs upward, but the term may still be applied if the relative densities of the dispersed and continuous phases are such that the concentrated emulsion settles downward.
   1972, *31*, 611

cream volume
The volume of *cream* formed in an *emulsion*.
   1972, *31*, 616

cresols
The monomethylphenols and their derivatives formed by substitution on the ring with substituents other than –OH.

[structure: benzene ring with OH and CH₃ substituents]

1995, *67*, 1328

critical embryo
See *embryo*.
1972, *31*, 608

critical energy (threshold energy)
A chemical reaction cannot occur, except by *quantum-mechanical tunnelling*, unless the total energy available is greater than a certain energy, which is known as the critical energy, or the threshold energy.
1996, *68*, 162

critical excitation energy (in *in situ* microanalysis), E_q
Minimum energy to excite a specific analytical signal. The term is most frequently used in connection with the generation of X-ray spectra, designating the *ionization* (binding) *energy* of the orbital on which the transition ends.
1983, *55*, 2027

critical micelle concentration (cmc)
There is a relatively small range of concentrations separating the limit below which virtually no *micelles* are detected and the limit above which virtually all additional surfactant molecules form micelles. Many properties of surfactant solutions, if plotted against the concentration, appear to change at a different rate above and below this range. By extrapolating the loci of such a property above and below this range until they intersect, a value may be obtained known as the critical micellization concentration (critical micelle concentration), symbol c_M, abbreviation cmc (or c.m.c.). As values obtained using different properties are not quite identical, the method by which the cmc is determined should be clearly stated.
See also *inverted micelle*.
1994, *66*, 1101

critical point
The temperature and pressure at which the liquid and vapour intensive properties (density, heat capacity, etc.) become equal. It is the highest temperature (critical temperature) and pressure (critical pressure) at which both a gaseous and a liquid phase of a given compound can coexist.
1990, *62*, 2183; 1993, *65*, 2399

critical pressure, p_c
The minimum pressure which would suffice to liquefy a substance at its critical temperature. Above the critical pressure, increasing the temperature will not cause a fluid to vaporize to give a two-phase system.
1993, *65*, 2399

critical quenching radius, r_0
See *Förster excitation transfer*.
1996, *68*, 2234

critical solution composition
See *critical solution point*.
1994, *66*, 580

critical solution point
The point, with coordinates critical solution temperature and critical composition, on a temperature-composition phase diagram at which the distinction between co-existent phases vanishes. In solid-solid, solid-liquid and liquid-liquid systems both upper and lower *critical solution temperatures* and corresponding *critical solution compositions* can occur. Synonymous with consolute point.
1994, *66*, 580

critical solution temperature
See *critical solution point*.
1994, *66*, 580

critical temperature, T_c
That temperature, characteristic of each gas, above which it is not possible to liquefy a given gas.
1990, *62*, 2183; 1993, *65*, 2399

critical thickness (of a film)
A film often thins gradually to a thickness at which it either ruptures or converts abruptly to an *equilibrium film*. This thickness is sometimes well enough defined statistically to be considered a critical thickness, t_c or h_c.
1972, *31*, 614

cross-conjugation
In a system XC_6H_4GY this is conjugation involving the substituent X, the benzene ring and the side-chain connective-plus-reaction site GY, i.e. either X is a +R group and GY is a –R group, or X is a –R group and GY is a +R group. In Hammett correlations this situation can lead to the need to apply exalted substituent constants σ^+ or σ^-, respectively, as in electrophilic or nucleophilic aromatic substitution, respectively. The term 'through resonance' is synonymous. Cross conjugation has also been used to describe the interactions occurring in 2-phenylallyl and and similar systems.
1994, *66*, 1101

crossed electric and magnetic fields (in mass spectrometry)
Electric and magnetic fields with the electric field direction at right angles to the magnetic field direction.
1991, *63*, 1544

crossed molecular beams
See *molecular beams*.
1996, *68*, 162

cross-flow filtration (in biotechnology)
Method of operating a filtration device where retained fluid is circulated over the membrane (filter) surface

thus preventing undue build-up of filtered material on membrane (filter).
1992, *64*, 151

crossing over (in biotechnology)
The usually reciprocal exchange of genetic material between *chromosomes*; part of natural genetic recombination.
1992, *64*, 151

crosslink
A small region in a *macromolecule* from which at least four *chains* emanate, and formed by reactions involving sites or groups on existing macromolecules or by interactions between existing macromolecules.
Notes:
1. The small region may be an atom, a group of atoms, or a number of branch points connected by bonds, groups of atoms, or oligomeric chains.
2. In the majority of cases, a crosslink is a covalent structure but the term is also used to describe sites of weaker chemical interactions, portions of crystallites, and even physical entanglements.
1996, *68*, 2298

cross-over concentration, c^*
The concentration range at which the sum of the volumes of the domains occupied by the solute molecules or particles in solution is approximately equal to the total volume of that solution. The term 'domain' refers to the smallest convex body that contains the molecule or particle in its average shape.
P.B. 57

cross reaction
The ability of substances other than the analyte to bind to the binding reagent and the ability of substances other than the binding reagent to bind the analyte in *competitive binding assays*.
1994, *66*, 2517

cross-section, activation
See *activation cross-section*.
1994, *66*, 2517

cross-section, capture
See *capture cross-section*.
1994, *66*, 2517

cross-section, collision
See *collision cross-section*.
1996, *68*, 158

cross-section, effective thermal
See *effective thermal cross-section*.
1994, *66*, 2517; O.B. 218

cross-section, macroscopic
See *macroscopic cross-section*.
1994, *66*, 2517; O.B. 218

cross-section, microscopic
See *microscopic cross-section*.
1994, *66*, 2517; O.B. 217

cross-section, reaction
See *reaction cross-section*.
1996, *68*, 183

cross-section, Westcott
See *effective thermal cross-section*.
1994, *66*, 2517; O.B. 218

crowding (in solvent extraction)
The displacement of an impurity from an extract phase by contact with a solution containing the main extractable solute. The main solute need not be present in a pure solution but should have a higher distribution ratio than the impurities present.
See also *scrubbing, exchange extraction*.
1993, *65*, 2386

crown
A *molecular entity* comprising a monocyclic ligand assembly that contains three or more *binding sites* held together by covalent bonds and capable of binding a *guest* in a central (or nearly central) position. The *adducts* formed are sometimes known as 'coronates'. The best known members of this group are macrocyclic polyethers, such as '18-crown-6', containing several repeating units $-CR_2-CR_2O-$ (where R is most commonly H), and known as crown ethers.

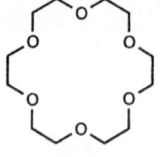
18-crown-6

See also *host*.
1994, *66*, 1101; see also 1995, *67*, 1328

crown conformation
A conformation of a saturated cyclic molecular entity, containing an even number (≥ 8) of atoms in the ring, in which these atoms lie alternately in each of two parallel planes and are symmetrically equivalent (D_{4d} for cyclooctane, D_{5d} for cyclodecane etc). It is analogous to the *chair* conformation of cyclohexane.

See also *tub conformation*.
1996, *68*, 2205

crud (in solvent extraction)
A deposit or emulsion at the interface between two partially settled phases.
Notes:
1. The phenomenon of crud formation arises from many causes and this definition does not imply any single one.
2. Other terms — some unprintable — have been used but crud is the generally accepted term.
1993, *65*, 2387

cryogenic
Low temperature processes, apparatus, etc.; usually applied to systems operated at the temperature of liquid nitrogen, helium or other condensed gas which boils at a very low temperature (at atmospheric pressure).

See *cryogenic sampling*.
1990, *62*, 2183

cryogenic sampling
The collection of trace compounds from gaseous media by cocondensation with a major constituent (e.g. water vapour, CO_2, N_2, Ar) of the matrix.
1990, *62*, 2212

cryptand
A *molecular entity* comprising a cyclic or polycyclic assembly of *binding sites* that contains three or more binding sites held together by *covalent bonds*, and which defines a molecular cavity in such a way as to bind (and thus 'hide' in the cavity) another molecular entity, the guest (a cation, an anion or a neutral species), more strongly than do the separate parts of the assembly (at the same total concentration of binding sites). The *adduct* thus formed is called a 'cryptate'. The term is usually restricted to bicyclic or oligocyclic molecular entities.
Example:

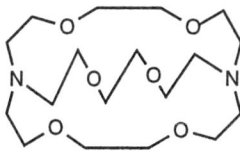

Corresponding monocyclic ligand assemblies (*crowns*) are sometimes included in this group, if they can be considered to define a cavity in which a guest can hide. The terms 'podand' and 'spherand' are used for certain specific ligand assemblies. Coplanar cyclic polydentate ligands, such as porphyrins, are not normally regarded as cryptands.
See also *host*.
1994, *66*, 1102; 1995, *67*, 1329

cryptate
See *cryptand*.
1994, *66*, 1102

crystal diffraction spectrometer
A *spectrometer*, in which diffraction by a crystal is used to obtain the energy spectra of electromagnetic *radiation* as well as of *slow neutrons*.
1982, *54*, 1552

crystal field splitting
The removal of a degeneracy of the energy levels of molecular entities or ions due to the lower site symmetry created by a crystalline environment. This term is sometimes incorrectly used synonymously with the term *ligand field splitting*.
1996, *68*, 2234

crystal laser
A *solid state laser* in which the active medium is composed of ionic species supported in a crystalline host. The first laser used a ruby crystal in which chromium ions as the active medium were supported in an aluminium oxide (sapphire) crystal. One of the most common crystal lasers involves neodymium atoms in a yttrium aluminium garnet host crystal (the YAG:Nd laser).
1995, *67*, 1920

crystalline electrodes
Electrodes which contain mobile ions of one sign and fixed sites of opposite sign. They may be homogeneous or heterogeneous.
1. Homogeneous membrane electrodes are ion-selective electrodes in which the membrane is a crystalline material prepared from either a single compound or a homogeneous mixture of compounds (i.e. Ag_2S, AgI/Ag_2S).
2. Heterogeneous membrane electrodes are formed when an active substance, or mixture of active substances, is mixed with an inert matrix, such as silicone rubber or PVC, or placed on hydrophobized graphite or conducting epoxy, to form the heterogeneous sensing membrane.
1994, *66*, 2534

crystalline polymer
A polymer showing crystallinity.
One or two-dimensional order leads to mesophase structure. The range of order may be as small as about 2 nm in one (or more) crystallographic direction(s) and is usually below 50 nm in at least one direction.
P.B. 74

crystallinity
The presence of three-dimensional order on the level of atomic dimensions.
Crystallinity may be detected by diffraction techniques, heat-of-fusion measurements, etc. The amount of disorder within the crystalline region is not incompatible with this concept.
P.B. 74

crystallization
The formation of a crystalline solid from a solution, melt vapour or a different solid phase, generally by the lowering of the temperature or by evaporation of a solvent.
1994, *66*, 580

CT
Abbreviation for *charge-transfer*.
1996, *68*, 2234

C-terminal residue (in a polypeptide)
See *amino-acid residue (in a polypeptide)*.
W.B. 48

cumulative double bonds
Those present in a chain in which at least three contiguous carbon atoms are joined by double bonds; non-cumulative double bonds comprise every other arrangement of two or more double bonds in a single structure. The generic name 'cumulene' is given to compounds containing three or more cumulative double bonds.
B.B. 20

cumulative fission yield
The fraction of *fissions* which have resulted in the production of a *nuclide* either directly or indirectly,

up to a specified time. If no time is specified, the yield is considered to be the asymptotic value.
1982, *54*, 1542

cumulative sample
A sample collected over an extended period of time.
1990, *62*, 2183

cumulenes
Hydrocarbons (and by extension, derivatives formed by substitution) having three or more *cumulative double bonds*, e.g. $R_2C=C=C=CR_2$.
See also *allenes, heterocumulenes*.
1995, *67*, 1329

cumulonimbus cloud (in atmospheric chemistry)
Heavy masses of dense cloud whose cumuliform summits rise in the forms of towers, the upper parts having a fibrous texture and often spreading out into the shape of an anvil; these clouds generally produce showers of rain and sometimes of snow, hail or soft hail, and often develop into thunderstorms; up to 12 000 m; summits may be as cold as –50 °C; strong convective motions with vertical velocities of 3 to greater than 30 m s^{-1}.
1990, *62*, 2180

cumulus cloud (in atmospheric chemistry)
Detached, dense clouds with a dome-shaped upper surface with sharp-edged, rounded protuberances and a nearly horizontal base; 600–6 000 m or more; convective motion with rising, large bubbles of warm air with vertical speeds of 1–5 m s^{-1}.
1990, *62*, 2180

cupola
A vertical shaft furnace used for melting metals; the melting of ores is accomplished in a blast furnace.
1990, *62*, 2183

curie
Non-SI unit of radioactivity, Ci ≈ 3.7 × 10^{10} Bq.
G.B. 113; 1996, *68*, 967; 1994, *66*, 2517

current, adsorption
See *adsorption current, limiting adsorption current*.
1985, *57*, 1494

current, alternating
See *alternating current*.
1985, *57*, 1494

current, alternating, amplitude of
See *amplitude (of alternating current)*.
1985, *57*, 1494

current, apex
See *apex current*.
1985, *57*, 1494

current, capacity
See *double-layer current*.
1985, *57*, 1494

current, catalytic
See *catalytic current*.
1985, *57*, 1494

current, charging
See *double-layer current*.
1985, *57*, 1495

current density
The current density j_B of a species B in a given point of the solution is obtained by multiplying the *flux density* of that species at the given point by the *Faraday constant* F and by the *charge number* z_B of the species:
$$j_B = z_B F N_B$$
where j_B is a vector which indicates the direction in which the charges transported by the species B flow and which gives the number of these charges going through a plane oriented perpendicular to the vector, divided by time and by area, and N_B is the flux density of a minor constituent of the solution with respect to a fixed frame of reference.
See also *electric current density*.
1981, *53*, 1833

current density (relative local)
See *current distribution*.
1981, *53*, 1836

current, diffusion (-controlled)
See *diffusion current*.
1985, *57*, 1495

current, direct
See *direct current*.
1985, *57*, 1495

current distribution
The ratio of *current density* at a point X on an interface to the average current density (j_x/j) is called the relative local current density. The current distribution is described by the function $j_x/j = f(x)$ (or more generally, $j_x/j = f(x, y, z)$ where x or (x, y, z) are the coordinates of the points of the electrode-solution interface.
The primary current distribution is that which establishes itself when the influence of overpotential is negligible.
The secondary current distribution is that which establishes itself when the influence of the overpotential cannot be neglected but concentration overpotential is negligible. The secondary distribution is often described in terms of dimensionless numbers of the form
$$Wa = \kappa (d\eta / dj) / l$$
where κ is the *conductivity* of the solution, $d\eta/dj$ the slope of the overpotential-current curve under the above conditions and l a characteristic length of the system, for instance the radius of a disc electrode. Wa is the Wagner number. It is a quantity which determines the throwing power and characterizes the equalizing influence of overpotential on the current distribution. In electroplating the throwing power is qualitatively defined as 'the ability of a solution to deposit metal uniformly upon a cathode of irregular shape'.

The tertiary current distribution is that which establishes itself when the influence of the overpotential (including concentration overpotential) cannot be neglected.
  1981, 53, 1836
current, double-layer
  See *double-layer current*.
  1985, 57, 1496
current efficiency
If several reactions take place simultaneously at the electrode a partial electrode *current density* (c.d.) j_k can be assigned to each reaction. It is given by the stoichiometry of the reaction and by the amount of substance of B reacting (per unit time and per unit electrode area) in the reaction considered. The current efficiency of reaction k, ε_k is defined as the ratio of j_k to the total c.d.:

$$\varepsilon_k = j_k / \sum_m j_m$$

Note that ε_k may be larger than one if cathodic and anodic reactions take place simultaneously at the same electrode. However, ε_k still gives correctly the product yield, which is the quantity of industrial interest. The product yield is the amount of substance of B produced per unit charge and is equal to $\varepsilon_k \nu_{B,k}/n_k F$ (in the absence of a *chemical reaction* which is consecutive to the *electrode reaction* and which consumes or produces species B). n_k is the charge number of electrode reaction k.
Note that in the case of simultaneous electrode reactions the distribution of the partial c.d. j_k may be different from that of the total c.d., i.e. the function $(j_k)_x/j = f_k(x)$ may be different from $j_x/j = f(x)$. In electroplating the term 'metal distribution' is sometimes used to designate the distribution $f_k(x)$ of the partial c.d. for metal deposition.
  1981, 53, 1836
current, faradaic
  See *faradaic current*.
  1985, 57, 1496
current, faradaic demodulation
  See *faradaic demodulation current*.
  1985, 57, 1496
current, faradaic rectification
  See *faradaic rectification current*.
  1985, 57, 1496
current, instantaneous
  See *instantaneous current*.
  1985, 57, 1496
current, kinetic
  See *kinetic current*.
  1985, 57, 1496
current, limiting
  See *limiting current*.
  1974, 37, 513; 1985, 57, 1497
current, limiting adsorption
  See *limiting adsorption current*.
  1985, 57, 1494

current, limiting catalytic
  See *limiting catalytic current*.
  1985, 57, 1495
current, limiting diffusion
  See *limiting diffusion current*.
  1985, 57, 1495
current, limiting kinetic
  See *limiting kinetic current*.
  1985, 57, 1497
current, limiting migration
  See *limiting migration current*.
  1985, 57, 1498
current migration
  See *migration current*.
  1985, 57, 1497
current, net faradaic
  See *faradaic current*.
  1985, 57, 1496
current, non-additive
  See *catalytic current*.
  1985, 57, 1498
current, peak
  See *peak current, apex current*.
  1985, 57, 1498
current, periodic
  See *alternating current*.
  1985, 57, 1499
current, regeneration
  See *catalytic current*.
  1985, 57, 1499
current, residual
  See *residual current*.
  1985, 57, 1499
current, square-wave
  See *square-wave current*.
  1985, 57, 1499
current, summit (in polarography)
  See *summit current (in polarography)*.
  1985, 57, 1499
current yield
  See *photocurrent yield*.
  1996, 68, 2234
Curtin–Hammett principle
In a *chemical reaction* that yields one product (X) from one conformational isomer (A′) and a different product (Y) from another conformational isomer (A″) (and provided these two isomers are rapidly interconvertible relative to the rate of product formation, whereas the products do not undergo interconversion) the product composition is not in direct proportion to the relative concentrations of the conformational isomers in the *substrate*; it is controlled only by the diference in standard free energies ($\delta\Delta^{\ddagger}G$) of the respective *transition states*.

$$X \xleftarrow{k_X} A' \underset{k_{-1}}{\overset{k_1}{\rightleftarrows}} A'' \xrightarrow{k_Y} Y$$

$$k_1 >> k_X \quad k_{-1} >> k_Y$$

It is also true that the product composition is formally related to the relative concentrations of the conformational isomers A′ and A″ (*i.e.* the conformational equilibrium constant) and the respective rate constants of their reactions; these parameters are generally — though not invariably — unknown.

The diagram below represents the energetic situation for transformation of interconverting isomers A and A′ into products X and Y.

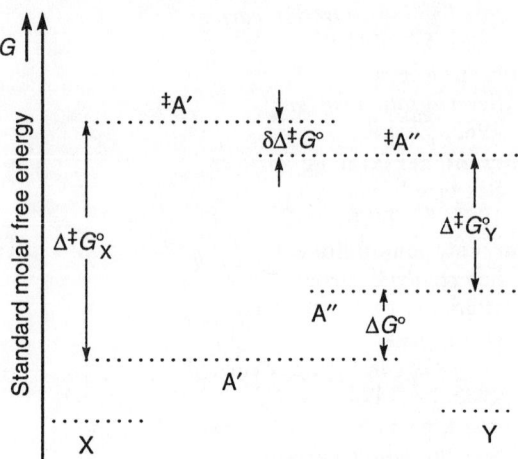

1994, *66*, 1102

cut off (in aerosol sizes)(in atmospheric chemistry)
The size of particles at which the retention efficiency of an instrument device drops below a specified value under defined conditions.
1990, *62*, 2183

cut-off filter
An optical device which only permits the transmission of radiation of wavelengths that are longer than or shorter than a specified wavelength. Usually, the term refers to devices which transmit radiation of wavelengths longer than the specified wavelength.
See *filter*.
1996, *68*, 2234

CVD diamond
See *diamond by CVD*.
1995, *67*, 487

CW
Abbreviation for continuous wave, i.e. non-pulsed source of electromagnetic radiation.
1996, *68*, 2234

cyanates
Salts and *esters* of cyanic acid, HOC≡N, e.g. KOCN potassium cyanate, PhOCN phenyl cyanate.
See *isocyanates*.
1995, *67*, 1329

cyanides
Salts and *C*-organyl derivatives of hydrogen cyanide, HC≡N, e.g. CH$_3$C≡N methyl cyanide (acetonitrile), NaCN sodium cyanide, PhC(=O)CN benzoyl cyanide.
See also *isocyanides, nitriles, carbonitriles*.
1995, *67*, 1329

cyanine dyes
Synthetic dyes with the general formula R$_2$N-[CH=CH]$_n$CH=N$^+$R$_2$ ↔ R$_2$N$^+$=CH[CH=CH]$_n$NR$_2$ (*n* is a small number) in which the nitrogen and part of the conjugated chain usually form part of a heterocyclic system, such as imidazole, pyridine, pyrrole, quinoline and thiazole, e.g.

1995, *67*, 1329

cyanogenic
Compounds able to produce cyanide; examples are the cyanogenic glycosides such as amygdalin in peach and apricot stones.
1993, *65*, 2035

cyanohydrins
Alcohols substituted by a cyano group, most commonly, but not limited to, examples having a cyano and a hydroxy group attached to the same carbon atom, formally derived from *aldehydes* or *ketones* by the addition of hydrogen cyanide. An individual cyanohydrin can systematically be named as a hydroxy nitrile, e.g. (CH$_3$)$_2$C(OH)C≡N 'acetone cyanohydrin' (2-hydroxy-2-methylpropanenitrile), HOCH$_2$CH$_2$C≡N 'ethylene cyanohydrin' (3-hydroxypropanenitrile).
See *halohydrins*.
1995, *67*, 1329

cybotactic region
That part of a solution in the vicinity of a solute molecule in which the ordering of the solvent molecules is modified by the presence of the solute molecule. The term solvent 'cosphere' of the solute has also been used.
See also *solvation*.
1994, *66*, 1103

cycles per second
See *hertz*.
1996, *68*, 967

cyclic acid anhydrides (cyclic anhydrides)
Acid anhydrides derived by loss of water between two oxoacid functions R$_k$E(=O)$_l$(OH)$_m$ (*l* ≠ 0) (carboxylic, sulfonic, etc.) in the same molecule so as to close a ring. Oxo oxygen replacement analogues are included, e.g.

See *thioanhydrides*.
1995, *67*, 1329

cyclitols
Hydroxylated *cycloalkanes* containing at least three hydroxy groups, each attached to a different ring carbon atom.
1995, *67*, 1330; W.B. 196

cyclization
Formation of a ring compound from a chain by formation of a new bond.
See also *annulation*.
1994, *66*, 1103

cyclo-
1. A prefix used in names to designate a ring structure, usually italicized in inorganic, but not in organic nomenclature.
Examples:

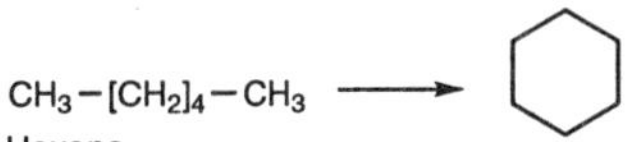

$CH_3-[CH_2]_4-CH_3$ → Cyclohexane
Hexane

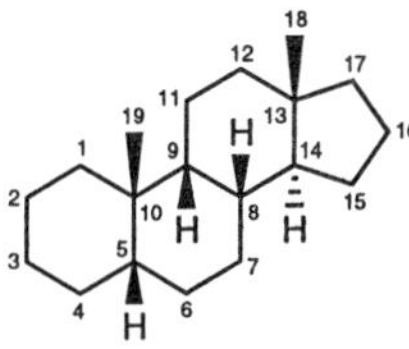

5β, 9β-Androstane

↓

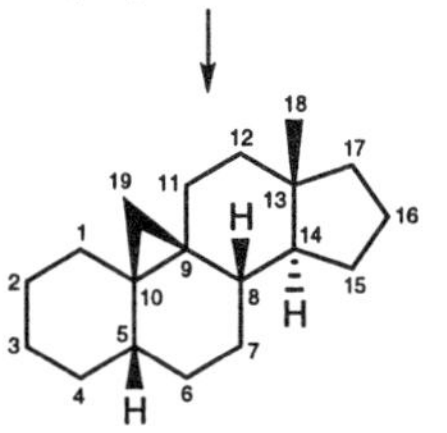

9,19-Cyclo-5β,9β-androstane

R.B. 245; B.B. 497; see also B.B.(G) 30
2. The prefix for an elementary reaction or some part thereof indicating that the bonds undergoing primitive changes form part of a ring in the transition state of a pericyclic process.
1989, *61*, 34

cycloaddition
A reaction in which two or more unsaturated molecules (or parts of the same molecule) combine with the formation of a cyclic *adduct* in which there is a net reduction of the bond multiplicity.
The following two systems of notations have been used for the more detailed specification of cycloadditions, of which the second, more recent system [described under (2)] is preferred:
(1) A $(i + j + ...)$ cycloaddition is a reaction in which two or more molecules (or parts of the same molecule), respectively, provide units of $i, j, ...$ linearly connected atoms: these units become joined at their respective termini by new σ-bonds so as to form a cycle containing $(i + j + ...)$ atoms. In this notation, (a) a Diels–Alder reaction is a (4+2) cycloaddition, (b) the initial reaction of ozone with an alkene is a (3+2) cycloaddition, and (c) the reaction shown below is a (2+2+2) cycloaddition. (N.B.: parentheses (...) are used in the description based on numbers of atoms.)

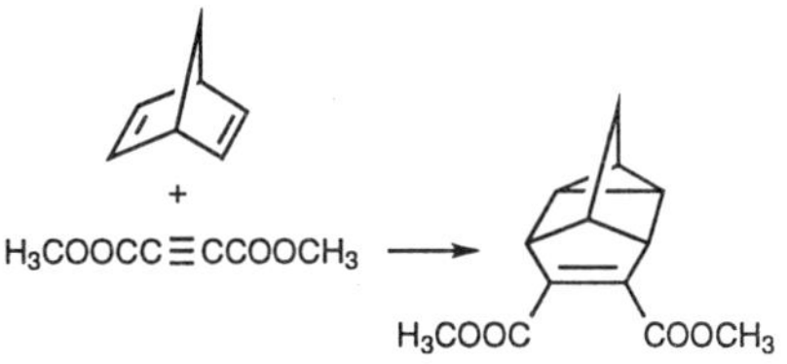

(2) The symbolism $[i + j + ...]$ for a cycloaddition identifies the numbers $i, j, ...$ of electrons in the interacting units that participate in the transformation of reactants to products. In this notation the reaction (a) and (b) of the preceding paragraph would both be described as [2+4] cycloadditions, and (c) as a [2+2+2] cycloaddition. The symbol a or s (a = antarafacial, s = suprafacial) is often added (usually as a subscript after the number to designate the stereochemistry of addition to each fragment. A subscript specifying the orbitals, *viz.*, σ, π (sigma, pi) with their usual significance) or n (for an orbital associated with a single atom only), may be added as a subscript before the number. Thus the normal Diels–Alder reaction is a $[4_s + 2_s]$ or $[_\pi 4_s + _\pi 2_s]$ cycloaddition, whilst the reaction:

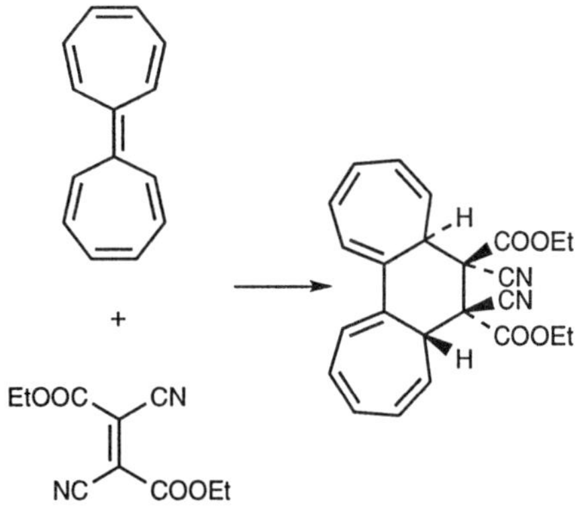

would be a $[14_a + 2_s]$ or $[_\pi 14_a + _\pi 2_s]$ cycloaddition. (N.B. Square brackets [...] are used in the descriptions based on numbers of electrons.)
Cycloadditions may be *pericyclic reactions* or (non-concerted) *stepwise reactions*. The term 'dipolar cycloaddition' is used for cycloadditions of 1,3-dipolar compounds.
See also *cheletropic reactions*.
1994, *66*, 1103

cycloalkanes
Saturated monocyclic hydrocarbons (with or without side chains), e.g. cyclobutane. Unsaturated monocyclic hydrocarbons having one endocyclic double or one triple bond are called cycloalkenes and cycloalkynes, respectively. Those having more than one such multiple bond are cycloalkadienes, cycloalkatrienes, etc. The inclusive terms for any cyclic hydrocarbons

having any number of such multiple bonds are cyclic *olefins* or cyclic *acetylenes*.

$$H_2C-CH_2$$
$$H_2C-CH_2$$
cyclobutane

See *alicyclic compounds*.
1995, *67*, 1330

cycloalkyl groups
Univalent groups derived from cycloalkanes by removal of a hydrogen atom from a ring carbon atom, e.g. 2-methylcyclopropyl.

CH₃–HC–CH–
    \ /
     C
     H₂

1995, *67*, 1330

cyclodepsipeptides
See *depsipeptides*.
1995, *67*, 1330

cyclodextrins
Cyclic oligoglucosides containing 5 to *ca.* 10 glucose residues in which an enclosed tubular space allows reception of a guest molecule to form a *clathrate*. The synonymous term Schardinger dextrins is not recommended. (α-Cyclodextrin has 6 glucose residues; β-cyclodextrin has 7.) Semi-systematically α-cyclodextrin is called cyclomaltohexaose.

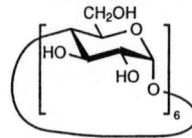

See also *dextrins*.
1995, *67*, 1330

cycloelimination
The reverse of *cycloaddition*. The term is preferred to the synonyms 'cycloreversion', 'retro-addition' and 'retrocycloaddition'.
1994, *66*, 1104

cyclohexadienyl cations
A subclass of *arenium ions*.
1995, *67*, 1330

cyclone (collector)(in atmospheric chemistry)
A dust, grit or droplet separator utilizing essentially the centrifugal force derived from the motion of the gas. The flow of gases containing suspended particles into the device is transformed into a confined vortex from which centrifugal forces tend to drive the suspended particles to the wall of the cyclone body. The agglomerated particles are subsequently removed from the cyclone by gravitational action.
1990, *62*, 2183

cyclophanes
The term originally applied to compounds having two *p*-phenylene groups held face to face by –[CH₂]ₙ– bridges. It now designates compounds having (i) *mancude-ring systems*, or assemblies of mancude-ring systems, and (ii) atoms and/or saturated or unsaturated chains as alternate components of a large ring. E.g. [2.2](1,4)(1,4)cyclophane [or 1(1,4),4(1,4)-dibenzenacyclohexaphane].

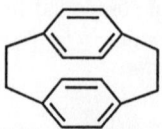

Many chemists include in this class any bridged aromatic system, irrespective of the attachment positions of the bridge.
1995, *67*, 1330

cyclopolymerization
A *polymerization* in which the number of cyclic structures in the *constitutional units* of the resulting *macromolecules* is larger than in the *monomer molecules*.
1996, *68*, 2306

+ cycloreversion
See *cycloelimination*.
1994, *66*, 1104

cyclosilazanes
Compounds having rings of alternating silicon and nitrogen atoms:

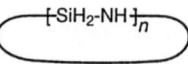

1995, *67*, 1330

cyclosiloxanes
Compounds having rings of alternating silicon and oxygen atoms, e.g. 2,2-dimethylcyclotrisiloxane.

$$H_2Si \overset{O}{\underset{6}{}} \overset{1}{} \overset{}{\underset{2}{}} Si(CH_3)_2$$
$$O \underset{5}{} \underset{4}{} \underset{3}{} O$$
$$\underset{\underset{H_2}{Si}}{}$$

1995, *67*, 1330

cyclotron
A particle accelerator in which the particles travel in a succession of semicircular orbits of increasing radii under the influence of a magnetic field and are accelerated at the beginning of each such orbit by traversing an electric field produced by a high-frequency generator.
1982, *54*, 1539

cytochromes
Conjugated proteins containing haem as the *prosthetic group* and associated with electron transport and with redox processes. The terminal electron transport chain of oxidative respiration contains at least five different cytochromes.
1992, *64*, 151

Δ (delta), Λ (lambda)
Designations of stereoisomers of tris(didentate) metal complexes and other octahedral complexes.
See *helicity*.
1996, *68*, 2205; R.B. 184

δ (delta)(in organic nomenclature)
Signifies, with its superscript, the number of skeletal *cumulative double bonds* terminating at a heteroatom in a cyclic compound.
B.B. 465; see also B.B. (G) 21; 1988, *60*, 1399

D, L, DL
See *Fischer–Rosanoff convention*.
1996, *68*, 2205

+ *d*, *l*, *dl*
Obsolete (usage strongly discouraged) alternatives for, respectively, the prefixes (+)- and (–)- [used to designate the sign of optical rotation of enantiomers under specified condition (and hence to distinguish the enantiomers)] and (±)- for a racemate.
1996, *68*, 2205

dalton
Non-SI unit of mass (symbol Da), equal to the *unified atomic mass unit* (atomic mass constant). It is often used in biochemistry and molecular biology although it was never approved by the Conférence Général des Poids et Mesures.
G.B. 111; 1996, *68*, 968

dark current
See *responsivity*.
1995, *67*, 1749

dark output
See *responsivity*.
1995, *67*, 1749

+ dark photochemistry (photochemistry without light)
Chemical reactions involving electronically excited molecular entities which are generated thermally rather than by *absorption* of electromagnetic radiation. The use of this term is discouraged.
1996, *68*, 2235

dark reaction (darkness reaction)
A *chemical reaction* that does not require or depend on the presence of light. Contrasts with a *photochemical* reaction which is initiated by light absorption by one or more of the reactants.
1990, *62*, 2183

dark resistance
See *responsivity*.
1995, *67*, 1749

Darlington phototransistor
See *phototransistor*.
1995, *67*, 1755

data reduction
The process of transforming the initial representation of a spectrometer output into a form which is amenable to interpretation; for example, a bar graph or a table of ion currents.
1991, *63*, 1544

dating, radioactive
See *radioactive dating*.
1982, *54*, 1539

+ dative bond
See *coordination, dipolar bond*.
1994, *66*, 1104

daughter ion (in mass spectrometry)
An electrically charged product of reaction of a particular parent ion. In general such ions have a direct relationship to a particular precursor ion and indeed may relate to a unique state of the precursor ion. The reaction need not necessarily involve fragmentation. It could, for example, involve a change in the number of charges carried. Thus, all fragment ions are daughter ions, but not all daughter ions are necessarily fragment ions.
1991, *63*, 1549

daughter product (in radiochemistry)
Any *nuclide* which follows a specified *radionuclide* in a *decay chain*.
1994, *66*, 2517

Davydov splitting (factor-group splitting)
The splitting of bands in the electronic or vibrational spectra of crystals due to the presence of more than one (interacting) equivalent molecular entity in the unit cell.
1996, *68*, 2235

day
Non-SI unit of time, d = 86 400 s. The day is not exactly defined in terms of the second since so-called leap seconds are added or subtracted from the day semiannually in order to keep the annual average occurrence of midnight at 24:00 on the clock.
G.B. 111; 1996, *68*, 968

deactivation
Any loss of energy by an excited molecular entity.
See *emission, energy transfer, internal conversion, radiationless deactivation (decay), radiationless transition, radiative transition*.
1996, *68*, 2235

dead time (in radioanalytical chemistry)
Of a *radiation counter*, the constant and known value imposed on the resolving time, usually in order to make the correction for resolving time losses more accurate.
1994, *66*, 2517

dead time (of an analyser)
The time which elapses between the moment at which a sudden change in concentration (or a new sample) is introduced and the moment at which the detector response indication reaches the measurement threshold of the analyser, a value conventionally fixed at 10% of the final change in indication.
1990, *62*, 2174

dead time correction (in radioanalytical chemistry)
Correction to be applied to the observed number of pulses in order to take into account the number of pulses lost during the resolving or *dead time*.
1994, *66*, 2518

+ dead-volume (in chromatography)
This term is used to express the *extra-column volume*. Strictly speaking, the term 'dead-volume' refers to volumes within the chromatographic system which are not swept by the mobile phase. On the other hand, mobile phase is flowing through most of the extra-column volumes. Due to this ambiguity the use of term 'dead-volume' is discouraged.
1993, *65*, 833; see also O.B. 101

debye
Non-SI unit of electric dipole moment. It is equal to the electric dipole moment for two charges of 10^{-10} franklin separated by 1 ångström, D = 10^{-18} Fr cm ≈ 3.335 64 × 10^{-30} C m.
G.B. 115

Debye–Hückel equation
The chemical potential or activity of ions cannot be determined on a purely thermodynamic basis. This is due to the fact that the effects of an ion cannot be separated from the effects of the accompanying counter-ion, or in other terms, the electrochemical potential of the ion cannot be separated into the chemical and the electrical component. Such a separation must necessarily be based on a non-thermodynamic convention. The present convention is based on the assumption that the molal activity coefficient of the chloride ion in dilute aqueous solutions ($I <$ 0.10 mol kg^{-1}) can be estimated by means of the Debye–Hückel equation:

$$-\lg \gamma_B = z_B^2 A I^{1/2}/(1 + å B I^{1/2})$$

where I is ionic strength, z is the charge number of the ion, $å$ is ion size parameter and A and B are temperature-dependent constants.
According to the Bates–Guggenheim convention $åB$ is taken to be 1.5 (mol kg^{-1})$^{1/2}$ at all temperatures and for all compositions of the solutions.
See also *pH*.
1984, *56*, 569

deca
SI prefix for 10 (symbol: da).
G.B. 74

decadic absorbance, *A*
The negative decadic logarithm of one minus *absorptance* as measured on a uniform sample:

$$A = -\lg (1 - \alpha)$$

1985, *57*, 116

decay chain
A series of *nuclides* in which each member transforms into the next through *nuclear decay* until a stable nuclide has been formed.

Synonymous with radioactive chain and radioactive series.
1994, *66*, 2518

decay constant
Proportionality constant relating the activity of a radioactive substance, A, to the number of decaying particles B, N_B, $A = \lambda N_B$. Probability of a nuclear decay within a time interval divided by that time interval.
Synonymous with *disintegration constant*.
G.B. 22; see also 1994, *66*, 2518; 1996, *68*, 968

decay curve
A graph showing the relative amount of *radioactive* substance remaining after any time interval.
1994, *66*, 2518

decay, nuclear
See *nuclear decay*.
1994, *66*, 2518

decay rate (in atmospheric chemistry)
The rate at which a pollutant is removed from the atmosphere either by reaction with reactive transient species such as the HO radical, O_3, etc., by photodecomposition initiated by light absorption by the impurity, or by loss at the surface of aerosols, the earth, etc. The decay rate as applied to radioactive materials is related to the radioactive half-life ($t_{1/2}$) of the particular isotopic species A and its concentration [A]$_t$, at the given time (t): Rate = [A]$_t$(ln 2)/$t_{1/2}$.
1990, *62*, 2183

decay rate (of a radioactive material)
See *activity (of a radioactive material)*.
1994, *66*, 2518

decay scheme (in radioanalytical chemistry)
A graphical representation of the energy levels of the members of a decay chain showing the way along which *nuclear decay* occurs.
1982, *54*, 1539

decay time (in heterogenous catalysis)
The time on stream during which the *rate constant* for the *catalytic reaction*, k, falls to a specified fraction of its original value, often 0.5.
1976, *46*, 83

deci
SI prefix for 10^{-1} (symbol: d).
G.B. 74

decibel
Non-SI unit for expressing power levels, dB = 0.115 1293 Np.
G.B. 79

decomposition
The breakdown of a single phase into two or more phases. The term applies also to other chemical entities such as a normal molecule and a reaction intermediate.
1994, *66*, 580

decomposition, chemical
See *chemical decomposition*.
1994, *66*, 1104

decontamination factor (in liquid-liquid distribution)
The ratio of the proportion of contaminant to product before treatment to the proportion after treatment. It is the reciprocal of the enrichment factor.
1993, *65*, 2381

DEDMR
See *ODMR*.
1996, *68*, 2235

+ de-electronation
See *oxidation (1)*.
1994, *66*, 1104

de-energization efficiency
See *collision efficiency*.
1996, *68*, 162

definitive method
A method of exceptional scientific status which is sufficiently accurate to stand alone in the determination of a given property for the certification of a reference material. Such a method must have a firm theoretical foundation so that systematic error is negligible relative to the intended use. Analyte masses (amounts) or concentrations must be measured directly in terms of the base units of measurements, or indirectly related through sound theoretical equations. Definitive methods, together with certified reference materials, are primary means for transferring accuracy, i.e. establishing *traceability*.
1995, *67*, 1701

deflection (for a precision balance)
(In terms of divisions of the pointer scale). The other point of reversal of an ideal (undamped) swing starting at the *zero point* of the pointer scale. Since the points of reversal of an ideal swing are located symmetrically about the *rest point*, the deflection is equal to twice the rest point.
O.B. 36

deflocculation (peptization)
The reversal of *coagulation* or *flocculation*, i.e. the *dispersion* of aggregates to form a colloidally stable *suspension* or *emulsion*.
1972, *31*, 610

degeneracy
Number of states having the same energy. Symbols: g, d, W, ω, β. Also called statistical weight.
G.B. 39

degenerate chain branching
Sometimes *chain branching* is brought about by an intermediate which has a long lifetime compared with an ordinary free radical. If this intermediate can break down in different ways, only one of which leads to branching, there may be a relatively slow increase in the number of radicals, and there is said to be degenerate chain branching.
1996, *68*, 162

degenerate chemical reaction
See *identity reaction*.
1983, *55*, 1307

degenerate rearrangement
A *molecular rearrangement* in which the principal product is indistinguishable (in the absence of isotopic labelling) from the principal reactant. The term includes both 'degenerate *intramolecular* rearrangements' and reactions that involve intermolecular transfer of atoms or groups ('degenerate *intermolecular* rearrangements'): both are degenerate *isomerizations*. The occurrence of degenerate rearrangements may be detectable by isotopic labelling or by dynamic NMR techniques. For example: the [3,3]*sigmatropic* rearrangement of hexa-1,5-diene (Cope rearrangement):

Synonymous but less preferable terms are 'automerization', 'permutational isomerism', 'isodynamic transformation', 'topomerization'.
See also *fluxional, molecular rearrangement, valence isomer*.
1994, *66*, 1104

degree (of arc)
Non-SI unit of plane angle, $1° = (\pi/180)$ rad ≈ 0.017 453 29 ... rad.
G.B. 113; 1996, *68*, 968

degree Celsius
SI derived unit of Celsius temperature equal to the kelvin (symbol: °C).
G.B. 72; 1996, *68*, 968

degree Fahrenheit
Non-SI unit of Fahrenheit temperature, °F = (5/9) K. The Fahrenheit temperature is related to the Celsius temperature by $\theta_F/°F = (9/5)(\theta/°C) + 32$.
G.B. 113

degree of activation
See *activation*.
1996, *68*, 162

degree of association (of a micelle)
The number of *surfactant* ions in the *micelle*. It does not concern the location of the counterions.
1972, *31*, 612

degree of crystallinity (of a polymer)
The fractional amount of crystallinity in the polymer sample (w_c for mass fraction; φ_c for volume fraction). Notes:
1. The assumption is made that the sample can be subdivided into a crystalline phase and an amorphous phase (the so-called two-phase model).
2. Both phases are assumed to have properties identical with those of their ideal states, with no influence of interfaces.
3. The degree of crystallinity may be expressed either as the mass fraction or as the volume fraction, the two quantities being related by

$$w_c = \varphi_c \rho_c / \rho$$

where ρ and ρ_c are the densities of the entire sample and of the crystalline fraction, respectively.

4. The degree of crystallinity can be determined by several experimental techniques; among the most commonly used are: (i) X-ray diffraction, (ii) calorimetry, (iii) density measurements, and (iv) infrared spectroscopy (IR). Imperfections in crystals are not easily distinguished from the amorphous phase. Also, the various techniques may be affected to different extents by imperfections and interfacial effects. Hence, some disagreement among the results of quantitative measurements of crystallinity by different methods is frequently encountered.
P.B. 75

degree of dissociation
Degree of reaction for a dissociation reaction.
G.B. 43; 1996, *68*, 968

degree of inhibition
See *inhibition*.
1996, *68*, 162

degree of ionization
Degree of reaction for an ionization reaction.
G.B. 43

degree of polymerization
The number of *monomeric units* in a *macromolecule* or *oligomer molecule*, a *block* or a *chain*.
1996, *68*, 2291

degree of reaction, α
Extent of reaction divided by the maximum extent of reaction.
G.B. 43

degrees of cistacticity and transtacticity
For a *regular polymer* containing double bonds in the main chain of the *constitutional repeating units*, these are the fractions of such double bonds that are in the *cis* and *trans* configurations, respectively.
P.B. 46

degrees of freedom, ν
A statistical quantity indicating the number of values which could be arbitrarily assigned within the specification of a system of observations. For simple replication, with n measurements and one estimated parameter (the mean), $\nu = n - 1$.
More generally, for multivariable computations, the number of degrees of freedom equals the number of observations minus the number of fitted parameters.
1994, *66*, 599

degrees of triad isotacticity, syndiotacticity, and heterotacticity
The fractions of *triads* in a regular vinyl *polymer* that are of the *mm*, *rr* and *mr* = *rm* types, respectively. In cases where triad analysis is not attainable, the diad isotacticity and diad syndiotacticity may be defined as the fractions of *diads* in a regular vinyl polymer that are of the *m* and *r* types, respectively. (*m* = *meso*; *r* = *racemic*).
P.B. 45

dehydroarenes
Species, usually transient, derived formally by the abstraction of a hydrogen atom from each of two ring atoms of an *arene*. The name for specific compounds requires the numerical prefix di-.
E.g. 1,8-didehydronaphthalene or naphthalene-1,8-diyl.

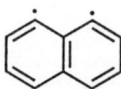

1,2-Didehydroarenes are called *arynes* and are commonly represented with a formal triple bond.
1995, *67*, 1331

dehydrobenzenes
A subclass of *dehydroarenes*.
See *benzynes*.
1995, *67*, 1331

delayed coincidence
The occurrence of two or more events separated by a short but measurable time interval.
1982, *54*, 1538

delayed coke
A commonly used term for a primary *carbonization* product (*green* or *raw coke*) from high-boiling hydrocarbon fractions (heavy residues of petroleum or coal processing) produced by the *delayed coking process*.
Notes:
Delayed coke has, with only a few exceptions, a better graphitizability than *cokes* produced by other coking processes even if the same feedstock is used. Delayed coke contains a mass fraction of volatile matter between 4 and 15 wt% which can be released during heat treatment.
1995, *67*, 486

delayed coking process
A thermal process which increases the molecular aggregation or association in petroleum-based residues or *coal tar pitches* leading to extended mesophase domains. This is achieved by holding them at an elevated temperature (usually 750–765 K) over a period of time (12 to 36 h). It is performed in a coking drum and is designed to ultimately produce *delayed coke*. The feed is rapidly pre-heated in a tubular furnace to about 760 K.
Notes:
Needle coke is the premium product of the delayed coking process. It is generally produced from highly aromatic residues from, for instance, the steam cracking of gas oil. Its appearance and preferred orientation of the *graphene layers* is a consequence of the evolved gaseous products percolating through the mesophase which must not have too high a viscosity. A close control of temperature, time and feedstock is essential. Lower grades, for instance isotropic cokes, are used for *carbon electrodes* applied, for example, in the production of aluminium.
1995, *67*, 486

delayed fluorescence
Three types of delayed fluorescence are known:
1. E-type delayed fluorescence: The process in which the first excited singlet state becomes populated by a thermally activated *radiationless transition* from the first excited triplet state. Since in this case the populations of the singlet and triplet states are in thermal equilibrium, the lifetimes of delayed fluorescence and the concomitant *phosphorescence* are equal.
2. P-type delayed fluorescence: The process in which the first excited singlet state is populated by interaction of two molecules in the triplet state (triplet-triplet *annihilation*) thus producing one molecule in the excited singlet state. In this biphotonic process the lifetime of delayed fluorescence is half the value of the concomitant phosphorescence.
3. Recombination fluorescence: The first excited singlet state becomes populated by recombination of *radical ions* with electrons or by recombination of radical ions of opposite charge.
See also *delayed luminescence*.
1984, *56*, 233; see also 1996, *68*, 2235; O.B. 185

delayed luminescence
Luminescence decaying more slowly than that expected from the rate of decay of the emitting state. The following mechanisms of luminescence provide examples:
1. triplet-triplet or singlet-singlet annihilation to form one molecular entity in its excited singlet state and another molecular entity in its electronic ground state (sometime referred to as P type),
2. thermally activated delayed fluorescence involving reversible intersystem crossing (sometimes referred to as E type), and
3. combination of oppositely charged ions or of an electron and a cation. For emission to be referred to in this case as delayed luminescence at least one of the two reaction partners must be generated in a photochemical process.
See also *delayed fluorescence*.
1988, *60*, 1065; see also O.B. 185

delayed neutrons
Neutrons emitted by *fission products* formed by *nuclear decay* (the observed delay is due to the preceding nuclear decay or decays).
1982, *54*, 1546

deliquescence
The process that occurs when the vapour pressure of the saturated aqueous solution of a substance is less than the vapour pressure of water in the ambient air. When water vapour is collected by the pure solid compound, a mixture of the solid and liquid or an aqueous solution of the compound forms until the substance is dissolved and is in equilibrium with its environment; at this time the vapour pressure of water over the aqueous solution will equal the partial pressure of water in the atmosphere in contact with it. A crystalline salt aerosol particle will deliquesce in the atmosphere when the relative humidity surpasses a characteristic value, the so-called *deliquescence point*.
1990, *62*, 2183

delocalization
A quantum mechanical concept most usually applied in organic chemistry to describe the π-bonding in a *conjugated system*. This bonding is not localized between two atoms: instead, each link has a 'fractional double bond character' or *bond order*. There is a corresponding 'delocalization energy', identifiable with the stabilization of the system compared with a hypothetical alternative in which formal (localized) single and double *bonds* are present. Some degree of delocalization is always present and can be estimated by quantum mechanical calculations. The effects are particularly evident in *aromatic* systems and in symmetrical *molecular entities* in which a lone pair of electrons or a vacant p-orbital is conjugated with a double bond (e.g. carboxylate ions, nitro compounds, enamines, the allyl cation). Delocalization in such species may be represented by partial bonds or as resonance (here symbolized by a two-headed arrow) between *contributing structures*.

These examples also illustrate the concomitant delocalization of charge in ionic conjugated systems. Analogously, delocalization of the spin of an unpaired electron occurs in conjugated *radicals*.
See also *mesomerism*.
1994, *66*, 1104

delta
For entries see δ, Δ (beginning of 'd').

demister
Apparatus made of wire mesh or glass fibre which is used to help remove acid mist as in the manufacture of sulfuric acid. Demisters are also components of wet arrestment plants.
1990, *62*, 2183

denaturation (of a macromolecule)
The process of partial or total alteration of the native structure of a macromolecule resulting from the loss of tertiary or tertiary and secondary structure that is a consequence of the disruption of stabilizing weak bonds. Denaturation can occur when proteins and *nucleic acids* are subjected to elevated temperature or to extremes of pH, or to non-physiological concentrations of salt, organic solvents, urea or other chemical agents.
1992, *64*, 151; 1994, *66*, 2593

denaturation (of alcohol)
Addition of methanol or acetone to alcohol to make it unfit for drinking.
1993, *65*, 2036

dendrite
A crystalline morphology produced by skeletal growth, leading to a 'tree-like' appearance.
P.B. 83

denitrification
The reduction of nitrates to nitrites, dinitrogen oxide (nitrous oxide) or dinitrogen catalysed by microoganisms, e.g. facultative aerobic soil bacteria under anaerobic conditions.
1992, *64*, 151

density, ρ
Mass of a sample or body divided by its volume.
G.B. 12; 1996, *68*, 968

density inversion (in solvent extraction)
The interchange of the denser and less dense phases due to changes in solute concentration. Phase inversion is often used in this context but is ambiguous.
See *inversion*.
1993, *65*, 2387

density, number
See *number density*.
1990, *62*, 2183

density of states, N_E, $r(E)$
1. Number of states within an energy interval divided by volume and that energy interval. Symbol: N_E.
2. Number of states within an energy interval divided by that energy interval. Symbol: $r(E)$.
G.B. 37, 39; 1996, *68*, 162

denticity
In a *coordination entity*, the number of donor groups from a given ligand attached to the same central atom.
See *chelation*.
R.B. 147

denuder system (tube or assembly)
An apparatus used to separate gases and aerosols (over a given diameter) which is based upon the difference in diffusion velocity between gases and aerosol particles. Usually a tube containing a selective internal wall coating which removes the gaseous compounds at the wall.
1990, *62*, 2184

deodorizer (in atmospheric chemistry)
Equipment for the removal of noxious gases and odours, which may consist of combustion, absorption, or adsorption units.
See *emission control equipment*.
1990, *62*, 2184

deoxyribonucleic acids (DNA)
High molecular weight, linear polymers, composed of *nucleotides* containing deoxyribose and linked by phosphodiester bonds; DNA contain the genetic information of organisms. The double-stranded form consists of a double helix of two complementary chains that run in opposite directions and are held together by hydrogen bonds between pairs of the complementary *nucleotides* and Hoogsteen (stacking) forces.
1992, *64*, 151

depolarization of scattered light
The phenomenon, due primarily to the anisotropy of the polarizability of the scattering medium, resulting from the fact that the electric vectors of the incident and scattered beams are not coplanar and that, therefore, light scattered from a vertically (horizontally) polarized incident beam contains a horizontal (vertical) component.
P.B. 67

+ depolarizer
Synonym for *electroactive substance* (preferred term).
1985, *57*, 1500; O.B. 58

depolymerization
The process of converting a *polymer* into a *monomer* or a mixture of monomers.
Note:
Unzipping is depolymerization occurring by a sequence of reactions, progressing along a *macromolecule* and yielding products, usually *monomer molecules* at each reaction step, from which macromolecules similar to the original can be regenerated.
1996, *68*, 2309

deposition (in atmospheric chemistry)
Deposition is normally considered to be one of two types: *dry deposition* is the process by which aerosols and gases in the air are deposited on the surface of the earth (soil, water, rock, plants, etc.); this is termed 'dry' deposition even when the receptor surface is moist. *Wet deposition* is that process which involves the transport of chemicals to the surface of the earth by water droplets or snow crystals which scavenge pollutants as they form and fall through the atmosphere.
1990, *62*, 2184

deposition velocity (in atmospheric chemistry)
The ratio of flux density (often given in units of g cm^{-2} s^{-1}) of a substance at a sink surface to its concentration in the atmosphere (corresponding units of g cm^{-3}). While the units of this ratio are clearly those of velocity (in this case cm s^{-1}), the ratio is not a flow velocity in the normal sense of the word.
1990, *62*, 2184

depsides
Intermolecular esters formed from two or more molecules of the same or different phenolic benzoic acids.

4-*O*-galloylgallic acid

Depending on the number of the component units, di-, tri-, etc. depsides may result, e.g. 4-*O*-galloylgallic acid.
1995, *67*, 1331

depsipeptides
Natural or synthetic compounds having sequences of amino and hydroxy carboxylic acid residues (usually α–amino and α–hydroxy acids), commonly but not necessarily regularly alternating. In cyclodepsipeptides, the residues are connected in a ring.

$$-\underset{\underset{R}{|}}{CH}-\underset{\underset{O}{\|}}{C}-NH-\underset{\underset{R}{|}}{CH}-\underset{\underset{O}{\|}}{C}-O-\underset{\underset{R}{|}}{CH}-\underset{\underset{O}{\|}}{C}-NH-\underset{\underset{R}{|}}{CH}-\underset{\underset{O}{\|}}{C}-O-$$

See *carboxylic acids*.
1995, *67*, 1331; see also 1984, *56*, 617

depth of penetration (of light)
The inverse of the *absorption coefficient*. If the decadic absorption coefficient, a, is used, the depth of penetration ($1/a$) is the distance at which the radiant power, P_λ, decreases to one tenth of its incident value, P_λ^0.
If the Naperian absorption coefficient, α, is used, the depth of penetration ($1/\alpha = \beta$ in this case) is the distance at which the radiant power decreases to $1/e$ of its incident value.
See *absorbance, attenuance*.
1996, *68*, 2235

depth profile
Dependence of concentration on depth perpendicular to the surface in a solid sample. It can be obtained by a simultaneous or sequential process of erosion and surface analysis or by measurement of the energy loss of primary backscattered particles produced by nuclear reactions.
1979, *51*, 2247

depth resolution
The distance between the 84 and 16 per cent levels of the *depth profile* of an element in a perfect sandwich sample with an infinitesimally small overlap of the components. These limits correspond to the 2σ-value of the Gaussian distribution of the signal at the interface.
1979, *51*, 2247

deracemization
See *asymmetric transformation*.
1996, *68*, 2205

derivative potentiometric titration
A *titration* that involves measuring, recording, or computing the first derivative of the potential of a single *indicator electrode* with respect to the volume or otherwise added amount of reagent.
O.B. 65

derivative spectroscopy
The first (second, ...) derivative absorption spectrum of a molecule is defined as the first (second, ...) derivative, $dA(\tilde{v})/d\tilde{v}$, $[dA(\tilde{v})/d\tilde{v}^2, ...]$ of the absorbance A as a function of wavenumber, $\tilde{v}$. Wavelengths may be used in place of wavenumbers but the shape of the derivative spectra will be slightly different.
When derivative spectra are obtained at low temperature, they are called first (second, ...) derivative *low temperature absorption spectra*, (specifying the solvent, temperature and solute concentration).
1988, *60*, 1455

derived coherent unit
A unit constructed exclusively from base units, e.g. mol kg^{-1}, kg m^{-3}.
1979, *51*, 2461

derived kind of quantity
A kind of quantity characterized by an equation between base kinds of quantities, e.g. mass concentration = mass (length)$^{-3}$.
1979, *51*, 2460

derived non-coherent unit
A unit constructed from base units and numerical factors, which may be named and symbolized, e.g. mg, mol l^{-1}.
1979, *51*, 2461

derived quantity
Quantity defined, in a system of quantities, as a function of base quantities of that system.
See *quantity, base quantity*.
1996, *68*, 969

derived unit (of measurement)
Unit of measurement of a derived quantity in a given system of quantities. Derived units are expressed algebraically in terms of base units by means of the mathematical symbols of multiplication and division.
See *unit (of measurement), derived quantity*.
1996, *68*, 969

descending elution (descending development) (in planar chromatography)
A mode of operation in which the mobile phase is supplied to the upper edge of the paper or plate and the downward movement is governed mainly by gravity.
1993, *65*, 829

deshielding
See *shielding*.
1994, *66*, 1105

desiccant
1. Drying agent.
2. In agriculture, a substance used for drying up plants and facilitating their mechanical harvesting.
1993, *65*, 2037

desolvation (in flame spectroscopy)
The process of evaporation of the solvent from an aerosol of fine droplets (the droplets being the result of nebulizing a solution). Desolvation converts the aerosol into a dry aerosol consisting of a suspension of solid or molten particles of the solute.
O.B. 165

desorption
The converse of *adsorption*, i.e. the decrease in the amount of adsorbed substance.
1972, *31*, 584; see also 1990, *62*, 2184

desorption (by displacement)
The process in which expulsion of a previously adsorbed component from the *interfacial layer* is effected by subsequent stronger *adsorption* of another component.
1972, *31*, 585

desulfurization
The process by which sulfur is removed from a material such as coal or oil. It may involve one of many techniques including elutriation, froth flotation, laundering, magnetic separation, chemical treatment, etc.
1990, *62*, 2184

desymmetrization
The modification of an object which results in the loss of one or more symmetry elements, such as those which preclude *chirality* (mirror plane, centre of inversion, rotation-reflection axis), as in the conversion of a *prochiral* molecular entity into a chiral one.
1996, *68*, 2205

desymmetrization step
The removal of the smallest possible number of symmetry elements from a molecule with or without the restriction to pointwise substitution.
1996, *68*, 2205

detachment
The reverse of an *attachment*.
See also *electron attachment*.
1994, *66*, 1105

detailed balancing (principle of)
When equilibrium is reached in a reaction system (containing an arbitrary number of components and reaction paths), as many atoms, in their respective *molecular entities*, will pass forward, as well as backwards, along each individual path in a given finite time interval. Accordingly, the reaction path in the reverse direction must in every detail be the reverse of the reaction path in the forward direction (provided always that the system is at equilibrium).
The principle of detailed balancing is a consequence for macroscopic systems of the principle of *microscopic reversibility*.
1994, *66*, 1105; see also 1996, *68*, 162

detection efficiency (in nuclear analytical chemistry)
The ratio between the number of *particles* or *photons* detected and the number of similar particles or photons emitted by the *radiation source*.
1982, *54*, 1539

detection limit
See *absolute detection limit; relative detection limit*.
1979, *51*, 2247

detection limit (in analysis)
The minimum single result which, with a stated probability, can be distinguished from a suitable blank value.
The limit defines the point at which the analysis becomes possible and this may be different from the lower limit of the determinable analytical range.
See also *limit of detection, relative detection limit*.
1989, *61*, 1663; see also 1991, *63*, 1553

detector, chromatographic
See *chromatographic detector*.
O.B. 99; 1993, *65*, 829

detector, concentration-sensitive (in chromatography)
See *concentration-sensitive detector (in chromatography)*.
1993, *65*, 849

detector, differential (in chromatography)
See *differential detector (in chromatography)*.
1993, *65*, 849; O.B. 99

detector, diffused junction semiconductor
See *diffused junction semiconductor detector*.
1982, *54*, 1540

detector efficiency, intrinsic
See *intrinsic detector efficiency*.
1994, *66*, 2518

detector, integral (in chromatography)
See *integral detector (in chromatography)*.
1993, *65*, 849; O.B. 99

detector, mass-flow sensitive (in chromatography)
See *mass-flow sensitive detector (in chromatography)*.
1993, *65*, 849

detector, radiation
See *radiation detector*.
1982, *54*, 1540

detector, scintillation
See *scintillation detector*.
1982, *54*, 1540

detector, selective (in chromatography)
See *selective detector (in chromatography)*.
1993, *65*, 849

detector, semiconductor
See *semiconductor detector*.
1982, *54*, 1540

detector, specific (in chromatography)
See *specific detector (in chromatography)*.
1993, *65*, 849

detector, universal (in chromatography)
See *universal detector (in chromatography)*.
1993, *65*, 849

detergent
A *surfactant* (or a mixture containing one or more surfactants) having cleaning properties in *dilute solution* (soaps are surfactants and detergents).
1972, *31*, 612

detoxification
1. Process, or processes, of chemical modification which make a toxic molecule less toxic.
2. Treatment of patients suffering from poisoning in such a way as to promote physiological processes which reduce the probability or severity of harmful effects.
1993, *65*, 2038

deuteriation
A term used to represent replacement of protium by deuterium. The use of the word deuteration is also acceptable.
1988, *60*, 1115; 1991, *63*, 1556

deuteride
See *deuterium*.
1988, *60*, 1116

deuterio
See *deuterium*.
1988, *60*, 1116

deuterium
A specific name for the atom ^{2}H. The cation ^{2}H$^+$ is a *deuteron*, the species ^{2}H$^-$ is a *deuteride* anion and ^{2}H is the *deuterio* group.
1988, *60*, 1116

deuteron
Nucleus of the ^{2}H (deuterium) atom.
G.B. 93; 1988, *60*, 1116

deviation, *d*
The difference between an observed value and the arithmetic mean of the set to which it belongs. It can be calculated by the formula:
$$d_i = x_i - \bar{x}$$
1994, *66*, 599; O.B. 4; see also VIM

devolatilizer
Material added to a sample to decrease its volatilization or that of some component of it.
O.B. 159

dew point (in atmospheric chemistry)
That temperature at which water vapour present in the atmosphere is just sufficient to saturate it. When air is cooled below the dew point, the excess of water vapour appears as tiny droplets or crystals of ice depending on the temperature of the air mass. This term is also applicable to acid gases as in the problem of acid corrosion and acid smut formation.
1990, *62*, 2184

dew point hygrometer (cooled surface condensation)
Instrument in which the sample is passed over a cooled surface. The temperature at which dew forms on the cooled surface is a function of the water content of the gas passing over the surface.
1990, *62*, 2195

Dexter excitation transfer (electron exchange excitation transfer)
Excitation transfer occurring as a result of an electron exchange mechanism. It requires an overlap of the wavefunctions of the energy donor and the energy acceptor. It is the dominant mechanism in triplet-triplet energy transfer. The transfer rate constant, k_{ET}, is given by:
$$k_{ET} \propto [h/(2\pi)]P^2J \exp[-2r/L]$$
where r is the distance between donor (D) and acceptor (A), L and P are constants not easily related to experimentally determinable quantities, and J is the spectral overlap integral. For this mechanism the spin conservation rules are obeyed.
See also *radiative energy transfer*.
1996, *68*, 2235

dextrans
Branched poly-α-D-glucosides of microbial origin having glycosidic bonds predominantly C-1 → C-6.
See *glycosides*.
1995, *67*, 1331

dextrins
Poly-α-D-glucosides of intermediate chain length derived from starch components (amylopectins) by the action of amylases (starch hydrolysing enzymes).
See also *cyclodextrins, dextrans*.
1995, *67*, 1331

DFDMR
See *ODMR*.
1996, *68*, 2235

diabatic coupling
Energy coupling between two potential-energy surfaces is sometimes known as diabatic coupling.
1996, *68*, 163

diabatic electron transfer
Electron transfer process in which the reacting system has to cross over between different electronic surfaces in passing from reactants to products. For diabatic electron transfer the electronic transmission factor is << 1 (see *Marcus equation*.) The term non-adiabatic electron transfer has also been used and is in fact more widespread, but should be discouraged because it contains a double negation.
See also *adiabatic electron transfer*.
1996, *68*, 2236

diabatic photoreaction
Within the Born–Oppenheimer approximation, a reaction beginning on one excited state 'potential-energy surface' and ending, as a result of radiationless transition, on another surface, usually that of the ground state. Also called non-adiabatic.
See also *adiabatic photoreaction*.
1996, *68*, 2235

diacylamines
Compounds having two *acyl groups* substituted on ammonia or a primary amine acyl-NR-acyl. They are also known as secondary *amides* and, especially the cyclic examples derived from diacids, as *imides*, e.g. [RC(=O)]$_2$NR, RS(=O)$_2$NHC(=O)R.
1995, *67*, 1331

diads (in polymers)
See *constitutional sequence*.
1996, *68*, 2299

diagram level (in X-ray spectroscopy)
A level described by the removal of one electron from the configuration of the neutral ground state. These levels form a spectrum similar to that of a one-electron or hydrogen-like atom but, being single-valency levels, have the energy scale reversed relative to that of single-electron levels. Diagram levels may be divided into valence levels and core levels according to the nature of the electron vacancy. Diagram levels with orbital angular momentum different from zero occur in pairs and form spin doublets.
1991, *63*, 738

diagram line (in X-ray spectroscopy)
See *characteristic X-ray emission, X-ray satellite*.
1991, *63*, 737

+ dialysate
After (complete or incomplete) *dialysis* two solutions are obtained. The one free from *colloidal* material is called dialysate; the other one, containing the colloidal particles may be called retentate, dialysis residue, or simply residue, but should not be called dialysate.
1972, *31*, 608

dialysis
The process of separating a colloidal *sol* from a colloid-free solution by a membrane permeable to all components of the system except the *colloidal* ones, and allowing the exchange of the components of small molar mass to proceed for a certain time.
1972, *31*, 608

dialysis residue
See *dialysate*.
1972, *31*, 608

diamagnetic
Substances having a negative magnetic susceptibility are diamagnetic. They are repelled out of a magnetic field.
See also *paramagnetic*.
1994, *66*, 1105

diameter, equivalent
See *equivalent diameter*.
1990, *62*, 2184

diamidides
Analogues of acyclic carboxylic *acid anhydrides* in which =O has been replaced by =NR and -O- by -NR-: RC(=NR)NRC(=NR)R, *N*-imidoyl amidines. The name 1,3,5-triazapentadienes is sometimes applied as a class name to diamidides.
See *imidines*.
1995, *67*, 1331

diamond
An allotropic form of the element *carbon* with cubic structure (space group O_h^7-$Fd3m$) which is thermodynamically stable at pressures above 6 GPa at room temperature and metastable at atmospheric pressure. At low pressures diamond converts rapidly to *graphite* at temperatures above 1900 K in an inert atmosphere. The chemical bonding between the carbon atoms is covalent with sp^3 hybridization.
Notes:
There is also a hexagonal diamond-like structure of the element carbon (lonsdaleite).
1995, *67*, 487

diamond by CVD
Diamond by CVD (chemical vapour deposition) is formed as crystals or as films from various gaseous hydrocarbons or other organic molecules in the presence of activated, atomic hydrogen. It consists of sp^3-hybridized carbon atoms with the three-dimensional crystalline structure of the diamond lattice.
Notes:
'CVD diamond' or 'low-pressure diamond' are synonyms of the term diamond by CVD. Diamond by CVD can be prepared in a variety of ways. Deposition parameters are: total (low) pressure, partial hydrogen pressure, precursor molecules in the gas phase, temperature for activation of the hydrogen and that of the surface of the underlying substrate. The energy supply for the hydrogen activation may be, for instance: heat, radio frequency, microwave excitation (plasma deposition) or accelerated ions (e.g. Ar$^+$ ions). CVD diamond has also been obtained at atmospheric pressure from oxyacetylene torches and by other flame-based methods.
Often CVD carbon films consist of a mixture of sp^2- and sp^3-hybridized carbon atoms and do not have the three-dimensional structure of the diamond lattice. In this case they should be called *hard amorphous carbon* or *diamond-like carbon films*.
1995, *67*, 487

diamond-like carbon films
Diamond-like carbon (DLC) films are hard, amorphous films with a significant fraction of sp^3-hybridized carbon atoms and which can contain a significant amount of hydrogen. Depending on the deposition conditions, these films can be fully amorphous or contain *diamond* crystallites. These materials are not called diamond unless a full three-dimensional crystalline lattice of diamond is proven.
Notes:
Diamond-like films without hydrogen can be prepared by carbon ion beam deposition, ion-assisted sputtering from *graphite* or by laser ablation of graphite. Diamond-like carbon films containing significant contents of hydrogen are prepared by chemical vapour deposition. The hydrogen content is usually over 25 atomic %. The deposition parameters are (low) total pressure, hydrogen partial pressure, precursor molecules and plasma ionization. The plasma activation can be radio frequency, microwave or Ar$^+$ ions. High ionization favours amorphous films while high atomic hydrogen contents favour diamond crystallite formation. Because of the confusion about structure engendered by the term diamond-like carbon films,

the term *hard amorphous carbon* films has been suggested as a synonym.
1995, *67*, 487

dianions
Molecular entities bearing two negative charges, which may be located on a single atom or on different atoms or may be delocalized.
1995, *67*, 1331

diastereoisomer excess (diastereoisomeric excess)
This is defined by analogy with *enantiomer excess*, as $D_1 - D_2$ [and the percent diastereoisomer excess as $100(D_1 - D_2)$], where the mole fractions of two diastereoisomers in a mixture or the fractional yields of two diastereoisomers formed in a reaction are D_1 and D_2 $(D_1 + D_2 = 1)$. The term is not applicable if more than two diastereoisomers are present. Frequently this term is abbreviated to d.e.
See *stereoselectivity, diastereoisomerism*.
1996, *68*, 2205

diastereoisomeric excess
See *diastereoisomer excess*.
1996, *68*, 2205

diastereoisomeric units (in a polymer)
Two non-superposable *configurational units* that correspond to the same *constitutional unit* are considered to be diastereomeric if they are not mirror images.
P.B. 27

diastereoisomerization
The interconversion of *diastereoisomers*.
1996, *68*, 2205

diastereoisomerism
Stereoisomerism other than *enantiomerism*. Diastereoisomers (or diastereomers) are stereoisomers not related as mirror images. Diastereoisomers are characterized by differences in physical properties, and by some differences in chemical behaviour towards achiral as well as chiral reagents.
1996, *68*, 2205; 1994, *66*, 1105

diastereoisomers (diastereomers)
See *diastereoisomerism*.
1996, *68*, 2205

diastereomeric ratio
This is defined by analogy with *enantiomeric ratio* as the ratio of the percentage of one *diastereoisomer* in a mixture to that of the other.
1996, *68*, 2206

diastereomers
See *diastereoisomerism*.
1996, *68*, 2206

diastereomorphism
The relationship between objects (or models) analogous to that between *diastereoisomeric* molecular entities.
1996, *68*, 2206

diastereoselectivity
See *stereoselectivity*.
1996, *68*, 2206

diastereotopic
Constitutionally equivalent atoms or groups of a molecule which are not symmetry related. Replacement of one of two diastereotopic atoms or groups results in the formation of one of a pair of *diastereoisomers*. In the example below the two hydrogen atoms of the methylene group C-3 are diastereotopic.

See also *prochirality, enantiotopic, heterotopic*.
1996, *68*, 2206

diazanylidenes
An alternative term for *isodiazenes*.
1995, *67*, 1331

+ 1,1-diazenes
An incorrectly formed name for *isodiazenes*. Use of this name is discouraged.
See also *nitrenes*.
1995, *67*, 1332

diazenyl radicals
Radicals of structure RN=N·.
1995, *67*, 1332

diazoamino compounds
Compounds having the structure $RN=N-NR_2$ (not all R = H, and one R commonly aryl). In systematic nomenclature, diazoamino prefixed to the name of RH names the compound RN=NNHR' (R = R'), e.g. PhN=N–NPhMe *N*-methyldiazoaminobenzene.
See also *triazenes*.
1995, *67*, 1332

diazoates
Salts $RN=NO^-M^+$ (R commonly *aryl*) of the compounds RN=NOH, (hydrocarbon)diazohydroxides, e.g. $PhN=NO^-K^+$ potassium benzenediazoate. In the IUPAC 'Blue Book' these compounds are named as hydrocarbyldiazenolates and hydrocarbyldiazenols, respectively. E.g. PhN=NOH phenyldiazenol. $RN=NO^-M^+$ are also known by the non-IUPAC name diazotates.
1995, *67*, 1332

diazo compounds
Compounds having the divalent diazo group, $=N^+=N^-$, attached to a carbon atom, e.g. $CH_2=N_2$ diazomethane.
See also *dipolar compounds*.
1995, *67*, 1332

diazonium salts
Compounds of structure $RN_2^+ Y^-$, in which R is generally but not necessarily *aryl*, and the cations of which are usually formulated as $RN^+\equiv N$, E.g. $PhN^+\equiv N$ benzenediazonium chloride. They may also be named, from the canonical form $RN=N^+$, hydrocarbyldiazenylium salts.
1995, *67*, 1332

diazooxides
Diazocyclohexadienones, which may also be considered as dipolar diazonio phenoxides, obtained by diazotizing aromatic primary *amines* having a hydroxy group in an *ortho* or *para* position. Also known as diazophenols.

$$^-O-C_6H_4-N_2^+ \leftrightarrow O=C_6H_4=N_2$$

See also *diazo compounds, phenoxides, quinone diazides*.
1995, *67*, 1332

diazotates
See *diazoates*.
1995, *67*, 1332

dicarbenium ions
Species carrying two positive charges, formally located on tervalent carbon atoms, e.g. $^+CH_2-^+CH_2$ ethane-1,2-diyl dication, $^+CH_2CH_2-^+CH_2$ propane-1,3-bis(ylium).
1995, *67*, 1332

dichotomous sampler
Device for dividing a polydispersed aerosol particle population into two size fractions during sampling. The fractionation is based on the momentum differences of the particles which allow the larger particles to pass through a zone of stagnant gas.
1990, *62*, 2212

+ dielectric constant
See *relative permittivity*.
G.B. 14; see also 1994, *66*, 1105

dielectric polarization, P
The difference between the *electric displacement* vector and the product of the *electric field strength* with the permittivity of vacuum, $P = D - \varepsilon_0 E$.
G.B. 14

dienes
Compounds that contain two fixed double bonds (usually assumed to be between carbon atoms). Dienes in which the two double-bond units are linked by one single bond are termed conjugated, e.g. $CH_2=CH-CH=CH_2$ buta-1,3-diene. Dienes in which the double bonds are adjacent are called cumulative, e.g. $CH_3CH=C=CH_2$ buta-1,2-diene. Those in which one or more of the unsaturated carbon atoms is replaced by a heteroatom may be called heterodienes.
See *alkenes, olefins, diols, allenes, cumulenes*.
1995, *67*, 1332

dienophile
The olefin component of a Diels–Alder reaction.
See *cycloaddition*.
1994, *66*, 1105

difference absorption spectroscopy
A highly concentrated analyte in an analytical sample can be determined with better precision by replacing the *blank (reference) cell* by one containing a solution of the analyte or other absorber of known concentration; this is known as difference absorption spectroscopy. Difference spectra can also be obtained by computer or other subtraction methods.
1988, *60*, 1455

differential capacitance
Differential capacitance (per unit area of interphase or electrode; SI unit Fm^{-2}) is given by $C = (\partial Q/\partial E)_{T, p, \mu_i ...}$ where E is the potential of the electrode with respect to a *reference electrode*, μ_i is a set of *chemical potentials* which are held constant, T is the thermodynamic temperature, Q is the electric charge (per unit area of interphase or electrode), and p is the pressure.
1974, *37*, 509; 1983, *55*, 1261

differential detector (in chromatography)
A device which measures the instantaneous difference in the composition of the column effluent.
1993, *65*, 849; O.B. 99

differential diffusion coefficient
Defined by

$$D_i = -J_i/\text{grad } c_i$$

where J_i is the amount of species i flowing through unit area in unit time and $\text{grad } c_i$ is the concentration gradient of species i. Different diffusion coefficients may be defined depending on the choice of the frame of reference used for J_i and $\text{grad } c_i$. For systems with more than two components, the flow of any component and hence its diffusion coefficient depends on the concentration distribution of all components.

The limiting differential diffusion coefficient is the value of D_i extrapolated to zero concentration of the diffusing species:

$$[D_i] = \lim_{c_i \to 0} D_i$$

1972, *31*, 617

differential molar energy of adsorption
When the addition of a differential amount dn_i^σ or dn_i^s is effected at constant gas volume, the differential molar energy of adsorption of component i, $\Delta_a U_i^\sigma$ or $\Delta_a U_i^s$, is defined as:

$$\Delta_a U_i^\sigma = U_i^\sigma - U_i^g,$$

or

$$\Delta_a U_i^s = U_i^s - U_i^g$$

where the differential molar surface excess energy, U_i^σ, is given by

$$U_i^\sigma = \left(\frac{\partial U^\sigma}{\partial n_i^s}\right)_{T, m, n_j^\sigma} = \left(\frac{\partial U}{\partial n_i^\sigma}\right)_{T, m, V^g, p_i, n_j^\sigma}$$

and the differential molar interfacial energy, U_i^s, by

$$U_i^s = \left(\frac{\partial U}{\partial n_i^s}\right)_{T, m, V^g, p_i, n_j^\sigma}$$
$$= \left(\frac{\partial U}{\partial n_i^s}\right)_{T, m, V^g, V^s, p_i, n_j^s}$$

U_i^g is the differential molar energy of component i in the gas phase, i.e $(\partial U/\partial n_i^g)_{T,V,n_i^g}$.
1972, *31*, 603

differential molar interfacial energy
See *differential molar energy of adsorption*.
1972, *31*, 603

differential molar surface excess energy
See *differential molar energy of adsorption*.
1972, *31*, 603

differential scanning calorimetry (DSC)
A technique in which the difference in energy inputs into a substance (and/or its reaction product(s)) and a reference material is measured as a function of temperature whilst the substance and reference material are subjected to a controlled temperature programme.
1985, *57*, 1739; O.B. 44

differential thermal analysis (DTA)
A technique in which the temperature difference between a substance and a reference material is measured as a function of temperature, while the substance and reference material are subjected to a controlled temperature program.
O.B. 42

differential viscosity
The derivative η_Δ of *stress* with respect to the rate of shear at a given *shear rate*.
1979, *51*, 1217

diffraction
A modification which light (or electron, neutron beams, etc.) undergoes in passing by the edges of opaque bodies or through narrow slits or in being reflected from ruled surfaces (or crystalline materials). The light waves, owing to their wave-like nature appear to be deflected and produce fringes of parallel light and dark bands corresponding to regions of constructive reinforcement or destructive interference, respectively, of the waves.
1990, *62*, 2184

diffraction analysis
The application of diffraction techniques (X-rays, electrons, neutrons) which are sometimes used to identify the presence of certain solid aerosols and dust particles through the characteristic diffraction patterns which result from each unique crystal structure.
1990, *62*, 2184

diffused junction semiconductor detector
A *semiconductor detector* in which the p–n or n–p junction is produced by diffusion of donor or acceptor impurities.
1982, *54*, 1540

diffuse layer (in electrochemistry)
The region in which non-specifically adsorbed ions are accumulated and distributed by the contrasting action of the electric field and thermal motion.
Counter and co-ions in immediate contact with the surface are said to be located in the *Stern layer*. Ions farther away from the surface form the diffuse layer or Gouy layer.
See also *interfacial double layer*.
1983, *55*, 1261; 1972, *31*, 618

diffuser
A porous plate or tube, commonly made of carborundum, alundum or silica sand, through which air is forced and divided into minute bubbles for diffusion in liquids.
1990, *62*, 2184

diffusion
The spreading or scattering of a gaseous or liquid material. Eddy diffusion in the atmosphere is the process of transport of gases due to turbulent mixing in the presence of a composition gradient. Molecular diffusion is the net transport of molecules which results from their molecular motions alone in the absence of turbulent mixing; it occurs when the concentration gradient of a particular gas in a mixture differs from its equilibrium value. Eddy diffusion is the most important mixing process in the lower atmosphere, while molecular diffusion becomes significant at the lower pressures of the upper atmosphere.
1990, *62*, 2185

diffusional transition
A transition that requires the rearrangement of atoms, ions or molecules in a manner that cannot be accomplished by a cooperative atomic displacement; it may require the movement of atoms, ions or molecules over distances significantly larger than a unit cell.
Example: The transition of graphite (hexagonal sheets of three-coordinated carbon atoms) to diamond (infinite three-dimensional framework of four-coordinated carbon atoms) at high temperature and pressure.
1994, *66*, 581

diffusion battery
An aerosol sizing instrument for particles with diameters below 0.2 µm. The fractionation is based on different diffusivities of the small particles and their deposition on the walls of the long parallel or circular channels, formed by equally spaced plates, bundles of small bore parallel tubes or sets of stainless wire screens.
1990, *62*, 2185

diffusion coefficient, D
Proportionality constant, D, relating the *flux* of *amount* of entities B to their concentration gradient $J_n = -D \text{ grad } c_B$.
G.B. 65; see also 1996, *68*, 969

diffusion control
See *microscopic diffusion control*.
1996, *68*, 163

diffusion-controlled rate
See *encounter-controlled rate, macroscopic diffusion control, microscopic diffusion control*.
1983, *55*, 1308

diffusion current (or diffusion-controlled current)
A *faradaic current* whose magnitude is controlled by the rate at which a reactant in an electrochemical

process diffuses toward an electrode-solution interface (and, sometimes, by the rate at which a product diffuses away from that interface).
For the reaction mechanism

$$C \underset{k_-}{\overset{k}{\rightleftharpoons}} B \xrightarrow{+ne} B'$$

there are two common situations in which a diffusion current can be observed. In one, the rate of formation of B from electroinactive C is small and the current is governed by the rate of diffusion of B toward the electrode surface. In the other, C predominates at equilibrium in the bulk of the solution, but its transformation into B is fast; C diffuses to the vicinity of the electrode surface and is there rapidly converted into B, which is reduced.

1985, *57*, 1495

diffusion current constant (in polarography)
The empirical quantity defined by the equation

$$i = i_{d,l} / c_B m^{2/3} t_1^{1/6}$$

where $i_{d,l}$ = *limiting diffusion current*, c_B = bulk concentration of the substance B whose *reduction* or *oxidation* is responsible for the wave in question, m = average rate of (mass) flow of mercury (or other liquid metal), and t_1 = drop time.

1985, *57*, 1500

diffusion current, limiting
  See *limiting diffusion current*.
  1985, *57*, 1495

diffusion layer (concentration boundary layer)
The region in the vicinity of an electrode where the concentrations are different from their value in the bulk solution. The definition of the thickness of the diffusion layer is arbitrary because the concentration approaches asymptotically the value c_0 in the bulk solution (see diagram).

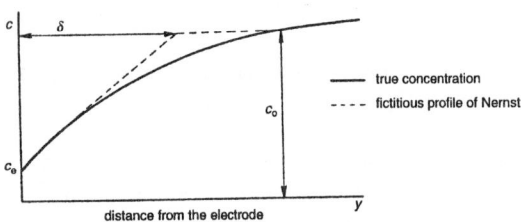

See also *Nernst's diffusion layer*.
  1981, *53*, 1837

diffusion layer thickness
Defined by the equation:

$$\delta = D/k_d = nFDcA/l_i$$

where D is the diffusion coefficient, k_d is the *heterogeneous diffusion rate constant*, F is the *Faraday constant*, c is the *concentration*, A is the geometric area of the electrode, l_i is the *limiting current*, and n is the *charge number of the cell reaction*.

1974, *37*, 513

diffusion layer, thickness of
  See *thickness of diffusion layer*.
  1982, *53*, 1837

diffusionless transition
A transition that does not involve long-range diffusion of atomic species over distances significantly greater than a typical interatomic distance.

1994, *66*, 581

diffusion potential
For an ideal *dilute solution*, $\Delta\Phi_d$ is the integral of $\nabla\Phi$ (given by the following equation) across the boundary between two regions of different concentrations.

$$\nabla\Phi = \frac{RT \Sigma D_i z_i \nabla c_i}{F \Sigma s_i^2 D_i c_i}$$

where D_i is the diffusion coefficient of species i, z_i is the *charge number* of species i, c_i is the *concentration* of species i, R is the gas constant, T is the thermodynamic temperature, and F is the *Faraday constant*.

1981, *53*, 1838

dihedral angle
The angle between two intersecting planes on a third plane normal to the intersection of the two planes.
  See *torsion angle*.
  1996, *68*, 2206

diisotactic polymer
An *isotactic polymer* that contains two chiral or *prochiral* atoms with defined stereochemistry in the main chain of the *configurational base unit*.
  P.B. 32

dilatancy
An increase in volume of the system which usually accompanies *shear thickening*. This term has formerly been used for the phenomenon of shear thickening.

1979, *51*, 1217

dilational (dilatational) transition
A transition in which the crystal structure is dilated (or compressed) along one (or more) crystallographic direction(s) while the symmetry about that direction is retained.
Examples:
i. The transition of a CsCl-type structure to a rock salt structure in which dilation occurs along the three-fold axis.
ii. The transition at T_D in quenched NiS in which volume expansion occurs without change of symmetry on going from a metallic state ($T > T_D$) to a semiconducting state ($T < T_D$).

1994, *66*, 581

dilational viscosity
  See *volume viscosity*.
  1979, *51*, 1216

diluent (in solvent extraction)
The liquid or homogeneous mixture of liquids in which extractant(s) and possible modifier(s) may be dissolved to form the solvent phase.

Notes:
i. The term carrier, which implies an inert diluent is not recommended.
ii. Although the diluent may well be a single liquid or even the major portion of the extracting phase, the term solvent should not be used in this sense as it has a much wider meaning in the context of liquid-liquid extraction, although the term cosolvent may be used in certain circumstances.
iii. The diluent by itself does not extract the main (extractable) solute appreciably.
1993, *65*, 2380

diluent gas
A gas of known quality introduced for analytical purposes so that it quantitatively lowers the concentration of the components of a gaseous sample; this may also be the complementary gas.
1990, *62*, 2185

dilute phase (in polymer chemistry)
See *polymer-poor phase*.
P.B. 68

diluter
A device used to add a measured volume or amount of the sample to a measured volume of a diluent.
1989, *61*, 1661

dilute solution
Solution in which the sum of amount fractions of all the solutes is small compared to 1.
Physical Chemistry Division, unpublished; 1994, *66*, 535; see also P.B. 57

dilution rate (in biotechnology), D
The ratio of the ingoing volume flow rate (dV/dt) and the culture volume (V). In continuous *fermentation*, a measure of a rate at which the existing medium is replaced with fresh medium, D is the reciprocal of the hydraulic retention time (HRT).
1992, *64*, 152

dimension (of a quantity)
Product of powers of the base dimensions (the dimensions attributed to the base quantities of a system) which parallels the expression of the quantity in terms of the base quantities. When all the exponents of the base dimensions are zero the quantity has the dimension one and is often called dimensionless.
Physical Chemistry Division, unpublished; 1996, *68*, 969

dimensionless quantities
Quantities of dimension one, or numerical quantities, are numbers which may be integers (or exceptionally half-integers) when obtained by counting (e.g. number of molecules, quantum numbers, ...), rational numbers such as ratios, factors and fractions when obtained as ratios of two quantities of the same kind or real numbers when obtained by taking logarithms (e.g. absorbance, power levels, ...).
G.B. 77; 1996, *68*, 969

dimeric ion (in mass spectrometry)
An ion formed either when a chemical species exists in the vapour as a dimer and can be detected as such, or when a molecular ion can attach to a neutral molecule within the ion source to form an ion such as $[M_2]^{+\cdot}$ where M represents the molecule.
1991, *63*, 1549

dimerization
The *transformation* of a *molecular entity* A to give a molecular entity A_2. For example:
$$H_3C\cdot + \cdot CH_3 \to CH_3CH_3$$
$$2\,CH_3COCH_3 \to (CH_3)_2C(OH)CH_2COCH_3$$
$$2\,RCOOH \to (RCOOH)_2$$
See also *association*.
1994, *66*, 1106

di-π-methane rearrangement
A *photochemical reaction* of a *molecular entity* comprising two π-systems, separated by a saturated carbon atom (a 1,4-diene or an allyl-substituted aromatic analog), to form an ene- (or aryl-) substituted cyclopropane. The rearrangement formally amounts to a 1,2 shift of one ene group (in the diene) or the aryl group (in the allyl-aromatic analog) and 'bond formation' between the lateral carbons of the non-migrating moiety.

See also *oxa-di-π-methane rearrangement*.
1996, *68*, 2236

Dimroth–Reichardt E_T parameter
A measure of the *ionizing power* (loosely *polarity*) of a solvent, based on the maximum wavenumber of the longest wavelength electronic absorption band of:

in a given solvent. E_T, called $E_T(30)$ by its originators, is given by:
$$E_T = 2.859 \times 10^{-3}\nu$$
$$= 2.859 \times 10^4 \lambda^{-1}$$
where E_T is in kcal mol^{-1}, ν is in cm^{-1} and λ is in nm.
The so-called normalized E_T^N scale is defined as:
$$E_T^N = \frac{E_T(\text{solvent}) - E_T(\text{SiMe}_4)}{E_T(\text{water}) - E_T(\text{SiMe}_4)}$$
$$= [E_T(\text{solvent}) - 30.7]/32.4$$
See also *Grunwald–Winstein equation*, *Z-value*.
1994, *66*, 1106

diode laser
A *solid state laser* in which the *active medium* is the p-n junction between p-type and n-type semiconductor host materials. This term is preferred over the sometimes-used term semiconductor laser.
1995, *67*, 1920; see also 1996, *68*, 2236

diols
Compounds that contain two hydroxy groups, generally assumed to be, but not necessarily, alcoholic. *Aliphatic diols* are also called *glycols*.
1995, *67*, 1332

diosphenols
Cyclic α-diketones, which exist predominantly in an enolic form.
1995, *67*, 1332

dioxin
Colloquial (short) name of a toxic by-product (and sometimes contaminant) of herbicides; the full name of this species is: 2,3,7,8-tetrachlorodibenzo[*b*,*e*][1,4]dioxin.
1990, *62*, 2185

dipolar aprotic solvent
A solvent with a comparatively high relative permittivity (or *dielectric constant*), greater than *ca.* 15, and a sizable permanent dipole moment, that cannot donate suitably labile hydrogen atoms to form strong hydrogen bonds, e.g. dimethyl sulfoxide. The term (and its alternative 'polar aprotic solvent') is a misnomer and is therefore discouraged. Such solvents are usually not *aprotic*, but *protophilic* (and at most weakly *protogenic*). In describing a solvent it is better to be explicit about its essential properties, e.g. dipolar and non-protogenic.
1994, *66*, 1106

dipolar bond
A *bond* formed (actually or conceptually) by *coordination* of two neutral moieties, the combination of which results in charge-separated structures, e.g.

$$R_3N: + \ddot{\underset{..}{O}}: \longrightarrow R_3N^+\!\!-\!\!\ddot{\underset{..}{O}}^-$$

The term is preferred to the obsolescent synonyms 'coordinate link', 'co-ordinate covalence', 'dative bond', 'semipolar bond'.
1994, *66*, 1106

dipolar compounds
Electrically neutral molecules carrying a positive and a negative charge in one of their major canonical descriptions. In most dipolar compounds the charges are delocalized; however the term is also applied to species where this is not the case. 1,2-Dipolar compounds have the opposite charges on adjacent atoms. The term 1,3-dipolar compounds is used for those in which a significant canonical resonance form can be represented by a separation of charge over three atoms (in connection with 1,3-dipolar cycloadditions). Subclasses of 1,3-dipolar compounds include:

i. Allyl type
$X=Y^+\!\!-\!Z^- \leftrightarrow {}^-X\!-\!Y^+\!\!=\!Z \leftrightarrow {}^+X\!-\!Y\!-\!Z^- \leftrightarrow {}^-X\!-\!Y\!-\!Z^+$
(X, Z = C, N, or O; Y = N or O)
See *azo imides, azomethine imides, azomethine ylides, azoxy compounds, carbonyl imides, carbonyl oxides, carbonyl ylides, nitrones, nitro compounds*.

ii. Propargyl type
$X\equiv N^+\!\!-\!Z^- \leftrightarrow {}^-X\!=\!N^+\!\!=\!Z \leftrightarrow {}^-X\!=\!N\!-\!Z^+ \leftrightarrow X\!-\!N\!=\!Z$
(X = C or O, Z = C, N, or O)
See *nitrile imides, nitrile oxides, nitrile ylides, nitrilium betaines, azides, diazo compounds*.

iii. Carbene type
$:X\!-\!C\!=\!Z \leftrightarrow {}^+X\!=\!C\!-\!Z^-$ (X = C or N; Z = C, N, or O)
See *acyl carbenes, imidoyl carbenes, vinyl carbenes*.
See *betaines*.
1995, *67*, 1333

dipolar cycloaddition
See *cycloaddition*.
1994, *66*, 1106

dipolar ions
See *zwitterionic compounds*.
1995, *67*, 1333

dipolar mechanism (of energy transfer)
Synonymous with *Förster excitation transfer*.
See also *energy transfer*.
1996, *68*, 2236

dipole–dipole excitation transfer
Synonymous with *Förster excitation transfer*.
See also *energy transfer*.
1996, *68*, 2236

dipole–dipole interaction
Intermolecular or intramolecular interaction between molecules or groups having a permanent electric dipole moment. The strength of the interaction depends on the distance and relative orientation of the dipoles. The term applies also to intramolecular interactions between bonds having permanent dipole moments.
See also *van der Waals forces*.
1994, *66*, 1106

dipole–induced dipole forces
See *van der Waals forces*.
1994, *66*, 1107

dipole length
Electric dipole moment divided by the *elementary charge*.
G.B. 24

dipole moment
See *electric dipole moment*.
G.B. 24

dipole moment per volume
See *dielectric polarization*.
G.B. 14

dipyrrins
Compounds containing two pyrrole rings linked through a methine, –CH=, group (see **I**).
1995, *67*, 1333

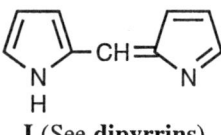

I (See **dipyrrins**)

dipyrromethenes
See *pyrromethenes*.
1995, *67*, 1333

diradicals
Molecular species having two unpaired electrons, in which at least two different electronic states with different multiplicities [electron-paired (singlet state) or electron-unpaired (triplet state)] can be identified, e.g. H$_2$Ċ–CH$_2$ĊH$_2$ propane-1,3-diyl (trimethylene).
See also *biradical*.
1995, *67*, 1333

direct amplification
A type of *amplification reaction*, where the constituent to be determined is amplified directly and it is this constituent which is finally measured.
1982, *54*, 2555

direct current
This term and its symbol, i_{dc} or I_{dc}, should be used (in preference to 'current' alone) only to denote the steady (time-dependent) component of a current that also has a periodic component.
1985, *57*, 1495

direct effect
See *field effect*.
1994, *66*, 1107

direct fission yield
The fraction of *fissions* giving rise to a particular *nuclide* before any *nuclear decay* has occurred.
1982, *54*, 1542

direct-injection burner (for analytical flame spectroscopy)
A burner in which oxidant and fuel emerge from separate ports and are mixed above the burner orifices through their turbulent motion.
O.B. 165

direct-injection enthalpimetry
An analytical method in which a reactant is injected into a calorimetric vessel containing another reagent. The enthalpy change of the ensuing reaction is measured and directly related to the amount of the limiting reagent (usually the analyte).
If the experiment determines information other than amounts of analytes, an acceptable synonym is *batch injection calorimetry*. Other terms which have been used in the literature but are not recommended include the *concentration thermometric technique* and the *direct thermometric method*.
1994, *66*, 2490

direct (radiochemical) isotope dilution analysis
Isotope dilution analysis used for the determination of a non-radioactive element with the aid of one of its *radionuclides*.
1994, *66*, 2521

direct reaction
A chemical process in which the reactive complex has a lifetime that is shorter than its period of rotation. In a *molecular-beam* experiment the reaction products of a direct reaction are scattered, with reference to the centre of mass of the system, in preferred directions rather than at random.
Some direct reactions are impulsive which means that there is an energy exchange that is very fast compared to the vibrational time scale.
A direct reaction is to be contrasted with an *indirect reaction*, also known as a *complex-mode reaction*.
1996, *68*, 163

+ direct thermometric method (in enthalpimetric analysis)
See *direct injection enthalpimetry*.
1994, *66*, 2490

direct transfer gas flow system (in spectrochemical analysis)
A system in which the purge gas transports the analyte directly to the sampling or excitation sources.
1992, *64*, 263

disaccharides
Compounds in which two *monosaccharides* are joined by a glycosidic bond.
See *saccharides, glycosides*.
1995, *67*, 1333

discomfort threshold (in atmospheric chemistry)
The lowest value (e.g. concentration of an impurity, etc.) at which a sensation of discomfort is perceived; a measure which varies from person to person.
1990, *62*, 2185

discontinuous analyser
An analyser with at least one discontinuous subassembly.
1990, *62*, 2174

discontinuous indication (of an analyser)
Indication related to the concentration during intervals of time which are not continuous.
1990, *62*, 2174

discontinuous measuring cell
A measuring cell which operates intermittently and not necessarily at fixed time intervals.
1990, *62*, 2178

discontinuous phase transition
See *first-order phase transition*.
1994, *66*, 581

discontinuous precipitation
A diffusional reaction in a multi-component system in which structural and compositional changes occur in regions immediately adjacent to the advancing interface. The parent phase remains unchanged until swept over by the interface; the transition is complete in regions over which the interface has passed.
Synonymous with *nucleation and growth*.
1994, *66*, 581

discontinuous simultaneous techniques (in analysis)
The application of coupled techniques to the same sample when sampling for the second technique is discontinuous, e.g. discontinuous thermal analysis and gas chromatography, when discrete portions of evolved volatile(s) are collected from the sample situation in the instrument used for the first technique.
O.B. 39

discrete transport
See *transport*.
1989, *61*, 1661

discriminator
A basic function unit comprising an electronic circuit which gives an output pulse for each input pulse whose amplitude lies above a given threshold value.
1982, *54*, 1540

disintegration constant
See *decay constant*.
G.B. 22; see also 1994, *66*, 2518; 1996, *68*, 968

disintegration energy, Q
Energy liberated in a decay process and carried as kinetic energy of the emitted alpha or beta particle and the recoil energy of the product atom.
G.B. 22

disintegration, nuclear
See *nuclear disintegration*.
1982, *54*, 1540

disintegration rate (of a radioactive material)
See *activity (of a radioactive material)*.
1994, *66*, 2518

disjoining pressure (for the attraction between two surfaces)
The force Π_d per unit area which can be obtained as the derivative of $-g_t$ with respect to the distance, where g_t is the total Gibbs energy of interaction per unit area of each of the two flat and parallel plates.
1972, *31*, 616

dismutation
See *disproportionation*.
1994, *66*, 1107

dispenser
A device used to deliver a measured amount of material.
See *sampler*.
1989, *61*, 1661

dispersion (for spectroscopic instruments)
Dispersion of a material = $dn/d\lambda$, where η = refractive index and λ = wavelength; angular dispersion = $d\Phi/d\lambda$, where Φ = angle; and linear dispersion = $dx/d\lambda$, where x = separation of spectral lines.
The reciprocal of the last-named quantity is more frequently used ($d\lambda/dx$), and is commonly expressed in Å/mm.
O.B. 101

dispersion (in atmospheric chemistry)
The dilution of a pollutant by spreading in the atmosphere due to diffusion or turbulent action (eddy diffusion).
1990, *62*, 2185

dispersion, energy (in emission spectrometry)
See *energy dispersion (in emission spectrometry)*.
O.B. 180

dispersion forces
See *London forces*.
1994, *66*, 1107

displacement chromatography
A procedure in which the mobile phase contains a compound (the displacer) more strongly retained than the components of the sample under examination. The sample is fed into the system as a finite slug.
1993, *65*, 824; O.B. 74

displacive transition
A transition in which a displacement of one or more kinds of atoms or ions in a crystal structure changes the lengths and/or directions of bonds, without severing the primary bonds.
Examples:
The transitions of the low-temperature polymorphs of SiO_2 (quartz, tridymite and cristobalite) to their respective high-temperature polymorphs, which involve distortions or rotations of the SiO_4 tetrahedra.
Also *Jahn–Teller* and *ferroic transitions*.
1994, *66*, 581

disproportionation
1. Any chemical reaction of the type $A + A \rightarrow A' + A''$, where A, A' and A'' are different chemical species. For example:

$$2\,ArH^+ \rightarrow ArH + ArH^{2+}$$

The reverse of disproportionation is called *comproportionation*.
A special case of disproportionation (or 'dismutation') is 'radical disproportionation', exemplified by:

$$\cdot CH_2CH_3 + \cdot CH_2CH_3 \rightarrow CH_2=CH_2 + CH_3CH_3$$

Reactions of the more general type:

$$R\dot{C}HCH_3 + R'\dot{C}HCH_3 \rightarrow$$
$$RCH=CH_2 + R'CH_2CH_3$$

are also loosely described as radical disproportionations.
1994, *66*, 1107

The following somewhat more restricted usage of the term prevails in inorganic chemistry.
1994, *66*, 581

2. A *reversible-* or *irreversible transition* in which species with the same oxidation state combine to yield one of higher oxidation state and one of lower oxidation state.
Example: $3Au^+ \rightarrow Au^{3+} + 2Au$
The term also applies to an internal oxidation-reduction process as occurs, for example, among the

iron atoms of $CaFeO_3$, where $2Fe^{4+} \rightarrow Fe^{(4-\delta)+} + Fe^{(4+\delta)+}$, at Fe subarrays on lowering the temperature.
1994, 66, 581

disrotatory
See *electrocyclic reaction*.
1994, 66, 1107

dissociation
1. The separation of a *molecular entity* into two or more molecular entities (or any similar separation within a polyatomic molecular entity). Examples include *unimolecular heterolysis* and *homolysis*, and the separation of the constituents of an *ion pair* into free ions.
2. The separation of the constituents of any aggregate of molecular entities.
In both senses dissociation is the reverse of *association*.
1994, 66, 1107

dissociation energy, E_d, D
Energy required to dissociate a molecule into two parts. Subscripts 0 and e, i.e. D_0 and D_e, are used to denote the dissociation from the ground state and potential energy minimum, respectively.
See also *bond-dissociation energy*.
G.B. 20

dissociative adsorption (dissociative chemisorption)
Adsorption with *dissociation* into two or more fragments, both or all of which are bound to the surface of the *adsorbent*. The process may be homolytic, as in the chemisorption of hydrogen:

$$H_2 + 2^* \rightarrow 2H^*$$

where * represents a surface site, or heterolytic, as in:

$$H_2 + M^{n+} + O^{2-} \rightarrow H^-M^{n+} + HO^-$$

where M^{n+} and O^{2-} are surface sites in which the ions are of lower coordination than the ions in the bulk phase.
Associative desorption is the reverse of dissociative adsorption.
1976, 46, 76, 84

dissociative ionization (in mass spectrometry)
An ionization process in which a gaseous molecule decomposes to form products, one of which is an ion.
1991, 63, 1547

dissociative surface reaction
Dissociative surface reaction and its reverse, associative surface reaction, are defined by:

$$2^* + C_2H_5^* \rightleftharpoons H^* + {}^*CH_2CH_2^*$$

where * represents a surface site.
1976, 46, 84

dissolution
The mixing of two phases with the formation of one new homogeneous phase (i.e. the solution).
1994, 66, 581; O.B. 83

+ dissymmetry (in stereochemistry)
Obsolescent synonym for *chirality*.
1996, 68, 2206

dissymmetry of scattering
The ratio of two Rayleigh ratios for different angles of observation, i.e. $z(\theta_1, \theta_2) = R(\theta_1)/R(\theta_2)$, $\theta_1 < \theta_2$. The angles must be specified; in light scattering it is customary to let $\theta_2 = 180° - \theta_1$ and, most frequently, $\theta_1 = 45°$ and $\theta_2 = 135°$.
P.B. 67; see also 1983, 55, 932

distonic radical cation
A radical cation in which charge and radical sites are separated.
1994, 66, 1107

distribuend
The substance that is distributed between two immiscible liquids or liquid phases.
O.B. 91

distribution
The apportionment of a solute between two phases. The term partition or extraction may also be used in this sense where appropriate.
1993, 65, 2377

distribution coefficient
See *distribution constant, distribution ratio*.
The term is not recommended as a synonym for the latter.
1993, 65, 2381

distribution constant
Synonymous with *partition ratio*.
1993, 65, 2381

distribution constant (in chromatography)
The concentration of a component in or on the stationary phase divided by the concentration of the component in the mobile phase. Since in chromatography a component may be present in more than one form (e.g. associated and dissociated forms), the analytical condition used here refers to the total amount present without regard to the existence of various forms. These terms are also called the distribution coefficients. However, the present term conforms more closely to the general usage in science.
1993, 65, 847; see also O.B. 88, 106; 1996, 68, 969

distribution function (in polymers)
A normalized function giving the relative amount of a portion of a polymeric substance with a specific value, or a range of values, of a random variable or variables.
Notes:
1. Distribution functions may be discrete, i.e. take on only certain specified values of the random variable(s), or continuous, i.e. take on any intermediate value of the random variable(s), in a given range. Most distributions in polymer science are intrinsically discrete, but it is often convenient to regard them as continuous or to use distribution functions that are inherently continuous.
2. Distribution functions may be integral (or cumulative), i.e. give the proportion of the population for which a random variable is less than or equal to a

given value. Alternatively they may be differential distribution functions (or probability density functions), i.e. give the (maybe infinitesimal) proportion of the population for which the random variable(s) is (are) within a (maybe infinitesimal) interval of its (their) range(s).

3. Normalization requires that: (i) for a discrete differential distribution function, the sum of the function values over all possible values of the random variable(s) be unity; (ii) for a continuous differential distribution function, the integral over the entire range of the random variable(s) be unity; (iii) for an integral (cumulative) distribution function, the function value at the upper limit of the random variable(s) be unity.
    P.B. 55

distribution isotherm
The relationship (algebraic or graphical) between the concentration of a solute in an extract and the corresponding concentration of the same solute in the other phase at equilibrium at a specified temperature. Alternative terms in common use are equilibrium line and in the appropriate contexts: extraction isotherm, scrubbing isotherm and stripping isotherm. Partition isotherm is not normal usage.
    1993, *65*, 2377

distribution ratio (in liquid-liquid distribution), D
The ratio of the total analytical concentration of a solute in the extract (regardless of its chemical form) to its total analytical concentration in the other phase.
Notes:
i. If there is possible confusion with the extraction factor or (mass) distribution ratio the term concentration distribution ratio (symbol D_c) should be used, but this is not common usage. This is reasonably compatible with chromatographic nomenclature.
ii. The terms distribution coefficient, extraction coefficient and, where appropriate, scrubbing coefficient, stripping coefficient are widely used alternatives but are not recommended. If they must be used in a given situation the term ratio is preferable to coefficient.
iii. In equations relating to aqueous/organic systems the organic phase concentration is, by convention, the numerator and the aqueous phase concentration the denominator. In the case of stripping ratio the opposite convention is sometimes used but should then be clearly specified.
iv. In the past there has been much confusion between the distribution ratio as defined above, the value of which varies with experimental conditions, e.g. pH, presence of complexing agents, extent of achievement of equilibrium etc. and the true partition coefficient which is by definition invariable or the partition coefficient or distribution constant which apply to a particular chemical species under specified conditions. For this reason the terms distribution constant, partition constant, partition coefficient, partition ratio and extraction constant should not be used in this context.

v. The use of the ratio: light phase concentration to heavy phase concentration is ambiguous and is not recommended.
vi. The distribution ratio is an experimental parameter and its value does not necessarily imply that distribution equilibrium between the phases has been achieved.
    1993, *65*, 2382; 1996, *68*, 970; O.B. 89

disyndiotactic polymer
A *syndiotactic polymer* that contains two chiral or *prochiral* atoms with defined stereochemistry in the main chain of the *configurational base unit*.
    P.B. 32

ditactic polymer
A *tactic polymer* that contains two sites of defined stereoisomerism in the main chain of the *configurational base unit*.
    P.B. 32

diterpenoids
Terpenoids having a C_{20} skeleton.
    1995, *67*, 1333

dithioacetals
    See *thioacetals*.
    1995, *67*, 1333

diurnal variation (in atmospheric chemistry)
Indicates variations which follow a distinctive pattern which recurs with a daily cycle.
    1990, *62*, 2185

dividing surface
A surface, usually taken to be a hyperplane constructed at right angles to the *minimum-energy path* on a *potential-energy surface*. In conventional *transition-state theory* it passes through the highest point on the *minimum-energy path*. In generalized versions of *transition-state theory* the dividing surface can be at other positions; in *variational transition-state theory* the position of the dividing surface is varied so as to get a better estimate of the rate constant.
    1996, *68*, 163

DM-interference
Interference in which the analytical signal is influenced by a different *mechanism* than the analyte e.g. the formation of non-dissociating species in emission spectroscopy.
    1983, *55*, 554

DNA
    See *deoxyribonucleic acids*.
    1992, *64*, 152

DNA probe
A short sequence of DNA labelled isotopically or chemically that is used for the detection of a complementary *nucleotide* sequence.
    1992, *64*, 152

Dobson unit (in atmospheric chemistry)
Unit sometimes used in the description of the total ozone in a column of air overhead. It is given as the thickness (in units of 10^{-3} cm or 10^{-5} m) of that layer which would be formed by the total ozone in a col-

umn reduced to the pressure of 760 Torr and temperature of 0 °C.
1990, *62*, 2185

dodecahedro-
An affix used in names to denote eight atoms bound into a dodecahedron with triangular faces.
R.B. 245; B.B. 464

dolichols
A group of *prenol* derivatives in which n in the general formula H-[CH$_2$-C(CH$_3$)=CH-CH$_2$]$_n$-OH is greater than 4 and in which the residue that carries the hydroxy group is saturated, i.e. 2,3-dihydropolyprenols.
Note:
The collective term prenol should not be used without qualification to include dolichols since these are derivatives of prenols.
W.B. 255

Donnan emf (Donnan potential)
The potential difference E_D at zero electric current between two identical salt bridges, usually saturated KCl bridges (conveniently measured by linking them to two identical electrodes) inserted into two solutions in *Donnan equilibrium*.
1972, *31*, 619

Donnan equilibrium
The equilibrium characterized by an unequal distribution of diffusible ions between two ionic solutions (one or both of the solutions may be gelled) separated by a membrane which is impermeable to at least one of the ionic species present, e.g. because they are too large to pass through the pores of the membrane. The membrane may be replaced by other kinds of restraint, such as gelation, the field of gravity, etc., which prevent some ionic components from moving from one phase to the other, but allow other components to do so.
1972, *31*, 619

Donnan pressure
See *colloid osmotic pressure*.
1972, *31*, 615

donor number (DN)
A quantitative measure of *Lewis basicity*.
1994, *66*, 1107

doping (in catalysis)
The action of adding a small amount of foreign atoms to form a solid solution in the lattice of a non-metallic catalyst.
1991, *63*, 1231

Doppler broadening (of a spectral line)
Broadening due to the random motion of the emitting or absorbing atoms. A Doppler broadened line has a Gaussian shape.
1985, *57*, 1463

Dorn effect
See *sedimentation potential difference*.
1972, *31*, 620

dosage (in atmospheric chemistry)
As applied to an air pollutant in an exposure chamber, dosage is commonly defined as the concentration of the pollutant times the duration of exposure.
1990, *62*, 2185

dose
The energy or amount of photons absorbed per unit area or unit volume by an irradiated object during a particular exposure time.
In medicine and in some other research areas (e.g. photopolymerization and water handling through irradiation) dose is used in the sense of *fluence*, i.e. the energy or amount of photons per unit area or unit volume received by an irradiated object during a particular exposure time. The SI units are J m^{-2} or J m^{-3} and mol m^{-2} or mol m^{-3}, respectively.
For special purposes it must be appropriately qualified, e.g. absorbed, maximum permissible, mean lethal.
See also *UV-dose, absorbed dose, dose equivalent*.
1996, *68*, 2237; O.B. 220; 1996, *68*, 970

dose (in radioanalytical chemistry)
A general term denoting the quantity of *radiation* or energy absorbed. For special purposes, it must be appropriately qualified e.g. absorbed, maximum permissible, mean lethal.
1982, *54*, 1540

dose, absorbed
See *absorbed dose*.
1996, *68*, 959; 1993, *65*, 2007; O.B. 220

dose equivalent, H
Product of the absorbed dose of ionizing radiation, a quality factor and a modifying factor to take into account the effect of radiation on different biological tissues.
G.B. 72; ISO 31-10: 1992; 1996, *68*, 970

double-beam spectrometer (for luminescence spectroscopy)
Double (spectral) beam spectrometers are used where two samples are to be excited by two different wavelengths.
A double- (synchronous) beam spectrometer is a luminescence spectrometer in which both the excitation and emission monochromators scan the excitation and emission spectra simultaneously, usually with a fixed wavelength difference between excitation and emission.
O.B. 191

double escape peak
In a gamma-ray spectrum, the peak due to *pair production* in the detector and escape, from the sensitive part of the detector, of two *photons* of 511 keV resulting from *annihilation*.
1982, *54*, 1541

double-focusing mass spectrograph
An instrument which uses both direction and velocity focusing, and therefore an ion beam initially diverg-

ing in direction and containing ions of different kinetic energies is separated into beams according to the quotient mass/charge, these beams being focused on to a photographic plate or film.
  1991, *63*, 1545

double-focusing mass spectrometer
An instrument which uses both direction and velocity focusing, and therefore an ion beam of a given mass/charge is brought to a focus when the ion beam is initially diverging and contains ions of the same mass and charge with different translational energies. The ion beam is measured electrically.
  1991, *63*, 1545; O.B. 201

double-layer
  See *layer*.
  1994, *66*, 1671

double-layer current
The non-faradaic current associated with the charging of the *electrical double layer* at an electrode-solution interface, given by:
$$i_{DL} = d(\sigma A)/dt$$
where σ = surface charge density of the double layer, A = area of the electrode-solution interface, and t = time.
Capital letters should be used as subscripts to avoid the possibility of confusing this symbol with that for the limiting diffusion current.
  1985, *57*, 1496

double-layer, interfacial
  See *interfacial double-layer*.
  1994, *66*, 1674

double-strand chain (in polymers)
A *chain* that comprises *constitutional units* connected in such a way that adjacent constitutional units are joined to each other through three or four atoms, two on one side and either one or two on the other side of each constitutional unit.
  1996, *68*, 2294

double-strand copolymer
A *copolymer* the *macromolecules* of which are *double-strand macromolecules*.
  1996, *68*, 2304

double-strand macromolecule
A *macromolecule* that comprises *constitutional units* connected in such a way that adjacent constitutional units are joined to each other through three or four atoms, two on one side and either one or two on the other side of each constitutional unit.
  1996, *68*, 2295

double-strand polymer
A *polymer* the *macromolecules* of which are *double-strand macromolecules*.
Notes:
1. A polymer, the macromolecules of which are *spiro macromolecules* is termed a *spiro polymer*.
2. A polymer, the macromolecules of which are *ladder macromolecules* is termed a *ladder polymer*.
  1996, *68*, 2304

doublet state
A state having a total electron spin quantum number equal to $1/2$.
  See *multiplicity*.
  1996, *68*, 2237

double-wavelength spectroscopy
The effect of spectral background due to impurities, solvent or radiation scattering may be reduced if the difference in the absorbances of a sample measured at two selected wavelengths is obtained. This is often achieved by repetitively switching from one wavelength to the other. Double-wavelength spectroscopy does this automatically by allowing two beams of radiation of different wavelengths to pass through the cell. One beam is fixed at a longer wavelength and the other measures absorbance while being scanned over a limited wavelength range at shorter wavelengths.
  1988, *60*, 1455

doubling time (in biotechnology), t_D
The time (min, h) required for a *cell* population to double either the number of cells or the active cell mass.
  1992, *64*, 152

downfield
  See *chemical shift*.
  1994, *66*, 1107

down-time (in analysis)
The loss of time that should be available for analysis. This might be due to breakdown, maintenance or other factors.
  1989, *61*, 1663

downwash (in atmospheric chemistry)
As applied to the action of chimney gases, it is the downward motion of the chimney gases brought on by eddies which form in the lee of a chimney when the wind is blowing. It may result in bringing the flue gases to the ground prematurely.
  1990, *62*, 2185

drift
A slow non-random change in signal with time.
  See also *noise*.
  1989, *61*, 1663

driving force (affinity) of a reaction, A
The decrease in Gibbs energy on going from the reactants to the products of a *chemical reaction* $(-\Delta G)$.
  1994, *66*, 1107; see also 1996, *68*, 2237

droplet (in atmospheric chemistry)
A small liquid particle. The size of droplets encountered in the atmosphere extends over a wide range; e.g. liquid aerosol solutions which make up the fine particle fraction of continental tropospheric aerosol are usually 2 μm in diameter. Cloud water droplets usually have diameters in the range of 5–70 μm, while rain droplets commonly have diameters ranging from 0.1–3 mm.
  1990, *62*, 2185

drop time (in polarography)
The time that elapses between the instants at which two successive drops of liquid metal are detached from the tip of the capillary.
1985, *57*, 1500

drug
Any substance which when absorbed into a living organism may modify one or more of its functions. The term is generally accepted for a substance taken for a therapeutic purpose, but is also commonly used for abused substances.
Synonymous with medicine, pharmaceutical.
1993, *65*, 2040

dry bulb temperature (in atmospheric chemistry)
In *psychrometry*, the temperature of the gas measured by a dry bulb thermometer.
1990, *62*, 2185

drying agent
A substance which removes water (liquid or gaseous).
1990, *62*, 2185

dual substituent-parameter equation
In a general sense, any equation which expresses substituent effects in terms of two parameters. However, in practice the term is used more specifically for an equation for summarizing the effects of *meta*- or *para*-substituents ($i = m$ or p) X on chemical reactivity, spectroscopic properties, etc. of a probe site Y in benzene or other aromatic system.

$$P^i = \rho_I^i \sigma_I + \rho_R^i \sigma_R$$

P is the magnitude of the property Y for substituent X, expressed relative to the property for X = H; σ_I and σ_R are inductive or polar and resonance substituent constants, respectively, there being various scales for σ_R; ρ_I and ρ_R are the corresponding regression coefficients.
See also *extended Hammett equation*.
1994, *66*, 1107

dust (in atmospheric chemistry)
Small, dry, solid particles projected into the air by natural forces, such as wind, volcanic eruption and by mechanical or manmade processes such as crushing, grinding, milling, drilling, demolition, shoveling, conveying, screening, bagging and sweeping. Dust particles are usually in the size range from about 1–100 µm in diameter and they settle slowly under the influence of gravity.
1990, *62*, 2185

dust collector (in atmospheric chemistry)
A device for monitoring dust emissions. Also the equipment used to remove and collect dust from process exhaust gases; this may employ simply sedimentation (dustfall jars, coated slides, papers, etc.), inertial separation (cyclones, impactors, impingers, etc.), precipitation (thermal and electrostatic) or filtration.
1990, *62*, 2185

dust fall (in atmospheric chemistry)
Solid particles in the air which fall to the ground under the influence of gravity.
1990, *62*, 2185

dye laser
A CW or pulsed source of coherent radiation in which the active medium is usually a solution of a fluorescent organic molecule (the dye) pumped with a suitable pump laser or with a flash lamp. These lasers can be tuned over a large part of the fluorescence band of the dye.
1996, *68*, 2237

dynamic field(s) mass spectrometer
A mass spectrometer in which the separation of a selected ion beam depends essentially on the use of fields, or a field, varying with time. These fields are generally electric.
1991, *63*, 1545; O.B. 202

dynamic interfacial tension
See *dynamic surface tension*.
1972, *31*, 597

dynamic quenching
See *quenching*.
1996, *68*, 2237

dynamic range (of an analyser)
The ratio between the maximum usable indication and the minimum usable indication (detection limit). A distinction may be made between the linear dynamic range, where the response is directly proportional to concentration, and the dynamic range where the response may be non-linear, especially at higher concentrations.
1990, *62*, 2185

dynamic surface tension
In certain circumstances, one may measure surface tensions that are different from the equilibrium value. Such a surface tension is called the dynamic surface (or interfacial) tension (γ^{dyn} or σ^{dyn}). The equilibrium value is then called the static surface (or interfacial) tension (γ^{st} or σ^{st}). The modifying signs may be omitted if there is no danger of ambiguity.
1972, *31*, 597

dynamic thermomechanometry
A technique in which the dynamic modulus and/or damping of a substance under oscillatory load is measured as a function of temperature while the substance is subjected to a controlled temperature program.
O.B. 45

dynamic viscosity, η
For a laminar flow of a fluid the ratio of the *shear stress* to the velocity gradient perpendicular to the plane of shear.
See also *viscosity*.
G.B. 13

dyne
cgs unit of force, dyn = 10^{-5} N.
G.B. 112

dyotropic rearrangement
An uncatalysed process in which two σ-bonds simultaneously migrate intramolecularly, e.g.

[structure: β-lactone with CH₃CH₂ and R substituents → γ-butyrolactone with CH₃ and R substituents, via MgBr₂]

1994, *66*, 1108

dypnones
1,3-Diphenylbut-2-en-1-one, PhC(=O)CH=C(CH$_3$)Ph, and its ring-substituted derivatives.
1995, *67*, 1333

η- (eta or hapto) (in inorganic nomenclature)
An affix giving a topological indication of the bonding between a π-electron ligand and the central atom in a coordination entity. A right superscript numerical index indicates the number of ligating atoms in the π-electron system of the ligand which bind to the central atom.
Examples:
i. [Fe(CO)$_3$(C$_4$H$_6$SO)], tricarbonyl(η^2-dihydrothiophene 1-oxide-κO)iron.
ii. [Cr(CO)$_4$(C$_4$H$_6$)], tetracarbonyl(η^4-2-methylene-1,3-propanediyl)chromium.
 R.B. 202; see also B.B. 465; 1994, *66*, 1120

E, Z
The approved *stereodescriptors* of stereoisomeric alkenes R^{1}R^{2}C=CR^{3}R^4 (R^1 ≠ R^2, R^3 ≠ R^4; neither R^1 nor R^2 need be different from R^3 or R^4), cumulenes R^{1}R^{2}C[=C=C]$_n$=CR^{3}R^4 and related systems e.g. R^{1}R^{2}C=NOH, HON=C{[CH$_2$]$_n$}$_2$C=NOH. The group of highest *CIP priority* attached to one of the terminal doubly bonded atoms of the alkene, oxime, *etc.* or cumulene (i.e. R^1 or R^2) is compared with the group of highest precedence attached to the other (i.e. R^3 or R^4). The stereoisomer is designated as *Z* (*zusammen* = together) if the groups lie on the same side of a reference plane passing through the double bond and perpendicular to the plane containing the bonds linking the groups to the double-bonded atoms; the other stereoisomer is designated as *E* (*entgegen* = opposite). The descriptors may be applied to structures with a fractional bond order between one and two; and to double bonds involving elements other than carbon. They are not used to describe ring substitution relationships.
 See also *cis-trans isomers*.
 1996, *68*, 2206

early-downhill surface
 See *attractive potential-energy surface*.
 1996, *68*, 163

eclipsed (eclipsing)
Two atoms or groups attached to adjacent atoms are said to be eclipsed if the *torsion angle* between the three bonds is zero (or approximately so).
 See *bisecting conformation, eclipsing strain*.
 1996, *68*, 2206

eclipsing conformation
 See *bisecting conformation*.
 1996, *68*, 2206

eclipsing strain
Intramolecular strain present due to *non-bonding interactions* between two *eclipsed* atoms or groups. It is, for example, one of the interactions responsible for the existence of a *rotational barrier* and hence *restricted rotation* about the C–C bond in ethane. It has also been called *Pitzer strain, torsional strain* and *bond opposition strain*.
 1996, *68*, 2206

ectohormone
 See *pheromone*.
 1993, *65*, 2083

eddy
In turbulent fluid motion, a blob of the fluid that has some definitive character and moves in some way differently from the main flow.
 1990, *62*, 2186

eddy dispersion (diffusion)
The process by which substances are mixed in the atmosphere or in any fluid system due to eddy motion.
 1990, *62*, 2186

+ educt
Term used mainly in the German literature for starting material (reactant). It should be avoided in English, because there it means 'something that comes out' and not 'something that goes in'. The German use of the term is in fact also incorrect.
 1994, *66*, 1108

effective cadmium cut-off energy (in nuclear analytical chemistry)
In a given experimental configuration, the *energy* value determined by the condition that the *detector* response would be unchanged if the cadmium cover surrounding the detector was replaced by a fictitious cover opaque to neutrons with energy below this value and transparent to neutrons with energy above this value.
 1982, *54*, 1540; O.B. 221

effective charge
Change in effective charge is a quantity obtained by comparison of the *polar effect* of substituents on the free energies of rate or equilibrium processes with that on a standard ionization equilibrium. Provided the effective charges on the states in the standard equilibrium are defined, then it is possible to measure effective charges for states in the reaction or equilibrium under consideration.
 1994, *66*, 1108

effective chimney height (stack height)
A height used for the purposes of calculating the dispersion of emitted gases from a chimney, and which differs from the real chimney height by an amount which depends on such factors as the exit velocity, buoyancy effects and wind speed; it may also be affected by the local topography. It denotes the maximum height of the centre of a plume path above the level of the ground. The effective height may be above or below the actual chimney height, although the former case is most common.
 1990, *62*, 2179

effective half life
For a *radioactive* substance, the time required for the amount of that substance in a biological system to be reduced to one half of its value by both *radioactive*

decay and biological processes, when the rate of removal is approximately exponential.
1994, *66*, 2520

effectively infinite thickness (in flame spectroscopy)
Solid or liquid samples of one gram or more are generally used in X-ray spectroscopy. If the sample is thick enough so that the intensity of characteristic fluorescent radiation is not significantly modified by an increase in thickness, the sample is described as a bulk sample of effectively infinite thickness.
O.B. 182

effective molarity (effective concentration)
The ratio of the first-order rate constant of an *intramolecular* reaction involving two functional groups within the same *molecular entity* to the second-order *rate constant* of an analogous *intermolecular elementary reaction*. This ratio has the dimension of concentration. The term can also apply to an equilibrium constant.
See also *intramolecular catalysis*.
1994, *66*, 1108

effectiveness
See *spectral (photon) effectiveness*.
1996, *68*, 2237

effective theoretical plate number (of a chromatographic column)
A number indicative of *column performance* when *resolution* is taken into account:
$$N = 16R_s^2/(1-\alpha)^2$$
where R_s is the peak resolution, and α is the *separation factor*.
See also *plate number (in chromatography)*.
O.B. 108

effective thermal cross-section
A fictitious *cross-section* for a specified reaction, which, when multiplied by the 2200 metre-per-second *flux density*, gives the correct reaction rate for thermal neutrons.
See also *microscopic cross-section*.
1994, *66*, 2517; O.B. 218

effective thickness of diffusion layer
See *Nernst's diffusion layer*.
1981, *53*, 1837

effector
A small molecule which increases (activator) or decreases (inhibitor) the activity of an (allosteric) protein by binding to the protein at the regulatory site (which is different from the substrate-binding catalytic site).
1992, *64*, 152

efficiency (of a step), η
The ratio between the useful energy delivered or bound and the energy supplied, i.e. energy output/energy input. It is also used in the sense of a quantitative measure of the relative rate of a given step involving a species with respect to the sum of the rates of all of the parallel steps which depopulate that species.
See also *quantum yield*.
1996, *68*, 2238

efficiency (of a counter)
See *counting efficiency*.
1994, *66*, 2518

efficiency, intrinsic
See *detector efficiency*.
1994, *66*, 2518

efficiency of atomization, local (in flame emission and absorption spectrometry)
See *local efficiency of atomization (in flame emission and absorption spectrometry)*.
1986, *58*, 1741; O.B. 169

efficiency of atomization (in analytical flame spectroscopy)
The ratio of the amount of analyte that passes through the flame cross-section at the observation height, as free neutral (or ionized) atoms, to the amount of analyte aspirated.
O.B. 124

efficiency of nebulization (in flame emission and absorption spectrometry), ε_n
The substance fraction of component entering the flame in the amount of component consumed.
The efficiency of nebulization is related to the amount of component and not to the amount of solvent. It cannot be calculated directly from the ratio: volume rate of sample drained from the spray chamber divided by volume rate of fluid consumed. Corrections usually have to be made to take account of differences in component concentrations in the drained and consumed solutions, respectively, because of evaporation of some solvent from mist droplets deposited on the walls.
1986, *58*, 1740; O.B. 168

efficiency spectrum
A plot of the efficiency of a step (η) against wavelength or photon energy.
See *action spectrum, conversion spectrum*.
See also *spectral effectiveness*.
1996, *68*, 2238

efflorescence
The reverse of deliquescence: the drying of a salt solution when the vapour pressure of water in the saturated solution of a substance is greater than the partial pressure of water in the ambient air. Also refers to the loss of water of crystallization from a solid salt such as $Na_2CO_3 \cdot 10H_2O$.
1990, *62*, 2186

effluent
Any spent liquors or other waste materials which are emitted by a source (waste from plating shops, pickling tanks, sewage treatment plants, chemical manufacturing plants, etc.).
1990, *62*, 2186

effluent (in chromatography)
The mobile phase leaving the column.
1993, *65*, 824

eicosanoids
See *icosanoids*.
1995, *67*, 1333

eighteen-electron rule
An electron-counting rule to which an overwhelming majority of stable diamagnetic transition metal complexes adhere. The number of nonbonded electrons at the metal plus the number of electrons in the metal-ligand bonds should be 18. The 18 electron rule in transition metal chemistry is a full analogue of the 'Lewis octet rule'.
1994, *66*, 1108

+ einstein
One mole of photons. Although widely used, it is not an IUPAC sanctioned unit. It is sometimes defined as the energy of one mole of photons. This use is discouraged.
1996, *68*, 2238; see also 1990, *62*, 2186

elastic collision
A particle (atom or molecule) can undergo a change in its state of excitation as a result of collisions with other particles. In an elastic collision an exchange only of kinetic energy takes place between the colliding species; in an inelastic collision there is an interchange between the kinetic energy and the internal energy of the particle.
1996, *68*, 159

elastic scattering (in reaction dynamics)
When in a molecular collision there is no transfer of energy among different degrees of freedom, there is said to be elastic scattering.
1996, *68*, 163; 1982, *54*, 1540

elastic scattering (in spectrochemistry)
Radiation may be scattered by its transmission through a medium containing particles. If the scatter results in no significant change in the wavelength relative to the primary radiation it is called elastic scattering. In cases where the scattering centres are small compared to the wavelength of the radiation the elastic scattering is called *Rayleigh scattering* and *Mie scattering* if this condition is not fulfilled.
1985, *57*, 1464

electrical arc
A self-sustaining electrical discharge between at least two electrodes, characterized by a comparatively small cathode fall voltage, a low burning voltage and a relatively high current density. The burning voltage of an arc is the voltage across the electrode gap during an arc discharge.
O.B. 123

electrical double-layer
See *interfacial double-layer*.
1994, *66*, 1674

electrical double layer, thickness of
See *thickness of electrical double layer*.
1972, *31*, 619

electrical hygrometer
A hygrometer whose sensitive element has electrical properties which vary with the humidity of the gas which traverses the hygrometer.
1990, *62*, 2195

electric capacitance
See *capacitance*.
1996, *68*, 970

electric charge, Q
Integral of the *electric current* over time. The smallest electric charge found on its own is the *elementary charge*, e, the charge of a proton.
G.B. 14; 1996, *68*, 970

electric coherence length (in thin films), ξ_E
See *coherence length*.
1994, *66*, 1675

electric conductance, G
Reciprocal of the *electric resistance*.
G.B. 15; 1996, *68*, 970

electric conductivity
See *conductivity*.
G.B. 15; 1996, *68*, 970

electric current, I
Base quantity in the system of quantities upon which *SI* is based.
G.B. 14; 1996, *68*, 970

electric current density, j
A vector quantity the scalar product of which with the cross sectional area vector is equal to the *electric current*. By magnitude it is the electric current divided by the cross sectional area.
G.B. 14; 1996, *68*, 971

electric dipole moment, p
Vector quantity, the vector product of which with the *electric field strength*, E, of a homogeneous field is equal to the *torque*. $T = p \times E$. The direction of the dipole moment is from the negative to the positive charge.
G.B. 24

electric displacement, D
Product of the *permittivity* and electric *field strength*.
G.B. 14

electric field (strength), E
Force acting on a charge divided by the charge.
G.B. 14; 1996, *68*, 971

electric mobility, u, μ
Speed of ions divided by the electric field strength in the phase concerned.
See also *mobility*.
G.B. 60; 1996, *68*, 982; see also 1996, *68*, 971

electric polarizability, α
Tensor quantity relating the induced *electric dipole moment*, p_i, to the applied *electric field strength*, E, $p_i = \alpha E$.
G.B. 22

electric potential difference (of a galvanic cell), ΔV
Difference in the potentials of electrodes on the right and left of a galvanic cell. When ΔV is positive, positive charge flows from left to right through the cell.
1996, 68, 971

electric potential, V
At a point, the work required to bring a charge from infinity to that point in the electric field divided by the charge.
G.B. 14; ISO 31-4: 1992

electric resistance, R
1. To direct current, the potential difference divided by the current when there is no *electromotive force* in the conductor.
2. To alternating current, the real part of *impedance*.
G.B. 15; ISO 31-4: 1992; *1996, 68, 971*

electric resistivity
See *resistivity*.
G.B. 15; *1996, 68, 971*

electric sector (in mass spectrometry)
An arrangement of two conducting sheets forming a capacitor and giving a radial electrostatic field which is used to deflect and focus ion beams of different energies. The capacitor may, for example, be cylindrical, spherical or toroidal.
1991, 63, 1544

electrified interphase
An *interphase* between phases containing free charged components which are usually accumulated or depleted in the surface region thus giving rise to net charges on the phase. This definition includes the special case when the net charge on each of the phases reduces to zero.
Charged components may or may not cross the interface between two phases, so that interphases may be divided into the limiting types unpolarizable and polarizable, respectively. *Ideally unpolarizable interphases* are those for which the exchange of common charged particles between the phases proceeds unhindered. *Ideally polarizable interphases* are those for which there are no common components between the phases or the exchange of these is hindered. This condition may arise as a result of the equilibrium conditions or from the kinetics of charge transfer and leads to an interphase which is impermeable to electric charge. *Real interphases* may approach more or less well one of these idealized cases. Polarisibility or non-polarisibility is not an absolute property of an interphase but depends on a number of conditions, e.g. time scale of the experiment.
1986, 58, 439

electroactive substance
In *voltammetry* and related techniques, a substance that undergoes a change of oxidation state, or the breaking or formation of chemical bonds, in a *charge-transfer step*.
If an electroactive substance B is formed, in the solution or electrode, by a *chemical reaction* from another substance C, the substance C should be called the precursor of B.
In *potentiometry* with *ion-selective electrodes*, a material containing, or in ion-exchange equilibrium with, the sensed ion. The electroactive substance is often incorporated in an inert matrix such as poly(vinyl chloride) or silicone rubber.
1985, 57, 1500

electrocapillarity
The dependence of the *interfacial tension* on the electrical state of the *interphase*.
1986, 58, 446

electrocapillary equation
A form of the Gibbs adsorption equation which includes an expression of the phenomenon of *electrocapillarity*:

$$sdT - \tau dp + d\gamma + \sigma^\alpha dE + \sum \Gamma_j d\mu_j = 0$$

where s is the surface excess of entropy of unit area of *interphase*, T is the temperature, τ is the thickness or excess volume of unit area of the interphase, p is the external pressure, γ is the interfacial tension, σ^α is the free surface charge density on phase α (areal amount of charge on the surface of phase α), E is the generalized potential, Γ_j is the surface excess, μ_j is the chemical potential and j is an electrically neutral component of one or other of the phases; the sum is over all the components but one in each phase.
1986, 58, 446

electrochemical detector (in gas chromatography)
An electrochemical cell which responds to certain substances in the carrier gas eluting from the column. The electrochemical process may be an oxidation, reduction or a change in conductivity.
1990, 62, 2191

electrochemical method of detection (in analysis)
A method in which either current or potential is measured during an electrochemical reaction. The gas or liquid containing the trace impurity to be analysed is sent through an electrochemical cell containing a liquid or solid electrolyte and in which an electrochemical reaction specific to the impurity takes place.
See also *amperometric detection method, coulometric detection method, potentiometric detection method*.
1990, 62, 2186

electrochemical potential
Of a substance in a specified phase, the *partial molar Gibbs energy* of the substance at the specified electric potential.
G.B. 59

electrochemiluminescence
See *electrogenerated chemiluminescence*.
1996, 68, 2238

electrochromic effect
See *Stark effect*.
1996, *68*, 2238

electrocyclic reaction
A *molecular rearrangement* that involves the formation of a σ-bond between the termini of a fully *conjugated* linear π-electron system (or a linear fragment of a π-electron system) and a decrease by one in the number of π-bonds, or the reverse of that process. For example:

The stereochemistry of such a process is termed 'conrotatory' or *antarafacial* if the substituents at the interacting termini of the conjugated system both rotate in the same sense, e.g.

or 'disrotatory' (or *suprafacial*) if one terminus rotates in a clockwise and the other in a counter-clockwise sense, e.g.

See also *pericyclic reaction*.
1994, *66*, 1108

electrode, auxiliary
See *auxiliary electrode*.
1985, *57*, 1500

electrodecantation (or electrophoresis convection)
Electrodialysis may lead to local differences in concentration and density. Under the influence of gravity these density differences lead to large scale separation of *sols* of high and of low (often vanishingly low) concentrations. This process is called electrodecantation.
1972, *31*, 609

electrode, counter
See *auxiliary electrode*.
O.B. 59

electrode current density, j
If the charging current is negligible, in the case of a single *electrode reaction*, the electrode current density (c.d.) of the *electric current* flowing through the electrode is related to the *flux density* of a species B by the equation:

$$j = n\nu_B^{-1} F(N_B)_e$$

where $(N_B)_e$ is the normal component of the vector N_B at the electrode-solution interface, n is the charge number of the electrode reaction, ν_B the *stoichiometric number* of species B. The ratio n/ν_B is to be taken as positive if the species B is consumed in a cathodic reaction or produced in an anodic reaction. Otherwise it is to be taken as negative. With the convention that the normal distance vector points into the electrolytic solution, a *cathodic current* is then negative, an *anodic current* positive.
1981, *53*, 1835

electrode, indicator
See *indicator electrode*.
1985, *57*, 1500

electrode memory
See *hysteresis*.
1994, *66*, 2530

electrodeposition
The deposition of dissolved or suspended material by an electric field on an electrode. (Includes electrocrystallization).
1972, *31*, 620

electrode potential, E
Electromotive force of a cell in which the electrode on the left is a standard hydrogen electrode and the electrode on the right is the electrode in question.
G.B. 61; 1996, *68*, 971

electrode potential, absolute
See *absolute electrode potential*.
1986, *58*, 957

electrode potential, relative
See *relative electrode potential*.
1985, *57*, 1501

electrode process
The totality of changes occurring at or near a single electrode during the passage of current.
1980, *52*, 235

electrode reaction
For an *electrode process* at a given potential the current is controlled by the kinetics of a number of steps which include the transport of reactants to and from the interface and the interfacial reaction itself. The latter, which is called the electrode reaction, must always include at least one elementary step in which charge is transferred from one phase to the other, but may also involve purely chemical steps within the *interfacial region*.
1980, *52*, 235

electrode reaction rate constants
The electrode reaction rate constants are related to the partial currents by

$$k_{ox} = I_a / nFA\Pi c_i^{\nu_i}$$
$$k_{red} = I_c / nFA\Pi c_i^{\nu_i}$$

where k_{ox} and k_{red} are the rate constants for the oxidizing (anodic) and reducing (cathodic) reactions respectively, n is the charge number of the cell reaction, F is the *Faraday constant*, A is the geometric area of the electrode, the product $\Pi c_i^{\nu_i}$ includes all the species i which take part in the partial reaction, c_i is the volume concentration of species i and v_i is the order of the reaction with respect to species i.

The conditional rate constant of an electrode reaction is the value of the electrode reaction rate constant at the *conditional (formal) potential* of the *electrode reaction*. When α the transfer coefficient is independent of potential,

$$k_c = k_{ox}/\exp[\alpha_a(E-E_c^{0\prime})\,nF/vRT]$$
$$= k_{red}/\exp[-\alpha_c(E-E_c^{0\prime})\,nF/vRT]$$

where α_a and α_c are the anodic and cathodic transfer coefficients respectively, E is the electric potential difference, $E_c^{0\prime}$ is the conditional (formal) potential, v is the *stoichiometric number*, R is the gas constant and T is the thermodynamic temperature.

Similar rate constants can be defined using activities in place of concentrations in the first two equations, and the standard electrode potential in place of the conditional potential in the latter two equations. This type of rate constant is called the standard rate constant of the electrode reaction.

The observable electrode rate constant is the constant of proportionality expressing the dependence of the rate of the electrode reaction on the *interfacial concentration* of the chemical species involved in the reaction.

1974, *37*, 515; 1980, *52*, 236

electrode, working
See *working electrode*.
O.B. 60

electrodialysis
Dialysis conducted in the presence of an electric field across the membrane(s).
1972, *31*, 609

electro-endosmosis
See *electro-osmosis*.
1972, *31*, 620

electrofuge
A *leaving group* that does not carry away the bonding electron pair. For example, in the nitration of benzene by NO_2^+, H^+ is the electrofuge. The adjective of electrofuge is electrofugal.
See also *electrophile, nucleofuge*.
1994, *66*, 1109

electrogenerated chemiluminescence (ECL)
Luminescence produced by electrode reactions. Also called electroluminescence or electrochemiluminescence.
1996, *68*, 2238

electrographite
A *synthetic graphite* made by electrical heating of *graphitizable carbon*.
1995, *67*, 488

electrokinetic potential, ζ
Potential drop across the mobile part of the double layer, that is responsible for electrokinetic phenomena. ζ is positive if the potential increases from the bulk of the liquid phase towards the interface. In calculating the electrokinetic potential from electrokinetic phenomena it is often asumed that the liquid adhering to the solid wall and the mobile liquid are separated by a sharp shear plane. As long as there is no reliable information on the values of the permittivity and the viscosity in the electrical double layer close to the interface, the calculation of the electrokinetic potential from electrokinetic experiments remains open to criticism. It is therefore essential to indicate in all cases which equations have been used in the calculation of ζ. It can be shown, however, that for the same assumptions about the permittivity and viscosity all electrokinetic phenomena must give the same value for the electrokinetic potential.
G.B. 60; 1996, *68*, 971; 1994, *66*, 894

electroluminescence
See *electrogenerated chemiluminescence*.
1996, *68*, 2238

electrolyte, base
See *base electrolyte*.
1985, *57*, 1501

electrolyte, indifferent
See *indifferent electrolyte*.
1985, *57*, 1501

electrolytic hygrometer
Hygrometer using a hygroscopic substance (for example, diphosphorus pentaoxide, P_2O_5) which is transformed into an electrolyte (phosphoric acid, H_3PO_4) in contact with the moisture in the gas. The electrolyte (phosphoric acid) is electrolysed continuously and the electrolysis current is measured. At a constant flow of the gas to be analysed, the electrolysis current is a linear function of the water concentration.
1990, *62*, 2195

electromeric effect
A molecular polarizability effect occurring by an intramolecular electron displacement (sometimes called the 'conjugative mechanism' and, previously, the 'tautomeric mechanism') characterized by the substitution of one electron pair for another within the same atomic octet of electrons. It can be indicated by curved arrows symbolizing the displacement of electron pairs, as in:

$$R_2N\!-\!C=\!C\!-\!C=\!O\!:$$

which represents the hypothetical electron shift

$$R_2N\!-\!C=\!C\!-\!C=\!O \quad \text{-----} \rightarrow \quad R_2N^+\!=\!C\!-\!C=\!C\!-\!O^-$$

The term has been deemed obsolescent or even obsolete (see *mesomeric effect, resonance effect*). It has long been custom to use phrases such as 'enhanced substituent resonance effect' which imply the operation of the electromeric effect, without using the term, and various modern theoretical treatments parametrize the response of substituents to 'electronic de-

mand', which amounts to considering the electromeric effect together with the *inductomeric effect*.
1994, *66*, 1109

electromotive force, *E*
Energy supplied by a source divided by the electric charge transported through the source. For a galvanic cell it is equal to the electric potential difference for zero current through the cell.
G.B. 58; ISO 31-4: 1992; 1996, *68*, 971

electron
Elementary particle not affected by the strong force having a spin quantum number $\frac{1}{2}$, a negative elementary charge and a rest mass of 0.000 548 579 903(13) u.
G.B. 93; see also 1982, *54*, 1540

electron acceptor
1. A substance to which an electron may be transferred; for example 1,4-dinitrobenzene or the dication 1,1′-dimethyl-4,4′-bipyridyldiium.
+ 2. A *Lewis acid*. This usage is discouraged.
1994, *66*, 1109

electron affinity, E_{ea}
Energy required to detach an electron from the singly charged negative ion (energy for the process $X^- \to X + e$). The equivalent more common definition is the energy released ($E_{initial} - E_{final}$) when an additional electron is attached to a neutral atom or molecule.
G.B. 20; 1994, *66*, 1109; 1991, *63*, 1546

electronation
See *electron attachment, reduction*.
1994, *66*, 1111

electron attachment
The transfer of an electron to a *molecular entity*, resulting in a molecular entity of (algebraically) increased negative charge. It is not an *attachment*.

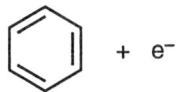 + e⁻

See also *oxidation (1), reduction*.
1994, *66*, 1109; 1991, *63*, 1547

electron capture
A *nuclear transformation* in which the nucleus captures an orbital *electron*. Often the shell, from which the electron is captured, is indicated. (K-, L-, etc.).
1982, *54*, 1537

electron capture detector (in gas chromatography)
A small radioactive source containing ^3H or ^{63}Ni ionizes the molecules of the carrier gas (nitrogen or argon–methane), and a potential difference creates a small current. This current is reduced when an electronegative substance (such as a halocarbon) is introduced. The reduction in current is a measure of the concentration of the electronegative substance. The detection limit (threshold) varies greatly according to the substances to be analysed and can reach a mixing ratio of 10^{-12}. The linear dynamic range may be 10^4 but the maximum measuring value generally lies below 1 ppmv.
1990, *62*, 2191

electron charge
The negative electric charge which appears on the electron or univalent ions [1.602×10^{-19} coulombs or 4.803×10^{-10} electrostatic units (esu)].
1990, *62*, 2186

electron configuration
See *configuration (electronic)*.
1994, *66*, 1110

electron correlation
The adjustment of electron motion to the instantaneous (as opposed to time-averaged) positions of all the electrons in a molecular entity.
See also *correlation energy*.
1996, *68*, 2238

electron-deficient bond
A single bond between adjacent atoms that is formed by less than two electrons, as in B_2H_6:

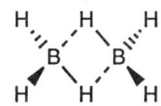

The B–H–B bonds are called a 'two-electron three-centre bonds'.
1994, *66*, 1110

electron density
If $P(x,y,z)$ dx dy dz is the probability of finding an electron in the volume element dx dy dz at the point of a molecular entity with coordinates x,y,z, then $P(x,y,z)$ is the electron density at this point. For many purposes (e.g. X-ray scattering, forces on atoms) the system behaves exactly as if the electrons were spread out into a continuously distributed charge. The term has frequently been wrongly applied to negative *charge population*.
See also *charge density*.
1994, *66*, 1110

electron detachment
The reverse of an *electron attachment*.
1994, *66*, 1110

electron donor
1. A *molecular entity* that can transfer an electron to another molecular entity, or to the corresponding chemical species.
+ 2. A *Lewis base*. This use is discouraged.
1994, *66*, 1110

+ electron-donor–acceptor complex
A term sometimes employed instead of *charge-transfer complex* or *Lewis adduct*.
See also *adduct, coordination*.
1994, *66*, 1110

electronegativity
Concept introduced by L. Pauling as the power of an atom to attract electrons to itself. There are several definitions of this quantity. According to Mulliken it is the average of the ionization energy and electron affinity of an atom, but more frequently a relative

scale due to Pauling is used where dimensionless relative electronegativity differences are defined on the basis of bond dissociation energies, E_d, expressed in electronvolts:

$$\chi_r(A) - \chi_r(B) = (eV)^{-1/2} \sqrt{E_d(AB) - 1/2[E_d(AA) + E_d(BB)]}$$

The scale is chosen so as to make the relative electronegativity of hydrogen $\chi_r = 2.1$. The sign of the square root was chosen intuitively by Pauling.

G.B. 20; 1994, *66*, 1111

electron energy (in mass spectrometry)
The potential difference through which electrons are accelerated before they are used to bring about electron ionization.

1991, *63*, 1547

electroneutrality principle
The principle expresses the fact that all pure substances carry a net charge of zero.

1994, *66*, 1111

electron exchange excitation transfer
Synonymous with *Dexter excitation transfer*.
See also *energy transfer*.

1996, *68*, 2238

electronically excited state
A state of an atom or molecular entity which has greater electronic energy than the ground state of the same entity.

1996, *68*, 2238

electronic configuration
See *configuration (electronic)*.

1996, *68*, 2238

electronic effect of substituents: symbols and signs
The *inductive effect* has universally been represented by the symbol I. This is now commonly taken to include both through-bonds and through-space transmission, but I is also used specifically for through-bonds transmission; through-space transmission is then symbolized as F (for *field effect*). The symbols for the influence of substituents exerted through electron delocalization have variously been M (*mesomeric*), E (*electromeric*), T (tautomeric), C (conjugative), K (konjugativ) and R (*resonance*).

Since the present fashion is to use the term *resonance effect*, R is the most commonly used symbol, although M is still seen quite often.

Both the possible sign conventions are in use. The Ingold sign convention associates *electronegativity* (relative to hydrogen atom) with a negative sign, electropositivity with a positive sign. Thus the nitro group is described as electronwithdrawing by virtue of its –I and –M effects; chloro is described as a –I, +M substituent, etc. For *correlation analysis* and *linear free-energy relationships* this convention has been found inconvenient, for it is in contradiction to the sign convention for polar substituent constants (σ-constants). Authors concerned with these fields often avoid this contradiction by adopting the opposite sign convention originally associated with Robinson, for electronic effects. This practice is almost always associated with the use of R for the electron delocalization effect: thus the nitro group is a +I, +R substituent; chloro a +I, –R substituent, etc.

1994, *66*, 1111

electronic energy migration (or hopping)
The movement of electronic excitation energy from one molecular entity to another of the same species, or from one part of a molecular entity to another of the same kind (e.g. excitation migration between the chromophores of an aromatic polymer). The migration can happen via radiative or radiationless processes.

1996, *68*, 2238

+ electron impact ionization (in mass spectrometry)
See *electron ionization*.

1991, *63*, 1547

electron ionization (in mass spectrometry)
This is the term used to describe ionization of any species by electrons. The process may, for example, be written:

$M + e \rightarrow M^{+\cdot} + 2e$ for atoms or molecules.

$M\cdot + e \rightarrow M^+ + 2e$ for radicals.

The term 'electron impact' should not be used.
See *radical ion* for a note concerning the symbolic representation of radical ions; see also *photoionization*.

1991, *63*, 1547

electron jump
The term can be applied to a reaction in which there is an electron transfer with the formation of an ionic intermediate.

1981, *53*, 767

electron kinetic energy
The difference between the energy of the exciting photon and the *ionization energy*.

1976, *45*, 223

electron microscopy
See *transmission electron microscopy (TEM)*.

1983, *55*, 2024

electron-pair acceptor
A synonym for *Lewis acid*.

1994, *66*, 1110

electron-pair donor
A synonym for *Lewis base*.

1994, *66*, 1110

electron paramagnetic resonance (EPR)
Electron paramagnetic resonance (EPR) and/or electron spin resonance (ESR) is defined as the form of spectroscopy concerned with microwave-induced transitions between magnetic energy levels of electrons having a net spin and orbital angular momentum. The term electron paramagnetic resonance and

the symbol EPR are preferred and should be used for primary indexing.
1989, *61*, 2196

electron probe microanalysis (EPMA)
General term for methods using bombardment of a solid specimen by electrons which generate a variety of signals providing the basis for a number of different analytical techniques.
1983, *55*, 2024

electron probe X-ray microanalysis (EPXMA)
Any analytical technique which uses the excitation and evaluation of the characteristic X-ray spectrum of a solid specimen by a focused electron beam (typically of a diameter of less than 1 μm). Qualitative and quantitative X-ray analysis is the main feature of EPXMA. For obtaining supplementary information the *secondary, back scattered* and *absorbed electrons* are also frequently observed. Qualitative and quantitative X-ray analysis is, however, the main feature of EPXMA.
1983, *55*, 2024

electron rest mass
Atomic fundamental physical constant used as *atomic unit* of mass, $m_e = 9.109\ 3897\ (54) \times 10^{-31}$ kg.
CODATA Bull., 1986, *63*, 1

electron spectroscopy for chemical analysis (ESCA)
Type of spectroscopy involving the measurement of the kinetic energy of electrons emitted by chemical substances usually as a result of excitation by monochromatic X-rays.
1976, *45*, 223

electron stopping power (in X-ray emission spectroscopy)
When a beam of electrons strikes a target or specimen there are three ways in which the electrons may lose energy (low energy collisions, X-ray production, and formation of a spectral continuum). The average energy loss per unit distance travelled along the electron path is called electron stopping power, dE/dx.
1980, *52*, 2545

electron transfer
The transfer of an electron from one *molecular entity* to another, or between two localized sites in the same *molecular entity*.
See also *inner sphere (electron transfer)*, *outer sphere (electron transfer)*, *Marcus equation*.
1994, *66*, 1110; 1996, *68*, 2239

electron-transfer catalysis
The term indicates a sequence of reactions such as shown in equations (1)–(3), leading from A to B:

$$A + e^- \rightarrow A^{\cdot -} \quad (1)$$
$$A^{\cdot -} \rightarrow B^{\cdot -} \quad (2)$$
$$B^{\cdot -} + A \rightarrow B + A^{\cdot -} \quad (3)$$

An analogous sequence involving radical cations ($A^{\cdot +}$, $B^{\cdot +}$) is also observed.
The most notable example of electron-transfer catalysis is the $S_{RN}1$ (or $T + D_N + A_N$) reaction of aromatic halides.
The term has its origin in a suggested analogy to acid-base catalysis, with the electron instead of the proton. However, there is a difference between the two catalytic mechanisms, since the electron is not a true catalyst, but rather behaves as the initiator of a *chain reaction*. 'Electron-transfer induced chain reaction' is a more appropriate term for the situation described by equations (1)–(3).
1994, *66*, 1110

electron transfer photosensitization
Photochemical process in which a reaction of a non-absorbing substrate is induced by *electron transfer* (not energy transfer) with an excited light-absorbing *sensitizer*. The overall process must be such that the sensitizer is recycled. Depending on the action of the excited sensitizer as electron donor or acceptor the sensitization is called reductive or oxidative.
See also *photosensitization*.
1996, *68*, 2239

electronvolt
Non-SI unit of energy, $\text{eV} \approx 1.602\ 177\ 33(49) \times 10^{-19}$ J.
G.B. 75; 1996, *68*, 971

electron work function, Φ
The minimum work needed to extract electrons from the *Fermi level* of a metal M across a surface carrying no net charge. It is equal to the sum of the potential energy and the kinetic *Fermi energy* taken with the reverse sign:

$$\Phi^M = -(V_e + \varepsilon_e^F)$$

where V_e is the potential energy for electrons in metals and ε_e^F is the kinetic energy of electrons at the *Fermi level*.
1986, *58*, 441

electro-osmosis
The motion of a liquid through a membrane (or plug or capillary) as a consequence of the application of an electric field across the membrane. The linear velocity of flow (per unit field strength) is called the electro-osmotic velocity (u).
The spelling of electro-osmosis with two o's is preferred to electrosmosis with one o and to the older term electro-endosmosis.
1972, *31*, 620

electro-osmotic pressure
The pressure difference Δp across the membrane, plug, etc., needed just to stop electro-osmotic volume flow. Δp is positive if the higher pressure is on the high potential side.
1972, *31*, 620

electro-osmotic velocity
See *electro-osmosis*.
1972, *31*, 620

electro-osmotic volume flow
The volume flow per unit time per unit field strength through the whole plug.
See also *electro-osmosis*.
1972, *31*, 620

electrophile (electrophilic)
An electrophile (or electrophilic reagent) is a reagent that forms a *bond* to its reaction partner (the *nucleophile*) by accepting both bonding electrons from that reaction partner.
An 'electrophilic *substitution reaction*' is a *heterolytic* reaction in which the reagent supplying the entering group acts as an electrophile. For example:

$$C_6H_6 + NO_2^+ \longrightarrow C_6H_5NO_2 + H^+$$
electrophile → electrofuge

Electrophilic reagents are *Lewis acids*. 'Electrophilic catalysis' is catalysis by Lewis acids.
The term 'electrophilic' is also used to designate the apparent polar character of certain *radicals*, as inferred from their higher relative reactivities with reaction sites of higher electron density.
See also *electrophilicity*.
1994, *66*, 1111

electrophilicity
1. The property of being electrophilic (see *electrophile*).
2. The relative reactivity of an electrophilic reagent. (It is also sometimes referred to as 'electrophilic power'.) Qualitatively, the concept is related to Lewis acidity. However, whereas Lewis acidity is measured by relative equilibrium constants, electrophilicity is measured by relative *rate constants* for reactions of different electrophilic reagents towards a common *substrate* (usually involving attack at a carbon atom).
See also *nucleophilicity*.
1994, *66*, 1111

electrophoresis
The motion of *colloidal* particles in an electric field. The term cataphoresis should be abandoned.
1972, *31*, 619

electrophoresis convection
See *electrodecantation*.
1972, *31*, 609

electrophoretic mobility, μ
The observed *rate of migration* of a component (v) divided by electric field strength (E) in a given medium.
Notes:
1. Mobilities are sometimes expressed with a negative sign, because migration of the solutes or particles generally occurs in the direction opposite to the electrophoretic field (which is taken as reference for that direction).
2. In a solid support medium, only apparent values can be determined.
1994, *66*, 894

electrophotography
Processes of photoimaging which are based on photo-induced changes of electric fields (photo-conductive or photo-electrostatic effects).
1996, *68*, 2239

+ electrosmosis
See *electro-osmosis*.
1972, *31*, 620

electrostatic filter
Filters for which an electrostatic charge is applied to the filter element. A fibrous filter material is often pleated in between V-shaped supports consisting of electrostatically charged, metal rods which are insulated from the supporting frame.
1990, *62*, 2186

electrostatic precipitator
A device which separates particles from a gas stream by passing the carrier gas between pairs of electrodes across which a unidirectional, high-voltage potential is placed. The particles are charged before passing through the field and migrate to an oppositely charged electrode. These devices are very efficient collectors of small particles, and their use in removing particles from power plant plumes and in other industrial applications is widespread.
1990, *62*, 2186

electrothermal atomizer (in spectrochemical analysis)
A device which is heated to the temperature required for analyte atomization by the passage of electrical current through its body.
In *open atomizers* the sample vapour, together with analyte atoms, leaving the atomizer surface may move freely into the surrounding space.
In *confined atomizers* the sample vapour is restricted by the atomizer wall except for the openings where the optical beam enters and leaves the atomizer and the sample introduction hole.
1992, *64*, 254

electroviscous effects
For *dispersions* of charged particles, these are those components of the *viscosity* connected with the charge of the particles.
1979, *51*, 1217

element
See *chemical element*.
Physical Chemistry Division, unpublished; R.B. 35

elementary charge
Electromagnetic fundamental physical constant equal to the charge of a proton and used as *atomic unit* of charge $e = 1.602\ 177\ 33\ (49) \times 10^{-19}$ C.
See also *electric charge*.
CODATA Bull., 1986, *63*, 1; G.B. 14; 1996, *68*, 972

elementary entity
Any countable object or event, but usually a molecule, an ion or a specified group of atoms.
Physical Chemistry Division, unpublished

elementary particle (in nuclear chemistry)
A particle in which, at the present, no structure can be observed at moderate energies.
1982, *54*, 1547

elementary reaction
A reaction for which no *reaction intermediates* have been detected or need to be postulated in order to describe the chemical reaction on a molecular scale. An elementary reaction is assumed to occur in a single step and to pass through a single *transition state*.
See also *stepwise reaction*.
1994, *66*, 1111; 1993, *65*, 2294; see also 1996, *68*, 163

element effect
The ratio of the rate constants of two reactions that differ only in the identity of the element of the atom in the *leaving group*, e.g. k_{Br}/k_{Cl}. As for *isotope effects*, a ratio of unity is regarded as a 'null effect'.
1994, *66*, 1111

element of chirality
See *chirality element*.
1996, *68*, 2206

elimination
The reverse of an *addition reaction* or *transformation*. In an elimination two groups (called eliminands) are lost most often from two different centres (1/2/elimination or 1/3/elimination, etc.) with concomitant formation of an unsaturation in the molecule (double bond, triple bond) or formation of a new ring.
If the groups are lost from a single centre (α-elimination, 1/1/elimination) the resulting product is a *carbene* or a 'carbene analogue'.
See also α-*elimination*.
1994, *66*, 1112

eluate
The effluent from a *chromatographic bed* emerging when *elution* is carried out.
O.B. 99

eluent
The liquid or gas entering a *chromatographic bed* and used to effect a separation by *elution*.
O.B. 97

elute
To chromatograph by *elution chromatography*. This term is preferred to the term 'develop', which has been used in paper chromatography and in *thin-layer chromatography*. The process of elution may continue until the components have left the *chromatographic bed*.
O.B. 95; 1993, *65*, 824; 1990, *62*, 2186

elution
See *selective elution, stepwise elution, gradient elution*.
O.B. 92

elution, anticircular (in planar chromatography)
See *anticircular elution (in planar chromatography)*.
1993, *65*, 830

elution, ascending (in planar chromatography)
See *ascending elution (in planar chromatography)*.
1993, *65*, 829

elution band (in chromatography)
Synonymous with *chromatographic peak*.
O.B. 96, 101

elution chromatography
A procedure for chromatographic separation in which the mobile phase is passed through the chromatographic bed after the application of the sample.
O.B. 92; 1993, *65*, 824

elution curve
A *chromatogram* or part of a chromatogram recorded when *elution* techniques are used.
O.B. 95

elution, descending (in planar chromatography)
See *descending elution (in planar chromatography)*.
1993, *65*, 829

elution, gradient
See *gradient elution (in chromatography)*.
1993, *65*, 826

elution, horizontal (in planar chromatography)
See *horizontal elution (in planar chromatography)*.
1993, *65*, 829

elution, radial (in planar chromatography)
See *radial elution (in planar chromatography)*.
1993, *65*, 829

elution, stepwise
See *stepwise elution (in chromatography)*.
1993, *65*, 827

elutriation
The process of separating the lighter particles of a powder from the heavier ones by means of an upward directed stream of fluid (gas or liquid).
1990, *62*, 2186

emanation thermal analysis
A thermoanalytical technique in which the release of radioactive emanation from a substance (and/or its reaction product(s)) is measured as a function of temperature whilst the substance is subjected to a controlled temperature programme.
1985, *57*, 1739

embryo
An *aggregate* of a small number of atoms, molecules or ions.
An embryo of such a size that the Gibbs energy at constant pressure and temperature is a maximum is called a critical embryo.
An embryo larger than a critical embryo is called a homogeneous nucleus.
Colloidal sols may be formed by condensation methods. When a condensation method is applied, molecules (or ions) are deposited on *nuclei* which may be of the same chemical species as the *colloid* (homoge-

neous nucleation) or different (heterogeneous nucleation).
1972, *31*, 608

emission
Radiative deactivation of an excited state; transfer of energy from a molecular entity to an electromagnetic field.
See also *fluorescence, luminescence, phosphorescence.*
1996, *68*, 2239

emission (in atmospheric chemistry)
The total rate at which a solid, liquid or gaseous pollutant is emitted into the atmosphere from a given source; usually expressed as mass per unit time. Primary emissions are those substances which are emitted directly to the atmosphere (e.g. NO, SO_2, etc.), while secondary emissions are formed from the primary emissions through thermal or photochemical reactions (e.g. ozone, aldehydes, ketones, sulfuric acid, nitric acid, etc.). The point or area from which the discharge takes place is called the source; the area in which the emission or its transformed products (e.g. in the case of aerosols, acidic deposition, etc.) may be deposited is called the *receptor* area or *sink*. *Emission* may be applied to noise, heat, etc., as well as pollutants.
1990, *62*, 2186

emission control equipment (in atmospheric chemistry)
Air pollution control equipment which either converts the pollutant chemically to a non-polluting substance or collects the pollutant by some means including gravity settling chambers, inertial separators, cyclonic separators, filters, electrical precipitators, scrubbers, (spray towers, jet scrubbers, Venturi scrubbers, inertial scrubbers, mechanical scrubbers and packed scrubbers). Certain gases and odoriferous compounds are controlled by combustion, absorption (spray chambers, mechanical contactors, bubble cap or sieve plate contactors and packed towers) and adsorption units (packed beds or fluidized beds).
1990, *62*, 2187

emission flux (in atmospheric chemistry)
The emission per unit area of the appropriate surface of an emitting source.
1990, *62*, 2187

emission spectrum
Plot of the emitted *spectral radiant power* (*spectral radiant exitance*) or of the emitted *spectral photon irradiance* (*spectral photon exitance*) against a quantity related to photon energy, such as frequency, ν, wavenumber, σ, or wavelength, λ. When corrected for wavelength-dependent variations in the equipment response, it is called a corrected emission spectrum.
1996, *68*, 2239; see also O.B. 196

emission spectrum, corrected
See *corrected emission spectrum.*
1984, *56*, 242

emittance, ε
Ratio of *radiant exitance* of a thermal radiator to that of a full radiator (black body) at the same temperature. In ISO this quantity is called emissivity.
G.B. 31; ISO 31-5: 1992; 1996, *68*, 2239; 1996, *68*, 972

empirical formula
Formed by juxtaposition of the atomic symbols with their appropriate subscripts to give the simplest possible formula expressing the composition of a compound.
R.B. 45

emulsifier
A *surfactant* which when present in small amounts facilitates the formation of an *emulsion*, or enhances its *colloidal* stability by decreasing either or both of the rates of *aggregation* and *coalescence*.
1972, *31*, 612

emulsion
A fluid *colloidal* system in which liquid droplets and/or liquid crystals are dispersed in a liquid. The droplets often exceed the usual limits for *colloids* in size. An emulsion is denoted by the symbol O/W if the continuous phase is an aqueous solution and by W/O if the continuous phase is an organic liquid (an 'oil'). More complicated emulsions such as O/W/O (i.e. oil droplets contained within aqueous droplets dispersed in a continuous oil phase) are also possible. Photographic emulsions, although colloidal systems, are not emulsions in the sense of this nomenclature.
1972, *31*, 606

enamines
Alkenylamines; by usage the term refers specifically to *vinylic amines*, which have the structure $R_2NCR=CR_2$.
1995, *67*, 1333

enantioconvergence
See *stereoconvergence.*
1996, *68*, 2206

enantioenriched
See *enantiomerically enriched.*
1996, *68*, 2206

enantiomer
One of a pair of molecular entities which are mirror images of each other and non-*superposable*.
See also *enantiomorph.*
1996, *68*, 2207; 1994, *66*, 1112

enantiomer excess (enantiomeric excess)
For a mixture of (+)- and (−)-enantiomers, with composition given as the mole or weight fractions $F_{(+)}$ and $F_{(-)}$ (where $F_{(+)} + F_{(-)} = 1$) the enantiomer excess is defined as $|F_{(+)} - F_{(-)}|$ (and the percent enantiomer excess by $100|F_{(+)} - F_{(-)}|$). Frequently this term is abbreviated as e.e.
See *optical purity.*
1996, *68*, 2207

enantiomerically enriched (enantioenriched)
A sample of a chiral substance whose enantiomeric ratio is greater than 50:50 but less than 100:0
1996, 68, 2207

enantiomerically pure (enantiopure)
A sample all of whose molecules have (within limits of detection) the same chirality sense. Use of homochiral as a synonym is strongly discouraged.
1996, 68, 2207

enantiomeric groups
Chiral groups that are mirror images of one another.
B.B. 480

enantiomeric purity
See *enantiomer excess*.
1996, 68, 2207

enantiomeric ratio
The ratio of the percentage of one *enantiomer* in a mixture to that of the other e.g. 70(+) : 30(–).
1996, 68, 2207

enantiomeric units (in a polymer)
Two *configurational units* that correspond to the same *constitutional unit* are considered to be enantiomeric if they are non-superposable mirror images.
P.B. 27

enantiomerization
The interconversion of *enantiomers*.
See *racemization*.
1996, 68, 2207

enantiomerism
The isomerism of *enantiomers*.
1996, 68, 2207

enantiomorph
One of a pair of *chiral* objects or models that are non-*superposable* mirror images of each other. The adjective enantiomorphic is also applied to mirror-image related groups within a molecular entity.
1996, 68, 2207

enantiomorphous structures (in polymers)
See *isomorphous structures (in polymers)*.
P.B. 43

enantiopure
See *enantiomerically pure*.
1996, 68, 2207

enantioselectivity
See *stereoselectivity*.
1996, 68, 2207; 1994, 66, 1112

enantiotopic
Constitutionally identical atoms or groups in molecules which are related by symmetry elements of the second kind only (mirror plane, inversion centre or rotation–reflection axis). For example the two groups c in a grouping Cabcc are enantiotopic. Replacement of one of a pair of enantiotopic groups forms one of a pair of enantiomers. Analogously, if complexation or addition to one of the two faces defined by a double bond or other molecular plane gives rise to a chiral species, the two faces are called enantiotopic.

See also *prochiral, diastereotopic*.
1996, 68, 2207

enantiotropic transition
See *polymorphic transition*.
Note:
In liquid crystal systems this term refers to a liquid crystal to *liquid crystal transition* that occurs above the melting point.
1994, 66, 581

encapsulation (in catalysis)
Enzymes or cells which are of relatively large size may be entrapped in a maze of polymeric molecules (a gel). This procedure is called *immobilization by inclusion*. When the biocatalyst is enclosed inside a semipermeable membrane, usually approximately spherical, the method is known as *encapsulation*. In the process of reticulation the primary biocatalyst particles (individual enzyme molecules, cofactors or individual cells) are covalently attached to each other by organic chains, into a three-dimensional network. The term grafting is also used in this context. Attachment to the support by adsorption forces is called *immobilization by adsorption*.
1991, 63, 1230

encounter
A set of collisions between *reactant* molecules in solids, liquids or dense gases, occurring in rapid succession as a result of the *cage effect*.
1996, 68, 163

encounter complex
A *complex* of *molecular entities* produced at an *encounter-controlled rate*, and which occurs as an intermediate in a reaction mechanism. When the complex is formed from two molecular entities it is called an 'encounter pair'. A distinction between encounter pairs and (larger) encounter complexes may be relevant in some cases, e.g. for mechanisms involving *pre-association*.
1994, 66, 1112; see also 1996, 68, 2239

encounter control
See *microscopic diffusion control*.
1996, 68, 164

encounter-controlled rate
A *rate of reaction* corresponding to the rate of encounter of the reacting *molecular entities*. This is also known as 'diffusion-controlled rate' since rates of encounter are themselves controlled by diffusion rates (which in turn depend on the viscosity of the *medium* and the dimensions of the reactant molecular entities). For a *bimolecular* reaction between solutes in water at 25 °C an encounter-controlled rate is calculated to have a second-order *rate constant* of about 10^{10} dm^3 mol^{-1} s^{-1}.
See also *microscopic diffusion control*.
1994, 66, 1112

encounter pair
See *encounter complex*.
1994, 66, 1112; see also 1996, 68, 2239

endergonic (or endoergic) reaction
This is usually taken to be a reaction for which the overall standard Gibbs energy change $\Delta G°$ is positive. Some workers use this term with respect to a positive value of $\Delta H°$ at the absolute zero of temperature.
 1996, *68*, 164

end-group
A *constitutional unit* that is an extremity of a *macromolecule* or *oligomer molecule*.
An end-group is attached to only one constitutional unit of a macromolecule or oligomer molecule.
 1996, *68*, 2294

endoenzymes
Enzymes that cut internal bonds of a polymer. Endonucleases are able to cleave phosphodiester bonds within a *nucleic acid* chain by hydrolysis either randomly or at specific base sequences (cf. *restriction enzymes*).
 1992, *64*, 152

endo, exo, syn, anti
Descriptors of the relative orientation of groups attached to non-bridgehead atoms in a bicyclo[*x.y.z*]alkane ($x \geq y > z > 0$).

If the group is orientated towards the highest numbered bridge (*z* bridge, e.g. C-7 in example below) it is given the description *exo*; if it is orientated away from the highest numbered bridge it is given the description *endo*. If the group is attached to the highest numbered bridge and is orientated towards the lowest numbered bridge (*x* bridge, e.g. C-2 in example below) it is given the description *syn*; if the group is orientated away from the lowest numbered bridge it is given the description *anti*.

2-*endo*-bromo-7-*anti*-fluoro-bicyclo [2.2.1] heptane

2-*exo*-bromo-7-*syn*-fluoro-bicyclo [2.2.1] heptane

See also *syn* and *anti*.
 1996, *68*, 2207

endothermic reaction
A reaction for which the overall standard enthalpy change $\Delta H°$ is positive.
 1996, *68*, 164

end-point
 See *titration*.
 O.B. 47

end-to-end distance (in polymers)
The length, *r*, of the end-to-end vector.
 P.B. 49

end-to-end vector (in polymers)
The vector, *r*, connecting the two ends of a linear polymer chain in a particular conformation.
 P.B. 49

ene reaction
The addition of a compound with a double bond having an allylic hydrogen (the 'ene') to a compound with a multiple bond (the 'enophile') with transfer of the allylic hydrogen and a concomitant reorganization of the bonding, as illustrated below for propene (the 'ene') and ethene (the 'enophile'). The reverse is a 'retro-ene' reaction.

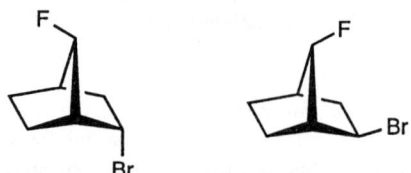

 1994, *66*, 1112

energized species
In an elementary process, a species having sufficient energy to become an *activated complex*, but which does not have the structure of the activated complex. If it is not inactivated by collisions it can become an activated complex and pass at once into products, after undergoing vibration.
 1996, *68*, 164

energy, *E*
In mechanics the sum of *potential energy* and *kinetic energy*.
In thermodynamics the *internal energy* or thermodynamic energy increase, ΔU, is the sum of *heat* and *work* brought to the system. Only changes in energy are measurable.
For photons $E = h\nu$, where h is the Planck constant and ν the frequency of radiation.
In relativistic physics $E = mc^2$, where c is the speed of light, and m the mass.
 G.B. 12

energy (of a radiation)
Energy of the individual particles or photons of which a *radiation* consists.
 1994, *66*, 2518

energy dispersion (in emission spectrometry)
The separation of characteristic photons according to their energy.
 O.B. 180.

energy dispersive detector
 See *radiation detector*.
 1995, *67*, 1748

energy dispersive X-ray fluorescence analysis
A method of *X-ray fluorescence analysis* where element specificity is obtained by measuring the energy *spectrum* of the induced *X-radiation*.
 1982, *54*, 1541

energy flux density
For mono-directional *radiation*, the *energy* traversing in a time interval over a small area perpendicular to the direction of the energy flow, divided by that time interval and by that area.
See also *intensity of radiation*.
1994, *66*, 2519

energy migration
See *electronic energy migration*.
1996, *68*, 2240

energy of activation
See *activation energy*.
G.B. 55; 1996, *68*, 164; 1994, *66*, 1112

energy of activation (of an electrode reaction)
This is defined by the equation:
$$U^{\ddagger} = -RT\,(\partial \ln I_0/\partial T^{-1})_{p,\,c_j\ldots}$$
where I_0 is the *exchange current*.
At any *overpotential* η it is defined by the equation:
$$U^{\ddagger}(\eta) = -R\,(\partial \ln |I|/\partial T^{-1})_{p,\,\eta,\,c_j\ldots}$$
where I is the current passing from the electrode into the electrolyte.
1974, *37*, 516

energy peak, full
See *full energy peak*.
1982, *54*, 1543

energy pooling
See *annihilation*.
1996, *68*, 2240

energy profile
See *Gibbs energy diagram, potential-energy profile*.
1994, *66*, 1112

energy resolution (in radiochemistry)
A measurement, at given *energy*, of the smallest difference between the energies of two particles or photons capable of being distinguished by a *radiation* spectrometer.
1994, *66*, 2519

energy storage efficiency, η
The rate of the Gibbs energy storage in an endothermic photochemical reaction divided by the incident irradiance.
See also *efficiency*.
1996, *68*, 2240

energy threshold (in radiochemistry)
The limiting kinetic *energy* of a incident particle or energy of an incident photon below which a specified process cannot take place.
1994, *66*, 2519

energy transfer (in photochemistry)
From a phenomenological point of view, the term is used to describe the process by which a molecular entity absorbs light and a phenomenon originates from the excited state of another molecular entity. In mechanistic photochemistry the term has been reserved for the photophysical process in which an excited state of one molecular entity (the donor) is deactivated to a lower-lying state by transferring energy to a second molecular entity (the acceptor) which is thereby raised to a higher energy state. The excitation may be electronic, vibrational, rotational or translational. The donor and acceptor may be two parts of the same molecular entity, in which case the process is called intramolecular energy transfer.
See also *Dexter excitation transfer, Förster excitation transfer, radiative energy transfer, spectral overlap*.
1996, *68*, 2240

energy transfer plot (in photochemistry)
A plot of the quenching rate constant of an excited molecular entity by a series of quenchers versus the excited state energy of the quenchers. Alternatively, a plot of the rate constant for the sensitization of a reaction versus the excited state energy of different sensitizers. This type of plot is used to estimate the energy of the excited molecular entity quenched (in the former case) or produced (in the latter case).
See also *Stern–Volmer kinetic relationships*.
1996, *68*, 2240

energy yield of luminescence
The ratio of the energy emitted as *luminescence* to the energy absorbed by a species.
1984, *56*, 243

enforced concerted mechanism
Variation of reaction parameters in a series of reactions proceeding in non-concerted steps may lead to a situation where the putative intermediate will possess a lifetime shorter than a bond vibration, so that the steps become concerted. The *transition state* structure will lie on the coordinate of the *More O'Ferrall–Jencks diagram* leading to that of the putative intermediate.
1994, *66*, 1112

enhanced phosphorescence analysis (in molecular luminescence spectroscopy)
A use of *luminescence quenching* effects to enhance *sensitivity*. The strong depopulation of the fluorescing singlet excited state by external heavy atom perturbers can lead to a large population of the phosphorescing triplet excited state.
1984, *56*, 235

enhancement reaction (in analytical chemistry)
A reaction which enables a favourable measurement to be made in which the constituent which is finally measured has not been magnified in any way (cf. *amplification reaction*).
1982, *54*, 2556

enhancer
A fluorescent compound which accepts energy and thus enhances or promotes the emission from a sample containing a chemically or enzymatically generated excited molecular entity.
See also *dark photochemistry*.
1996, *68*, 2240

enolates
Salts of *enols* (or of the tautomeric *aldehydes* or *ketones*), in which the anionic charge is delocalized over oxygen and carbon, or similar covalent metal derivatives in which the metal is bound to oxygen.

$$R_2C=C\begin{smallmatrix}O^-\\R\end{smallmatrix} \leftrightarrow R_2\bar{C}-C\begin{smallmatrix}O\\R\end{smallmatrix} \quad ; \quad R_2C=C\begin{smallmatrix}OM\\R\end{smallmatrix}$$

1995, 67, 1334

enols
Alkenols; the term refers specifically to vinylic alcohols, which have the structure HOCR'=CR$_2$. Enols are tautomeric with *aldehydes* (R' = H) or ketones (R' ≠ H).
See also *phenols*.
1995, 67, 1334

enophile
See *ene reaction*.
1994, 66, 1113

enoses
Monosaccharides having a carbon–carbon double bond anywhere in the backbone chain. Glycals (term not recommended) designates the enoses that are generated by formal elimination of the hemiacetal hydroxy group and an adjacent hydrogen atom. Thus glycals are cyclic *enol ethers*. Unsaturated hexoses, for example, are called hexenoses, e.g. the hex-1-enopyranose derived from D-glucopyranose:

1995, 67, 1334

enrichment factor (in liquid-liquid distribution), S
The factor by which the ratio of two substances in the feed must be multiplied to give their ratio after treatment.
$Q_A/Q_B = S_{A,B}(Q'_A/Q'_B)$ where Q_A and Q'_A are the final and initial amounts of species A and Q_B and Q'_B are the final and initial amounts of species B. Hence $S_{A,B} = E_A/E_B$ where E is the fraction extracted. In terms of D, n, r (where n is the number of stages and r the phase ratio),

$$S_{A,B} = \frac{1-(1+rD_A)^{-n}}{1-(1+rD_B)^{-n}}$$

1993, 65, 2382; O.B. 90

enrichment, isotopic
See *isotopic enrichment*.
1982, 54, 1541

ent
A prefix used to indicate the enantiomer for natural products and related molecules where the trivial name only refers to one *enantiomer*, e.g. *ent*-kaurene is the enantiomer of kaurene.
1996, 68, 2208

entering group
An atom or *group* that forms a *bond* to what is considered to be the main part of the *substrate* during a reaction. For example: the attacking *nucleophile* in a bimolecular nucleophilic *substitution reaction*.
1994, 66, 1113

enthalpimetric analysis
The generic designation for a family of analytical methods in which the enthalpy change of a chemical reaction is measured, directly or indirectly, in order to perform a quantitative determination of a reactant or catalyst. In general, at least one reactant is a liquid or solution. It is a subset of *calorimetry* in which the system is maintained at constant (usually atmospheric pressure), and the emphasis is on a quantitative determination in a reasonable length of time.
Analytical solution calorimetry is a synonymous designation that may be used but is not preferred. *Thermochemical analysis* has been used but is not recommended because of potential confusion and overlap with the methods of *thermal analysis*. The adjective *thermometric* is also associated with the names of some of these methods, but is not recommended as a general designation.
See also *thermal analysis*.
1994, 66, 2489

enthalpimetric flow injection analysis
See *flow injection enthalpimetry*.
1994, 66, 2491

enthalpimetric titration
See *thermometric titration*.
1994, 66, 2490

enthalpimetry
See *enthalpimetric analysis*.
1994, 66, 2489

enthalpimetry, continuous flow
See *continuous flow enthalpimetry*.
1994, 66, 2491

enthalpimetry, direct injection
See *direct injection enthalpimetry*.
1994, 66, 2490

enthalpimetry, flow
See *flow enthalpimetry*.
1994, 66, 2491

enthalpimetry, flow injection
See *flow injection enthalpimetry*.
1994, 66, 2491

enthalpimetry, peak
See *peak enthalpimetry*.
1994, 66, 2491

enthalpogram
A plot of temperature *versus* time or heat change *versus* time which is produced in *direct injection enthalpimetry*. Use of the term *thermogram* is not rec-

ommended because it is too general and overlaps with other fields.
1994, *66*, 2490

enthalpy, *H*
Internal energy of a system plus the product of pressure and volume. Its change in a system is equal to the heat brought to the system at constant pressure.
G.B. 48; 1996, *68*, 972; 1990, *62*, 2187

enthalpy of activation, $\Delta^{\ddagger}H^{o}$
The standard enthalpy of activation $\Delta^{\ddagger}H^{o}$ is the enthalpy change that appears in the thermodynamic form of the rate equation obtained from conventional *transition-state theory*. This equation is only correct for a first order reaction, for which the rate constant has the dimension reciprocal time. For a second order reaction, for which the rate constant has the dimension (reciprocal time) × (reciprocal concentration), the left hand side should be read as kc^{o}, where c^{o} denotes the standard concentration (usually 1 mol dm^{-3}).

$$k = (k_{B}T/h) \exp(\Delta^{\ddagger}S^{o}/R) \exp(-\Delta^{\ddagger}H^{o}/RT)$$

The quantity $\Delta^{\ddagger}S^{o}$ is the standard *entropy of activation*, and care must be taken with standard states. In this equation k_{B} is the Boltzmann constant, T the absolute temperature, h the Planck constant, and R the gas constant.
The enthalpy of activation is approximately equal to the *activation energy*; the conversion of one into the other depends on the molecularity.
The enthalpy of activation is always the standard quantity, although the word standard and the superscript ^{o} on the symbol are often omitted. The symbol is frequently (but incorrectly) written $\Delta H^{\ddagger}$, where the standard symbol is omitted and the $\ddagger$ is placed after the H.
1996, *68*, 164; 1994, *66*, 1113; G.B. 56; 1993, *65*, 2294

enthalpy of immersion
Synonymous with *enthalpy of wetting*.
1972, *31*, 605

enthalpy of wetting
Referred to unit of mass of the solid, the difference (at constant temperature) between the enthalpy of a solid completely immersed in a wetting liquid, and that of the solid and the liquid taken separately. It must be specified whether the solid in the initial state is in contact with vacuum or with the vapour of the liquid at a given partial pressure. Measurements of the enthalpy of wetting of a solid equilibrated with varying relative pressures of the vapour of a pure wetting liquid may be used to derive the differential enthalpy of adsorption of the vapour.
1972, *31*, 605

entitic
Modifier used to denote divided by number of entities.
1996, *68*, 972

entitic quantity
One which refers to a single entity (particle) of a system.
1986, *58*, 441

entrainment (in atmospheric chemistry)
The act of forming a mist or fog: droplets of a liquid carried off by the vapours of a boiling liquid or from a liquid through which bubbles of gas or vapour are passing rapidly.
1990, *62*, 2187

entrance channel
The region of a *potential-energy surface* or hypersurface that corresponds to molecular configurations that are closer in geometry to those of the reactants than to those of the products.
1996, *68*, 165

entropy, *S*
Quantity the change in which is equal to the heat brought to the system in a reversible process at constant temperature divided by that temperature. Entropy is zero for an ideally ordered crystal at 0 K. In statistical thermodynamics $S = k \ln W$, where k is the Boltzmann constant and W the number of possible arrangements of the system.
G.B. 48; 1990, *62*, 2187; 1996, *68*, 972

entropy of activation, $\Delta^{\ddagger}S^{o}$
The entropy change that appears in the thermodynamic form of the rate equation obtained from conventional transition-state theory (see *enthalpy of activation*).
1996, *68*, 165; 1994, *66*, 1113

entropy unit
Non-SI unit of molar entropy, e.u. = 4.184 J K^{-1} mol^{-1}.
G.B. 113

envelope conformation
The conformation (of symmetry group C_{s}) of a five-membered ring in which four atoms are coplanar, and one atom (the flap) projects out of the plane.

1996, *68*, 2208

enzyme activator
See *activator*.
1993, *65*, 2292

enzyme activity
See *catalytic activity*.
1996, *68*, 972

enzyme, immobilized
See *immobilized enzyme*.
1994, *66*, 2593

enzyme induction
The process whereby an (inducible) enzyme is synthesized in response to a specific molecule (cf. *inducer*). The inducer molecule (often a *substrate* that needs the catalytic activity of the inducible enzyme for its *metabolism*) combines with a *repressor* and

thereby prevents the blocking of an *operator* by the repressor.
1992, *64*, 152

enzyme inhibitor
See *inhibitor*.
1993, *65*, 2295

enzyme repression
The mode by which the synthesis of an enzyme is prevented by *repressor* molecules. In many cases, the end product of a synthesis chain (e.g. an amino acid) acts as a feed-back corepressor by combining with an intracellular aporepressor protein, so that this complex is able to block the function of an operator. As a result, the whole *operon* is prevented from being transcribed into mRNA, and the expression of all enzymes necessary for the synthesis of the end product amino acid is abolished.
1992, *64*, 152

enzymes
Macromolecules, mostly of protein nature, that function as *(bio)catalysts* by increasing the reaction rates. In general, an enzyme catalyses only one reaction type (reaction specificity) and operates on only one type of *substrate* (substrate specificity). Substrate molecules are attacked at the same site (regiospecificity) and only one or preferentially one of the enantiomers of chiral substrates or of racemic mixtures is attacked (stereospecificity).
1992, *64*, 152; 1993, *65*, 2295; 1994, *66*, 2593

enzyme substrate electrode
A sensor in which an *ion-selective electrode* is covered with a coating containing an enzyme which causes the reaction of an organic or inorganic substance (substrate) to produce a species to which the electrode responds. Alternatively, the sensor could be covered with a layer of substrate which reacts with the enzyme, co-factor or inhibitor to be assayed.
1994, *66*, 2535

+ enzyme thermistor
See *flow injection enthalpimetry*.
1994, *66*, 2491

enzymic decomposition
Decomposition of organic materials (e.g. starch, sugars, proteins, etc.) can be achieved by enzymic decomposition, in which the enzyme converts a high molecular mass compound into lower molecular mass species. The process can be regarded as an example of enzymic degradation.
1988, *60*, 1469

epicadmium neutrons
Neutrons of kinetic *energy* greater than the *effective cadmium cut-off energy*.
1994, *66*, 2522

epigenetic
Descriptive term for processes that change the *phenotype* without altering the *genotype*.
1992, *64*, 153

epihalohydrins
Compounds having the (halomethyl)oxirane skeleton:

$$\underset{CH_2-CH}{\overset{O}{\triangle}}-CH_2X$$

1995, *67*, 1334

epimerization
Interconversion of *epimers*.
1996, *68*, 2208; 1994, *66*, 1114

epimers
Diastereoisomers that have the opposite *configuration* at only one of two or more tetrahedral *stereogenic* centres present in the respective molecular entities.
1996, *68*, 2208; 1994, *66*, 1113

epi-phase
The less dense phase in a distribution system. The term is often used when two non-aqueous phases are present or when the solvent is an aqueous solution.
See also *hypo-phase*.
1993, *65*, 2380

episulfonium ions
Ions derived from thiiranes, in which a trivalent sulfur atom bears a positive charge:

$$\underset{R_2C-CR_2}{\overset{R}{\underset{|}{S^+}}}$$

1995, *67*, 1334

epithermal neutrons
Neutrons of kinetic *energy* greater than that of thermal agitation. The term is often restricted to energies just above thermal.
1994, *66*, 2522

epitope
Any part of a molecule that acts as an antigenic determinant. A macromolecule can contain many different epitopes each capable of stimulating production of a different specific *antibody*.
1992, *64*, 153

epoxides
See *epoxy compounds*.
1995, *67*, 1334

epoxy compounds
Compounds in which an oxygen atom is directly attached to two adjacent or non-adjacent carbon atoms of a carbon chain or ring system; thus cyclic *ethers*. The term epoxides represents a subclass of epoxy compounds containing a saturated three-membered cyclic ether; thus oxirane derivatives, e.g. 1,2-epoxypropane, or 2-methyloxirane (an epoxide); 9,10-epoxy-9,10-dihydroanthracene (an epoxy compound).

$$CH_3-\overset{O}{\overset{\triangle}{CH-CH_2}}$$
1,2-epoxypropane

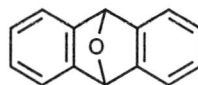
9,10-epoxy-9,10-dihydroanthracene
1995, *67*, 1334

equatorial
1. For the use of this term in the context of cyclohexane see *axial, equatorial*.
2. For the use of this term in the context of a bipyramidal structure see *apical, basal, equatorial*.
1996, *68*, 2208

equilibration
The operation by which a system of two or more phases is brought to a condition where further changes with time do not occur. This term is not synonymous with pre-equilibrium and should not be used in that sense.
1993, *65*, 2377

equilibrium, chemical
See *chemical equilibrium*.
1994, *66*, 1114

equilibrium constant
Quantity characterizing the equilibrium of a chemical reaction and defined by an expression of the type $K_x = \Pi_B x_B^{\nu_B}$, where ν_B is the *stoichiometric number* of a reactant (negative) or product (positive) for the reaction and x stands for a quantity which can be the equilibrium value either of pressure, fugacity, amount concentration, amount fraction, molality, relative activity or reciprocal absolute activity defining the pressure based, fugacity based, concentration based, amount fraction based, molality based, relative activity based or *standard equilibrium constant* (then denoted $K°$), respectively.
G.B. 50; see also 1990, *62*, 2187; 1996, *68*, 972

equilibrium control
See *thermodynamic control*.
1994, *66*, 1114

equilibrium dialysate
The colloid-free solution obtained at equilibrium in *dialysis*.
Its composition approaches that of the *dispersion* medium (more precisely, the limit to which the composition of the dispersion medium tends at large distances from the particles). In the dialysis equilibrium an osmotic pressure difference exists between *sol* and equilibrium dialysate.
1972, *31*, 608

equilibrium distance (in a molecule)
Internuclear distance at the minimum of the electronic (or potential) energy surface.
G.B. 24

equilibrium film
A liquid film of thickness, or thicknesses, at which it is stable or *metastable* with respect to small thickness changes. Unless the area of the film is small, its composition may not be the same over its area and the (metastable) equilibrium thickness may be characteristic of the local condition only.
1972, *31*, 614

equilibrium isotope effect
See *isotope effect*.
1994, *66*, 1114

equilibrium, radioactive
See *radioactive equilibrium*.
1994, *66*, 2519

equilibrium reaction
When the reactants in a chemical reaction are initially in a Boltzmann distribution the reaction is referred to as an equilibrium reaction.
See also *non-equilibrium reaction*.
1996, *68*, 165

equilibrium, secular
See *secular equilibrium*.
1982, *54*, 1541

equilibrium sedimentation
A method by which the distribution of the concentration of the solute or dispersed component in a dilute solution or dispersion along the centrifuge cell is measured at sedimentation equilibrium and the results are interpreted in terms of molar masses or their distribution, or both.
P.B. 62

equilibrium solution
See *coacervation*.
1972, *31*, 611

equivalence point
See *titration*.
O.B. 47

equivalence postulate (in polymer chemistry)
The working hypothesis that the chain monomeric units are geometrically equivalent.
P.B. 78

equivalent chain (in polymers)
A hypothetical freely jointed chain with the same mean-square end-to-end distance and contour length as an actual polymer chain in a theta state.
P.B. 50

equivalent diameter
The diameter of a spherical particle which will give identical geometric, optical, electrical or aerodynamic behaviour to that of the particle (non-spherical) being examined; sometimes referred to as the Stokes diameter for particles in non-turbulent flows.
1990, *62*, 2184

equivalent entity
Entity corresponding to the transfer of a H^+ ion in a neutralization reaction, of an electron in a redox reaction, or to a magnitude of charge number equal to 1 in ions. Examples: $\frac{1}{2}H_2SO_4$, $\frac{1}{5}KMnO_4$, $\frac{1}{3}Fe^{3+}$.
Physical Chemistry Division, unpublished

erg
cgs unit of energy, erg = 10^{-7} J.
G.B. 112; 1996, *68*, 973

error (of measurement)
Result of a measurement minus the true value of the *measurand*. Since a true value cannot be determined, in practice the *conventional true value* is used.
  See also measurement result.
  VIM; O.B. 6; see also 1994, *66*, 601

error, base-line (in spectrochemical analysis)
  See *base-line error (in spectrochemical analysis)*.
  1988, *60*, 1456

error, cell (in spectrochemical analysis)
  See *cell error (in spectrochemical analysis)*.
  1988, *60*, 1457

error, fluorescence (in spectrochemical analysis)
  See *fluorescence error (in spectrochemical analysis)*.
  1988, *60*, 1457

error, inhomogeneity (in spectrochemical analysis)
  See *inhomogeneity error (in spectrochemical analysis)*.
  1988, *60*, 1456

error, non-linearity (in spectrochemical analysis)
  See *non-linearity error (in spectrochemical analysis)*.
  1988, *60*, 1456

error, optical-beam (in spectrochemical analysis)
  See *optical-beam-error (in spectrochemical analysis)*.
  1988, *60*, 1457

error, polarization
  See *polarization error (in spectrochemical analysis)*.
  1988, *60*, 1456

error, random
  See *random error*.
  1990, *62*, 2188

error, sample
  See *sample error (in spectrochemical analysis)*.
  1988, *60*, 1457

error, sampling
  See *sampling error*.
  1990, *62*, 1201

error, scattering (in spectrochemical analysis)
  See *scattering error (in spectrochemical analysis)*.
  1988, *60*, 1457

error, settling (in spectrochemical analysis)
  See *settling error (in spectrochemical analysis)*.
  1988, *60*, 1456

error, spectral bandwidth (in spectrochemical analysis)
  See *spectral bandwidth error (in spectrochemical analysis)*.
  1988, *60*, 1456

error, stray radiation (in spectrochemical analysis)
  See *stray radiation error (in spectrochemical analysis)*.
  1988, *60*, 1456

error, systematic
  See *systematic error*.
  1990, *62*, 2188

error, wavelength (in spectrochemical analysis)
  See *wavelength error (in spectrochemical analysis)*.
  1988, *60*, 1456

erythro **structures (in a polymer)**
The *relative configuration* at two contiguous carbon atoms in the main chain bearing, respectively, substituents a and b (a ≠ b), is designated by the prefix *erythro* or *threo*, as appropriate, by analogy with the terminology for carbohydrate systems in which the substituents are OH.
Examples:

$$\begin{array}{c} \text{H} \quad \text{H} \\ | \quad | \\ -\text{C}-\text{C}- \\ | \quad | \\ \text{a} \quad \text{b} \end{array}$$
erythro

$$\begin{array}{c} \text{H} \quad \text{b} \\ | \quad | \\ -\text{C}-\text{C}- \\ | \quad | \\ \text{a} \quad \text{H} \end{array}$$
threo

Similar systems in which a higher level of substitution exists may be treated analogously if the *erythro* or *threo* designation is employed to denote the relative placements of those two substituents, one for each backbone carbon atom, which rank highest according to the Sequence Rule.
  P.B. 36

erythro, threo
Descriptors of the diastereoisomers of an acyclic structure or partial structure having two stereogenic centres. This notation is derived from carbohydrate nomenclature. The extension of this system has given rise to conflicting interpretations of these prefixes. It is recommended that for such cases the l,u or R^*,S^* system should be used.
  1996, *68*, 2208

ESCA
  See *photoelectron spectroscopy*.
  1996, *68*, 2240

escape depth (for surface analysis techniques)
The distance into the sample measured from the physical surface from which all but a fraction $1/e$ of the particles or *radiation* detected have originated.
  1979, *51*, 2247

escape peak, double
  See *double escape peak*.
  1982, *54*, 1541

escape peak, single
  See *single escape peak*.
  1982, *54*, 1541

escape peak, X-ray
See *X-ray escape peak*.
1982, *54*, 1541

Esin and Markov coefficient
The left-hand side of one of the various cross-differential relationships that can be obtained from the Gibbs adsorption equation when only one chemical potential (μ) is considered as variable, viz.

$$\left(\frac{\partial E}{\partial \mu}\right)_{T, p, \sigma} = -\left(\frac{\partial \Gamma}{\partial \sigma}\right)_{T, p, \mu}$$

where E is the potential difference, T is the temperature, p is the pressure, Γ is the surface excess and σ is the charge density.
1986, *58*, 446

esters
Compounds formally derived from an *oxoacid* $R_kE(=O)_l(OH)_m$ ($l \neq 0$), and an alcohol, phenol, heteroarenol, or enol by linking with formal loss of water from an acidic hydroxy group of the former and a hydroxy group of the latter. By extension *acyl* derivatives of alcohols, etc. Acyl derivatives of chalcogen analogues of alcohols (thiols, selenols, tellurols) etc. are included. E.g. R'C(=O)OR, R'C(=S)OR, R'C(=O)SR, R'S(=O)$_2$OR, (HO)$_2$P(=O)OR, (R'S)$_2$C(=O), ROCN (but not R–NCO) (R $\neq$ H).
Note:
O-Alkyl derivatives of other acidic compounds [see *amides (1)*] may be named as esters but do not belong to the class esters proper. E.g. (Ph)$_2$POCH$_3$ methyl diphenylphosphinite.
See also *acylals, ortho esters, depsides, depsipeptides, glycerides, lactides, lactones, macrolides*.
1995, *67*, 1334

eta
See η (*eta or hapto*), at the beginning of 'e'.
R.B. 202

ethers
Compounds ROR (R $\neq$ H). (Compounds R$_3$SiOR, silicon analogues of ethers, are trialkylsilyloxy compounds). E.g. CH$_3$CH$_2$OCH$_2$CH$_3$,

 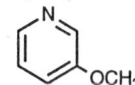

See also *acetals, epoxy compounds, ortho esters*.
1995, *67*, 1334

E_T-value
See *Dimroth–Reichardt E_T parameter, Z-value*.
1994, *66*, 1108

euatmotic reaction
An isothermal, reversible reaction between two (or more) solid phases during the heating of a system as a result of which a single vapour phase is produced.
1994, *66*, 582

eupeptide bond
See *peptide*.
W.B. 48

eutectic reaction
An isothermal, reversible reaction between two (or more) solid phases during the heating of a system, as a result of which a single liquid phase is produced.
1994, *66*, 582

evaluation function
The inverse of the *calibration function*.
1995, *67*, 1704

evaporation
The physical process by which a liquid substance is converted to a gas or vapour. This may occur at or below the normal boiling point of the liquid (the temperature at which a liquid boils at 1 atmosphere pressure) and the process is endothermic.
1990, *62*, 2188

even-electron ion
An ion containing no unpaired electrons, for example CH$_3^+$ in its ground state.
1991, *63*, 1549

evolved gas analysis (EGA)
A technique in which the nature and/or amount of volatile product(s) released by a substance subjected to a controlled temperature program is (are) determined. The method of analysis should always be clearly stated.
O.B. 42

evolved gas detection (EGD)
A technique in which the evolution of gas from a substance is detected as a function of temperature while the substance is subjected to a controlled temperature program.
O.B. 41

+ Ewens–Bassett number
Usage no longer recommended.
See *oxidation number, charge number*.
R.B. 66

exa
SI prefix for 10^{18} (symbol: E).
G.B. 74

EXAFS
See *X-ray absorption*.
1991, *63*, 738

excess acidity
See *Bunnett–Olsen equations, Cox–Yates equation*.
1994, *66*, 1114

excess mass (at a solid/liquid interface)
See *excess volume*.
1986, *58*, 972

excess Rayleigh ratio
The difference between the *Rayleigh ratio* for a dilute solution and for pure solvent.
If the scattering intensity is not reduced to the Rayleigh ratio, the difference between the scattering

intensities for a dilute solution and that for pure solvent is named 'excess scattering'.
P.B. 66

excess volume (at a solid/liquid interface)
For a pure liquid, despite its low compressibility, the variation of density near a solid surface can be detected and measured. The total volume V of a system consisting of solid and pure liquid is different from (usually less than) that calculated assuming a constant liquid density. If the densities of bulk solid (ρ^{sol}) and liquid (ρ^l) are known then an excess volume (usually negative) can be defined as:

$$V^\sigma = V - V^{sol} - V^o$$
$$= V - m^{sol}/\rho^{sol} - m^l/\rho^l$$

where m^{sol} is the mass of solid, V^{sol} its volume calculated from the bulk density, V^o is the initial volume of liquid and m^l is the mass of liquid.
The *excess mass* is given by:

$$m^\sigma = m^l - (V - V^{sol})\rho^l$$

1986, *58*, 972

exchange current (of an electrode reaction)
The common value I_0 of the anodic and cathodic partial currents when the reaction is at equilibrium

$$I_0 = I_a = -I_c$$

For an electrode at equilibrium at which only one reaction is significant $I = 0$. When more than one reaction is significant at a given electrode, subscripts to I_0 may be used to distinguish exchange currents. I is not usually zero when only one of these reactions is at equilibrium.
1974, *37*, 513

exchange extraction
An extraction operation or process in which a metal from one phase is exchanged with the equivalent amount of a second metal from the other phase.
Notes:
i. This term may be used in connection with any step (e.g. loading, scrubbing or stripping in a process).
ii. This applies also to organic or molecular species.
See also *crowding*.
1993, *65*, 2387

exchange-inversion transition
A transition between antiferromagnetic and ferromagnetic coupling between two subarrays of magnetic atoms.
Example:
Ordered FeRh changes from antiferromagnetic to ferromagnetic coupling with the simple cubic Fe array in a *first-order transition*
1994, *66*, 582

exchange labelling
Labelling of a substance by *isotope exchange*.
1994, *66*, 2521

excimer
An electronically excited dimer, 'non-bonding' in the ground state. For example, a complex formed by the interaction of an excited *molecular entity* with a ground state partner of the same structure.
See also *exciplex*.
1996, *68*, 2240; 1994, *66*, 1114

excimer laser
A source of pulsed coherent radiation obtained from an exciplex. The proper name is exciplex laser. Typical lasing species are noble gas halides (XeCl, KrF, etc.) emitting in the UV domain.
See *gas lasers*.
1996, *68*, 2241; 1995, *67*, 1920

excimer-luminescence (in luminescence quenching)
An effect accompanying concentration *quenching* in some cases, namely the formation of a new bimolecular species which is capable of emission (excimer- and exciplex-luminescence).
1984, *56*, 235

excipient (in toxicology)
Any more or less inert substance added to a drug to give suitable consistency or form to the drug.
1993, *65*, 2046

exciplex
An electronically excited complex of definite stoichiometry, 'non-bonding' in the ground state. For example, a complex formed by the interaction of an excited *molecular entity* with a ground state counterpart of a different structure.
See also *excimer*.
1994, *66*, 1114; see also 1996, *68*, 2241

exciplex-luminescence (in luminescence quenching)
See *excimer-luminescence (in luminescence quenching)*.
1984, *56*, 235

excitation
Process causing the transition of a system from one state to another of higher energy.
1982, *54*, 1541

excitation-emission spectrum
The three-dimensional spectrum generated by scanning the emission spectrum at incremental steps of *excitation* wavelength (x axis = emission wavelength, y axis = excitation wavelength, z axis = emission flux) is called a (*fluorescence, phosphorescence*) excitation-emission spectrum (or EES).
The spectra are particularly useful for investigating samples containing more than one emitting species. Corrected EES are obtained if (a) the emission is corrected for instrumental response with wavelength, and (b) the exciting radiation flux in photons s^{-1} is held constant for all excitation wavelengths.
1984, *56*, 242

excitation energy
The minimum energy required to bring a system to a specified higher energy level.
O.B. 223

excitation energy (in *in situ* microanalysis)
Sometimes called acceleration energy or initial energy. Kinetic energy of bombarding particles at the surface of the solid.
  1983, *55*, 2027

excitation level (in X-ray spectroscopy)
X-ray levels have various degrees of ionization — single, double or higher — and may in some cases also be electrically neutral. An excitation level (exciton) is an electrically neutral X-ray level with an expelled electron bound in the field of a core electron vacancy.
  1991, *63*, 738

excitation spectrum
A plot of the *spectral radiant exitance* or of the *spectral photon exitance* against the frequency (or wavenumber, or wavelength) of excitation. When corrected for wavelength dependent variations in the excitation radiant power this is called a corrected excitation spectrum.
  See also *emission spectrum*.
  1996, *68*, 2241; see also O.B. 197

excitation spectrum, corrected
  See *corrected excitation spectrum*.
  1984, *56*, 242

excitation spectrum, measured
  See *measured excitation spectrum*.
  1984, *56*, 242

excitation transfer
Synonymous with *energy transfer*.
  1996, *68*, 2241

excited state
State of a system with energy higher than that of the ground state. This term is most commonly used to characterize a molecule in one of its electronically excited states, but can also refer to vibrational and/or rotational excitation in the electronic ground state.
  1994, *66*, 1114; 1996, *68*, 2241

exciton
In some applications it is useful to consider electronic excitation as if a quasi-particle, capable of migrating, were involved. This is termed an exciton. In organic materials two models are used: the band or wave model (low temperature, high crystalline order) and the hopping model (higher temperature, low crystalline order or amorphous state). Energy transfer in the hopping limit is identical with energy migration.
  See *electronic energy migration*.
  1996, *68*, 2241

excluded volume of a macromolecule (in polymers)
The volume from which a macromolecule in a dilute solution of a macromolecule effectively excludes all other macromolecules.
The excluded volume of a macromolecule depends on the Gibbs and Helmholtz energies of mixing of solvent and polymer, i.e. on the thermodynamic quality of the solvent, and is not a measure of the geometrical volume of that macromolecule.
  P.B. 58

excluded volume of a segment (in polymers)
The volume from which a segment of a macromolecule in solution effectively excludes all other segments, i.e. those belonging to the same macromolecule as well as those belonging to other macromolecules.
The excluded volume of a segment depends on the Gibbs and Helmholtz energies of mixing of solvent and polymer, i.e. on the thermodynamic quality of the solvent, and is not a measure of the geometrical volume of that segment.
  P.B. 58

exclusion chromatography
Chromatography in which separation is based mainly on exclusion effects, such as differences in molecular size and/or shape or in charge. The term size-exclusion chromatography may also be used when separation is based on molecular size. The terms gel filtration and gel-permeation chromatography (GPC) were used earlier to describe this process when the stationary phase is a swollen gel. The term ion-exclusion chromatography is specifically used for the separation of ions in an aqueous phase.
  1993, *65*, 826

exergonic (exoergic) reaction
This expression is often applied to a reaction for which the overall standard Gibbs energy change ΔG^o is negative.
Some workers use this term with reference to a negative value of ΔH^o at the absolute zero of temperature.
  1996, *68*, 165

exfoliated graphite
The product of very rapid heating (or flash heating) of graphite intercalation compounds, such as graphite hydrogensulfate of relatively large particle diameter (flakes). The vaporizing intercalated substances force the graphite layers apart. The exfoliated graphite assumes an accordion-like shape with an apparent volume often hundreds of times that of the original graphite flakes.
Notes:
Exfoliated graphite is usually prepared from well-crystallized natural flake *graphite*. It is used for the production of graphite foils. Exfoliated graphite is different from the deflagration product of graphite oxide (graphitic acid).
  1995, *67*, 488

exitance
  See *radiant exitance*.
  1996, *68*, 2241

exit channel
This expression refers to the region of a *potential-energy surface* or hypersurface that corresponds to mo-

lecular configurations that are closer to those of the *products* than to those of the reactants.
1996, *68*, 165

exo
See *endo, exo, syn, anti*.
1996, *68*, 2208

exoenzymes
Enzymes that cleave monomers (sometimes also di- or oligomers) from one end of a polymer chain. Exonucleases are able to cleave *nucleotides*, one by one, from either the 5'- or the 3'- (or both) ends of *nucleic acids*.
1992, *64*, 153

exon
A section of DNA which carries the coding sequence for a protein or part of it. Exons are separated by intervening, non-coding sequences (cf. *intron*). In *eukaryotes* most *genes* consist of a number of exons.
1992, *64*, 153

exothermic reaction
A reaction for which the overall standard enthalpy change ΔH^o is negative.
1996, *68*, 165

expansion factor (in polymers)
The ratio of a dimensional characteristic of a macromolecule in a given solvent at a given temperature to the same dimensional characteristic in the θ state at the same temperature. The most frequently used expansion factors are: expansion factor of the mean-square end-to-end distance, $\alpha_r = (<r^2>/<r^2>_0)^{1/2}$; expansion factor of the radius of gyration $\alpha_s = (<s^2>/<s^2>_0)^{1/2}$; viscosity expansion factor $\alpha_\eta = ([\eta]/[\eta]_\theta)^{1/3}$ where $[\eta]$ and $[\eta]_\theta$ are the intrinsic viscosity in a given solvent and in the θ state at the same temperature, respectively.
Expansion factors defined by different dimensional characteristics are not exactly equal, nor need they have a constant ratio as a function of relative molecular mass.
P.B. 59

expectation value
See *measurement result*.
1995, *67*, 1705

expected value
See *measurement result*.
1995, *67*, 1705

experimental perturbational calculation
To assess the uncertainty of derived final results from a least-squares procedure one can investigate the *sensitivity* of output data for errors in the input data by means of this calculation. One performs the calculations many times with perturbed input data and studies the relation between the changes (perturbations) in the input data and the changes in the output data. The perturbations are preferably generated using random numbers with a distribution function chosen to mimic the distribution of uncertainties in the primary results.

See also *least squares technique*.
1981, *53*, 1823

explosivity limits (explosion limits)
The concentration limits, usually of a substance in air, between which combustion will be self-sustaining.
1990, *62*, 2188

exponential decay
Variation of a quantity (generally the *activity* of a quantity of a radionuclide) according to the law $A = A_0 e^{-\lambda t}$ where A and A_0 are the values of the quantity being considered at time t and zero respectively, and is an appropriate constant.
1982, *54*, 1541

exposure
For X- or *gamma radiation* in air: the sum of the electrical charges of all the *ions* of one sign produced when all *electrons* liberated by *photons* in a suitably small element of volume of air completely stopped, divided by the mass of the air in the volume element.
1982, *54*, 1541

expression (in biotechnology)
1. The cellular production of the protein encoded by a particular *gene*. The process includes *transcription* of DNA, processing of the resulting mRNA product and its *translation* into an active protein.
2. A recombinant gene inserted into a *host* cell by means of a *vector* is said to be expressed if the synthesis of the encoded protein can be demonstrated.
1992, *64*, 153

extended-chain crystal (in polymers)
A polymer crystal in which the chains are in an essentially fully extended conformation.
P.B. 84

extended Hammett equation
This term applies in a general way to any multiparametric extension of the *Hammett equation*. It is sometimes used specifically for a form of *dual substituent-parameter equation* in which the actual value of the correlated property P under the influence of the substituent X is used, rather than the value relative to that for X = H. An intercept term h corresponding to the value of P for X = H is introduced, e.g.

$$P = \alpha \sigma_I + \beta \sigma_R + h$$

The equation may be applied to systems for which the inclusion of further terms to represent other effects, e.g. steric, is appropriate.
1994, *66*, 1114

extended X-ray absorption fine structure (EXAFS)
Structure in X-ray absorption spectra remote from the edge and extending to higher energies, due mainly to the scattering of the expelled electron by neighbouring atoms. Structure closer to the absorption edge is

sometimes called near-edge X-ray absorption fine structure (NEXAFS).
1991, *63*, 738

extensive quantity
Physical quantity whose magnitude is additive for subsystems
G.B. 7; 1986, *58*, 1407

extent of an interface (surface)
A quantity measured by its area, A (or S). For solids, a real (true, actual) and a geometric surface (interface) area, A_r and A_g, respectively, may be defined if asperities are present whose height is orders of magnitude greater than the atomic or molecular size. The geometric surface is the projection of the real surface on a plane parallel to the macroscopic, visible phase boundary. If asperities are of the order of the atomic size, the surface of the solid may be described as stepped. (High index faces of crystals are *stepped surfaces* but may be ideally smooth in the sense of the *roughness factor*.)
1986, *58*, 439

extent of reaction, ξ
Extensive quantity describing the progress of a chemical reaction equal to the number of chemical transformations, as indicated by the reaction equation on a molecular scale, divided by the Avogadro constant (it is essentially the *amount* of chemical transformations). The change in the extent of reaction is given by $d\xi = dn_B/\nu_B$, where ν_B is the *stoichiometric number* of any reaction entity B (reactant or product) and n_B is the corresponding amount.
G.B. 43; 1996, *68*, 165; 1996, *68*, 973; 1992, *64*, 1572; 1993, *65*, 2295

+ external compensation
The absence of *optical activity* in a *racemate*. (Usage strongly discouraged).
1996, *68*, 2208

external heavy atom effect
See *heavy atom effect*.
1996, *68*, 2241

external ion-pair return
See *ion-pair return*.
1994, *66*, 1127

external ion return
See *ion-pair return*.
1994, *66*, 1127

external photoelectric effect
See *photoemissive detector*
1995, *67*, 1752

external return
See *ion-pair return*.
1994, *66*, 1114

external standard (in chromatography)
A compound present in a standard sample of known concentration and volume which is analysed separately from the unknown sample under identical conditions. It is used to facilitate the qualitative indentification and/or quantitative determination of the sample components. The volume of the external standard (standard sample) need not to be known if it is identical to that of the unknown sample.
1993, *65*, 837

external surface
When a porous solid consists of discrete particles it is inconvenient to describe the outer boundary of the particles as the external surface.
1972, *31*, 585; 1976, *46*, 79

+ exterplex
Termolecular analogue of an *exciplex*. Use of this term is discouraged.
See also *exciplex*.
1996, *68*, 2242

+ extinction
This term, sometimes used as equivalent to *absorbance*, is no longer recommended.
See *attenuance*.
G.B. 32; 1996, *68*, 2242; 1996, *68*, 973

+ extinction coefficient
See *attenuation coefficient*.
G.B. 32; 1996, *68*, 2242; 1996, *68*, 973

extra-column volume (in chromatography)
The volume between the effective injection point and the effective detection point, excluding the part of the column containing the stationary phase. It is composed of the volumes of the injector, connecting lines and detector.
1993, *65*, 832

extract (n.)
The separated phase (often but not necessarily organic) that contains the material extracted from the other phase.
Where appropriate the term 'loaded solvent' may be used, but is not recommended.
1993, *65*, 2380; O.B. 91

extract (v.)
To transfer a solute from a liquid phase to another immiscible or partially miscible liquid phase in contact with it. The term is also applied to the dissolution of material from a solid phase with a liquid in which it is not wholly soluble (i.e. leaching).
See also *solvent extraction*.
1993, *65*, 2377; O.B. 95

extractability (in solvent extraction)
A property which qualitatively indicates the degree to which a substance is extracted. The term is imprecise and generally used in a qualitative sense. It is not a synonym for fraction extracted.
1993, *65*, 2382

extractant
The active component(s) primarily responsible for transfer of a solute from one phase to the other.
Notes:
i. The term extracting agent is a synonym but solvent and ligand should not be used in this context.
ii. Certain extractants that consist of liquids immiscible with water (e.g. tributyl phosphate or certain

ketones) might comprise the only component of the initial organic phase but extractant(s) can also be dissolved in diluent.
1993, 65, 2380; O.B. 91

extracting agent
See *extractant*.
1993, 65, 2380

extraction
Distribution and *partition* are often used as synonyms for the general phenomenon of extraction where appropriate.
See *liquid-liquid extraction*.
1993, 65, 2377

extraction (in process liquid-liquid distribution)
In connection with processes, this term often refers to the initial transfer step whereby the main solute, often together with impurities, is transferred from feed to solvent.
Notes:
i. Partition and distribution are not synonymous in this specific instance.
ii. The term extraction may be used in a more general sense.
See also *loading*.
1993, 65, 2387

+ extraction coefficient
This term is not recommended as a synonym for *distribution ratio*.
1993, 65, 2382; O.B. 89

extraction (equilibrium) constant
The extraction constant at zero ionic strength, K_{ex}^o, is the equilibrium constant of the distribution reaction expressed in terms of the reacting species. Thus, for the gross reaction:

$$M^{n+}_{aq} + nHL_{org} \rightleftharpoons ML_{n,\,org} + nH^+_{aq}$$

in which the reagent HL initially dissolved in an organic phase reacts with a metal ion M^{n+} in aqueous solution to form a product ML_n which is more soluble in the organic phase than in water,

$$K_{ex}^o = \frac{a_{ML_n,\,org} \times a^n_{H^+,\,aq}}{a_{M^{n+},\,aq} \times a^n_{HL,\,org}}$$

Notes:
i. When concentrations are used instead of activities or mixed terms are employed as when H^+ and/or M^{n+} are measured with an electrode, the appropriate name is extraction constant, symbol K_{ex}, accompanied by a careful definition. K_{ex}^o may be termed the thermodynamic extraction constant.
ii. The extraction constant is related to other terms relevant to such systems by:

$$K_{ex} = \frac{D_{ML_n}\beta_n K_a^n}{D_{HL}^n}$$

where β_n is the overall formation constant of ML_n and K_a is the dissociation constant of HL. Where the reagent HL is more soluble in water than the other immiscible phase it may be more convenient to define a special equilibrium constant in terms of HL_{aq}:

$$K_{ex} = D_{ML_n}\beta_n K_a^n$$

iii. In distribution equilibria involving non-aqueous systems, e.g. liquid SO_2, molten salts and metals, the mass action equilibrium constant for the relevant extraction process can be identified with K_{ex} which should be explicitly defined in this context.
iv. In actual practice, it may be necessary to include other terms to take into account other complexes formed by auxiliary reagents and the solvation and/or polymerization of the various species. In such cases, K_{ex} must be defined with reference to the relevant explicit chemical equation. An example is complex formation between the metal ion and an uncharged crown ether or cryptand molecule followed by ion-pair extraction:

$$M^{n+}_{aq} + L_{org} + nA^-_{aq} \rightleftharpoons (ML^{n+} \cdot A_n^{n-})_{org}$$

$$K_{ex} = \frac{[ML^{n+} A_n^{\,n-}]_{org}}{[M^{n+}]_{aq}[L]_{org}[A^-]_{aq}^n}$$

v. Use of Ringbom's 'conditional extraction constant',

$$K_{ex}^{eff} = \frac{a_{H^+}^n \,^n[ML_n']_{org}}{[M']_{aq}[HL']_{org}^n}$$

in conjunction with alpha coefficients is useful.
vi. The phases can also be specified by the formula of the solvent or by other symbols (preferably Roman numerals) or by overlining formulae referring to one phase, usually the less polar one. The subscript aq (or w) is often omitted; aq is preferable to w as the latter is appropriate only in English and German.
vii. The qualification 'equilibrium' is often omitted.
viii. The terms partition constant and distribution constant must not be used in this sense.
1993, 65, 2383; O.B. 89

extraction factor, D_m
The ratio of the total mass of a solute in the extract to that in the other phase.
Notes:
i. It is the product of the (concentration) distribution ratio and the appropriate phase ratio.
ii. It is synonymous with the concentration factor or mass distribution ratio, this latter term being particularly apt.
iii. The term concentration factor is often employed for the overall extraction factor in a process or process step.
1993, 65, 2384

extraction fractionation (of polymers)
A process in which a polymeric material, consisting of macromolecules differing in some characteristic affecting their solubility, is separated from a polymer-rich phase into fractions by successively increasing the solution power of the solvent, resulting in the repeated formation of a two-phase system in which

extraction isotherm
See *distribution isotherm*.
1993, *65*, 2387

extraction process (in catalysis)
See *abstraction process (in catalysis)*.
1976, *46*, 85

extrapolated range (in radiochemistry)
The distance from a *radiation source*, calculated by extrapolation to zero of the *flux density*, of the tangent to the flux density versus distance curve, taken at the point where the flux density has decreased to one half of its initial value.
1982, *54*, 1541

extremophiles
Organisms which require extreme physico-chemical conditions for their optimum growth and proliferation. Extremophilic microorganisms are e.g. *thermophiles* or psychrophiles, *halophiles*, alkalophiles or acidophiles, osmophiles and barophiles, based on their growth at extremes of temperature, salt concentration, pH, osmolarity or pressure, respectively.
1992, *64*, 153

extrusion transformation
A *transformation* in which an atom or *group* Y connected to two other atoms or groups X and Z is lost from a molecule, leading to a product in which X is bonded to Z, e.g.

X–Y–Z → X–Z + Y

or

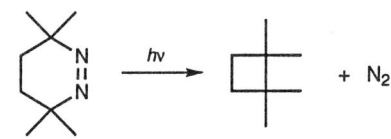

The reverse of an extrusion is called *insertion*.
See *cheletropic reaction*.
1994, *66*, 1114

fabric filter
　See *bag filter*.
　1990, *62*, 2188

fac-
An affix used in names to designate three groups occupying the corners of the same face of an octahedral coordination sphere, not now generally recommended for precise nomenclature purposes.
　R.B. 245; B.B. 464

factor
Proportionality constant between two quantities of the same dimension.
　ISO 31-0: 1992; 1996, *68*, 973

factor-group splitting
　See *Davydov splitting*.
　1996, *68*, 2242

fallout (in atmospheric chemistry)
A measurement of air contamination consisting of the mass rate at which solid particles deposit from the atmosphere.
　See *dust fall*.
　1990, *62*, 2188

fallout, radioactive
　See *radioactive fallout*.
　1982, *54*, 1541

fall time
　See *response time*.
　1995, *67*, 1751; see also 1990, *62*, 2174

fanning (in atmospheric chemistry)
In blast furnace operation, the idling period between the blowing periods when the blast pressure is reduced to a minimum. This also applies to plume behaviour under inversion conditions.
　See *gas plume*.
　1990, *62*, 2188

farad
SI derived unit of electric capacitance, $F = C\ V^{-1} = m^{-2}\ kg^{-1}\ s^4\ A^2$.
　G.B. 72; 1996, *68*, 973

faradaic current
A current corresponding to the reduction or oxidation of some chemical substance. The net faradaic current is the algebraic sum of all the faradaic currents flowing through an *indicator* or *working electrode*.
　1985, *57*, 1496

faradaic current density
Measures the rate of the interfacial reaction, the proportionality constant being the charge number of the reaction multiplied by the *Faraday constant*.
　1980, *52*, 236

faradaic demodulation current
Component of the current that is due to the demodulation associated with an *electrode reaction* and that appears if an *indicator* or *working electrode* is subjected to the action of two intermodulated *applied potentials* of different frequency.
　1985, *57*, 1496

faradaic rectification current
A component of the current that is due to the rectifying properties of an *electrode reaction* and that appears if an *indicator* or *working electrode* is subjected to any periodically varying *applied potential* while the mean value of the applied potential is controlled.
　1985, *57*, 1496

Faraday constant
Fundamental physical constant representing molar elementary charge: $F = 9.648\ 5309\ (29) \times 10^{-27}$ C mol^{-1}.
　CODATA Bull., 1986, *63*, 1; 1996, *68*, 973

Faraday cup (or cylinder) collector
A hollow collector, open at one end and closed at the other, used to collect beams of ions.
　1991, *63*, 1553; O.B. 204

fast atom bombardment ionization
This term is used to describe ionization of any species by causing interaction of the sample (which may be dissolved in a solvent matrix) and a beam of neutral atoms having high translational energy.
　See also *secondary ionization*.
　1991, *63*, 1547

fast-atom bombardment (FAB) mass spectroscopy
A method in which ions are produced in a mass spectrometer from nonvolatile or thermally fragile organic molecules by bombarding the compound in the condensed phase with energy-rich neutral particles.
　1994, *66*, 1115

fast neutrons
Neutrons of kinetic *energy* greater than some specified value. This value may vary over a wide range and will be dependent upon the application.
　1994, *66*, 2522

fatty acids
Aliphatic monocarboxylic acids derived from or contained in esterified form in an animal or vegetable fat, oil or wax. Natural fatty acids commonly have a chain of 4 to 28 carbons (usually unbranched and even-numbered), which may be saturated or unsaturated. By extension, the term is sometimes used to embrace all acyclic aliphatic *carboxylic acids*.
　1995, *67*, 1335; see also W.B. 180

feedback (in analysis)
The process whereby the output of a device is used to modify the operation of an analytical instrument.
　1989, *61*, 1659

feedback (in kinetics)
　See *composite reaction*.
　1996, *68*, 166

feed-back inhibition (end product inhibition)(in biotechnology)
A metabolic control mechanism in which the end product of a biochemical sequence is able to inhibit the activity of an early *enzyme* in the sequence, thereby controlling the metabolic flux through this

pathway. As an example, isoleucine controls its own synthesis by inhibiting threonine deaminase; adenosine 5′-triphosphate (ATP) and citrate control glycolysis by inhibiting phosphofructokinase.
1992, *64*, 154

feed rate (in catalysis)
The amount of reactant fed per unit time to the inlet of the reactor.
1976, *46*, 82

femto
SI prefix for 10^{-15} (symbol: f).
G.B. 74

fenestranes
Compounds of the general formula shown below. They may be considered to be *spiro compounds* having bridges of carbon atoms connecting the α and α′ positions.

1995, *67*, 1335

fermentation
1. In *metabolism*, the dehydrogenating degradation of organic substance by organisms or *cells* under anaerobic conditions in which electrons are transferred to *metabolites* which accumulate and are excreted in reduced form. Fermentation is only possible if the organism is able to gain energy by this process.
2. In microbiology, the process in which cells (*microorganisms*, plant or animal cells) are cultured in a *bioreactor* in liquid or solid medium to convert organic substances into *biomass* (growth) or into products.
1992, *64*, 154

fermenter
A *bioreactor* which enables optimal *fermentation* conditions to be maintained, allowing addition of nutrients, removal of products and insertion of measuring and/or control probes as well as other necessary equipment (e.g. for heating, cooling, aeration, agitation, sterilization, etc.) under sterile conditions.
1992, *64*, 154

fermi
Non-SI unit of length, f = 10^{-15} m, used in nuclear physics.
G.B. 110

Fermi energy
The total energy of an electron in an uncharged metal at the *Fermi level*.
1986, *58*, 440

Fermi level, E_F
The chemical potential of electrons in a solid (metals, semiconductors or insulators) or in an electrolyte solution.

See *bandgap energy, conduction band, valence band*.
1996, *68*, 2242; see also 1986, *58*, 440

fermion
Particle of half-integer spin quantum number following Fermi–Dirac statistics.
Physical Chemistry Division, unpublished

feromone
See *pheromone*.
1993, *65*, 2083

ferrimagnetic transition
See *ferroic transition, magnetic transitions*.
1994, *66*, 582

ferrocenophanes
Compounds in which the two ring components of ferrocene are linked by one or more bridging chains.

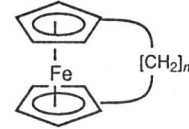

See also *cyclophanes, metallocenes, sandwich compounds*.
1995, *67*, 1335

ferroelastic transition
A transition in which a crystal switches from one stable orientation state into another that is equally stable by the application of a mechanical stress along an appropriate direction.
See *ferroic transition*.
1994, *66*, 582

ferroelectric (antiferroelectric) transition
A transition from a ferroelectric to either another ferroelectric, or a paraelectric, or an antiferroelectric state.
Example: The transition of the low-temperature, cubic paraelectric $BaTiO_3$ to the high-temperature, tetragonal, ferroelectric form at 393 K.
Notes:
1. In an antiferroelectric transition individual dipoles become arranged antiparallel to adjacent dipoles with the result that the net spontaneous polarization is zero.
2. Ferroelectric/antiferroelectric transitions also occur in the liquid-crystal state. These states are dependent on the alternating nature of dipoles between layers in the *smectic state*.
See *ferroic transition*.
1994, *66*, 582

ferroic transition
A general term for ferroelastic, ferroelectric, antiferroelectric, ferromagnetic, antiferromagnetic and ferrimagnetic transitions in which a suitable driving force switches a ferroic crystal from one orientation state, or domain state, to another.
1994, *66*, 582

ferromagnetic transition
See *ferroic transition, magnetic transition*.
1994, *66*, 583

fertile (in radioanalytical chemistry)
1. Of a *nuclide*: capable of being transformed, directly or indirectly, into a *fissile* nuclide by neutron *capture*.
2. Of a material: containing one or more fertile nuclides.
1982, *54*, 1542

***f*-functional branch point**
See *branch point*.
1996, *68*, 2297

***f*-functional branch unit**
See *branch unit*.
1996, *68*, 2297

fibrous activated carbon
An *activated carbon* in the form of fibres, filaments, yarns or rovings and fabrics or felts. Such fibres differ from *carbon fibres* used for reinforcement purposes in composites by their high surface area, high porosity and low mechanical strength.
Notes:
Sometimes fabrics of fibrous activated carbon are named *charcoal* cloths; a more precise term is 'activated carbon cloth'.
1995, *67*, 488

fibrous carbon
See *filamentous carbon*.
1995, *67*, 488

fibrous crystal (in polymers)
A type of crystal significantly longer in one dimension than in either of the other two.
Fibrous crystals may comprise essentially extended chains parallel to the fibre axis; however, macroscopic polymer fibres containing chain-folded crystals are also known.
P.B. 83

fiducial group
See *helicity, torsion angle*.
1996, *68*, 2208

field desorption (in mass spectrometry)
A term used to describe the formation of ions in the gas phase from a material deposited on a solid surface (known as an 'emitter') in the presence of a high electrical field. 'Field desorption' is an ambiguous term because it implies that the electric field desorbs a material as an ion from some kind of emitter on which the material is deposited. There is growing evidence that some of the ions formed are due to thermal ionization and some to field ionization of vapour evaporated from material on the emitter. Because there is generally little or no ionization unless the emitter is heated by an electric current, 'field desorption' is a misnomer. The term is, however, firmly implanted in the literature and most users undertand what is going on regardless of the implications of the term. In addition, no better simple term has been suggested to take its place and so, reluctantly, it is recommended that it be retained.
1991, *63*, 1547

field effect
An experimentally observable effect symbolized by F (on reaction rates, etc.) of *intramolecular* coulombic interaction between the centre of interest and a remote unipole or dipole, by direct action through space rather than through bonds. The magnitude of the field effect (or 'direct effect') depends on the unipolar charge/dipole moment, orientation of dipole, shortest distance between the centre of interest and the remote unipole or dipole, and on the effective *dielectric constant*.
See also *electronic effect, inductive effect, polar effect*.
1994, *66*, 1115

field effect phototransistor
See *phototransistor*.
1995, *67*, 1755

field ionization (in mass spectrometry)
The removal of electrons from any species by interaction with a high electrical field.
See also *field desorption*.
1991, *63*, 1548; O.B. 203

field level
Logarithm of the ratio of two values of a field quantity, usually amplitude. Levels are expressed in different ways: $L_F = \ln(F/F_0)$ Np, where Np is the symbol for the unit *neper* coherent with the SI, or $L_F = 20 \lg(F/F_0)$ dB, where dB is the symbol for the unit *decibel*.
G.B. 79

filamentous carbon
A carbonaceous deposit from gaseous *carbon* compounds, consisting of filaments grown by the catalytic action of metal particles.
Notes:
In general, such deposits are obtained at pressures of < kPa in the temperature region 600–1300 K on metals such as iron, cobalt or nickel. Typical filaments consist of a duplex structure, a relatively oxidation-resistant skin surrounding a more easily oxidizable core, with a metal particle located at the growing end of the filament. They generally range from 0.01 to 0.5 μm in diameter and up to 10 μm in length. In some systems, the metal particles are located in the middle of the filaments, and there are also examples where several filaments originate from a single particle. The filaments may be produced in different conformations, such as helical, twisted and straight.
1995, *67*, 488

filler
Filler (also called *grist*) is a petroleum- or coal-based coke fraction of a green, carbon mix or formulation. Coarse particles, > 0.425 mm, are sometimes referred to as tailings; fine particles, < 0.074 mm, are referred to as flour. *Graphite* flour is also used as a filler.
1995, *67*, 489

filler coke
The main constituent of a carbon artifact, introduced as solid component (predominantly in the form of

particulate carbon) into the '*carbon mix*' from which *polygranular carbon* and *graphite* materials are obtained by heat treatment.
Note:
Filler coke is not necessarily the only, but it is commonly the most important *filler* material used in a 'carbon mix' which consists of filler and binder.
1995, *67*, 489

filling solution (of a reference electrode)
Solution containing the anion to which the *reference electrode* of the *operational pH cell* is reversible, e.g. chloride for silver-silver chloride electrode. In the absence of a *bridge solution*, a high concentration of filling solution comprising cations and anions of almost equal mobility is employed as a means of maintaining the *liquid junction* potential small and approximately constant on substitution of *test solution* for *standard solution(s)*.
1985, *57*, 541

film
A generic term referring to condensed matter restricted in one dimension.
1994, *66*, 1671

film, asymmetric
See *asymmetric film*.
1994, *66*, 1671

film badge (in radioanalytical chemistry)
A package containing one or more small photographic films for approximate measurement of *radiation exposure*, used for the purpose of monitoring personnel.
1982, *54*, 1542

film, closed
See *open film*.
1994, *66*, 1671

film element
A small homogeneous part of a film including the two *interfaces* and any fluid between them.
1972, *31*, 613

film, macroscopic
See *macroscopic film*.
1994, *66*, 1671

film, metastable
See *stable film*.
1972, *31*, 614

film, microscopic
See *microscopic film*.
1994, *66*, 1671

film, nanoscopic
See *nanoscopic film*.
1994, *66*, 1671

film, open
See *open film*.
1994, *66*, 1671

film, partly open
See *partly open film*.
1994, *66*, 1671

film, planar
See *planar film*.
1994, *66*, 1671

film, soap
See *soap film*.
1972, *31*, 613

film, stable
See *stable film*.
1972, *31*, 614

film, symmetrical
See *symmetrical films*.
1994, *66*, 1671

film tension
The contractile force per unit length, Σ_f, exerted by an *equilibrium film* in contact with the bulk phase.
1972, *31*, 614

film, thick
See *thick film*.
1994, *66*, 1672

film, thin
See *thin film*.
1994, *66*, 1672

filter
A porous material on which solid particles present in air or other fluid which flows through it are largely caught and retained. Filters are made with a variety of materials: cellulose and derivatives, glass fibre, ceramic, synthetic plastics and fibres. Filters may be naturally porous or be made so by mechanical or other means. Membrane/ceramic filters are prepared with highly controlled pore size in a sheet of suitable material such as polyfluoroethylene, polycarbonate or cellulose esters. Nylon mesh is sometimes used for reinforcement. The pores constitute 80–85% of the filter volume commonly and several pore sizes are available for air sampling (0.45–0.8 μm are commonly employed).
1990, *62*, 2188

filter (of a radiation)
Material interposed in the path of *radiation* to modify the spectral distribution of the radiation.
1994, *66*, 2519

filter, optical
See *optical filter*.
1996, *68*, 2242

filter spectrometer
A *spectrometer* which has one or more spectral filters for isolating one or more spectral bands.
1995, *67*, 1729

filtration
The process of segregation of phases; e.g. the separation of suspended solids from a liquid or gas, usually by forcing a carrier gas or liquid through a porous medium.
1990, *62*, 2188

fine structure constant, α
Atomic fundamental physical constant, $\alpha = \mu_0 e^2 c_0 / 2h = 7.297\ 353\ 08\ (33) \times 10^{-3}$, where μ_0 is

the permeability of vacuum, *e* the elementary charge, *h* the Planck constant, and c_0 the speed of light.
CODATA Bull., 1986, *63*, 1

first-order phase transition
A transition in which the molar Gibbs energies or molar Helmholtz energies of the two phases (or chemical potentials of all components in the two phases) are equal at the transition temperature, but their first derivatives with respect to temperature and pressure (for example, specific enthalpy of transition and specific volume) are discontinuous at the transition point, as for two dissimilar phases that coexist and that can be transformed into one another by a change in a field variable such as pressure, temperature, magnetic or electric field.
Example:
The transition on heating CsCl to 752 K at which it changes from the low-temperature, CsCl-type structure to the high-temperature NaCl-type structure.
Synonymous with discontinuous phase transition.
1994, *66*, 583

Fischer projection (Fischer–Tollens projection)
A *projection formula* in which vertically drawn bonds are considered to lie below the projection plane and horizontal bonds to lie above that plane. Thus for the molecule Cabcd:

$$d \blacktriangleright \underset{c}{\overset{a}{\text{C}}} \blacktriangleleft b \equiv d \text{—}\underset{c}{\overset{a}{\text{|}}}\text{—} b$$

1996, *68*, 2208; B.B. 484; see also W.B. 128

Fischer–Rosanoff convention (or Rosanoff convention)
An arbitrary convention according to which (+)-glyceraldehyde, now known to be (*R*)-2,3-dihydroxypropanal, was named D-glyceraldehyde (with the enantiomer L-glyceraldehyde and its racemate DL-glyceraldehyde) and taken to have the absolute configuration represented by the *Fischer projection formula* shown in the diagram.

$$\text{H} \blacktriangleright \underset{\text{CH}_2\text{OH}}{\overset{\text{CHO}}{\text{C}}} \blacktriangleleft \text{OH} \equiv \text{H} \text{—}\underset{\text{CH}_2\text{OH}}{\overset{\text{CHO}}{\text{|}}}\text{—} \text{OH}$$

D-glyceraldehyde

$$\text{HO} \blacktriangleright \underset{\text{CH}_2\text{OH}}{\overset{\text{CHO}}{\text{C}}} \blacktriangleleft \text{H} \equiv \text{HO} \text{—}\underset{\text{CH}_2\text{OH}}{\overset{\text{CHO}}{\text{|}}}\text{—} \text{H}$$

L-glyceraldehyde

The atom numbered 1 according to normal nomenclature rules is conventionally placed at the top of the main chain, which is drawn vertically and other groups are drawn on either side of that main chain.

The convention is still in use for α-amino acids and for sugars.
1996, *68*, 2208

Fischer–Tollens projection
See *Fischer projection*.
1996, *68*, 2206

fissile (in radioanalytical chemistry)
1. Of a *nuclide*: capable of undergoing *fission* by interaction with (slow) neutrons.
2. Of a material: containing one or more fissile nuclides.
1982, *54*, 1542

fissionable
Synonymous with *fissile*.
1982, *54*, 1542

fission fragment ionization (in mass spectrometry)
See *plasma desorption ionization*.
1991, *63*, 1548

fission fragments
Nuclei resulting from *fission* and possessing kinetic energy acquired from the fission process.
1982, *54*, 1542

fission neutrons
Neutrons originating in the fission process which have retained their original energy.
1982, *54*, 1546

fission, nuclear
See *nuclear fission*.
1994, *66*, 2519

fission products
Nuclides produced by *fission* and the *daughter products* of these nuclides.
1982, *54*, 1542

fission yield
The fraction of *fissions* leading to *fission products* of a given type.
1982, *54*, 1542

fixed ions
In an ion exchanger, the non-exchangeable ions which have a charge opposite to that of the counterions.
1993, *65*, 854

fixed neutral loss (gain) scan (in mass spectrometry)
A scan which determines, in a single experiment, all the parent ion mass-to-charge ratios which react to the loss or gain of a selected neutral mass.
1991, *63*, 1551

fixed neutral loss (gain) spectrum (in mass spectrometry)
A spectrum obtained when data are acquired that determine all the parent ion mass-to-charge ratios that react by the loss (gain) of a selected neutral mass.
1991, *63*, 1551

fixed precursor ion scan (in mass spectrometry)
A scan which determines, in a single experiment, all the daughter ion mass-to-charge ratios that are pro-

duced by the reaction of a selected parent ion mass-to-charge ratio.
1991, *63*, 1551

fixed product ion scan (in mass spectrometry)
A scan which determines, in a single experiment, all the parent ion mass-to-charge ratios that react to produce a selected daughter ion mass-to-charge ratio (sometimes called a 'parent ion scan')
1991, *63*, 1551

fixed product ion spectrum (in mass spectrometry)
A spectrum obtained when data are acquired that determine all the parent ion mass-to-charge ratios that react to produce a selected daughter ion mass-to-charge ratio.
1991, *63*, 1551

flagpole
See *bowsprit*.
1996, *68*, 2209

flame-in-tube atomizer (in spectrochemical analysis)
An atomizer in which hydrogen, used as carrier gas, is partially combusted in the inlet arm of a T-shaped tube furnace by introducing a limited amount of oxygen or air.
1992, *64*, 2647

flame ionization detector (in gas chromatography)
The gases emerging from the column are fed into a hydrogen flame across which an electrical potential is placed. Certain molecules ionize easily in the flame and the current produced is proportional to the instantaneous flow rate of the eluted component. The detector is relatively insensitive to inorganic molecules and is most used for organic compounds. Concentrations below 1 ppmv are easily detected. The linear dynamic range is between 10^3 and 10^5.
1990, *62*, 2191

flame photometric detector (in gas chromatography)
The eluent from the column is fed into a hydrogen-rich flame and produces light emission. Optical filters are used to select the wavelength range of the emission which is characteristic of specific atoms (usually sulfur or phosphorus). The detector is very specific, depending on the choice of optical filters. It can detect the S- and P-containing compounds down to 10^{-3} ppmv, but the detector is non-linear.
1990, *62*, 2191

flame photometry
The use of emission spectroscopy in the ultraviolet and visible regions to identify and estimate the amounts of various elements which are excited in a flame, an arc or high voltage spark.
1990, *62*, 2188

flame temperature, local (in flame emission and absorption spectrometry)
See *local flame temperature (in flame emission and absorption spectrometry)*.
1986, *58*, 1741

flammable limits
See *explosivity limits*.
1990, *62*, 2188

flare, solar
See *solar flare*.
1990, *62*, 2188

flash fluorimetry (phosphorimetry)
The term used when *decay times* of *luminescence* are measured using a pulsed source of *radiation*. It is often necessary to separate the signal due to the light flash from luminescence emission signal by a deconvolution technique in order to obtain the correct *decay curve* for emission. Decay times corrected for this effect are termed corrected decay times of *fluorescence* or *phosphorescence*.
1984, *56*, 242

flash lamps
Lamps which contain an inert gas which can be rapidly pulsed.
1984, *56*, 238

flash photolysis
A technique of transient spectroscopy and transient kinetic studies in which a light pulse is used to produce transient species. Commonly, an intense pulse of short duration is used to produce a sufficient concentration of a transient species suitable for spectroscopic observation.
1996, *68*, 2242

flash point
The lowest temperature at which a substance, e.g. fuel oil, will give off a vapour that will flash or burn momentarily when ignited.
1990, *62*, 2188

flash vacuum pyrolysis (FVP)
Thermal reaction of a molecule by exposing it to a short thermal shock at high temperature, usually in the gas phase.
1994, *66*, 1115; 1993, *65*, 2407

flat band potential (at a semiconductor/solution interphase)
If effects of *surface states* are negligible, $\Delta\varphi^{sc}$, the potential drop associated with the space charge in the semiconductor, vanishes as the charge on the semiconductor becomes zero. The potential of the semiconductor corresponding to this condition is the flat band potential, which plays the same role as the potential of zero charge for metals.
1986, *58*, 444

flat bands
See *bending of energy bands*.
1986, *58*, 443

flavins
Derivatives of the dimethylisoalloxazine (7,8-dimethylbenzo[g]pteridine-2,4(3H,10H)-dione) skeleton, with a substituent on the 10 position. (Riboflavin, having a 10-D-ribityl group, is the most prominent member).

riboflavin

1995, *67*, 1335

flavonoids (isoflavonoids and neoflavonoids)
Flavonoids, isoflavonoids and neoflavonoids are natural products derived from 2-phenylchromen-4-one (flavone), 3-phenylchromen-4-one and 4-phenylcoumarin, respectively. Derivations include reduction of the 2(3) carbon-carbon double bond (flavanones), reduction of the keto group (flavanols), and hydroxylation at various positions.

2-phenylchromen-4-one   3-phenylchromen-4-one

4-phenylcoumarin

See also *anthocyanins, chalcones*.
1995, *67*, 1335

flavoproteins
Flavins tightly bound or covalently attached to a protein chain, commonly through the 8-methyl carbon atom.
1995, *67*, 1335

flavylium salts
See *anthocyanidins*.
1995, *67*, 1336

floating monolayer
See *monolayer*.
1994, *66*, 1672

floc
When a *sol* is colloidally unstable the formation of *aggregates* is called *coagulation* or *flocculation*. Some authors distinguish between coagulation and flocculation, the latter implying the formation of a loose or open network which may or may not separate macroscopically. The loose structure formed in this way is called a floc.
1972, *31*, 610

flocculation
A process of contact and adhesion whereby the particles of a dispersion form larger-size clusters.
Synonymous with *agglomeration* and *coagulation*.
1990, *62*, 2188

floccule
A small loosely aggregated mass of material suspended in or precipitated from a liquid; a cluster of particles.
1990, *62*, 2188

Flory–Huggins theory
A thermodynamic theory of polymer solutions, first formulated independently by Flory and by Huggins, in which the thermodynamic quantities of the solution are derived from a simple concept of combinatorial entropy of mixing and a reduced Gibbs energy parameter, the 'χ parameter'.
P.B. 59

flotation
The removal of matter by entrainment at an *interface*.
1972, *31*, 609

flow analysis
The generic name for all analytical methods that are based on the introduction and processing of test samples in flowing media.
1994, *66*, 2496; 1990, *62*, 2189

flow birefringence
See *streaming birefringence*.
P.B. 61

+ flow enthalpimetry
See *flow injection enthalpimetry*
1994, *66*, 2491

flow injection
The introduction of a sample or reagent into a continuous stream by use of a rapid delivery device.
1989, *61*, 1661

flow injection enthalpimetry
A term used to describe several related methods in which a transient temperature change in a flowing liquid stream, caused by a chemical reaction, is used to quantitatively determine an analyte. The analyte is introduced as a discrete liquid sample (i.e. test portion) into the flowing stream.
This definition limits the scope to experiments with primarily an analytical emphasis, performed in a flow injection calorimeter. This maintains a clear distinction from classical flow calorimetric methods where reagents are combined continuously in flowing streams, or where the heat effect is measured *via* the flowing of a fluid over a reaction vessel.
An acceptable synonym is enthalpimetric flow injection analysis. Other non-recommended terms which have been used in the literature include peak enthalpimetry and enzyme thermistor. Flow enthalpimetry has been used to describe the flow injection technique, but its use for this purpose is discouraged.
1994, *66*, 2491

flow-programmed chromatography
A procedure in which the rate of flow of the *mobile phase* is changed systematically during a part or the whole of the separation.
O.B. 92

flow rate (in chromatography)
The volume of mobile phase passing through the column in unit time. The flow rate is usually measured at column outlet, at ambient pressure (p_a) and temperature (T_a, in K); this value is indicated with the symbol F. If a water-containing flowmeter was used for the measurement (e.g. the so-called soap bubble flowmeter) then F must be corrected to dry gas conditions in order to obtain the mobile phase flow rate at ambient temperature (F_a):
$$F_a = F(1 - p_w/p_a)$$
where p_w is the partial pressure of water vapour at ambient temperature.
In order to specify chromatographic conditions in column chromatography, the flow-rate (mobile phase flow rate at column temperature, F_c) must be expressed at T_c (kelvin), the column temperature:
$$F_c = F_a(T_c/T_a)$$
1993, *65*, 839

flow rate (in flame emission and absorption spectrometry)
The volume rate of one component, X (e.g. air, O_2), of the unburnt gas mixture (such as C_2H_2, O_2, etc.) in the burner tip. Ambient temperature and pressure must be stated.
1986, *58*, 1742

flow rate (of a quantity)
Quantity X (e.g. heat, amount, mass, volume, ...) transferred in a time interval divided by that time interval. General symbols: q_X, X.
G.B. 65

flow rate of unburnt gas mixture (in flame emission and absorption spectrometry)
The volume rate of the gas mixture of the burner tip. It depends on operating conditions such as gas pressure; the ambient temperature and pressure should also be stated.
1986, *58*, 1742

flow reactor
A reactor through which the reactants pass while *catalysis* is in progress.
1976, *46*, 80

flow resistance parameter (in chromatography), Φ
A term used to compare packing density and permeability of columns packed with different particles; it is dimensionless.
$$\Phi = d_p^2/B_0$$
where d_p is the average particle diameter. In open-tubular columns $\Phi = 32$.
1993, *65*, 833

flow system, mercury
See *mercury flow system*.
1992, *64*, 263

flue gas (in atmospheric chemistry)
Waste gas from the combustion process.
1990, *62*, 2189

flue gas scrubber (in atmospheric chemistry)
Equipment for removing fly ash and other objectionable materials from the products of combustion by means of sprays or wet baffles.
1990, *62*, 2189

fluence, F, Ψ, H_0
At a given point in space, the radiant energy incident on a small sphere divided by the cross-sectional area of that sphere. It is used in photochemistry to specify the energy delivered in a given time interval (for instance by a laser pulse).
See also *dose, photon fluence*.
G.B. 31; ISO 31-5: 1992; see also 1996, *68*, 2242

fluid coke
The *carbonization* product of high-boiling hydrocarbon fractions (heavy residues of petroleum or coal processing) produced by the fluid coking process.
Notes:
Fluid coke consists of spherulitic grains with a spherical layer structure and is generally less graphitizable than *delayed coke*. Therefore, it is not suitable as *filler coke* for *polygranular graphite* products and is also less suitable for polycrystalline carbon products. Because of its isotropy it is less suitable to produce an anisotropic *synthetic graphite*. All *cokes* contain a fraction of matter that can be released as volatiles during heat treatment. This mass fraction, the so-called volatile matter, is in the case of fluid coke about 6 wt.%.
1995, *67*, 489

fluidity, φ
Reciprocal of *dynamic viscosity*.
G.B. 13

fluidized bed
A *catalyst* bed in which the flow of gases is sufficient to cause the finely divided particles to behave like a fluid.
1976, *46*, 80

fluoresceins
See *xanthene dyes*.
1995, *67*, 1374

fluorescence
Luminescence which occurs essentially only during the *irradiation* of a substance by electromagnetic *radiation*.
1994, *66*, 2519; 1984, *56*, 233

fluorescence analysis, energy dispersive X-ray
See *energy dispersive X-ray fluorescence analysis*.
1982, *54*, 1541

fluorescence error (in spectrochemical analysis)
Some samples are excited to fluorescence by the incident radiation beam. When added to the transmitted radiation beam this fluorescence results in an erroneous reading called the *fluorescence error*. It is particularly noticeable if the photodetector is situated close to the sample.
1988, *60*, 1457

fluorescence excitation (of X-rays)
See *characteristic X-ray emission*.
1991, *63*, 737

fluorescence yield
For a given transition from an excited state of a specified atom, the ratio of the number of excited atoms which emit a photon to the total number of excited atoms.
1994, *66*, 2526; see also O.B. 174

fluorimeter
An instrument used to measure the intensity and the wavelength distribution of the light emitted as fluorescence from a molecule excited at a specific wavelength or wavelengths within the absorption band of a particular compound. Characteristic fluorescence bands may be used to identify specific pollutants such as the polynuclear aromatic hydrocarbons. Excitation spectra of impurities can be observed by scanning the wavelength of the excitation light which is incident on the sample over a range of wavelengths and observing the relative intensity of the fluorescence emitted at a given wavelength. These spectra are also characteristic of the impurity.
1990, *62*, 2189

fluorocarbons
Compounds consisting wholly of fluorine and carbon.
1995, *67*, 1336

fluorohydrins
See *halohydrins*
1995, *67*, 1336

flux (of a quantity), J_X
Flow rate of X through a cross-section perpendicular to the flow divided by the cross-sectional area.
G.B. 65; 1996, *68*, 974

flux depression
The lowering of the particle (or photon) *flux density* in the neighbourhood of an object due to absorption of particles (or photons) in the object.
1994, *66*, 2519

fluxional
A *chemical species* is said to be fluxional if it undergoes rapid *degenerate rearrangements* (generally detectable by methods which allow the observation of the behaviour of individual nuclei in a rearranged chemical species, e.g. NMR, X-ray).
Example: bullvalene (1 209 600 interconvertible arrangements of the ten CH groups).
The term is also used to designate positional change among ligands of complex compounds and or-

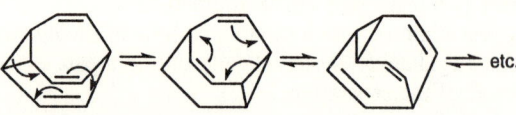

ganometallics. In these cases, the change is not necessarily degenerate.
See also *valence tautomerization*.
1994, *66*, 1115

flux, radiant energy
See *radiant energy flux, radiant power*.
1996, *68*, 2243

fly ash (in atmospheric chemistry)
Finely divided particles of ash entrained in flue gases resulting from the combustion of fuel.
1990, *62*, 2190

f number
See *oscillator strength*.
1996, *68*, 2243

foam
A dispersion in which a large proportion of gas by volume in the form of gas bubbles, is dispersed in a liquid, solid or gel. The diameter of the bubbles is usually larger than 1 μm, but the thickness of the lamellae between the bubbles is often in the usual *colloidal* size range.
The term froth has been used interchangeably with foam. In particular cases froth may be distinguished from foam by the fact that the former is stabilized by solid particles (as in *froth flotation* q.v.) and the latter by soluble substances.
1972, *31*, 606

foam fractionation
A method of separation in which a component of the bulk liquid is preferentially adsorbed at the liquid/vapour (L/V) *interface* and is removed by foaming.
1972, *31*, 609

foaming agent
A *surfactant* which when present in small amounts facilitates the formation of a *foam*, or enhances its *colloidal* stability by inhibiting the *coalescence* of bubbles.
1972, *31*, 612

fog
A general term applied to a suspension of droplets in a gas. In meteorology, it refers to a suspension of water droplets resulting in a visibility of less than 1 km.
1990, *62*, 2190; 1972, *31*, 606

fog horizon (in atmospheric chemistry)
The top of a fog layer which is confined by a low-level temperature inversion and gives the appearance of the horizon (which it actually obscures) when viewed from above, against the sky.
1990, *62*, 2190

folates

A group of heterocyclic compounds based on the pteroic acid skeleton (below) conjugated with one or more L-glutamate units.

W.B. 266

fold (in polymer crystals)

The loop connecting two different stems in a folded chain.
P.B. 83

fold domain (in polymer crystals)

A portion of a polymer crystal wherein the fold planes have the same orientation.
The sectors of lamellar crystals frequently represent fold domains.
P.B. 83

folded-chain crystal (in polymers)

A polymer crystal consisting predominantly of chains that traverse the crystal repeatedly by folding as they emerge at its external surfaces.
The re-entry of the chain into the crystal is assumed to be adjacent or near-adjacent within the lattice.
P.B. 84

fold plane (in polymer crystals)

A crystallographic plane defined by a large number of stems that are connected by chain folds.
P.B. 83

fold surface (in polymer crystals)

A surface approximately tangential to the folds.
P.B. 83

foot

Non-SI unit of length, ft = 12 in = 0.3048 m.
G.B. 110

forbidden line (in X-ray spectroscopy)

See *selection rule*.
1991, 63, 738

force, F

Derivative of momentum with respect to time.
G.B. 12; 1996, 68, 974

force constants

The coefficients in an expansion of the intramolecular potential function in terms of a definite set of coordinates, the values of which define the deformation of the molecule away from its equilibrium configuration. In order to define force constants, one has to indicate both the definition of each coordinate and the expansion of the potential.
1978, 50, 1709

force-field calculation

See *molecular mechanics calculation*.
1994, 66, 1115

foreign-gas broadening

See *collisional broadening (of a spectral line)*.
O.B. 122

formamidine disulfides

The compound $H_2NC(=NH)SSC(=NH)NH_2$ and its derivatives formed by substitution at nitrogen.
1995, 67, 1336

formation constant

For a mononuclear binary complex, if a central atom (central group) M (the 'metal') and a ligand L have been defined, then in the following expressions K_n is the stepwise formation constant, and β_n is the cumulative formation constant for the complex ML_n. They can both be referred to as stability constants (stepwise and cumulative).

$$K_n = K(ML_{n-1} + L \rightleftharpoons ML_n)$$
$$\beta_n = K(M + nL \rightleftharpoons ML_n)$$

O.B. 11

formazans

The parent compound $H_2NN=CHN=NH$ and its derivatives formed by substitution at carbon and/or nitrogen.
1995, 67, 1336

Förster cycle

An indirect method of determination of excited state equilibria, such as $pK^{*,a}$ values, based on ground state thermodynamics and electronic transition energies. This cycle considers only the difference in molar enthalpy change ($\Delta\Delta H$) of reaction of ground and excited states, neglecting the difference in molar entropy change of reaction of those states ($\Delta\Delta S$).
1996, 68, 2243

Förster excitation transfer (dipole–dipole excitation transfer)

A mechanism of excitation transfer which can occur between molecular entities separated by distances considerably exceeding the sum of their van der Waals radii. It is described in terms of an interaction between the transition dipole moments (a dipolar mechanism). The transfer rate constant $k_{D \to A}$ is given by:

$$k_{D \to A} = \frac{K^2 J \; 8.8 \times 10^{-28} \, \text{mol}}{n^4 \tau_0 r^6}$$

where K is an orientation factor, n the refractive index of the medium, τ_0 the radiative lifetime of the donor, r the distance (cm) between donor (D) and acceptor (A), and J the spectral overlap (in coherent units $cm^6 \, mol^{-1}$) between the absorption spectrum of the acceptor and the fluorescence spectrum of the donor. The critical quenching radius, r_0, is that distance at which $k_{D \to A}$ is equal to the inverse of the radiative lifetime.

See also *Dexter excitation transfer, energy transfer, radiative energy transfer*.
1996, 68, 2243

forward scattering

See *backscatter*.
1983, 55, 932

fossil fuel
A fuel such as coal, oil and natural gas which has formed over many years through the decomposition of deposited vegetation which was under extreme pressure of an overburden of earth.
1990, *62*, 2190

fouling agent (in catalysis)
Fouling agents (or mechanical inhibitors) are inhibitory substances bound by neither covalent nor other strong bonds to the active centres: the interaction is usually of the van der Waals, H-bond or sometimes ionic, type. They form protective layers or block pores, thus physically impeding access of reactants to the active centres. The fouling agents which cause real problems are those which have a long standing effect and do not disappear spontaneously. Carbon deposits act, partially or totally, this way (see *coking*). Other examples are vanadium and nickel sulfide deposits in hydrotreating catalysts.
1991, *63*, 1244

Fourier transform ion cyclotron resonance (FT-ICR) mass spectrometer
A high-frequency mass spectrometer in which the cyclotron motion of ions, having different mass/charge ratios, in a constant magnetic field is excited essentially simultaneously and coherently by a pulse or a radio-frequency electric field applied perpendicular to the magnetic field. The excited cyclotron motion of the ions is subsequently detected on so-called receiver plates as a time domain signal that contains all the cyclotron frequencies that have been excited. Fourier transformation of the time domain signal results in the frequency domain FT-ICR signal which, on the basis of the inverse proportionality between frequency and the mass/charge ratio, can be converted into a mass spectrum. The term is sometimes contracted to Fourier transform mass spectrometer (FT-MS).
See also *ion cyclotron resonance (ICR) mass spectrometer*.
1991, *63*, 1545

Fourier transform spectrometer
A scanning interferometer, containing no principal dispersive element, which first splits a beam into two or more components, then recombines these with a phase difference. The spectrum is obtained by a Fourier transformation of the output of the interferometer.
See also *multiplex spectrometer*.
1996, *68*, 2243

fraction
In general, a ratio of two quantities of the same kind, the numerator quantity applying to one constituent (or part) of the system and the denominator to the sum of quantities for all constituents (parts) of the system. When applied to mixtures fractions represent a group of 3 quantities: *mass fraction*, *volume fraction* and *amount fraction* (or mole fraction equal to the number fraction).
G.B. 41; 1996, *68*, 974

fractional change of a quantity
A term which may be expressed infinitesimally at time t by the differential $dQ(t)/Q(t)$. For a finite time interval the quotient is

$$\Delta Q(t_1; t_2)/Q(t_1) = [Q(t_2) - Q(t_1)]/Q(t_1)$$

The quantities $Q(t_1)$ and $Q(t_2)$ are of the same kind and have the same type of component. Fractional change has dimension one.
Examples are: mass fractional change, $dm(t)/m(t)$; amount of substance fractional change, $dn(t)/n(t)$.
1992, *64*, 1571

fractional selectivity (in catalysis)
The term *selectivity* (S) is used to describe the relative rates of two or more competing reactions on a *catalyst*. Such *competition* includes cases of different reactants undergoing simultaneous reactions or of a single reactant taking part in two or more reactions. For the latter case, S may be defined in two ways.
The first of these defines a fractional selectivity, S_F, for each product by the equation

$$S_F = \dot\xi_i / \Sigma \dot\xi_i$$

The second defines relative selectivities, S_R, for each pair of products by

$$S_R = \dot\xi_i / \dot\xi_j$$

In each case, $\dot\xi_i$ and $\dot\xi_j$ are the rates of increase of the *extent of reactions* i and j respectively, i.e.

$$\dot\xi_i = d\xi_i/dt \quad \text{and} \quad \dot\xi_j = d\xi_j/dt$$

where ξ_i and ξ_j are the extents of reactions i and j respectively.
1976, *46*, 81

fractionation (of polymers)
A process by means of which macromolecular species differing in some characteristic (chemical composition, relative molecular mass, branching, stereoregularity, etc.) are separated from each other.
P.B. 68

fractionation factor, isotopic
See *isotopic fractionation factor*.
1994, *66*, 1115

fraction atomized, local (in flame emission and absorption spectrometry)
See *local fraction atomized (in flame emission and absorption spectrometry)*.
1986, *58*, 1741; O.B. 168

fraction, bound
See *bound fraction*.
1994, *66*, 2519

fraction collector (in chromatography)
A device for recovering fractional volumes of the column effluent.
1993, *65*, 829

fraction desolvated, local (in flame emission and absorption spectrometry)
See *local fraction desolvated (in flame emission and absorption spectrometry)*.
1986, *58*, 1740; O.B. 168

fraction extracted, E
The fraction of the total quantity of a substance extracted (usually by the solvent) under specified conditions, i.e. $E_A = Q_A/Q'_A$ where Q_A is the mass of A extracted and Q'_A is the total mass of A present at the start.
Notes:
i. E may be expressed as a percentage, $\%E$.
ii. The term extractability is qualitative and should not be used as a synonym for fraction extracted.
iii. If the aqueous phase is extracted with n successive portions of solvent, the phase volume ratio (solvent/feed) being r each time, the fraction extracted is given by:

$$E_n = 1 - (rD + 1)^{-n}$$

If $n = r = 1$, $E_1 = D/(1 + D)$ this expression is a concept of value in chromatography theory.
iv. The fraction extracted is also known as the recovery factor, especially for a multistage process.
1993, *65*, 2384

fraction, free
See *free fraction*.
1994, *66*, 2519

fraction volatilized, local (in flame emission and absorption spectrometry)
See *local fraction volatilized (in flame emission and absorption spectrometry)*.
1986, *58*, 1741; O.B. 168

fragmentation
1. The *heterolytic* cleavage of a molecule according to the general reaction:

$$a–b–c–d–X \rightarrow (a–b)^+ + c=d + X^-$$

where a–b is an *electrofuge* and X is a *nucleofuge* (which may emerge from the reaction in combined form), and the middle group affords the unsaturated fragment c=d. For example,

$$Ph_3C–CO_2H + H^+ \rightarrow Ph_3C^+ + C=O + H_2O$$

2. The breakdown of a *radical* into a diamagnetic molecule or ion and a smaller radical, e.g.

$$(CH_3)_3C–O· \rightarrow (CH_3)_2C=O + H_3C·$$

$$[ArBr]·^- \rightarrow Ar· + Br^- \quad (\text{solution})$$

3. The breakdown of a *radical ion* in a mass spectrometer or in solution, forming an ion of lower molar mass and a radical, e.g.

$$[(CH_3)_3C–OH]·^+ \rightarrow (CH_3)_2C=O^+H + H_3C·$$
(mass spectrometer)

See also *fragmentation reaction (in mass spectrometry)*.
1994, *66*, 1116

fragmentation reaction (in mass spectrometry)
A fragmentation reaction may be written:

$$M_1^+ \rightarrow M_2^+ + M_3$$

The decomposition of a metastable ion of mass-to-charge ratio m_1/z_1 into an ion of mass-to-charge ratio m_2/z_2 after electric acceleration of the metastable ion and before magnetic deflection gives rise to a peak in the mass spectrum at an apparent mass, $m_2^2 z_1/(m_1 z_2^2)$. The symbol m^* should be used to indicate the apparent mass of the product ions giving rise to this peak. Traditionally, the peak itself has been called a *metastable peak* and this should preferably be expressed as *metastable ion peak*. It is nevertheless recommended that the former term should be retained. The word metastable should never be used as a noun. Its use as an adjective should be limited to such terms as metastable ion, where it is used correctly, and to the special case of metastable peak discussed above. It should never be used in such terms as metastable reaction, metastable decomposition, metastable studies, etc. It should be clear that the metastable ion is the ion that undergoes fragmentation; it is not detected. When a reaction is written with an asterisk above the arrow as shown:

$$M_1^+ \xrightarrow{*} M_2^+ + M_3$$

this means that the reaction has been confirmed by the observation of a metastable peak.
The textual description of such a process may be written as, for example, (m^*, 43 $\rightarrow$ 28, calcd. 18.2, obsd. 18.3) meaning 'for the fragmentation m/z 43 $\rightarrow$ m/z 28 a metastable peak at m/z 18.3 (calculated 18.2) has been observed'.
1991, *63*, 1557

fragment ion (in mass spectrometry)
An electrically charged dissociation product of an ionic fragmentation. Such an ion may dissociate further to form other electrically charged molecular or atomic moieties of successively lower formula weight.
See also *daughter ion*.
1991, *63*, 1549; O.B. 205; 1990, *62*, 2190

frame-shift mutation
A point *mutation* involving either the deletion or insertion of a *nucleotide* in a *gene*. By the frame-shift mutation, the normal reading frame used when decoding the *nucleotide* triplets in the gene is altered to another reading frame.
1992, *64*, 154

Franck–Condon principle
Classically, the Franck–Condon principle is the approximation that an electronic transition is most likely to occur without changes in the positions of the nuclei in the molecular entity and its environment. The resulting state is called a Franck–Condon state, and the transition involved, a vertical transition.
The quantum mechanical formulation of this principle is that the intensity of a vibronic transition is propor-

tional to the square of the overlap integral between the vibrational wavefunctions of the two states that are involved in the transition.
1996, *68*, 2243

Franck–Condon state
See *Franck–Condon principle*.
1996, *68*, 2244

franklin
esu unit of electric charge, Fr ≈ 3.335 64 × 10^{-10} C.
G.B. 114

free charge density on the interface
The physical *charge density* believed to occur on either side of the *electrical double layer*.
1974, *37*, 508

free electron laser
Source of coherent radiation in which the active medium is an electron beam moving at speeds close to the speed of light in the spatially periodic magnetic field produced by an array of magnets (the wiggler). The emitted wavelength, λ_L, is approximately given by $\lambda_\omega/(4E^2)$, with λ_ω being the wiggler period and E the electron's energy in MeV.
See *laser*.
1996, *68*, 2244

free energy
See *Gibbs energy (function)*.
G.B. 48; 1996, *68*, 975; 1990, *62*, 2192

freely draining
An adjective referring to a chain macromolecule the segments of which produce such small frictional effects when moving in a medium that the hydrodynamic field in the vicinity of a given segment is not affected by the presence of other segments. Thus, the solvent can flow virtually undisturbed through the domain occupied by a freely draining macromolecule.
P.B. 61

freely jointed chain (in polymers)
A hypothetical linear chain molecule consisting of infinitely thin rectilinear segments uniform in length; each segment can take all orientations in space with equal probability, independently of its neighbours.
For models in which the segments are not all uniform in length, the name 'random-walk chain' has been used.
P.B. 50

freely rotating chain (in polymers)
A hypothetical linear chain molecule, free from short-range and long-range interactions, consisting of infinitely thin rectilinear segments (bonds) of fixed length, jointed at fixed bond angles; the torsion angles of the bonds can assume all values with equal probability.
P.B. 50

free radical
See *radical*.
1994, *66*, 1116; 1990, *62*, 2190

free rotation (hindered rotation, restricted rotation)
In a stereochemical context the rotation about a bond is called 'free' when the *rotational barrier* is so low that different *conformations* are not perceptible as different chemical species on the time scale of the experiment.
The inhibition of rotation of groups about a bond due to the presence of a sufficiently large *rotational barrier* to make the phenomenon observable on the time scale of the experiment is termed hindered rotation or restricted rotation.
1996, *68*, 2209

free-running laser
A term applied to a pulsed laser, meaning that the laser emission lasts as long as the pumping process is sufficient to sustain lasing conditions. Typical pulse durations are in the μs–ms range, depending on the pumping source. When the operation mode of a pulsed laser is not specified as mode-locked, Q-switched, or anything else, it must be considered as free-running.
1996, *68*, 2244

free spiro union
See *spiro union*.
B.B. 37

freezing
The conversion of a liquid to a solid by lowering the temperature and/or the application of pressure.
1994, *66*, 583

freezing out (in atmospheric chemistry)
A term used in combustion for the analysis of incinerator flue gas components (largely the organic fraction) in which a series of traps at progressively lower temperatures is employed. In more general use, the term implies the removal of a condensable gas or liquid by condensation in a trap at low temperatures.
1990, *62*, 2190

frequency, f, ν
Reciprocal of the *period*.
G.B. 11; see also 1996, *68*, 974; 1996, *68*, 2244

frequency distribution
If the range of observed measurement values is subdivided into a regular sequence of smaller intervals, this distribution is a plot of the frequency of occurrence of values falling into each interval.
1990, *62*, 2190

frequency doubling
See *harmonic frequency generation, non-linear optical effects*.
1996, *68*, 2244

frictional coefficient (in polymer chemistry), f
A tensor correlating the frictional force, F, opposing the motion of a particle in a viscous fluid and the velocity u of this particle relative to the fluid.
In the case of an isolated spherical particle in a viscous isotropic fluid, f is a constant and $F = fu$.
P.B. 60

friction coefficient
See *friction factor*.
1996, *68*, 974

friction factor, μ
Quotient of frictional force and normal force, for a sliding body. Formerly called frictional coefficient or coefficient of friction.
1996, *68*, 974

fringed-micelle model (in polymer crystals)
A model of crystallinity in which the crystallized segments of a macromolecule belong predominantly to different crystals.
P.B. 84

frontal chromatography
A procedure in which the sample (liquid or gas) is fed continuously into the chromatographic bed. In frontal chromatography no additional mobile phase is used.
1993, *65*, 824

frontier orbitals
The *H*ighest-energy *O*ccupied *M*olecular *O*rbital (HOMO) (filled or partly filled) and *L*owest-energy *U*noccupied *M*olecular *O*rbital (LUMO) (completely or partly vacant) of a *molecular entity*. Examination of the mixing of frontier molecular orbitals of reacting molecular entities affords an approach to the interpretation of reaction behaviour; this constitutes a simplified perturbation *molecular orbital* theory of chemical behaviour.
See also *SOMO, subjacent orbital*.
1994, *66*, 1116

fronting (in chromatography)
Asymmetry of a peak such that, relative to the baseline, the front is less steep than the rear. In paper chromatography and thin-layer chromatography, it refers to the distortion of a spot showing a diffuse region in front of the spot in the direction of flow.
1993, *65*, 837; O.B. 96

front surface geometry (in luminescence)
A system where excitation and observation are from the same face of the sample ($\alpha \angle 90°, \beta \angle 180°$ where α = angle of incidence of the exciting beam on the plane surface of the sample, and β = angle between the exciting direction and observation direction.)
1984, *56*, 244

frost point hygrometer
Instrument in which the sample is passed over a cooled surface. The temperature at which frost forms on it is a function of the water content of the gas passing over the surface.
1990, *62*, 2195

froth flotation
the removal of particulate matter by foaming.
1972, *31*, 609

Frumkin effect
In an *electrode reaction*, when reactants or *intermediates* are adsorbed, the *rate of reaction* may no longer be related to the *concentration* by a simple law. The deviation may be due to either entropic or energetic effects or both. The situation best understood is that where a reactant is non-specifically adsorbed in the outer Helmholtz plane (inner boundary of the *diffuse layer*). The effect of such *adsorption* on electrode kinetics is usually termed the Frumkin effect. *Rate constants*, transfer coefficients etc. corrected for this effect are frequently called 'true' rate constants etc. It would be preferable to describe them as 'corrected for the Frumkin effect', but in any case, if such a correction is carried out, the basis on which it is made should be clearly described.
1980, *52*, 239

fucolipid
See glycolipids.
W.B. 184

fuel cycle (nuclear)
The sequence of steps, such as utilization, reprocessing, and refabrication, through which *nuclear fuel* may pass.
1982, *54*, 1543

fuel element (nuclear)
The smallest structurally discrete part of a *reactor* which has fuel as its principal constituent.
1982, *54*, 1543

fuel reprocessing (nuclear)
The processing of *nuclear fuel*, after its use in a *reactor*, to remove *fission products* and recover *fissile* and *fertile* material.
1982, *54*, 1543

fugacity, f, $\tilde{p}$
Of a substance B, f_B or $\tilde{p}_B$, in a gaseous mixture is defined by $f_B = \lambda_B \lim_{p \to 0} (p_B/\lambda_B)$, where p_B is the partial *pressure* of B and λ_B its *absolute activity*.
G.B. 50; see also 1994, *66*, 539; 1984, *56*, 571

fugacity coefficient, φ
Ratio of *fugacity* to the partial *pressure* of a gaseous constituent.
G.B. 50; 1994, *66*, 539

fulgides
Dialkylidenesuccinic anhydrides (generally photochromic).

1995, *67*, 1336

full energy peak
Of a *spectrum* of *radiation*, the part of the spectral response curve corresponding to the total *absorption* in a detecting material of the energy of the detected radiation.
1982, *54*, 1543

fullerenes
Compounds composed solely of an even number of carbon atoms, which form a cage-like fused-ring

polycyclic system with twelve five-membered rings and the rest six-membered rings. The archetypal example is [60]fullerene, where the atoms and bonds delineate a truncated icosahedron. The term has been broadened to include any closed cage structure consisting entirely of three-coordinate carbon atoms.
1995, *67*, 1336

fulminates
1. Compounds having the structure RON=C:. So called because fulminic acid (actually HC≡N⁺–O⁻ formonitrile oxide) was previously considered to be HON=C:.
2. Salts of fulminic acid, e.g. Na⁺[–C≡N⁺–O⁻].
1995, *67*, 1336

fulvalenes
The hydrocarbon fulvalene and its derivatives formed by substitution (and by extension, analogues formed by replacement of one or more carbon atoms of the fulvalene skeleton by a heteroatom).

1995, *67*, 1336

fulvenes
The hydrocarbon fulvene and its derivatives formed by substitution (and by extension, analogues formed by replacement of one or more carbon atoms of the fulvene skeleton by a heteroatom).

1995, *67*, 1336

fume (in atmospheric chemistry)
Fine solid particles (aerosol), predominantly less than 1 μm in diameter, which result from the condensation of vapour from some types of chemical reaction. Usually this is formed from the gaseous state generally after volatilization from melted substances and often accompanied by chemical reactions such as oxidation.
1990, *62*, 2190

fumes
In popular usage, a term often taken to mean airborne effluents, unpleasant and malodorous, which might arise from chemical processes.
See also *smoke*.
1990, *62*, 2191

fumigation (in atmospheric chemistry)
An atmospheric phenomenon in which pollution, retained by an inversion layer near its level of emission, is brought rapidly to ground level as the inversion breaks up. This term also applies to the exposure of material (e.g. grain) to chemicals to kill insects, etc.
1990, *62*, 2191

functional class name
A name that expresses the characteristic group as a class term written as a separate word following the name of a parent structure or a name derived from a parent structure. In the latter case, when the derived name is that for a substituent group (formerly called a 'radical'), the method has been called 'radicofunctional nomenclature'.
B.B.(G) 14

functional group
Organic compounds are thought of as consisting of a relatively unreactive backbone, for example a chain of sp³ hybridized carbon atoms, and one or several functional groups. The functional group is an atom, or a group of atoms that has similar chemical properties whenever it occurs in different compounds. It defines the characteristic physical and chemical properties of families of organic compounds.
1994, *66*, 1116

functional parent
A structure the name of which implies the presence of one or more *characteristic groups* and which has one or more hydrogen atoms attached to at least one of its skeletal atoms or one of its characteristic groups, or in which at least one of its characteristic groups can form at least one kind of functional modification.
Note:
A parent hydride bearing a characteristic group denoted by a suffix, for example, cyclohexanol, is not considered to be a functional parent, but may be described as a 'functionalized parent hydride'.
B.B.(G) 13

fungicide
See *antimycotic*.
1993, *65*, 2015

furanocoumarins
An alternative name for furocoumarins.
1995, *67*, 1336

furanoses
Cyclic hemiacetal forms of monosaccharides in which the ring is five-membered (i.e. a tetrahydrofuran skeleton).

1995, *67*, 1336

furnace black
A type of *carbon* that is produced industrially in a furnace by incomplete combustion in an adjustable and controllable process that yields a wide variety of properties within the product.
Note:
The most widely employed industrial process for *carbon black* production is the furnace process.
See also *gas black*.
1995, *67*, 490

furnace pyrolysis (in spectrochemical analysis)
A flowing stream of gas (hydrogen, oxygen, nitrogen, chlorine, etc.) required to produce volatile species of the elements being determined, is passed over the *test sample* in a heated furnace. The analytes leave the

furocoumarins

furnace in the gas stream, or are entrained by a carrier gas. The analytes in the gas stream may be collected in an absorbing solution, on a carbon or other filter or by condensation on a cool surface. In the case of mercury this can also be done by amalgamation with a noble metal. The analytes may then be swept and released from the trap, by heating, into a sampling source for analysis.
1988, *60*, 1470

furocoumarins
Derivatives (mostly of natural origin) of the linear furocoumarin skeleton, psoralen, or its angular isomer, angelicin, variously substituted with hydroxy, methoxy, alkyl, or hydroxymethyl groups, among others.

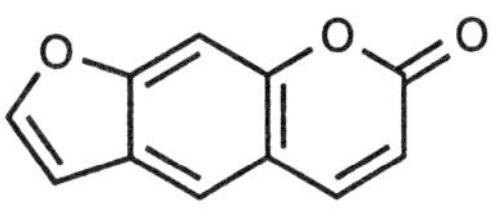

psoralen 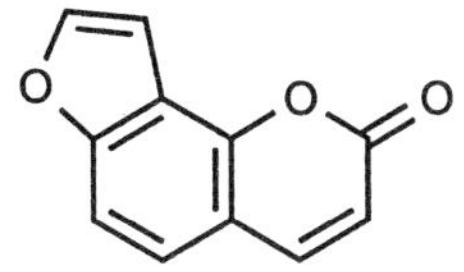   angelicin

1995, *67*, 1337

fusion (in biotechnology)
The amalgamation of two distinct *cells* or macromolecules into a single integrated unit.
1992, *64*, 154

fusion name
A composite name for a polycyclic parent structure having the maximum number of non-cumulative double bonds and at least one *ortho* fusion. Name formation involves the dissection of the structure into contiguous components having recognized trivial or semisystematic names, one of which is selected as the 'base component'. Attachment of the other components is described by prefixes.
B.B.(G) 14

FWHM (Full Width at Half Maximum)
See *half-width (of a band)*, *line width*.
1996, *68*, 2244; G.B. 31; see also O.B. 226

γ–cascade
Two or more different gamma rays emitted successively from one *nucleus* when it passes through one or more energy levels.
1982, *54*, 1543

γ-quantum
A *photon* of *γ-radiation*.
1982, *54*, 1543

γ-radiation
Electromagnetic *radiation* emitted in the process of *nuclear transformation* or particle *annihilation*.
1982, *54*, 1543

γ-radiation, capture
See *capture γ-radiation*.
1982, *54*, 1543

γ-ray spectrometer
A measuring assembly for determining the energy *spectrum* of γ-rays.
1982, *54*, 1543

G value (in nuclear chemistry)
The number of specified chemical events in an irradiated substance produced per 100 eV of energy absorbed from *ionizing radiation*.
1982, *54*, 1544

gain (of a photomultiplier), G
The gain of a *photomultiplier* is given by the equation $G = k\sigma^n$, where k is the efficiency of collection of photoelectrons on the first dynode, σ is the secondary emission ratio, i.e. the number of secondary electrons emitted for each electron incident on the dynode and n is the number of dynodes.
1995, *67*, 1753

galvanic corrosion
See *contact corrosion*.
1989, *61*, 22

Galvani potential difference
An electric potential difference between points in the bulk of two phases. It is measurable only when the two phases have identical composition (e.g. two copper wires). It is the difference of *inner electric potentials* in two phases. Symbol: $\Delta_\alpha^\beta \varphi$.
G.B. 59

gamma
See entries under γ (beginning of 'g').

gas analysis installation (in atmospheric chemistry)
An assembly for the purpose of determining one or more components in a gaseous mixture. It generally comprises the following elements: sample probe, region of primary treatment of the gases, region of secondary treatment of gases, points for venting to the atmosphere and for calibration sample injection, transfer line and the analytical unit.
1990, *62*, 2191

gas black (carbon black, channel black, furnace black)
Finely divided carbon (graphite) produced by incomplete combustion or thermal decomposition of natural gas.
1990, *62*, 2191

gas chromatography (GC)
A separation technique in which the mobile phase is a gas. Gas chromatography is always carried out in a column.
1993, *65*, 825; O.B. 93; 1990, *62*, 2191

gas constant
Fundamental physical constant, R = 8.314 510 (79) J K^{-1} mol^{-1}.
CODATA Bull., 1986, *63*, 1; 1996, *68*, 974

gaseous diffusion separator (in atmospheric chemistry)
Instrumentation to separate a gas mixture into its components by means of differences in the diffusion rates of the component molecules. This method has been used in separation of various isotopes of uranium (by means of UF$_6$) and hydrogen (H$_2$O, HDO, D$_2$O, HTO, etc.)
1990, *62*, 2192

gas-filled phototube
Similar in construction to a *vacuum phototube* except that it is filled with a noble gas (usually Ar) at a pressure of about 10 Pa. Photoelectrons accelerated by the anode electric potential ionize gas atoms. The additional electrons provide a substantial intrinsic gain.
1995, *67*, 1753

gas-filled X-ray detector
A cylindrical cathode with a window, an axial wire anode and an ionizable gas. The gas may be continuously replenished giving a flow-through detector or the detector may be sealed. Following an original ionizing event, electron multiplication occurs through a process of gas amplification in the high electric field surrounding the anode wire. The gain of this process is defined as the number of electrons collected on the anode wire for each primary electron produced. For X-rays having energies higher than the excitation potential of the detector gas, the *spectral responsivity function* has a second peak in addition to the main peak that is called the escape peak. The escape peak has a mean pulse height proportional to the difference between the photon energy of the incident X-rays and of the spectral characteristic line of the detector gas. A quenching gas, a molecular gas, is added to the detector gas in order to neutralize the detector gas ions and to absorb secondary electrons as well as UV radiation resulting from neutralization of detector gas ions. According to the potential applied to the anode, the detector can work as an ionization chamber, proportional counter, or Geiger counter.
1995, *67*, 1753

gas flow system, collection (in spectrochemical analysis)
See *collection gas flow system (in spectrochemical analysis)*.
1992, *64*, 263

gas flow system, direct transfer (in spectrochemical analysis)
See *direct transfer gas flow system (in spectrochemical analysis)*.
1992, *64*, 263

gas laser
A *laser* in which the *active medium* is a gas containing a laser *active species*. It is typically composed of a plasma discharge tube containing a gas that can be excited with an electric discharge.
1995, *67*, 1919; see also 1996, *68*, 2244

gas-liquid chromatography
Comprises all gas-chromatographic methods in which the *stationary phase* is a liquid dispersed on a solid support. Separation is achieved by partition of the components of a sample between the phases.
O.B. 93

gas-phase acidity
The negative of the Gibbs energy (ΔG_r^o) change for the reaction:

A–H $\rightarrow$ A$^-$ + H$^+$

in the gas phase.
1994, *66*, 1116

gas-phase basicity
The negative of the Gibbs energy (ΔG_r^o) change associated with the reaction:

B + H$^+$ $\rightarrow$ BH$^+$

in the gas phase. Also called absolute or intrinsic basicity.
See also *proton affinity*.
1994, *66*, 1116

gas-phase-grown carbon fibres
Carbon fibres grown in an atmosphere of hydrocarbons with the aid of fine particulate solid catalysts such as iron or other transition metals and consisting of *graphitizable carbon*.
Notes:
Gas-phase-grown carbon fibres transform during *graphitization heat treatment* into *graphite fibres*. These show a very high degree of preferred orientation and are particularly suitable for intercalation treatments. The term 'vapour-grown carbon fibres' alternatively used in the literature is acceptable. The use of the term 'CVD fibres' is not recommended as an alternative for gas-phase-grown carbon fibres since the term 'CVD fibres' also describes fibres grown by a chemical vapour deposition (CVD) process on substrate fibres.
1995, *67*, 490

gas sensing electrode
A sensor composed of an indicator and a reference electrode in contact with a thin film of solution which is separated from the bulk of the sample solution by a gas-permeable membrane or an air gap. This intermediate solution interacts with the gaseous species (penetrated through the membrane or an air gap) in such a way as to produce a change in a measured constituent (e.g. the H$^+$ activity) of the intermediate solution. This change is then sensed by the *ion-selective electrode* and is related to the partial pressure of the gaseous species in the sample. [Note: In electrochemical literature the term gas electrode is used for the classical, redox-equilibrium-based gas electrodes as well, such as the hydrogen or the chlorine gas electrodes (Pt(s)|H$_2$(g) |H$^+$(aq) or Pt(s) |Cl$_2$(g) |Cl$^-$(aq)]. These electrodes respond both to the partial pressure of the gas (H$_2$ or Cl$_2$) and to the ionic activities (H$^+$ or Cl$^-$). The Clark oxygen electrode fits under this classification although, in contrast to other gas sensors, it is an amperometric and not a potentiometric device.
1994, *66*, 2534

gas-solid chromatography
Comprises all gas chromatographic methods in which the stationary phase is an active solid (e.g. charcoal, molecular sieves). Separation is achieved by adsorption of the components of a sample.
In gas chromatography the distinction between gas-liquid and gas-solid may be obscure because liquids are used to modify solid stationary phases, and because the solid supports for liquid stationary phases affect the chromatographic process. For classification by the phases used, the term relating to the predominant effect should be chosen.
O.B. 93

gated photodetector
Where photodetectors are switched on (or off) usually in a repetitive manner employing electronic switches, they are described as gated.
O.B. 193

gauche
Synonymous with a synclinal alignment of groups attached to adjacent atoms.
See *torsion angle*.
1996, *68*, 2209; B.B. 484

gauche conformation (in polymers)
See cis *conformation (in polymers)*.
P.B. 41

gauche effect
1. The stabilization of the gauche (synclinal) conformation in a two carbon unit bonded vicinally to electronegative elements e.g. 1,2-difluoroethane.
2. The destabilization of the gauche (synclinal) conformation in a two carbon unit bonded vicinally to large, soft and polarizable elements such as sulfur and bromine.
1996, *68*, 2209

gauss
emu unit of magnetic flux density, G = 10^{-4} T.
G.B. 115

Gaussian band shape
A band shape described by the Gaussian function:
$$F(v - v_0) = (a/\sqrt{\pi}) \exp[-a^2(v - v_0)^2]$$
In this equation, a^{-1} is proportional to the width of the band, and v_0 is the frequency of the band maximum.
See also *Lorentzian band shape*.
1996, *68*, 2244; see also 1990, *62*, 2192

Geiger counter
A gas-filled X-ray detector in which gas amplification reaches saturation and proportionality no longer exists. The output signal does not depend on the incident energy. The time taken for the counter to recover from the saturation is called *dead time*.
1995, *67*, 1754

Geiger–Muller counter tube
A *counter tube* operated under such conditions that the magnitude of each pulse is independent of the amount of energy deposited in it.
1982, *54*, 1538

gel
A *colloidal* system with a finite, usually rather small, *yield stress*.
1972, *31*, 606

gel permeation chromatography
See *permeation chromatography*.
O.B. 94

geminate pair
Pair of *molecular entities* in close proximity in solution within a solvent cage and resulting from reaction (e.g. bond scission, electron transfer, group transfer) of a precursor that constitutes a single kinetic entity.
See also *ion pair, radical pair*.
1994, *66*, 1117

geminate recombination
This expression refers to the reaction, with each other, of two *transient species* produced from a common precursor in solution. If reaction occurs before any separation by diffusion has occurred, this is termed primary geminate recombination. If the mutually reactive entities have been separated, and come together by diffusion, this is termed secondary geminate recombination. This is illustrated in the reaction diagram below:

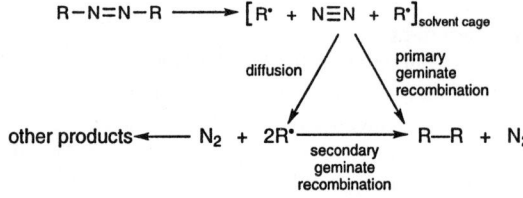

See also *cage*.
1996, *68*, 166; 1994, *66*, 1117

gene (cistron)
Structurally, a basic unit of hereditary material; an ordered sequence of nucleotide bases that encodes one polypeptide chain (via mRNA). The gene includes, however, regions preceding and following the coding region (leader and trailer) as well as (in *eukaryotes*) intervening sequences (*introns*) between individual coding segments (*exons*). Functionally, the gene is defined by the *cis-trans* test that determines whether independent *mutations* of the same *phenotype* occur within a single gene or in several genes involved in the same function.
1992, *64*, 154

gene amplification
An increase in the number of copies of a specific *gene* in an organism. This can lead to the production of a corresponding protein at elevated levels.
1992, *64*, 154

gene library
A collection of cloned DNA fragments from a variety of species.
1992, *64*, 155

gene manipulation
The use of *in vitro* techniques to produce DNA molecules containing novel combinations of *genes* or altered sequences, and the insertion of these into *vectors* that can be used for their incorporation into *host* organisms or *cells* in which they are capable of continued propagation of the modified genes.
1992, *64*, 155

general acid–base catalysis
Catalysis by acids or bases in solution is said to be general when it is possible to detect *catalysis* brought about by species other than the ions formed from the solvent itself (e.g. when water is the solvent, by species other than H^+ and OH^- ions).
See *catalysis, general acid catalysis, general base catalysis*.
1996, *68*, 150

general acid catalysis
The *catalysis* of a chemical reaction by a series of *Brønsted acids* (which may include the solvated hydrogen ion) so that the rate of the catalysed part of the reaction is given by $\Sigma k_{HA}[HA]$ multiplied by some function of *substrate* concentrations. (The acids HA are unchanged by the overall reaction.) General catalysis by acids can be experimentally distinguished from *specific catalysis* by hydrogen cations (*hydrons*) by observation of the *rate of reaction* as a function of buffer concentration.
See also *catalysis, catalytic coefficient, intramolecular catalysis, pseudo-catalysis, specific catalysis*.
1994, *66*, 1117

general base catalysis
The catalysis of a *chemical reaction* by a series of *Brønsted bases* (which may include the *lyate* ion) so that the rate of the catalysed part of the reaction is given by $\Sigma k_B[B]$ multiplied by some function of *substrate* concentration.
See also *general acid catalysis*.
1994, *66*, 1117

general force field (gff)
A force field expressed in terms of $3N - 6$ basis coordinates:
$$V = \frac{1}{2}\sum f_{ij}(\text{or}\, F_{ij})S_i S_j$$
where f_{ij} (or F_{ij}) are force constants and the basis coordinates S (or sometimes s) may be internal symmetry coordinates, local symmetry coordinates or any others suitable to the problem, but the number of the coordinates has to be reduced to $3N - 6$ ($3N - 5$ for linear molecules), N being the number of atoms in the molecule.
1978, *50*, 1709

generalized transition-state theory
Any form of *transition-state theory* (TST), such as *microcanonical variational TST*, *canonical variational TST*, and *improved canonical variational TST*, in which the transition state is not necessarily located at the saddle point, is referred to as generalized transition-state theory.
1996, *68*, 166

generally labelled tracer
A *tracer* in which the position of the *label* is not defined.
1994, *66*, 2526

generation time (in biotechnology), t_G
The average time required by a *cell* for the completion of one *cell cycle*.
1992, *64*, 155

genetic code
The set of rules which governs the relationship between the linear order of the *nucleotides* in an mRNA molecule and the sequence of the amino acids in the protein which it encodes. The genetic code is a triplet code, nearly universal. It is not overlapping: a *mutation* which alters only a single nucleotide in a gene can change only one amino acid in the encoded protein. It is degenerate: it assigns each of the 20 amino acids to one or more of 61 of the 64 possible nucleotide triplets (cf. *codon*) that can be constructed from four nucleotides. The remaining three triplets are used to signal the termination of a genetic message.
1992, *64*, 155

genome
The complete set of chromosomal and extrachromosomal *genes* of an organism, a *cell*, an *organelle* or a *virus*; the complete DNA component of an organism.
1992, *64*, 155

genotype
The genetic constitution of an organism as revealed by genetic or molecular analysis, i.e. the complete set of *genes*, both dominant and recessive, possessed by a particular *cell* or organism.
1992, *64*, 155

geometrical equivalence (in polymers)
The symmetry correspondence among units belonging to the same chain. The symmetry elements always bear a special relationship to the chain axis.
See also *line repetition groups*.
P.B. 77

geometric attenuation
The reduction of a *radiation* quantity due to the effect only of the distance between the point of interest and the source, excluding the effect of any matter present.
1982, *54*, 1543

+ geometric isomerism
Obsolete synonym for cis-trans *isomerism*. (Usage strongly discouraged).
1996, *68*, 2209

geometric (logarithmic) mean, $\bar{x}_g$
The nth root of the product of the absolute values of the observations, taken with the proper sign. It can be calculated with the formula:
$$\bar{x}_g = (\Pi\, |x_i|)^{1/n}$$
The Π product is taken from $i = 1$ to n.
Comment:
This quantity is often calculated directly from experimental measurements (e.g. determination of concentrations by electrode potential measurements, or pH), although its significance may not always have been recognized. The problem is that the average value of a variable (such as pH) that is a function of concentration is not the same as the value of the function at the average concentration. In the case of electrode potentials, the average potential is equivalent to the geometric mean concentration. The correct procedure is to transform to units of concentration before averaging. There is one notable case where the geometric mean is appropriate, namely, when the analyte itself is distributed in a log-normal fashion, as in certain environmental and geological samples.
1994, *66*, 602

geometric surface (interface) area
See *extent of an interface (surface)*.
1986, *58*, 439

geometry (counting) (in radioanalytical chemistry)
A term used colloquially to signify the arrangement in space of the various components in an experiment, particularly the *source* and the *detector* in *radiation* measurements.
1994, *66*, 2520

geometry factor (in radioanalytical chemistry)
The average solid angle in steradians at *source* subtended by the aperture or sensitive volume of the detector, divided by 4π.
1994, *66*, 2520

+ germylenes
See *germylidenes*.
1995, *67*, 1337

germylidenes
Carbene analogues having the structure $R_2Ge{:}$. The older synonym germylenes is no longer recommended.
1995, *67*, 1337

Gibbs adsorption
The surface excess amount or Gibbs adsorption of component i, n_i^σ, which may be positive or negative, is defined as the excess of the amount of this component actually present in the system over that present in a reference system of the same volume as the real system and in which the bulk concentrations in the two phases remain uniform up to the Gibbs dividing surface. That is

$$n_i^\sigma = n_i - V^\alpha c_i^\alpha - V^\beta c_i^\beta$$

where n_i is the total amount of the component i in the system, c_i^α and c_i^β are the concentrations in the two bulk phases α and β, and V^α and V^β are the volumes of the two phases defined by the *Gibbs surface*.
1972, *31*, 588

Gibbs dividing surface
See *Gibbs surface*.
1972, *31*, 588

Gibbs energy (function), G
Enthalpy minus the product of thermodynamic temperature and entropy. It was formerly called free energy or free enthalpy.
G.B. 48; 1996, *68*, 975; 1990, *62*, 2192

Gibbs energy diagram
A diagram showing the relative standard Gibbs energies of reactants, *transition states*, reaction *intermediates* and products, in the same sequence as they occur in a chemical reaction. These points are often connected by a smooth curve (a 'Gibbs energy profile', commonly still referred to as a 'free energy profile') but experimental observation can provide information on relative standard Gibbs energies only at the maxima and minima and not at the configurations between them. The abscissa expresses the sequence of reactants, products, reaction intermediates and transition states and is usually undefined or only vaguely defined by the *reaction coordinate* (extent of bond breaking or bond making). In some adaptations the abscissas are however explicitly defined as *bond orders*, Brønsted exponents, etc.

Contrary to statements in many text books, the highest point on a Gibbs energy diagram does not necessarily correspond to the transition state of the *rate-limiting step*. For example, in a *stepwise reaction* consisting of two reaction steps:

(1) $A + B \rightleftarrows C$

(2) $C + D \to E$

one of the transition states of the two reaction steps must (in general) have a higher standard Gibbs energy than the other, whatever the concentration of D in the system. However, the value of that concentration will determine which of the reaction steps is rate-limiting. If the particular concentrations of interest, which may vary, are chosen as the standard state, then the rate-limiting step is the one of highest Gibbs energy.

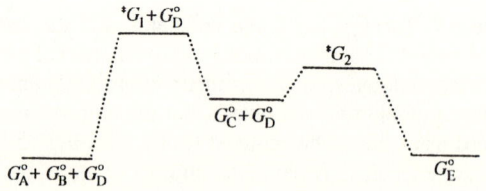

Gibbs energy diagram

See also *potential energy profile, potential energy (reaction) surface, reaction coordinate*.
1994, *66*, 1117; 1996, *68*, 167

Gibbs energy of activation (standard free energy of activation), $\Delta^\ddagger G^o$
The standard Gibbs energy difference between the *transition state* of a reaction (either an *elementary reaction* or a *stepwise reaction*) and the ground state of the reactants. It is calculated from the experimental rate constant k via the conventional form of the absolute rate equation:

$$\Delta^\ddagger G = RT\,[\ln(k_B/h) - \ln(k/T)]$$

where k_B is the Boltzmann constant and h the Planck constant ($k_B/h = 2.08358 \times 10^{10}$ K^{-1} s^{-1}). The values of the rate constants, and hence Gibbs energies of activation, depend upon the choice of concentration units (or of the thermodynamic standard state).

See also *enthalpy of activation, entropy of activation*.
1994, *66*, 1118; 1996, *68*, 166

Gibbs energy of repulsion
Indicated by G_r or G_{el} if the repulsion is due to electric effects (g_r or g_{el} is taken for unit area of each of two flat and parallel surfaces). G_r (or G_{el}) is defined as

$$G_r \text{ (or } G_{el}) = \left[\int_{\text{final distance}}^{\infty} \text{Force}.d\,(\text{distance})\right]_{T,p}$$

1972, *31*, 615

Gibbs energy profile
See *Gibbs energy diagram, transition-state theory*.
1994, *66*, 1117; 1996, *68*, 167, 190

Gibbs film elasticity
Pertains to a film element of a *soap film* changing in area at constant mass and is the differential change of its *surface tension* σ with relative change in area A,

$$E = A(\partial\sigma/\partial A)_{T,p,n_i}$$

where T is the thermodynamic temperature, p is the pressure, and n_i is the amount of substance of the species i.
1972, *31*, 615

Gibbs surface
A geometrical surface chosen parallel to the *interface* and used to define the volumes of the bulk phases in the calculation of the extent of *adsorption*, and of other surface excess properties. Also referred to as Gibbs dividing surface.
1972, *31*, 588

giga
SI prefix for 10^9 (symbol: G).
G.B. 74

glass electrode error
Deviation of a glass electrode from the hydrogen-ion response function. An example often encountered is the error due to sodium ions at alkaline pH values, which by convention is regarded as positive.
1985, *57*, 540

glass laser
A *solid state laser* in which the *active medium* is a glass host that supports the active species. The neodymium-doped phosphate glass laser is one example.
1995, *67*, 1920

glass-like carbon
An agranular *non-graphitizable carbon* with a very high isotropy of its structural and physical properties and with a very low permeability for liquids and gases. The original surfaces and the fracture surfaces have a pseudo-glassy appearance.
Notes:
The often used synonyms 'glassy carbon' and 'vitreous carbon' have been introduced as trademarks and should not be used as terms. From a scientific viewpoint, all synonymous terms suggest a similiarity with the structure of silicate glasses which does not exist in glass-like carbon, except for the pseudo-glassy appearance of the surface.
Glass-like carbon cannot be described as *amorphous carbon* because it consists of two-dimensional structural elements and does not exhibit 'dangling' bonds.
See *agranular carbon*.
1995, *67*, 490

glass transition
A *second-order transition* in which a supercooled melt yields, on cooling, a glassy structure. Below the glass-transition temperature the physical properties vary in a manner similar to those of the crystalline phase.
Example:
Lithium disilicate crystals melt at 1305 K; the melt can be supercooled to the *glass-transition temperature* at approximately 773 K below which the viscous liquid freezes to a rigid amorphous glass.
1994, *66*, 583

glass-transition temperature
The temperature at which the viscosity of the glass is 10^{13} dPa s.
1994, *66*, 583

globular-chain crystal (in polymers)
A type of crystal comprised of macromolecules having globular conformations.
Globular-chain crystals usually occur with globular proteins.
P.B. 84

glove box
An enclosure in which material may be manipulated in isolation from the operator's environment. This manipulation is effected by gauntlet gloves or flexible plastic devices fixed to ports in the walls of the box.
1982, *54*, 1544

glycals
See *enoses*.
1995, *67*, 1337

glycans
Synonymous with *polysaccharides*. Glycans composed of a single type of monosaccharide residue (homopolysaccharide, synonym homoglycan) are named by replacing the ending '-ose' of the sugar by '-an', e.g. mannans, fructans, xylans, arabinans. Dextrans and dextrins belong to the class glucans.
1995, *67*, 1337

+ glycaric acids
An obsolescent synonym for *aldaric acids*.
1995, *67*, 1337

glycerides
Esters of glycerol (propane-1,2,3-triol) with *fatty acids*, widely distributed in nature. They are by long-established custom subdivided into triglycerides, 1,2- or 1,3-diglycerides, and 1- or 2-monoglycerides, according to the number and position of acyl groups (not, as one might suppose, the number of glycerol residues). The recommended method for naming individual glycerides is mono-, di- or tri-*O*-acylglycerol, as appropriate.

$$\underset{\text{(a triglyceride)}}{\overset{R}{\underset{O}{\overset{\|}{C}}}-O-CH_2-CH-CH_2-O-\overset{R}{\underset{O}{\overset{\|}{C}}}}$$

with $\overset{O}{\underset{}{\overset{\|}{C}}}-R$ on center carbon

1995, *67*, 1337; see also W.B. 181

glycerophospholipid
Any derivative of glycerophosphoric acid that contains at least one *O*-acyl, or *O*-alkyl, or *O*-(1-alkenyl) group attached to the glycerol residue.
W.B. 184

+ glycitols
An obsolescent synonym for *alditols*.
1995, *67*, 1337

glyco-amino-acid
A saccharide attached to a single amino acid by any kind of covalent bond.
A glycosyl-amino-acid is a compound consisting of saccharide linked through a glycosyl linkage (*O*-, *N*-, or *S*-) to an amino acid.

Note:
The hyphens are needed to avoid imply that the carbohydrate is necessarily linked to the amino group.
W.B. 85

glycoconjugate
A type of compound consisting of carbohydrate units covalently linked with other types of chemical constituent.
1988, *60*, 1390; W.B. 84

glycoglycerolipid
See *glycolipids*.
W.B. 129

glycolipids
Naturally occurring 1,2-di-*O*-acylglycerols joined at oxygen 3 by a glycosidic linkage to a carbohydrate part (usually a mono-, di- or tri-saccharide). Some substances classified as bacterial glycolipids have the sugar part acylated by one or more *fatty acids* and the glycerol part may be absent.

R' = H or glycosyl

See also *glycosides, lipids, lipopolysaccharides*.
1995, *67*, 1337; see also W.B. 187

glycols
Dihydric alcohols, also known as *diols*, in which the two hydroxy groups are on different carbon atoms, usually but not necessarily adjacent. E.g. $HOCH_2CH_2OH$ 'ethylene glycol' (ethane-1,2-diol), $HO[CH_2]_4OH$ butane-1,4-diol.
1995, *67*, 1337

+ glyconic acids
An obsolescent synonym for aldonic acids.
1995, *67*, 1337

glycopeptides (glycoproteins)
Compounds in which a *carbohydrate* component is linked to a peptide/protein component.
1995, *67*, 1337; see also W.B. 84

glycoproteins
See *glycopeptides*.
1995, *67*, 1338

glycosamines
A term synonymous with *amino sugars*; now restricted to some trivial names.
1995, *67*, 1338

glycosaminoglycan
The carbohydrate units in *proteoglycans* are polysaccharides that contain amino sugars; these polysaccharides are known as glycosaminoglycans.
W.B. 85

glycoses
A less frequently used term for *monosaccharides*.
1995, *67*, 1338

glycosides
Originally mixed *acetals* resulting from the attachment of a *glycosyl group* to a non-acyl group RO– (which itself may be derived from a *saccharide*) and chalcogen replacements thereof (RS–, RSe–).

m and n may be 0, 1, 2, etc.; and x usually 2 or 3

The bond between the *glycosyl group* and the OR group is called a glycosidic bond. By extension, the terms *N*-glycosides and *C*-glycosides are used as class names for *glycosylamines* and for compounds having a glycosyl group attached to a hydrocarbyl group respectively.

4-β-D-glucopyranosylbenzoic acid, a *C*-glycosyl compound

These terms are misnomers and should not be used. The preferred terms are *glycosylamines* and *C*-glycosyl compounds, respectively.
See *osides, glycosylamines*.
1995, *67*, 1338; W.B. 136

glycosphingolipid
See *glycolipids*.
W.B. 187

glycosylamines
Compounds having a glycosyl group attached to an amino group, NR_2; less elegantly called *N*-glycosides, e.g. *N,N*-dimethyl-β-D-glucopyranosylamine.

See *glycosides*.
1995, *67*, 1338

glycosyl-amino-acid
See *glyco-amino-acid*.
W.B. 85

glycosyl group
The structure obtained by removing the hydroxy group from the *hemiacetal* function of a *monosaccharide* and, by extension, of a lower *oligosaccharide*.
1995, *67*, 1338

+ glycuronic acids
An obsolescent synonym for *uronic acids*.
1995, *67*, 1338; see also W.B. 174

Gouy layer (in electrochemistry)
See *diffuse layer (in electrochemistry)*.
1972, *31*, 618

grab sampling
The taking of a sample (often in an evacuated bulb) in a very short time; preferred terms are *instantaneous sampling* or *spot sampling*.
1990, *62*, 2212

gradient
A differential ratio; the change in a quantity such as the mixing ratio of an impurity in air, the temperature of the air, etc., with height or distance.
1990, *62*, 2192

gradient elution (in chromatography)
A procedure in which the composition of the mobile phase is changed continuously or stepwise during the elution process.
1993, *65*, 826; O.B. 93

gradient layer (in chromatography)
The chromatographic bed used in thin-layer chromatography in which there is a gradual transition in some property.
1993, *65*, 830; see also O.B. 98

gradientless reactor (in catalysis)
Stirred flow or recirculation reactors, characterized ideally by very small concentration and temperature gradients within the *catalyst* region.
1976, *46*, 80

gradient packing
Synonymous with *gradient layer*.
O.B. 79

gradual (sudden) potential-energy surface
These terms relate to the steepness of the energy rise along the *reaction path* of a *potential-energy surface*.
1996, *68*, 167

graft copolymer
A *copolymer* that is a *graft polymer*.
Note:
In the constituent macromolecules of a graft copolymer, adjacent blocks in the main chain or side-chains, or both, are constitutionally different, i.e. adjacent blocks comprise *constitutional units* derived from different species of *monomer* or from the same species of monomer but with a different composition or sequence distribution of constitutional units.
1996, *68*, 2303

graft copolymerization
Polymerization in which a *graft copolymer* is formed.
P.B. 19

grafting (in catalysis)
See *encapsulation (in catalysis)*.
1991, *63*, 1230; 1996, *68*, 975

graft macromolecule
A *macromolecule* with one or more species of *block* connected to the *main chain* as *side chains*, these side chains having constitutional or configurational features that differ from those in the main chain.
1996, *68*, 2293

graft polymer
A polymer composed of *graft macromolecules*.
1996, *68*, 2303

gram
cgs base unit of mass, g = 10^{-3} kg.
G.B. 111

Gram stain
A set of two stains that are used to stain (crystal violet and iodine) and counterstain (fuchsin) *bacteria*. Gram-negative bacteria are completely decolorized after treating the stained bacteria with ethanol or acetone and can be visualized by the counterstain. They have a complex cell wall in which the peptidoglycan layer is covered by lipopolysaccharides in an outer membrane structure. Gram-positive bacteria resist decolorization by ethanol or acetone, they retain the initial Gram stain. They have a cell wall consisting predominantly of peptidoglycan not overlayed by an outer membrane.
1992, *64*, 155

granular carbon
The term granular carbon is equivalent to coarse *particulate carbon*. This is a *carbon material* consisting of separate particles or grains which are monolithic, on the average larger than about 100 µm in diameter, but smaller than about 1 cm.
Notes:
Although limits of size cannot be exactly defined, coke grains obtained by grinding belong to coarse particulate carbon for grain sizes above *ca.* 100 µm, or to fine particulate carbon for grain sizes below *ca.* 100 µm. Colloidal graphite obtained by grinding of *natural graphite* is a typical extra fine particulate carbon. Industrial carbon materials (such as electrodes) are made with *fillers* composed of coarse particulate carbon (*coke* grains) and fine particulate carbon (flour), and sometimes even *colloidal carbon* (*carbon blacks* or *soot*). They are therefore polygranular materials.
1995, *67*, 491

graphene layer
A single carbon layer of the graphite structure, describing its nature by analogy to a polycyclic aromatic hydrocarbon of quasi infinite size.
Notes:
Previously, descriptions such as graphite layers, carbon layers or carbon sheets have been used for the term graphene. Because *graphite* designates that modification of the chemical element *carbon*, in which planar sheets of carbon atoms, each atom bound to three neighbours in a honeycomb-like structure, are stacked in a three-dimensional regular order, it is not correct to use for a single layer a term which includes the term graphite, which would imply a three-dimensional structure. The term graphene should be used only when the reactions, structural relations or other properties of individual layers are discussed.
1995, *67*, 491

graphite
An allotropic form of the element *carbon* consisting of layers of hexagonally arranged carbon atoms in a

planar condensed ring system (*graphene layers*). The layers are stacked parallel to each other in a three-dimensional crystalline long-range order. There are two allotropic forms with different stacking arrangements, hexagonal and rhombohedral. The chemical bonds within the layers are covalent with sp^2 hybridization and with a C–C distance of 141.7 pm. The weak bonds between the layers are metallic with a strength comparable to van der Waals bonding only.
Notes:
The term graphite is also used often but incorrectly to describe *graphite materials*, i.e. materials consisting of *graphitic carbon* made from carbon materials by processing to temperatures greater than 2500 K, even though no perfect graphite structure is present.
1995, *67*, 491

graphite electrode
See *carbon electrode*.
1995, *67*, 492

graphite fibres
Carbon fibres consisting mostly of *synthetic graphite* for which three-dimensional crystalline order is confirmed by X-ray diffraction.
Notes:
Graphite fibres can be obtained by *graphitization heat treatment* of carbon fibres if these consist mostly of *graphitizable* carbon. If the *h,k,l* diffraction lines are difficult to recognize because they are of minor intensity, the mean interlayer spacing *c*/2 can be used as indication for the presence of a graphitic structure. The *c*/2 value of 0.34 nm is generally considered as an upper limit for *synthetic graphite*.
1995, *67*, 492

graphite material
A material consisting essentially of *graphitic carbon*.
Notes:
The use of the term *graphite* as a short term for material consisting of graphitic *carbon* is incorrect. The term graphite can only be used in combination with other nouns or clarifying adjectives for special types of graphite materials (graphite electrodes, *natural graphite* and others). The use of the term graphite without a noun or clarifying adjective should be restricted to the allotropic form of the element carbon.
1995, *67*, 492

graphite whiskers
Thin, approximately cylindrical filaments in which *graphene layers* are arranged in a scroll-like manner. There is, at least in part, a regular stacking of the layers as in the *graphite* lattice, giving rise to *h,k,l*, X-ray reflections. The physical properties of graphite whiskers approach, along the cylinder axis, those of graphite.
Notes:
If there is, due to misalignment of the layers caused by their bending, no three-dimensional stacking order as in graphite, the term *carbon whiskers* should be used. Graphite whiskers and carbon whiskers should be distinguished from more disordered *filamentous carbon*.
1995, *67*, 492

graphitic carbon
All varieties of substances consisting of the element carbon in the allotropic form of *graphite* irrespective of the presence of structural defects.
Notes:
The use of the term graphitic carbon is justified if three-dimensional hexagonal crystalline long-range order can be detected in the material by diffraction methods, independent of the volume fraction and the homogeneity of distribution of such crystalline domains. Otherwise, the term *non-graphitic carbon* should be used.
1995, *67*, 493

graphitizable carbon
A *non-graphitic carbon* which upon *graphitization heat treatment* converts into *graphitic carbon*.
Notes:
If it is preferred to define the characterizable state of material instead of its behaviour during subsequent treatment, the term 'pregraphitic carbon' could be considered.
1995, *67*, 493

graphitization
A solid-state transformation of thermodynamically unstable *non-graphitic carbon* into *graphite* by means of heat treatment.
Notes:
Graphitization is also used for the transformation of metastable *diamond* into graphite by heat treatment, as well as in metallurgy for the formation of graphite from thermodynamically unstable carbides by thermal decomposition at high temperatures. Such uses of the term graphitization are in line with the above definition. The use of the term graphitization to indicate a process of thermal treatment of *carbon materials* at $T > 2500$ K regardless of any resultant crystallinity is incorrect.
1995, *67*, 493

graphitization heat treatment
A process of heat treatment of a *non-graphitic carbon*, industrially performed at temperatures in the range between 2500 and 3300 K, to achieve transformation into *graphitic carbon*.
Notes:
The term graphitization heat treatment does not include information as to the crystallinity achieved by the heat treatment, that is the extent of transformation into graphitic carbon or the degree of *graphitization*. Only for such a transformation into graphitic carbon should the term graphitization be used. Consequently: the common use of the term graphitization for the heat treatment process only, regardless of the resultant crystallinity, is incorrect and should be avoided.
1995, *67*, 493

graphitized carbon
A *graphitic carbon* with more or less perfect three-dimensional hexagonal crystalline order prepared from *non-graphitic carbon* by *graphitization heat treatment*.
Notes:
Non-graphitizable carbons do not transform into graphitic carbon on heat treatment at temperatures above 2500 K and therefore are not graphitized carbons.
1995, *67*, 494

gravimetric method
A determination by weight; e.g. in the older method of Cl^- ion determination, a weighed amount of sample is dissolved in water, $AgNO_3$ solution added, AgCl is precipitated, dried and weighed. From the known mass fraction of silver in AgCl, the weights of the initial sample and that of the AgCl precipitated, the percentage of chlorine in the sample can be calculated readily.
1990, *62*, 2192

gravitational constant
Universal fundamental physical constant in Newton's law of gravitation, $G = 6.672\,59\,(85) \times 10^{-11}$ m^3 kg^{-1} s^{-2}.
CODATA Bull., 1986, *63*, 1

gray
SI derived unit of energy imparted to an element of matter by ionizing radiation divided by the mass of that element (absorbed dose of radiation) equal to one joule per kilogram and admitted for reasons of safeguarding human health, Gy = J kg^{-1}.
G.B. 72; 1996, *68*, 975

green coke
Green coke (*raw coke*) is the primary solid *carbonization* product from high boiling hydrocarbon fractions obtained at temperatures below 900 K. It contains a fraction of matter that can be released as volatiles during subsequent heat treatment at temperatures up to approximately 1600 K. This mass fraction, the so-called volatile matter, is in the case of green coke between 4 and 15 wt.%, but it depends also on the heating rate.
Note:
Raw coke is an equivalent term to green coke although it is now less frequently used. The so-called volatile matter of green coke depends on temperature and time of coking, but also on the method for its determination.
1995, *67*, 494

greenhouse effect (in atmospheric chemistry)
Heating effect produced by certain gases (e.g. CO_2, O_3, etc.), which by virtue of their characteristic infrared absorption, lower the earth to space transmission of long wavelength radiation but allow transmission of shorter wavelength radiation inward from the sun to the earth.
1990, *62*, 2192

Grignard reagents
Organomagnesium halides, RMgX, having a carbon–magnesium bond (or their equilibrium mixtures in solution with $R_2Mg + MgX_2$).
1995, *67*, 1338

grist
See *filler*.
1995, *67*, 489

grit (in atmospheric chemistry)
Airborne solid particles in the atmosphere which are of natural or manmade origin and which remain in suspension for some time; in the United Kingdom the size of the grit particles is defined as greater than 75 μm in diameter (retained on a 200 mesh British Standard sieve).
1990, *62*, 2193

ground level concentration (in atmospheric chemistry)
The concentration of a chemical species, normally a pollutant, in air; usually measured at a specific height above the ground.
1990, *62*, 2181

ground level inversion (in atmospheric chemistry)
The inversion of the normal temperature gradient in the atmosphere; the temperature of the air increases with increasing height of the air above the ground. This leads to poor mixing of gases released below the inversion.
1990, *62*, 2193

ground state
The state of lowest Gibbs energy of a system.
See also *excited state*.
1994, *66*, 1118; see also 1996, *68*, 2245; O.B. 227

group
A defined linked collection of atoms or a single atom within a *molecular entity*. This use of the term in physical organic and general chemistry is less restrictive than the definition adopted for the purpose of nomenclature of organic compounds.
See also *characteristic group (in organic nomenclature), principal group (in organic nomenclature), substituent*.
1994, *66*, 1118; see also B.B. 82, 85

group electronegativity
Synonymous with substituent electronegativity.
See *electronegativity*.
1994, *66*, 1169

group preconcentration (in trace analysis)
An operation (process) as a result of which several microcomponents are isolated in one step. This may be achieved in one of two ways, by transportation either of the matrix or of the microcomponents into a second phase. The first method is widely used in the analysis of relatively simple substances and materials.
1979, *51*, 1198

growth curve (of activity)
Curve giving the *activity* of a *radioactive nuclide* as a function of time and showing the increase of the activity through the *decay* of the *precursor* or as a result of *activation*.
  1994, *66*, 2520

growth rate (in biotechnology)
The measure (h^{-1}, d^{-1}) of the rate of growth or multiplication of an organism or a culture, usually expressed as specific growth rate (the increase of mass or cell number per time unit referred to the unit of mass, $d\ln X/dt$).
  1992, *64*, 156

Grunwald–Winstein equation
The linear free energy relation:
$$\log(k_s/k_0) = mY$$
expressing the dependence of the rate of solvolysis of a substrate on ionizing power of the solvent. The rate constant k_0 applies to the reference solvent (ethanol–water, 80:20, v/v) and k_s to the solvent s, both at 25 °C. The parameter m is characteristic of the substrate and is assigned the value unity for *tert*-butyl chloride. The value Y is intended to be a quantitative measure of the ionizing power of the solvents. The equation was later extended to the form:
$$\log(k_s/k_0) = mY + lN$$
where N is the *nucleophilicity* of the solvent and l its susceptibility parameter. The equation has also been applied to reactions other than solvolysis.

See also *Dimroth–Reichardt E_T parameter, polarity, Z-value*.
  1994, *66*, 1118

guest
An organic or inorganic ion or molecule that occupies a cavity, cleft or pocket within the molecular structure of a *host molecular entity* and forms a *complex* with it or that is trapped in a cavity within the crystal structure of a host.
  See also *crown, cryptand, inclusion compound*.
  1994, *66*, 1118

Guinier plot
A diagrammatic representation of data on scattering from large particles, obtained at different angles but at the same concentration, constructed by plotting log $\Delta R(\theta)$ or log $P(\theta)$ *versus* $\sin^2(\theta/2)$ or q^2 and (usually) used for the evaluation of the radius of gyration. $\Delta R(\theta)$ is the excess Rayleigh ratio, $P(\theta)$ the particle scattering function, θ the scattering angle and q the length of the scattering vector.
  P.B. 67

gustiness (in atmospheric chemistry)
Intensity of turbulence; the ratio of the root mean square of wind velocity fluctuations to the mean wind velocity.
  1990, *62*, 2193

haem
An alternative spelling for *heme*.
1995, *67*, 1338

half-chair
The *conformation* of a six-membered ring structure in which four contiguous atoms are in a plane and the other two atoms lie on opposite sides of the plane.

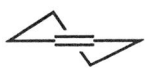

See also *chair, boat, twist*.
1996, *68*, 2209

half life, $t_{1/2}$
For a given reaction the half life $t_{1/2}$ of a *reactant* is the time required for its concentration to reach a value that is the arithmetic mean of its initial and final (equilibrium) values. For a reactant that is entirely consumed it is the time taken for the reactant concentration to fall to one half its initial value:
The half life of a reaction has meaning only in special cases:
1. For a first-order reaction, the half life of the reactant may be called the half life of the reaction.
2. For a reaction involving more than one reactant, with the concentrations of the reactants in their stoichiometric ratios, the half life of each reactant is the same, and may be called the half life of the reaction.
If the concentrations of reactants are not in their stoichiometric ratios, there are different half lives for different reactants, and one cannot speak of the half life of the reaction.
See also *lifetime*.
1996, *68*, 167; 1994, *66*, 1119; G.B. 55; 1996, *68*, 975

half life (of a radionuclide)
For a single *radioactive decay* process, the time required for the activity to decrease to half its value by that process.
1994, *66*, 2520

half life, biological
See *biological half life*.
1994, *66*, 2520

half life, effective
See *effective half life*.
1994, *66*, 2520

half-peak potential
In linear-sweep *voltammetry*, triangular-wave voltammetry, cyclic triangular-wave voltammetry, and similar techniques, the potential of the *indicator electrode* at which the difference between the total current and the *residual current* is equal to one-half of the *peak current*. This potential is attained in the interval in which the rate of the charge-transfer process, and hence the (absolute value of the) current, increase monotonically with time.
1985, *57*, 1502

half thickness (in radiochemistry)
The thickness of a specified substance which, when introduced into the path of a given beam of *radiation*, reduces the value of a specified radiation quantity by one half.
1994, *66*, 2520

half-wave potential
The potential of a polarographic or voltammetric *indicator electrode* at the point, on the rising part of a polarographic or voltammetric wave, where the difference between the total current and the *residual current* is equal to one-half of the *limiting current*. The quarter-wave potential, the three-quarter-wave potential, etc., may be similarly defined.
1985, *57*, 1502

half-width (of a band)
The full width of a spectral band at a height equal to half of the height at the band maximum. Also known as *full width at half maximum* (*FWHM*). The dimension of band width should be either inverse length (wavenumbers) or inverse time (frequencies) so that the values give an indication of the energies. Note the hyphen in half-width. Half bandwidth has the meaning of half-width at half maximum.
1996, *68*, 2245; see also O.B. 226

halirenium ions
Cyclic cations having the structure:

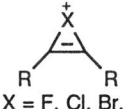

X = F, Cl, Br, I

1995, *67*, 1338

halochromism
The colour change which occurs on addition of *acid* (or *base*, or a salt) to a solution of a compound. A *chemical reaction* (e.g. ion formation) transforms a colourless compound into a coloured one.
1994, *66*, 1119

haloforms
Trihalomethanes CHX_3.
1995, *67*, 1338

halohydrins
A traditional term for alcohols substituted by a halogen atom at a saturated carbon atom otherwise bearing only hydrogen or hydrocarbyl groups (usually used to mean β-halo alcohols). E.g. $BrCH_2CH_2OH$ 'ethylene bromohydrin' (2-bromoethanol), $ClCH_2CH_2CH_2OH$ 'trimethylene chlorohydrin' (3-chloropropan-1-ol), $PhCH(OH)CH_2Cl$ 'styrene chlorohydrin' (2-chloro-1-phenylethanol).
1995, *67*, 1338

halonium ions
Ions of the form R_2X^+, where X is any halogen (X = Br^+, bromonium ions; X = Cl^+, chloronium ions; X

= F$^+$, fluoronium ions; X = I$^+$, iodonium ions). They may be open-chain or cyclic.

$$\underset{R_2C-CR_2}{\overset{\overset{+}{X}}{}} \qquad \underset{R_2C-CR_2}{\overset{R_2C\overset{\overset{+}{X}}{\diagup\diagdown}CR_2}{}}$$

1995, *67*, 1339

halophiles
Organisms which require a minimum concentration of sodium chloride in their environment (cf. *extremophiles*).
  1992, *64*, 156

hamiltonian operator, $\hat{H}$
Quantum mechanical operator of total energy.
  G.B. 16

Hammett acidity function
  See *acidity function*.
  1994, *66*, 1119

Hammett equation (Hammett relation)
The equation in the form:

$$\lg (k/k_0) = \rho\sigma$$

or

$$\lg (K/K_0) = \rho\sigma$$

applied to the influence of *meta*- or *para*-substituents X on the reactivity of the functional group Y in the benzene derivative *m*- or *p*-XC$_6$H$_4$Y. *k* or *K* is the rate or equilibrium constant, respectively, for the given reaction of *m*- or *p*-XC$_6$H$_4$Y; k_0 or K_0 refers to the reaction of C$_6$H$_5$Y, i.e. X = H; σ is the substituent constant characteristic of *m*- or *p*-X: ρ is the reaction constant characteristic of the given reaction of Y. The equation is often encountered in a form with lg k_0 or lg K_0 written as a separate term on the right hand side, e.g.

$$\lg k = \rho\sigma + \lg k_0$$

or

$$\lg K = \rho\sigma + \lg K_0$$

It then signifies the intercept corresponding to X = H in a regression of lg *k* or lg *K* on σ.
  See also ρ-*value*, σ-*constant*, *Taft equation*, *Yukawa–Tsuno equation*.
  1994, *66*, 1119

Hammond–Herkstroeter plot
  See *energy transfer plot*.
  1996, *68*, 2245

Hammond principle (Hammond postulate)
The hypothesis that, when a *transition state* leading to an unstable *reaction intermediate* (or product) has nearly the same energy as that intermediate, the two are interconverted with only a small reorganization of molecular structure. Essentially the same idea is sometimes referred to as 'Leffler's assumption', namely, that the transition state bears the greater resemblance to the less stable species (reactant or reaction intermediate/product). Many text books and physical organic chemists, however, express the idea in Leffler's form, but attribute it to Hammond.
As a corollary, it follows that a factor stabilizing a reaction *intermediate* will also stabilize the *transition state* leading to that intermediate.
The acronym 'Bemahapothle' (Bell, Marcus, Hammond, Polanyi, Thornton, Leffler) is sometimes used in recognition of the principal contributors towards expansion of the original idea of the Hammond postulate.
  See also *More O'Ferrall–Jencks diagram*.
  1994, *66*, 1119

handedness
This term has been used in two ways, either *chirality* or *chirality sense*.
  1996, *68*, 2209

Hansch constant
A measure of the capability of a solute for *hydrophobic (lipophilic)* interaction based on the partition coefficient *P* for distribution of the solute between octan-1-ol and water. The most general way of applying *P* in *correlation analysis*, QSAR, etc. is as log *P*, but the behaviour of substituted benzene derivatives may be quantified by a substituent constant scale, π, which is defined in a way analogous to the Hammett σ scale. There are various π scales, depending on the substrate series used as reference.
  1994, *66*, 1119

Hantzsch–Widman name
A name for a heteromonocyclic parent hydride having no more than ten ring members formed by the citation of 'a' prefixes denoting the heteroatoms followed by an ending (the 'stem') defining the size of the ring.
  B.B.(G) 14

hapten
A low-molecular weight molecule which contains an antigenic determinant but which is not itself antigenic unless complexed with an anitgenic carrier. Examples of haptens are dinitrophenols, phosphorylcholine and dextran.
  1992, *64*, 156

hapto
  See η (*eta or hapto*), under 'e'.

hard acid
A *Lewis acid* with an acceptor centre of low *polarizability*. Other things being approximately equal, complexes of *hard acids* and *bases* or *soft acids* and bases have an added stabilization (sometimes called 'HSAB' rule). For example the hard O- (or N-) bases are preferred to their S- (or P-) analogues by hard acids. Conversely a 'soft acid' possesses an acceptor centre of high polarizability and exhibits the reverse preference for coordination of a soft base. These preferences are not defined in a quantitative sense.
  See also *class (a) metal ion, hard base*.
  1994, *66*, 1120

hard amorphous carbon films
A synonym for *diamond-like carbon films*.
1995, *67*, 494

hard base
A *Lewis base* with a donor centre (e.g. an oxygen atom) of low *polarizability;* the converse applies to 'soft bases'.
See also *hard acid*.
1994, *66*, 1120

hard-sphere collision cross section
See *reaction probability*.
1981, *53*, 766

harmonic frequency generation
Production of coherent radiation of frequency $k\nu$ ($k = 2, 3, \ldots$) from coherent radiation of frequency ν. In general, this effect is obtained through the interaction of laser light with a suitable optical medium with non-linear polarizability. The case $k = 2$ is referred to as frequency doubling, $k = 3$ is frequency tripling and $k = 4$ is frequency quadrupling. Even higher integer values of k are possible.
1996, *68*, 2245

harmonic mean
The number of observations, divided by the sum of reciprocals of the observations. Symbol: $\bar{x}_h$. It can be calculated with the formula:

$$\bar{x}_h = \frac{n}{\sum x_j^{-1}}$$

Comment:
As in the case of the geometric mean, this quantity is sometimes directly (but inappropriately) calculated, for example, when evaluating kinetic analytical results where the reaction time is inversely proportional to concentration.
1994, *66*, 602

harpoon mechanism
Reaction sequence (thermal or photoinduced) between neutral molecular or atomic entities in which long-range *electron transfer* is followed by a considerable reduction of the distance between donor and acceptor sites as a result of the electrostatic attraction in the *ion pair* created.
1996, *68*, 2245

hartree
Atomic unit of energy, $E_h \approx 4.359\,7482(26) \times 10^{-18}$ J.
G.B. 76

Hartree energy
Atomic fundamental physical constant used as *atomic unit* of energy: $E_h = \hbar^2/m_e a_0^2 = 4.359\,7482\,(26) \times 10^{-18}$ J, where $\hbar$ is the Planck constant divided by 2π, m_e the electron rest mass and a_0 the Bohr radius.
CODATA Bull., 1986, *63*, 1

Haworth representation
The Haworth representation of the cyclic forms of monosaccharides can be derived from the *Fischer projection*, as follows. The monosaccharide is depicted with the carbon-chain horizontal and in the plane of the paper, the potential carbonyl group being to the right. The oxygen bridge is then depicted as being formed behind the plane of the paper.
The heterocyclic ring is therefore located in a plane approximately perpendicular to the plane of the paper and the groups attached to the carbon atoms of that ring are above and below the ring. The carbon atoms of the ring are not shown.
Groups that appear to the right of the vertical chain in the Fischer projection (structures A, D) then appear below the plane of the ring in the Haworth representation (structures B, C, E). However, at the asymmetric carbon atom (C-5 in A; C-4 in D) involved via oxygen in ring formation with the carbon atom of the carbonyl group a formal double inversion must be envisaged to obtain the correct Haworth representation.
In the pyranose forms of D-aldohexoses C-6 will always be above the plane. In the furanose forms of D-aldohexoses the position of C-6 will depend on the configuration at C-4; it will, for example, be above the plane in D-glucofuranoses (e.g. C) but below the plane in D-galactofuranoses (e.g. E).

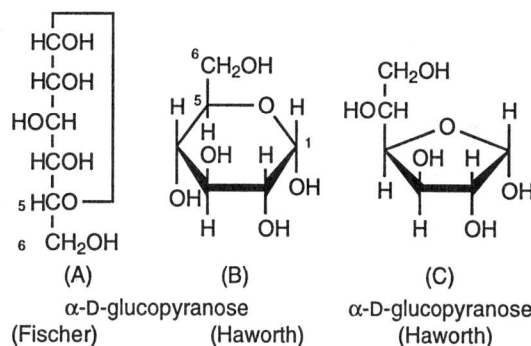

(A) α-D-glucopyranose (Fischer) (B) α-D-glucopyranose (Haworth) (C) α-D-glucopyranose (Haworth)

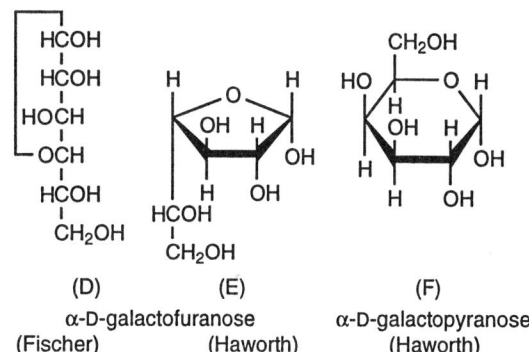

(D) α-D-galactofuranose (Fischer) (E) (Haworth) (F) α-D-galactopyranose (Haworth)

W.B. 128

haze (in atmospheric chemistry)
A state of reduced visibility (1–2 km) resulting from the increased light scatter due to the presence of fine dust or aerosol particles (H_2SO_4, NH_4HSO_4, products of the ozone-terpene reactions, etc.).
1990, *62*, 2193

haze horizon (in atmospheric chemistry)
The top of a haze layer which is confined by a low-level temperature inversion so that it gives the appearance of the horizon which it may obscure.
1990, *62*, 2193

heat, *q*, *Q*
Energy transferred from a hotter to a cooler body due to a temperature gradient.
G.B. 48

heat capacity, *C*
Heat brought to a system to increase its temperature divided by that temperature increase. At constant volume $C_v = (\partial U/\partial T)_v$ at constant pressure $C_p = (\partial H/\partial T)_p$, where U is the *internal energy* and H the *enthalpy* of the system.
G.B. 48; 1996, *68*, 975

heat capacity of activation, $\Delta^{\ddagger}C_p^{\circ}$
A quantity related to the temperature coefficient of $\Delta^{\ddagger}H$ (*enthalpy of activation*) and $\Delta^{\ddagger}S$ (*entropy of activation*) according to the equations:

$$\Delta^{\ddagger}C_p = (\partial \Delta^{\ddagger}H/\partial T)_p = T(\partial \Delta^{\ddagger}S/\partial T)_p$$

If the rate constant is expressible in the form $\ln k = a/T + b + c \ln T + dT$, then:

$$\Delta^{\ddagger}C_p = (c - 1)R + 2dRT$$

SI unit: J mol^{-1} K^{-1}.
1994, *66*, 1120; 1996, *68*, 168

heat flux, J_q
Heat transferred through a cross section perpendicular to the flow in a small time interval divided by that time interval and the cross sectional area.
See also *flux*.
G.B. 65

heavy atom effect
The enhancement of the rate of a spin-forbidden process by the presence of an atom of high atomic number, which is either part of, or external to, the excited molecular entity. Mechanistically, it responds to a spin-orbit coupling enhancement produced by a heavy atom.
1996, *68*, 2245

heavy atom isotope effect
An *isotope effect* due to isotopes other than those of hydrogen.
1994, *66*, 1130

heavy water
Water containing a significant fraction (up to 100 per cent) of deuterium in the form of D$_2$O or HDO.
1982, *54*, 1544

hecto
SI prefix for 10^2 (symbol: h).
G.B. 74

height equivalent to an effective theoretical plate (h.e.e.t.p.) (in chromatography)
The column length divided by the effective *theoretical plate number*.
O.B. 108

height equivalent to a theoretical plate (h.e.t.p.) (in chromatography)
The column length divided by the *theoretical plate number*.
O.B. 108

helicenes
ortho-Fused polycyclic *aromatic* or *heteroaromatic* compounds in which all rings (minimum five) are angularly arranged so as to give helically shaped molecules, which are thus chiral, e.g. hexahelicene.

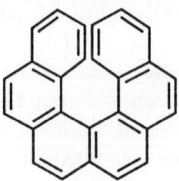

hexahelicene

1995, *67*, 1339

helicity
The *chirality* of a helical, propeller or screw-shaped molecular entity. A right-handed helix is described as *P* (or plus), a left-handed one as *M* (or minus).

M or Λ   *P* or Δ

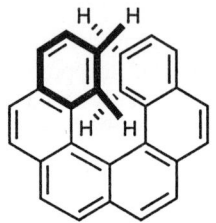

(*M*)-hexahelicene

The application of this system to the description of conformations considers the *torsion angle* between two specified (fiducial) groups that are attached to the atoms linked by that bond. The sign of the smaller torsion angle between the fiducial groups defines the chirality sense of the helix. Rules for the selection of fiducial groups according to priority are given by R.S. Cahn, C.K. Ingold and V. Prelog, *Angew. Chem.* 78, 413–447 (1966), *Angew. Chem. Internat. Ed. Eng.* 5, 385–415, 511 (1966).
See also *axial chirality*; Δ (*delta*), Λ (*lambda*).
1996, *68*, 2209

helion
Nucleus of the ^{3}He atom.
G.B. 93

helium–cadium laser
A continuous wave laser emitting mainly at 325.0 and 441.6 nm from singly ionized cadmium.

See *gas lasers*.
1996, *68*, 2246

helium dead-space (in colloid and surface chemistry)
The position of the *Gibbs surface* is often defined experimentally as that surface which encloses the volume of space from which the solid excludes helium gas (the so-called helium dead-space), and is associated with the assumptions that the volume of the solid is unaffected by the *adsorption* of component i, and that helium is not adsorbed by the solid. This requires that the measurement of the helium dead-space be made at a sufficiently high temperature.
1972, *31*, 595

helium ionization detector (in gas chromatography)
A weak beta source and a high potential raise the helium atom of the carrier gas to a metastable state. All other substances having an ionization potential lower than 18 eV are ionized, and the current that results is used to measure the components. The detector is usually employed to measure inorganic compounds at concentrations between 0.1 and 10 ppmv. It has a linear range of about 10^2 but is somewhat unstable and requires great care to ensure the helium purity and to eliminate all leaks in the system.
1990, *62*, 2191

helium–neon laser
A continuous wave laser emitting mainly at 632.8, 1152.3, and 3391.3 nm from excited neutral Ne atoms.
See *gas lasers*.
1996, *68*, 2246

helix
The molecular conformation of a spiral nature, generated by regularly repeating rotations around the backbone bonds of a macromolecule.
P.B. 78

helix residue (in a polymer)
The smallest set of one or more successive configurational base units that generates the whole chain through helical symmetry.
P.B. 79

helix sense
The right-handed sense of a helix traces out a clockwise rotation moving away from the observer; the left-handed sense of a helix traces out a counterclockwise rotation moving away from the observer, e.g. the ...TG⁺TG⁺TG⁺... helix of isotactic poly(propylene) is left-handed.
P.B. 43

Helmholtz energy (function), A
Internal energy minus the product of *thermodynamic temperature* and entropy. It was formerly called free energy.
G.B. 48

hemes (heme derivatives)
Complexes consisting of an iron ion coordinated to a *porphyrin* acting as a tetradentate ligand, and to one or two axial ligands.

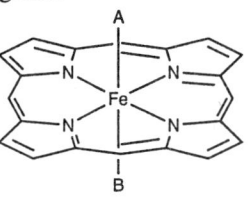

1995, *67*, 1339; see also W.B. 311, 316

hemiacetals
Compounds having the general formula $R_2C(OH)OR'$ ($R' \neq H$).
See also *lactols, hemiketals*.
1995, *67*, 1339

hemiaminals
α-Amino alcohols, improperly called *carbinolamines* (the adducts of ammonia, or of primary or secondary amines to the carbonyl group of *aldehydes* and *ketones*) $R_2C(OH)(NR_2)$. Compounds of structure $R_2C(OR')NR_2$ ($R' \neq H$) are hemiaminal ethers, or α-amino ethers.
See *alcohols, aminals, ethers*.
1995, *67*, 1339

hemiketals
Hemiacetals having the structure $R_2C(OH)OR$ ($R \neq H$), derived from ketones by formal addition of an alcohol to the carbonyl group. This term, once abandoned, has been reinstated as a subclass of hemiacetals.
1995, *67*, 1339

+ hemileptic
See *homoleptic*.
1994, *66*, 1120

hemins
Chloro(porphyrinato)iron(III) complexes.
See also *hemes and heme derivatives*.
1995, *67*, 1339; see also W.B. 316

hemochromes
Iron–*porphyrin* complexes, having one or two basic ligands (e.g. piperidine, amines).
See also *hemes (heme derivatives)*.
1995, *67*, 1339; see also W.B. 316

hemoglobins
Heme derivatives having a *protein* chain as axial ligand.
1995, *67*, 1339

Henderson–Hasselbach equation
An equation of the form:
$$\text{pH} = pK_a - \lg([HA]/[A^-])$$
for the calculation of the pH of solutions where the ratio $[HA]/[A^-]$ is known.
1994, *66*, 1120

henry
SI derived unit of inductance, H = V A⁻¹ s = m² kg s⁻² A⁻².
G.B. 72; see also 1996, *68*, 975

Henry's law
The partial pressure (fugacity) of a solute (B) in a solution is directly proportional to the rational chemical activity (a_x) of the solute; this relationship is called Henry's law:

$$p_B = a_{x,B} / \alpha_{x,B}^\infty$$

where $\alpha_{x,B}^\infty$ is the rational solubility coefficient for infinite dilution, i.e. for pure solvent. For the solvent (A) the relationship is called Raoult's law, and the proportionality factor is the fugacity of the pure solvent, $\tilde{p}_A^*$:

$$p_A = \tilde{p}_A^* a_A$$

1984, *56*, 571; see also 1990, *62*, 2193; 1994, *66*, 544

Herkstroeter plot
See *energy transfer plot*.
1996, *68*, 2246

hertz
SI derived unit of frequency equal to one cycle per second, Hz = s^{-1}.
G.B. 72; 1996, *68*, 976

Herz compounds
1,2λ^4,3-Benzodithiazolium chlorides (formed in the reaction of aniline and derivatives thereof with disulfur dichloride).

1995, *67*, 1339

hetarenes
Synonymous with *heteroarenes*.
1995, *67*, 1340

hetaryl groups
Synonymous with *heteroaryl groups*.
1995, *67*, 1340

hetarynes
Synonymous with *heteroarynes*.
1995, *67*, 1340

heteroalkenes
Analogues of alkenes in which a doubly bonded carbon atom is replaced by a heteroatom, e.g. H$_2$Si=CH$_2$ methylidenesilane (silene less preferred), MeN=CH$_2$ N-methylmethanimine.
1995, *67*, 1340

heteroarenes
Heterocyclic compounds formally derived from *arenes* by replacement of one or more methine (–C=) and/or vinylene (–CH=CH–) groups by trivalent or divalent heteroatoms, respectively, in such a way as to maintain the continuous π-electron system characteristic of aromatic systems and a number of out-of-plane π-electrons corresponding to the Hückel rule (4n + 2); an alternative term is hetarenes.

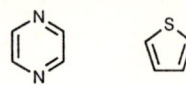

1995, *67*, 1340

heteroaryl groups
The class of *heterocyclyl groups* derived from *heteroarenes* by removal of a hydrogen atom from any ring atom; an alternative term is hetaryl. E.g.

2-pyridyl (pyridin-2-yl)   indol-1-yl

1995, *67*, 1340

heteroarynes
Compounds derived from *heteroarenes* by replacement of a formal carbon–carbon double bond by a formal triple bond (with loss of two hydrogen atoms). Also known as hetarynes and as 1,2-didehydroheteroarenes. E.g.

See *arynes*.
1995, *67*, 1340

heterobimetallic complex
A metal complex having two different metal atoms.
1994, *66*, 1120

heterochain polymer
A class of polymer in which the main chain is constructed from atoms of two or more elements.
Heterochain polymers are named by placing the names or symbols of all the elements in the main chain, immediately before the expresssion '-chain polymer', e.g. (oxygen, carbon)-chain polymer or (O, C)-chain polymer; (oxygen, nitrogen, carbon)-chain polymer or (O, N, C)-chain polymer.
P.B. 146

heteroconjugation
+ 1. Association between a base and the conjugate acid of a different base through a *hydrogen bond* (B' ... HB$^+$ or A'H ... A$^-$). The term has its origin in the *conjugate acid–base pair* and is in no way related to *conjugation* of *orbitals*. Heteroassociation is a more appropriate term.
+ 2. Some authors refer to *conjugated systems* containing a heteroatom, e.g. pyridine, as 'heteroconjugated systems'. This usage is discouraged since it inappropriately suggests an analogy to *homoconjugation* (2), and conflicts with the currently accepted definition of that term.
1994, *66*, 1120

heterocumulenes
Cumulenes in which one or more carbon atoms of the cumulative bond system have been replaced by heteroatoms. E.g. O=C=C=C=O, but not CH$_2$=C=O, ketene, nor O=C=O, carbon dioxide, which are heteroallenes.
1995, *67*, 1340

heterocyclic compounds
Cyclic compounds having as ring members atoms of at least two different elements, e.g. quinoline, 1,2-thiazole, bicyclo[3.3.1]tetrasiloxane.
See *homocyclic compounds, carbocyclic compounds.*
1995, *67*, 1340

heterocyclyl groups
Univalent groups formed by removing a hydrogen atom from any ring atom of a *heterocyclic compound.* E.g.

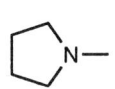 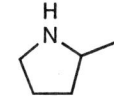

pyrrolidin-1-yl   pyrrolidin-2-yl

See *organyl.*
1995, *67*, 1340

heterodetic cyclic peptide
A *peptide* consisting only of *amino-acid residues*, but in which the linkages forming the ring are not solely *eupeptide* bonds; one or more is an *isopeptide*, disulfide, ester, or other bond.
W.B. 59

heterodisperse
Describes a *colloidal* system in which all the particles are of different sizes.
1972, *31*, 607

heteroexcimer
Synonymous with *exciplex.*
1996, *68*, 2246

heterogeneity (in analytical chemistry)
See *homogeneity.*
1990, *62*, 1201

heterogeneous catalysis
See *catalyst.*
1993, *65*, 2293

heterogeneous diffusion rate constant (in electrochemistry)
Defined by the equation:
$$k_d = I_l/nFcA$$
where the *limiting current* I_l is assumed to be due to the diffusion species of *concentration c* and of diffusion coefficient *D*. *n* is the *charge number of the cell reaction* written so that the stoichiometric coefficient of this species is unity. *A* is usually taken as the geometric area of the electrode, and *F* is the *Faraday constant.*
1974, *37*, 513

heterogeneous membrane electrode
See *crystalline electrodes.*
1994, *66*, 2534

heterogeneous nucleation
See *embryo, nucleation.*
1972, *31*, 608

heteroleptic
Transition metal of Main Group compounds having more than one type of ligand.

See also *homoleptic.*
1994, *66*, 1120

heterolysis (heterolytic)
The cleavage of a *covalent bond* so that both bonding electrons remain with one of the two fragments between which the bond is broken, e.g.

$$A\!-\!B \longrightarrow A^+ + B^-$$

Heterolytic bond fission is a feature of many *bimolecular reactions* in solution (e.g. *electrophilic* substitution, *nucleophilic* substitution).
See also *homolysis, heterolytic bond-dissociation energy.*
1994, *66*, 1120

heterolytic bond-dissociation energy
The energy required to break a given *bond* of some specific compound by *heterolysis*. For the *dissociation* of a neutral molecule AB in the gas phase into A^+ and B^- the heterolytic *bond-dissociation energy* $D(A^+B^-)$ is the sum of the bond dissociation energy, $D(A\!-\!B)$, and the adiabatic ionization energy of the radical A· minus the electron affinity of the radical B·.
1994, *66*, 1121

heterolytic dissociative adsorption
Relates in the usual sense to the formal nature of the cleavage of a single bond. If in the bond of the *adsorptive* A:B, A or B retains the electron pair, the *adsorption* is heterolytic dissociative adsorption.
1976, *46*, 76

heteropolysaccharide (heteroglycan)
The class name for *polysaccharides* composed of two or more different kinds of monomeric units (i.e. monosaccharides).
W.B. 174

heterotactic polymer
See *triads (in polymers).*
P.B. 40

heterotactic triads (in polymers)
See *triads (in polymers).*
P.B. 40

heterotopic
See *stereoheterotopic.*
1996, *68*, 2210

heterotrophic (organisms)
Descriptive of organisms which are not able to synthesize *cell* components from carbon dioxide as sole carbon source. Heterotrophic organisms use preformed oxidizable organic *substrates* such as glucose as carbon and energy sources.
1992, *64*, 156

heterovalent hyperconjugation
See *hyperconjugation.*
1994, *66*, 1121

hexagonal graphite
The thermodynamically stable form of *graphite* with an ABAB stacking sequence of the *graphene layers*. The exact crystallographic description of this allo-

tropic form is given by the space group d_{6h}^4-$P6_3/mmc$ (unit cell constants: $a = 245.6$ pm, $c = 670.8$ pm). Hexagonal graphite is thermodynamically stable below approximately 2600 K and 6 GPa.
Note:
The use of the term graphite instead of the more exact term hexagonal graphite may be tolerated in view of the minor importance of *rhombohedral graphite*, the other allotropic form.
    1995, *67*, 494

hexahedro- (in inorganic nomenclature)
An affix used in names to denote eight atoms bound into a hexahedron (e.g. cube).
    R.B. 245; B.B. 464

hexaprismo- (in inorganic nomenclature)
An affix used in names to denote twelve atoms bound into a hexagonal prism.
    R.B. 245; B.B. 464

hidden return
    See *ion-pair return*.
    1994, *66*, 1121

higher-order transition
A general term used to describe a transition in which the first and second derivatives of the molar *Gibbs energy* or molar *Helmholtz energy* (or chemical potential) with respect to temperature and pressure are continuous, but derivatives of some higher order are discontinuous at the transition point.
Synonymous with smooth transition.
    1994, *66*, 583

highly oriented pyrolytic graphite
Highly oriented pyrolytic graphite (HOPG) is a *pyrolytic graphite* with an angular spread of the c-axes of the crystallites of less than 1 degree.
Note:
Commercial HOPG is usually produced by stress annealing at approximately 3300 K.
    1995, *67*, 495

high-pressure graphitization
A term which refers to the solid-state transformation of *non-graphitic carbon* into *graphite* by heat treatment under elevated pressure (e.g. 100 to 1000 MPa) so that a definitely higher degree of *graphitization* is achieved at lower temperature and/or for a shorter heat treatment time than in heat treatment of the same non-graphitic material at atmospheric pressure.
    1995, *67*, 495

high-pressure mercury lamp (arc)
Radiation source containing mercury at a pressure of *ca*. 8 MPa (*ca*. 80 bar) or higher which emits lines over a background continuum between about 200 and 1400 nm.
    See *lamp*.
    1996, *68*, 2246

high resolution energy loss spectroscopy (HRELS)
    See *reflection electron energy loss spectroscopy (REELS)*.
    1983, *55*, 2025

Hildebrand parameter
A parameter measuring the cohesion of a solvent (energy required to create a cavity in the solvent).
    1994, *66*, 1121

hindered rotation
    See *free rotation, hindered rotation, restricted rotation*.
    1996, *68*, 2210

histones
A class of basic proteins associated with *DNA* in the *chromosomes* of eukaryotic *cells* forming the nucleosome as the basic subunit of chromatin. They contain an unusually large proportion of the basic amino acids arginine and lysine.
    1992, *64*, 156

Hofmann rule
"The principal olefin formed in the decomposition of quaternary ammonium hydroxides that contain different primary alkyl groups is always ethylene, if an ethyl group is present." Originally given in this limited form by A. W. Hofmann, the rule has since been extended and modified as follows: "When two or more alkenes can be produced in a β-elimination reaction, the alkene having the smallest number of alkyl groups attached to the double bond carbon atoms will be the predominant product." This orientation described by the Hofmann rule is observed in elimination reactions of quaternary ammonium salts and tertiary sulfonium salts, and in certain other cases.
    See also *Zaitsev rule*.
    1994, *66*, 1121

hold-back carrier
A *carrier* used to prevent a particular species from following other species in a chemical or physical operation.
    1994, *66*, 2516

hold-up volume (in chromatography)
The volume of *eluent* required to *elute* a component, the concentration of which in the *stationary phase* is negligible compared to that in the *mobile phase*.
    O.B. 100

hold-up volume (time) (in column chromatography), V_M, t_M
The volume of the mobile phase (or the corresponding time) required to elute a component the concentration of which in the stationary phase is negligible compared to that in the mobile phase. In other words, this component is not retained at all by the stationary phase. Thus, the hold-up volume (time) is equal to the retention volume (time) of an unretained compound. The hold-up volume (time) includes any volumes contributed by the sample injector, the detector, and connectors.

$$t_M = V_M/F_c$$
In gas chromatography this term is also called the gas hold-up volume (time). The corrected gas hold-up volume (V_M^o) is the gas hold-up volume multiplied by the compression (compressibility) correction factor (j):
$$V_M^o = V_M j$$
Assuming that the influence of extra column volume on V_M is negligible,
$$V_M^o = V_G$$
1993, *65*, 841

hole burning
The photo*bleaching* of a feature, normally a narrow range, within an inhomogeneous broader absorption or emission band. The holes are produced by the disappearance of resonantly excited molecules as a result of photophysical or photochemical processes. The resulting spectroscopic technique is site-selection spectroscopy.
1996, *68*, 2246

hole transfer
Charge migration process in which the majority carriers are positively charged.
1996, *68*, 2246

holoenzyme
An active *enzyme* consisting of the *apoenzyme* and *coenzyme*.
1994, *66*, 2593

+ Holtsmark broadening (of a spectral line)
See *collisional broadening (of a spectral line)*.
1985, *57*, 1463

homo
1. An acronym for *Highest Occupied Molecular Orbital* (HOMO).
    See *frontier orbitals*.
2. A prefix (consisting of lower case letters, homo), used to indicate a higher homologue of a compound.
    1994, *66*, 1121
3. homo-
Affix used to denote a ring expansion by inclusion of a methylene group into a *ring-sector*.
    B.B. 500

homoaromatic
Whereas in an *aromatic* molecule there is continuous overlap of p-orbitals over a cyclic array of atoms, in a homoaromatic molecule there is a formal discontinuity in this overlap resulting from the presence of a single sp^3 hybridized atom at one or several positions within the ring; p-orbital overlap apparently bridges these sp^3 centres, and features associated with aromaticity are manifest in the properties of the compound. Pronounced homoaromaticity is not normally associated with neutral molecules, but mainly with species bearing an electrical charge, e.g. the 'homotropylium' cation, C$_8$H$_9^+$:

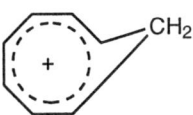

In bis, tris, (etc.) homoaromatic species, two, three, (etc.) single sp^3 centres separately interrupt the π-electron system.
See also *homoconjugation (2)*.
1994, *66*, 1121

homochain polymer
A class of polymer in which the main chain is constructed from atoms of a single element.
Homochain polymers are named by placing the name or symbol of the element in the main chain immediately before the expression '-chain polymer', e.g. carbon-chain polymer or C-chain polymer; sulfur-chain polymer or S-chain polymer.
P.B. 146

homochiral
See *enantiomerically pure (enantiopure)*.
1996, *68*, 2210

homoconjugation
+ 1. Association between a base and its *conjugate acid* through a *hydrogen bond* (B ... HB$^+$ or AH ... A$^-$). Homoassociation is a more appropriate term for this phenomenon.
2. The orbital overlap of two π-systems separated by a non-conjugating group, such as CH$_2$.
    See also *conjugate acid–base pair, conjugated system, homoaromatic*.
1994, *66*, 1121

homocyclic compounds
Cyclic compounds having as ring members atoms of the same element only, e.g. benzene, pentazole, cyclohexasilane.
1995, *67*, 1340

homodetic cyclic peptide
A cyclic peptide in which the ring consists solely of *amino-acid residues* in *eupeptide* linkage.
W.B. 59

homogeneity (in analytical chemistry)
The degree to which a property or a constituent is uniformly distributed throughout a quantity of material.
A material may be homogeneous with respect to one analyte or property but heterogeneous with respect to another. The degree of heterogeneity (the opposite of homogeneity) is the determining factor of sampling error.
1990, *62*, 1201

homogeneous catalysis
See *catalyst*.
1993, *65*, 2293

homogeneous membrane electrode
See *crystalline electrodes*.
1994, *66*, 2534

homogeneous nucleation
See *embryo, nucleation*.
1972, *31*, 608

homogeneous nucleus
See *embryo*.
1972, *31*, 608

homoleptic
Transition metal or Main Group compounds having only one type of *ligand* are said to be homoleptic, e.g. $TaMe_5$.
See also *heteroleptic*.
1994, *66*, 1121

homolysis (homolytic)
The cleavage of a *bond* ('homolytic cleavage' or 'homolytic fission') so that each of the molecular fragments between which the bond is broken retains one of the bonding electrons. A *unimolecular* reaction involving homolysis of a bond (not forming part of a cyclic structure) in a molecular entity containing an even number of (paired) electrons results in the formation of two radicals:

$$A\!-\!B \longrightarrow A^\bullet + B^\bullet$$

It is the reverse of *colligation*. Homolysis is also commonly a feature of *bimolecular substitution reactions* (and of other reactions) involving radicals and molecules.
See also *bond dissociation energy, heterolysis*.
1994, *66*, 1122

homolytic dissociative adsorption
Relates in the usual sense to the formal nature of the cleavage of a single bond. If the electron pair in the bond of the *adsorptive* A:B is divided in the course of its *dissociative adsorption*, the *adsorption* is homolytic dissociative adsorption.
1976, *46*, 76

homomorphic
Superposable ligands are called homomorphic.
1996, *68*, 2210

homopolymer
A *polymer* derived from one species of (real, implicit or hypothetical) *monomer*.
Notes:
1. Many polymers are made by the mutual reaction of complementary monomers. These monomers can readily be visualized as reacting to give an 'implicit monomer', the homopolymerization of which would give the actual product, which can be regarded as a homopolymer. Common examples are poly(ethylene terephthalate) and poly(hexamethylene adipamide).
2. Some polymers are obtained by the chemical modification of other polymers such that the structure of the macromolecules that constitute the resulting polymer can be thought of as having been formed by the homopolymerization of a hypothetical monomer. These polymers can be regarded as homopolymers. Example: poly(vinyl alcohol).
1996, *68*, 2300

homopolymerization
Polymerization in which a *homopolymer* is formed.
1996, *68*, 2306

homopolysaccharide (homoglycan)
The class name for *polysaccharides* composed of only one kind of monosaccharide.
W.B. 174

homotopic
Atoms or groups of a molecule which are related by an *n*-fold rotation axis ($n = 2, 3$, etc.) are called homotopic. For example, chiral tartaric acid (C_2 axis), chloroform (C_3 axis) and cyclohexaamylose (α-cyclodextrin, C_6 axis) have respectively two homotopic carboxyl groups, three homotopic chlorine atoms and six homotopic D-glucose residues.
See *prochirality*.
1996, *68*, 2210

horizontal elution (horizontal development) (in planar chromatography)
A mode of operation in which the paper or plate is in a horizontal position and the mobile-phase movement along the plane depends on capillary action.
1993, *65*, 829

host
1. A *molecular entity* that forms *complexes* with organic or inorganic guests, or a chemical species that can accommodate guests within cavities of its crystal structure. Examples include *cryptands* and *crowns* (where there are ion-dipole attractions between heteroatoms and positive ions), *hydrogen-bonded* molecules that form 'clathrates' (e.g. hydroquinone and water), and host molecules of *inclusion compounds* (e.g. urea or thiourea). *van der Waals forces* and *hydrophobic interactions* bind the guest to the host molecule in clathrates and inclusion compounds.
1994, *66*, 1122
2. (in biotechnology)
A *cell* whose *metabolism* is used for growth and reproduction of a *virus*, *plasmid* or other form of foreign DNA.
1992, *64*, 156

host-vector system
A compatible combination of *host* (e.g. *bacteria*) and *vector* (e.g. *plasmid*) that allows propagation of DNA.
1992, *64*, 156

hot atom
An atom in an excited energy state or having kinetic energy above the ambient thermal level, usually as a result of nuclear processes.
1994, *66*, 2520

hot cell
A heavily shielded enclosure for highly *radioactive* materials. It may be used for their handling or processing by remote means or for their storage.
1994, *66*, 2520

hot ground state reaction
A hot state reaction of the ground electronic state.
1996, *68*, 2246

+ hot quartz lamp
A term sometimes used to describe a high-pressure mercury lamp. The use of this term is not recommended.
1996, *68*, 2246

hot state reaction
A reaction proceeding from an ensemble of molecular entities possessing a higher average vibrational, rotational or translational energy than they would at thermal equilibrium with the surrounding medium.
1996, *68*, 2246

hour
Non-SI unit of time, h = 3600 s.
G.B. 111; 1996, *68*, 976

Hückel ($4n + 2$) rule
Monocyclic planar (or almost planar) systems of trigonally (or sometimes digonally) hybridized atoms that contain ($4n + 2$) π-electrons (where n is a nonnegative integer) will exhibit *aromatic* character. The rule is generally limited to $n = 0$–5.
This rule is derived from the Hückel MO calculation on planar monocyclic conjugated hydrocarbons $(CH)_m$ where m is an integer equal to or greater than 3 according to which ($4n + 2$) π-electrons are contained in a closed-shell system. Examples of systems that obey the Hückel rule include:

cyclopropenyl cation   cyclopentadienyl anion
($m = 3, n = 0$)   ($m = 5, n = 1$)

benzene
($m = 6, n = 1$)

Systems containing $4n$ π-electrons (such as cyclobutadiene and the cyclopentadienyl cation) are 'antiaromatic'.
See also *conjugation*, *Möbius aromaticity*.
1994, *66*, 1122

Huggins coefficient, k_H
A parameter in the *Huggins equation*.
P.B. 64

Huggins equation
The equation describing the dependence of the *reduced viscosity*, η_i/c, on the mass concentration of a polymer, c, for dilute polymer solutions of the form:

$$\eta_i/c = [\eta] + k_H[\eta]^2 c$$

where k_H is the Huggins coefficient and $[\eta]$ is the *intrinsic viscosity*.
P.B. 63

humidity (in atmospheric chemistry)
A general term referring to the water content of a gas.
See *relative humidity*.
1990, *62*, 2194

Hund rules
1. Of the different multiplets resulting from different configurations of electrons in degenerate orbitals of an atom those with greatest multiplicity have the lowest energy (multiplicity rule).
2. Among multiplets having the same multiplicity, the lowest-energy one is that with the largest total orbital angular momentum (angular momentum rule) (valid if the total orbital angular momentum is a constant of motion).
3. In configurations containing shells less than half full of electrons, the term having the lowest total angular momentum J lies lowest in energy, whereas in those with shells more than half filled, the term having the largest value of J lies lowest (fine structure rule).
Hund rules apply if the 'Russell–Saunders' coupling scheme is valid. Sometimes the first rule is applied with questionable validity to molecules.
1996, *68*, 2246

Hush model
See *Marcus–Hush relationship*.
1996, *68*, 2247

HWHM (Half Width at Half Maximum)
See *line width*.
G.B. 31

hybridization
1. Linear combination of *atomic orbitals* on an atom. Hybrid orbitals are often used in organic chemistry to describe the bonding molecules containing tetrahedral (sp^3), trigonal (sp^2) and digonal (sp) atoms.
1994, *66*, 1122
2. The formation of stable duplexes of two DNA and/or RNA (complementary) strands via Watson-Crick *base pairing* used for locating or identifying *nucleotide* sequences and to establish the effective transfer of *nucleic acid* material to a new *host*.
3. The formation of a novel diploid organism either by sexual processes or by *protoplast fusion*.
1992, *64*, 156

hybridoma
A hybrid *cell* line resulting from the *fusion* of a specific *antibody*-producing spleen cell (lymphocyte) with a *myeloma* cell, which has the growth characteristics of the myeloma component and the antibody-secreting characteristics of the lymphocyte, and will multiply to become a source of pure *monoclonal antibody*.
1992, *64*, 156

hydration
Addition of water or of the elements of water (i.e. H and OH) to a *molecular entity*. For example, hydration of ethene:

$$H_2C=CH_2 + H_2O \rightarrow CH_3CH_2OH$$

The term is also used in a more restricted sense for the process:

A (gas) → A (aqueous soln.)

cf. the use of the term in inorganic/physical chemistry to describe the state of the ions of an electrolyte in aqueous solution.

See also *aquation, solvation*.
1994, *66*, 1122

hydrazide hydrazones
See *hydrazidines*.
1995, *67*, 1341

hydrazide imides
See *amidrazones*.
1995, *67*, 1341

hydrazides
Compounds derived from oxoacids $R_kE(=O)_l(OH)_m$ ($l \neq 0$) by replacing –OH by –NRNR$_2$ (R groups are commonly H), as in carbohydrazides, RC(=O)NHNH$_2$, sulfonohydrazides, RS(=O)$_2$NHNH$_2$, and phosphonic dihydrazides, RP(=O)(NHNH$_2$)$_2$.
1995, *67*, 1340

hydrazidines
Compounds, RC(NHNH$_2$)=NNH$_2$, derived from *carboxylic acids* by replacing –OH by –NHNH$_2$ (or *N*-substituted analogues) and =O by =NNH$_2$ (or substituted analogues). A specific hydrazidine is named as a hydrazide hydrazone, e.g. hexanohydrazide hydrazone.
1995, *67*, 1341

hydrazines
Hydrazine (diazane), H$_2$NNH$_2$, and its *hydrocarbyl* derivatives. When one or more substituents are *acyl groups*, the compound is a hydrazide. *N*-Alkylidene derivatives are *hydrazones*.
See *azines, hydrazo compounds*.
1995, *67*, 1341

hydrazinylidenes
Synonymous with *isodiazenes*.
1995, *67*, 1341

hydrazo compounds
Compounds containing the divalent hydrazo group, –NHNH–, such as hydrazoarenes (1,2-diarylhydrazines or 1,2-diaryldiazanes, usually with both aryl groups the same) and their *N*-substituted derivatives ArNRNRAr.
See also *hydrazines*.
1995, *67*, 1341

hydrazones
Compounds having the structure R$_2$C=NNR$_2$, formally derived from aldehydes or ketones by replacing =O by =NNH$_2$ (or substituted analogues).
1995, *67*, 1341

hydrazonic acids
Compounds derived from *oxoacids* $R_kE(=O)_l(OH)_m$ ($l \neq 0$) by replacing a double-bonded oxygen atom by =NNR$_2$, as in carbohydrazonic acids, RC(OH)=NNH$_2$ and sulfonohydrazonic acids, RS(=O)(=NNH$_2$)OH.
1995, *67*, 1341

hydride
See *hydron*.
1988, *60*, 1116

hydro
See *hydron*.
1988, *60*, 1116

hydrocarbon cracking
See *cracking*.
1990, *62*, 2194

hydrocarbons
Compounds consisting of carbon and hydrogen only.
1995, *67*, 1341

hydrocarbylene groups
Divalent groups formed by removing two hydrogen atoms from a *hydrocarbon*, the free valencies of which are not engaged in a double bond, e.g. 1,3-phenylene, –CH$_2$CH$_2$CH$_2$– propane-1,3-diyl, –CH$_2$–methylene.
1995, *67*, 1341

hydrocarbyl groups
Univalent groups formed by removing a hydrogen atom from a *hydrocarbon*, e.g. ethyl, phenyl.
See *heterocyclyl, organoheteryl, organyl groups*.
1995, *67*, 1341

hydrocarbylidene groups
Divalent groups, R$_2$C=, formed by removing two hydrogen atoms from the same carbon atom of a *hydrocarbon*, the free valencies of which are engaged in a double bond.
1995, *67*, 1341

hydrocarbylidyne groups
Trivalent groups, RC≡, formed by removing three hydrogen atoms from the same carbon atom of a hydrocarbon, the free valencies of which are engaged in a triple bond.
1995, *67*, 1341

hydrocarbylsulfanyl nitrenes
Nitrenes substituted with hydrocarbylsulfanyl groups, RS–N: ↔ RS$^+$=N$^-$ ↔ RS≡N. Sulfenyl nitrenes is a older synonymous term. E.g. MeSN methylsulfanylnitrene or methylthionitrene. The synonymous term thiazynes (from the third canonical form; confusable with the hetarynes derivable from 1,2- and 1,4-thiazine) is best avoided.
1995, *67*, 1341

hydrocracking unit
Used in the thermal decomposition of heavy (high molecular weight) hydrocarbons to smaller (low molecular weight) hydrocarbons; high pressures of hydrogen and a special catalyst are employed. Sulfur compounds in the fuel are reduced to H$_2$S, and the final hydrocarbon product can be obtained relatively sulfur-free.
1990, *62*, 2195

hydrodynamically equivalent sphere (in polymers)
A hypothetical sphere, impenetrable to the surrounding medium, displaying in a hydrodynamic field the same frictional effect as an actual polymer molecule. The size of a hydrodynamically equivalent sphere may be different for different types of motion of the

macromolecule, e.g. for diffusion and for viscous flow.
P.B. 60

hydrodynamic volume (in polymers)
The volume of a *hydrodynamically equivalent sphere*.
P.B. 60

hydrogen
The general name for the atom H, without regard for its nuclear mass, either for hydrogen in its natural abundance or where it is not desired to distinguish between the isotopes. The systematic name for atomic hydrogen is monohydrogen.
See also *hydron*.
1988, *60*, 1116

hydrogen bond
A form of *association* between an electronegative atom and a hydrogen atom attached to a second, relatively electronegative atom. It is best considered as an electrostatic interaction, heightened by the small size of hydrogen, which permits proximity of the interacting dipoles or charges. Both electronegative atoms are usually (but not necessarily) from the first row of the Periodic Table, i.e. N, O or F. Hydrogen bonds may be intermolecular or intramolecular. With a few exceptions, usually involving fluorine, the associated energies are less than 20–25 kJ mol^{-1} (5–6 kcal mol^{-1}).
1994, *66*, 1123

hydrogen gas electrode
A thin foil of platinum electrolytically coated with a finely divided deposit of platinum or palladium metal, which catalyses the electrode reaction:-

$$H^+ + e \rightarrow \tfrac{1}{2}H_2$$

in solutions saturated with hydrogen gas.
1985, *57*, 540

hydrolases
Enzymes that catalyse the cleavage of C–O, C–N, C–C and other bonds by reactions involving the addition or removal of water.
1992, *64*, 156

hydrolysis
Solvolysis by water.
1994, *66*, 1123

hydrometeor (in atmospheric chemistry)
Any condensed water particle in the atmosphere of sufficient size to be potentially capable of undergoing precipitation (in fogs, clouds, some hazes, raindrops, snowflakes, etc.).
1990, *62*, 2195

hydron
The general name for the cation H$^+$; the species H$^-$ is the *hydride* anion and H is the *hydro* group. These are general names to be used without regard to the nuclear mass of the hydrogen entity, either for hydrogen in its natural abundance or where it is not desired to distinguish between the isotopes.
1988, *60*, 1116; 1994, *66*, 1123; R.B. 103

hydroperoxides
Monosubstitution products of hydrogen peroxide (dioxidane), HOOH, having the skeleton ROOH, in which R is any *organyl group*. Compounds in which R is acyl are known as *peroxy acids*.
1995, *67*, 1342

hydrophilic
'Water loving'. The capacity of a *molecular entity* or of a *substituent* to interact with polar solvents, in particular with water, or with other polar groups.
1994, *66*, 1123

hydrophobic interaction
The tendency of hydrocarbons (or of *lipophilic* hydrocarbon-like groups in solutes) to form *intermolecular* aggregates in an aqueous *medium*, and analogous intramolecular interactions. The name arises from the attribution of the phenomenon to the apparent repulsion between water and hydrocarbons. However, the phenomenon ought to be attributed to the effect of the hydrocarbon-like groups on the water-water interaction. The misleading alternative term 'hydrophobic bond' is discouraged.
1994, *66*, 1123

hydropolysulfides
Compounds having the structures RS$_2$H, RS$_3$H ... RS$_n$H, in which S$_n$ is a chain of sulfur atoms, and R is *hydrocarbyl*. Some people exclude hydrodisulfides from the class hydropolysulfides.
1995, *67*, 1342

hydrosphere (in atmospheric chemistry)
The gaseous, liquid and solid water of the earth (oceans, ice caps, lakes, rivers, etc.) as distinguished from the lithosphere and the atmosphere.
1990, *62*, 2195

hydrosulfides
Term used in radicofunctional nomenclature of *thiols*. Hydrosulfides is not commonly used as a class name for thiols.
1995, *67*, 1342

hydroxamic acids
Compounds, RC(=O)NHOH, derived from *oxoacids* R$_k$E(=O)$_l$(OH)$_m$ ($l \neq 0$) by replacing –OH by –NHOH, and hydrocarbyl derivatives thereof. Specific examples are preferably named as *N*-hydroxy amides.
1995, *67*, 1342

hydroximic acids
Compounds derived from *oxoacids* R$_k$E(=O)$_l$(OH)$_m$ ($l \neq 0$) by replacing =O by =NOH (=NOR), as in carbohydroximic acids, RC(OH)=NOH, and sulfonohydroximic acids, RS(=O)(=NOH)OH.
1995, *67*, 1342

hydroxylamines
Hydroxylamine, H$_2$N–OH, and its hydrocarbyl derivatives.
See also *hydroxamic acids, oximes*.
1995, *67*, 1342

hygrometer
Instrument used to measure the water vapour content of the atmosphere.
  See *capacitance hygrometer, dew point hygrometer, electrical hygrometer, electrolytic hygrometer, frost point hygrometer, mechanical hygrometer, psychrometric hygrometer.*
  1990, *62*, 2195

hygrometer, capacitance
  See *capacitance hygrometer.*
  1990, *62*, 2195

hygrometer, dew point (cooled surface condensation)
  See *dew point hygrometer (cooled surface condensation).*
  1990, *62*, 2195

hygrometer, electrical
  See *electrical hygrometer.*
  1990, *62*, 2195

hygrometer, electrolytic
  See *electrolytic hygrometer.*
  1990, *62*, 2195

hygrometer, frost point
  See *frost point hygrometer.*
  1990, *62*, 2195

hygrometer, mechanical
  See *mechanical hygrometer.*
  1990, *62*, 2195

hygrometer, psychrometric
  See *psychrometric hygrometer.*
  1990, *62*, 2195

hygrometry (moisture analysis)
The measurement or indication of the water content of the ambient air or of a sample of gas.
  1990, *62*, 2195

hyperchromic effect
The increase in the intensity of a spectral band due to substituents or interactions with the molecular environment.
  See also *auxochrome, hypochromic effect.*
  1996, *68*, 2247

hyperconjugation
In the formalism that separates bonds into σ and π types, hyperconjugation is the interaction of σ-bonds (e.g. C–H, C–C, etc.) with a π network. This interaction is customarily illustrated by *contributing structures*, e.g. for toluene (below), sometimes said to be an example of 'heterovalent' or 'sacrificial hyperconjugation', so named because the contributing structure contains one two-electron bond less than the normal *Lewis formula* for toluene:

At present, there is no evidence for sacrificial hyperconjugation in neutral hydrocarbons.
The concept of hyperconjugation is also applied to *carbenium* ions and *radicals*, where the interaction is now between σ-bonds and an unfilled or partially filled π- or p-orbital. A contributing structure illustrating this for the *tert*-butyl cation is:

This latter example is sometimes called an example of 'isovalent hyperconjugation' (the contributing structure containing the same number of two-electron bonds as the normal Lewis formula).
Both structures shown on the right hand side are also examples of 'double bond-no-bond resonance'.
The interaction between filled π- or p- orbitals and adjacent antibonding σ^* orbitals is referred to as 'negative hyperconjugation', as for example in the fluoroethyl anion:

  See also *sigma, pi, n-σ^* delocalization.*
  1994, *66*, 1123

hypersatellite (in X-ray spectroscopy)
  See *X-ray satellite.*
  1991, *63*, 739

hypho-
An affix used in names to designate an open structure, especially a boron skeleton, more closed than a *klado* structure but more open than an *arachno* structure.
  R.B. 245

hypochromic effect
The opposite of *hyperchromic effect* (i.e. a decrease in intensity).
  1996, *68*, 2247

hypo-phase
The denser phase in an extraction system. This term is often used when two non-aqueous phases are present or when the solvent is an aqueous phase.
  See also *epi-phase.*
  1993, *65*, 2381

hypsochromic shift
Shift of a spectral band to higher frequency or shorter wavelength upon substitution or change in medium. It is informally referred to as blue shift.
  See also *bathochromic shift.*
  1994, *66*, 1124; 1996, *68*, 2247

hysteresis
1. Regarding a material quantity or instrument's reading; the dependence of a value on the direction of change from a previous characteristic value. It may be quantified by the difference between the upscale and downscale variation starting from fixed lower and upper measurement points (inversion).
  1990, *62*, 2196
2. (in solid-state transitions) The difference in temperature or pressure for the transition of one phase to

another in the forward and reverse directions. It also refers to the corresponding difference in magnetic, electric or stress field in reversing the sense of magnetic, electric or strain polarization in ferromagnetic, ferroelectric or ferroelastic materials.

1994, *66*, 583

3. (in electroanalytical chemistry) Hysteresis (electrode memory) occurs when there is a difference between the emf first observed in a solution containing a concentration of A and a second observation of the emf in the same solution after exposing the electrode to a different concentration of A. The systematic error is generally in the direction of the concentration of the solution in which the electrode was previously immersed. Hysteresis is thought to be a kinetic process. Normal, reversible responses are expected when sufficient time is allowed for the system to return to its initial condition.

See also *reproducibility (in electrochemical analysis), electrode memory.*

1994, *66*, 2530

icosahedro-
An affix used in names to denote twelve atoms bound into a triangular icosahedron.
R.B. 245; B.B. 465

icosanoids
Unsaturated C_{20} fatty acids and skeletally related compounds. The spelling icosanoids is preferred over the spelling eicosanoids for consistency with icosanoic acid. Specialists working in this field commonly use eicosanoids, however.
See *prostanoids, leukotrienes*.
1995, *67*, 1342

ideal adsorbed state
The adsorbed state in a system following Langmuir's isotherm.
1976, *46*, 78

ideal dilute solution
Dilute solution in which the solute may be regarded as obeying *Henry's law*, so that all the solute *activity coefficients* may be approximated to 1.
Physical Chemistry Division, unpublished; see also 1994, *66*, 545

ideal gas
Gas which obeys the equation of state $pV = nRT$ (the ideal gas law; p is the pressure, V the volume, n the amount of molecules, R the gas constant and T the thermodynamic temperature). For an ideal gas the *fugacity*, f, of each constituent B is equal to its partial pressure, p_B, $f_B = p_B = x_B p$, where x_B is the amount fraction of B.
Physical Chemistry Division, unpublished; see also 1990, *62*, 2196

ideally polarized (electrified) interphase
See *electrified interphase*.
1986, *58*, 439

ideally unpolarized (electrified) interphase
See *electrified interphase*.
1986, *58*, 439

ideal mixture
A mixture of substances B, C, ... is called an ideal mixture when $a_B = x_B$, $a_C = x_C$, ... or $f_B = 1, f_C = 1,$ ... , where a_B and f_B are the *relative activity* and the *activity coefficient*, respectively, of a substance B in a liquid or solid mixture.
1994, *66*, 543

identity period (of a polymer)
See *(chain) identity period (of a polymer)*.
P.B. 77

identity reaction
A *chemical reaction* whose products are chemically identical with the reactants, for example the bimolecular self exchange reaction of CH_3I with I^-.
See also *degenerate rearrangement*.
1994, *66*, 1124

illuminance, E, E_v
Luminous flux received by a surface divided by the area of that surface.
G.B. 72; 1996, *68*, 976; see also 1990, *62*, 2196

image converter tube
An electron tube that produces on its fluorescent screen an image of the irradiation pattern of its photosensitive input surface. An image converter which produces an image with enhanced radiance is sometimes called an image intensifier.
1995, *67*, 1758

image dissection tube
A two-dimensional *radiation detector* in which the electron image produced by a photo-emitting surface, usually a photocathode, is focused in the plane of a defining aperture. Magnetic or electric fields scan this image across the defining aperture.
1995, *67*, 1756

imaging (photoimaging)
The use of a photosensitive system for the capture, recording and retrieval of information associated with an object using electromagnetic energy.
1996, *68*, 2247

imbalance
The situation in which *reaction* parameters that characterize different bond forming or bond breaking processes in the same reaction have developed to different extents as the *transition state* is approached along some arbitrarily defined reaction coordinate. For example, in the nitroalkane anomaly, the Brønsted β exponent for proton removal is smaller than the Brønsted α for the nitroalkane, because of imbalance between the amount of bond breaking and resonance delocalization in the transition state. Imbalance is common in reactions such as elimination, addition and other complex reactions that involve proton (hydron) transfer.
See also *synchronous, synchronization (principle of imperfect synchronization)*.
1994, *66*, 1124

imbibition (in colloid chemistry)
The uptake of a liquid by a *gel* or porous substance. It may or may not be accompanied by swelling.
1972, *31*, 615

+ imenes
See *nitrenes*.
1994, *66*, 1124

imides
1. Diacyl derivatives of ammonia or primary *amines*, especially those cyclic compounds derived from diacids, e.g.

2. In additive nomenclature, in which imide is analogous to oxide, the term is used to name compounds of the type $R_3Y^+\text{–}N^-R$ (Y = N, P) and $R_2Z^+\text{–}N^-R$ (Z = O, S, Se, Te), which are the products of formal attachment of an RN= group to N, P, O, S, Se, Te. E.g. amine imides, azomethine imides.

3. Salts having the anion RN^{2-}.
1995, *67*, 1342

imidic acids
Compounds derived from *oxoacids* R$_k$E(=O)$_l$(OH)$_m$ ($l \neq 0$) by replacing =O by =NR; thus tautomers of *amides*. In organic chemistry an unspecified imidic acid is generally a carboximidic acid, RC(=NR)OH. E.g. RS(=NH)$_2$OH a sulfonodiimidic acid.
See *imino acids*.
1995, *67*, 1342

imidines
Analogues of cyclic *acid anhydrides* in which =O has been replaced by =NR and –O– by –NR–.

See *diamidides*.
1995, *67*, 1342

+ imidogens
A non-IUPAC term for *nitrenes* used in the Chemical Abstracts Service index nomenclature.
1993, *65*, 1357

+ imidonium ions
A term, which is not recommended, for *nitrenium ions*.
1995, *67*, 1343

imidoyl carbenes
Carbenes having the structure RC(=NR)C̈–R. Imidoyl is a shortened but imprecise term for carboximidoyl, RC(=NH)–.
1995, *67*, 1343

imidoyl nitrenes
Nitrenes having the structure RC(=NR)N: ↔ RC(–N$^-$R)=N$^+$.
See *imidoyl carbenes*.
1995, *67*, 1343

+ imin
See *nitrenes*.
1994, *66*, 1124

+ imine radical
See *nitrenes*.
1994, *66*, 1124

imines
1. Compounds having the structure RN=CR$_2$ (R = H, hydrocarbyl). Thus analogues of *aldehydes* or *ketones*, having NR doubly bonded to carbon; aldimines have the structure RCH=NR, ketimines have the structure R′$_2$C=NR (R′ ≠ H). Imines include *azomethines* and *Schiff bases*. Imine is used as a suffix in systematic nomenclature to denote the C=NH group excluding the carbon atom.
+ 2. An obsolete term for azacycloalkanes.
See also *azomethines, Schiff bases*.
1995, *67*, 1343

iminium compounds
Salts in which the cation has the structure R$_2$C=N$^+$R$_2$. Thus *N*-hydronated *imines* and their *N*-substituted derivatives. The synonymous terms imonium compounds and immonium compounds are irregularly formed and should not be used.
See *quaternary ammonium compounds*.
1995, *67*, 1343

imino acids
+ 1. An obsolete term, which should not be used, for *imidic acids*.
2. Any *carboxylic acid* having an imino substituent, HN=, replacing two hydrogens. A shortened form of imino carboxylic acid.
3. Obsolescent term for azaalkanoic acids and azacycloalkane-2-carboxylic acids, e.g. proline:

1995, *67*, 1343

imino carbenes
This term systematically means *carbenes* bearing an imino or *N*-substituted imino group, RN=, somewhere in the molecule. E.g. R–C̈–CH$_2$C(=NR)R. It is listed here in order to warn against its misuse for alkylideneamino carbenes.
See *carbenes, nitrile ylides*.
1995, *67*, 1343

+ iminooxy (iminoxy radicals)
This term has improperly been used for *alkylideneaminoxyl radicals*, also called *iminoxyl radicals* R$_2$C=N–O·. Its use is strongly discouraged.
See *aminoxyl radicals, iminyl radicals*.
1995, *67*, 1343

iminoxyl radicals
Synonymous with *alkylideneaminoxyl radicals*. The term can be regarded as a contraction of iminyloxyl radicals.
1995, *67*, 1343

iminyl carbenes
See *nitrile ylides*.
1995, *67*, 1343

iminylium ions
Cations having the structure R$_2$C=N$^+$. A contraction of alkaniminylium ions. Alkylideneaminylium ions is a synonymous term. A subclass of nitrenium ions.
1995, *67*, 1344

iminyl radicals
Radicals having the structure R$_2$C=N·. A contraction of alkaniminyl radicals, e.g. MeCH=N· ethaniminyl. A synonymous term is alkylideneaminyl radicals.
See *aminyl radicals*.
1995, *67*, 1343

immersional wetting
A process in which a solid or liquid, β, is covered with a liquid, α, both of which were initially in contact with a gas or liquid, δ, without changing the area of the αδ-interface.
1972, *31*, 598

immission (in atmospheric chemistry)
A Germanic term, pronounced in English, 'eye-mission'; the transfer of pollutants from the atmosphere to a 'receptor'; for example, pollutants retained by the lungs. It does not have the same meaning as ground level concentration, but is the opposite in meaning to emission. This term has not been used commonly in the English language.
1990, *62*, 2196

immission dose (in atmospheric chemistry)
The integral of the immission flow into the receptor over the exposure period.
1990, *62*, 2196

immission flux (in atmospheric chemistry)
The immission rate divided by the unit surface area of the receptor.
1990, *62*, 2196

immission rate (in atmospheric chemistry)
The mass (or other physical quantity) of pollutant transferring per unit time into a receptor.
1990, *62*, 2196

immobile adsorption
A situation in which the freedom of molecules of *adsorbate* to move about the surface is limited. *Adsorption* is immobile when kT is small compared to ΔE, the energy barrier separating adjacent sites. The adsorbate has little chance of migrating to neighbouring sites and such adsorption is necessarily localized.
1976, *46*, 76

immobilized phase (in chromatography)
A stationary phase in which is immobilized on the support particles, or on the inner wall of the column tubing e.g. by *in situ* polymerization (cross-linking) after coating.
1993, *65*, 823

immobilization (in biotechnology)
The technique used for the physical or chemical fixation of *cells, organelles, enzymes,* or other proteins (e.g. *monoclonal antibodies*) onto a solid support, into a solid matrix or retained by a membrane, in order to increase their stability and make possible their repeated or continued use. The principle is also used for *affinity chromatography*.
1992, *64*, 157

immobilization by adsorption (in catalysis)
See *encapsulation (in catalysis)*.
1991, *63*, 1230

immobilization by inclusion (in catalysis)
See *encapsulation (in catalysis)*.
1991, *63*, 1230

immobilized enzyme
A soluble *enzyme* bound to an insoluble organic or inorganic matrix, or encapsulated within a membrane in order to increase its stability and make possible its repeated or continued use.
1994, *66*, 2593

immune response
Selective reaction of the body to substances that are foreign to it, or that the immune system identifies as foreign, shown by the production of antibodies and antibody-bearing cells or by a cell-mediated hypersensitivity reaction.
See also *antibody*.
1993, *65*, 2058

immunization
The administration of an *antigen* to an animal in order to stimulate the production of *antibodies* by that organism. Also, the administration of antigens, antibodies or lymphocytes to an animal to generate the corresponding active, passive or adaptive immunity.
1992, *64*, 157

immunoassay
A ligand-binding assay that uses a specific *antigen* or *antibody*, capable of binding to the analyte, to identify and quantify substances. The antibody can be linked to a radiosotope (radioimmunoassay, RIA), or to an *enzyme* which catalyses an easily monitored reaction (enzyme-linked immunosorbent assay, ELISA), or to a highly fluorescent compound by which the location of an antigen can be visualized (immunofluorescene).
1992, *64*, 157; 1994, *66*, 2520

immunochemistry
Study of biochemical and molecular aspects of immunology, especially the nature of antibodies, antigens and their interactions.
1993, *65*, 2058

immunogen
A substance that elicits a cellular immune response and/or *antibody* production (cf. *antigen*).
1992, *64*, 157

immunoglobulin (Ig)
A protein of the globulin-type found in serum or other body fluids that possesses *antibody* activity. An individual Ig molecule is built up from two light (L) and two heavy (H) polypeptide chains linked together by disulfide bonds. Igs are divided into five classes based on antigenic and structural differences in the H chains.
1992, *64*, 157

immunoradiometric assay
An *assay* based on the reversible and non-covalent binding of an *antigen* by a specific *antibody* labelled with a *radioactive nuclide* as a *tracer*.
1994, *66*, 2520

immunosuppression
Reduction in the functional capacity of the immune response; may be due to:
1. Inhibition of the normal response of the immune system to an antigen.
2. Prevention, by chemical or biological means, of the production of an antibody to an antigen by inhibition of the processes of transcription, translation or formation of tertiary structure.
1993, *65*, 2058

impaction
A forcible contact of particles of matter with a surface.
1990, *62*, 2196

impactor, cascade
See *cascade impactor*.
1990, *62*, 2196

impact parameter, *b*
In simple hard sphere collision theory the distance of closest approach that would result for two particles in a collision if the particle trajectories were undeflected by the collision.
G.B. 56; 1991, *63*, 1555; 1996, *68*, 168

impedance, Z
Complex representation of potential difference divided by the complex representation of the current.
G.B. 15; ISO 31-4: 1992; 1996, *68*, 976

impingement
Equivalent to *impaction;* often refers to impaction on a liquid surface.
1990, *62*, 2196

impinger
A sampling instrument employing impingement for the collection of particulate matter. Common types are: (i) the midget impinger employing impingement in 1–10 cm^3 water, (ii) the standard impinger employing impingement in 75 cm^3 water and (iii) dry impingers. Impingers are also suitable for sampling certain gases and vapours.
1990, *62*, 2196

imprecision (in analysis)
Variation of the result in a set of replicate measurements. This can be expressed, e.g. as the *standard deviation* or coefficient of variation (relative standard deviation). This term may have a more general meaning, e.g. if the replicates constitute a batch or involve different instruments, laboratories and analyst.
See *precision*.
1989, *61*, 1663; 1995, *67*, 1706

impregnation (in chromatography)
The modification of the separation properties of the chromatographic bed used in planar chromatography by appropriate additives.
1993, *65*, 830

improved canonical variational transition-state theory (ICVTST)
A modification of *canonical variational transition-state theory* in which, for energies below the *threshold energy*, the position of the *dividing surface* is taken to be that of the microcanonical threshold energy. This forces the contributions to *rate constants* to be zero below the threshold energy. A compromise dividing surface is then chosen so as to minimize the contributions to the rate constant made by *reactants* having higher energies.
1996, *68*, 168

impulsive reaction
See *direct reaction*.
1996, *68*, 168

inaccuracy (in analysis)
A quantitative term to describe the (lack of) accuracy of a *chemical measurement process*; comprises the *imprecision* and the *bias*. Inaccuracy must be viewed as a two-component quantity (vector); imprecision and bias should never be combined to give a scalar measure for chemical measurement process inaccuracy. (One or the other component may, however, be negligible under certain circumstances.) Inaccuracy should not be confused with uncertainty. Inaccuracy (imprecision, bias) is characteristic of the measurement process, whereas error and uncertainty are characteristics of a result. (The latter characteristic, of course, derives from the imprecision and bounds for bias of the chemical measurement process.)
Note:
The resultant bias and imprecision for the overall measurement process generally arise from several individual components, some of which act multiplicatively (e.g. *sensitivity*), and some of which act additively (e.g. the blank).
1995, *67*, 1706; 1989, *61*, 1663

inch
Non-SI unit of length, in = 0.0254 m.
G.B. 110

incinerator
Equipment in which solid, liquid or gaseous combustible wastes are ignited and burned. Types include flue-fed and multiple-chamber incinerators with several stages of combustion.
1990, *62*, 2196

inclusion compound (inclusion complex)
A *complex* in which one component (the *host*) forms a cavity or, in the case of a crystal, a crystal lattice containing spaces in the shape of long tunnels or channels in which molecular entities of a second *chemical species* (the *guest*) are located. There is no covalent bonding between guest and host, the attraction being generally due to *van der Waals forces*. If the spaces in the host lattice are enclosed on all sides so that the guest species is 'trapped' as in a cage, such compounds are known as *clathrates* or 'cage' compounds'.
See also *crown compounds, cryptands, cryptates, intercalation compounds*.
1994, *66*, 1124; 1995, *67*, 1344

incoherent radiation
Not having the properties of *coherent radiation*.
1996, *68*, 2247

incoherent scattering
See *coherent scattering*.
1983, *55*, 932

incoherent structure
See *coherent structure*.
1976, *46*, 76

incongruent reaction
See *peritectic reaction*.
1994, *66*, 583

increment (for bulk materials and large units)
An individual portion of material collected by a single operation of a sampling device.
Increments may be reduced individually or tested either (a) individually or (b) combined with other increments with the resulting composite reduced in size and tested as a single unit. Increments are created by the sampling operation and are usually taken from parts of a lot separated in time or space. Increments of a bulk population correspond to units of a packaged population.
1990, *62*, 1205

indicated hydrogen
Under certain circumstances it is necessary to indicate in the name of a ring, or ring system, containing the maximum number of non-cumulative double bonds, one or more positions where no multiple bond is attached. This is done by specifying the presence of an 'extra' hydrogen atom at such positions by citation of the appropriate numbered locant followed by an italicized capital *H*.
Example:

3*H*-Pyrrole

A second type of indicated hydrogen (sometimes referred to as 'added hydrogen') describes hydrogen atoms added to a specific structure as a consequence of the addition of a suffix or a prefix describing a structural modification. This type of indicated hydrogen is normally cited in parentheses after the locant of the additional feature.
Example:

Phosphinine   Phosphinin-2(1*H*)-one

B.B.(G) 34

indicator (visual)
See *acid-base indicator, adsorption indicator, mixed indicator, colour indicator*.
O.B. 48

indicator electrode
An electrode that serves as a transducer responding to the excitation signal (if any) and to the composition of the solution being investigated, but that does not affect an appreciable change of bulk composition within the ordinary duration of a measurement.
O.B. 59

indicator reaction
See *coupled (indicator) reaction (in analysis)*.
1993, *65*, 2294

indifferent absorbing ion
An ion which absorbs through Coulomb forces only and which is hence only repelled by surfaces of like sign, attracted by surfaces of opposite sign, and is not absorbed on an uncharged surface.
See also *specifically absorbing ion*
1991, *63*, 899

indifferent electrolyte
Synonymous with supporting electrolyte.
1985, *57*, 1501

indirect amplification
A type of *amplification reaction* where the constituent required is associated with some other constituent which is then amplified and measured.
1982, *54*, 2555

indirect reaction
A chemical process in which the reactive complex has a lifetime that is longer than its period of rotation is called an indirect reaction (in contrast to a *direct reaction*). In a *molecular-beam* experiment the *products* of such reactions are scattered at random with reference to the centre of mass of the system.
Indirect reactions are also called *complex-mode reactions*. (They are sometimes called complex reactions, but this usage is not recommended in view of the danger of confusion with composite reactions, which occur in more than one step).
1996, *68*, 168

individual isotherm (in surface chemistry)
See *partial isotherm (in surface chemistry)*.
1972, *31*, 594

individual particle analysis
See *selective micro-sample*.
1988, *60*, 1467

individual perception threshold (IPT) (in atmospheric chemistry)
A term used in odour testing which signifies the lowest concentration of a particular species at which a subject indicates both an initial positive and repeated response.
1990, *62*, 2196

induced radioactivity
Radioactivity induced by *irradiation*.
1982, *54*, 1544

induced reaction
A reaction which occurs or is accelerated only if another reaction is simultaneously occurring in the same system. The substance, which causes an induced reaction, is termed *inductor*. The inductor, unlike the catalyst, is used up irreversibly during the process.
1993, *65*, 2295

inducer (in enzyme catalysis)
A small molecule that triggers *gene transcription* by binding to a regulator protein (cf. *enzyme induction*). It acts by combining with the corresponding *repressor* protein to bring about an allosteric change so

that the repressor is made incapable of combining with the *operator* of the system.
1992, *64*, 157

induction (in enzyme catalysis)
1. An increase in the rate of synthesis of an (inducible) *enzyme* in response to the action of an *inducer* or environmental conditions. An inducer is often the *substrate* of the inducible enzyme or a structurally similar substance (gratuitous inducer) that is not metabolized.
2. The experimental elicitation of lytic growth by a *prophage* from *lysogenic bacteria*.
1992, *64*, 157; 1994, *66*, 2593

induction period
The initial slow phase of a *chemical reaction* which later accelerates. Induction periods are often observed with radical reactions, but they may also occur in other systems (for example before steady-state concentration of the reactants is reached).
1994, *66*, 1124; 1996, *68*, 169

induction, stereochemical
See *asymmetric induction*.
1996, *68*, 2200

inductive effect
In strict definition, an experimentally observable effect (on rates of reaction, etc.) of the transmission of charge through a chain of atoms by electrostatic induction. A theoretical distinction may be made between the *field effect*, and the inductive effect as models for the Coulomb interaction between a given site within a *molecular entity* and a remote unipole or dipole within the same entity. The experimental distinction between the two effects has proved difficult, except for molecules of peculiar geometry, which may exhibit 'reversed field effects'. Ordinarily the inductive effect and the field effect are influenced in the same direction by structural changes in the molecule and the distinction between them is not clear. This situation has led many authors to include the field effect in the term 'inductive effect'. Thus the separation of σ values into inductive and *resonance* components does not imply the exclusive operation of a through-bonds route for the transmission of the non-conjugative part of the substituent effect. To indicate the all-inclusive use of the term inductive, the phrase 'so-called inductive effect' is sometimes used. Certain modern theoretical approaches suggest that the 'so-called inductive effect' reflects a field effect rather than through-bonds transmission.
See also *field effect, mesomeric effect, polar effect*.
1994, *66*, 1124

inductomeric effect
A molecular polarizability effect occurring by the inductive mechanism of electron displacement. The consideration of such an effect and the descriptive term have been regarded as obsolescent or even obsolete, but in recent years theoretical approaches have reintroduced substituent polarizability as a factor governing reactivity, etc. and its parametrization has been proposed.
1994, *66*, 1125

inductor
See *induced reaction*.
1993, *65*, 2295

inelastic collision
See *elastic collision*.
1996, *68*, 159

inelastic scattering
If in a molecular collision there is transfer of energy among degrees of freedom, but no chemical reaction occurs, there is said to be inelastic scattering.
1996, *68*, 169

inert
Stable and *unreactive* under specified conditions.
1994, *66*, 1125

inert gas
A non-reactive gas under particular conditions. For example, nitrogen at ordinary temperatures and the noble gases (helium, argon, krypton, xenon and radon) are unreactive toward most species.
1990, *62*, 2196

inertial defect
Difference of principal *moments of inertia* $\Delta = I_C - I_A - I_B$. For planar molecules it is close to zero and for non-planar molecules it is negative.
G.B. 23

inertial separator
Any dry type collector which utilizes the relatively greater inertia of particles to effect their removal from a gas stream; e.g. cyclonic and impingement separators, gravity settling chambers and high-velocity gas reversal chambers.
1990, *62*, 2196

infinite source thickness
For a specified *radiation*, the minimum thickness of a flat preparation of a radioactive material where the *intensity* of the specified radiation at the surface does not increase when the thickness is increased by adding more of similar radioactive material.
1982, *54*, 1544

infinite thickness, effectively (in flame spectroscopy)
See *effectively infinite thickness (in flame spectroscopy)*.
O.B. 182

information theory
As applied in reaction dynamics, information theory analyses the deviations of experimentally-determined energy distributions from predictions made on the basis of equal *a priori* probabilities for all possible outcomes.
1996, *68*, 169

inherent viscosity (of a polymer)
The ratio of the natural logarithm of the relative viscosity, η_r, to the mass concentration of the polymer, c, i.e.

$$\eta_{inh} \equiv \eta_{ln} = (\ln \eta_r)/c$$

The quantity η_{ln}, with which the inherent viscosity is synonymous, is the logarithmic viscosity number.
Notes:
1. The unit must be specified; $cm^3\ g^{-1}$ is recommended.
2. These quantities are neither viscosities nor pure numbers. The terms are to be looked on as traditional names. Any replacement by consistent terminology would produce unnecessary confusion in the polymer literature.
P.B. 63

inhibition
The decrease in *rate of reaction* brought about by the addition of a substance (*inhibitor*), by virtue of its effect on the concentration of a reactant, *catalyst* or *reaction intermediate*. For example, molecular oxygen and *p*-benzoquinone can react as 'inhibitors' in many reactions involving *radicals* as *intermediates* by virtue of their ability to act as *scavengers* toward these radicals.
If the rate of a reaction in the absence of inhibitor is v_0 and that in the presence of a certain amount of inhibitor is v, the degree of inhibition (i) is given by:

$$i = (v_0 - v)/v_0$$

See also *mechanism based inhibition, inhibitor*.
1994, *66*, 1125; 1996, *68*, 169; see also 1992, *64*, 157; 1993, *65*, 2295

inhibitor
A substance that diminishes the rate of a chemical reaction; the process is called *inhibition*. Inhibitors are sometimes called negative catalysts, but since the action of an inhibitor is fundamentally different from that of a *catalyst*, this terminology is discouraged. In contrast to a catalyst, an inhibitor may be consumed during the course of a reaction. In enzyme-catalysed reactions an inhibitor frequently acts by binding to the enzyme, in which case it may be called an *enzyme inhibitor*.
See also *effector*.
1993, *65*, 2295; see also 1991, *63*, 1244

inhibitory concentration (IC)
Concentration of a substance that causes a defined inhibition of a given system: IC_{50} is the median concentration that causes 50% inhibition.
See also *lethal concentration*.
1993, *65*, 2060

inhibitory dose (ID)
Dose of a substance that causes a defined inhibition of a given system: ID_{50} is the median dose that causes 50% inhibition.
See also *lethal dose*
1993, *65*, 2060

inhomogeneity error (in spectrochemical analysis)
An error due to inhomogenous distribution of absorbing species in the analytical sample. It may be observed by probing the sample volume with an incident beam of radiation of reduced cross-section.
1988, *60*, 1456

initial energy (in *in situ* microanalysis)
See *excitation energy (in* in situ *microanalysis)*.
1983, *55*, 2027

initial rate method
A mode of measurement in a kinetic method of analysis, in which the initial reaction rate is determined (usually by the slope method) and utilized for the measurement of concentration.
1993, *65*, 2296

initial (final) state correlations
These are correlations between the states of the products of a reaction and the initial states of the reactants. The correlations may relate to energy levels, quantum numbers, and symmetries.
1996, *68*, 169

initiation
A reaction or process generating free *radicals* (or some other *reactive reaction intermediates*) which then induce a *chain reaction*. For example, in the chlorination of alkanes by a radical mechanism the initiation step is the *dissociation* of molecular chlorine.
1994, *66*, 1125; 1996, *68*, 169

initiator
A substance introduced into a reaction system in order to bring about an *initiation* reaction.
1996, *68*, 170; 1976, *46*, 74

injection temperature (in chromatography)
The temperature at the point of injection.
O.B. 100

in-laboratory processing (in analytical chemistry)
The selection, removal and preparation of the test (or analytical) portions from the laboratory sample.
The processing may include a reduction in the size of the unit(s) (division) and particle size (reduction), as well as mixing to achieve homogeneity.
1990, *62*, 1204

inner electric potential, φ
Quantity the gradient of which is equal to the negative of the electric field strength within the phase concerned. It can be calculated from the model of the phase. It is sometimes called the Galvani potential.
G.B. 59

inner filter effect
1. In an emission experiment, an apparent decrease in emission quantum yield and/or distortion of bandshape as a result of reabsorption of emitted radiation.
2. During a light irradiation experiment, absorption of incident radiation by a species other than the intended primary absorber.
1996, *68*, 2247

inner Helmholtz plane (IHP)
The locus of the electrical centres of specifically adsorbed ions.

See also *outer Helmholtz plane*.
1991, 63, 899; 1986, 58, 448

inner layer (compact layer) (in electrochemistry)
The region between the *outer Helmholtz plane* and the *interface*.
1983, 55, 1261

inner orbital X-ray emission spectra
X-ray *spectral lines* generated by the transition of electrons between 'inner orbitals'. The term 'inner orbital' serves to designate the sharp, not degenerate electronic levels of atoms in a solid and therefore is a general term for all orbitals except the valence band.
1983, 55, 2026

inner salts
See *zwitterionic compounds*.
1995, 67, 1344

inner-sphere electron transfer
Historically an electron transfer between two metal centres sharing a ligand or atom in their respective coordination shells. The definition has more recently been extended to any situation in which the interaction between the donor and acceptor centres in the *transition state* is significant (>20 kJ mol^{-1}).
See also *outer-sphere electron transfer*.
1994, 66, 1125; 1996, 68, 2247

inoculation
The introduction of a small sample of organisms (microbial, plant or animal *cells*) into a culture medium to act as a seed for the production of large numbers of the same organism by growth and propagation.
1992, 64, 158

inositols
Cyclohexane-1,2,3,4,5,6-hexols.
1995, 67, 1344; see also W.B. 151

in-out isomerism
Isomerism found in bicyclic systems having long enough bridges to allow the bridgehead exocyclic bond or lone pair of electrons to point either inside the structure or outside.

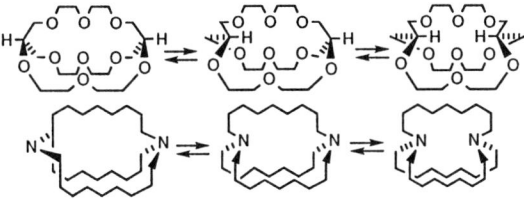

1996, 68, 2210

input rate (in analysis)
The number of samples that are processed by the instrument divided by time of operation. It should be stated if the residence time is included or not.
See *residence time*.
1989, 61, 1663

insert (in biotechnology)
A sequence of foreign *DNA* introduced into a restriction site (insertion site, cloning site) of a *vector* DNA.
1992, 64, 158

insertion
1. A *chemical reaction* or *transformation* of the general type:

$$X–Z + Y \rightarrow X–Y–Z$$

in which the connecting atom or *group* Y replaces the bond joining the parts X and Z of the reactant XZ. An example is the carbene insertion reaction:

$$R_3C–H + H_2C: \rightarrow R_3C–CH_3$$

The reverse of an insertion is called an *extrusion*.
See also α-*addition*.
1994, 66, 1125

2. A general term given to a reaction involving the transfer of a guest atom, ion or molecule into a host crystal lattice.
See *intercalation reaction, topochemical reaction, topotactic reaction*.
1994, 66, 583

***in situ* microanalysis**
Term used for direct analytical investigation of the microstructural domains of a solid by focused beams of particles and radiation. Analytical characterization includes obtaining information about type, quantity and distribution of the elements, their state of chemical bonding, morphology and crystalline (geometric) and electronic structure of the individual phases. The combination of this information serves as a basis for the property-relevant characterization of solids.
1983, 55, 2024

***in situ* micro-X-ray diffraction (Kossel-technique)**
Any technique which utilizes the diffraction of X-rays generated in a microstructural domain of a solid under bombardment with a finely focused electron beam, thus providing an X-ray diffraction pattern of this microstructural domain. The pattern can be recorded with a film either on the reflection or transmission side of the specimen (in the latter case the crystalline sample has to be a thin film or a small particle).
1983, 55, 2025

instability (with reference to instrumentation)
A change which takes place in an instrument reading over a stated period of unattended operation for a given value of the air quality characteristic. It can be characterized by the variation with time of its mean, specifying the drift and by the dispersion. *Span instability* is the change which takes place in instrument span over a stated period of unattended operation. *Zero instability* is the change in instrument reading in response to a zero sample over a stated period of unattended operation.
1990, 62, 2197

instantaneous current
At a dropping electrode, the total current that flows at the instant when a time t has elapsed since the fall of the preceding drop.
At any other electrode, the total current that flows at the instant when a time t has elapsed since the beginning of an *electrolysis*.
The instantaneous current is usually time-dependent and may have the character of an *adsorption, catalytic, diffusion, double-layer*, or *kinetic current*, and may include a *migration current*. A plot of the dependence of instantaneous current on time is commonly called an '*i-t* curve'.
1985, *57*, 1496

instantaneous rate of flow (in polarography)
The rate of increase of the mass of a drop at a particular instant t seconds after it has begun to form.
1985, *57*, 1504

instantaneous (spot) sampling (in atmospheric chemistry)
Obtaining a sample of the atmosphere in a period which is short compared with the duration of the sampling exercise. Such samples are often called 'grab' samples, a term not recommended. These are useful for the analysis of hydrocarbons and other complicated mixtures of trace gases which are relatively stable in a stainless steel canister or tank and can be transported back to the laboratory for chromatographic or other analysis which cannot be done satisfactorily in the field.
1990, *62*, 2197

instrumental activation analysis
A kind of *activation analysis* in which element specificity is obtained by using appropriate *irradiation* conditions, *radiation* measurement techniques and mathematical techniques for the interpretation of the measurement results.
1994, *66*, 2515

instrumental dependability
This relates to the frequency of failures which interrupt the operation of an instrument.
It can be quantified, e.g. by the infrequency of breakdowns or by the availability of the instrument for use when required.
1989, *61*, 1663

instrumental indication (for a precision balance)
The observed *deflection* or *rest point* multiplied by the value of the division for the load in question.
O.B. 36

integral capacitance (of an electrode)
Integral capacitance (per unit area of electrode) is given by

$$K = Q/(E - E_{Q=0})$$

where Q is the electric charge (per unit area of electrode), E is the potential of the electrode with respect to a *reference electrode*, and $E_{Q=0}$ is the *potential at the point of zero charge*.
1974, *37*, 509

integral detector (in chromatography)
A device which measures the accumulated quantity of sample component(s) reaching the detector.
1993, *65*, 849

integrating sphere
A hollow sphere having a highly reflecting inside surface used as a device to collect, with very high efficiency, light scattered or emitted from a sample contained in it or located outside and near one of the ports. Small ports allow the entrance of light and access to a detector.
1996, *68*, 2248

intended crossing (of potential-energy surfaces)
Synonymous with *avoided crossing*. The term 'intended' should not be used in this context since it is an anthropomorphic term.
1996, *68*, 2248

intensity
1. The traditional term for *photon flux, fluence* rate, *irradiance* or *radiant power* (*radiant flux*). In terms of an object exposed to radiation, the term should now be used only for qualitative descriptions.
1996, *68*, 2248; see also O.B. 228; G.B. 31; 1990, *62*, 2197
2. The magnitude of a particular feature in a spectrum.
1996, *68*, 2248

intensity (relative to base peak) (in mass spectrometry)
The ratio of intensity of a separated ion beam to the intensity of the separated ion beam which has the greatest intensity. This ratio is generally equated to the normalized ratio of the heights of the respective peaks in the mass spectrum, with the height of the base peak being taken as 100.
1991, *63*, 1554; O.B. 206

intensive quantity
Physical quantity whose magnitude depends on the extent of the system.
G.B. 7

interaction distance
The farthest distance of approach of two particles at which it is discernible that they will not pass at the impact parameter.
1991, *63*, 1555

intercalation compounds
Compounds resulting from reversible inclusion, without covalent bonding, of one kind of molecule in a solid matrix of another compound, which has a laminar structure. The host compound, a solid, may be macromolecular, crystalline or amorphous.
See also *inclusion compounds*.
1995, *67*, 1344

intercalation reaction
A reaction, generally reversible, that involves the introduction of a guest species into a host structure

without a major structural modification of the host. In the strictest sense, intercalation refers to the insertion of a guest into a two-dimensional host; however, the term also now commonly refers to one-dimensional and three-dimensional host structures.
Example:
The insertion of lithium into layered TiS_2:

$x\,Li + TiS_2 \rightarrow Li_xTiS_2\ (0 \leq x \leq 1)$

Synonymous with *insertion reaction*.
1994, *66*, 583; R.B. 79

interchromophoric radiationless transition
See *radiationless transition*.
O.B. 187

intercomparison, analytical
See *analytical intercomparison*.
1982, *54*, 1544

interconal region
The region of a Bunsen flame where the combustion zone has the form of a cone.
O.B. 166

interconvertible enzyme
An enzyme existing in at least two well defined, reversibly convertible forms, produced by covalent modifications of amino acid side chains under biological conditions. Covalent modifications that occur as intermediates in the catalytic process are not included in this definition.
W.B. 94

interface
The plane ideally marking the boundary between two phases.
Note:
The term should not be used synonymously with *interphase*.
1986, *58*, 439; see also 1972, *31*, 583; O.B. 251

interfacial concentration (in electrochemistry), c
The interfacial concentration $c_{B,e}$ (or simply c_e) of a species B is the concentration of that species at the boundary of the *electric double layer* facing the solution, i.e. just outside the region where the departures from the electroneutrality of the solution are significant. This concept is mainly used in the usual case where the thickness of the electric double layer is very small as compared to that of the *diffusion layer*. It is often calculated from theory or derived from measurements of the *limiting current*.
See *mass transfer coefficient (in electrochemistry)*.
1981, *53*, 1837; see also 1980, *52*, 236

interfacial double-layer
The Coulombic interaction of interfacial charges (e.g. ions) and the magnetic or electrostatic interaction of interfacial molecules lead to particularly complex interfacial structures. Complex interfacial profiles that can be approximated by two distinct sub-layers with different physical properties (e.g. structure and/or nature and/or composition), are referred to as interfacial double-layers. Examples of such approximated complex profiles are: the electrical double-layer consisting of a surface charge layer (i.e. a two dimensional distribution of one type of ions) and a diffuse charge layer (counter-ions distributed over the space region next to the surface); the approximated profile of the orientation angle of anisotropic liquid molecules within a 'double-layer' consisting of a distribution of so-called anchored molecules which are perturbed (strongly bound and orientated) by the surface, and the adjacent, so-called, transition layer, i.e. the region where the surface perturbation is damped.
1994, *66*, 1674

interfacial layer
The inhomogeneous space region intermediate between two bulk phases in contact, and where properties are significantly different from, but related to, the properties of the bulk phases. Examples of such properties are: compositions, molecular density, orientation or conformation, charge density, pressure tensor, electron density, etc. The interfacial properties vary in the direction normal to the surface. Complex profiles of interfacial properties occur in the case of multicomponent systems with coexisting bulk phases where attractive/repulsive molecular interactions involve adsorption or depletion of one or several components.
This interfacial region may also be regarded as a distinct, though not autonomous, phase and be called the *interphase*.
1994, *66*, 1673; 1986, *58*, 439

interfacial layer width (in thin films), σ_S
The *characteristic length* defined by the variance of the Gaussian function fitted to the gradient of the electron density profile established by X-ray specular reflectivity studies of the liquid interface.
1994, *66*, 1674

interfacial region
See *interfacial layer*.
1994, *66*, 1673; 1986, *58*, 439

interfacial tension
See *surface tension*.
G.B. 12; 1972, *31*, 596

interference (in analysis)
A systematic error in the measure of a signal caused by the presence of *concomitants* in a sample.
1983, *55*, 554

interference filter
See *filter*.
1996, *68*, 2248

interference, transport (in flame spectroscopy)
See *transport interference (in flame spectroscopy)*.
O.B. 170

interfering lines
Radiation from nearby *spectral lines* may perturb the measurement of the *intensity* of the line wanted. These lines are termed interfering lines.
O.B. 159

interfering substance (in electroanalytical chemistry)
Any substance, other than the ion being measured, whose presence in the sample solution affects the measured emf of a cell. Interfering substances fall into two classes: 'electrode/electrochemical' interferences and 'chemical' interferences. Examples of the first class include:
1. Those substances which give a similar response to the ion being measured and whose presence generally results in an apparent increase in the activity (or concentration) of the ion to be determined (e.g. Na^+ for the Ca^{2+} electrode).
2. Electrolytes present at a high concentration that give rise to appreciable liquid junction potential differences or results in a significant activity coefficient decrease, or incipient Donnan exclusion failure.
The second class of substances that should be recognized as chemical interferences includes:
3. Species that interact with the ion being measured so as to decrease its activity or apparent concentration. The electrode continues to report the true activity (e.g. CN^- present in the measurement of Ag^+), but a considerable gap will occur between the activity and concentration of the ions even in very dilute solutions. Under these circumstances the determination of ionic concentration may be problematic.
4. Substances interacting with the membrane itself, blocking the surface or changing its chemical composition [i.e. organic solvents for the liquid or poly(vinyl chloride) (PVC) membrane electrodes] are grouped as interferences or electrode poisons.
1994, *66*, 2530

interferometer
See *Fourier transform spectrometer*.
1996, *68*, 2248

interferons
A class of glycoproteins (with sugar groups attached at specific locations) important in immune function. They are able to inhibit the multiplication of *viruses* in *cells*.
1992, *64*, 158

intermediate
A *molecular entity* with a *lifetime* appreciably longer than a molecular vibration (corresponding to a local potential energy minimum of depth greater than RT) that is formed (directly or indirectly) from the reactants and reacts further to give (either directly or indirectly) the products of a *chemical reaction*; also the corresponding *chemical species*.
See *reaction step, elementary reaction, stepwise reaction*.
1994, *66*, 1126; see also 1996, *68*, 170

intermediate neutrons
Neutrons of kinetic energy between the energies of *slow* and *fast neutrons*. In reactor physics, the range might be 1 eV to 0.1 MeV.
1982, *54*, 1547

intermolecular
1. Descriptive of any process that involves a transfer (of atoms, *groups*, electrons, etc.) or interactions between two or more *molecular entities*.
2. Relating to a comparison between different molecular entities.
See also *intramolecular*.
1994, *66*, 1126

intermolecular radiationless transition
See *radiative transition*.
O.B. 185

internal absorptance, α_i
Absorptance when surface effects and effects of the cuvette such as reflection losses are excluded.
G.B. 32

+ internal compensation
Usage strongly discouraged.
See *meso-compound*.
1996, *68*, 2210

internal conversion
A photophysical process. An isoenenergetic radiationless transition between two electronic states of the same multiplicity. When the transition results in a vibrationally excited molecular entity in the lower electronic state, this usually undergoes deactivation to its lowest vibrational level, provided the final state is not unstable to dissociation.
1996, *68*, 2248; see also O.B. 187, 216

internal energy, U
Quantity the change in which is equal to the sum of heat, q, brought to the system and work, w, performed on it, $\Delta U = q + w$. Also called thermodynamic energy.
G.B. 48; 1996, *68*, 976

internal energy, quantized
See *quantized internal energy*.
1988, *60*, 1452

internal filling solution (of a glass electrode)
Aqueous electrolyte solution, which may be gelled, containing fixed concentration of hydrogen ions, e.g. HCl or a buffer solution, and a fixed concentration of the ion to which the *internal reference electrode* is reversible, e.g. chloride ion in the case of silver-silver chloride, or calomel electrodes.
1985, *57*, 540

internal reference electrode
An 'inner' reference electrode which is contained inside an *ion-selective electrode* assembly.
Comment: The system frequently consists of a silver/silver chloride electrode in contact with an appropriate solution containing fixed concentrations of chloride and the ion for which the membrane is selective. This 'inner' filling solution is in contact with the ion-selective membrane.
1994, *66*, 2533

internal return
See *ion-pair return*.
1994, *66*, 1126

internal standard (in chromatography)
A compound added to a sample in known concentration to facilitate the qualitative indentification and/or quantitative determination of the sample components.
1993, *65*, 837

internal surface
With respect to porous solids, the surface associated with pores communicating with the outside space. Since the accessibility of pores may depend on the size of the fluid molecules, the extent of the accessible internal surface may depend on the size of the molecules comprising the fluid, and may be different for the various components of a fluid mixture (*molecular sieve effect*).
1976, *46*, 79; 1972, *31*, 585

internal transmittance
See *transmittance*.
1996, *68*, 2248

internal valence force field (IVFF)
A force field expressed in terms of bond-stretching, angle-bending, torsional and other displacements directly connected to the structural parameters of the molecule:

$$V = \frac{1}{2}\sum k_{ij} \text{ (or } K_{ij}) R_i R_j$$

where k (or K) are the *force constants* and R the internal valence coordinates.
1978, *50*, 1709

international calorie
See *calorie*.
G.B. 112; 1996, *68*, 976

International system of units
See *SI*.
G.B. 69

+ international unit
The unit of enzyme activity proposed by the International Union of Biochemistry in 1964. Specifically, it is the amount of enzyme that catalyses the conversion of one micromole of substrate per minute under the specified conditions of the assay method.
Note:
This unit is no longer recommended because the term does not indicate what physical quantity it refers to, and because the minute is not the SI unit of time.
1994, *66*, 2593

interparticle porosity (in chromatography), ε
The interparticle volume of a packed column per unit column volume:

$$\varepsilon = V_0/V_c$$

It is also called the interstitial fraction of the column.
1993, *65*, 832

interparticle volume of the column (in chromatography), V_0
The volume occupied by the mobile phase between the particles in the packed section of a column. It is also called the interstitial volume or the void volume of the column. In *liquid chromatography*, the interparticle volume is equal to the mobile-phase holdup volume (V_M) in the ideal case, neglecting any extra-column volume. In gas chromatography, the symbol V_G may be used for the interparticle volume of the column. In the ideal case, neglecting any extra-column volume, V_G is equal to the corrected gas hold-up volume (V_M^o).

$$V_G = V_M^o = V_M j$$

interpenetrating polymer network (IPN)
A *polymer* comprising two or more *networks* which are at least partially interlaced on a molecular scale but not covalently bonded to each other and cannot be separated unless chemical bonds are broken.
A mixture of two or more preformed polymer networks is not an IPN.
1996, *68*, 2305

interphase
See *interfacial layer*.
1986, *58*, 439; 1994, *66*, 1673

interphase transition
A transition that occurs at boundaries between phases.
Example:
The precipitation of second phases or the initiation of new phase growth at an interface.
1994, *66*, 584

interstitial fraction (in chromatography)
The *interstitial volume* per unit volume of a packed column. Defined by:

$$\varepsilon_I = V_I/X$$

where V_I is the interstitial volume and X the *column volume*.
O.B. 100

interstitial velocity (in chromatography)
The linear velocity of the *mobile phase* inside a packed column calculated as the average over the entire cross section. This quantity can, under idealized conditions, be calculated from the equation:

$$u = F/\varepsilon_I$$

where F is the *nominal linear flow* and ε_I the *interstitial fraction*.
O.B. 102

interstitial volume (in gas chromatography)
The volume V_G of the column not occupied by the liquid phase and its solid support, or by the active solid. It does not include any volume external to the column, such as the volume of the sample injector or of the detector.
O.B. 99

intersystem crossing
A photophysical process. An isoenergetic radiationless transition between two electronic states having different multiplicities. It often results in a vibrationally excited molecular entity in the lower electronic state, which then usually deactivates to its lowest vibrational level.
1996, *68*, 2248; see also O.B. 187

interval analysis
A method to estimate uncertainties in the results of a *least-squares* procedure. The input data are primary values plus limits of intervals expressing their uncertainty. The output consists of derived values plus limits of intervals expressing a pessimistic estimate of maximal uncertainty.
1981, *53*, 1823

intervalence charge transfer
Electron transfer (thermal or photoinduced) between two metal sites differing only in oxidation state. Quite often such electron transfer reverses the oxidation states of the sites. The term is frequently extended to the case of *metal-to-metal charge transfer* between non-equivalent metal centres.
1996, *68*, 2248

interzonal region
The region of the Bunsen flame confined by the inner and outer zones, where in many instances the conditions for flame analysis are optimum.
O.B. 166

intimate ion pair
See *ion pair*.
1994, *66*, 1126

intra- (in organic reaction mechanisms)
The prefix for an elementary reaction or some part thereof indicating that the bond(s) undergoing primitive changes form part of a ring in the transition state of the elementary reaction. This prefix is used for nonpericyclic processes only and is replaced by 'cyclo' for pericyclic processes.
1989, *61*, 34

intrachromophoric radiationless transition
See *radiationless transition*.
O.B. 187

intramolecular
1. Descriptive of any process that involves a transfer (of *atoms, groups, electrons*, etc.) or interactions between different parts of the same *molecular entity*.
2. Relating to a comparison between atoms or groups within the same *molecular entity*.
See also *intermolecular*.
1994, *66*, 1126

intramolecular catalysis
The acceleration of a chemical transformation at one site of a *molecular entity* through the involvement of another *functional* ('catalytic') *group* in the same molecular entity, without that group appearing to have undergone change in the reaction product. The use of the term should be restricted to cases for which analogous *intermolecular catalysis* by *chemical species* bearing that catalytic group is observable. Intramolecular catalysis can be detected and expressed in quantitative form by a comparison of the reaction rate with that of a comparable model compound in which the catalytic group is absent, or by measurement of the *effective molarity* of the catalytic group.

See also *effective molarity, neighbouring group participation*.
1994, *66*, 1126

intramolecular isotope effect
A kinetic *isotope effect* observed when a single substrate, in which the isotopic atoms occupy equivalent reactive positions, reacts to produce a non-statistical distribution of *isotopomeric* products. In such a case the isotope effect will favour the pathway with lower force constants for displacement of the isotopic nuclei in the *transition state*.
1994, *66*, 1130

intraphase transition
A transition that occurs within a phase.
Example:
Initiation of phase growth at an intragranular feature, such as twin planes and dislocations in steels.
1994, *66*, 584

intrinsic activation energy, $E_{a,i}$
If $E_{a,1}$ and $E_{a,-1}$ are the *activation energies* for a reaction in forward and reverse directions, the lesser of the two is sometimes called the intrinsic activation energy. It is the activation energy for the reaction in the exothermic direction.
1996, *68*, 170

intrinsic barrier
The Gibbs energy of activation ($\Delta^{\ddagger}G$) in the limiting case where $\Delta G^{\circ} = 0$, i.e. when the effect of thermodynamic driving force is eliminated. According to the *Marcus equation*, the intrinsic barrier is related to the *reorganization energy*, λ, of the reaction by the equation:

$$\Delta^{\ddagger}G = \lambda/4$$

1994, *66*, 1126

intrinsic detector efficiency
The ratio of the number of particles or photons detected to the number of similar particles or photons which have struck the envelope limiting the sensitive volume of a *radiation detector*.
1994, *66*, 2518

intrinsic efficiency
See *detector efficiency*.
1994, *66*, 2518

intrinsic full energy peak efficiency
The *detector efficiency* when considering only events where the total *energy* of the *radiation* is absorbed in the sensitive volume of the detector.
1982, *54*, 1543

intrinsic photopeak efficiency
The *detector efficiency* when only counts resulting from the *photoelectric effect* are considered.
1982, *54*, 1548

intrinsic viscosity (of a polymer)
The limiting value of the *reduced viscosity*, η_i/c, or the *inherent viscosity*, η_{inh}, at infinite dilution of the polymer, i.e.

$$[\eta] = \lim_{c \to 0} (\eta_i/c) = \lim_{c \to 0} \eta_{\text{inh}}$$

Notes:
1. This term is also known in the literature as the Staudinger index.
2. The unit must be specified; $cm^3\ g^{-1}$ is recommended.
3. This quantity is neither a viscosity nor a pure number. The term is to be looked on as a traditional name. Any replacement by consistent terminology would produce unnecessary confusion in the polymer literature.
Synonymous with limiting viscosity number.
P.B. 63

intron
An intervening section of DNA occuring almost exclusively within a eukaryotic *gene* but which is not translated to amino acid sequences in the gene product. The introns are removed from the premature mRNA through a process called *splicing* to form an active mRNA.
1992, *64*, 158

inverse isotope effect
A kinetic *isotope effect* in which $k^l/k^h < 1$, i.e. the heavier substrate reacts more rapidly than the lighter one, as opposed to the more usual 'normal' isotope effect, in which $k^l/k^h > 1$. The isotope effect will normally be 'normal' when the frequency differences between the isotopic *transition states* are smaller than in the reactants. Conversely, an inverse isotope effect can be taken as evidence for an increase in the corresponding force constants on passing from the reactant to the transition state.
1994, *66*, 1130

inverse kinetic isotope effect
See *isotope effect*.
1994, *66*, 1126

inverse micelle
See *inverted micelle*.
1994, *66*, 1126

inverse square law (in radiation chemistry)
The intensity of *radiation* from a point source in free space is inversely proportional to the square of the distance from the source. Presence of *absorbers* necessitates corrections to this law.
1982, *54*, 1544

inversion
1. See *Walden inversion*.
2. See *pyramidal inversion*.
3. See *ring inversion*.
4. A symmetry operation involving a centre of inversion (*i*).
1996, *68*, 2210

inversion height (in atmospheric chemistry)
The height above ground level at which there is change in sign of the normal temperature 'lapse rate', dT/dz (the rate of change of the temperature with height). Several temperature inversions may be present in the air over a given site at different altitudes as result of various meteorological factors.
1990, *62*, 2197

inversion, phase
See *phase inversion*.
1993, *65*, 2387

inversion point (in phase transitions)
The temperature (pressure) at which one polymorph of a solid phase changes into another at constant pressure (temperature).
See *polymorphic transition*.
1994, *66*, 584

inversion, temperature (in atmospheric chemistry)
See *temperature inversion*.
1990, *62*, 2197

inverted micelle
The reversible formation of association colloids from surfactants in non-polar solvents leads to aggregates termed inverted (or inverse, reverse or reversed) *micelles*. Such association is often of the type:

Monomer $\rightleftarrows$ Dimer $\rightleftarrows$ Trimer $\rightleftarrows$ ... *n*-mer

and the phenomenon of *critical micelle concentration* (or an analogous effect) is consequently not observed. In an inverted micelle the polar *groups* of the surfactants are concentrated in the interior and the *lipophilic* groups extend towards and into the non-polar solvent.
1994, *66*, 1126

inverted region (for electron transfer)
In plots relating rate constants to charges in standard Gibbs energy (ΔG^o) for *electron transfer* a region may occur in which the rate constants decrease as the exergonicity of the reaction increases. This region is often referred to as the inverted region and its presence is predicted by the theory developed for *outer sphere electron transfer* for the case $-\Delta G^o > \lambda$ in the *Marcus equation*, λ being the *reorganization energy*. Note the similarity to the energy gap law for radiationless conversion of an excited state.
See *normal region*.
1996, *68*, 2249

in vitro
In glass, referring to a study in the laboratory usually involving isolated organ, tissue, cell, or biochemical systems.
See also *in vivo*.
1993, *65*, 2061

in vivo
In the living body, referring to a study performed on a living organism.
See also *in vitro*.
1993, *65*, 2061

iodohydrins
See *halohydrins*.
1995, *67*, 1344

iodometric titration
See *titration*.
O.B. 47

iodonium ions
See *halonium ions*.
1995, *67*, 1344

ion
An atomic or molecular *particle* having a net electric charge.
1982, *54*, 1545

ion, adduct
See *adduct ion*.
1991, *63*, 1549

ion, cluster
See *cluster ion*.
1991, *63*, 1549

ion collector (in mass spectrometry)
Device for the capture of selected ions such as a Faraday cup collector (with a d.c. amplifier) or an electron multiplier.
1990, *62*, 2200

ion cyclotron resonance (ICR) mass spectrometer
A high-frequency *mass spectrometer* in which the ions to be detected, with a selected value of the quotient mass/charge, absorb maximum energy through the effect of a high-frequency electric field and a constant magnetic field perpendicular to the electric field. Maximum energy is gained by the ions which satisfy the cyclotron resonance condition and as a result they are separated from ions of different mass/charge.
1991, *63*, 1545; O.B. 202

ion, daughter
See *daughter ion*.
1991, *63*, 1549

ion, dimeric
See *dimeric ion*.
1991, *63*, 1549

ion energy loss spectra (in mass spectrometry)
Spectra that show the loss of translational energy of ions involved in ion/neutral species reactions.
1991, *63*, 1555

ion, even-electron
See *even-electron ion*.
1991, *63*, 1549

ion exchange
1. In surface chemistry, if the *adsorption* of one or several ionic species is accompanied by the simultaneous *desorption* (displacement) of an equivalent amount of one or more other ionic species, this process is called ion exchange.
1972, *31*, 585
2. The process of exchanging ions between a solution and an *ion exchanger*.
1993, *65*, 854

ion-exchange chromatography
Chromatography in which separation is based mainly on differences in the ion-exchange affinities of the sample components.
Present day ion-exchange chromatography on small particle high efficiency columns and usually utilizing conductometric or spectroscopic detectors is often referred to as ion chromatography (IC).
1993, *65*, 826

ion-exchange isotherm
The concentration of a counter-ion in the ion exchanger expressed as a function of its concentration in the external solution under specified conditions and at constant temperature.
1993, *65*, 854

ion-exchange membrane
A thin sheet or film of ion-exchange material which may be used to separate ions by allowing the preferential transport or either cations (in the case of a cation-exchange membrane) or anions (in the case of an anion-exchange membrane). If the membrane material is made from only ion-exchanging material, it is called a homogeneous ion-exchange membrane. If the ion-exchange material is embedded in an inert binder, it is called a heterogeneous ion-exchange membrane.
1993, *65*, 856

ion exchanger
A solid or liquid, inorganic or organic substance containing exchangeable ions with others of the same charge, present in a solution in which the ion exchanger is considered to be insoluble. It is recognized that there are cases where liquid exchangers are employed and where it may be difficult to distinguish between the separation process as belonging to ion exchange or liquid-liquid distribution, but the broad definition given here is regarded as that which is most appropriate. A monofunctional ion exchanger contains only one type of ionogenic group, a bifunctional ion exchanger two types and a polyfunctional ion exchanger more than one type. In a macroporous ion exchanger the pores are large compared to atomic dimensions.
1993, *65*, 855

ion, fragment
See *fragment ion*.
1991, *63*, 1549

ion-free layer
An *interphase* in which only non-specifically adsorbed ions are present.
1983, *55*, 1261

ionic concentration
See *concentration*.
1996, *68*, 977

ionic conductivity
Defined for ionic species B by

$$\lambda = |z_B| F u_B$$

where z_B is the charge number of the ionic species B, F is the *Faraday constant*, and u_B is the *electric mobility* of species B. In most current practice z_B is taken as unity, i.e. ionic conductivity is taken as that of species such as Na^+, $Ca^{2+}/2$, $La^{3+}/3$ etc. To avoid

ambiguity the species considered should be clearly stated, e.g. as $\lambda(Ca^{2+}/2)$.
1974, *37*, 512

ionic copolymerization
A *copolymerization* which is an *ionic polymerization*.
1996, *68*, 2308

ionic dissociation (in mass spectrometry)
The decomposition of an ion into another ion of lower mass and one or more neutral species.
1991, *63*, 1547

ionic polymerization
A *chain polymerization* in which the kinetic-chain carriers are ions or ion pairs.
Note:
Usually, the growing chain ends are ions.
1996, *68*, 2308

ionic spectral lines
See *atomic spectral lines*.
O.B. 118

ionic strength, I
On a molality basis, $I_m = \frac{1}{2}\Sigma m_B z_B^2$ where the sum goes over all the ions B. z_B is the charge number of ion B. The ionic strength on a concentration basis is defined analogously $I_c = \frac{1}{2}\Sigma c_B z_B^2$.
G.B. 50; 1996, *68*, 977

ionic transport number, t
Quotient of the current carried by an ionic component and the total current.
1996, *68*, 977

ion, isotopic
See *isotopic ion*.
1991, *63*, 1549

ionization
The generation of one or more ions. It may occur, e.g. by loss of an electron from a neutral *molecular entity*, by the *unimolecular heterolysis* of such an entity into two or more ions, or by a heterolytic substitution reaction involving neutral molecules, such as:

$$CH_3CO_2H + H_2O \rightarrow H_3O^+ + CH_3CO_2^-$$
$$Ph_3CCl + AlCl_3 \rightarrow Ph_3C^+ + AlCl_4^-$$
(electrophile assisted)
$$Ph_3CCl \rightarrow Ph_3C^+Cl^-$$
(ion pair, in benzene)

The loss of an electron from a singly, doubly, etc. charged cation is called second, third, etc. ionization. This terminology is used especially in mass spectroscopy.
See also *dissociation, ionization energy*.
1994, *66*, 1127

ionization, adiabatic
See *adiabatic ionization*.
1991, *63*, 1547

ionization, associative
See *associative ionization*.
1991, *63*, 1547

ionization, atmospheric pressure
See *atmospheric pressure ionization*.
1991, *63*, 1547

ionization, auto-
See *auto-ionization*.
1991, *63*, 1547

ionization buffer (in flame spectroscopy)
A buffer added to increase the free-electron concentration by the flame gases, thus repressing and stabilizing the degree of *ionization*.
O.B. 172

ionization by sputtering
Ionization by bombardment of a solid specimen with accelerated ions or electrons or fast neutrals.
O.B. 204

ionization chamber
A *radiation detector* which employs an electric field for the collection at the electrodes of charges associated with the *ions* produced in the sensitive volume by *ionizing radiation*, without charge multiplication. The solid angle (2Π, 4π) subtended when using the detector, is sometimes specified.
1982, *54*, 1545

ionization, charge exchange (charge transfer)
See *charge exchange (charge transfer) ionization*.
1991, *63*, 1547

ionization, chemi-
See *chemi-ionization*.
1991, *63*, 1547

ionization, chemical
See *chemical ionization*.
1991, *63*, 1547

ionization cross-section
A measure of the probability that a given ionization process will occur when an atom or molecule interacts with an electron or a photon.
1991, *63*, 1548

ionization, dissociative
See *dissociative ionization*.
1991, *63*, 1547

ionization efficiency
The ratio of the number of ions formed to the number of electrons or photons used in an ionization process.
1991, *63*, 1548

ionization efficiency curve (in mass spectrometry)
The number of ions produced as a function of the energy of the electrons or photons used to produce ionization.
1991, *63*, 1548; O.B. 206

ionization, electron
See *electron ionization*.
1991, *63*, 1547

ionization energy, E_i
Minimum energy required to eject an electron out of a neutral atom or molecule in its ground state. The adiabatic ionization energy refers to the formation of the molecular ion in its ground vibrational state and the vertical ionization energy applies to the transition to the molecular ion without change in geometry. This quantity was formerly called ionization potential. The

second ionization energy of an atom is the energy required to eject the second electron from the atom (energy for the process $X^+ \rightarrow X^{2+} + e$).
G.B. 20; 1994, *66*, 1128

ionization, fast atom bombardment
See *fast atom bombardment ionization*.
1991, *63*, 1547

ionization, field
See *field ionization*.
1991, *63*, 1548

ionization, fission fragment
See *plasma desorption ionization*.
1991, *63*, 1548

ionization, laser
See *laser ionization*.
1991, *63*, 1548

ionization, multiphoton
See *multiphoton ionization*.
1991, *63*, 1548

ionization, Penning
See *Penning ionization*.
1991, *63*, 1548

ionization, photo
See *photoionization*.
1991, *63*, 1548

ionization, plasma desorption
See *plasma desorption ionization*.
1991, *63*, 1548

+ ionization potential
See *ionization energy*.
G.B. 20

ionization, secondary
See *secondary ionization*.
1991, *63*, 1548

ionization, spark (source)
See *spark (source) ionization*.
1991, *63*, 1548

ionization, surface
See *surface ionization*.
1991, *63*, 1548

ionization, thermal
See *thermal ionization*.
1991, *63*, 1548

ionization, vertical
See *vertical ionization*.
1991, *63*, 1548

ionizing collision (in mass spectrometry)
An ion/neutral species interaction in which an electron or electrons are stripped from the ion and/or the neutral species in the collision. Generally, this term has come to be used to describe collisions of fast-moving ions with a neutral species in which the neutral species is ionized with no change in the number of charges carried by the ion. Care should be taken when this term is used to emphasize if charge stripping of the ion has taken place.
1991, *63*, 1555

ionizing power
A term to denote the tendency of a particular solvent to promote *ionization* of an uncharged or, less often, charged solute. The term has been used both in a kinetic and in a thermodynamic context.
See also *Dimroth–Reichardt E_T parameter, Grunwald–Winstein equation, Z-value*.
1994, *66*, 1128

ionizing radiation
Any *radiation* consisting of directly or indirectly ionizing particles or a mixture of both, or photons with *energy* higher than the energy of photons of ultraviolet light or a mixture of both such particles and photons.
1994, *66*, 2520

ionizing voltage
The voltage difference through which electrons are accelerated before they are used to bring about electron impact ionization. To obtain the true ionizing voltage, corrections for any contact or surface potentials must be made.
Note:
The term electron energy is frequently used in place of ionizing voltage.
O.B. 203

ion kinetic energy spectrum (in mass spectrometry)
A spectrum obtained when a beam of ions is separated according to the translational energy-to-charge ratios of the ionic species contained within it. A radial electric field (E) achieves separation of the various ionic species in this way.
1991, *63*, 1550

ion laser
A *gas laser* in which the *active species* is an ion formed in the electric discharge. Most practically important lasers of this type have continuous wave (CW) output. Examples are the argon ion, krypton ion and helium-cadmium (HeCd) lasers.
1995, *67*, 1920

ion, metastable
See *metastable ion*.
1991, *63*, 1549

ion microscopy
Use of the secondary ion mass spectrometry (SIMS) technique to obtain micrographs of the elemental (or isotopic) distribution at the surface of a sample with a spatial resolution of 2 µm or better.
O.B. 249

ion, molecular
See *molecular ion*.
1991, *63*, 1549

ion/molecule reaction (in mass spectrometry)
An ion/neutral species reaction in which the neutral species is a molecule. The use of ion-molecule reaction is not recommended; the hyphen suggests a re-

action of a species that is both an ion and a molecule and is not the intended meaning.
1991, *63*, 1556; O.B. 206

ion, negative
See *negative ion*.
1991, *63*, 1549

ion/neutral species exchange reaction (in mass spectrometry)
In this reaction an association reaction is accompanied by the subsequent or simultaneous liberation of a different neutral species as product.
1991, *63*, 1556

ion/neutral species reaction (in mass spectrometry)
A process wherein a charged species interacts with a neutral reactant to produce either chemically different species or changes in the internal energy of one or both of the reactants. The contrasting expression ion/neutral reaction is not ideal, simply because the word neutral is not a noun. However, any alternatives such as ion/neutral species reaction are so clumsy as to mitigate against their general acceptance
1991, *63*, 1556

ion, odd-electron
See *odd-electron ion*.
1991, *63*, 1549

ionogenic groups
The fixed groupings in an ion exchanger which are either ionized or capable of dissociation into fixed ions and mobile counter-ions.
1993, *65*, 854

ionomer
A polymer composed of *ionomer molecules*.
1996, *68*, 2305

ionomer molecule
A *macromolecule* in which a small but significant proportion of the *constitutional units* have ionizable or ionic groups, or both.
Some protein molecules may be classified as ionomer molecules.
1996, *68*, 2299

ion pair
A pair of oppositely charged ions held together by Coulomb attraction without formation of a *covalent bond*. Experimentally, an ion pair behaves as one unit in determining conductivity, kinetic behaviour, osmotic properties, etc.
Following Bjerrum, oppositely charged ions with their centres closer together than a distance:

$$q = 8.36 \times 10^6 z^+ z^- / (\varepsilon_r T) \text{ pm}$$

are considered to constitute an ion pair ('Bjerrum ion pair'). [z^+ and z^- are the charge numbers of the ions, and ε_r is the relative permittivity (or dielectric constant) of the medium.]
An ion pair, the constituent ions of which are in direct contact (and not separated by an intervening solvent or other neutral molecule) is designated as a 'tight ion pair' (or 'intimate' or 'contact ion pair'). A tight ion pair of X^+ and Y^- is symbolically represented as X^+Y^-.
By contrast, an ion pair whose constituent ions are separated by one or several solvent or other neutral molecules is described as a 'loose ion pair', symbolically represented as $X^+||Y^-$. The members of a loose ion pair can readily interchange with other free or loosely paired ions in the solution. This interchange may be detectable (e.g. by isotopic labelling) and thus afford an experimental distinction between tight and loose ion pairs.
A further conceptual distinction has sometimes been made between two types of loose ion pairs. In 'solvent-shared ion pairs' the ionic constituents of the pair are separated by only a single solvent molecule, whereas in 'solvent-separated ion pairs' more than one solvent molecule intervenes. However, the term 'solvent-separated ion pair' must be used and interpreted with care since it has also widely been used as a less specific term for 'loose' ion pair.
See also *common-ion effect, dissociation, ion-pair return, special salt effect*.
1994, *66*, 1126; 1995, *67*, 1344

ion-pair formation (in mass spectrometry)
Involves an ionization process in which a positive fragment ion and a negative fragment ion are among the products.
1991, *63*, 1548

ion pair return
The recombination of a pair of ions R^+ and Z^- formed from ionization of RZ.
If the ions are paired as a tight *ion pair* and recombine without prior separation into a loose ion pair this is called 'internal ion-pair return':

R⁺Z⁻ ⇌ RZ
tight ion pair   covalent molecule

It is a special case of 'primary *geminate recombination*'.
If the ions are paired as a loose ion pair and form the covalent chemical species *via* a tight ion pair, this is called 'external ion-pair return':

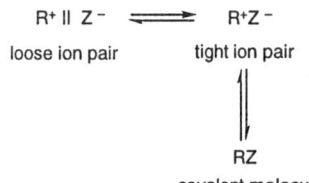

It is a special case of 'secondary geminate recombination'.
When the covalent molecule RZ is reformed without direct evidence of prior partial racemization or without other direct evidence of prior formation of a tight ion pair, (e.g. without partial racemization if the group

R is suitably chiral) the internal ion-pair return is sometimes called a 'hidden return'.

External (unimolecular) ion-pair return is to be distinguished from 'external (bimolecular) ion return', the (reversible) process whereby dissociated ions are converted into loose ion pairs:

$$R^+ + Z^- \rightleftharpoons R^+ \| Z^-$$

1994, *66*, 1127

ion, parent
See *parent ion*.
1991, *63*, 1549

ion, positive
See *positive ion*.
1991, *63*, 1549

ion, precursor
See *precursor ion*.
1991, *63*, 1549

ion, principal
See *principal ion*.
1991, *63*, 1550

ion probe microanalysis (IPMA)
Any technique in which the specimen is bombarded by a focused beam of (primary) ions (diameter less than 10 μm) and the (secondary) ions ejected from the specimen are detected after passage through a *mass spectrometer*.
1983, *55*, 2025; 1979, *51*, 2246

ion, product
See *product ion*.
1991, *63*, 1550

ion, quasi-molecular
See *quasi-molecular ion*.
1991, *63*, 1550

ion radical
See *radical ion*.
1994, *66*, 1156

ion, rearrangement
See *rearrangement ion*.
1991, *63*, 1550

ion scattering spectrometry (ISS)
Any technique using low energy (10 keV) ions in which the bombarding particles scattered by the sample are detected and recorded as a function of energy and/or angle. This technique is used mainly for determining the composition and structure of the first few atomic layers of a sample.
1979, *51*, 2246

ion-selective electrode (ISE)
An electrochemical sensor, based on thin films or selective membranes as recognition elements, and an electrochemical half-cell equivalent to other half-cells of the zeroth (inert metal in a redox electrolyte), 1st, 2nd and 3rd kinds. These devices are distinct from systems that involve redox reactions (electrodes of zeroth, 1st, 2nd and 3rd kinds), although they often contain a 2nd kind electrode as the 'inner' or 'internal' reference electrode. The potential difference response has, as its principal component, the Gibbs energy change associated with permselective mass transfer (by ion-exchange, solvent extraction or some other mechanism) across a phase boundary. The ion-selective electrode must be used in conjunction with a reference electrode (i.e. 'outer' or 'external' reference electrode) to form a complete electrochemical cell. The measured potential differences (ion-selective electrode vs. outer reference electrode potentials) are linearly dependent on the logarithm of the activity of a given ion in solution.

Comment: The term 'ion-specific electrode' is not recommended. The term 'specific' implies that the electrode does not respond to additional ions. Since no electrode is truly specific for one ion, the term 'ion-selective' is recommended as more appropriate. 'Selective ion-sensitive electrode' is a little-used term to describe an ion-selective electrode. 'Principal' or 'primary' ions are those which an electrode is designed to measure. It is never certain that the 'principal' ion is most sensitively measured, e.g. nitrate ion-selective electrodes.
1994, *66*, 2531

ion-selective electrode cell
An ion-selective electrode in conjunction with a reference electrode. Generally, the cell contains two reference electrodes, internal and external, and the thin film or membrane recognition-transduction element. However, besides this conventional type of cell (with solution contact on both sides of the membrane) there are cell arrangements with wire contact to one side of the membrane (all solid state and coated wire types).
1994, *66*, 2531

ions, isotopically enriched
See *isotopically enriched ions*.
1991, *63*, 1549

ion source (in mass spectrometry)
Generally an assembly composed of: (i) an ionization chamber in which a stream of electrons flows from a hot filament across a stream of gas to collector (the potential between filament and collector is usually between 50 and 70 V); (ii) a device for the acceleration of these ions.
1990, *62*, 2200

ion, stable
See *stable ion*.
1991, *63*, 1550

ion trap mass spectrometer
An arrangement in which ions with a desired range of quotients mass/charge are first made to describe stable paths under the effect of a high-frequency electric quadrupole field, and are then separated and presented to a detector by adjusting the field so as to selectivity induce path instability according to their respective mass/charge ratios.
1991, *63*, 1545

ion, unstable
See *unstable ion*.
1991, *63*, 1550

ipso-attack
The attachment of an entering group to a position in an aromatic compound already carrying a *substituent group* (other than hydrogen). The entering group may displace that substituent group but may also itself be expelled or migrate to a different position in a subsequent step. The term '*ipso*-substitution' is not used, since it is synonymous with substitution.
For example:

E$^+$ + ⟨benzene-Z⟩ ⟶ ⟨cyclohexadienyl cation with Z and E⟩

where E$^+$ is an *electrophile* and Z is a substituent (other than hydrogen).
See also *cine-substitution, tele-substitution*.
1994, *66*, 1128

iridescent layers
See *Schiller layers*.
1972, *31*, 611

iridoids
Cyclic monoterpenoids having the iridane skeleton (1-isopropyl-2,3-dimethylcyclopentane), the traditional, but not undisputed, numbering of which is shown in the diagram.

1995, *67*, 1344

irradiance, *E*
Radiant power received by a surface divided by the area of that surface. For collimated beams this quantity is sometimes called intensity and given the symbol *I*.
See also *photon irradiance, spectral irradiance*.
G.B. 31; 1996, *68*, 2249; 1996, *68*, 977

irradiation
Exposure to *ionizing radiation*.
1994, *66*, 2520

irregular macromolecule
A *macromolecule* the structure of which essentially comprises the repetition of more than one type of *constitutional unit* or a macromolecule the structure of which comprises constitutional units not all connected identically with respect to directional sense.
1996, *68*, 2289

irregular polymer
A polymer composed of *irregular macromolecules*.
1996, *68*, 2302

irreversible transition
A transition that changes the state of a system which cannot be readily reversed to restore the system to its original state.
Example:
The conversion of TiO_2 (anatase) to TiO_2 (rutile).
1994, *66*, 584

+ isoabsorption point
The use of this term, equivalent to *isosbestic point*, is not recommended.
1996, *68*, 2249

isobar (in atmospheric chemistry)
Lines on a plot joining points of equal barometric pressure in the atmosphere.
1990, *62*, 2198

isobaric mass-change determination
A technique in which the equilibrium mass of a substance at constant partial pressure of the volatile product(s) is measured as a function of temperature while the substance is subjected to a controlled temperature program.
The record is the isobaric mass-change curve. The mass should be plotted on the ordinate decreasing downwards and temperature on the abscissa increasing from left to right.
O.B. 41

isobaric separation
Chromatographic separation carried out using constant inlet and outlet pressure conditions.
1993, *65*, 2401

isobars
Different nuclides of equal mass number.
Physical Chemistry Division, unpublished

isoclined structures (in polymers)
See *isomorphous structures (in polymers)*.
P.B. 43

isoclinic point
A wavelength, wavenumber, or frequency at which the first derivative of an absorption spectrum of a sample does not change upon a chemical reaction or physical change of the sample.
1996, *68*, 2249

+ isoconfertic separation
See *isopycnic separation*.
1993, *65*, 2401

isocoumarins
Isocoumarin (1*H*-isochromen-1-one) and its derivatives formed by substitution.

See *coumarins*.
1995, *67*, 1344

isocratic analysis (in chromatography)
A procedure in which the composition of the mobile phase remains constant during the elution process.
1993, *65*, 826

isocyanates
The isocyanic acid tautomer, HN=C=O, of cyanic acid, HOC≡N and its hydrocarbyl derivatives RN=C=O.
1995, *67*, 1344

isocyanides
The isomer $HN^+{\equiv}C^-$ of hydrocyanic acid, $HC{\equiv}N$, and its *hydrocarbyl* derivatives RNC ($RN^+{\equiv}C^-$).
1995, *67*, 1344

isocyclic compounds
A less preferred synonym for *homocyclic compounds*.
1995, *67*, 1344

isodesmic reaction
A reaction (actual or hypothetical) in which the types of bonds that are made in forming the products are the same as those which are broken in the reactants, e.g.

(a) $PhCOOH + p\text{-}ClC_6H_4COO^-$
$\rightarrow PhCOO^- + p\text{-}ClC_6H_4COOH$

(b) $ClCH{=}CH_2 + ClCH_2CH_2Cl$
$\rightarrow CH_2{=}CH_2 + Cl_2CHCH_2Cl$

Such processes have advantages for theoretical treatment. The *Hammett equation* as applied to equilibria [cf. (a)] essentially deals with isodesmic processes.
1994, *66*, 1128

isodiazenes
Compounds having the structure $R_2NN{:} \leftrightarrow R_2N^+{=}N^-$. These compounds can also be called *diazanylidenes* or *hydrazinylidenes*. They have also been called by the trivial name azamines and the incorrectly formed name 1,1-*diazenes*. They should not be called aminonitrenes.
See *carbenes*.
1995, *67*, 1344

isoelectric
A macro-ion of a polyampholyte (in particular a protein) is said to be isoelectric if it exhibits no *electrophoresis*.
1972, *31*, 618

isoelectric point (in electrophoresis)
The pH value at which the net electric charge of an elementary entity is zero. pI is a commonly used symbol for this kind-of-quantity. It should be replaced by pH(I) because it is a pH determined under that particular condition.
1994, *66*, 894; 1972, *31*, 618

isoelectronic
Two or more *molecular entities* are described as isoelectronic if they have the same number of valence electrons and the same structure, i.e. number and *connectivity* of atoms, but differ in some of the elements involved. Thus:
CO, N_2 and NO^+ are isoelectronic.
$CH_2{=}C{=}O$ and $CH_2{=}N{=}N$ are isoelectronic.
CH_3COCH_3 and $CH_3N{=}NCH_3$ have the same number of electrons, but have different structures, hence they are not described as isoelectronic.
1994, *66*, 1128

isoemissive point
Synonymous with *isostilbic point*.
1996, *68*, 2249

isoentropic
A reaction series is said to be isoentropic if the individual reactions of the series have the same standard *entropy of activation*.
1994, *66*, 1128

isoenzyme
One of a group of related *enzymes* catalysing the same reaction but having different molecular structures and characterized by varying physical, biochemical and immunological properties.
1994, *66*, 2594; 1992, *64*, 158; see also W.B. 93

isoequilibrium relationship
A relationship analogous to the *isokinetic relationship* but applied to equilibrium data. The equation defining the isoequilibrium temperature β is:

$$\Delta_r H - \beta \Delta_r S = \text{constant}$$

where ΔH and ΔS are enthalpy and entropy of reaction, respectively.
See also *isokinetic relationship*.
1994, *66*, 1128

isoflavonoids
See *flavonoids*.
1995, *67*, 1345

isoionic
A macro-ion of a polyampholyte (in particular a protein) is said to be isoionic if besides the polyampholyte and H^+ or OH^- ions (in general ions of the solvent) no other ions are present in the system.
1972, *31*, 618

isoionic point (in electrophoresis)
The pH value at which the net electric charge of an elementary entity in pure water equals zero.
1994, *66*, 895

isokinetic line (in atmospheric chemistry)
A line in a given surface connecting points with equal wind speed; also called isotach or isovel.
1990, *62*, 2198

isokinetic relationship
When a series of structurally related substrates undergo the same general reaction or when the reaction conditions for a single substrate are changed in a systematic way, the *enthalpies* and *entropies of activation* sometimes satisfy the relation:

$$\Delta^{\ddagger} H - \beta \Delta^{\ddagger} S = \text{constant}$$

where the parameter β is independent of temperature. This equation (or some equivalent form) is said to represent an 'isokinetic relationship'. The temperature $T = \beta$ (at which all members of a series obeying the isokinetic relationship react at the same rate) is termed the 'isokinetic temperature'.

Supposed isokinetic relationships as established by direct correlation of $\Delta^{\ddagger} H$ with $\Delta^{\ddagger} S$ are often spurious and the calculated value of β is meaningless, because errors in $\Delta^{\ddagger} H$ lead to compensating errors in $\Delta^{\ddagger} S$. Satisfactory methods of establishing such relationships have been devised.

See also *compensation effect, isoequilibrium relationship, isoselective relationship.*
1994, *66*, 1129

isokinetic sampling (in atmospheric chemistry)
A technique for collecting airborne particulate in which the sampling device has a collection efficiency of unity for all sizes of particles in sampled air, regardless of wind velocity and direction of the instrument. The air stream entering the collector has a velocity (speed and direction) equal to that of the air in the gas stream just ahead of the sampling port of the collector.
1990, *62*, 2198

isolated double bonds
Those bonds which are neither *conjugated* nor *cumulative* as in

or the B ring of

B.B. 69

isolobal
The term is used to compare molecular fragments with each other and with familiar species from organic chemistry. Two fragments are isolobal if the number, symmetry properties, approximate energy and shape of the *frontier orbitals* and the number of electrons in them are similar.
See *isoelectronic.*
1994, *66*, 1129

isomer
One of several species (or *molecular entities*) that have the same atomic composition (molecular formula) but different *line formulae* or different *stereochemical formulae* and hence different physical and/or chemical properties.
1994, *66*, 1129; 1996, *68*, 2210

isomerases
Enzymes that catalyse intramolecular rearrangements. Isomerases are classified into racemases and epimerases, *cis-trans* isomerases, intramolecular lyases and other isomerases.
1992, *64*, 158

isomeric
Adjective derived from *isomer.*
1996, *68*, 2210

isomeric state (in nuclear chemistry)
A nuclear state having a *mean life* long enough to be observed.
1982, *54*, 1545

isomeric transition (in nuclear chemistry)
A spontaneous transition between two *isomeric states* of a *nucleus.*
1982, *54*, 1545

isomerism
The relationship between *isomers.*
1996, *68*, 2210

isomerization
A *chemical reaction*, the principal product of which is isomeric with the principal reactant. An *intramolecular* isomerization that involves the breaking or making of bonds is a special case of a *molecular rearrangement.*
Isomerization does not necessarily imply *molecular rearrangement* (e.g. in the case of the interconversion of conformational isomers).
1994, *66*, 1129

isomer shift (in Mössbauer spectroscopy)
Measure of the energy difference between the source (E_s) and the absorber (E_a) transition. The measured Doppler velocity shift, δ, is related to the energy difference by

$$E_a - E_s = \delta E_\gamma / c$$

where E_γ is the Mössbauer gamma energy and c is the speed of light in vacuum.
1976, *45*, 214

isomers, nuclear
See *nuclear isomers.*
1982, *54*, 1545

isometric
Two *molecular entities* that are superposable or can be made superposable by reflection of one of them in a mirror are isometric; otherwise they are anisometric.
1996, *68*, 2210

isomorphous structures (in polymers)
In the crystalline state, polymer chains are generally parallel to one another but neighbouring chains of equivalent conformation may differ in chirality and/or orientation.
Chains of identical *chirality* and *conformation* are isomorphous. Chains of opposite chirality but equivalent conformation are enantiomorphous. For example, two ...TG⁺TG⁺TG⁺... helices of isotactic poly(propylene) are isomorphous. Isotactic poly(propylene) chains of the ...TG⁺TG⁺TG⁺... and ...G⁻TG⁻TG⁻T... types are mutually enantiomorphous.
With regard to orientation, consider a repeating side group originating at atom A^i_1, the first atom of the side group being B^i_α. For certain chain symmetries (helical, for instance) the bond vectors $\vec{b}(A^i_1, B^i_\alpha)$ have the same components (positive or negative) $\vec{b} \cdot \vec{c} / |\vec{c}|$ along the c axis for every i.
Two equivalent (isomorphous or enantiomorphous) chains in the crystal lattice, having identical components of the bond vectors along c, both positive or both negative, are designated isoclined; two equivalent chains having bond vectors along c of the same

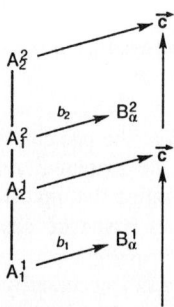

magnitude but opposite sign are designated anticlined.

P.B. 43

+ isonitriles
An obsolete term, which should not be used, for *isocyanides*.

1995, *67*, 1345

+ isonitroso compounds
An obsolete term for *oximes*, based on the fact that compounds with a nitroso group, –N=O, bonded to a –CR$_2$H group readily tautomerize to oximes.

1995, *67*, 1345

isooptoacoustic point
A wavelength, wavenumber, or frequency at which the total energy emitted by a sample as heat does not change upon a chemical reaction or physical change of the sample. Its position depends on the experimental conditions. The spectral differences between the *isosbestic points* and the isooptoacoustic points are the result of the non-linear relationship between the molar absorption coefficient and the photoacoustic signal.

1996, *68*, 2249

isopeptide bond
See *peptides*.
W.B. 48

isopotential point
For an ion-selective electrode cell, there is often a particular activity of the measured ion for which the emf of the cell is independent of temperature. That activity, and the corresponding potential difference, defines the isopotential point. The specification of the *ion-selective electrode* and outer reference electrode must be described.

Comment: When an isothermal cell is used with identical reference electrodes, the isopotential point is the activity of sensed species that gives zero net *membrane potential*, e.g. sensed activity is the same in the inner and outer (test) solution. Calibration lines for different cell temperatures have different slopes, but intersect at a common activity point. Cells with temperature gradients are not recommended.

1994, *66*, 2531

isoprenes
See *prenols*.
W.B. 252

isoprenoids
Compounds formally derived from isoprene (2-methylbuta-1,3-diene), the skeleton of which can generally be discerned in repeated occurrence in the molecule. The skeleton of isoprenoids may differ from strict additivity of isoprene units by loss or shift of a fragment, commonly a methyl group. The class includes both *hydrocarbons* and oxygenated derivatives.

See also *carotenoids, steroids, terpenes, terpenoids, prenols*.
1995, *67*, 1345; see also W.B. 255

isopycnic
An adjective describing components of a multicomponent system with equal partial specific volumes.
P.B. 60

isopycnic separation
Chromatographic separations carried out using constant density conditions. The temperature and pressure may be altered during the run (originally the term isoconfertic separation was used but this term is not recommended).

1993, *65*, 2401

isorefractive
An adjective describing components of a multicomponent system having zero *refractive index* increments with respect to each other.
P.B. 68

isosbestic point
This term is usually employed with reference to a set of absorption spectra, plotted on the same chart for a set of solutions in which the sum of the concentrations of two principal absorbing components, A and B, is constant. The curves of absorbance against wavelength (or frequency) for such a set of mixtures often all intersect at one or more points, called isosbestic points.

Isosbestic points are commonly met when electronic spectra are taken (a) on a solution in which a *chemical reaction* is in progress (in which case the two absorbing components concerned are a reactant and a product, A + B), or (b) on a solution in which the two absorbing components are in equilibrium and their relative proportions are controlled by the concentration of some other component, typically the concentration of hydrogen ions, e.g. an *acid–base indicator* equilibrium.

$$A \rightleftharpoons B + H^+_{aq}$$

The effect may also appear (c) in the spectra of a set of solutions of two unrelated non-interacting components having the same total concentration. In all these examples, A (and/or B) may be either a single *chemical species* or a mixture of chemical species present in invariant proportion.

If A and B are single chemical species, isosbestic points will appear at all wavelengths at which their molar absorption coefficients (formerly called extinction coefficients) are the same. (A more involved identity applies when A and B are mixtures of constant proportion.)

If absorption spectra of the types considered above intersect not at one or more isosbestic points but over progressively changing wavelength, this is prima facie evidence in case (a) for the formation of a *reaction intermediate* in substantial concentration (A → C → B), in case (b) for the involvement of a third absorbing species in the equilibrium, e.g.

$$A \rightleftharpoons B + H^+_{aq} \rightleftharpoons C + 2H^+_{aq}$$

or in case (c) for some interaction of A and B, e.g.

$$A + B \rightleftharpoons C$$

1994, *66*, 1129

isoselective relationship
A relationship analogous to the *isokinetic relationship*, but applied to *selectivity* data of reactions. At the isoselective temperature, the selectivities of the series of reactions following the relationship are identical
See also *isoequilibrium relationship, isokinetic relationship*.
1994, *66*, 1130

isoselenocyanates
Selenium analogues of *isocyanates* RN=C=Se.
1995, *67*, 1345

isosteric enthalpy of adsorption
When the addition of the differential amount of component i dn_i^σ or dn_i^s is effected at constant pressure p, the *differential molar enthalpy of adsorption*, $\Delta_a H_i^\sigma$ or $\Delta_a H_i^s$ also called the isosteric enthalpy of adsorption (q^{st}) is defined as:

$$\Delta_a H_i^\sigma = -q^{st,\sigma} = U_i^\sigma - H_i^g$$

$$\Delta_a H_i^s = -q^{st,\sigma} = H_i^\sigma - H_i^g$$

where $H_i^s = (\partial H^s/\partial n_i^s)_{T,p,m,n_i^s}$ and H_i^g is the partial molar enthalpy of component i in the gas phase, i.e. $(\partial H^g/\partial n_i^g)_{T,p,n_i^g}$

1972, *31*, 603

isostilbic point
The wavelength at which the intensity of emission of a sample does not change during a *chemical reaction* or physical change. The term derives from the Greek word for 'same luminescence.' The terms isoemissive and isolampsic are sometimes used.
See *isosbestic point*.
1996, *68*, 2250

isotactic macromolecule
A *tactic macromolecule,* essentially comprising only one species of *configurational base unit*, which has chiral or prochiral atoms in the *main chain* in a unique arrangement with respect to its adjacent *constitutional units*.
Notes:
1. In an isotactic macromolecule, the *configurational repeating unit* is identical with the configurational base unit.
2. An isotactic macromolecule consists of *meso diads*.
1996, *68*, 2292

isotactic polymer
A polymer composed of *isotactic macromolecules*.
1996, *68*, 2302

isotactic triads (in polymers)
See *triads (in polymers)*.
1996, *68*, 2299

isotherm (in atmospheric chemistry)
Lines joining points of equal temperature in the atmosphere.
1990, *62*, 2198

isothermal chromatography
A procedure in which the temperature of the column is kept constant during the separation.
1993, *65*, 827

isothiocyanates
Sulfur analogues of *isocyanates* RN=C=S.
1995, *67*, 1345

isotones
Nuclides having the same *neutron number* but different *atomic numbers*.
1982, *54*, 1545

isotope dilution
Mixing of a given *nuclide* with one or more of its *isotopes*.
1994, *66*, 2521

isotope dilution analysis
A kind of quantitative analysis based on the measurement of the isotopic abundance of a *nuclide* after *isotope dilution* with the test portion.
1994, *66*, 2521

isotope dilution analysis, direct (radiochemical)
See *direct (radiochemical) isotope dilution analysis*
1994, *66*, 2521

isotope dilution analysis, reversed (radiochemical)
See *reversed (radiochemical) isotope dilution analysis*.
1994, *66*, 2521

isotope dilution analysis, substoichiometric
See *substoichiometric isotope dilution analysis*.
1994, *66*, 2521

isotope effect
The effect on the rate or equilibrium constant of two reactions that differ only in the isotopic composition of one or more of their otherwise chemically identical components is referred to as a *kinetic isotope effect* or a *thermodynamic* (or *equilibrium*) *isotope effect*, respectively.
1994, *66*, 1130; 1994, *66*, 2521; see also 1996, *68*, 170

isotope exchange
A *chemical reaction* in which the reactant and product *chemical species* are chemically identical but have different isotopic composition. In such a reaction the isotope distribution tends towards equilibrium (as expressed by fractionation factors) as a result of trans-

fers of isotopically different atoms or groups. For example:

$$DCl + \underset{R}{\underset{|}{R}}\underset{R}{\overset{H}{\bigcirc}}\underset{R}{R} \longrightarrow \underset{R}{\underset{|}{R}}\underset{R}{\overset{D}{\bigcirc}}\underset{R}{R} + HCl$$

1994, 66, 1132; 1994, 66, 2521

isotope exchange analysis
A kind of quantitative analysis based on the *isotope exchange* between *isotopes* of the element to be determined and other isotopes of this element in different valency states or in different molecules.
1994, 66, 2521

isotope pattern (in mass spectrometry)
Set of peaks related to ions with the same chemical formula but containing different isotopes; e.g. the 16 and 17 mass/charge peaks in a CH_4 sample arising from $^{12}CH_4^+$ and $^{13}CH_4^+$ ions
1990, 62, 2198

isotopes
Nuclides having the same atomic number but different *mass numbers*.
1994, 66, 2520; R.B. 35; see also O.B. 230

isotopic abundance
The relative number of *atoms* of a particular *isotope* in a mixture of the isotopes of an element, expressed as a fraction of all the atoms of the element.
1982, 54, 1535

isotopic abundance, natural
See *natural isotopic abundance*.
1982, 54, 1535

isotopically deficient
An *isotopically labelled* compound may be designated as isotopically deficient when the isotopic content of one or more elements has been depleted, i.e. when one or more *nuclide(s)* is (are) present in less than a natural ratio.
B.B. 518; 1981, 53, 1896

isotopically enriched ions
When the abundance of a particular *nuclide* is increased above the level at which it occurs in nature and is incorporated in a molecule, the term 'isotopically enriched ion' is used to describe any ion enriched in the isotope.
1991, 63, 1549

isotopically labelled
Describes a mixture of an *isotopically unmodified* compound with one or more analogous *isotopically substituted* compound(s).
B.B. 514; 1981, 53, 1891

isotopically modified
Describes a compound that has a macroscopic composition such that the isotopic ratio of *nuclides* for at least one element deviates measurably from that occurring in nature. It is either an *isotopically substituted* compound or an *isotopically labelled* compound.
B.B. 514; 1981, 53, 1889

isotopically substituted
Describes a compound that has a composition such that essentially all the molecules of the compound have only the indicated *nuclide(s)* at each designated position. For all other positions, the absence of nuclide indication means that the nuclide composition is the natural one.
B.B. 514; 1981, 53, 1890

isotopically unmodified
Describes a compound that has a macroscopic composition such that its constituent *nuclides* are present in the proportions occurring in nature.
B.B. 514; 1981, 53, 1889

isotopic carrier
A *carrier* which differs only in isotopic composition from the trace it has to carry.
1994, 66, 2516

isotopic enrichment
Any process by which the *isotopic abundance* of a specified *isotope* in a mixture of isotopes of an element is increased.
1982, 54, 1541

isotopic enrichment factor
For a material enriched in a specified *isotope*, the ratio between the *isotopic abundance* and the *natural abundance* of that isotope.
1982, 54, 1541

isotopic fractionation factor
The ratio $(x_1/x_2)_A/(x_1/x_2)_B$, where x is the abundance, expressed as the atom fraction of the isotope distinguished by the subscript numeral, when the two isotopes are equilibrated between two different *chemical species* A and B (or between specific sites A and B in the same or different chemical species). The term is most commonly met in connection with deuterium solvent *isotope effects*, when the fractionation factor Φ expresses the ratio:

$$\Phi = (x_D/x_H)_{solute}/(x_D/x_H)_{solvent}$$

for the exchangeable hydrogen atoms in the chemical species (or sites) concerned. The concept is also applicable to *transition states*.
1994, 66, 1115

isotopic ion
Any ion containing one or more of the less abundant naturally occurring isotopes of the elements that make up its structure, e.g. CH_2D^+.
1991, 63, 1549; O.B. 205

isotopic labelling
Labelling in which the resulting product is only different from the initial one by its isotopic composition.
1994, 66, 2521

isotopic molecular ion
A molecular ion containing one or more of the less abundant naturally occurring isotopes of the atoms that make up the molecular structure.
Thus, for ethyl bromide there exist isotopic molecular ions such as: $^{13}CCH_5Br^{+\cdot}$, $C_2H_4DBr^{+\cdot}$, $C_2H_5{}^{81}Br^{+\cdot}$, $^{13}C_2H_5{}^{81}Br^{+\cdot}$, etc.
1991, *63*, 1549

isotopic perturbation, method of
See *method of isotopic perturbation*.
1994, *66*, 1132

isotopic scrambling
The achievement, or the process of achieving, an equilibrium distribution of isotopes within a specified set of atoms in a *chemical species* or group of chemical species (for an example, see diagram).

(a) [reaction scheme: iodobenzene (labeled *) + KNH₂ → aniline (labeled *) + aniline (labeled *) + KI]

(b) $(Ph-\overset{*}{N}\equiv N)^+ \longrightarrow (Ph-\overset{*}{N}\equiv N)^+ + (Ph-N\equiv \overset{*}{N})^+$

See also *fractionation factor*.
1994, *66*, 1132; see also 1991, *63*, 1556

isotopic separation
An operation for the purpose of modifying an *isotopic abundance* in a mixture of isotopes.
1982, *54*, 1545

isotopic tracer
A *tracer* which only differs in isotopic composition from the substance to be traced.
1994, *66*, 2526

isotopologue
A *molecular entity* that differs only in isotopic composition (number of isotopic substitutions), e.g. CH_4, CH_3D, CH_2D_2.
1994, *66*, 1132

isotopomer
Isomers having the same number of each isotopic atom but differing in their positions. The term is a contraction of 'isotopic isomer'.
Isotopomers can be either constitutional isomers (e.g. $CH_2DCH=O$ and $CH_3CD=O$) or isotopic stereoisomers [e.g. (*R*)- and (*S*)-CH_3CHDOH or (*Z*)- and (*E*)-$CH_3CH=CHD$].
1994, *66*, 1132; 1996, *68*, 2211

isotropic
A quantity which is independent of direction. *Anisotropic* and *nonisotropic* refer to quantities which are direction dependent.
1990, *62*, 2198

isotropic carbon
A monolithic *carbon material* without preferred crystallographic orientation of the microstructure.
Notes:
Isotropic carbon can also be a *graphite material*. The isotropy can be gross (bulk), macroscopic or microscopic, depending on the structural level at which isotropy is obtained. This word is widely used today and its meaning covers all the above levels. For example, the aerospace graphites have isotropy built in by random grain orientation. Some *nuclear graphites* are isotropic at the crystalline (sub-grain) level.
1995, *67*, 495

isotropic pitch-based carbon fibres
Carbon fibres obtained by *carbonization* of isotropic pitch fibres after these have been stabilized (i.e. made non-fusible).
Notes:
During fabrication of isotropic pitch-based carbon fibres no means (neither mechanical nor chemical) are applied to achieve preferred orientation of the polyaromatic molecules in the fibre direction. They belong to the *carbon fibres type LM* (low modulus), and because of the relatively low values of strength and Young's modulus this *pitch-based carbon fibre* type is not used for high-performance reinforcement purposes.
1995, *67*, 495

isoureas
The imidic acid tautomer of urea, $H_2NC(=NH)OH$, and its hydrocarbyl derivatives.
1995, *67*, 1345

isovalent hyperconjugation
See *hyperconjugation*.
1994, *66*, 1132

isozyme
Synonymous with *isoenzyme*.
W.B. 93

j-value (in atmospheric chemistry)
An effective first order rate constant for a photochemical reaction (photodecomposition, etc.) of light-absorbing impurities in the atmosphere.
1990, *62*, 2198

Jablonski diagram
Originally, a diagram showing that the fluorescent state of a *molecular entity* is the lowest *excited state* from which the transition to the *ground state* is allowed, whereas the phosphorescent state is a metastable state below the fluorescent state, which is reached by *radiationless transition*. In the most typical cases the fluorescent state is the lowest singlet excited state and the phosphorescent state the lowest triplet state, the ground state being a singlet. Presently, modified Jablonski diagrams are frequently used and are actually state diagrams in which molecular electronic states, represented by horizontal lines displaced vertically to indicate relative energies, are grouped according to multiplicity into horizontally displaced columns. *Excitation* and *relaxation* processes that interconvert states are indicated in the diagram by arrows. Radiative transitions are generally indicated with straight arrows (→), while radiationless transitions are generally indicated with wavy arrows ⤳ .
1996, *68*, 2250

Jahn–Teller effect
For non-linear *molecular entities* in a geometry described by a point symmetry group possessing degenerate irreducible representations there always exists at least one non-totally symmetric vibration that makes electronically degenerate states unstable at this geometry. The nuclei are displaced to new equilibrium positions of lower symmetry causing a splitting of the originally degenerate states. This effect is due to the odd terms in the vibronic perturbation expansion.
See also *Renner–Teller effect*.
1996, *68*, 2250

Jahn–Teller transition
A displacive crystal distortion to lower symmetry that cooperatively removes a localized-electron orbital degeneracy at an array of like atoms so as to leave the atoms in the centre of symmetry of their distorted sites.
Example: The crystallographic distortion in Mn_3O_4.
1994, *66*, 584

joule
SI derived unit of energy, $J = N\ m = kg\ m^2\ s^{-2}$.
G.B. 72; 1996, *68*, 977

junction point
See *branch point*.
1996, *68*, 2297

junction unit
A non-repeating atom or non-repeating group of atoms between blocks in a *block macromolecule*.
1996, *68*, 2293

κ (kappa) (in inorganic nomenclature)
An affix in the name of a polydentate *chelate* complex indicating single ligating atom attachments of a polyatomic *ligand* to a coordination centre. A right superscript numerical index indicates the number of such attachments.
Example: [Ni{(CH$_3$)$_2$PCH$_2$CH$_2$P(CH$_3$)$_2$}Br$_2$], dibromobis[1,2-ethanediylbis(dimethylphosphine)-κ^2P]nickel(II).
R.B. 174

Kamlet–Taft solvent parameters
Parameters of the Kamlet–Taft solvatochromic relationship which measure separately the hydrogen bond donor (α), *hydrogen bond acceptor (β)*, and *dipolarity/polarizability (π*)* properties of solvents as contributing to overall solvent *polarity*.
1994, *66*, 1132

kappa
See κ (beginning of 'k').
R.B. 174

Kaptein–Closs rules
Rules used to predict the sign of *CIDNP* effects.
1996, *68*, 2250

Kasha rule
Polyatomic *molecular entities* luminescence with appreciable yield only from the lowest excited state of a given multiplicity. There are exceptions to this rule.
1996, *68*, 2250

Kasha–Vavilov rule
The quantum yield of luminescence is independent of the wavelength of exciting radiation. There are exceptions to this rule.
1996, *68*, 2250

katal
Unit for catalytic activity coherent with the SI, equal to the catalytic activity that catalyses a reaction rate of one mole per second in an assay system. 1 kat = 1 mol s^{-1}. The katal is recommended for use in the specific context of *enzymes* and clinical chemistry.
1996, *68*, 977; see also 1994, *66*, 2595

Kekulé structure (for aromatic compounds)
A representation of an *aromatic molecular entity* (such as benzene), with fixed alternating single and double bonds, in which interactions between multiple bonds are assumed to be absent.
For benzene:

 and

are the Kekulé structures.
1994, *66*, 1132

kelvin
SI base unit of thermodynamic temperature (symbol: K). The kelvin is the fraction 1/273.16 of the thermodynamic temperature of the triple point of water.
G.B. 70; 1996, *68*, 977

kephalins
An alternative spelling of *cephalins*.
1995, *67*, 1345

ketals
Acetals derived from *ketones* by replacement of the oxo group by two hydrocarbyloxy groups R$_2$C(OR)$_2$ (R ≠ H). This term, once abandoned, has been reinstated as a subclass of acetals.
1995, *67*, 1345

ketazines
Azines of *ketones* R$_2$C=NN=CR$_2$.
1995, *67*, 1345

ketenes
Compounds in which a carbonyl group is connected by a double bond to an alkylidene group R$_2$C=C=O.
1995, *67*, 1345

ketenimines
Compounds having the structure R$_2$C=C=NR. Thus imino analogues of *ketenes*.
1995, *67*, 1345

ketides
See *polyketides*.
1995, *67*, 1345

ketimines
Compounds having the structure R$_2$C=NR' (R ≠ H).
See *imines*.
1995, *67*, 1345

keto
See *oxo compounds*.
1995, *67*, 1345

ketoaldonic acids
Monosaccharides in which a structure containing a keto group and a carboxylic acid group is in equilibrium with a *hemiacetal* structure. Specific compounds are named using the -ulosonic acid suffix.
E.g. D-*arabino*-hexulosonic acid:

```
    O   OH
     \\ /
      C
      |
      C=O
      |
  HO-C-H
      |
   H-C-OH
      |
   H-C-OH
      |
      CH₂-OH
```

See also *oxo compounds*.
1995, *67*, 1346

ketoaldoses
Monosaccharides which contain both an aldehydic and a ketonic carbonyl group in equilibrium with intramolecular hemiacetal forms.
See also *oxo compounds*.
1995, *67*, 1346

keto carbenes
The term has imprecisely been used to designate carboxylic *acyl carbenes*. Keto carbenes are carbenes bearing an oxo function at an unspecified site.
1995, *67*, 1345

ketones

Compounds in which a carbonyl group is bonded to two carbon atoms $R_2C=O$ (neither R may be H).
Note:
Compounds of structure such as $R_3SiC(=O)R$ are not ketones but acyl derivatives of substituted *silanes*.
1995, *67*, 1346

ketoses

Ketonic parent *sugars* (polyhydroxy ketones H–[CHOH]$_n$–C(=O)[CHOH]$_m$–H with three or more carbon atoms) and their intramolecular *hemiacetals*. The oxo group is usually at C-2. E.g. D-fructose:

See also *monosaccharides*.
1995, *67*, 1346; W.B. 129

ketoximes

Oximes of *ketones*, $R_2C=NOH$ (R ≠ H).
1995, *67*, 1346

ketyls

Radical anions (or the corresponding salts) derived from *ketones* by addition of an electron $R_2\dot{C}-O^- \leftrightarrow R_2C^--O\cdot$.
Note:
Ketyls produce two types of conjugate acids: $R_2\dot{C}-OH$ and $R_2CH-O\cdot$. The former are α-hydroxyalkyl radicals and the latter are alkoxyl *radicals*, but they have also been called ketyls in photochemistry.
1995, *67*, 1346

kilo

SI prefix for 10^3 (symbol: k).
G.B. 74

kilogram

SI base unit for mass (symbol: kg). The kilogram is equal to the mass of the international prototype of the kilogram.
G.B. 70; 1996, *68*, 978

kind-of-property

An attribute of phenomena, bodies or substances that may be distinguished qualitatively.
Examples:
Colour (value: green, blue), transparency, length (value: long, short; 2 m, 5 m), amount-of-substance (value: 2 mol, 5 mol).
Kind-of-property includes the concept *kind-of-quantity*. All kinds-of-property may be related to nominal (e.g. green, blue) and ordinal scales (e.g. small, large), but kinds-of-quantity are generally related to difference [e.g. 10 °C (i.e. 10 °C more than an arbitrary zero)] or ratio scales (length 2 m or 5 m).
1995, *67*, 1565

kind-of-quantity

Abstract property common to several real quantities.
See *quantity*.
1996, *68*, 978

kinematics

The study of the properties of motion that are independent of the nature of the forces.
In *reaction dynamics* the term is often applied in particular to effects that occur when atomic masses are changed, the forces remaining the same.
1996, *68*, 170

kinematic viscosity, ν

Dynamic viscosity divided by the *density* of the fluid.
G.B. 13; 1996, *68*, 978

kinetic activity factor

A factor involving activity coefficients that appears as a multiplier in the rate equation. For a bimolecular reaction the factor is $y_A y_B / y_\ddagger$, where y_A, y_B and $y_\ddagger$ are the activity coefficients for the reactants A and B and the activated complex, respectively. For a unimolecular reaction the factor is $y_A / y_\ddagger$, and for a trimolecular reaction it is $y_A y_B y_C / y_\ddagger$.
1996, *68*, 171

kinetic ambiguity

See *kinetic equivalence*.
1994, *66*, 1133

kinetic control (of product composition)

The term characterizes conditions (including reaction times) that lead to reaction products in a proportion governed by the relative rates of the parallel (forward) reactions in which the products are formed, rather than by the respective overall equilibrium constants.
See also *thermodynamic control*.
1994, *66*, 1133

kinetic current

A *faradaic current* that corresponds to the reduction or oxidation of an electroactive substance formed by a prior *chemical reaction* from another substance that is not electroactive, and that is partially or entirely controlled by the rate of that reaction. The reaction may be heterogeneous, occurring at an electrode-solution interface (surface reaction), or it may be homogeneous, occurring at some distance from the interface (volume reaction).
O.B. 55

kinetic current, limiting

See *limiting kinetic current*.
1985, *57*, 1497

kinetic electrolyte effect (kinetic ionic-strength effect)

The general effect of an added electrolyte (i.e. an effect other than, or in addition to, that due to its possible involvement as a reactant or catalyst) on the observed rate constant of a reaction in solution. At low concentrations (when only long-range coulombic forces need to be considered) the effect on a given reaction is determined only by the *ionic strength* of the solution and not by the chemical identity of the ions. For practical purposes, this concentration range is roughly the same as the region of validity of the Debye–Hückel limiting law for activity coefficients. At higher concentrations, the effect of an added electrolyte depends also on the chemical identity of the ions. Such specific action can usually be interpreted as the incursion of a *reaction path* involving an ion of the electrolyte as reactant or catalyst, in which case the action is not properly to be regarded just as a kinetic electrolyte effect.

Kinetic electrolyte effects are usually (too restrictively and therefore incorrectly) referred to as 'kinetic salt effects'.

A kinetic electrolyte effect ascribable solely to the influence of the ionic strength on activity coefficients of ionic reactants and transition states is called a 'primary kinetic electrolyte effect'.

A kinetic electrolyte effect arising from the influence of the ionic strength of the solution upon the *pre-equilibrium* concentration of an ionic species that is involved in a subsequent *rate-limiting step* of a reaction is called a 'secondary kinetic electrolyte effect'. A common case encountered in practice is the effect on the concentration of hydrogen ion (acting as catalyst) produced from the ionization of a weak acid in a buffer solution.

Synonymous with *kinetic equivalence*.
See also *common-ion effect, order of reaction, kinetic ambiguity*.
1994, *66*, 1133

kinetic energy, E_k
Energy of motion: for a body of mass m, $E_k = mv^2/2$, where v is the speed.
G.B. 12; 1996, *68*, 978

kinetic equivalence
Two reaction schemes are kinetically equivalent if they imply the same *rate law*.
For example, consider the two schemes (i) and (ii) for the formation of C from A:

(i) $A \underset{k_{-1},\ OH^-}{\overset{k_1,\ OH^-}{\rightleftarrows}} B \xrightarrow{k_2} C$

Providing that B does not accumulate as a *reaction intermediate*.

$$\frac{d[C]}{dt} = \frac{k_1 k_2 [A][OH^-]}{k_2 + k_{-1}[OH^-]} \quad (1)$$

(ii) $A \underset{k_{-1}}{\overset{k_1}{\rightleftarrows}} B \xrightarrow[OH^-]{k_2} C$

Providing that B does not accumulate as a reaction intermediate:

$$\frac{d[C]}{dt} = \frac{k_1 k_2 [A][OH^-]}{k_{-1} + k_2 [OH^-]} \quad (2)$$

Both equations for d[C]/dt are of the form:

$$\frac{d[C]}{dt} = \frac{r[A][OH^-]}{1 + s[OH^-]} \quad (3)$$

where r and s are constants (sometimes called 'coefficients in the rate equation'). The equations are identical in their dependence on concentrations and do not distinguish whether OH^- catalyses the formation of B, and necessarily also its reversion to A, or is involved in its further transformation to C. The two schemes are therefore kinetically equivalent under conditions to which the stated provisos apply.
1994, *66*, 1133

kinetic ionic-strength effect
See *kinetic electrolyte effect*.
1994, *66*, 1133

kinetic isotope effect
The effect of isotopic substitution on a rate constant is referred to as a kinetic isotope effect.
For example, in the reaction:

$$A + B \rightarrow C$$

the effect of isotopic substitution in reactant A is expressed as the ratio of rate constants k^l/k^h, where the superscripts l and h represent reactions in which the molecules A contain the light and heavy isotopes, respectively.
Within the framework of *transition state theory* in which the reaction is rewritten as:

$$A + B \rightleftarrows [TS]^{\ddagger} \rightarrow C$$

and with neglect of isotopic mass on *tunnelling* and the transmission coefficient, k^l/k^h can be regarded as if it were the equilibrium constant for an isotope exchange reaction between the transition state $[TS]^{\ddagger}$ and the isotopically substituted reactant A, and calculated from their vibrational frequencies as in the case of a *thermodynamic isotope effect*.

Isotope effects like the above, involving a direct or indirect comparison of the rates of reaction of *isotopologues*, are called 'intermolecular', in contrast to *intramolecular isotope effects*, in which a single substrate reacts to produce a non-statistical distribution of *isotopomeric* product molecules.
See also *isotope effect*.
1994, *66*, 1130

kinetic method (of analysis)
An analytical method in which the rate of a reaction or a related quantity is measured and utilized to determine concentrations.
1993, *65*, 2296

kinetic resolution
The achievement of partial or complete resolution by virtue of unequal rates of reaction of the *enantiomers* in a *racemate* with a *chiral* agent (reagent, catalyst, solvent, etc.).
1996, *68*, 2211

+ kinetic synergist
This term is not recommended for *catalyst* or *accelerator*.
1993, *65*, 2381

kinetic theory of collisions
See *collision theory*.
1996, *68*, 171

klado-
An affix used in names to designate a very open polyboron structure.
R.B. 245

Koopmans' theorem
See *photoelectron spectroscopy*.
1996, 68, 2250

Koppel–Palm solvent parameters
Parameters to measure separately the ability of a solvent to enter into non-specific solvent–solute interactions (permittivity ε and refractive index η_D) and specific solvent–solute interaction (solvent basicity or nucleophilicity B and solvent acidity or electrophilicity E) as contributing to overall solvent polarity.
1994, *66*, 1134

Kosower Z-value
See *Z-value*.
1994, *66*, 1134

Kováts (retention) index
See *retention index*.
1993, *65*, 844

Krafft point
The temperature (more precisely, narrow temperature range) above which the solubility of a *surfactant* rises sharply. At this temperature the solubility of the surfactant becomes equal to the *critical micelle concentration*. It is best determined by locating the abrupt change in slope of a graph of the logarithm of the solubility against t or $1/T$.
1972, *31*, 613

Kratky plot
A diagrammatic representation of scattering data on large particles, obtained at different angles but at the same concentration, constructed by plotting $\sin^2(\theta/2) \cdot \Delta R(\theta)$ vs. $\sin^2(\theta/2)$, or $q^2 \Delta R(\theta)$ vs. q and used for the determination of molecular shape. $\Delta R(\theta)$ is the excess Rayleigh ratio, $P(\theta)$ the particle scattering function, θ the scattering angle and q the length of the scattering vector.
P.B. 67

krypton ion laser
A CW or pulsed laser emitting lines from 337 to 859 nm from singly ionized krypton. Principal emissions are at 530.9, 568.2, 647.1 and 752.5 nm.
See *gas lasers*.
1996, *68*, 2250

λ (lambda)
Signifies, with its superscript, the *bonding number*, i.e. the sum of the number of skeletal bonds to an atom and the number of hydrogen atoms associated with it in a parent hydride, real or hypothetical. Its usage in names of compounds is generally restricted to non-standard valence states.
1984, *56*, 774; see also B.B. 465 and R.B. 245

Λ (lambda)
See Δ *(delta)*.
1996, *68*, 2211

λ-transition (lambda-transition)
A *second-order* or *higher-order transition*, in which the heat capacity shows either a discontinuity (second-order) or a vertex (higher-order) at the transition temperature.
Example: The transition at 846 K, accompanied by a finite change in specific heat capacity at constant pressure C_p, of the low-temperature polymorph of quartz to the high-temperature polymorph.
1994, *66*, 584

L
See D, L, DL.
1996, *68*, 2211

l
1. An obsolete prefix for (−)-, See *d, l, dl*.
2. See *l,u* for like and unlike diastereoisomers.
1996, *68*, 2211

l, u
Stereodescriptors of *diastereoisomers* with two *chirality elements*. When the two chirality elements are both *R* or both *S* the *molecular entity* is described as having an *l* relationship (*l* = like), when one chirality element is *R* and the other *S* the molecular entity is described as having a *u* relationship (*u* = unlike). The notation can be extended to molecules with more than two chirality elements whereby the *l,u* relationships of the chiral elements are considered in pairs.
1996, *68*, 2211

label
A marker, tag or indicator distinguishable by the observer but not by the system and used to identify a *tracer*.
1994, *66*, 2521

labelling
Providing a substance with a *label*.
1994, *66*, 2521

labelling, conjugation
See *conjugation labelling*.
1994, *66*, 2521

labelling, exchange
See *exchange labelling*.
1994, *66*, 2521

labelling, isotopic
See *isotopic labelling*.
1994, *66*, 2521

labelling, non-isotopic
See *non-isotopic labelling*
1994, *66*, 2521

labelling, recoil
See *recoil labelling*.
1994, *66*, 2521

labelling, Wilzbach
See *Wilzbach labelling*.
1994, *66*, 2521

labile
The term has loosely been used to describe a relatively *unstable* and *transient chemical species* or (less commonly) a relatively stable but reactive species. It must therefore not be used without explanation of the intended meaning.
See also *inert, persistent, reactive, unreactive*.
1994, *66*, 1134

laboratory sample
The sample or subsample(s) sent to or received by the laboratory.
When the laboratory sample is further prepared (reduced) by subdividing, mixing, grinding or by combinations of these operations, the result is the test sample. When no preparation of the laboratory sample is required, the laboratory sample is the test sample. A test portion is removed from the test sample for the performance of the test or for analysis. The laboratory sample is the final sample from the point of view of sample collection but it is the initial sample from the point of view of the laboratory. Several laboratory samples may be prepared and sent to different laboratories or to the same laboratory for different purposes. When sent to the same laboratory, the set is generally considered as a single laboratory sample and is documented as a single sample.
1990, *62*, 1206; 1988, *60*, 1465

lachrymator
A substance which produces a flow of tears in a person or animal (e.g. acetyl chloride, acrylaldehyde, etc.).
1990, *62*, 2199

lactams
Cyclic *amides* of amino carboxylic acids, having a 1-azacycloalkan-2-one structure, or analogues having unsaturation or heteroatoms replacing one or more carbon atoms of the ring. E.g.

1995, *67*, 1346

lactides
Cyclic *esters* derived by multiple esterification between two (usually) or more molecules of lactic acid or other hydroxy *carboxylic acids*. They are designated as dilactides, trilactides, etc. according to the

number of hydroxy acid residues, e.g. a dilactide (a 1,4-dioxane-2,5-dione):

1995, *67*, 1346

lactims
Tautomeric forms of *lactams*, having an endocyclic carbon–nitrogen double bond. Thus, cyclic *carboximidic acids*, e.g.

1995, *67*, 1346

lactols
1. Cyclic *hemiacetals* formed by intramolecular addition of a hydroxy group to an aldehydic or ketonic carbonyl group. They are thus 1-oxacycloalkan-2-ols or unsaturated analogues, e.g.

+ 2. This term has also been used for hydroxy *lactones*, but such use is not recommended.
1995, *67*, 1347

lactones
Cyclic *esters* of hydroxy *carboxylic acids*, containing a 1-oxacycloalkan-2-one structure, or analogues having unsaturation or heteroatoms replacing one or more carbon atoms of the ring.

1995, *67*, 1347

ladder chain
A *double-strand chain*, consisting of an uninterrupted sequence of rings, with adjacent rings having two or more atoms in common.

Alternatively, a ladder chain is a double-strand chain with adjacent *constitutional units* joined to each other through four atoms, two on one side and two on the other side of each constitutional unit.
1996, *68*, 2295

ladder macromolecule
A *double-strand macromolecule* consisting of an uninterrupted sequence of rings, with adjacent rings having two or more atoms in common.

Alternatively, a ladder macromolecule is a double-strand macromolecule with *adjacent constitutional units* joined to each other through four atoms, two on one side and two on the other side of each constitutional unit.
1996, *68*, 2295

ladder polymer
See *double-strand polymer*.
1996, *68*, 2304

lag phase (in biotechnology)
The growth interval (adaption phase) between *inoculation* and start of the *exponential phase* during which there is little or no growth.
1992, *64*, 158

lambda
For entries see λ, Λ (beginning of 'l').

Lambert law
The fraction of light absorbed by a system is independent of the incident radiant power (P_λ^0). This law holds only if P_λ^0 is small, scattering is negligible, and multiphoton processes, excited state populations, and photochemical reactions are negligible.
See *absorbance, Beer–Lambert law*.
1996, *68*, 2251

lamellar crystal
A type of crystal with a large extension in two dimensions and a uniform thickness.

A lamellar crystal is usually of a thickness in the 5–50 nm range, and it may be found individually or in aggregates. The parallel chain stems intersect the lamellar plane at an angle between 45° and 90°. The lamellae often have pyramidal shape owing to differences in the fold domains; as a result, one can deduce different fold planes and fold surfaces from the lamellar morphology.
P.B. 81

lamp
A source of incoherent radiation.
See *high-pressure mercury lamp (arc), medium-pressure mercury lamp (arc), low-pressure mercury lamp (arc), antimony–xenon lamp (arc), quartz–iodine lamp, tungsten–halogen lamp, resonance lamp, xenon lamp*.
1996, *68*, 2251

lamp black
A special type of *carbon black* produced by incomplete combustion of a fuel rich in aromatics that is burned in flat pans. Lamp black is characterized by a relatively broad particle size distribution.
1995, *67*, 496

land fill, sanitary
See *sanitary land fill*.
1990, *62*, 2212

Landolt reaction
A slow *chemical reaction*, in which the formation of a product is delayed by a suitable reagent added for the purpose.
1993, *65*, 2296

Langmuir–Hinshelwood mechanism
A mechanism for *surface catalysis* in which the reaction occurs between species that are adsorbed on the surface is often known as a Langmuir–Hinshelwood mechanism.
1996, *68*, 171

+ Langmuir monolayer
See *monolayer*.
1994, *66*, 1672

Langmuir–Rideal (Rideal–Eley) mechanism
A mechanism for *surface catalysis* in which the reaction occurs between a reactant molecule in the gas phase and one that is adsorbed on the surface.
1996, *68*, 171

Laporte rule
This rule states that for monophotonic radiative transitions in centrosymmetric systems, the only non-vanishing electric-dipole transition moments are those which connect an even term (*g*) with an odd term (*u*).
1996, *68*, 2251

lapse rate (in atmospheric chemistry)
The variation of an atmospheric variable with height; unless otherwise stated the variable is temperature.
1990, *62*, 2199

lapse rate, temperature (in atmospheric chemistry)
See *temperature lapse rate*.
1990, *62*, 2199

large particle (in radiation scattering)
A particle with dimensions comparable with the wavelength of the radiation in the medium or larger. In practice a particle must be treated as large if its largest dimension exceeds about one-twentieth of the wavelength employed.
P.B. 65

lariat ethers
Crown *ethers* (see *crown*) having a side chain that holds one or more additional coordinating sites.
1995, *67*, 1347; see also 1988, *60*, 461

laser
A source of ultraviolet, visible, or infrared radiation which produces light amplification by stimulated emission of radiation from which the acronym is derived. The light emitted is coherent except for superradiance emission. The essential elements of a laser are: 1, an *active medium*; 2, a pumping process to make a population inversion; and 3, suitable geometry of optical feedback elements. The active medium consists of a host material (gas, liquid or solid) containing an *active species*.
See *argon ion laser, atomic laser, crystal laser, glass laser, ion laser, molecular laser, organic dye laser, helium–cadmium laser, chemical laser, CO_2 laser, copper vapour laser, diode lasers, dye laser, excimer laser, free electron laser, free-running laser, gas lasers, helium–neon laser, krypton ion laser, mode-locked laser, neodymium laser, nitrogen laser, Q-switched laser, solid-state lasers, ruby laser*.
See also *lasing*.
1996, *68*, 2251; 1995, *67*, 1915

laser, atomic
See *atomic laser*.
1995, *67*, 1919

laser beam ionization
Ionization by *irradiation* of a specimen with a laser beam.
O.B. 204

laser, chemical
See *chemical laser*.
1988, *60*, 1062

laser, crystal
See *crystal laser*.
1995, *67*, 1920

laser, diode
See *diode laser*.
1995, *67*, 1920

laser, excimer
See *excimer laser*.
1995, *67*, 1920

laser, gas
See *gas laser*.
1995, *67*, 1919.

laser, glass
See *glass laser*.
1995, *67*, 1920

laser, ion
See *ion laser*.
1995, *67*, 1920

laser ionization (in mass spectrometry)
Occurs when a sample is irradiated with a laser beam. In the irradiation of gaseous samples, ionization occurs *via* a single- or multi-photon process. In the case of solid samples, ionization occurs *via* a thermal process.
1991, *63*, 1548; O.B. 204

laser, liquid
See *liquid laser*.
1995, *67*, 1921

laser, liquid excimer
See *liquid excimer laser*.
1995, *67*, 1921

laser, liquid ion
See *liquid ion laser*.
1995, *67*, 1921

laser micro emission spectroscopy (LAMES)
Any technique in which a specimen is bombarded with a finely focused laser beam (diameter less than 10 μm) in the ultraviolet or visible range under conditions of vapourization and thermal excitation of electronic states of sample material and in which the photon emission spectrum is observed.
O.B. 250

laser micro mass spectrometry (LAMMS)
Any technique in which a specimen is bombarded with a finely focused laser beam (diameter less than 10 μm) in the ultraviolet or visible range under conditions of vapourization and ionization of sample material and in which the ions generated are recorded with a time-of-flight *mass spectrometer*.
O.B. 250

laser, molecular
　See *molecular laser*.
　1995, *67*, 1920

laser, organic dye
　See *organic dye laser*.
　1995, *67*, 1921

laser Raman microanalysis (LRMA)
Any technique in which a specimen is bombarded with a finely focused laser beam (diameter less than 10 µm) in the ultraviolet or visible range and the intensity versus wavelength function of the Raman radiation is recorded yielding information about vibrational states of the excited substance and therefore also about functional groups and chemical bonding.
　O.B. 249

lasers, solid state
　See *solid state lasers*.
　1995, *67*, 1920

lasing
The process of light amplification by stimulated emission of radiation.
　See *laser*.
　1996, *68*, 2251

late-downhill surface
　See *repulsive potential-energy surface*.
　1996, *68*, 185

latent image
The primary result of radiation absorption in a photoimaging system which is susceptible to development.
　1996, *68*, 2251

lateral order (in a polymer)
Order in the side-by-side packing of the molecules of a linear *polymer*.
　P.B. 46

lateral resolution (in *in situ* microanalysis)
For qualitative analysis, this should be defined as the minimum distance of two points (areas) on the specimen corresponding to signal levels of 16 and 84%. For electron signals originating from top surface layers like *secondary* or *Auger electrons* and secondary ions the lateral resolution for qualitative purposes corresponds to the *beam diameter*. For signals originating in a greater depth of the sample (*back scattered electrons*, X-rays) the lateral resolution is worse than the corresponding value of the beam diameter due to the diffusion of the *primary electrons*. The lateral resolution may be determined exactly with a sandwich specimen or a sharp edge specimen, or approximately with a specimen showing a regular microstructure of known dimensions.
Recommended abbreviation: Lat. Res. (qualitative); unit: m; range: nm or µm.
For quantitative analysis, this should be defined as the minimum distance of two points (areas) on the specimen corresponding to signal ratios of 10^4. This condition assures that the value for the lateral resolution defines the diameter of the analytical area which yields the total analytical signal. Due to the effect of electron diffusion in a solid and 'tail effects' in ion beam analysis the quantitative lateral resolution is significantly larger than the beam diameter.
Recommended abbreviation: Lat. Res. (quantitative); unit: m; range: nm or µm.
　1983, *55*, 2027

latex
An *emulsion* or *sol* in which each *colloidal* particle contains a number of macromolecules.
　1972, *31*, 606

lath crystal
A lamellar crystal prevailingly extended along one lateral dimension.
　P.B. 81

lattice distortion
Structural disorder resulting from misalignment of the unit cells within the crystals.
　P.B. 80

laws of distribution (in precipitation)
During the formation of a mixed crystal from a solution containing two components 'A' and 'B', the latter may be distributed according to the equation

$$K_{A,B} = b(a_0 - a)/a(b_0 - b).$$

In this homogeneous distribution, a_0 and b_0 are the respective concentrations in the solution before crystallization and a and b are the respective concentrations in the solution after crystallization. $K_{A,B}$ is usually called the separation factor. The term homogeneous distribution coefficient is not recommended.
Alternatively the distribution of the micro-component may follow the equation of Doerner and Hoskins

$$\ln(a_0/a) = \lambda \ln(b_0/b)$$

(logarithmic distribution) where λ is usually called the logarithmic distribution coefficient, the meaning of the other symbols remaining the same.
Exactly homogeneous or logarithmic distributions are extreme cases and very seldom encountered.
　O.B. 85

layer
Any conceptual region of space restricted in one dimension, within or at the surface of a condensed phase or a *film*. The usage of the term 'film' for an adsorption layer is confusing and is discouraged. The term *double-layer* applies to layers approximated by two 'distinct' sublayers.
　1994, *66*, 1671

layer equilibrium (in chromatography)
Saturation of the *stationary phase* with the *mobile phase* via the vapour phase.
　O.B. 99

leaching, microbial
　See *microbial leaching*.
　1992, *64*, 158

leader sequence (in biotechnology)
1. A polynucleotide region between *promoter* and structural *gene*, necessary for the correct *transcription* of DNA into mRNA.
2. An N-terminal extension of *polypeptides* (also known as the signal sequence) which is necessary for the transport of the *protein* into or through the membrane or for its secretion into the extracellular medium.
 1992, *64*, 158

least-squares technique
A procedure for replacing the discrete set of results obtained from an experiment by a continuous function. It is defined by the following.
For the set of variables $y, x_0, x_1, \ldots$ there are n measured values such as $y_i, x_{0i}, x_{1i}, \ldots$ and it is decided to write a relation:

$$y = f(a_0, a_1, \ldots, a_K; x_0, x_1, \ldots)$$

where $a_0, a_1, \ldots, a_K$ are undetermined constants. If it is assumed that each measurement y_i of y has associated with it a number w_i^{-1} characteristic of the uncertainty, then numerical estimates of the $a_0, a_1, \ldots, a_K$ are found by constructing a variable S, defined by

$$S = \sum_i [w(y-f)]_i^2,$$

and solving the equations obtained by writing

$$(\partial S/\partial a_j)\tilde{a}_j = 0, \quad (\tilde{a}_j = \text{all } a \text{ except } a_j).$$

If the relations between the a and y are linear, this is the familiar least-squares technique of fitting an equation to a number of experimental points. If the relations between the a and y are non-linear, there is an increase in the difficulty of finding a solution, but the problem is essentially unchanged.
 1981, *53*, 1822

leaving group
An atom or *group* (charged or uncharged) that becomes detached from an atom in what is considered to be the residual or main part of the *substrate* in a specified reaction.
For example, in the heterolytic *solvolysis* of benzyl bromide in acetic acid:

$$PhCH_2Br + AcOH \rightarrow PhCH_2OAc + HBr$$

the leaving group is Br^-; in the reaction:

$$MeS^- + PhCH_2N^+Me_3 \rightarrow MeSCH_2Ph + NMe_3$$

the leaving group is NMe_3; in the *electrophilic* nitration of benzene, it is H^+. The term has meaning only in relation to a specified reaction. The leaving group is not, in general, the same as the *substituent group* present in the substrate (e.g. bromo and trimethylammonio in the substrates of the first two examples above.)
A slightly different usage of the term prevails in the (non-mechanistic) naming of transformations, where the actual substituent group present in the substrate (and also in the product) is referred to as the leaving group.
 See also *electrofuge, entering group, nucleofuge*.
 1994, *66*, 1134

lecithins
Choline *esters* of *phosphatidic acids*. Specific compounds should be named systematically.

 1995, *67*, 1347; see also W.B. 184

lectins
Glycoproteins isolated from plants but recently found also in animals and microorganisms that react specifically with terminal glycosidic residues of other molecules (e.g. cell wall polysaccharides); some causing cells to agglutinate (cf. *agglutination*).
 1992, *64*, 158

Leffler's assumption
 See *Hammond principle*.
 1994, *66*, 1134

left-to-right convention
Arrangement of the structural formulae of the reactants so that the bonds to be made or broken form a linear array in which the electrons move from left to right.
 1994, *66*, 1134

length, l
Base quantity in the system of quantities upon which SI is based. Special symbols are used for different types of lengths.
 G.B. 11; 1996, *68*, 978

lethal concentration
Concentration of a potentially toxic substance in an environmental medium that causes death following a certain period of exposure (denoted by LC).
 See also *lethal dose*.
 1993, *65*, 2063

lethal dose
Amount of a substance or physical agent (radiation) that causes death when taken into the body by a single absorption (denoted by LD).
 See also *lethal concentration*.
 1993, *65*, 2063

lethal synthesis
Metabolic formation of a highly toxic compound from one that is relatively non-toxic (bioactivation), often leading to death of affected cells.
Synonymous with suicide metabolism.
 1993, *65*, 2063

+ leuco bases
Colourless compounds formed by reduction of triphenylmethane dyes. Generally, they are amino or hydroxy derivatives of triphenylmethane. Use of this class name, which is more or less parochial to the

technology of dyes, is not endorsed. E.g. (p-Me$_2$NC$_6$H$_4$)$_2$CHPh, the leuco base of Malachite Green.

1995, 67, 1347

+ leuco compounds

Dihydroxy polycyclic *aromatic compounds* which on oxidation form polycyclic quinone dyes (intermediates in vat dyeing). Use of this class name, which is more or less parochial to the technology of dyes, is not endorsed.

1995, 67, 1347

leukotrienes

Linear C$_{20}$ endogenous metabolites of arachidonic acid (icosa-5,8,11,14-tetraenoic acid) containing a terminal carboxy function and four or more double bonds (three or more of which are conjugated) as well as other functional groups. A subclass of *icosanoids*.

1995, 67, 1347

level

Logarithm of the ratio of a quantity and a reference value of that quantity.

See *field level* and *power level*.

G.B. 79

levelling effect

The tendency of a solvent to make all *Brønsted acids* whose *acidity* exceeds a certain value appear equally acidic. It is due to the complete transfer to a *protophilic* solvent of a hydron from a dissolved acid stronger than the *conjugate acid* of the solvent. The only acid present to any significant extent in all such solutions is the *lyonium ion*. For example, the solvent water has a levelling effect on the acidities of HClO$_4$, HCl and HI: aqueous solutions of these acids at the same (moderately low) concentrations have the same acidities. A corresponding levelling effect applies to strong bases in *protogenic* solvents.

1994, 66, 1135

level width, Γ

Planck constant over 2π divided by the mean life.

G.B. 22

Lewis acid

A *molecular entity* (and the corresponding *chemical species*) that is an electron-pair acceptor and therefore able to react with a *Lewis base* to form a *Lewis adduct*, by sharing the electron pair furnished by the Lewis base. For example:

$$\text{Me}_3\text{B} + :\text{NH}_3 \longrightarrow \text{Me}_3\overset{-}{\text{B}}-\overset{+}{\text{NH}}_3$$

Lewis acid   Lewis base   Lewis adduct

See also *coordination, dipolar bond*.

1994, 66, 1135

Lewis acidity

The thermodynamic tendency of a substrate to act as a *Lewis acid*. Comparative measures of this property are provided by the equilibrium constants for Lewis adduct formation of a series of Lewis acids with a common reference *Lewis base*.

See also *acceptor number (AN), electrophilicity*.

1994, 66, 1135

Lewis adduct

The *adduct* formed between a *Lewis acid* and a *Lewis base*.

1994, 66, 1135

Lewis base

A *molecular entity* (and the corresponding *chemical species*) able to provide a pair of electrons and thus capable of *coordination* to a *Lewis acid*, thereby producing a *Lewis adduct*.

1994, 66, 1135

Lewis basicity

The thermodynamic tendency of a substance to act as a *Lewis base*. Comparative measures of this property are provided by the equilibrium constants for *Lewis adduct* formation for a series of Lewis bases with a common reference *Lewis acid*.

See also *donor number (DN), nucleophilicity*.

1994, 66, 1135

Lewis formula (electron dot or Lewis structure)

Molecular structure in which the valency electrons are shown as dots so placed between the bonded atoms that one pair of dots represents two electrons or one covalent (single) bond, e.g.

$$\text{H}:\overset{..}{\underset{..}{\text{Cl}}}:$$

A double bond is represented by two pairs of dots, etc. Dots representing non-bonded outer-shell electrons are placed adjacent to the atoms with which they are associated, but not between the atoms. Formal charges (e.g. +, −, 2+, etc.) are attached to atoms to indicate the difference between the positive nuclear charge (atomic number) and the total number of electrons (including those in the inner shells), on the formal basis that bonding electrons are shared equally between atoms they join. (Bonding pairs of electrons are usually denoted by lines, representing *covalent bonds*, as in *line formulae*.)

1994, 66, 1135

LIDAR (in atmospheric chemistry)

An acronym for Light Detection And Ranging. It is a real time remote sensing technique for the determination of aerosol and trace gas concentrations in air by measurement of scattered laser radiation.

1990, 62, 2199

lifetime (mean lifetime), τ

The lifetime of a *chemical species* which decays in a first-order process is the time needed for a concentration of this species to decrease to 1/e of its original value. Statistically, it represents the mean life expectancy of an excited species. In a reacting system in which the decrease in concentration of a particular chemical species is governed by a first-order *rate law*, it is equal to the reciprocal of the sum of the (pseudo)unimolecular rate constants of all processes which cause the decay. When the term is used for processes which are not first order, the lifetime depends on the initial concentration of the species, or of a quencher, and should be called apparent lifetime instead.

See also *chemical relaxation, half-life, rate of reaction*.
1994, *66*, 1135; see also 1996, *68*, 2251; 1990, *62*, 2199; 1996, *68*, 978; G.B. 22

lifetime (of luminescence)
The time required for the *luminescence* intensity to decay from some initial value to 1/e of that value (e = 2.718 . . .). Lifetimes can be measured by phase fluorimetry (phosphorimetry) where the phase shift between the sinusoidally modulated exciting light and the emitted light is measured.
1984, *56*, 242

ligand field splitting
The removal of a degeneracy of atomic or molecular levels in a molecule or ion with a given symmetry, induced by the attachment or removal of ligands to produce reduced symmetries.

See *crystal field splitting*.
1996, *68*, 2251

ligands
1. In an inorganic coordination entity, the atoms or groups joined to the central atom.
R.B. 146
2. In biochemistry: if it is possible or convenient to regard part of a polyatomic molecular entity as central, then the atoms, groups or molecules bound to that part are called ligands. Biochemical usage is thus wider, in that the central entity can be polyatomic. Thus H^+ may be a ligand for proteins and for citrate as well as for O^{2-}. It may even be a ligand for a univalent entity such as acetate: in other circumstances, AcO^- may be the ligand for H^+, since the definition makes it clear that the view of which entity is central may change for convenience. Thus, four calcium ions are ligands for calmodulin, when the protein is regarded as central: four carboxylate groups of calmodulin ligate (are ligands of) each calcium ion when this ion is regarded as central. It is the ligand that is said to ligate the central entity, which is said to be ligated. When the hormone binding to a receptor is called a ligand, the receptor is thus regarded as the central entity. Biochemists should bear in mind that the usage in inorganic chemistry has been that ligands bind only single atoms, so they should be cautious in fields such as bioinorganic chemistry where confusion may be possible.
W.B. 335; 1994, *66*, 1136

ligand to ligand charge transfer (LLCT) transition
An electronic transition of a metal complex that corresponds to excitation populating an electronic state in which considerable *electron transfer* between two ligands has occurred.
1996, *68*, 2252

ligand to metal charge transfer (LMCT) transition
An electronic transition in a metal complex that corresponds to excitation populating an electronic state in which considerable *electron transfer* from a ligand to a metal centre has occurred.

See also *metal to ligand charge transfer transition*.
1996, *68*, 2252

ligases (synthetases)
Enzymes that catalyse the ligation of two molecules with concomitant hydrolysis of the pyrophosphate bond in adenosine 5′-triphosphate (ATP) or a similar triphosphate, forming C–C, C–O, C–S, P–O or C–N bonds. In *recombination* DNA technology, ligases covalently join together two sequences of DNA (e.g. *host* DNA and foreign DNA) by a phosphodiester bond.
1992, *64*, 159

ligate
To coordinate to a central atom as a ligand.
R.B. 146

light-atom anomaly
A dynamical effect that arises for a process
A + B–C → A–B + C when the species A is light (e.g. a hydrogen atom) compared to B and C. The vibrational excitation of the product A–B is low, since the light atom A approaches to within the bonding distance of BC before the C atom retreats. The energy of reaction is therefore released as repulsion between A–B and C, with the result that there is translational excitation of the products.
1996, *68*, 171

light polarization
When the end point of the electric vector of a polarized light beam is viewed along the direction of light propagation, it moves along a straight line if the light is linearly polarized, along a circle if it is circularly polarized, and along an ellipse if it is elliptically polarized.
1996, *68*, 2252

light scattering
The redirection of a light beam due to interactions with molecules (Rayleigh and Raman scattering) and aerosols (Mie scattering). Scattered light received at any point in the atmosphere (sometimes called *sky radiation*) is a very important component of the total radiation received from the sun; it is dependent on the solar zenith angle, elevation aerosol concentration etc.

See also *elastic scattering*.
1990, *62*, 2199

light source
See *lamp, laser*.
1996, *68*, 2252

lignans
Plant products of low molecular weight formed primarily from oxidative coupling of two *p*-propylphe-

nol moieties at their β carbon atoms; products with units coupled in other ways are neolignans, e.g.

galbacin, a lignan

(−)-eusiderin, a neolignan

1995, *67*, 1347

lignins
Macromolecular constituents of wood related to *lignans*, composed of phenolic propylbenzene skeletal units, linked at various sites and apparently randomly.
1995, *67*, 1348

limiting adsorption current
The potential-independent value that is approached by an *adsorption current* as the rate of reduction or oxidation of the electroactive substance is increased by varying the applied potential.
The terms adsorption current and limiting adsorption current should not be applied to *faradaic currents* that have been increased or decreased by adding a non-electroactive *surfactant* to a solution containing an electroactive substance, nor to apparent waves resulting from the effect of *adsorption* or *desorption* on *double-layer currents*.
1985, *57*, 1494

limiting catalytic current
The potential-independent value that is approached by a *catalytic current* as the rate of the charge-transfer process is increased by varying the *applied potential*.
1985, *57*, 1495

limiting condition of operation
Range of physical and operational parameters in which the method meets given values of performance characteristics with 95% probability.
1990, *62*, 2199

limiting current
The limiting value of a *faradaic current* that is approached as the rate of the charge-transfer process is increased by varying the potential. It is independent of the *applied potential* over a finite range, and is usually evaluated by subtracting the appropriate *residual current* from the measured total current. A limiting current may have the character of an *adsorption*, catalytic, diffusion, or kinetic current, and may include a *migration current*.
1974, *37*, 513; 1985, *57*, 1497

limiting differential diffusion coefficient
See *differential diffusion coefficient*.
1972, *31*, 617

limiting diffusion current
The potential-independent value that is approached by a *diffusion current* as the rate of the charge-transfer process is increased by varying the *applied potential*.
1985, *57*, 1495

limiting kinetic current
The potential-independent value that is approached by a *kinetic current* as the rate of the charge-transfer process is increased by varying the applied potential.
1985, *57*, 1497

limiting mean
See *measurement result*.
1995, *67*, 1705

limiting migration current
The limiting value of a *migration current*, which is approached as the rate of the charge-transfer process is increased by varying the *applied potential*.
1985, *57*, 1498

limiting sedimentation coefficient
See *sedimentation coefficient*.
1972, *31*, 616

limiting viscosity number
See *intrinsic viscosity*.
P.B. 63

limit of detection (in analysis)
The limit of detection, expressed as the *concentration*, c_L, or the quantity, q_L, is derived from the smallest measure, x_L, that can be detected with reasonable certainty for a given analytical procedure. The value of x_L is given by the equation

$$x_L = \bar{x}_{bi} + ks_{bi}$$

where $\bar{x}_{bi}$ is the mean of the *blank measures*, s_{bi} is the *standard deviation* of the blank measures, and k is a numerical factor chosen according to the confidence level desired.
O.B. 5

limit test (in toxicology)
Acute toxicity test in which, if no ill-effects occur at a pre-selected maximum dose, no further testing at greater exposure levels is required.
See also *fixed dose test*.
1993, *65*, 2063

linear absorption coefficient
See *absorption coefficient*.
1996, *68*, 978

linear (decadic) absorption coefficient (in optical spectroscopy)
See *absorption coefficient*.
G.B. 32; see also 1990, *62*, 2169; 1988, *60*, 1058; 1996, *68*, 959

linear attenuation coefficient
See *attenuation coefficient*.
1996, *68*, 978

linear (decadic) attenuation coefficient (in optical spectroscopy)
See *attenuation coefficient*.
G.B. 32; see also O.B. 212; 1996, *68*, 963

linear chain
A *chain* with no *branch points* intermediate between the boundary units.
1996, *68*, 2293

linear copolymer
A *copolymer* composed of *linear macromolecules*.
1996, *68*, 2303

linear dispersion
See *dispersion (for spectroscopic instruments)*.
O.B. 158

linear electron accelerator
An evacuated metal tube in which *electrons* pass through a series of small gaps (usually) in the form of cavity resonators in the high frequency range) so arranged and spaced that, at a specific excitation frequency, the stream of electrons on passing through successive gaps gains additional energy from the electric field in each gap.
1982, *54*, 1545

linear energy transfer
The average energy locally imparted to a medium by a charged particle of specified energy, per unit distance traversed.
1982, *54*, 1545

linear free-energy relation
A linear correlation between the logarithm of a rate constant or equilibrium constant for one series of reactions and the logarithm of the rate constant or equilibrium constant for a related series of reactions. Typical examples of such relations (also known as linear Gibbs energy relations) are the *Brønsted relation*, and the *Hammett equation* (see also σ-value).
The name arises because the logarithm of an equilibrium constant (at constant temperature and pressure) is proportional to a standard free energy (Gibbs energy) change, and the logarithm of a rate constant is a linear function of the free energy (Gibbs energy) of activation.
It has been suggested that this name should be replaced by *linear Gibbs energy relation*, but at present there is little sign of acceptance of this change.
The area of physical organic chemistry which deals with such relations is commonly referred to as 'Linear Free-Energy Relationships'.
1994, *66*, 1136

linear Gibbs energy relation
See *linear free-energy relation*.
1994, *66*, 1136

linearity of responsivity (of a radiation detector)
The extent to which the output of the *detector* is directly proportional to the incident radiant power at a given wavelength and at constant irradiation geometry.
1995, *67*, 1750

linear macromolecule
A *macromolecule* the structure of which essentially comprises the multiple repetition in linear sequence of units derived, actually or conceptually, from molecules of low relative molecular mass.
1996, *68*, 2290

linear polarizer
An optical device which allows the transmission of *radiation* of which the electric vector is restricted to one plane resulting in linearly polarized radiation.
1984, *56*, 238

linear polymer
A *polymer* composed of *linear macromolecules*.
1996, *68*, 2303

linear pulse amplifier
A *pulse amplifier* which, within the limits of its normal operating characteristics, delivers an output pulse of amplitude proportional to that of the input pulse.
1982, *54*, 1535

linear range
Concentration range over which the intensity of the signal obtained is directly proportional to the concentration of the species producing the signal.
1990, *62*, 2199

linear solvation energy relationships
Equations involving the application of *solvent parameters* in linear or multiple (linear) regression expressing the solvent effect on the rate or equilibrium constant of a reaction.
See *Dimroth–Reichardt E_T parameter, Kamlet–Taft solvent parameter, Koppel–Palm solvent parameter, Z-value*.
1994, *66*, 1136

linear strain, ε, e
Change of length divided by the original length.
G.B. 12

line formula
A two-dimensional representation of *molecular entities* in which atoms are shown joined by lines representing single or multiple bonds, without any indication or implication concerning the spatial direction of bonds. For example, methanol is represented as:

$$\begin{array}{c} \text{H} \\ | \\ \text{H}-\text{C}-\text{O}-\text{H} \\ | \\ \text{H} \end{array}$$

(The term should not be confused with the representation of chemical formulae by the 'Wiswesser line notation', a method of string notation. Formulae in this notation are also known as 'Wiswesser line formulae'.)
1994, *66*, 1136

lineic
Attribute to a physical quantity obtained by division by length. Lineic mass of a thread is the mass of the thread divided by its length.
ISO 31-0: 1992; 1996, *68*, 978

line-of-centres model
A form of the collision theory of *chemical reactions* in which the assumption is made that reaction can only occur if on collision the component of energy along the line of centres of the masses of the colliding species is greater than a particular *threshold energy*.
1996, *68*, 171

line repetition groups
The possible symmetries of arrays extending in one direction with a fixed repeating distance.
Linear polymer chains in the crystalline state must belong to one of the line repetition groups. Permitted symmetry elements are: the identity operation (symbol l); the translation along the chain axis (symbol t); the mirror plane orthogonal to the chain axis (symbol m) and that containing the chain axis (symbol d); the glide plane containing the chain axis (symbol c); the inversion centre, placed on the chain axis (symbol i); the two-fold axis orthogonal to the chain axis (symbol 2); the helical, or screw, symmetry where the axis of the helix coincides with the chain axis. In the latter case, the symbol is s($A*M/N$), where s stands for the screw axis, A is the class of the helix, * and / are separators, and M is the integral number of residues contained in N turns, corresponding to the identity period (M and N must be prime to each other). The class index A may be dropped if deemed unnecessary, so that the helix may also be simply denoted as s(M/N).
P.B. 79

line-shape analysis
Determination of rate constants for a chemical exchange from the shapes of spectroscopic lines of dynamic processes. The method is most often used in nuclear magnetic resonance spectroscopy.
1994, *66*, 1136

Lineweaver–Burk plot
A plot of the reciprocal of velocity of an enzyme-catalysed reaction (ordinate) versus the reciprocal of substrate concentration (abscissa). The plot is used to graphically define the maximum velocity of an enzyme-catalysed reaction and the *Michaelis constant* for the enzyme.
See also *Michaelis–Menten kinetics*.
1994, *66*, 2593

line width
Width of a spectral line in terms of *wavelength, frequency* or *wavenumber*. Usually defined as full or half width of the line at half of the maximum intensity (*FWHM* or *HWHM*, respectively).
G.B. 31

line width (in Mössbauer spectroscopy)
The full width at half maximum peak-height.
The natural linewidth is the theoretical value of the full width at half maximum of the *nuclear transition*, usually calculated from lifetime data.
1976, *45*, 213

linked scan (in mass spectrometry)
A scan, in an instrument comprising two or more analysers, in which two or more of the analyser fields are scanned simultaneously so as to preserve a predetermined relationship between parameters characterizing these fields. Often, these parameters are the field strengths, but they may also be the frequencies in the case of analysers in which alternating fields are employed.
1991, *63*, 1551

lipid film
Established term for W/O/W films (i.e. oil film in water).
1972, *31*, 613

lipids
A loosely defined term for substances of biological origin that are soluble in nonpolar solvents. They consist of saponifiable lipids, such as *glycerides* (fats and oils) and *phospholipids*, as well as nonsaponifiable lipids, principally *steroids*.
1995, *67*, 1348; see also W.B. 180

lipophilic
Literally 'fat-loving'. Applied to *molecular entities* (or parts of molecular entities) having a tendency to dissolve in fat-like (e.g. hydrocarbon) solvents.
See also *hydrophilic, hydrophobic interaction, lyophilic*.
1994, *66*, 1136; 1972, *31*, 607

lipophobic
Literally, 'fat-rejecting'; the converse of *lipophilic*.
1972, *31*, 607

lipopolysaccharides
Natural compounds consisting of a trisaccharide repeating unit (two heptose units and octulosonic acid) with oligosaccharide side chains and 3-hydroxytetradecanoic acid units (they are a major constituent of the cell walls of Gram-negative bacteria).
1995, *67*, 1348

lipoproteins
Clathrate complexes consisting of a *lipid* enwrapped in a *protein* host without covalent binding in such a way that the complex has a hydrophilic outer surface consisting of all the protein and the polar ends of any *phospholipids*.
1995, *67*, 1348

liposome
Originally a lipid droplet in the endoplasmic reticulum of a fatty liver. Now applied to an artificially formed lipid droplet, small enough to form a relatively stable suspension in aqueous media and with potential use in drug delivery.
1993, *65*, 2064

Lippman's equation
An equation which gives the electric charge per unit area of an interface (electrode):

$$(\partial \gamma / \partial E_A)_{T, p, \mu_i \neq \mu} = - Q_A$$

where γ is the interfacial tension, E_A is the potential of a cell in which the reference electrode has an interfacial equilibrium with one of the ionic components of A, Q_A is the charge on unit area of the interface, μ_i is the chemical potential of the combination of species i whose net charge is zero, T is the thermodynamic temperature and p is the external pressure.
Since more than one type of reference electrode may be chosen, more than one quantity Q may be obtained. Consequently Q cannot be considered as equivalent to the physical charge on a particular region of the interphase. It is in fact an alternative way of expressing a *surface excess* or combination of surface excess of charged species.
1974, *37*, 508; 1986, *58*, 445

liquid chromatography (LC)
A separation technique in which the mobile phase is a liquid. Liquid chromatography can be carried out either in a column or on a plane.
Present-day liquid chromatography generally utilizing very small particles and a relatively high inlet pressure is often characterized by the term high-performance (or high-pressure) liquid chromatography, and the acronym HPLC.
1993, *65*, 825

liquid-crystal transitions
A liquid crystal is a molecular crystal with properties that are both solid- and liquid-like. Liquid crystals are composed predominantly of rod-like or disc-like molecules, that can exhibit one or more different, ordered fluid phases as well as the isotropic fluid; the translational order is wholly or partially destroyed but a considerable degree of orientational order is retained on passing from the crystalline to the liquid phase in a *mesomorphic transition*.
1. Transition to a *nematic phase*.
A *mesomorphic transition* that occurs when a molecular crystal is heated to form a nematic phase in which the mean direction of the molecules is parallel or antiparallel to an axis known as the director.
2. Transition to a *cholesteric phase*.
A *mesomorphic transition* that occurs when a molecular crystal is heated to form a cholesteric phase in which there is simply a spiralling of the local orientational order perpendicular to the long axes of the molecules.
3. Transition to a *smectic state*.
A *mesomorphic transition* that occurs when a molecular crystal is heated to yield a smectic state in which there is a one-dimensional density wave which produces very soft/disordered layers.
1994, *66*, 584

liquid excimer laser
A *liquid laser* in which excimer species in liquids are used as the *active medium*.
1995, *67*, 1921

liquid-gel chromatography
Includes *gel-permeation* and *ion-exchange chromatography*. Liquid is the *mobile phase*, gel the *stationary phase*.
O.B. 93

liquid ion exchange
A term used to describe a *liquid-liquid extraction* process that involves a transfer of ionic species from the extractant to the aqueous phase in exchange for ions from the aqueous phase. The term does not imply anything concerning the nature of the bonding in the extracted complex. The term solvent ion exchange is not recommended.
1993, *65*, 2378

liquid ion laser
A *liquid laser* in which rare earth ions in solution are used as the laser *active medium*. An example is KrF in liquid Kr laser.
1995, *67*, 1921

liquid junction
Any junction between two electrolyte solutions of different composition. Across such a junction there arises a potential difference, called the liquid junction potential. In the *operational pH cell* the junction is between the *test*, or pH standard, *solution* and the *filling solution* or the *bridge solution* of the *reference electrode*.
1985, *57*, 541

liquid laser
A *laser* in which a liquid solution is the *active species*.
1995, *67*, 1921

liquid-liquid distribution (extraction) (partition)
The process of transferring a dissolved substance from one liquid phase to another (immiscible or partially miscible) liquid phase in contact with it. Although extraction, partition and distribution are not synonymous, extraction may replace distribution where appropriate.
1993, *65*, 2378; O.B. 87

liquid-liquid extraction
This term may be used in place of liquid-liquid distribution when the emphasis is on the analyte(s) being distributed (or extracted).
Note:
The distinction between the distribution constant (K_D) and the partition constant (K_D^o) or the concentration distribution ratio (D_c) is reaffirmed and it is recommended that the terms partition constant, partition coefficient and extinction constant should not be used as synonyms for the (analytical) *distribution ratio*, D_c. A distinction is drawn between the terms solvent and diluent, and the term extractant is now restricted to the active substance in the solvent (i.e. the homogeneous 'organic phase' which comprises the extractant,

the diluent and/or the modifier) which is primarily responsible for the transfer of solute from the 'aqueous' to the 'organic' phase.
O.B. 88

liquid-phase loading (in chromatography)
A term used in *partition chromatography* to express the relative amount of the liquid stationary phase in the column packing. It is equal to the mass fraction (%) of liquid stationary phase in the total packing (liquid stationary phase plus support).
1993, *65*, 831

liquid scintillation detector
A *scintillation detector* in which the test portion is mixed with a liquid *scintillator*.
1994, *66*, 2522

liquidus
A line on a binary phase diagram (or surface on a ternary phase diagram) that indicates the temperature at which *solidification* begins on cooling or at which *melting* is completed on heating under equilibrium conditions.
1994, *66*, 585

liquid volume (in gas chromatography)
That volume occupied by the *liquid phase* in the *column*. Defined by

$$V_L = w_L/\rho_L$$

where w_L is the weight of the liquid in the column, and ρ_L is the density at the column temperature.
O.B. 99

lithometeor (in atmospheric chemistry)
A particle of dry substance in the atmosphere, as contrasted to a *hydrometeor*.
1990, *62*, 2199

lithosphere (in atmospheric chemistry)
The crust of the earth, usually thought of as discrete from and in contact with the hydrosphere and the atmosphere.
1990, *62*, 2199

litre
Non-SI unit of volume, L = 10^{-3} m^3. The other allowed symbol is l.
G.B. 111; see also 1996, *68*, 979; 1986, *58*, 1407

live time
For a measurement, the time during which a *radiation* measuring assembly is capable of processing events occurring in the radiation detector. It equals the clock time minus the integrated resolving or *dead time* (to be distinguished from 'life time').
1994, *66*, 2522

living copolymerization
A *copolymerization* which is a *living polymerization*.
1996, *68*, 2309

living polymerization
A chain polymerization from which chain transfer and chain termination are absent.

In many cases, the rate of chain initiation is fast compared with the rate of chain propagation, so that the number of kinetic-chain carriers is essentially constant throughout the polymerization.
1996, *68*, 2308

load (on a precision balance)
The total weight acting, after counterbalancing, upon the terminal bearing which carries the object being weighed.
O.B. 35

loading capacity (in solvent extraction)
The maximum concentration of solute(s) that a solvent can contain under specified conditions.
Notes:
1. The terms maximum loading, saturation capacity and saturation loading are synonymous.
2. All the above terms should clearly be distinguished from ultimate capacity.
1993, *65*, 2384

local conformation (of a polymer)
The *conformation* of a *macromolecule* at the scale of the *constitutional units*.
In the polymer literature, local conformation is sometimes referred to as microconformation.
P.B. 77

local current density
See *average current density*.
1981, *53*, 1836

local efficiency of atomization (in flame emission and absorption spectrometry), ε_a
The substance fraction of atomized component in the component consumed.
The efficiency of atomization is measured is measured in a given part of the flame, usually the observation space; $\varepsilon_a = \varepsilon_n \chi_s \chi_v \chi_a$. The signal is a function of the product $q_v \varepsilon_a$, but ε_a is also a function of q_v, usually decreasing at high volume rates.
1986, *58*, 1741; O.B. 169

local flame temperature (in flame emission and absorption spectrometry), T_1
The effective thermodynamic temperature in the observation space as measured by a specific sensor for a specified element (in K).
The temperature of a flame (or other plasma) is not homogeneous. It is usually lower at the borders of the flame. It is therefore appropriate to speak of an effective temperature which represents an average value of all temperatures throughout the observation space.
The flame temperature depends on several factors such as: kind of plasma, kind of gas or gas mixture and concentration gradient of the thermometric species in the observation space.
1986, *58*, 1741

local fraction atomized (in flame emission and absorption spectrometry), χ_a, β_a
The substance fraction of the atomized component in the total volatilized component.

This quantity is measured in a defined part of the flame, usually the observation space. The fraction atomized is the result of chemical reactions in the gaseous state. It depends on the bond strength of the compounds that the component may form within the flame and on the composition and temperature of the flame. When analysing elements that tend to become oxidized in the flame, it may be advisable to use as fuel gas mixtures with a reducing component such as C_2H_2 or N_2O.
1986, 58, 1741; O.B. 168

local fraction desolvated (in flame emission and absorption spectrometry), χ_s, β_s
The amount-of-substance fraction of component in a desolvated state in the amount of component entering the flame.
This quantity is measured in a defined part of the flame, usually the observation space. Because it varies with height in the flame as a result of progressive evaporation of aerosol droplets, it is appropriate to term the expression local. The fraction desolvated does not account for losses by incomplete volatilization of the dry aerosol (which largely depends upon the nature and concentration of the component). Such losses are described by local fraction volatilized, which usually depends on the solute. Since χ_s varies markedly with the height in the flame, its observed value represents an average. Local fraction desolvated depends on the solvent, the temperature of the flame and the time the component takes to travel from the tip of the burner to the height in the flame considered.
1986, 58, 1740; O.B. 168

local fraction volatilized (in flame emission and absorption spectrometry), χ_v, β_v
The substance fraction of the volatilized component in the total desolvated component. The gaseous state includes free atoms, molecules and radicals.
This quantity is measured in a defined part of the flame, usually the observation space. The fraction volatilized varies inversely with the size of the desolvated particles. Since χ_v varies markedly with the height in the flame, its observed value represents an average.
1986, 58, 1741; O.B. 168

localized adsorption
See *mobile adsorption, immobile adsorption*.
1976, 46, 76

localized-itinerant transition
A transition of either of two types:
(a) a *Mott transition* (single-valent systems).
(b) a small-polaron to itinerant-electron transition (mixed-valent systems).
1994, 66, 585

lock-in state
See *commensurate-incommensurate transition*.
1994, 66, 585

logarithmic distribution coefficient (in precipitation)
See *laws of distribution (in precipitation)*.
O.B. 85

logarithmic normal distribution (of a macromolecular assembly)
A continuous distribution with the differential mass-distribution function of the form:

$$f_w(x)\,dx = \frac{1}{a\sqrt{\pi}x}\exp(-\frac{1}{a^2}\ln^2\frac{x}{b})\,dx$$

where x is a parameter characterizing the chain length, such as relative molecular mass or degree of polymerization and a and b are positive adjustable parameters.
P.B. 57

logarithmic viscosity number
See *inherent viscosity*.
P.B. 63

logit
In *competitive binding assays*, the logit-log dose relationship, in which the response is defined by: $R =$ logit $(y) = \log [y/(1 - y]$ where $y = b/b_0$ with $b =$ fraction of *tracer* bound and $b_0 =$ value of b with no unlabelled *ligand* in the system. Logit transformed assay data frequently yield straight-line dose-response curves, amenable to statistical analysis.
1994, 66, 2522

log-normal distribution
A distribution function $F(y)$, in which the logarithm of a quantity is normally distributed, i.e. $F(y) = f_{\text{gauss}}(\ln y)$ where $f_{\text{gauss}}(x)$ is a Gaussian distribution. The size distributions of atmospheric aerosols are often described using this distribution function, although the term also applies to gaseous pollutants.
1990, 62, 2199

London–Eyring–Polanyi (LEP) method
A semiempirical method of calculating a *potential-energy surface*, based on a simplified quantum-mechanical equation (the London equation).
1996, 68, 171

London–Eyring–Polanyi–Sato (LEPS) method
A modification of the *London–Eyring–Polanyi* method proposed in 1955 by S. Sato.
1996, 68, 171

London forces
Attractive forces between apolar molecules, due to their mutual polarizability. They are also components of the forces between polar molecules. Also called 'dispersion forces'.
See also *van der Waals forces*.
1994, 66, 1136

lone (electron) pair
Two paired electrons localized in the valence shell on a single atom. Lone pairs should be designated with two dots. The term 'nonbonding electron pair' is more appropriate, and is found in many modern text books.
1994, 66, 1137

long chain
A *chain* of high relative molecular mass.
See *macromolecule (1)*.
1996, *68*, 2294

long-chain branch
See *branch*.
1996, *68*, 2297

longitudinal order (in a polymer)
Order in the atomic positions along the chains of a linear *polymer*.
P.B. 46

long-lived collision complex
When a potential-energy surface is such that the system performs a number of vibrations and rotations in the region of a *col*, the corresponding *intermediate* is known as a long-lived collision complex, and the reaction is said to be indirect.
1981, *53*, 767

long-range intramolecular interaction (in polymers)
An interaction between segments, widely separated in sequence along the chain, that occasionally approach one another during molecular flexing.
This type of interaction is closely related to the excluded volume of a segment, the latter quantity reflecting interactions involving segments and solvent molecules. If no confusion can occur, the word 'intramolecular' may be omitted.
P.B. 48

long spacing (in polymer crystals)
The average separation between stacked lamellar crystals, usually measured by small-angle X-ray or neutron diffraction.
P.B. 82

loose end
A *chain* only one point of which is attached to a *network*.
1996, *68*, 2298

loose ion pair
See *ion pair*.
1994, *66*, 1137

+ Lorentz broadening (of a spectral line)
See *collisional broadening (of a spectral line)*.
1985, *57*, 1463

Lorentzian band shape
This band shape is described by the function:

$$F(v - v_0) = (1/\pi) \gamma [(v - v_0)^2 + \gamma^2]^{-1}$$

where v_0 is the mean band position, γ is the half band width at half maximum, and $F(v - v_0)$ is the frequency distribution function.
See also *Gaussian band shape*.
1996, *68*, 2252

Lorenz–Mie theory
Theory of light scattering by *isotropic*, homogeneous spheres.
1983, *55*, 935

lot (in analytical chemistry)
A quantity of material which is assumed to be a single population for sampling purposes.
1990, *62*, 1201; 1988, *60*, 1463

low energy electron diffraction (LEED)
Any technique which measures the angular intensity distribution of electrons reflected from a crystalline surface under bombardment with low energy electrons ($E_0 < 500$ eV) in larger angles of incidence. The diffraction pattern also provides very surface sensitive information on the atomic arrangement of the top layers of a solid.
1983, *55*, 2025

lowest lethal concentration found
See *minimum lethal concentration*.
1993, *65*, 2071

lowest-observed-adverse-effect-level (LOAEL)
Lowest concentration or amount of a substance, found by experiment or observation, which causes an adverse alteration of morphology, functional capacity, growth, development, or life span of a target organism distinguishable from normal (control) organisms of the same species and strain under defined conditions of exposure.
See also *lowest-observed-effect-level, no-observed-adverse-effect-level, no-observed-effect-level*.
1993, *65*, 2065

lowest-observed-effect-level (LOEL)
Lowest concentration or amount of a substance, found by experiment or observation, that causes any alteration in morphology, functional capacity, growth, development, or life span of target organisms distinguishable from normal (control) organisms of the same species and strain under the same defined conditions of exposure .
See also *lowest-observed-adverse-effect-level, no-observed-adverse-effect-level, no-observed-effect-level*.
1993, *65*, 2065

low-pressure diamond
See *diamond by CVD*.
1995, *67*, 487

low pressure electrical discharge
Low pressure electrical discharge sources are radiation sources in which radiation is produced by electrical discharges in gases at low pressures, i.e. pressures from 10^2 to 10^{-3} Pa.
O.B. 145

low-pressure mercury lamp (arc)
A type of resonance lamp which contains mercury vapour at pressures of about 0.1 Pa (0.75×10^{-3} Torr). At 250 °C, such a lamp emits mainly at 253.7 and 184.9 nm. Other terms used for such a lamp are germicidal, cold and hot cathode, Wood lamp.
1996, *68*, 2252

low temperature UV–VIS absorption spectroscopy
The measurement of the absorption of UV–VIS radiation by a solute in a highly viscous or solid matrix at low temperature.
1988, *60*, 1452

lumen
SI derived unit of luminous flux, lm = sd sr^{-1}.
G.B. 72; 1996, *68*, 979

luminance
Photometric counterpart of radiance, producing the visual sensation called brightness. Typical units are: candela m^{-2} (nit), candela cm^{-2} (stilb), foot lambert (2.426 nit). As with all photometric quantities, luminance does not refer to a specific wavelength, but applies to light emitted by a standard source (formerly a 'standard international candle', now a blackbody radiator emitting at the temperature of solidifying platinum, 2042 K). Conversion from photometric units to radiometric units (e.g. J s^{-1}) requires convolution over wavelength of the relative spectral response of the human eye (photopic response tables).
1990, *62*, 2199

luminescence
Spontaneous emission of radiation from an electronically or vibrationally excited species not in thermal equilibrium with its environment.
See also *bioluminescence, chemiluminescence, electro-generated chemiluminescence, fluorescence, phosphorescence, photoluminescence, radioluminescence, sonoluminescence, thermoluminescence, triboluminescence*.
1996, *68*, 2252; 1994, *66*, 2522; see also O.B. 184, 231; 1990, *62*, 2200

luminescence emission polarization spectrum, corrected
The corrected luminescence polarization spectrum of an analyte is obtained when the polarization is measured as a function of the excitation wavelength. Since this spectrum may depend on the emission wavelength monitored, this wavelength should be specified. The polarization is usually given as r or P.
O.B. 198

luminescence excitation polarization spectrum, corrected
The (fluorescence, phosphorescence) spectrum observed when the polarization r (or P) is measured as a function of emission wavelength using a fixed and specified excitation wavelength.
O.B. 198

luminescence quenching
Radiationless redistribution of *excitation energy* via interaction (electronic energy or charge transfer) between an emitting species and the quencher.
1984, *56*, 235

luminescence spectrometer
The *instrument* used to measure *luminescence* emission spectra.
1984, *56*, 238

luminous flux
Of a source of *luminous intensity I* in an element of solid angle dΩ is given by d$\Phi = I$ dΩ
G.B. 72; 1996, *68*, 979

luminous intensity, I_v
Base quantity in the system of quantities upon which SI is based.
G.B. 4; 1996, *68*, 979

luminous quantities
Group of quantities characterizing visible light in terms of the appearance to the human eye. The symbols of luminous quantities may be distinguished by adding a subscript v. The units for luminous quantities are derived from the SI base unit *candela*.
See also *radiant quantities, photon quantities*.
G.B. 30

lumiphore (luminophore)
A part of a molecular entity (or atom or group of atoms) in which electronic excitation associated with a given emission band is approximately localized. (Analogous to *chromophore* for absorption spectra).
1996, *68*, 2253

LUMO
See *frontier orbitals*.
1994, *66*, 1137

lux
SI derived unit of illuminance, lx = lm m^{-2} = cd sr^{-1} m^{-2}.
G.B. 72; 1996, *68*, 979

lyases
Enzymes cleaving C–C, C–O, C–N and other bonds by other reactions than by hydrolysis or oxidation. Lyases catalyse the addition of molecules to unsaturated compounds or the elimination of molecules creating an unsaturated residue.
1992, *64*, 159

lyate ion
The *anion* produced by *hydron* removal from a solvent molecule. For example, the hydroxide ion is the lyate ion of water.
1994, *66*, 1137

lyonium ion
The *cation* produced by hydronation of a solvent molecule. For example, $CH_3OH_2^+$ is the lyonium ion of methanol.
See also *onium ion*.
1994, *66*, 1137

lyophilic
The terms lyophilic (*hydrophilic, lipophilic, oleophilic*, etc.) and *lyophobic* (*lipophobic*, etc.) may be used to describe the character of interaction of a particular atomic group with the medium. In this usage the terms have the relative qualitative meaning of 'solvent preferring' (water-preferring, fat-preferring etc.) and 'solvent rejecting' (water-rejecting, fat-rejecting, etc.) respectively.
1972, *31*, 607

lyophilic sols
These comprise both *association colloids* in which *aggregates* of small molecules are formed reversibly and macromolecules in which the molecules themselves are of *colloidal* size.
 1972, *31*, 607

lyophobic
 See *lyophilic*.
 1972, *31*, 607

lysimeter
Laboratory column of selected representative soil or a protected monolith of undisturbed field soil with which it is possible to sample and monitor the movement of water and substances.
 1993, *65*, 2065

μ- (mu) (in inorganic nomenclature)
An affix used in names to signify that a group so designated bridges two or more centres of coordination.
  R.B. 245; B.B. 465

M
  See *helicity*.
  1996, *68*, 2211

+ machine (in analysis)
Usage not recommended.
  See *analytical instrument*.
  1989, *61*, 1659

macrocycle
A cyclic *macromolecule* or a macromolecular cyclic portion of a macromolecule.
Notes:
1. See *chain (2)*.
2. In the literature, the term 'macrocycle' is sometimes used for molecules of low molecular mass that would not be considered 'macromolecules' as specified in the definition given in this book.
  1996, *68*, 2298

macrolides
Macrocyclic *lactones* with a ring of twelve or more members.
  1995, *67*, 1348

macrometeorology (in atmospheric chemistry)
Study of the largest-scale aspects of the atmosphere, e.g. general global circulation.
  1990, *62*, 2200

macromolecular
  See *macromolecule*.
  1996, *68*, 2289

macromolecular isomorphism
Statistical co-crystallization of different *constitutional repeating units*, which may either belong to the same copolymer chains (copolymer isomorphism) or originate from different homopolymer chains (homopolymer isomorphism).
Isomorphism is a general term: in the strict sense, the crystal structure is essentially the same throughout the range of compositions; in isodimorphism or isopolymorphism, there are two or more crystal structures, respectively, depending on composition.
  P.B. 80

macromolecule (polymer molecule)
A molecule of high relative molecular mass, the structure of which essentially comprises the multiple repetition of units derived, actually or conceptually, from molecules of low relative molecular mass.
Notes:
1. In many cases, especially for synthetic polymers, a molecule can be regarded as having a high relative molecular mass if the addition or removal of one or a few of the units has a negligible effect on the molecular properties. This statement fails in the case of certain macromolecules for which the properties may be critically dependent on fine details of the molecular structure.
2. If a part or the whole of the molecule has a high relative molecular mass and essentially comprises the multiple repetition of units derived, actually or conceptually, from molecules of low relative molecular mass, it may be described as either *macromolecular* or *polymeric*, or by polymer used adjectivally.
  1996, *68*, 2289

macromonomer
A polymer composed of *macromonomer molecules*.
  1996, *68*, 2304

macromonomeric unit (macromonomer unit)
The largest *constitutional unit* contributed by a single *macromonomer molecule* to the structure of a *macromolecule*.
  1996, *68*, 2291

macromonomer molecule
A *macromolecule* that has one end-group which enables it to act as a *monomer molecule*, contributing only a single *monomeric unit* to a chain of the final macromolecule.
  1996, *68*, 2290

macromonomer unit
  See *macromonomeric unit*.
  1996, *68*, 2291

macropore (in catalysis)
Pore with width exceeding about 0.05 μm or 50 nm (500 Å).
Cf. *mesopore (in catalysis)*, *micropore (in catalysis)*.
  1972, *31*, 585; 1976, *46*, 79

macroradical
A *macromolecule* which is also a free radical.
  1996, *68*, 2290

macroscopic cross-section
The *cross-section* per unit volume of a given material for a specified process. For a pure *nuclide*, it is the product of the *microscopic cross-section* and the number of target nuclei per unit volume; for a mixture of nuclides, it is the sum of such products.
  See also *microscopic cross-section*.
  1994, *66*, 2517; O.B. 218

+ macroscopic diffusion control
This terminology is not recommended.
  See *mixing control*.
  1996, *68*, 171

macroscopic film
A *film* which has lateral dimensions above 100 μm.
  1994, *66*, 1671

macroscopic kinetics
A term applied to a kinetic study in which the interest is in the behaviour of bulk systems, e.g. in changes in concentrations of reactants and products.
  1996, *68*, 172

magic acid
  See *superacid*.
  1994, *66*, 1137

magnetic coherence length (in thin films), ξ_M
See *coherence length*.
1994, *66*, 1675

magnetic deflection (in mass spectrometry)
The deflection of an ion beam as a result of the motion of the ions in a magnetic field (magnetic sector). Generally the direction of motion of the ions is at right angles to the direction of the magnetic field, and the motion is uniform.
1991, *63*, 1544; O.B. 201

magnetic equivalence
Nuclei having the same resonance frequency in nuclear magnetic resonance spectroscopy and also identical spin-spin interactions with the nuclei of a neighbouring group are magnetically equivalent. The spin-spin interaction between magnetically equivalent nuclei does not appear, and thus has no effect on the multiplicity of the respective NMR signals. Magnetically equivalent nuclei are necessarily also chemically equivalent, but the reverse is not necessarily true.
1994, *66*, 1137

magnetic field (B) scan (in mass spectrometry)
The usual method of producing a momentum (mass) spectrum, by varying the strength of the magnetic field.
1991, *63*, 1550

magnetic field strength, H
Magnetic flux density divided by the permeability.
G.B. 14

magnetic flux, Φ
Scalar product of magnetic flux density and area.
G.B. 14

magnetic flux density (in Mössbauer spectroscopy)
Magnetic flux density at the nucleus (from experiment) in those cases in which the magnetic hyperfine interaction can be described by an effective field. In other cases the vector components of the magnetic hyperfine interaction should be reported if possible.
1976, *45*, 214

magnetic flux density, B
Magnetic induction is a vector characterizing a magnetic field. The force acting on a moving charge in a magnetic field is the charge multiplied by the vector product of the velocity and magnetic flux density. This quantity is sometimes loosely called the magnetic field.
See also *flux*.
G.B. 14

magnetic induction
See *magnetic flux density*
G.B. 14

magnetic moment, m, μ
Vector quantity, the vector product of which with the magnetic flux density of a homogeneous field is equal to the torque.
G.B. 21

magnetic susceptibility, χ
Relative permeability minus one.
G.B. 15

magnetic transition
A transition between disordered and ordered arrays of atomic magnetic moments. Where the ordered phase has a net spontaneous magnetization, M_S, the magnetic-ordering temperature is called a Curie temperature, T_C; where the net spontaneous magnetization of the ordered phase remains zero, the ordering temperature is called a Néel temperature, T_N. The temperature at which the two ferromagnetic subarrays of a ferrimagnet just cancel each other is called the compensation point.
Materials exhibit *ferromagnetic* behaviour when unpaired electron spins are aligned in parallel, *antiferromagnetic* behaviour when the alignment is antiparallel, and *ferrimagnetic* behaviour if the alignment of the spins is antiparallel with unequal numbers in the two orientations or if the spins are canted. Therefore, ferromagnetic, ferrimagnetic and weak ferromagnetic transitions involve a net magnetic moment change, whereas antiferromagnetic transitions have zero net magnetic moment change.
Note:
Antiferromagnetic order below T_N may be complex; for example, canting of spins as in $GdFeO_3$, spiral-spin configuration as may occur in MnO_2, and amplitude modulation, as in some rare-earth metals.
1994, *66*, 585

magnetizability, ξ
Tensor quantity relating the induced *magnetic moment*, m_i, to the applied *magnetic flux density*, B, $m_i = \xi B$.
G.B. 21

magnetization transfer
NMR method for determining kinetics of chemical exchange by perturbing the magnetization of nuclei in a particular site or sites and following the rate at which magnetic equilibrium is restored. The most common perturbations are saturation and inversion, and the corresponding techniques are often called 'saturation transfer' and 'selective inversion-recovery'.
See also *saturation transfer*.
1994, *66*, 1137

magnetogyric ratio, γ
Ratio of the *magnetic moment* to the *angular momentum*. It is often misleadingly called the gyromagnetic ratio.
G.B. 21

main chain (backbone) (of a polymer)
That *linear chain* to which all other *chains, long* or *short* or both, may be regarded as being pendant.

Note:
Where two or more chains could equally be considered to be the main chain, that one is selected which leads to the simplest representation of the molecule.
1996, *68*, 2294

mancude-ring systems
Rings having (formally) the maximum number of noncumulative double bonds, e.g. benzene, indene, indole, 4*H*-1,3-dioxine. Also called *mancunide-ring systems*.
1995, *67*, 1348

mancunide-ring systems
See *mancude-ring systems*.
1995, *67*, 1348

manipulator
A hand-operated or -controlled device for remotely handling *radioactive materials*.
1982, *54*, 1546

manual (in analysis)
Refers to physical human intervention in an analytical procedure.
1989, *61*, 1659

mapping (in biotechnology)
The determination of the relative positions of *genes* within the *chromosomes* or of restriction sites along a DNA molecule.
1992, *64*, 159

Marangoni effect
Motions of the surface of a liquid are coupled with those of the subsurface fluid or fluids, so that movements of the liquid normally produce stresses in the surface and vice versa. The movement of the surface and of the entrained fluid(s) caused by surface tension gradients is called the Marangoni effect.
1979, *51*, 1218

Marcus–Coltrin path
A path (*trajectory*) over a *potential-energy surface* which corresponds to the outermost vibrational turning points for the molecular species involved. This path was suggested as a device for calculating the probability of quantum-mechanical *tunnelling*: the trajectory for the system is considered to follow this path.
1996, *68*, 172

Marcus equation
A general expression which correlates the Gibbs energy of activation ($\Delta^\ddagger G$) with the driving force ($\Delta_r G^{o\prime}$) of the reaction:
$$\Delta^\ddagger G = (\lambda/4)(1 + \Delta_r G^{o\prime}/\lambda)^2$$
where λ is the *reorganization energy* and $\Delta_r G^{o\prime}$ is the standard free energy of the reaction corrected for the electrostatic work required to bring the reactants together. $\lambda/4$ is the *intrinsic barrier of the reaction*. Originally developed for *outer-sphere electron transfer* reactions, the Marcus equation has later been applied also to atom and group transfer reactions.
1994, *66*, 1137; see also 1996, *68*, 2153

Marcus–Hush relationship
Relationship between the barrier ($\Delta G^\ddagger$) to thermal *electron transfer*, the energy of a corresponding optical *charge transfer transition* (ΔE_{op}), and the overall change in standard Gibbs energy accompanying thermal electron transfer (ΔG^o). Assuming a quadratic relation between the energy of the system and its distortions from equilibrium (harmonic oscillator model) the expression obtained is:
$$\Delta G^\ddagger = \Delta E_{op}^2/4(\Delta E_{op} - \Delta G^o)$$
The simplest form of this expression obtains for degenerate electron transfer (ΔG^o) in e.g. symmetrical mixed valence systems:
$$\Delta G^\ddagger = \Delta E_{op}/4$$
Note that for this situation the *Marcus equation* reads:
$$\Delta G^\ddagger = \lambda/4$$
1996, *68*, 2253

Marcus inverted region (for electron transfer)
See *inverted region*.
1996, *68*, 2253

marker
1. (DNA) A fragment of known size used as reference for analytical purposes.
2. (genetic) A *gene* with known *phenotype* and mapped position.
3. (chromatography) A reference substance co-chromatographed with the sample to assist in identifying the components.
1992, *64*, 159; 1993, *65*, 837; O.B. 97

Mark–Houwink equation
The equation describing the dependence of the *intrinsic viscosity* of a polymer on its relative molecular mass (molecular weight) and having the form:
$$[\eta] = K \cdot M_r^a$$
where $[\eta]$ is the intrinsic viscosity, K and a are constants the values of which depend on the nature of the polymer and solvent as well as on temperature and M_r is usually one of the relative molecular mass averages.
Notes:
1. The use of this equation with the relative molecular mass (molecular weight) is recommended, rather than with molar mass (which has the dimension of mass divided by amount of substance), since in the latter case the constant K assumes awkward and variable dimensions owing to the fractional and variable nature of the exponent a.
2. Kuhn and Sakurada have also made important contributions and their names are sometimes included, as, for example, in the Kuhn–Mark–Houwink–Sakurada equation.
P.B. 64

Markownikoff rule
'In the addition of hydrogen halides to unsymmetrically constituted [unsaturated] hydrocarbons, the halogen atom becomes attached to the carbon bearing the lesser number of hydrogen atoms.' Originally for-

mulated by Markownikoff (Markovnikov) to generalize the orientation in additions of hydrogen halides to simple alkenes, this rule has been extended to polar *addition reactions* as follows. 'In the *heterolytic* addition of a polar molecule to an alkene or alkyne, the more electronegative (nucleophilic) atom (or part) of the polar molecule becomes attached to the carbon atom bearing the smaller number of hydrogen atoms.' This is an indirect statement of the common mechanistic observation, that the more electropositive (electrophilic) atom (or part) of the polar molecule becomes attached to the end of the multiple bond that would result in the more stable *carbenium ion* (whether or not a carbenium ion is actually formed as a *reaction intermediate* in the addition reaction). Addition in the opposite sense is commonly called 'anti-Markovnikov addition'.
1994, 66, 1137

martensitic transition
A *diffusionless transition* (first studied in the steel alloy, Martensite), at constant composition, generated by coordinated atomic, ionic or molecular displacements over distances smaller than interatomic distances in the parent phase. The cooperative rearrangements of the crystal structure generally take place progressively by the movement of a two-dimensional interface through the solid.
Examples:
The face-centred-cubic to body-centred-tetragonal transition of iron containing some carbon; the transition of tetragonal ZrO_2 to monoclinic ZrO_2.
1994, 66, 585

mass, m
Base quantity in the system of quantities upon which SI is based.
G.B. 12; 1996, 68, 979

mass analysis (in mass spectrometry)
A process by which a mixture of ionic or neutral species is identified according to the mass-to-charge (m/z) ratios (ions) or their aggregate atomic masses (neutrals). The analysis may be qualitative and/or quantitative.
1991, 63, 1545

mass-average velocity (in electrolytes)
Mass-average velocity:
$$v_b = \rho^{-1} \Sigma\, C_i M_i v_i$$
Cf. molar average velocity:
$$v_m = c_t^{-1} \Sigma\, c_i v_i$$
with $c_t = \Sigma c_i$ where M_i = *molar mass*, c_t = total concentration (mol m^{-1}), ρ = density of the solution (kg m^{-3}).
1981, 53, 1831

mass balance (in atmospheric chemistry)
Summation of the masses of a given element in its various compounds before and after reaction (changes) in the atmosphere; provides a test of the completeness of the accounting of the various reaction paths for this element which can be had through the compounds which have been analysed.
1990, 62, 2200

mass concentration, γ, ρ
Mass of a constituent divided by the volume of the mixture.
See *concentration*.
G.B. 42; 1996, 68, 979

mass density
See *density*.
G.B. 12; see also 1996, 68, 980

mass density gradient, grad ρ
Change in mass density in a small distance divided by that distance.
1996, 68, 980

mass-distribution function
A distribution function in which the relative amount of a portion of a substance with a specific value, or a weight distribution range of values, of the random variable(s) is expressed in terms of mass fraction.
P.B. 56

mass distribution ratio
See *extraction factor*.
1993, 65, 2384

mass distribution ratio (in chromatography)
The fraction $(1 - R)$ of a component in the stationary phase divided by the fraction (R) in the mobile phase:
$$D_m = \frac{\text{amount of substance in the stationary phase}}{\text{amount of component in the mobile phase}}$$
This term is recommended in preference to the term capacity factor frequently used in the chromatographic literature.
See also *extraction factor*.
O.B. 107; 1993, 65, 2384

mass excess, Δ
Mass of an atom minus the product of mass number and atomic mass constant.
G.B. 20

mass flow rate, q_m
Mass of a material crossing a surface divided by the time.
1996, 68, 980

mass-flow sensitive detector (in chromatography)
A device the response of which is proportional to the amount of sample component reaching the detector in unit time.
1993, 65, 849

mass fraction, w
Mass of a constituent divided by the total mass of all constituents in the mixture.
G.B. 41; 1996, 68, 980

massic
Attribute to a physical quantity obtained by division by mass; synonymous with *specific*.
ISO 31-0: 1992; 1996, 68, 980

massive transition
A transition in which a crystal structure is changed into a new structure during cooling or heating, predominantly by interface-controlled reactions and the usual *nucleation and growth* characteristics associated with thermal diffusion.
1994, *66*, 585

mass-law effect
At equilibrium, the product of the activities (or concentrations) of the reacting species is constant. Thus for the equilibrium:

$$\alpha A + \beta B \rightleftharpoons \gamma C + \delta D$$

$$K = [C]^\gamma [D]^\delta / [A]^\alpha [B]^\beta$$

See also *common-ion effect, equilibrium*.
1994, *66*, 1138

mass number, A
Total number of heavy particles (protons and neutrons jointly called nucleons) in the atomic nucleus. Also called nucleon number. Symbol m in mass spectrometry.
G.B. 20; R.B. 35; 1996, *68*, 980; 1991, *63*, 1545

mass (weight) of the stationary phase (in chromatography), W_s
The mass (weight) of the liquid stationary phase or the active solid in the *column*. The mass (weight) of any solid support is not included. In the case of *partition chromatography* with a liquid stationary phase it is identical to the liquid phase mass (weight).
1993, *65*, 833

mass peak (in mass spectrometry)
A record of the ion current, at a specific mass to charge ratio, received by the collector.
1990, *62*, 2200

mass range (in mass spectrometry)
The range of mass numbers which can be characterized by a *mass spectrometer* with sufficient resolution to differentiate adjacent peaks.
1990, *62*, 2200

mass resolving power (in mass spectrometry)
Commonly and also acceptably defined in terms of the overlap (or 'valley') between two peaks. Thus for two peaks of equal height, masses m_1 and m_2, when there is overlap between the two peaks to a stated percentage of either peak height (10% is recommended), then the resolving power is defined as $m_1/(m_1 - m_2)$. The percentage overlap (or 'valley') concerned must always be stated.
1978, *50*, 72

mass spectrograph
An instrument in which beams of ions are separated (analysed) according to the quotient mass/charge, and in which the deflection and intensity of the beams are recorded directly on photographic plate or film.
1991, *63*, 1545

mass spectrometer
An instrument in which beams of ions are separated (analysed) according to the quotient mass/charge, and in which the ions are measured electrically. This term should also be used when a scintillation detector is employed.
1991, *63*, 1545; O.B. 201; see also 1990, *62*, 2200

mass spectrometer (operating on the linear accelerator principle)
A mass spectrometer in which the ions to be separated absorb maximum energy through the effect of alternating electric fields which are parallel to the path of the ions. These ions are then separated from other ions with different mass/charge by an additional electric field.
1991, *63*, 1545

mass spectrometer, double focusing
See *double focusing mass spectrometer*.
1991, *63*, 1545

mass spectrometer, dynamic field(s)
See *dynamic field(s) mass spectrometer*.
1991, *63*, 1545

mass spectrometer focusing system (deflection system)
An assembly permitting the separation of ions according to their mass to charge ratio.
1990, *62*, 2200

mass spectrometer, Fourier transform ion cyclotron resonance (FT-IR)
See *Fourier transform ion cyclotron resonance (FT-IR) mass spectrometer*.
1991, *63*, 1545

mass spectrometer, ion cyclotron resonance (ICR)
See *ion cyclotron resonance (ICR) mass spectrometer*.
1991, *63*, 1545

mass spectrometer, ion trap
See *ion trap mass spectrometer*.
1991, *63*, 1545

mass spectrometer, prolate trochoidal
See *prolate trochoidal mass spectrometer*.
1991, *63*, 1545

mass spectrometer, single-focusing
See *single-focusing mass spectrometer*.
1991, *63*, 1546

mass spectrometer, static fields
See *static fields mass spectrometer*.
1991, *63*, 1546

mass spectrometer, tandem
See *tandem mass spectrometer*.
1991, *63*, 1546

mass spectrometer, time-of-flight
See *time-of-flight mass spectrometer*.
1991, *63*, 1546

mass spectrometric detector (in gas chromatography)
A mass spectrometer used as a detector to give qualitative and quantitative data on the various eluted substances. The mass spectrum of the eluted compounds provides evidence beyond the elution time as to the chemical nature of the species.
1990, 62, 2191

mass spectrometry
The branch of science dealing with all aspects of mass spectroscopes and the results obtained with these instruments.
1991, 63, 1546; O.B. 201

+ mass spectroscope
A term (now essentially obsolete) which may refer to either a mass spectrometer or a mass spectrograph.
1991, 63, 1546

mass spectroscopy
The study of systems by causing the formation of gaseous ions, with or without fragmentation, which are then characterized by their mass-to-charge ratios and relative abundances.
1991, 63, 1546

mass spectrum
A spectrum which is obtained when a beam of ions is separated according to the mass-to-charge (m/z) ratios of the ionic species contained within it.
1991, 63, 1550; see also O.B. 204; 1990, 62, 2200

2E mass spectrum
Processes of the partial charge transfer type:
$$M^{2+} + X \rightarrow M^+ + X^+$$
occuring in a collision cell (containing a gas, X) located in a field-free region preceding in a magnetic and electric sector combination placed in either order, can be detected as follows.
If the instrument slits are wide, and if the electric sector field E is set to twice the value required to transmit the main ion beam, the only ions to be transmitted will be those with a kinetic energy/charge ratio twice, or almost exactly twice, that of the main ion beam. The product ions of the process shown fulfil this condition. If the magnetic field B is scanned, the mass spectrum of such singly-charged product ions, and thus of their doubly-charged precursors, is obtained. Such a spectrum is called a 2E mass spectrum.
1991, 63, 1552

$E/2$ mass spectrum
Processes of the charge-stripping type:
$$M^+ + X \rightarrow M^{2+} + X + e$$
occuring in a collision cell (containing a gas, X) located in a field-free region preceding a magnetic and electric sector combination placed in either order, may be detected as follows.
If the instrument slits are wide and if the electric sector field E is set to half the value required to transmit the main ion-beam, the only ions to be transmitted will be those with a kinetic energy/charge ratio half, or almost exactly half, that of the main ion beam. The product ions of the charge-stripping process fulfil this condition. If the magnetic field B is scanned, a mass spectrum of such doubly-charged product ions, and thus of their singly-charged precursors, is obtained. Such a spectrum is called an $E/2$ mass spectrum. Interference from product ions from processes of the type:
$$M_1^+ + X \rightarrow M_2^+ + X + M_3$$
where m_1, m_2, and $(m_1 - m_2)$ are the masses of M_1^+, M_2^+ and M_3, respectively, and where $m_2 = 0.5 m_1$, can arise in $E/2$ mass spectra.
1991, 63, 1552

mass-to-charge ratio (in mass spectrometry), m/z
The abbreviation m/z is used to denote the dimensionless quantity formed by dividing the mass number of an ion by its charge number. It has long been called the mass-to-charge ratio although m is not the ionic mass nor is z a multiple or the elementary (electronic) charge, e. The abbreviation m/e is, therefore, not recommended. Thus, for example, for the ion $C_7H_7^{2+}$, m/z equals 45.5.
1991, 63, 1544

mass transfer (in biotechnology)
Spontaneous (irreversible) process of transfer of mass across non-homogeneous fields. The driving force can be difference in concentration (in liquids) or partial pressure (in gases) of the component. In fluids, mass transfer may be enhanced by turbulent flow. In biological systems, mass transfer through membranes may result from normal diffusion, facilitated diffusion or *active transport*.
1992, 64, 159

mass transfer coefficient (in electrochemistry)
The *heterogeneous diffusion rate constant* defined for the case of the *limiting current*. A more general definition in terms of the *electrode current densities* is:
$$k_d = jv/nF(c_e - c_0)$$
or
$$k_d = jv(1 - t_B n v^{-1} z_B^{-1})/nF(c_e - c_0)$$
where j is the *electrode current density*, v is the *stoichiometric number*, n is the *charge number of the electrode reaction*, F is the *Faraday constant*, c_e is the *interfacial concentration*, c_0 is the *bulk concentration*, t_B is the *transport number* of species B, and z_B is the charge number of species B.
1981, 53, 1839

mass-transfer-controlled electrolyte rate constant
In controlled-potential coulometry and related techniques, the empirically evaluated constant of proportionality defined by the equation
$$s_B = -(1/c_B)(dc_B/dt)$$
where c_B is the *bulk concentration* of the substance B, and dc_B/dt is the rate of change of that concentra-

tion, resulting from the consumption of B by reduction or oxidation at the *working electrode.*
1985, *57*, 1501

matched cells (in spectrochemical analysis)
A pair of cells with closely similar optical properties are called *matched cells.* One cell is the *sample cell* while the other, the *reference (or blank) cell* contains the solvent or a reference solution. In *double beam spectrometers*, radiation is passed either simultaneously or alternately through the cells. In single beam instruments the cells are moved sequentially into the radiation beam.
1988, *60*, 1453

material safety data sheet (MSDS)
Compilation of information required under the US OSHA Hazard Communication Standard on the identity of hazardous substances, health and physical hazards, exposure limits and precautions.
1993, *65*, 2066

matrix (in analysis)
The components of the sample other than the analyte.
1989, *61*, 1660

matrix effect
1. (in analytical chemistry) The combined effect of all components of the sample other than the analyte on the measurement of the quantity.
If a specific component can be identified as causing an effect then this is referred to as interference.
See *matrix.*
1989, *61*, 1660
2. (in surface analysis) Effects which cause changes in Auger-electron, photoelectron, secondary ion yield, or scattered ion intensity, the energy or shape of the signal of an element in any environment as compared to these quantities in a pure element.
(a) Chemical matrix effects: changes in the chemical composition of the solid which affect the signals as described above.
(b) Physical matrix effects: topographical and/or crystalline properties which affect the signal as described above.
1979, *51*, 2247

matrix isolation
A term which refers to the isolation of a reactive or unstable species by dilution in an inert matrix (argon, nitrogen, etc.), usually condensed on a window or in an optical cell at low temperature, to preserve its structure for identification by spectroscopic or other means.
1994, *66*, 1138; 1990, *62*, 2200

Mattauch–Herzog geometry
An arrangement for a double-focusing mass spectrograph in which a deflection of $\pi/(4\sqrt{2})$ radians in a radial electrostatic field is followed by a magnetic deflection of $\pi/2$ radians.
1991, *63*, 1546

maximum allowable concentration (in atmospheric chemistry)
The maximum concentration of a pollutant which is considered harmless to healthy adults during their working hours, assuming they breathe uncontaminated air at all other times.
1990, *62*, 2200; see also 1993, *65*, 2066

maximum emission concentration (in atmospheric chemistry)
Standards for maximum concentration of air pollutant emission from stationary or mobile sources.
1990, *62*, 2200

maximum loading
See *loading capacity.*
1993, *65*, 2384

maximum permissible daily dose
Maximum daily dose of substance whose penetration into a human body during a lifetime will not cause diseases or health hazards that can be detected by current investigation methods and will not adversely affect future generations.
1993, *65*, 2067

maximum permissible level (MPL)
Level, usually a combination of time and concentration, beyond which any exposure of humans to a chemical or physical agent in their immediate environment is unsafe.
See also *maximum allowable concentration.*
1993, *65*, 2067

maximum storage life
Period during which there is no change in concentration in excess of the value of the uncertainty of the concentration.
1990, *62*, 2201

maximum tolerable concentration (MTC)
Highest concentration of a substance in an environmental medium that does not cause death of test organisms or species (denoted by LC_0).
1993, *65*, 2067

maximum tolerable dose (MTD)
Highest amount of a substance that, when introduced into the body, does not kill test animals (denoted by LD_0).
1993, *65*, 2067

maximum tolerable exposure level (MTEL)
Maximum amount or concentration of a substance to which an organism can be exposed without leading to an adverse effect after prolonged exposure time.
1993, *65*, 2068

maximum tolerated dose (MTD)
High dose used in chronic toxicity testing that is expected on the basis of an adequate subchronic study to produce limited toxicity when administered for the duration of the test period. It should not induce:
(a) overt toxicity, for example appreciable death of cells or organ dysfunction, or (b) toxic manifestations that are predicted materially to reduce the life span of the animals except as the result of neoplastic devel-

opment or (c) 10% or greater retardation of body weight gain as compared with control animals. In some studies, toxicity that could interfere with a carcinogenic effect is specifically excluded from consideration.
1993, *65*, 2068

McLafferty rearrangement (in mass spectrometry)
β-Cleavage with concomitant specific transfer of a γ-hydrogen atom in a six-membered transition state in mono-unsaturated systems, irrespective of whether the rearrangement is formulated by a radical or an ionic mechanism, and irrespective of the position of the charge.
O.B. 207

mean (average), $\bar{x}$
The sum of a series of observations divided by the number of observations in the series.
O.B. 4

mean activity of an electrolyte in solution
Defined by the equation:
$$a_{\pm} = \exp[(\mu_B - \mu_B^{\ominus})\nu RT]$$
where μ_B is the *chemical potential* of the solute B in a solution containing B and other species. The nature of B must be clearly stated: it is taken as a group of ions of two kinds carrying an equal number of positive and negative charges, e.g. $Na^+ + NO_3^-$ or $Ba^{2+} + 2Cl^-$ or $2Al^{3+} + 3SO_4^{2-}$. ν is the total number of ions making up the group i.e. 2, 3 and 5 respectively in the above examples. $\mu_B^{\ominus}$ is the chemical potential of B in its standard state, usually the hypothetical ideal solution of concentration 1 mol kg^{-1} and at the same temperature and pressure as the solution under consideration.
See also *activity*.
1974, *37*, 510

mean catalytic activity rate, $\Delta z/\Delta t$
The catalytic activity of a specified component in or moved from a system divided by the time during which the component was changed or moved. This kind of quantity should not be confused with rate of reaction.
1979, *51*, 2475

mean current density
Defined as
$$j = I/A$$
where I is the *electric current* and A is usually taken as the geometric area of the electrode. However the nature of the area used in this calculation must be clearly stated. In interpreting the mean current density it is important to know whether the current is uniformly distributed over the electrode interface.
1974, *37*, 513

mean exchange current density
Defined by
$$j_0 = I_0/A$$
where I_0 is the *exchange current of the electrode reaction* and A is usually taken as the geometric area of the electrode.
1974, *37*, 516

mean free path, λ
The average distance a molecule travels between collisions. For a molecule, $\lambda = \{(2)^{1/2}\pi n d_m^2\}^{-1}$, where n is the number of molecules per unit volume and d_m is their mean diameter. For O_2 at one atmosphere and 25 °C, this distance is only 9.7×10^{-6} cm; at 10^{-6} atmospheres and 25 °C it is 9.7 cm. For an aerosol particle, the mean free path, λ_B in the Stokes region (see Stokes law) is given by: $\lambda_B = (3kT/m)^{1/2}mB$ where m is the mass of the particle, k is the Boltzmann constant (1.381×10^{-23} J K^{-1}), T is the temperature (K) and B is the mobility.
1990, *62*, 2201; G.B. 56

mean interstitial velocity of the carrier gas (in chromatography)
The *interstitial velocity*, u of the *carrier gas* multiplied by the *pressure-gradient correction-factor, j*:
$$\bar{u} = uj$$
$$\bar{u} = Fj/\varepsilon_I$$
where F is the *nominal linear flow*, and ε_I is the *interstitial fraction*.
O.B. 103

mean life, τ
See *lifetime*.
1994, *66*, 1138

mean lifetime, τ
See *lifetime*.
1994, *66*, 1138

mean linear range (in nuclear chemistry)
In a given material, for specified charged particles of a specified energy, the average displacement of the particles before they stop.
1982, *54*, 1550

mean mass range (in nuclear chemistry)
The *mean linear range* multiplied by the mass density of the material.
1982, *54*, 1550

mean mass rate, $\Delta m/\Delta t$
The mass of a specified component changed in or moved to or from a system divided by the time during which the component was changed or moved.
1979, *51*, 2474

mean residence time (of adsorbed molecules)
The mean time during which the molecules remain on the surface of the *adsorbent*, i.e. the mean time interval between impact and *desorption*.
1976, *46*, 79

mean substance rate, $\Delta n/\Delta t$
The amount of substance of a specified component changed in or moved to or from a system divided by the time during which the component was changed or

moved. This kind of quantity usually should be preferred to mean mass rate or mean volume rate.
1979, *51*, 2475

mean volume rate, $\Delta V/\Delta t$
The volume of a specified component changed in or moved to or from a system divided by the time during which the component was changed or moved.
See also *clearance*.
1979, *51*, 2474; 1996, *68*, 980

measurable quantity
An attribute of a substance which may be distinguished qualitatively and determined quantitatively. In the context of analytical chemistry, the attribute may refer to a physical quantity such as X- or γ-ray energy, or it may refer to a measure of amount such as mass or concentration.
VIM; 1995, *67*, 1701

measurand
Particular quantity subject to measurement.
VIM; 1996, *68*, 980

measured excitation spectrum
The spectrum observed by measuring the variation of the *luminescence* flux from an analyte as a function of the excitation wavelength.
1984, *56*, 242

measured quantum yield (of luminescence)
See *quantum yield (of luminescence)*.
1984, *56*, 243

measured spectrum
Synonymous with *spectrogram*.
1982, *54*, 1546

measured value (in analysis)
The observed value of weight, volume, meter-reading or other quantity, found in the analysis of a material.
1994, *66*, 597

measurement
A description of a property of a system by means of a set of specified rules, that maps the property onto a scale of specified values, by direct or 'mathematical' comparison with specified reference(s).

The demand for rules makes 'measurement' a scientific concept in contrast to the mere colloquial sense of 'description'. However, in the present definition, 'measurement' has a wider meaning than given in elementary physics. Even a very incomplete description of, for instance, a patient (at a stated time) has to be given by a set of measurements, that are easier to manage and grasp.
1995, *67*, 1565

measurement resolution (in atmospheric trace component analysis)
The minimum value above which the difference of two values of air quality characteristic can be distinguished with 95% probability.
1990, *62*, 2201

measurement result
The outcome of an analytical measurement (application of the *chemical measurement process*), or value attributed to a *measurand*. This may be the result of direct observation, but more commonly it is given as a statistical estimate derived from a set of observations. The distribution of such estimates (estimator distribution) characterizes the chemical measurement process, in contrast to a particular estimate, which constitutes an experimental result.

Additional characteristics become evident if we represent $\hat{x}$ as follows:

$$\hat{x} = \tau + e = \tau + \left[\begin{array}{c} e \\ \Delta \\ \mu \end{array} \right] + \delta = \mu + \delta$$

The *true value*, τ, is the value x that would result if the *chemical measurement process* were error-free.

The *error*, e, is the difference between an observed (estimated) value and the true value; i.e. $e = \hat{x} - \tau$ (signed quantity). The total error generally has two components, *bias* (Δ) and *random error* (δ), as indicated above.

The *limiting mean*, μ, is the asymptotic value or population mean of the distribution that characterizes the measured quantity; the value that is approached as the number of observations approaches infinity. Modern statistical terminology labels this quantity the *expectation value* or *expected value*, $E(\hat{x})$.

The *bias*, Δ, is the difference between the limiting mean and the true value; i.e. $\Delta = \mu - \tau$ (signed quantity).

The *random error*, δ, is the difference between an observed value and the limiting mean; i.e. $\delta = \hat{x} - \mu$ (signed quantity).
1995, *67*, 1705

measurement solution (in analysis)
The solution that is presented to the measuring device. It comprises the test solution, or an aliquot thereof, after it has undergone any treatment required prior to the presentation to the measuring device.
1989, *61*, 1660

measurement threshold (of an analyser)
The minimum concentration of a substance which produces a quantifiable signal with a given uncertainty.
1990, *62*, 2174

measurement, upper limit of (in atmospheric trace component analysis)
See *upper limit of measurement (in atmospheric trace component analysis)*.
1990, *62*, 2201

mechanical entrapment
1. The process of random incorporation of comparatively small quantities of other phases (e.g. water, dust, particles, etc.) in the bulk of a precipitate during its formation.

2. The deliberate capture of small quantities of such phases by the deliberate addition of solids to a liquid phase. (The term inclusion is not recommended).
O.B. 85

mechanical hygrometer
An apparatus containing an element (hair, goldbeater's skin, carbon-film) whose dimension or mass varies as a function of its water content.
1990, *62*, 2195

mechanical inhibitor (in catalysis)
See *fouling agent (in catalysis)*.
1991, *63*, 1244

mechanization (in analysis)
The use of devices to replace, refine, extend or supplement human effort. The corresponding verb is to mechanize.
1989, *61*, 1659

mechanism (of a reaction)
A detailed description of the process leading from the reactants to the products of a reaction, including a characterization as complete as possible of the composition, structure, energy and other properties of *reaction intermediates*, products and *transition states*. An acceptable mechanism of a specified reaction (and there may be a number of such alternative mechanisms not excluded by the evidence) must be consistent with the reaction stoichiometry, the *rate law* and with all other available experimental data, such as the stereochemical course of the reaction. Inferences concerning the electronic motions which dynamically interconvert successive species along the *reaction path* (as represented by curved arrows, for example) are often included in the description of a mechanism.
It should be noted that for many reactions all this information is not available and the suggested mechanism is based on incomplete experimental data. It is not appropriate to use the term mechanism to describe a statement of the probable sequence in a set of stepwise reactions. That should be referred to as a reaction sequence, and not a mechanism.
See also *Gibbs energy diagram*.
1994, *66*, 1138

mechanism-based inhibition
Irreversible *inhibition* of an enzyme due to its catalysis of the reaction of an artificial substrate. Also called 'suicide inhibition'.
1994, *66*, 1138

median
Depending on whether the number of observations is even or odd, the median can be estimated as follows:
(a) If $n = 2m + 1$: The middle value of a series of observations, arranged in increasing or decreasing order.
(b) If $n = 2m$: The arithmetic mean of the two middle values of a series of observations, arranged in increasing or decreasing order.

Comment:
The use of the median when reporting results of chemical analysis is generally not recommended, because its statistical efficiency is less than that of the mean. In certain cases, however, especially when treating small sets of data, the median may offer advantages because it is a so-called 'robust statistic', i.e. it offers considerable resistance to the effects of isolated outliers.
1994, *66*, 603

median effective concentration (EC_{50})
Statistically derived concentration of a substance in an environmental medium expected to produce a certain effect in 50% of test organisms in a given population under a defined set of conditions.
1993, *65*, 2068

median effective dose (ED_{50})
Statistically derived dose of a chemical or physical agent (radiation) expected to produce a certain effect in 50% of test organisms in a given population or to produce a half-maximal effect in a biological system under a defined set of conditions.
1993, *65*, 2068

median lethal concentration (LC_{50})
Statistically derived concentration of a substance in an environmental medium expected to kill 50% of organisms in a given population under a defined set of conditions.
1993, *65*, 2068

median lethal dose (LD_{50})
Statistically derived dose of a chemical or physical agent (radiation) expected to kill 50% of organisms in a given population under a defined set of conditions.
1993, *65*, 2068

median lethal time (TL_{50})
Statistically derived average time interval during which 50% of a given population may be expected to die following acute administration of a chemical or physical agent (radiation) at a given concentration under a defined set of conditions.
1993, *65*, 2068

median narcotic concentration (NC_{50})
Statistically derived concentration of a substance in an environmental medium expected to cause narcotic conditions in 50% of a given population under a defined set of conditions.
1993, *65*, 2068

median narcotic dose (ND_{50})
Statistically derived dose of a substance expected to cause narcosis in 50% of test animals under a defined set of conditions.
1993, *65*, 2068

medium
The phase (and composition of the phase) in which *chemical species* and their reactions are studied in a particular investigation.
1994, *66*, 1138

medium effect
The medium effect on ionic species B due to transfer from solvent S_1 to solvent S_2 (number) is defined by
$$RT\ln \gamma_{S_1}^{S_2}(B) = \mu_B^{o, S_2} - \mu_B^{o, S_1}$$
where R is the gas constant, T is the thermodynamic temperature and μ_B^{o, S_i} is the standard *chemical potential* of B in solvent S_i (where $i = 1$ or 2), the reference states being the same in both solvents. $\gamma_{S_1}^{S_2}(B)$ is not an exactly measurable quantity.
1974, *37*, 508

medium-pressure mercury lamp (arc)
Radiation source containing mercury vapour at pressures ranging from 100 to several hundred kPa. Emits mostly from 310 to 1000 nm with the most intense lines at 300, 303, 313, 334, 366, 405, 436, 546, and 578 nm.
See *lamp*.
1996, *68*, 2254

mega
SI prefix for 10^6 (symbol: M).
G.B. 74

meiosis
The reductive *cell* division which results in daughter cells containing one copy of each of the *chromosomes* of the parent. The entire meiotic process involves two separate divisions (meiosis I and meiosis II). The first division is a true reductive division with the chromosome number being halved, whereas the second division resembles mitosis in many ways. Thus, a diploid parental cell will give rise to haploid daughter cells (cf. *ploidy*).
1992, *64*, 159

Meisenheimer adduct
A cyclohexadienyl derivative formed as *Lewis adduct* from a *nucleophile (Lewis base)* and an *aromatic* or heteroaromatic compound, also called Jackson–Meisenheimer adduct. In earlier usage the term 'Meisenheimer complex' was restricted to the typical Meisenheimer alkoxide *adducts* of nitro-substituted aromatic ethers, e.g.

Analogous cationic adducts, such as:

considered to be *reaction intermediates* in *electrophilic* aromatic *substitution reactions*, are called 'Wheland intermediates', and sometimes, inappropriately, σ-complexes.
See also σ-adduct.
1994, *66*, 1138; see also 1995, *67*, 1348

Meker burner
See *Bunsen burner*.
O.B. 166

melting
The conversion of a solid to a liquid by the application of heat and/or pressure.
Note:
For a liquid crystal, melting occurs when the three-dimensional periodic structure of a solid collapses to give a liquid-crystal phase in which the molecules have orientational order, but not long range translational order.
1994, *66*, 585

melting point (corrected/uncorrected)
The term originally signified that a correction was made (not made) for the emergent stem of the thermometer. In current usage it often means that the accuracy of the thermometer was (was not) verified. This current usage is inappropriate and should be abandoned.
1994, *66*, 1139

membrane (in an ion-selective electrode)
A general term which refers to a continuous layer, usually consisting of a semi-permeable material, with controlled permeability covering a structure, such as carbon or an inert metal, or separating two electrolyte solutions. This latter case is the most general form of an *ion-selective electrode*. The membrane separates the internal components of the ion-selective electrode from the test solution. The membrane of an ion-selective electrode is responsible for the EMF response and selectivity of the entire electrode.
Comment:
Membranes of sensor electrodes are thought to be practically homogeneous, but an actual membrane may contain inhomogeneous regions, often at surfaces, and connected with materials and preparation methods used. Inhomogeneous regions include low dielectric polymer regions with few charge sites and regions with high local site densities. Surface regions of plasticized liquid membranes often are low in sites and high in plasticizer or exuded impurities.
1994, *66*, 2531

membrane emf
The potential difference E_m between two saturated KCl bridges inserted into two solutions separated by a membrane. The solutions need not be in equilibrium with one another and need not contain any *colloidal* material.
1972, *31*, 619

membrane potential
Synonymous with *membrane emf*.
1972, *31*, 619

membrane sites (in an ion-selective electrode)
Membranes frequently contain built-in 'fixed' charged sites (e.g. immobile $-SO_3^-$ in polystyrenesulfonate), or intentionally added, hydrophobically trapped, 'mobile' sites [e.g. tetraphenylborate in plas-

ticized poly(vinyl chloride)]. Such membranes with charged sites are named sited membranes. Ions of opposite sign in the membrane are '*counter-ions*'. Ions of the same sign as sites are not present in significant quantities, and are *co-ions*. Sited membranes are 'permselective' to counter-ions i.e. only counterions exchange into the membrane and therefore have some mobility in the membrane bulk.
   1994, *66*, 2532

memory effect (in atmospheric chemistry)
In instruments used for atmospheric trace component analysis, the dependence of an instrument reading on one or several previous sample(s).
   1990, *62*, 2201

***mer-* (in inorganic nomenclature)**
An affix used in names to denote meridional; three groups occupying vertices of an octahedral coordination sphere in such a relationship that one is *cis* to the two others which are themselves *trans*, not now recommended for precise nomenclature purposes.
   R.B. 245

mer (in polymer chemistry)
   See *monomeric unit*.
   1996, *68*, 2290

+ mercaptals
A term once used for dithioacetals derived from *aldehydes*; its use is discouraged.
   See *thioacetals*.
   1995, *67*, 1348

+ mercaptans
A traditional term abandoned by IUPAC, synonymous with thiols. This term is still widely used.
   1995, *67*, 1348

+ mercaptides
An obsolescent synonym for thiolates.
   1995, *67*, 1348

+ mercaptoles
An archaic term for dithioketals.
   See *thioacetals*.
   1995, *67*, 1348

mercury flow system (in spectrochemical analysis)
For mercury released directly as atomic vapour, different direct transfer systems are used. In an *open dynamic system* the liberated analyte is transported by a carrier gas through the sampling or excitation source and swept away.
In an *open static system* the equilibrated gaseous phase is forced into the absorption cell by displacement, e.g. by water. During measurement the gaseous phase is thus static.
In a *closed system* the analyte and carrier gas are circulated through the absorption cell and the generator vessel until equilibrium between the liquid and gaseous phases is established.
   1992, *64*, 263

mercury–xenon lamp (arc)
An intense source of ultraviolet, visible, and near infrared radiation produced by an electrical discharge in a mixture of mercury vapour and xenon under high pressure.
   See *lamp*.
   1996, *68*, 2254

mero
A prefix meaning part, partial or fragment, used in formation of compound words, e.g. merocyanines are compounds related to cyanines but have a nitrogen atom at only one end of the chromogenic system instead of at both ends.

$$R-N-C(=CHCH)_n^{[X]_n}=C-C=O^{[X']_n}$$

   1995, *67*, 1348

merry-go-round reactor (turntable reactor)
An apparatus in which several samples are rotated around a radiation source in order to expose each to equal amounts of radiation.
   1996, *68*, 2254

meso
A prefix to indicate the presence of a symmetry element of the second kind (see *chirality*) in a chemical species which is usually one of a set of *diastereoisomers* that also includes a chiral member.
   See meso-*compound*.
   1996, *68*, 2211

meso-**compound**
A term for the achiral member(s) of a set of *diastereoisomers* which also includes one or more chiral members. For example:

   1996, *68*, 2211

mesogenic monomer
A *monomer* which can impart the properties of liquid crystals to the *polymers* formed by its *polymerization*.
   1996, *68*, 2304

mesogenic pitch
A *pitch* with a complex mixture of numerous essentially aromatic hydrocarbons. It does not contain anisotropic particles detectable by optical microscopy. Mesogenic pitch is low in quinoline-insoluble fractions and capable of transforming into *mesophase pitch* during continuous heat treatment above 750 K

by the formation of optically detectable *carbonaceous mesophase*.
1995, *67*, 496

mesoionic compounds
Dipolar five- (possibly six-) membered heterocyclic compounds in which both the negative and the positive charge are delocalized, for which a totally covalent structure cannot be written, and which cannot be represented satisfactorily by any one polar structure. The formal positive charge is associated with the ring atoms, and the formal negative charge is associated with ring atoms or an exocyclic nitrogen or chalcogen atom. Mesoionic compounds are a subclass of *betaines*.
See also *munchnones, sydnones*.
1995, *67*, 1349

mesolytic cleavage
Cleavage of a bond in a *radical ion* whereby a *radical* and an ion are formed. The term reflects the mechanistic duality of the process, which can be viewed as homolytic or heterolytic depending on how the electrons are attributed to the fragments.
1994, *66*, 1139

mesomeric effect
The effect (on reaction rates, ionization equilibria, etc.) attributed to a substituent due to overlap of its p- or π-orbitals with the p- or π-orbitals of the rest of the *molecular entity*. *Delocalization* is thereby introduced or extended, and electronic charge may flow to or from the substituent. The effect is symbolized by M. Strictly understood, the mesomeric effect operates in the ground electronic state of the molecule. When the molecule undergoes electronic excitation or its energy is increased on the way to the *transition state* of a *chemical reaction*, the mesomeric effect may be enhanced by the *electromeric effect*, but this term is not much used, and the mesomeric and electromeric effects tend to be subsumed in the term *resonance effect* of a *substituent*.
See also *electronic effect, field effect, inductive effect*.
1994, *66*, 1139

mesomerism
Essentially synonymous with *resonance*. The term is particularly associated with the picture of π-electrons as less localized in an actual molecule than in a *Lewis formula*. The term is intended to imply that the correct representation of a structure is intermediate between two or more Lewis formulae.
See also *aromatic (2), delocalization*.
1994, *66*, 1139

mesomorphic phase
Often formed by concentrated systems of *surfactants*. These are states of matter in which anisometric molecules (or particles) are regularly arranged in one (nematic state) or two (smectic state) directions, but randomly arranged in the remaining direction(s).
1972, *31*, 613

mesomorphic transition
A transition that occurs between a fully-ordered crystalline solid and an isotropic liquid. *Mesomorphic transitions* can occur (i) from a crystal to a liquid crystal (ii) from a liquid crystal to another liquid crystal and (iii) from a liquid crystal to an isotropic liquid.
See *liquid-crystal transitions*.
1994, *66*, 586

mesopause (in atmospheric chemistry)
That region of the atmosphere between the *mesosphere* and the *thermosphere* at which the temperature is a minimum.
1990, *62*, 2201

mesophase
The phase of a liquid crystalline compound between the crystalline and the isotropic liquid phase.
1994, *66*, 1139

mesophase pitch
A *pitch* with a complex mixture of numerous essentially aromatic hydrocarbons containing anisotropic liquid-crystalline particles (*carbonaceous mesophase*) detectable by optical microscopy and capable of coalescence into the *bulk mesophase*.
Notes:
The carbonaceous mesophase particles are formed from the aromatics of high molecular mass in *mesogenic pitch*, which have not yet been aggregated to particles detectable by optical microscopy within the apparently isotropic *pitch* matrix. The carbonaceous mesophase is insoluble in quinoline and pyridine, but the amount of mesophase measured from microscopical observation appears somewhat higher because parts of the carbonaceous mesophase can be extracted by the solvents.
1995, *67*, 496

mesophase pitch-based carbon fibres
Mesophase pitch-based carbon fibres (MPP-based carbon fibres) are *carbon fibres* obtained from *mesogenic pitch* after it has been transformed into *mesophase pitch* (MPP) at least during the process of spinning, and after the spun mesophase pitch fibres have been made non-fusible (stabilized) and carbonized.
1995, *67*, 496

mesophiles
Microorganisms which grow at moderate temperatures in the range 20–45 °C, and which have an optimum growth temperature in the range 30–39 °C.
1992, *64*, 160

mesopore (in catalysis)
Pore of intermediate size. [Those with widths exceeding about 0.05 µm or 50 nm (500 Å) are called macropores; those with widths not exceeding about 2 nm (20 Å) are called micropores].
1972, *31*, 585; 1976, *46*, 79

mesoscale
In meteorology, the size or scale of phenomena smaller than ordinary cyclones or weather systems but

larger than such microscale phenomena as the thickness of the boundary layer, the wakes of objects, etc. Thunderstorms involve mesoscale processes, and other meteorological events the size of cities are usually mesoscale processes.
1990, *62*, 2201

mesosphere
That region of the atmosphere which lies above the *stratopause* (about 4752 km) and below the *mesopause* (about 8090 km) and in which temperature decreases with increasing height; this is the region in which the lowest temperatures of the atmosphere occur.
1990, *62*, 2201

meso structures (in polymers)
Relative configurations of consecutive, but not necessarily contiguous, constitutionally equivalent carbon atoms that have a symmetrically-constituted connecting group (if any) are designated 'meso' or 'racemo', as appropriate.
Examples:

```
    a           a
    |           |
 — C —〰〰— C —    abbreviation m
    |           |
    b           b
          meso

    a           b
    |           |
 — C —〰〰— C —    abbreviation r
    |           |
    b           a
         racemo
```

(The symbol —〰— represents a symmetrically-constituted connecting group, such as $-CH_2-$, $-CH_2-CH_2-$, or $-CR_2-CH_2-CR_2-$.)
P.B. 37

messenger RNA (mRNA)
An RNA molecule that transfers the coding information for protein synthesis from the *chromosomes* to the *ribosomes*. mRNA is formed from a DNA template by *transcription*. It may be a copy of a single *gene* or of several adjacent genes (polycistronic mRNA). On the ribosome, the sequence is converted into the programmed amino acid sequence through *translation*.
1992, *64*, 160

metabolism
The entire physical and chemical processes involved in the maintenance and reproduction of life in which nutrients are broken down to generate energy and to give simpler molecules (*catabolism*) which by themselves may be used to form more complex molecules (*anabolism*). In case of *heterotrophic organisms*, the energy evolving from catabolic processes is made available for use by the organism.
1992, *64*, 160

metabolite
Any intermediate or product resulting from *metabolism*.
1992, *64*, 160

metal–carbene complexes
Metal complexes of the type R_2CML_n (M = metal, L = ligand) in which formally a carbene is coordinated to a metal, e.g.

$$(CO)_5W=C\begin{matrix}OR\\Ph\end{matrix}$$

1995, *67*, 1349

metal–carbyne complexes
Metal complexes of the type $RCML_n$ (M = metal, L = ligand) in which formally a carbyne is coordinated to a metal, e.g.

$$\begin{matrix}&L&\\OC&|&\\&M\equiv CPh\\Cl&|&\\&L&\end{matrix}$$

1995, *67*, 1349

metal distribution
See *current efficiency*.
1981, *53*, 1837

metal–insulator transition
A transition characterized by a sudden change in electrical transport properties (conductivity) due to a reversible change from localized to itinerant behaviour of the electrons.
Example:
The transition at 339 K in VO_2 where it changes from a high-temperature metallic behaviour to a low-temperature semiconductor behaviour.
Synonymous with metal–nonmetal transition.
See *localized-itinerant transition*.
1994, *66*, 586

metallacycloalkanes
Monocyclic compounds containing a metal atom and saturated carbon atoms as ring members, e.g.:

L is a ligand, e.g. triphenylphosphane

1995, *67*, 1349

metallocenes
Organometallic coordination compounds in which one atom of a transition metal such as iron, ruthenium or osmium is bonded to and only to the face of two cyclopentadienyl [η^5-(C_5H_5)] ligands which lie in parallel planes. The term should not be used for analogues having rings other than cyclopentadienyl as ligands.

1995, *67*, 1349

metallurgical coke
Metallurgical coke is produced by *carbonization* of coals or coal blends at temperatures up to 1400 K to

produce a macroporous *carbon material* of high strength and relatively large lump size.
Notes:
Metallurgical cokes must have a high strength to support heavy loads in the blast furnace without disintegration. Metallurgical coke is also used as *filler coke* for *polygranular carbon* products.
1995, *67*, 497

metal–nonmetal transition
See *metal–insulator transition*.
1994, *66*, 586

metal to ligand charge transfer (MLCT) transition
An electronic transition of a metal complex that corresponds to excitation populating an electronic state in which considerable *electron transfer* from the metal to a ligand has occurred.
See also *ligand to metal charge transfer transition*.
1996, *68*, 2254

metal to metal charge transfer (MMCT) transition
An electronic transition of a bi- or poly-nuclear metal complex that corresponds to excitation populating an electronic state in which considerable *electron transfer* between two metal centres has occurred.
See also *intervalence charge transfer*.
1996, *68*, 2254

metamagnetic transition
An *antiferromagnetic* to *ferromagnetic transition* for $H_{applied} > H_c$ and/or $T > T_t$ where H_c and T_t refer to the critical magnetic field and transition temperature, respectively.
1994, *66*, 586

metastability (of a phase)
A term that describes the state of a phase in which an energy barrier considerably higher than kT must be surmounted before this phase can transform to a phase of lower molar *Gibbs energy* and molar *Helmholtz energy*, where k is the *Boltzmann constant* and T the *thermodynamic temperature*.
Note:
In a thermodynamic sense, the equilibrium state is the state with the lowest molar Gibbs energy; a metastable state corresponds to a relative minimum in the molar Gibbs energy.
1994, *66*, 586

metastable
See *metastable state (in nuclear chemistry)*, *metastable state (in spectrochemistry)*, *metastable ion (in mass spectrometry)*, *transient (chemical) species*.
1972, *31*, 609; 1982, *54*, 1546; 1985, *57*, 1463; 1994, *66*, 1139; 1991, *63*, 1549

metastable film
See *stable film*.
1972, *31*, 614

metastable ion (in mass spectrometry)
An *ion* which is formed with sufficient excitation to dissociate spontaneously during its flight from the ion source to the detector.
1991, *63*, 1549

metastable state (in nuclear chemistry)
An *isomeric state* in an energy state higher than the *ground state*.
1982, *54*, 1546

metastable state (in spectrochemistry)
Any *excited state* which in principle, by virtue of the selection rules, cannot radiatively combine with any lower state. It usually has a considerably longer lifetime than an ordinary excited state.
1985, *57*, 1463

metatectoid reaction
See *peritectoid reaction*.
1994, *66*, 586

metathesis
A *bimolecular* process formally involving the exchange of a bond (or bonds) between similar interacting *chemical species* so that the bonding affiliations in the products are identical (or closely similar) to those in the reactants. For example:

$$\begin{array}{c} RCH=CHR \\ + \\ R'CH=CHR' \end{array} \longrightarrow \begin{array}{c} RCH \\ \| \\ R'CH \end{array} + \begin{array}{c} CHR \\ \| \\ CHR' \end{array}$$

(The term has its origin in inorganic chemistry with a different meaning, but this older usage is not applicable in physical organic chemistry.)
1994, *66*, 1139

methanogens
Strictly anaerobic *archaebacteria*, able to use only a very limited *substrate* spectrum (e.g. molecular hydrogen, formate, methanol, methylamine, carbon monoxide or acetate) as electron donors for the reduction of carbon dioxide to methane.
1992, *64*, 160

method of isotopic perturbation
NMR shift difference measurement of the *isotope effect* on a fast (degenerate) equilibrium between two species which are equivalent except for isotopic substitution. This can be used to distinguish a rapidly equilibrating mixture with time-averaged symmetry from a single structure with higher symmetry.
1994, *66*, 1132

methylene
See *carbene*.
1994, *66*, 1139

methylidyne
See *carbyne*.
1994, *66*, 1140

methylotrophic microorganisms
Organisms that use, as *substrates* for growth, C_1 compounds containing carbon in a lower oxidation state than in carbon dioxide.
1992, *64*, 160

metre
SI base unit for length (symbol: m). The metre is the length of path travelled by light in vacuum during a time interval of 1/299 792 458 of a second.
G.B. 70; 1996, *68*, 981

micellar catalysis
The acceleration of a *chemical reaction* in solution by the addition of a surfactant at a concentration higher than its *critical micelle concentration* so that the reaction can proceed in the environment of surfactant aggregates (*micelles*). (Rate enhancements may be due, for example, to higher concentration of the reactants in that environment, more favourable orientation and solvation of the species, or enhanced rate constants in the micellar pseudophase of the surfactant aggregate.) Micelle formation can also lead to a decreased reaction rate.
   See also *catalyst*.
1994, *66*, 1140

micellar mass
The term usually refers to a neutral *micelle* and therefore includes an equivalent amount of *counter-ions* with the *surfactant* ions.
1972, *31*, 612

micellar solubilization
In a system formed by a solvent, an association *colloid* and at least one other component (the solubilizate), the incorporation of this other component into or on the *micelles* is called micellar solubilization, or, briefly, solubilization.
1972, *31*, 612

micellar weight
   See *relative micellar mass*.
1972, *31*, 612

micelle
Surfactants in solution are often association colloids, that is, they tend to form aggregates of colloidal dimensions, which exist in equilibrium with the molecules or ions from which they are formed. Such aggregates are termed micelles.
   See also *inverted micelle*.
1994, *66*, 1140

micelle charge
Usually understood to include the net charge of the *surfactant* ions and the counterions bound to the *micelle*.
1972, *31*, 612

Michaelis constant, K_m
Substance concentration of *substrate* at which the rate of reaction is equal to one half of the limiting rate (maximum rate).
Notes
1. Also called the Michaelis concentration
2. The Michaelis constant (Michaelis concentration) may be used only when *Michaelis–Menten kinetics* is obeyed.
1996, *68*, 981

Michaelis–Menten kinetics
The dependence of an initial *rate of reaction* upon the concentration of a *substrate* S that is present in large excess over the concentration of an enzyme or other *catalyst* (or reagent) E with the appearance of saturation behaviour following the Michaelis–Menten equation:

$$v = V[S]/(K_m + [S]),$$

where v is the observed initial rate, V is its limiting value at substrate saturation (i.e. $[S] \gg K_m$), and K_m the substrate concentration when $v = V/2$. The definition is experimental, i.e. it applies to any reaction that follows an equation of this general form. The symbols V_{max} or v_{max} are sometimes used for V.
The parameters V and K_m (the 'Michaelis constant') of the equation can be evaluated from the slope and intercept of a linear plot of v^{-1} vs. $[S]^{-1}$ (a 'Lineweaver–Burk plot') or from slope and intercept of a linear plot of v vs. $v/[S]$ ('Eadie–Hofstee plot').
A Michaelis–Menten equation is also applicable to the condition where E is present in large excess, in which case the concentration [E] appears in the equation instead of [S].
The term has sometimes been used to describe reactions that proceed according to the scheme:

$$E + S \underset{k_{-1}}{\overset{k_1}{\rightleftharpoons}} ES \xrightarrow{k_{cat}} \text{Products}$$

in which case $K_m = (k_{-1} + k_{cat})/k_1$ (Briggs–Haldane conditions). It has more usually been applied only to the special case in which $k_{-1} \gg k_{cat}$ and $K_m = k_{-1}/k_1 = K_s$; in this case K_m is a true dissociation constant (Michaelis–Menten conditions).
   See also *rate-determining step*.
1994, *66*, 1140; 1996, *68*, 172

Michaelis–Menten mechanism
The Michaelis–Menten mechanism is the simplest mechanism that will explain *Michaelis–Menten kinetics*. According to the mechanism, a substrate A first combines with a molecule of enzyme E, and this process is followed by a step in which the enzyme-substrate complex EA breaks down (sometimes with the participation of the solvent) into enzyme and reaction products:

$$E + A \underset{k_{-1}}{\overset{k_1}{\rightleftharpoons}} EA$$

$$EA \xrightarrow{k_2} E + \text{products}$$

If, as is usual, the substrate A is present in great excess of the enzyme it can be shown that steady-state conditions apply, and that the rate equation is:

$$v = \frac{k_2 [E]_0 [A]}{(k_{-1} + k_1)/k_1 + [A]}$$

where $[E]_0$ is the total concentration of enzyme. This equation is of the form of the Michaelis–Menten equation.

Other, more complicated, mechanisms lead to the Michaelis–Menten equation, adherence to which therefore does not require that the Michaelis–Menten mechanism applies.
    1996, *68*, 172

micro
SI prefix for 10^{-6} (symbol: μ).
    G.B. 74

microbial leaching
A process for the solubilization of metals, mostly from poor ores, by lithotrophic *bacteria*.
    1992, *64*, 158

microcanonical rate constant
A rate constant calculated by *microcanonical variational transition-state theory*, and therefore a rate constant that relates to a fixed energy.
    1996, *68*, 173

microcanonical variational transition-state theory (μVTST)
A development of *transition-state theory* in which the *dividing surface* is varied so as to minimize the rate calculated for a fixed energy.
The rate expressions obtained in a microcanonical treatment can be integrated over the energy, taking into account the statistical distribution over energy states, so as to give the canonical, or thermal, rates.
    1996, *68*, 173

microcarrier (in biotechnology)
A small, beaded material, derived from silica, glass, dextran or similar materials, used for the *immobilization* of *biocatalysts*, or as a support for the culture of anchorage-dependent animal cell lines.
    1992, *64*, 160

microclimatology
The science that deals with the climate of restricted areas and investigates their phenomena and causes.
    1990, *62*, 2201

microelectrophoresis
The *electrophoresis* technique involving the movement of a mass of particles on a small scale (e.g. paper electrophoresis).
    1972, *31*, 620

microgel
A *network* of microscopic dimensions.
    P.B. 52

microheterogeneity (in biochemistry)
A particular *glycoprotein* may occur in forms that differ in the structure of one or more or its carbohydrate units, a phenomenon known as microheterogeneity.
    1988, *60*, 1391; W.B. 86

micrometeorology
The study of the meteorological processes on scales from a millimeter or less up to tens or hundreds of metres; e.g. meteorology of a local site that is usually small and often is confined to a shallow layer of air next to the ground.
    1990, *62*, 2202

micro-network
A highly ramified *macromolecule* that contains cyclic structures and is of colloidal dimensions.
    1996, *68*, 2298

microphotometer
The optical instrument used to measure *transmittance* in atomic emission spectroscopy.
    O.B. 160

micropore (in catalysis)
Pore with width not exceeding about 2.0 nm (20 Å).
Cf. *macropore (in catalysis)*, *mesopore (in catalysis)*.
    1972, *31*, 585; 1976, *46*, 79

micropore filling (in catalysis)
The process in which molecules are adsorbed in the adsorption space within *micropores*.
    1972, *31*, 588; 1976, *46*, 76

micropore volume (in catalysis)
Conventionally measured by the volume of the adsorbed material which completely fills the *micropores*, expressed in terms of bulk liquid at atmospheric pressure and at the temperature of measurement.
    1972, *31*, 588; 1976, *46*, 76

microporous carbon
A porous *carbon material*, usually a *char* or *carbon fibres*, which may or may not have been subjected to an activation process to increase its adsorptive properties. A microporous carbon is considered to have a major part of its porosity in pores of less than 2 nm width and to exhibit apparent surface areas usually higher than 200 to 300 $m^2\ g^{-1}$.
Notes:
The surface areas determined by the Brunauer–Emmett–Teller (BET) method are apparent surface areas only since the BET adsorption equation is, in principle, not valid when micropore filling occurs. The determination of the true surface area in the micropores depends on the method used for the evaluation of the adsorption isotherms and on the model used for the shape of the micropores (cylindrical, slit-shaped or other).
See *micropore*.
    1995, *67*, 497; see also 1972, *31*, 518

microscopic chemical event
See *chemical reaction*, *molecularity*.
    1994, *66*, 1140

microscopic cross-section, σ
A measure of the probability of a specified interaction or reaction between an incident *radiation* and a target particle or system of particles. It is the reaction rate per target particle for a specified process divided by the *flux* of the incident radiation. In general, a specification is added of the type of radiation (e.g. neutron, photon), the *energy* of the incident radiation (e.g. thermal, epithermal, fast) and the type of interaction of reaction (e.g. *activation*, *fission*, scattering).
See also *macroscopic cross-section*.
    1994, *66*, 2517; O.B. 217

microscopic diffusion control (encounter control)
The observable consequence of the limitation that the rate of a bimolecular *chemical reaction* in a homogeneous medium cannot exceed the rate of encounter of the reacting *molecular entities*.

If (hypothetically) a *bimolecular* reaction in a homogeneous medium occurred instantaneously when two reactant molecular entities made an encounter, the *rate of reaction* would be an *encounter-controlled rate*, determined solely by rates of diffusion of reactants. Such a hypothetical 'fully diffusion controlled rate' is also said to correspond to 'total microscopic diffusion control', and represents the asymptotic limit of the rate of reaction as the rate constant for the chemical conversion of the encounter pair into product (or products) becomes large relative to the rate constant for separation (or dissociation) of the encounter pair.

'Partial microscopic diffusion control' is said to operate in a homogeneous reaction when the rates of chemical conversion and of separation are comparable. (The degree of microscopic diffusion control cannot usually be determined with any precision.)

See also *mixing control*.
1994, *66*, 1140; see also 1996, *68*, 173

microscopic electrophoresis
A technique in which the *electrophoresis* of individual particles is observed with the aid of a microscope or ultra-microscope.
1972, *31*, 620

microscopic film
A *film* which has lateral dimensions in the range 0.1–100 μm.
1994, *66*, 1671

microscopic kinetics
This term applies to a kinetic study in which the emphasis is on the interpretation of the behaviour in terms of elementary reactions.
See *macroscopic kinetics, molecular kinetics*.
1996, *68*, 173

microscopic reversibility at equilibrium
The principle of microscopic reversibility at equilibrium states that, in a system at equilibrium, any molecular process and the reverse of that process occur, on the average, at the same rate.
This definition corresponds to the statement of the principle that was given by R.C. Tolman in 1924. However, many workers have interchanged the meanings of microscopic reversibility and *detailed balance*, and it seems best now to regard the two, which are closely related, as synonymous.
1996, *68*, 173

microsome
Artefactual spherical particle, not present in the living cell, derived from pieces of the endoplasmic reticulum present in homogenates of tissues or cells: microsomes sediment from such homogenates when centrifuged at 10^6 g and higher: the microsomal fraction obtained in this way is often used as a source of mono-oxygenase enzymes.
1993, *65*, 2070

middle atmosphere
The combined *stratosphere* and *mesosphere* in the atmosphere.
1990, *62*, 2202

Mie scattering
The scattering of electromagnetic radiation by spherical particles of any size r, relative to the wavelength, λ. Since the cases $r \ll \lambda$ and $r \gg \lambda$ are covered by Rayleigh (dipole) scattering and geometric scattering theories, respectively, Mie scattering often refers to the case of $r \approx \lambda$.
1990, *62*, 2202

migration
1. The (usually *intramolecular*) transfer of an atom or *group* during the course of a *molecular rearrangement*.
2. The movement of a *bond* to a new position, within the same *molecular entity*, is known as 'bond migration'.
Allylic rearrangements, e.g.

$$RCH=CHCH_2X \longrightarrow RCHCH=CH_2$$
$$\qquad\qquad\qquad\qquad\qquad\quad |$$
$$\qquad\qquad\qquad\qquad\qquad\;\; X$$

exemplify both types of migration.
1994, *66*, 1141

migration current
The difference between the current that is actually obtained, at any particular value of the potential of the *indicator* or *working electrode*, for the reduction or oxidation of an ionic *electroactive substance* and the current that would be obtained, at the same potential, if there were no transport of that substance due to the electric field between the electrodes. The sign convention regarding current is such that the migration current is negative for the reduction of a cation or for the oxidation of an anion, and positive for the oxidation of a cation or the reduction of an anion. Hence the migration current may tend to either increase or decrease the total current observed. In any event the migration current approaches zero as the *transport number* of the electroactive substance is decreased by increasing the concentration of the supporting *electrolyte*, and hence the *conductivity*.
1985, *57*, 1497

migration current, limiting
See *limiting migration current*.
1985, *57*, 1498

migratory aptitude
The term is applied to characterize the relative tendency of a group to participate in a rearrangement. In nucleophilic rearrangements (*migration* to an electron-deficient centre), the migratory aptitude of a group is loosely related to its capacity to stabilize a partial positive charge, but exceptions are known, and

the position of hydrogen in the series is often unpredictable.
1994, *66*, 1141

migratory insertion
A combination of *migration* and *insertion*. The term is mainly used in organometallic chemistry.

$$\begin{array}{c} M-R \\ | \\ CO \end{array} \longrightarrow M-CO-R \qquad M = metal$$

1994, *66*, 1141

milli
SI prefix for 10^{-3} (symbol: m).
G.B. 74

milligram equivalent of readability (of a precision balance)
The product of *readability* and the value of the scale division (in mg per division).
O.B. 36

millimetre of mercury
Non-SI unit for pressure; mmHg ≈ 133.322 Pa. This old unit is not recognized by CIPM or ISO, but it is used for blood pressure.
1996, *68*, 981

milling (grinding)
The mechanical reduction of the particle size of a sample by attribution (friction), impact or cutting.
The required particle size of a sample is related to the size of the test portion and the number of particles required to ensure homogeneity among test portions. The reduction in particle size may sometimes result in particles of different hardness and density, which produces inhomogeneity during the preparation of the test sample or during the withdrawal of the test portion.
1990, *62*, 1205

minimum consumption time (in flame emission and absorption spectrometry), t_{min}
The time during which nebulization must be carried out in order to perform one analysis with a given precision.
The minimum consumption time is equal to the minimum volume consumed of the prepared sample divided by the rate of fluid consumption in order to obtain a given precision. It can be reduced by means of integrating the signal or measuring the signal increments (kinetic methods).
1986, *58*, 1740

minimum density of states criterion
This is a criterion used in *Rice–Ramsperger–Kassel–Marcus (RRKM) theory* for locating the transition state. The reaction coordinate *r* at the transition state, according to this criterion, is given by $\partial N(\varepsilon_r)/\partial r = 0$, where $N(\varepsilon_r)$ is the density of states at an internal energy ε_r.
1996, *68*, 174

minimum-energy reaction path
The path corresponding to the steepest descent from the *col* of a *potential-energy surface* into the two valleys. The reaction coordinate corresponds to this minimal path. Some workers refer to the minimum-energy reaction path as simply the reaction path but this is not recommended as it leads to confusion.
1996, *68*, 174

minimum lethal concentration (LC_{min})
Lowest concentration of a toxic substance in an environmental medium that kills individual organisms or test species under a defined set of conditions.
Synonymous with lowest lethal concentration found.
1993, *65*, 2071

minimum lethal dose (LD_{min})
Lowest amount of a substance that, when introduced into the body, may cause death to individual species of test animals under a defined set of conditions.
1993, *65*, 2071

minus
See *plus, minus*.
1996, *68*, 2211

minute (of arc)
Non-SI unit of plane angle.
$1' = (1/60)° = 2.908\,882\,\ldots \times 10^{-4}$ rad.
G.B. 113; 1996, *68*, 981

mist (in atmospheric chemistry)
A qualitative term applied to a suspension of droplets in a gas. In the atmosphere a mist produces a generally thin, greyish veil over the landscape. It reduces visibility to a lesser extent than fog but somewhat more than haze (visibility of less than 2 km but greater than 1 km).
1990, *62*, 2202

mitochondria
Organelles appearing in all eukaryotic cells which produce ATP as useful energy for the cell by oxidative phosphorylation. The proteins for the adenosine 5′-triphosphate (ATP)-generating electron transport of the respiration chain are located in the inner mitochondrial membrane. Mitochondria contain many *enzymes* of the citric acid cycle and for fatty acid acyle and for fatty acid β-oxidation. Many of them are coded for by nuclear DNA.
1992, *64*, 160; 1993, *65*, 2071

mitosis
The process whereby a *cell* nucleus divides into two daughter nuclei, each having the same genetic complement as the parent cell.
1992, *64*, 161

mixed control
See *transport control*.
1981, *53*, 1838

mixed crystal (solid solution)
A crystal containing a second constituent which fits into and is distributed in the lattice of the host crystal. (The use of 'solid solution' for amorphous materials is not recommended.)
O.B. 85

mixed energy release
This is an effect found in a process A + BC when the attacking species A is heavy. Because A is heavy the A–B distance continues to decrease while the repulsive energy of B–C is being released.
1996, *68*, 174

mixed indicator
An indicator containing a supplementary dye selected to heighten the overall colour change.
O.B. 48

mixed labelled
A *specifically labelled* compound is referred to as mixed labelled when the *isotopically substituted* compound has more than one kind of modified atom e.g.
$CH_3–CH_2–[^{18}O][^2H]$.
B.B. 515; 1981, *53*, 1893

mixed potential
The value of the potential of a given electrode with respect to a suitable *reference electrode* when appreciable contributions to the total anodic and/or cathodic partial currents are made by species belonging to two or more different couples, but the total current is zero, i.e.
$$I = \sum_i I_i = 0 \quad (I_i \neq 0)$$
where I_i is the partial current of reaction i
$$I_i = I_{i,a} + I_{i,c}.$$
1974, *37*, 513

mixing (in analytical chemistry)
The combining of components, particles or layers into a more homogeneous state.
The mixing may be achieved manually or mechanically by shifting the material with stirrers or pumps or by revolving or shaking the container. The process must not permit segregation of particles of different size or properties.
Homogeneity may be considered to have been achieved in a practical sense when the *sampling error* of the processed portion is negligible compared to the total error of the *measurement* system.
1990, *62*, 1204

mixing control
The experimental limitation of the *rate of reaction* in solution by the rate of mixing of solutions of the two reactants. It can occur even when the reaction rate constant is several powers of 10 less than that for an *encounter-controlled* rate. Analogous (and even more important) effects of the limitation of reaction rates by the speed of mixing are encountered in heterogeneous (solid/liquid, solid/gas, liquid/gas) systems.
See also *microscopic diffusion control, stopped flow.*
1994, *66*, 1141; 1996, *68*, 174

mixing height (in atmospheric chemistry)
The height to which significant mixing of added pollutants occurs within the atmosphere. In reference to stack gases, it is considered the height at which stack effluent begins mixing with the atmosphere as it leaves the stack.
1990, *62*, 2202

mixing ratio (in atmospheric chemistry)
In meteorology, the dimensionless ratio of the mass of a substance (such as water vapour) in an air parcel to the mass of the remaining substances in the air parcel. For trace substances, this is approximated by the ratio of the mass of the substance to the mass of air. However, in the case of water vapour the mass of dry air is used. In atmospheric chemistry, mixing ratios (molecular, molar, by volume, as well as by weight) are used to describe relative concentrations of atmospheric trace gases and impurities.
See *concentration*.
1990, *62*, 2202

mixture
Portion of matter consisting of two or more chemical substances called *constituents*.
Physical Chemistry Division, unpublished; see also 1994, *66*, 535

MLCT
Abbreviation for *Metal to Ligand Charge Transfer*.
1988, *60*, 1078

mobile adsorption
A situation in which molecules of *adsorbate* are free to move about the surface. Mobility of the adsorbate will increase with temperature and mobile adsorption may be either localized or non-localized. In localized mobile adsorption the adsorbate spends most of the time on the adsorption sites but can migrate or be desorbed and readsorbed elsewhere. In non-localized adsorption the mobility is so great that a small fraction of the adsorbed species is on the desorption sites and a large fraction at other positions on the surface.
1976, *46*, 76

mobile phase (in chromatography)
A fluid which percolates through or along the stationary bed, in a definite direction. It may be a liquid (*liquid chromatography*) or a gas (*gas chromatography*) or a supercritical fluid (*supercritical-fluid chromatography*). In gas chromatography the expression carrier gas may be used for the mobile phase. In elution chromatography the expression 'eluent' is also used for the mobile phase.
1993, *65*, 824; 1990, *62*, 2202; O.B. 97

mobile phase velocity, reduced
See *reduced mobile phase velocity (in chromatography)*.
1993, *65*, 840

mobile-phase velocity (in chromatography), u
The linear velocity of the mobile phase across the average cross-section of the chromatographic bed or column.
1993, *65*, 840

mobility, μ
Drift velocity of electrons or holes divided by the electric field strength.
See also *electric mobility*.
G.B. 37

mobility (in aerosol physics)
The velocity of a particle per unit applied force.
1990, *62*, 2202

Möbius aromaticity
A monocyclic array of orbitals in which there is a single out-of-phase overlap (or, more generally, an odd number of out-of-phase overlaps) reveals the opposite pattern of *aromatic* character to Hückel systems; with $4n$ electrons it is stabilized (aromatic), whereas with $4n + 2$ it is destabilized (antiaromatic). In the excited state $4n + 2$ Möbius π-electron systems are stabilized, and $4n$ systems are destabilized. No examples of ground-state Möbius π-systems are known, but the concept has been applied to *transition states* of *pericyclic reactions* [see *aromatic* (3)].
The name is derived from the topological analogy of such an arrangement of orbitals to a Möbius strip.
See also *Hückel (4n + 2) rule*.
1994, *66*, 1141

mode
The value of the variable occurring with the greatest frequency in the series of observations. A *bimodal* distribution is one which contains two maxima. The use of mode when reporting results of chemical analysis is generally not recommended.
1994, *66*, 603; 1990, *62*, 2202

mode-locked laser
A laser in which many resonant modes are coupled in phase, to yield a train of very short pulses (e.g. ps pulses). The coupling of the modes is obtained by modulation of the gain in the resonator, and can be active (electro-optic modulation of the losses or of the pump intensity), or passive (with a saturable absorber).
See also *free-running laser*.
1996, *68*, 2254

moderation (in nuclear chemistry)
Reduction of the *neutron* energy by *scattering* without appreciable *capture*.
1982, *54*, 1546

moderator
A material used to reduce the neutron *energy* by scattering without appreciable *capture*.
1994, *66*, 2522

modified active solid (in chromatography)
An active solid the sorptive properties of which have been changed by some treatment.
1993, *65*, 830; O.B. 98

modified Arrhenius equation
This is an extension of the simple *Arrhenius equation* in which the *pre-exponential factor* is proportional to T^n where T is the temperature and n a constant:

$$k = BT^n \exp(-E_a/RT)$$

B is a temperature-independent constant.
1996, *68*, 174

modified sample
A sample or a known fraction of the parent population in which the analyte has been isolated or (usually) concentrated before being submitted to the laboratory. If the isolation or concentration occurs in the laboratory, the procedure is usually considered part of the preparation of the test sample from the laboratory sample (in-laboratory processing).
1990, *62*, 1204

modifier (in solvent extraction)
A substance added to a solvent to improve its properties e.g. by increasing the solubility of an extractant, changing interfacial parameters or reducing adsorption losses. Additives used to enhance extraction rates should be called accelerators or catalysts.
1993, *65*, 2381

modulus of elasticity (Young's modulus), E
The normal stress divided by linear strain.
G.B. 12

Mohr amplification process (in analysis)
The process in which iodide is oxidized to iodate to achieve a more favourable measurement.
1982, *54*, 2555

moiety
In physical organic chemistry moiety is generally used to signify part of a molecule, e.g. in an ester R^1COOR^2 the alcohol moiety is R^2O. The term should not be used for a small fragment of a molecule.
1994, *66*, 1141

molal
See *molality*.
G.B. 42; 1996, *68*, 982

molality, m, b
Amount of entities of a *solute* divided by the mass of the *solvent*.
G.B. 42; 1996, *68*, 982

molar
Adjective before the name of an extensive quantity generally meaning division by *amount of substance* to make it intensive. Molar volume is volume divided by amount of substance. In a few special cases molar is used to denote division by *amount concentration* (e.g. molar absorption coefficient).
G.B. 7; 1996, *68*, 982

molar absorption coefficient
See *absorption coefficient*.
1996, *68*, 982

molar absorptivity
See *absorption coefficient*.
1996, *68*, 982

molar activity (in radiochemistry)
For a specified *isotope*, the *activity* of the compound divided by the amount of the material in moles: $A_m = A/n$.
1994, *66*, 2515

molar average velocity (in electrolytes)
See *mass average velocity (in electrolytes)*.
1981, *53*, 1831

molar conductivity
See *conductivity*.
1974, *37*, 512

molarity
Term sometimes used for *amount concentration*.
G.B. 42

molar-mass exclusion limit (in polymers)
The maximum value of the molar mass or molecular weight of molecules or particles, in a specific polymer solvent system, that can enter into the pores of the porous non-adsorbing material used in size-exclusion chromatography.
For particles with molar mass or molecular weight larger than the exclusion limit the separation effect of the size-exclusion chromatography vanishes.
P.B. 69

molar refraction, R
Function of the *refractive index*, n, given by $R = V_m (n^2 - 1)/(n^2 + 2)$, where V_m is the molar volume.
G.B. 33

mole
SI base unit for the *amount of substance* (symbol: mol). The mole is the amount of substance of a system which contains as many elementary entities as there are atoms in 0.012 kilogram of carbon-12. When the mole is used, the elementary entities must be specified and may be atoms, molecules, ions, electrons, other particles, or specified groups of such particles.
G.B. 70; 1996, *68*, 983

molecular anion
See *molecular ion*.
O.B. 205; 1991, *63*, 1549

molecular beams
A molecular beam is produced by allowing a gas at higher pressure to expand through a small orifice into a container at lower pressure. The result is a beam of particles (atoms, free radicals, molecules or ions) moving at approximately equal velocities, with few collisions occurring between them.
In a crossed molecular-beam experiment a reaction is studied using collimated beams of reactant molecules. For a bimolecular reaction, beams of the two reactants are caused to impinge on one another, often at an angle of 90°.
In a beam-gas scattering experiment a collimated beam is introduced into a gas, and the scattering patterns are observed.
1996, *68*, 175

molecular cation
See *molecular ion*.
O.B. 205; 1991, *63*, 1549

molecular conformation (of a polymer)
The conformation of the *macromolecule* as a whole. In the polymer literature, molecular conformation is sometimes referred to as macroconformation. In molecular conformations involving parallel stems, the latter may be confined to the same crystal or may also extend over several crystals.
P.B. 77

molecular dynamics
Synonymous with *reaction dynamics*.
1996, *68*, 175

molecular entity
Any constitutionally or isotopically distinct *atom*, *molecule*, ion, *ion pair*, *radical*, *radical ion*, *complex*, conformer etc., identifiable as a separately distinguishable entity.
Molecular entity is used in this Compendium as a general term for singular entities, irrespective of their nature, while *chemical species* stands for sets or ensembles of molecular entities. Note that the name of a compound may refer to the respective molecular entity or to the chemical species, e.g. methane, may mean a single molecule of CH_4 (molecular entity) or a molar amount, specified or not (chemical species), participating in a reaction.
The degree of precision necessary to describe a molecular entity depends on the context. For example 'hydrogen molecule' is an adequate definition of a certain molecular entity for some purposes, whereas for others it is necessary to distinguish the electronic state and/or vibrational state and/or nuclear spin, etc. of the hydrogen molecule.
1994, *66*, 1142

molecular formula
For compounds consisting of discrete molecules, a formula according with the relative molecular mass (or the structure).
R.B. 45

molecular ion (in mass spectrometry)
An ion formed by the removal from (positive ions) or addition to (negative ions) a molecule of one or more electrons without fragmentation of the molecular structure. The mass of this ion corresponds to the sum of the masses of the most abundant naturally occuring isotopes of the various atoms that make up the molecule (with a correction for the masses of the electron(s) lost or gained). For example, the mass of the molecular ion of ethyl bromide $C_2H_5{}^{79}Br$ will be 2×12 plus 5×1.0078246 plus 78.91839 minus the mass of the electron (m_e). This is equal to 107.95751 u $- m_e$, u being the unified atomic mass unit based on the standard that the mass of the isotope $^{12}C = 12$ u exactly.
1991, *63*, 1549; O.B. 205

molecularity
The number of reactant *molecular entities* that are involved in the 'microscopic chemical event' constituting an *elementary reaction*. (For reactions in solution this number is always taken to exclude molecular entities that form part of the *medium* and which are involved solely by virtue of their solvation of solutes.) A reaction with a molecularity of one is called 'uni-

molecular kinetics
molecular', one with a molecularity of two 'bimolecular' and of three 'termolecular'.
See also *chemical reaction, order of reaction*.
1994, *66*, 1143; 1996, *68*, 175; 1993, *65*, 2296

molecular kinetics
This term is applied to a kinetic study in which there is emphasis on the behaviour of the system at the molecular level.
See *macroscopic kinetics*.
1996, *68*, 175

molecular laser
A *gas laser* which makes use of electronic, vibrational or rotational energy levels in molecules. The molecules can be excited by an electric discharge, as in the case of the N_2 or CO_2 laser, by chemical reactions, as in the HF or DF lasers, or by another laser, as in the CH_3F far IR laser.
1995, *67*, 1920

molecular mass, relative
See *relative molecular mass*.
G.B. 41

molecular mechanics calculation
An empirical calculational method intended to give estimates of structures and energies for *conformations* of molecules. The method is based on the assumption of 'natural' bond lengths and angles, deviation from which leads to strain, and the existence of torsional interactions and attractive and/or repulsive *van der Waals* and dipolar forces between non-bonded atoms. The method is also called '(empirical) force-field calculations'.
1994, *66*, 1142

molecular metal
A non-metallic material whose properties resemble those of metals, usually following oxidative doping; e.g. polyacetylene following oxidative doping with iodine.
1994, *66*, 1142

molecular nucleation (in polymers)
The initial crystallization of a small portion of a macromolecule, after which further crystallization is thermodynamically favoured.
Molecular nucleation may give rise to a new crystal or increase the size of a pre-existing one.
P.B. 85

molecular orbital
A one-electron wavefunction describing an electron moving in the effective field provided by the nuclei and all other electrons of a *molecular entity* of more than one atom. Such molecular orbitals can be transformed in prescribed ways into component functions to give 'localized molecular orbitals'. Molecular orbitals can also be described, in terms of the number of nuclei (or 'centres') encompassed, as two-centre, multi-centre, etc. molecular orbitals, and are often expressed as a linear combination of *atomic orbitals*. An orbital is usually depicted by sketching contours on which the wavefunction has a constant value (contour map) or by indicating schematically the envelope of the region of space in which there is an arbitrarily fixed high (say 96%) probability of finding the electron occupying the orbital, giving also the algebraic sign (+ or –) of the wavefunction in each part of that region.
1994, *66*, 1142; G.B. 18

molecular rearrangement
The term is traditionally applied to any reaction that involves a change of connectivity (sometimes including hydrogen), and violates the so-called 'principle of minimum structural change'. According to this oversimplified principle, *chemical species* do not isomerize in the course of a *transformation*, e.g. *substitution*, or the change of a functional *group* of a chemical species into a different functional group is not expected to involve the making or breaking of more than the minimum number of bonds required to effect that transformation. For example, any new substituents are expected to enter the precise positions previously occupied by displaced groups.

The simplest type of rearrangement is an *intramolecular* reaction in which the product is isomeric with the reactant (one type of 'intramolecular isomerization'). An example is the first step of the Claisen rearrangement:

The definition of molecular rearrangement includes changes in which there is a *migration* of an atom or bond (unexpected on the basis of the principle of minimum structural change), as in the reaction:

$CH_3CH_2CH_2Br + AgOAc$
$\rightarrow (CH_3)_2CHOAc + AgBr$

where the *rearrangement stage* can formally be represented as the '1,2-shift' of hydride between adjacent carbon atoms in the carbocation:

$CH_3CH_2CH_2^+ \rightarrow CH_3CH^+CH_3$

Such migrations occur also in radicals, e.g.:

The definition also includes reactions in which an *entering group* takes up a different position from the *leaving group*, with accompanying bond migration.

An example of the latter type is the 'allylic rearrangement':

$(CH_3)_2C=CHCH_2Br + OH^-$
$\rightarrow (CH_3)_2C(OH)CH=CH_2 + Br^-$

A distinction is made between 'intramolecular rearrangements' (or 'true molecular rearrangements') and '*intermolecular* rearrangements' (or 'apparent rearrangements'). In the former case the atoms and groups that are common to a reactant and a product never separate into independent fragments during the rearrangement stage (i.e. the change is intramolecular), whereas in an 'intermolecular rearrangement' a migrating group is completely free from the parent molecule and is re-attached to a different position in a subsequent step, as in the Orton reaction:

PhNCOCH$_3$ + HCl $\longrightarrow$ PhNHCOCH$_3$ + Cl$_2$
|
Cl

$\longrightarrow$ (*o*- and *p*-) ClC$_6$H$_4$NHCOCH$_3$ + HCl

1994, *66*, 1142

molecular sieve effect
With respect to porous solids, the surface associated with pores communicating with the outside space may be called the *internal surface*. Because the accesibility of ores may depend on the size of the fluid molecules, the extent of the internal surface may depend on the size of the molecules comprising the fluid, and may be different for the various components of a fluid mixture. This effect is known as the molecular sieve effect.

1972, *31*, 585

molecular spectra
Spectra formed by bands consisting of rotational lines originating from rotational, vibrational and electronic transitions of molecules. These may be emission or absorption spectra.

1985, *57*, 1461

molecular weight
See *relative molecular mass*.
G.B. 41; 1996, *68*, 983

molecular-weight exclusion limit (in polymers)
See *molar-mass exclusion limit*.
P.B. 69

molecule
An electrically neutral entity consisting of more than one atom ($n > 1$). Rigorously, a molecule, in which $n > 1$ must correspond to a depression on the potential energy surface that is deep enough to confine at least one vibrational state.
See also *molecular entity*.
1994, *66*, 1143

mole fraction
See *amount fraction*.
G.B. 41

molozonides
1,2,3-Trioxolanes, the primary products of the reaction of ozone at a carbon–carbon double bond.

$$R_2C-CR_2$$ with O-O-O bridge

See *ozonides*.
1995, *67*, 1349

moment of a force, *M*
About a point is the vector product of the radius vector from this point to a point on the line of action of the force and the force, $M = r \times F$.
G.B. 12; ISO 31-2: 1992

moment of inertia, *I, J*
Of a body about an axis is the sum of the products of its mass elements and the squares of their distances to the axis. *Principal moments of inertia* of a molecule are chosen so that $I_A \leq I_B \leq I_C$.
G.B. 12; 1996, *68*, 983

momentum, *p*
Vector quantity equal to the product of *mass* and *velocity*.
G.B. 12

momentum spectrum
A spectrum which is obtained when a beam of ions is separated according to the momentum-to-charge ratios of the ionic species contained within it. A sector magnetic field achieves separation of the various ionic species in this way. If the ion beam is homogeneous in translational energy, as is the case with sector instruments, separation according to the *m/z* ratios is also achieved and a mass spectrum is obtained.

1991, *63*, 1551

monochromator
Spectrographic *instrument* where the focal plane is obstructed except for one slit. The bandwidth of the spectrum which emerges through the exit slit depends inter alia upon the widths of the entrance and exit slits.
O.B. 156

monoclonal antibodies (MAbs)
A single species of *immunoglobulin* molecules produced by culturing a single *clone* of a *hybridoma* cell. MAbs recognize only one chemical structure, i.e. they are directed against a single *epitope* of the antigenic substance used to raise the antibody.

1992, *64*, 161

monodisperse medium
A *colloidal* system in which all particles are of (nearly) the same size.

1972, *31*, 607

monodisperse polymer
See *uniform polymer*.
1996, *68*, 2301

mono-energetic radiation
Radiation consisting of *particles* of a single kinetic energy or *photons* of a single energy.

1982, *54*, 1546

monoisotopic mass spectrum
A spectrum containing only ions made up of the principal isotopes of atoms making up the original molecule.
1991, *63*, 1556

monolayer
A single, closely packed *layer* of atoms or molecules. The term floating monolayer is used for certain spread monolayers or *films* which are in the condensed — often solid — state. The term Langmuir monolayer has recently been coined for spread monolayers. This usage is not recommended.
1994, *66*, 1672; see also O.B. 251

monolayer capacity
For *chemisorption* the amount of adsorbate which is needed to occupy all adsorption sites as determined by the structure of the adsorbent and by the chemical nature of the absorptive. For *physisorption*, the amount needed to cover the surface with a complete monolayer of atoms or molecules in close-packed array, the kind of close packing having to be stated explicitly when necessary.
O.B. 252; 1972, *31*, 587; 1976, *46*, 75

monomer
A substance composed of *monomer molecules*.
1996, *68*, 2290

monomeric unit (monomer unit, mer)
The largest *constitutional unit* contributed by a single monomer molecule to the structure of a *macromolecule* or *oligomer molecule*.
Note:
The largest constitutional unit contributed by a single monomer molecule to the structure of a macromolecule or oligomer molecule may be described as either *monomeric*, or by *monomer* used adjectivally.
1996, *68*, 2290

monomer molecule
A molecule which can undergo *polymerization* thereby contributing *constitutional units* to the essential structure of a macromolecule.
1996, *68*, 2289

monomer unit
See *monomeric unit*.
1996, *68*, 2290

monosaccharides
A term which includes *aldoses, ketoses* and a wide variety of derivatives. Derivation includes oxidation, deoxygenation, introduction of other substituents, alkylation and acylation of hydroxy groups, and chain branching. E.g.

See also *aldoketoses, uronic acids, amino sugars, saccharides, glycosides*.
1995, *67*, 1349

monotectic reaction
The *reversible transition*, on cooling, of a liquid to a mixture of a second liquid and a solid:

$$\text{liquid}_1 \rightleftarrows \text{liquid}_2 + \text{solid}.$$

1994, *66*, 586

monotectoid reaction
A reaction in a system containing two solid solution phases, α' and α'' in which α' decomposes into α'' and a new phase β:

$$\alpha' \rightleftarrows \alpha'' + \beta$$

1994, *66*, 586

monotectoid temperature
The maximum temperature at which a *monotectoid reaction* can occur.
1994, *66*, 586

monoterpenes
See *terpenes*.
1995, *67*, 1349

monoterpenoids
Terpenoids having a C_{10} skeleton.
1995, *67*, 1349

monothioacetals
See *thioacetals*.
1995, *67*, 1349

monotropic transition
The *irreversible transition* from a metastable polymorphic form to the stable polymorph.
Example: The transition of metastable $CaCO_3$ (aragonite-type) to the stable $CaCO_3$ (calcite-type).
Note:
For liquid crystals the term refers to a liquid crystal to *liquid-crystal transition* that occurs below the melting point and is revealed by supercooling of the crystal.
1994, *66*, 586

mordant
Substance that fixes a dyestuff in or on a material by combining with the dye to form an insoluble compound, used to fix or intensify stains in a tissue or cell preparation.
1993, *65*, 2072

More O'Ferrall–Jencks diagram
Visualization of the potential energy surfaces for a reacting system, as a function of two chosen coordinates. It is particularly useful to discuss structural effects on the *transition state* geometry for processes occurring either by stepwise or concerted routes. The use of such diagrams, first suggested for elimination

reactions, was later extended to acid–base catalysis and to certain other reactions.
1994, 66, 1143

Morin transition
A transition specific to α-Fe_2O_3 in which there is a change in the direction of the atomic magnetic moments in the antiferromagnetic state from parallel to perpendicular to the c-axis.
Synonymous with spin-flop transition.
1994, 66, 587

morphotropic transition
An abrupt change in the structure of a solid solution with variation in composition.
1994, 66, 587

Mössbauer effect
Resonance *absorption* of γ-*radiation* by nuclei arranged in a crystal lattice in such a way that the recoil momentum is shared by many atoms.
1982, 54, 1546

Mössbauer thickness (in Mössbauer spectroscopy)
The effective thickness of a source (t_s) or absorber (t_a) in the optical path.
1976, 45, 214

most probable distribution (in macromolecular assemblies)
A discrete distribution with the differential mass-distribution function of the form:

$$f_w(x) = a^2 x(1 - a)^{x - 1}$$

where x is a parameter characterizing the chain length, such as relative molecular mass or degree of polymerization and a is a positive adjustable parameter.
For large values of x, the most probable distribution converges to the particular case of the *Schulz–Zimm distribution* with $b = 1$. In the literature, this distribution is sometimes referred to as the Flory distribution or the Schulz–Flory distribution.
P.B. 56

Mott–Hubbard transition
See *Mott transition*.
1994, 66, 587

Mott transition
A transition occurring only in 'single-valent' systems from strongly correlated ($U > W$) to weakly correlated ($U < W$) electrons as a result of a change of bandwidth W. W reflects the strength of the interatomic interactions in a periodic array of like atoms and U is a measure of the intra-atomic interactions, i.e. the electrostatic energy involved in the creation of polar states by transferring an electron from one atom to the next, as shown schematically by the *disproportionation* reaction:

$$M^{n+} + M^{n+} \rightarrow M^{(n + 1)+} + M^{(n - 1)+}$$

A $U > W$ produces a magnetic semiconductor; a $U < W$ gives rise to a metal (normally nonmagnetic). Note also that U and W can be altered by changes in interatomic distances, brought about through temperature or pressure variations or by introducing an alloying element.
Synonymous with Mott–Hubbard transition.
1994, 66, 587

MPP-based carbon fibres
See *mesophase pitch-based carbon fibres*.
1995, 67, 497

mRNA
See *messenger RNA*.
1992, 64, 161

mu
See entries under μ (beginning of 'm').

mucopolysaccharides
Polysaccharides composed of alternating units from *uronic acids* and glycosamines, and commonly partially esterified with sulfuric acid.
1995, 67, 1350; see also W.B. 86

multi-centre bond
Representation of some *molecular entities* solely by localized two-electron two-centre *bonds* appears to be unsatisfactory. Instead, multi-centre bonds have to be considered in which electron pairs occupy orbitals encompassing three or more atomic centres. Examples include the three-centre bonds in diborane, the delocalized π-bonding of benzene and *bridged carbocations*.
1994, 66, 1144

+ multi-centre reaction
A synonym for *pericyclic reaction*. The number of 'centres' is the number of atoms not bonded initially, between which single bonds are breaking or new bonds are formed in the *transition state*. This number does not necessarily correspond to the ring size of the transition state for the pericyclic reaction. Thus, a Diels–Alder reaction is a 'four-centre reaction'. This terminology has largely been superseded by the more detailed one developed for the various pericyclic reactions.
See *cycloaddition*, *sigmatropic rearrangement*.
1994, 66, 1144

multi-channel pulse height analyser
A *pulse amplitude analyser* which includes a storage function to record the number of pulses received per channel.
1982, 54, 1548

multident
See *ambident*.
1994, 66, 1144

multienzyme
A protein possessing more than one catalytic function contributed by distinct parts of a polypeptide chain ('domains'), or by distinct subunits, or both.
W.B. 107

multienzyme complex
A multienzyme with catalytic domains on more than one type of polypeptide chain.
W.B. 107

multienzyme polypeptide
A *polypeptide* chain containing at leat two types of catalytic domains.
W.B. 107

multilayer
A system of adjacent *layers* or *monolayers*. The term bilayer applies to the particular case of a multilayer two monolayers thick. Monolayers and multilayers may be alternatively named '*films*' provided boundaries can be defined for them.
1994, *66*, 1672

multilayer adsorption
Adsorption such that the adsorption space accommodates more than one layer of molecules, and not all adsorbed molecules are in contact with the *surface layer* of the *adsorbent*.
1972, *31*, 587; 1976, *46*, 75

multilayer aggregate (in polymer crystals)
A stack of *lamellar crystals* generated by spiral growth at one or more screw dislocations.
The axial displacement over a full turn of the screw (Burgers vector) is usually equal to one lamellar thickness.
P.B. 82

multiphoton absorption
See *multiphoton process*.
See also *biphotonic excitation*.
1996, *68*, 2255

multiphoton ionization (in mass spectrometry)
Occurs when an atom or molecule and their concomitant ions have energy states whereby the energy in two or more photons is absorbed.
1991, *63*, 1548

multiphoton process
A process involving interaction of two or more photons with a molecular entity.
See *biphotonic process, two-photon process*.
1996, *68*, 2255

multiple ionization satellite (in X-ray spectroscopy)
See *X-ray satellite*.
1991, *63*, 739

multiple-pass cell (in spectrochemical analysis)
Permits multiple passage of radiation to increase the absorption pathlength. It is constructed in such a way that mirrors either form part of the cell or are mounted in the sample cell holder.
1988, *60*, 1454

multiple peak scanning (in mass spectrometry)
An alternative to multiple ion detection which emphasizes that monitoring of peaks rather than detection of ions is the important feature.
1978, *50*, 69

multiple scattering
Successive rescattering of radiation within the scattering medium.
See also *light scattering*.
1983, *55*, 932

multiplex spectrometer
A *spectrometer* in which a single photodetector simultaneously receives signals from different spectral bands which are specifically encoded. In the case of frequency multiplexing, each spectral band is modulated at a specific frequency. Decoding is achieved by filtering out, by electronic means, the corresponding signals.
Frequency multiplexing may be realized by changing the path difference between the two interfering beams at a uniform rate. Fourier transform of the interferogram so obtained yields the spectrum. This method is called *Fourier transform spectrometry* (*FTS*).
1995, *67*, 1729

multiplication, neutron
See *neutron multiplication*.
1982, *54*, 1546

multiplicative name
A name that expresses the multiple occurrence of identical parent structures, two or more of which are connected by a symmetrical structure expressible by means of a multivalent simple or composite prefix.
B.B. (G) 17

multiplicity (spin multiplicity)
The number of possible orientations, calculated as $2S + 1$, of the spin angular momentum corresponding to a given total spin quantum number (S), for the same spatial electronic wavefunction. A state of singlet multiplicity has $S = 0$ and $2S + 1 = 1$. A doublet state has $S = ½$, $2S + 1 = 2$, etc. Note that when $S > L$ (the total orbital angular momentum possible) there are only $2L + 1$ orientations of total angular momentum possible.
1996, *68*, 2255

multiply labelled
A *selectively labelled* compound may be multiply labelled when in the unmodified compound there is more than one atom of the same element at the position where the isotopic modification occurs, e.g. H in CH_4, or there are several atoms of the same element at different positions where the isotopic modification occurs, e.g. C in C_4H_8O.
B.B. 515; 1981, *53*, 1893

multipole line (in X-ray spectroscopy)
See *selection rule*.
1991, *63*, 738

multistage sampling
Samples taken in a series of steps with the sampling portions constituting the sample (units or increments) at each step being selected from the larger or greater number of portions of the previous step, or from a primary or composite sample.
The first set of portions (units or increments) taken from the population available for sampling is the primary sample. The subsequent samples (secondary, tertiary, etc.) are the sets of subsamples, units, items, individuals or increments taken from the preceding

step. The units may be different steps of multistage sampling (e.g. pallets, cases, packages).
1990, *62*, 1203

multi-strand chain (in polymers)
A *chain* that comprises *constitutional units* connected in such a way that adjacent constitutional units are joined to each other through more than four atoms, more than two on at least one side of each constitutional unit.
Note:
A chain that comprises constitutional units joined to each other through *n* atoms on at least one side of each constitutional unit is termed an n-*strand chain*, e.g. three-strand chain. If an uncertainty exists in defining *n*, the highest possible number is selected.
1996, *68*, 2295

multi-strand macromolecule
A *macromolecule* that comprises *constitutional units* connected in such a way that adjacent constitutional units are joined to each other through more than four atoms, more than two on at least one side of each constitutional unit.
Note:
A macromolecule that comprises constitutional units joined to each other through *n* atoms on at least one side of each constitutional unit is termed an n-*strand macromolecule*, e.g. three-strand macromolecule. If an ambiguity exists in defining *n*, the highest possible number is selected.
1996, *68*, 2296

munchnones
Mesoionic compounds having a 1,3-oxazole skeleton bearing an oxygen atom attached to the 5-position with the following delocalized structure:

$$\begin{array}{c}\text{[structure diagram]}\end{array}$$

1995, *67*, 1350

muonium
Atom-like particle consisting of a positive muon and an electron.
G.B. 93

+ mustard oils
An archaic term for *isothiocyanates*, RN=C=S.
1995, *67*, 1350

mustards
Compounds having two β-haloalkyl groups bound to a sulfur atom, as in $(XCH_2CH_2)_2S$, and their analogues, the nitrogen (and phosphorus) mustards, $(XCH_2CH_2)_2NR$. Compounds having one β-haloalkyl group and one β-hydroxyalkyl group on S are termed hemi- or semi-mustards by some chemists.
1995, *67*, 1350

mutagenesis
The introduction of permanant heritable changes i.e. *mutations* into the DNA of an organism.
1992, *64*, 161

mutarotation
The change in optical rotation accompanying *epimerization*. In sugar chemistry this term usually refers to epimerization at the hemiacetal carbon atom.
1996, *68*, 2212

mutation
A heritable change in the *nucleotide* sequence of genomic DNA (or RNA in RNA *viruses*), or in the number of *genes* or *chromosomes* in a *cell*, which may occur spontaneously or be brought about by chemical mutagens or by radiation (induced mutation).
1992, *64*, 161

mutation rate (in biotechnology)
The frequency (h^{-1}) with which a mutation occurs within a organism or *gene*. In general, rates of spontaneous mutation vary between one in 10^4 and one in 10^8 per gene per generation, and can be considerably increased by mutagens.
1992, *64*, 161

mutual inductance, *M*
For two thin conducting loops, the *magnetic flux* through one loop, caused by an electric current in the other loop, divided by that current.
G.B. 15; ISO 31-4: 1992

Myelin cylinders
Birefringent cylinders which form spontaneously from lipid-containing material in contact with water.
1972, *31*, 613

n → π* state
An excited state related to the ground state by a *n → π* transition*.
1996, *68*, 2255

n → π* transition
An electronic transition described approximately as promotion of an electron from a 'non-bonding' (lone-pair) n orbital to an 'antibonding' π orbital designated as π*.
1996, *68*, 2255

n-σ* delocalization (or n-σ* no bond resonance)
Delocalization of a free electron pair (n) into an antibonding σ-orbital (σ*).
See *hyperconjugation, resonance*.
1994, *66*, 1146

n → σ* transition
An electronic transition described approximately as promotion of an electron from a 'non-bonding' (lone-pair) n orbital to an 'antibonding' σ orbital designated as σ*. Such transitions generally involve high transition energies and appear close to or mixed with *Rydberg transitions*.
1996, *68*, 2255

nano
SI prefix for 10^{-9} (symbol: n).
G.B. 74

nanoscopic film
A *film* which has lateral dimensions in the range of 0.1–100 nm.
1994, *66*, 1671

+ naphthenes
Cycloalkanes especially cyclopentane, cyclohexane and their alkyl derivatives. The term seems to be obsolescent, except in the petrochemical industry.
1995, *67*, 1350

+ naphthenic acids
Acids, chiefly monocarboxylic, derived from *naphthenes*. The term seems to be obsolescent, except in the petrochemical industry.
1995, *67*, 1350

+ narcissistic reaction
A *chemical reaction* that can be described as the conversion of a reactant into its mirror image, without rotation or translation of the product, so that the product enantiomer actually coincides with the mirror image of the reactant molecule. Examples of such reactions are cited under the entries *fluxional* and *degenerate rearrangement*.
1994, *66*, 1144

natural broadening (of a spectral line)
Broadening which has its origin in the finite optical lifetime of one or both levels involved in the transition.
O.B. 121; 1985, *57*, 1463

natural graphite
A mineral found in nature. It consists of *graphitic carbon* regardless of its crystalline perfection.
Notes:
Some natural graphites, often in the form of large flakes, show very high crystalline perfection. Occasionally, they occur as single crystals of *graphite*. The use of the term natural graphite as a synonym for the term 'graphite single crystal' is incorrect and should be avoided. Varieties of natural graphite with lower structural perfection are classified as 'microcrystalline natural graphite'. Commercial natural graphite is often contaminated with other minerals, e.g. silicates, and may contain *rhombohedral graphite* due to intensive milling.
1995, *67*, 497

natural isotopic abundance
Of a specified isotope of an element, the *isotopic abundance* in the element as found in nature.
1982, *54*, 1535

+ natural lifetime
Synonymous with *radiative lifetime*. The use of this term is discouraged.
1996, *68*, 2255

natural radiation
Radiation originating from *natural radioactivity*.
1982, *54*, 1549

natural radioactivity
Radioactivity of naturally occurring *nuclides* in materials where the *isotopic abundance* of that nuclide is natural.
1982, *54*, 1546

neat soap
See *soap*.
1972, *31*, 612

nebulization, efficiency of (in flame emission and absorption spectroscopy)
See *efficiency of nebulization (in flame emission and absorption spectroscopy)*.
1986, *58*, 1740; O.B. 168

needle coke
The commonly used term for a special type of *coke* with extremely high graphitizability resulting from a strong preferred parallel orientation of its turbostratic layer structure and a particular physical shape of the grains.
Notes:
Needle coke is derived mainly from clean (i.e. lacking hetero atoms and solids) and highly aromatic (i.e. several condensed rings per cluster) feedstocks with a very low concentration of insolubles. Upon solidification a material with a distinctive streaked or flow-like macroscopic appearance is produced. Upon grinding the coke breaks up first into macroscopic needles and then, after further grinding, into microplatelets. Sometimes the word 'acicular' is used as a synonym for needle-like.
1995, *67*, 497

negative adsorption
The depletion of one or more components in an *interfacial layer*.
1972, *31*, 584

negative feedback
See *composite mechanism*.
1996, *68*, 161

negative ion (in mass spectrometry)
An atom, radical, molecule or molecular moiety in the vapour phase which has gained one or more electrons thereby acquiring an electrically negative charge. The use of the term anion as an alternative is not recommended because of its connotations in solution chemistry.
See *positive ion*.
1991, *63*, 1549; O.B. 204

negaton
See *electron*.
O.B. 221

neighbouring group participation
The direct interaction of the reaction centre (usually, but not necessarily, an incipient *carbenium* centre) with a lone pair of electrons of an atom or with the electrons of a σ- or π-bond contained within the parent molecule but not conjugated with the reaction centre. A distinction is sometimes made between n-, σ- and π-participation.
A rate increase due to neighbouring group participation is known as 'anchimeric assistance'. 'Synartetic acceleration' is the special case of anchimeric assistance ascribed to participation by electrons binding a substituent to a carbon atom in a β-position relative to the leaving group attached to the α-carbon atom. According to the underlying model, these electrons then provide a three-centre bond (or 'bridge') 'fastening together' (as the word 'synartetic' is intended to suggest) the α- and β-carbon atoms between which the charge is divided in the intermediate *bridged ion* formed (and in the *transition state* preceding its formation). The term synartetic acceleration is not widely used.
See also *intramolecular catalysis, multi-centre bond*.
1994, *66*, 1145

nematic phase
See *liquid-crystal transitions*.
1994, *66*, 587

nematic state
See *mesomorphic phase*.
1972, *31*, 613

neodymium laser
A CW pulsed laser emitting radiation from excited Nd^{3+} principally occurring around 1.06 μm (the precise position depends on the matrix). The Nd^{3+} is present as a dopant in suitable crystals (e.g. yttrium–aluminium garnet, YAG) or in suitable glasses (phosphate, silicate, etc.).
See *solid state lasers*.
1996, *68*, 2256

neoflavonoids
See *flavonoids*.
1995, *67*, 1350

neolignans
See *lignans*.
1995, *67*, 1350

neper
Unit, coherent with the SI, for expressing field levels, Np = 1.
G.B. 78

nephelometry
Analytical methods which depend on the measurement of the intensity of scattered light emanating from an illuminated volume of an aerosol. The ratio of scattered intensity to illuminating intensity is compared with a standard of known properties.
1990, *62*, 2202

Nernst's diffusion layer
A fictitious layer corresponding to the dotted straight lines of the diagram which shows the concentration profile along the direction perpendicular to an electrode surface. The thickness δ of this layer is called the effective (or equivalent) thickness of the *diffusion layer*. Its definition is apparent from the figure. It is the thickness which the diffusion layer would have if the concentration profile were a straight line coinciding with the tangent to the true concentration profile at the *interface*, and that straight line were extended up to the point where the bulk concentration is reached. δ has a formal significance only. It is simply another way of writing the *mass transfer coefficient* k_d defined in terms of a resistivity instead of a conductivity.

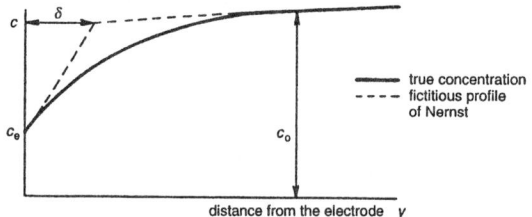

1981, *53*, 1837

net (in surface chemistry)
A two-dimensional lattice (in a particular range of surface coverage and temperature) into which the adsorbate is ordered, especially in some cases of localized adsorption.
1976, *46*, 76

net current
The sum of cathodic and anodic partial currents.
1980, *52*, 236

net electric charge (of a particle) (in electrophoresis), z
The algebraic sum of the charges present at the surface of the particle divided by the elementary charge

of the proton. The symbol z is also used for charge number of an ion.
1994, *66*, 895

network (in polymer chemistry)
A highly ramified *macromolecule* in which essentially each *constitutional unit* is connected to each other constitutional unit and to the macroscopic phase boundary by many permanent paths through the macromolecule, the number of such paths increasing with the average number of intervening bonds; the paths must on the average be co-extensive with the macromolecule.
Notes:
1. Usually, and in all systems that exhibit rubber elasticity, the number of distinct paths is very high, but, in most cases, some constitutional units exist which are connected by a single path only.
2. If the permanent paths through the structure of a network are all formed by covalent bonds, the term *covalent network* may be used.
3. The term *physical network* may be used if the permanent paths through the structure of a network are not all formed by covalent bonds but, at least in part, by physical interactions, such that removal of the interactions leaves individual macromolecules or a macromolecule that is not a network.
1996, *68*, 2298

network polymer
A polymer composed of one or more networks.
1996, *68*, 2305

neutral-density filter
See *attenuance filter*.
1996, *68*, 2256

neutrino (electron neutrino)
An elementary particle of negligible mass and zero electric charge. There are also a muon neutrino and tau neutrino.
G.B. 93

neutron
Nuclear particle of zero charge, spin quantum number ½ and a mass of 1.008 664 904(14) u.
G.B. 93

neutron density
The number of free neutrons divided by the containing volume. Partial densities may be defined for neutrons characterized by such parameters as *energy* and directions.
1994, *66*, 2522

neutron multiplication
The process in which a neutron produces on the average more than one neutron in a medium containing *fissile* material.
1982, *54*, 1546

neutron number, N
Number of neutrons in an atomic nucleus.
G.B. 20

neutron rest mass
Atomic fundamental physical constant m_n = 1.674 9286 (10) × 10^{-27} kg.
CODATA Bull., 1986, *63*, 1

neutrons, cold
See *cold neutrons*.
1982, *54*, 1546

neutrons, delayed
See *delayed neutrons*.
1982, *54*, 1546

neutrons, epicadmium
See *epicadmium neutrons*.
1994, *66*, 2522

neutrons, epithermal
See *epithermal neutrons*.
1994, *66*, 2522

neutrons, fast
See *fast neutrons*.
1994, *66*, 2522

neutrons, fission
See *fission neutrons*.
1982, *54*, 1546

neutrons, intermediate
See *intermediate neutrons*.
1982, *54*, 1547

neutrons, prompt
See *prompt neutrons*.
1982, *54*, 1547

neutrons, resonance
See *resonance neutrons*.
1994, *66*, 2522

neutrons, slow
See *slow neutrons*.
1982, *54*, 1547

neutrons, thermal
See *thermal neutrons*.
1994, *66*, 2522

neutron temperature
The temperature assigned to a population of neutrons when this population is approximated by a Maxwellian distribution.
1982, *54*, 1547

neutrophilic organisms
Organisms preferring a neutral medium for growth.
1992, *64*, 161

Newman projection
A *projection formula* representing the spatial arrangement of bonds on two adjacent atoms in a molecular entity. The structure appears as viewed along the bond between these two atoms, and the bonds from them to other groups are drawn as projections in the plane of the paper. The bonds from the atom nearer to the observer are drawn so as to meet at the centre of a

circle representing that atom. Those from the further atom are drawn as if projecting from behind the circle.

1996, 68, 2212

newton
SI derived unit of force, N = kg m s^{-2}.
G.B. 72; 1996, 68, 984

Newton black film
In soap films, two types of *equilibrium film* are often observed, sometimes successively in the same system: one characterized by thicknesses of the order of 7 nm or more which varies significantly with minor changes in composition such as ionic strength, and the other having a lesser thickness relatively independent of such changes. When a distinction is needed, the former are designated as *common black films*, and the latter as Newton black films. The current use of first or secondary for the common black film and of second, primary or Perrin's for the Newton black film is discouraged.
1972, 31, 614

Newton diagram
A vector diagram used to show the relationship between initial and final velocities or momenta in a two-particle scattering process. It is commonly used to relate laboratory and centre-of-mass coordinates.
1996, 68, 175

Newtonian fluid
A fluid in which the components of the stress tensor are linear functions of the first spatial derivatives of the velocity components. These functions involve two material parameters (taken as constants throughout the fluid, although depending on ambient temperature and pressure).
1979, 51, 1216

NHOMO
See *subjacent orbital*.
1994, 66, 1145

nido-
An affix used in names to denote a nest-like structure, especially a boron skeleton that is almost closed.
R.B. 245; B.B. 465

Nier–Johnson geometry
An arrangement for a double-focusing mass spectrometer in which a deflection of $\pi/2$ radians in a radial electrostatic field analyser is followed by a magnetic deflection of $\pi/3$ radians. The electrostatic analyser uses a symmetrical object-image arrangement and the magnetic analyser is used asymmetrically.
1991, 63, 1546; O.B. 201

NIH shift
The *intramolecular* hydrogen *migration* which can be observed in enzymatic and chemical hydroxylations of aromatic rings. It is evidenced by appropriate deuterium labelling, i.e.

In enzymatic reactions the NIH shift is generally thought to derive from the rearrangement of arene oxide intermediates, but other pathways have been suggested. (NIH stands for U.S. National Institutes of Health where the shift was discovered.)
1994, 66, 1145

nimbostratus cloud (in atmospheric chemistry)
An amorphous, dark grey, rainy cloud layer reaching almost to the ground; 300–600 m; vertical velocities of 0.05–0.2 m s^{-1}.
1990, 62, 2180

nitramines
Amines substituted at N with a nitro group (a contracted form of *N*-nitroamines); they are thus amides of nitric acid, and the class is composed of nitramide, O_2NNH_2, and its derivatives formed by substitution.
1995, 67, 1350

nitrenes
1. The neutral compound HN: having univalent nitrogen, and its derivatives RN:. Aminylenes is a recognized but less widely used synonym. Other names that have been used include aminediyls, *imidogens*, azenes. They may exist in either a singlet or a triplet electronic state (four spin-paired electrons, or two spin-paired and two with parallel spins, respectively), e.g. CH_3N: methylnitrene.
2. Until the 1960s, nitrenes had a totally different meaning: analogues of *nitrones* in which the double-bonded oxygen is replaced by double-bonded carbon, thus *azomethine ylides*.
See *carbenes*.
1995, 67, 1350; 1993, 65, 1357; see also 1994, 66, 1145

nitrenium ions
The cation H_2N:$^+$ and its *N*-hydrocarbyl derivatives R_2N:$^+$, in which the nitrogen has a positive charge, and two unshared electrons. A synonymous but less widely used term is aminylium ions. The alkylidene derivatives of H_2N:$^+$, $R_2C=N$:$^+$, still belong to the class nitrenium ions, but are more precisely designated by the term *iminylium ions*.
1995, 67, 1350; see also 1994, 66, 1145

nitrification
Sequential oxidation of ammonium salts to nitrite and nitrate by microorganisms.
1993, 65, 2075

nitrile imides
Ylides having the structure RC≡N$^+$–N$^-$–R ↔ RC$^+$=NN$^-$–R ↔ RC$^-$=N$^+$=NR. Also known as nitrile imines or nitrilimines, especially in the German literature.

See also *dipolar compounds, ylides*.
1995, *67*, 1351

nitrile imines
See *nitrile imides, imides (2)*.
1995, *67*, 1351

nitrile oxides
Ylides having the structure $RC{\equiv}N^+{-}O^- \leftrightarrow RC^-{=}N^+{=}O$.
See also *dipolar compounds*.
1995, *67*, 1351

nitriles
Compounds having the structure $RC{\equiv}N$; thus *C*-substituted derivatives of hydrocyanic acid, $HC{\equiv}N$. In systematic nomenclature, the suffix nitrile denotes the triply bound $\equiv N$ atom, not the carbon atom attached to it.
See also *cyanides, isocyanides, carbonitriles*.
1995, *67*, 1350

nitrile sulfides
Sulfur analogues of *nitrile oxides* $RC{\equiv}N^+{-}S^- \leftrightarrow RC^-{=}N^+{=}S$.
See also *dipolar compounds*.
1995, *67*, 1351

nitrile ylides
1,3-*Dipolar compounds* having the structure
$RC{\equiv}N^+{-}C^-R_2 \leftrightarrow RC^-{=}N^+{=}CR_2 \leftrightarrow RC^+{=}NC^-R_2 \leftrightarrow R\ddot{C}{-}N{=}CR_2$. The uncharged canonical form $R\ddot{C}{-}N{=}CR_2$ is called an alkylidene- (or hydrocarbylidene-) amino carbene; the name iminyl carbene is incorrect because iminyl is not a recognized prefix for $-N{=}CR_2$.
See also *carbenes, dipolar componds*.
1995, *67*, 1351

nitrilimines
See *nitrile imides*.
1995, *67*, 1351

nitrilium betaines
Derivatives of *nitriles* having the general structure $R{-}C{\equiv}N^+{-}Y^-$. A subclass of 1,3-*dipolar compounds* including *nitrile imides, nitrile oxides, nitrile sulfides* and *nitrile ylides*.
1995, *67*, 1351

nitrilium ions
Cations derived formally by attachment of one hydron to the nitrogen atom of a nitrile and hydrocarbyl derivatives thereof. E.g. $PhC{\equiv}N^+H \leftrightarrow PhC^+{=}NH$ benzonitrilium.
1995, *67*, 1351

nitrimines
Compounds having the structure $O_2NN{=}CR_2$ (also called *N*-nitroimines).
1995, *67*, 1351

nitro compounds
Compounds having the nitro group, $-NO_2$ (free valence on nitrogen), which may be attached to carbon, nitrogen (as in *nitramines*), or oxygen (as in nitrates), among other elements (in the absence of specification, *C*-nitro compounds are usually implied).

See also *dipolar compounds*.
1995, *67*, 1351

nitrogen fixation
The assimilation of atmospheric nitrogen by microbial reduction to ammonia and conversion into organonitrogen compounds such as amino acids. Only a limited number of *microrganisms* are able to fix nitrogen.
1992, *64*, 161

nitrogen laser
A source of pulsed semi-coherent *superradiance* mainly around 337 nm. The lasing species is molecular nitrogen.
See *gas lasers*.
1996, *68*, 2256

nitrogen mustards
See *mustards*.
1995, *67*, 1351

nitrogen ylides
See *ylides*.
1995, *67*, 1351

nitrolic acids
Compounds having the structure $RC({=}NOH)NO_2$.
1995, *67*, 1351

nitrones
The *N*-oxides of *imines*, that have the structure $R_2C{=}N^+(O^-)R'$ ($R' \neq H$). Synonymous with *azomethine oxides*. *N*-Oxides $R_2C{=}N^+(O^-)H$ may be included.
See also *dipolar compounds*.
1995, *67*, 1351

nitronic acids
See *azinic acids*.
1995, *67*, 1351

nitrosamides
Although this term has, regrettably, been used to mean the *N*-nitroso derivatives of *amides*, logically it refers to amides of nitrous acid; the unstable parent compound H_2NNO is, in fact, named nitrosamide, but its derivatives formed by substitution with *hydrocarbyl* groups are by long custom called *nitrosamines*.
See also *nitramines*.
1995, *67*, 1352

nitrosamines
N-Nitroso *amines* compounds of the structure R_2NNO. Compounds RNHNO are not ordinarily isolable, but they, too, are nitrosamines. The name is a contraction of *N*-nitrosoamine and, as such, does not require the *N* locant.
See also *nitrosamides*.
1995, *67*, 1352

nitrosimines
N-Nitroso *imines* $O{=}NN{=}CR_2$.
1995, *67*, 1352

nitroso compounds
Compounds having the nitroso group, $-NO$, attached to carbon, or to another element, most commonly nitrogen or oxygen.
See *nitrosamines*.
1995, *67*, 1352

nitrosolic acids
Compounds having the structure RC(=NOH)NO.
1995, *67*, 1352

nitroxides
Nitroxide is the parent name used by Chemical Abstracts Service for $H_2N-O·$, e.g. $(ClCH_2)_2N-O·$ bis(chloromethyl) nitroxide. The IUPAC name is bis(chloromethyl)aminoxyl. Nitroxides should not be used as a class name for aminoxyl radicals.
See *aminoxyl radicals*.
1995, *67*, 1352

nitroxyl radicals
See *aminoxyl radicals*.
1995, *67*, 1352

no-bond resonance
See *hyperconjugation*.
1994, *66*, 1145

no carrier added
A preparation of a *radioactive isotope* which is essentially free from stable *isotopes* of the element in question.
1994, *66*, 2522

noise
The random fluctuations occurring in a signal that are inherent in the combination of instrument and method.
1989, *61*, 1663; see also 1995, *67*, 1750

no-load indication (for a precision balance)
The deflection or *rest point* (no-load reading) multiplied by the value of the division for zero load (rider at zero); symbol i_0.
O.B. 36

nominal linear flow (in chromatography), *F*
The volumetric flowrate of the *mobile phase* divided by the area of the cross section of the column (cm min^{-1}) i.e. the linear flowrate in a part of the column not containing *packing*.
O.B. 102

nominally labelled tracer
A *tracer* in which the *label* is present mainly in a specified position.
1994, *66*, 2526

nonadiabatic
See *adiabatic*.
1996, *68*, 175

nonadiabatic coupling
This is momentum coupling between two adiabatic potential-energy surfaces.
See also *adiabatic*.
1996, *68*, 175

nonadiabatic electron transfer
See *diabatic electron transfer*.
1996, *68*, 2255

+ nonadiabatic photoreaction
Synonymous with *diabatic photoreaction*. Use of the double negative is discouraged.
1996, *68*, 2256

nonbonded interactions
Intramolecular attractions or repulsions between atoms that are not directly linked to each other, affecting the thermodynamic stability of the chemical species concerned.
See also *eclipsing strain, transannular strain*.
1996, *68*, 2212

non-calorimetric thermophysical measurements
These include (p, V, T) measurements, saturated density measurements, and any other measurements which give information on the (p, V, T) surface of a fluid.
1981, *53*, 1818

nonclassical carbocation
A *carbocation* the ground state of which has delocalized (bridged) bonding π- or σ-electrons. (N.B. Allylic and benzylic carbocations are not considered nonclassical.)
1994, *66*, 1145

non-coherent source (in spectrochemistry)
See *coherent source (in spectrochemistry)*.
1985, *57*, 1456

non-crystalline electrodes
Electrodes in which a support matrix, containing an ion exchanger (either cationic or anionic), a plasticizer solvent, and possibly an uncharged, selectivity-enhancing species, forms the ion-selective membrane which is usually interposed between two aqueous solutions. The support used can be either macroporous [e.g. poly(propylene carbonate) filter, glass frit, etc.] or microporous (e.g. 'thirsty' glass or inert polymeric material such as PVC) yielding with the ion-exchanger and the solvent a 'solidified' homogeneous mixture. These electrodes exhibit a response due to the presence of the ion-exchange material in the membrane. The solvent-polymeric-membrane is an example.
1994, *66*, 2534

non-destructive activation analysis
An *activation analysis* procedure in which, after the *irradiation*, no chemical and physical operations are applied which cause a change of any properties of the sample.
1982, *54*, 1535

non-diagram line (in X-ray spectroscopy)
See *X-ray satellite*.
1991, *63*, 739

non-dissociative chemisorption
See *chemisorption*.
1976, *46*, 76

non-draining
An adjective describing a chain macromolecule that behaves in a hydrodynamic field as though the solvent within the domain of the macromolecule were virtually immobilized with respect to the macromolecule.
P.B. 61

non-equilibrium reaction
If the reactants in a chemical reaction are not initially present in a Boltzmann distribution the reaction is referred to as a non-equilibrium reaction.
See *equilibrium reaction*.
1996, *68*, 175

non-graphitic carbon
All varieties of solids consisting mainly of the element *carbon* with two-dimensional long-range order of the carbon atoms in planar hexagonal networks, but without any measurable crystallographic order in the third direction (*c*-direction) apart from more or less parallel stacking.
Notes:
Some varieties of non-graphitic carbon convert on heat treatment to *graphitic carbon* (*graphitizable carbon*) but some others do not (*non-graphitizable carbon*).
See also *amorphous carbon*.
1995, *67*, 498

non-graphitizable carbon
A *non-graphitic carbon* which cannot be transformed into *graphitic carbon* solely by high-temperature treatment up to 3300 K under atmospheric pressure or lower pressure.
Notes:
The term non-graphitizable is limited to the result of heat treatment without additional influence of foreign matter or neutron radiation. Non-graphitizable carbon can be transformed into graphitic carbon by a high-temperature process via intermediate dissolution in foreign matter and precipitation under high pressure or by radiation damage.
1995, *67*, 498

non-isotopic labelling
Labelling in which the resulting product has a different chemical composition from the initial one.
1994, *66*, 2521

non-linearity error (in spectrochemical analysis)
An error caused by any deviation from linearity of the response of the detector to the measured *radiant power*.
1988, *60*, 1456

non-linear optical effect
An effect brought about by electromagnetic radiation the magnitude of which is not proportional to the irradiance. Non-linear optical effects of importance to photochemistry are *harmonic frequency generation*, lasers, Raman shifting, *upconversion*, and others.
1996, *68*, 2256

non-localized adsorption
See *mobile adsorption*.
1976, *46*, 76

non-polarized interphases
Interphases for which the exchange of common charged components between the phases proceeds unhindered.
1983, *55*, 1253

non-radiative decay
The disappearance of an excited species due to a radiationless transition.
1996, *68*, 2256

nonselective detector
See *radiation detector*.
1995, *67*, 1748

nonselectively labelled
An *isotopically labelled* compound is designated as nonselectively labelled when the position(s) and the number of the labelling *nuclide(s)* are both undefined.
B.B. 517; 1981, *53*, 1896

nonselective quantum counter
See *radiation detector*.
1995, *67*, 1748

non-specific adsorption
Ions approach an *interface* differently depending on the forces in play. Ions are non-specifically adsorbed (positively or negatively) when they are subjected in the *interphase* only to long-range coulombic interactions (attraction or repulsion). They are believed to retain their *solvation* shell, and in the position of closest approach to the interface they are separated from it by one or more molecular layers.
See also *specific adsorption*.
1986, *58*, 447

non-uniform corrosion
Corrosion is non-uniform if the time average of the corrosion current through a unit area depends on its position on the surface.
Non-uniform corrosion can be due to inhomogeneities of structure or of composition of the corroding material, or to inhomogeneities of the enviroment. Special cases of non-uniform corrosion such as pitting or intergranular corrision are sometimes called localized corrosion.
1989, *61*, 21

non-uniform polymer
A *polymer* comprising molecules non-uniform with respect to relative molecular mass or constitution or both.
See *uniform polymer (3)*.
1996, *68*, 2302

non-vertical energy transfer
An energy transfer process which has a low Franck–Condon factor.
See *Franck–Condon principle*.
1996, *68*, 2256

no-observed-adverse-effect-level (NOAEL)
Greatest concentration or amount of a substance, found by experiment or observation, which causes no detectable adverse alteration of morphology, functional capacity, growth, development, or life span of the target organism under defined conditions of exposure.
1993, *65*, 2076

no-observed-effect-level (NOEL)
Greatest concentration or amount of a substance, found by experiment or observation, that causes no alterations of morphology, functional capacity, growth, development, or life span of target organisms distinguishable from those observed in normal (control) organisms of the same species and strain under the same defined conditions of exposure.
1993, 65, 2076

nor-
Affix used to denote the elimination of one methylene group from a side chain of a *parent* structure (including a methyl group).
B.B. 497

normal
The term 'normal' in e.g. 'normal boiling temperature' means the value at a pressure of 101325 Pa.
1994, 66, 537

normal distribution
See *Gaussian band shape*.
1990, 62, 2203

normal kinetic isotope effect
See *isotope effect*.
1994, 66, 1145

normal-phase chromatography
An elution procedure in which the stationary phase is more polar then the mobile phase. This term is used in liquid chromatography to emphasize the contrast to reversed-phase chromatography.
1993, 65, 826

normal region (for electron transfer)
In plots relating rate constants for *electron transfer*, or quantities related to it, with the standard Gibbs energy for the reaction (ΔG^o), the region for which the rate constants increase with increasing exergonicity of the reaction is called the normal region. This region is predicted by the Marcus theory for *outer-sphere electron transfer* for the case of $\Delta G^o \leq \lambda$ in the *Marcus equation*.
1996, 68, 2255

normal stress, σ
Force acting normally to a surface divided by the area of the surface.
G.B. 12

normal X-ray level
Synonymous with *diagram level*.
1991, 63, 738

Norrish type II photoreaction
Intramolecular abstraction of a γ-hydrogen by an excited carbonyl compound to produce a 1,4-biradical as a primary photoproduct, e.g.

1996, 68, 2256

Norrish Type I photoreaction
α-Cleavage of an excited carbonyl compound leading to an acyl-alkyl radical pair (from an acyclic carbonyl compound) or an acyl-alkyl biradical (from a cyclic carbonyl compound) as a primary photoproduct; e.g.

1996, 68, 2256

n-star macromolecule
See *star macromolecule*.
1996, 68, 2296

n-strand chain
See *multi-strand chain*.
1996, 68, 2295

n-strand molecule
See *multi-strand macromolecule*.
1996, 68, 2296

N-terminal residue (in a polypeptide)
See *amino-acid residue (in a polypeptide)*.
W.B. 80

nth order phase transition
A transition in which the molar *Gibbs energy* (or chemical potential) and its $(n - 1)$th-order derivatives are continuous, whereas the nth-order derivatives with respect to temperature and pressure are discontinuous at the transition point.
See *first-order transition, second-order transition*.
1994, 66, 587

nuclear atom
See *coordination entity*.
R.B. 145; B.B. 337

nuclear chemistry
The part of chemistry which deals with the study of nuclei and nuclear reactions using chemical methods.
1994, 66, 2516

nuclear decay
A spontaneous nuclear transformation.
1994, 66, 2518

nuclear disintegration
Nuclear decay involving a splitting into more nuclei or the emission of particles.
1982, 54, 1540

nuclear fission
The division of a nucleus into two or more parts with masses of equal order of magnitude, usually accompanied by the emission of neutrons, gamma radiation and, rarely, small charged nuclear fragments.
1994, 66, 2519

nuclear fuel
Material containing fissile nuclides, which when placed in a *reactor*, enables a *chain reaction* to be achieved.
1982, 54, 1543

nuclear fusion
See *nuclear fusion reaction*.
1982, 54, 1547

nuclear fusion reaction
A reaction between two light nuclei resulting in the production of a nuclear species heavier than either initial nucleus.
1982, *54*, 1543

nuclear graphite
A *polygranular graphite* material for use in nuclear reactor cores consisting of *graphitic carbon* of very high chemical purity. High purity is needed to avoid absorption of low-energy neutrons and the production of undesirable radioactive species.
Notes:
Apart from the absence of neutron-absorbing impurities, modern reactor graphites are also characterized by a high degree of *graphitization* and no preferred bulk orientation. Such properties increase the dimensional stability of the nuclear graphite at high temperatures and in a high flux of neutrons. The term nuclear graphite is often, but incorrectly, used for any *graphite material* in a nuclear reactor, even if it serves only for structural purposes.
1995, *67*, 498

nuclear isomers
Nuclides having the same *mass number* and *atomic number*, but occupying different nuclear energy states.
1982, *54*, 1545

nuclear level
One of the energy values at which a *nucleus* can exist for an appreciable time ($> 10^{-22}$ s).
1982, *54*, 1547

nuclear magneton
Electromagnetic fundamental physical constant $\mu_N = (m_e/m_p)\mu_B = 5.050\ 7866\ (17) \times 10^{-27}$ J T^{-1}, where m_e is the electron rest mass, m_p the proton rest mass and μ_B the Bohr magneton.
CODATA Bull., 1986, *63*, 1

nuclear particle
A *nucleus* or any of its constituents in any of their energy states.
1982, *54*, 1547

nuclear quadrupole moment (spectroscopic)
A parameter which describes the effective shape of the equivalent ellipsoid of the nuclear charge distribution, $Q\ 0$ for prolate (e.g. ^{57}Fe, ^{197}Au); $Q < 0$ for oblate (e.g. ^{119}Sn, ^{129}I) nuclei.
1976, *45*, 214

nuclear reactor
A device in which a self-sustaining *nuclear fission* chain reaction can be maintained and controlled. The term is sometimes applied to a device in which a *nuclear fusion* reaction can be produced and controlled.
1982, *54*, 1550

nuclear transformation
The change of one *nuclide* into another with a different *proton number* or *nucleon number*.
1982, *54*, 1553

nuclear transition
For a *nucleus* a change from one quantized energy state into another or a *nuclear transformation*.
1982, *54*, 1553

nucleating agent
A material either added to or present in a system, which induces either homogeneous or heterogeneous *nucleation*.
1972, *31*, 608

nucleation (in colloid chemistry)
The process by which nuclei are formed in solution. The condensation of a single chemical compound is called homogeneous nucleation. The simultaneous condensation of more than one compound is called simultaneous nucleation. The condensation of a compound on a foreign substance is called heterogeneous nucleation.
O.B. 84; see also 1972, *31*, 608

nucleation and growth
A process in a *phase transition* in which nuclei of a new phase are first formed, followed by the propagation of the new phase at a faster rate.
See *continuous precipitation, discontinuous precipitation*.
1994, *66*, 587

nucleic acids
Macromolecules, the major organic matter of the nuclei of biological cells, made up of *nucleotide* units, and hydrolysable into certain *pyrimidine* or *purine bases* (usually adenine, cytosine, guanine, thymine, uracil), D-ribose or 2-deoxy-D-ribose and phosphoric acid.

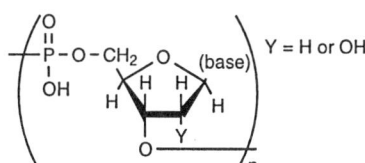

See *nucleotides, ribonucleic acids, deoxyribonucleic acids*.
1995, *67*, 1352; see also W.B. 110; 1992, *64*, 161

nucleofuge
A *leaving group* that carries away the bonding electron pair. For example, in the *hydrolysis* of an alkyl chloride, Cl$^-$ is the nucleofuge. The tendency of atoms or groups to depart with the bonding electron pair is called nucleofugality. The adjective is nucleofugal.
See also *electrofuge, nucleophile*.
1994, *66*, 1145

nucleon
Heavy nuclear particle: *proton* or *neutron*.
Physical Chemistry Division, unpublished

nucleon number
Synonymous with *mass number*.
G.B. 20

nucleophile (nucleophilic)
A nucleophile (or nucleophilic reagent) is a reagent that forms a bond to its reaction partner (the *electrophile*) by donating both bonding electrons.
A 'nucleophilic substitution reaction' is a *heterolytic* reaction in which the reagent supplying the entering group acts as a nucleophile. For example:

$$MeO^- + Et-Cl \longrightarrow MeO-Et + Cl^-$$
nucleophile $\quad\quad\quad\quad\quad\quad\quad\quad\quad$ nucleofuge

The term 'nucleophilic' is also used to designate the apparent polar character of certain *radicals*, as inferred from their higher relative reactivity with reaction sites of lower electron density.
Nucleophilic reagents are *Lewis bases*.
1994, 66, 1146

nucleophilic catalysis
Catalysis by a *Lewis base*, involving formation of a *Lewis adduct* as a *reaction intermediate*. For example, the hydrolysis of acetic anhydride in aqueous solution catalysed by pyridine:

$$C_5H_5N + (CH_3CO)_2O$$
$$\rightarrow [C_5H_5NCOCH_3]^+ + CH_3CO_2^-$$
$$[C_5H_5NCOCH_3]^+ + H_2O$$
$$\rightarrow C_5H_5N + CH_3CO_2H + H^+_{aq}$$

See also *electrophilic, nucleophilicity*.
1994, 66, 1146

nucleophilicity
1. The property of being *nucleophilic*.
2. The relative reactivity of a nucleophilic reagent. (It is also sometimes referred to as 'nucleophilic power'.) Qualitatively, the concept is related to *Lewis basicity*. However, whereas Lewis basicity is measured by relative equilibrium constants:

$$B: + A \xrightleftharpoons{K} B^+\text{-}A^-$$

nucleophilicity of a Lewis base is measured by relative *rate constants* of different nucleophilic reagents towards a common *substrate*, most commonly involving formation of a bond to carbon:

$$B: + A\text{-}Z \rightarrow B^+\text{-}A^- + Z:^-$$

See also *electrophilicity, Ritchie equation, Swain–Scott equation*.
1994, 66, 1146

nucleoproteins
Proteins having *nucleic acids* as prosthetic groups, and thus yielding *nucleic acids* (or their cleavage products) as well as amino acids on hydrolytic cleavage.
1995, 67, 1352

nucleosides
Ribosyl or deoxyribosyl derivatives (rarely, other glycosyl derivatives) of certain *pyrimidine* or *purine* bases. They are thus *glycosylamines* or N-*glycosides* related to *nucleotides* by the lack of phosphorylation. It has also become customary to include among nucleosides analogous substances in which the *glycosyl group* is attached to carbon rather than nitrogen ('C-nucleosides').
See also *nucleic acids*.
1995, 67, 1352; see also W.B. 110; 1992, 64, 161

nucleotide bases
The heterocyclic pyrimidine and purine compounds which are constituents of all *nucleic acids*. Adenine (A), guanine (G) and cytosine (C) are found in both *DNA* and *RNA*, thymine (T) is found (primarily) in DNA and uracil (U) only in RNA.
1992, 64, 147

nucleotides
Compounds formally obtained by esterification of the 3′ or 5′ hydroxy group of *nucleosides* with phosphoric acid. They are the monomers of *nucleic acids* and are formed from them by hydrolytic cleavage.

Y = H, a deoxyribonucleotide
Y = OH, a ribonucleotide
1995, 67, 1352; see also W.B. 109, 111; 1992, 64, 162

nucleus
The positively charged central portion of an *atom*, excluding the orbital *electrons*.
1982, 54, 1547

nuclide
A species of atom, characterized by its mass number, atomic number and nuclear energy state, provided that the mean life in that state is long enough to be observable.
1994, 66, 2522; R.B. 35; O.B. 233; 1982, 54, 1547

nuclidic mass
The rest mass of a *nuclide* expressed in *atomic mass units*.
1982, 54, 1547

nuisance threshold (in atmospheric chemistry)
That concentration of an *air pollutant* that is considered objectionable. In the case of a substance with an objectionable odour, it is the smallest concentration of the substance which can be detected by a human being (nose).
1990, 62, 2203

number concentration, C, n
Number of entities of a constituent in a mixture divided by the volume of the mixture.
See *concentration*.
G.B. 39; 1996, 68, 984

number content, N_B/m
Number of defined particles, or *elementary entities*, of a component in a system divided by the mass of that system.
 1996, *68*, 984

number density, n
Number of particles divided by the volume they occupy.
 G.B. 39; 1990, *62*, 2183

number-distribution function
A distribution function in which the relative amount of a portion of a substance with a specific value, or a range of values, of the random variable(s) is expressed in terms of mole fraction.
 P.B. 56

number flow rate, q_N
Number of defined particles, or *elementary entities* of a defined component, crossing a cross-section divided by the time.
 See *mass flow rate, volume flow rate*.
 1996, *68*, 984

number fraction
 See *amount fraction*.
 G.B. 41; 1996, *68*, 984

number of entities, N
Integer number obtained by counting of entities, which are usually *molecules*, *atoms* or *ions*.
 G.B. 39; 1996, *68*, 985

number of moles
 See *amount of substance*.
 G.B. 46

number of theoretical plates
 See *plate number (in chromatography)*.
 1993, *65*, 847

numerical value (of a quantity)
Quotient of the value of a quantity and the unit used in its expression.
 See *value (of a quantity)*.
 1996, *68*, 985

observation height (in flame emission and absorption spectrometry), h_{obs}
The difference in height between the axis of the observed space (optical axis) and the burner tip (in mm). The optical axis of the instrument should be the same as the optical axis of the observation space.
Other definitions which express the observation height as a fraction of the whole flame height are not recommended because the height to the tip of the flame is not well defined. The observation height should be stated in the method.
1986, *58*, 1741

observation pathlength (in flame emission and absorption spectrometry), l_{obs}
The intersection of the optical axis and the observation space (in mm).
1986, *58*, 1741

observation space (in flame emission and absorption spectrometry)
The intersection of the optical beam and that part of the flame where the net signal is at least half of the maximum net signal.
The characteristics of the observation space depend on the temperature of the flame, the stoichiometry of the gases and the properties of the processed fluid (presence of proteins for instance). The observation space in atomic absorption is analogous to the observation space of a cuvette in molecular absorption spectrometry.
See also *observation volume*.
1986, *58*, 1741

observation volume (in flame emission and absorption spectrometry), V_{obs}
The volume of that part of the flame that is observed through the optical device (in μl).
See also *observation space*.
1986, *58*, 1741

observed rate coefficient
See *order of reaction*.
1994, *66*, 1147

occlusion (molecular)
The process of incorporation of foreign substances as molecular species within *precipitates* as they are formed.
O.B. 85

***octahedro-* (in inorganic nomenclature)**
An affix used in names to denote six atoms bound into an octahedron.
R.B. 245; B.B.465

odd-electron ion
Synonymous with *radical ion*.
1991, *63*, 1549

ODMR (Optically Detected Magnetic Resonance)
A double resonance technique in which transitions between spin sublevels are detected by optical means. Usually these are sublevels of a triplet and the transitions are induced by microwaves. For different types of optical detection the following expressions are used: ADMR (absorption), DEDMR (delayed emission, non-specified), DFDMR (delayed fluorescence), FDMR (fluorescence), PDMR (phosphorescence). If a reaction yield is followed the expression RYDMR (reaction yield detected magnetic resonance) is used.
1996, *68*, 2257

odour threshold (in atmospheric chemistry)
The concentration of a compound which produces an odour which is detectable by a human being (nose). For certain compounds this threshold is very low (e.g. 1 part in 10^9 for certain sulfides).
1990, *62*, 2204

ohm
SI derived unit of electrical resistance, $\Omega = V A^{-1} = m^2 kg s^{-3} A^{-2}$.
G.B. 72; 1996, *68*, 985

olefins
Acyclic and cyclic *hydrocarbons* having one or more carbon–carbon double bonds, apart from the formal ones in *aromatic compounds*. The class olefins subsumes *alkenes* and cycloalkenes and the corresponding polyenes.
See also *cycloalkanes*.
1995, *67*, 1353

oligo
A prefix meaning 'a few', and used for compounds with a number of repeating units intermediate between those in monomers and those in high polymers. The limits are not precisely defined, and in practice vary with the type of structure being considered, but are generally from 3 to 10, e.g. oligopeptides, oligosaccharides.
1995, *67*, 1353

oligomer
A substance composed of *oligomer molecules*.
An oligomer obtained by telomerization is often termed a *telomer*.
1996, *68*, 2289

oligomeric
See *oligomer molecule*.
1996, *68*, 2289

oligomerization
The process of converting a *monomer* or a mixture of monomers into an *oligomer*.
An oligomerization by chain reaction carried out in the presence of a large amount of *chain-transfer* agent, so that the *end groups* are essentially fragments of the chain transfer agent, is termed *telomerization*.
1996, *68*, 2306

oligomer molecule
A molecule of intermediate relative molecular mass, the structure of which essentially comprises a small plurality of units derived, actually or conceptually, from molecules of lower relative molecular mass.

Notes:
1. A molecule is regarded as having an intermediate relative molecular mass if it has properties which do vary significantly with the removal of one or a few of the units.
2. If a part or the whole of the molecule has an intermediate relative molecular mass and essentially comprises a small plurality of units derived, actually or conceptually, from molecules of lower relative molecular mass, it may be described as oligomeric, or by oligomer used adjectivally.
   1996, 68, 2289

oligonucleotides
   See *oligo, nucleotides*.
   1995, 67, 1353; see also W.B. 111

oligopeptides
   See *oligo, peptides*.
   1995, 67, 1353; see also W.B. 48

oligosaccharides
   See *oligo, saccharides*.
   1995, 67, 1353

one-colour indicator
   See *colour indicator*.
   O.B. 48

onium compounds
1. *Cations* (with their counter-ions) derived by addition of a *hydron* to a mononuclear *parent hydride* of the nitrogen, chalcogen and halogen families.

H_4N^+ ammonium	H_3Se^+ selenonium
H_3O^+ oxonium	H_2Br^+ bromonium
H_2F^+ fluoronium	H_4Sb^+ stibonium
H_4P^+ phosphonium	H_3Te^+ telluronium
H_3S^+ sulfonium	H_2I^+ iodonium
H_2Cl^+ chloronium	H_4Bi^+ bismuthonium
H_4As^+ arsonium	

2. Derivatives formed by substitution of the above parent ions by univalent groups. The number of substituted hydrogen atoms is, especially in the case of *hydrocarbyl* substituents, indicated by the adjectives primary, secondary, tertiary or quaternary. E.g. Cl_2F^+ dichlorofluoronium, $(CH_3)_2S^+H$ dimethylsulfonium (a secondary sulfonium ion), $Cl(CH_3)_3P^+$ chlorotrimethylphosphonium, $(CH_3CH_2)_4N^+$ tetraethylammonium (a quaternary ammonium ion).

   See also *arsonium compounds, halonium ions, oxonium ions, phosphonium compounds, quaternary ammonium compounds, stibonium compounds, sulfonium compounds*.

3. Derivatives formed by substitution of the above parent ions by groups having two or three free valencies on the same atom. Such derivatives are, where possible, designated by a specific class name. E.g. $RC\equiv O^+$ hydrocarbylidyne oxonium ions, $R_2C=N^+H_2\ X^-$ iminium compounds, $RC\equiv NH^+$ nitrilium ions.
   1995, 67, 1353; 1994, 66, 1146

open atomizer
   See *electrothermal atomizer*.
   1992, 64, 254

open dynamic system (in spectrochemical analysis)
   See *mercury flow system (in spectrochemical analysis)*.
   1992, 64, 263

open film
A *film* in which mass transfer can occur between the film and the coexisting bulk phases, for all the components. The term partly open film applies to a film in which mass transfer can occur only for certain components. The term closed film applies to those films with fixed mass.
   1994, 66, 1671

open hearth furnace (in atmospheric chemistry)
Reverberatory furnace, containing a basin-shaped hearth, for melting and refining suitable types of pig iron, iron ore, and scrap for steel production. A large amount of dust from ore and other materials and splashings from slag are carried away by the waste gases; a supplementary chamber is commonly used for collecting slag and dust.
   1990, 62, 2204

open static system (in spectrochemical analysis)
   See *mercury flow system (in spectrochemical analysis)*.
   1992, 64, 263

open-tubular column (in chromatography)
A column, usually having a small diameter, in which either the inner tube wall, or a liquid or active solid held stationary on the tube wall acts as the stationary phase and there is an open, unrestricted path for the mobile phase.
   1993, 65, 831; O.B. 97

operational pH cell
An electrochemical cell which is the basis of practical pH measurement consisting of a hydrogen ion-responsive electrode (hydrogen gas, or glass) and a *reference electrode* immersed in the *test solution*.
   See *pH*.
   1985, 57, 540

operational pH standard
Certain substances which meet the criteria of (i) preparation in highly pure state reproducibly and (ii) stability of solution over a reasonable period of time are designated as operational standards in aqueous solution of specified concentration. Their number is in principle unlimited but values are available now for 15 solutions. The values of operational pH standard, i.e. pH(OS), are assigned by comparison with the *reference value pH standard*, i.e. pH(RVS), in cells with *liquid junction*, the operational cells, where the liquid junctions are formed within vertical 1 mm tubes.

$$Pt\ |\ H_2\ |\ OS\ \|\ KCl\ \|\ RVS\ |\ H_2\ |\ Pt\ (Pd).$$
$$> 3.5\ mol\ dm^{-3}$$

O.B. 21
operator gene
See *operon*.
1992, *64*, 162

operon
A functional unit consisting of a promoter, an operator and a number of structural *genes*, found mainly in prokaryotes. The structural genes commonly code for several functionally related *enzymes*, and although they are transcribed as one (polycistronic) mRNA each is independently translated. In the typical operon, the operator region acts as a controlling element in switching on or off the synthesis of mRNA.
1992, *64*, 162

opposing reactions
Composite reactions, occurring in forward and reverse directions:

$$A + B \underset{-1}{\overset{1}{\rightleftharpoons}} Z$$

See also *composite reaction*.
1993, *65*, 2296; 1994, *66*, 1146

optical activity
A sample of material able to rotate the plane of polarization of a beam of transmitted plane-polarized light is said to possess optical activity (or to be optically active). This optical rotation is the classical distinguishing characteristic (sufficient but not necessary) of systems containing unequal amounts of corresponding *enantiomers*. An enantiomer causing rotation in a clockwise direction (when viewed in the direction facing the oncoming light beam) under specified conditions is called dextrorotatory and its chemical name or formula is designated by the prefix (+)-; one causing rotation in the opposite sense is laevorotatory and designated by the prefix (–)-.
Materials with optical activity also exhibit other *chiroptic* phenomena.
1996, *68*, 2212

+ optical antipodes
Obsolete synonym for *enantiomers*. (Usage strongly discouraged).
1996, *68*, 2212

optical-beam error (in spectrochemical analysis)
For wavelength-scanning spectrometers, an error may be introduced if the image position on the photodetector changes. Reflection of the incident radiation, e.g. between the cell walls, is another source of error and results in a measured absorbance slightly higher than the true absorbance. These errors are collectively termed *optical-beam errors*.
1988, *60*, 1457

+ optical density
Synonymous with *absorbance*. The use of the term optical density is discouraged.
1996, *68*, 2257

optical filter
A device which reduces the spectral range (bandpass, cut-off, and interference filter) or radiant power of incident radiation (neutral density or attenuance filter) upon transmission of radiation.
1996, *68*, 2257

+ optical isomers
Obsolescent synonym for *stereoisomers* with different optical properties. They should be described as diastereoisomers or enantiomers. (Usage strongly discouraged).
1996, *68*, 2212

+ optically labile
A term describing a system in which stereoisomerization results in a change of optical rotation with time. (Usage strongly discouraged).
1996, *68*, 2212

optical purity
The ratio of the observed optical rotation of a sample consisting of a mixture of enantiomers to the optical rotation of one pure enantiomer.
See *enantiomeric excess*.
1996, *68*, 2212

+optical resolution
Usage strongly discouraged.
See *resolution*.
1996, *68*, 2212

optical rotation
See *optical activity*.
1996, *68*, 2212

optical rotatory power
Angle of optical rotation divided by the optical path length through the medium and by either the mass concentration of the substance giving the specific optical rotatory power $[\alpha]_\lambda^\theta$ or by amount concentration, giving the molar optical rotatory power, α_m.
G.B. 33

optical spectroscopy
The study of systems by the electromagnetic radiation with which they interact or that they produce. By convention, spectroscopic properties are expressed in terms of wavelength *in vacuo*, not the wavelength within the medium being studied.
1985, *57*, 107

optical yield
In a *chemical reaction* involving *chiral* reactants and products, the ratio of the *optical purity* of the product to that of the precursor, reactant or catalyst. This should not be confused with *enantiomeric excess*. The optical yield is in no way related to the chemical yield of the reaction.
See *stereoselectivity*.
1994, *66*, 1146; 1996, *68*, 2212

optoacoustic spectroscopy
Synonymous with *photoacoustic spectroscopy*.
1996, *68*, 2257

orbital
See *atomic orbital, molecular orbital*.
1994, *66*, 1147

orbital energy
Eigenvalue of a one-electron effective hamiltonian belonging to an orbital.
G.B. 18

orbital steering
A concept expressing that the stereochemistry of approach of two reacting species is governed by the most favourable overlap of their appropriate *orbitals*.
1994, *66*, 1147

orbital symmetry
The behaviour of an atomic or localized *molecular orbital* under molecular symmetry operations characterizes its orbital symmetry. For example, under a reflection in an appropriate symmetry plane, the phase of the orbital may be unchanged (symmetric), or it may change sign (antisymmetric), i.e. the positive and negative lobes are interchanged.

A principal context for the use of orbital symmetry is the discussion of chemical changes that involve 'conservation of orbital symmetry'. If a certain symmetry element (e.g. the reflection plane) is retained along a reaction pathway, that pathway is 'allowed' by orbital symmetry conservation if each of the occupied orbitals of the reactant(s) is of the same symmetry type as a similarly (e.g. singly or doubly) occupied orbital of the product(s). This principle permits the qualitative construction of correlation diagrams to show how molecular orbitals transform (and how their energies change) during idealized chemical changes (e.g. *cycloadditions*).

An idealized single bond is a σ-bond, i.e. it has cylindrical symmetry, whereas a p-orbital or π-bond orbital has π-symmetry, i.e. it is antisymmetric with respect to reflection in a plane passing through the atomic centres with which it is associated. In ethene, the π-bonding orbital is symmetric with respect to reflection in a plane perpendicular to and bisecting the C–C bond, whereas the π^*-antibonding orbital is antisymmetric with respect to this operation.

Considerations of orbital symmetry are frequently grossly simplified in that, for example, the p-orbitals of a carbonyl group would be treated as having the same symmetry as those of ethene, and the fact that the carbonyl group in, for example, camphor, unlike that in formaldehyde, has no mirror planes would be ignored. These simplified considerations nevertheless afford the basis of one approach to the understanding of the rules which indicate whether *pericyclic reactions* are likely to occur under thermal or photochemical conditions.
See also *sigma, pi*.
1994, *66*, 1147

order-disorder transition
A transition in which the degree of order of the system changes. Three principal types of disordering transitions may be distinguished: (i) positional disordering in a solid, (ii) orientational disordering which may be static or dynamic and (iii) disordering associated with electronic and nuclear spin states.

Examples:
(i) The transition of $LiFeO_2$, with a tetragonal unit cell, in which the Li^+ and Fe^{3+} cations are perfectly ordered on crystallographically non-equivalent octahedral sites to cubic $LiFeO_2$ in which the Li^+ and Fe^{3+} cations are distributed randomly over all the octahedral sites.
(ii) The transition of orthorhombic KCN to cubic KCN in which the CN^- ions become oriented in any of the eight [111] directions.
(iii) A *superconducting transition*.
1994, *66*, 587

order of reaction, *n*
If the macroscopic (observed, empirical or phenomenological) *rate of reaction* (*v*) for any reaction can be expressed by an empirical differential rate equation (or rate law) which contains a factor of the form $k[A]^\alpha[B]^\beta$... (expressing in full the dependence of the rate of reaction on the concentrations [A], [B] ...) where α, β are constant exponents (independent of concentration and time) and k is independent of [A] and [B] etc. (rate constant, rate coefficient), then the reaction is said to be of order α with respect to A, of order β with respect to B, ... , and of (total or overall) order $n = \alpha + \beta + ...$ The exponents $\alpha, \beta, ...$ can be positive or negative integral or rational nonintegral numbers. They are the reaction orders with respect to A, B, ... and are sometimes called 'partial orders of reaction'. Orders of reaction deduced from the dependence of initial rates of reaction on concentration are called 'orders of reaction with respect to concentration'; orders of reaction deduced from the dependence of the rate of reaction on time of reaction are called 'orders of reaction with respect to time'.

The concept of order of reaction is also applicable to chemical rate processes occurring in systems for which concentration changes (and hence the rate of reaction) are not themselves measurable, provided it is possible to measure a *chemical flux*. For example, if there is a dynamic equilibrium according to the equation:

$$aA \rightleftharpoons pP$$

and if a chemical flux is experimentally found, (e.g. by NMR *line shape analysis*) to be related to concentrations by the equation:

$$\varphi_{-A}/\alpha = k[A]^\alpha[L]^\lambda$$

then the corresponding reaction is of order α with respect to A ... and of total (or overall) order n (= $\alpha + \lambda + ...$). The proportionality factor k above is called the (*n*th order) 'rate coefficient'. Rate coefficients referring to (or believed to refer to) *elementary reactions* are called 'rate constants' or, more appropriately 'microscopic' (hypothetical, mechanistic) rate constants.

The (overall) order of a reaction cannot be deduced from measurements of a 'rate of appearance' or 'rate of disappearance' at a single value of the concentration of a species whose concentration is constant (or effectively constant) during the course of the reaction. If the overall rate of reaction is, for example, given by:

$$v = k [A]^\alpha [B]^\beta$$

but [B] stays constant, then the order of the reaction (with respect to time), as observed from the concentration change of A with time, will be α, and the rate of disappearance of A can be expressed in the form:

$$v_A = k_{obs} [A]^\alpha$$

The proportionality factor k_{obs} deduced from such an experiment is called the 'observed rate coefficient' and it is related to the $(\alpha + \beta)$th order rate coefficient k by the equation:

$$k_{obs} = k[B]^\beta$$

For the common case when $\alpha = 1$, k_{obs} is often referred to as a 'pseudo-first order rate coefficient' (k_ψ).
For a simple (*elementary*) *reactions* a partial order of reaction is the same as the stoichiometric number of the reactant concerned and must therefore be a positive integer (see *rate of reaction*). The overall order is then the same as the *molecularity*. For *stepwise reactions* there is no general connection between stoichiometric numbers and partial orders. Such reactions may have more complex rate laws, so that an apparent order of reaction may vary with the concentrations of the *chemical species* involved and with the progress of the reaction: in such cases it is not useful to speak of orders of reaction, although apparent orders of reaction may be deducible from initial rates. In a stepwise reaction, orders of reaction may in principle always be assigned to the elementary steps.
See also *kinetic equivalence*.
1994, *66*, 1147; 1993, *65*, 2296; 1996, *68*, 176; G.B. 55

order parameter
A normalized parameter that indicates the degree of order of a system. An order parameter of 0 indicates disorder; the absolute value in the ordered state is 1.
1994, *66*, 587

organelles
Separated compartments within a *cell* with specialized functions, e.g. nuclei (containing most of the genetic material), *mitochondria* (respiratory energy supply for the cell), chloroplasts (location of *photosynthesis*) etc.
1992, *64*, 162

organic dye laser
A *liquid laser* in which the active medium is an organic dye in a solvent.
1995, *67*, 1921

organo-
See *organometallic compounds, organoheteryl groups*.
1995, *67*, 1353

organoheteryl groups
Univalent groups containing carbon, which are thus organic, but which have their free valence at an atom other than carbon. This collective term is seldom used; specific subclasses are more frequently encountered (organothio or organylthio, organogermanium or organylgermanium groups). The synonymous term organoelement groups is occasionally encountered. E.g. phenoxy, acetamido, pyridinio ($C_5H_5N^+$), thiocyanato ($N\equiv C-S^-$), trimethylsilyl; but not hydroxyphenyl, aminoacetyl.
See *organyl*.
1995, *67*, 1353

organometallic compounds
Classically compounds having bonds between one or more metal atoms and one or more carbon atoms of an organyl group. Organometallic compounds are classified by prefixing the metal with organo-, e.g. organopalladium compounds. In addition to the traditional metals and semimetals, elements such as boron, silicon, arsenic and selenium are considered to form organometallic compounds, e.g. organomagnesium compounds MeMgI iodo(methyl)magnesium, Et_2Mg diethylmagnesium; an organolithium compound BuLi butyllithium; an organozinc compound $ClZnCH_2C(=O)OEt$ chloro(ethoxycarbonylmethyl)zinc; an organocuprate $Li^+[CuMe_2]^-$ lithium dimethylcuprate; an organoborane Et_3B triethylborane. The status of compounds in which the canonical anion has a delocalized structure in which the negative charge is shared with an atom more electronegative than carbon, as in enolates, may vary with the nature of the anionic moiety, the metal ion, and possibly the medium; in the absence of direct structural evidence for a carbon–metal bond, such compounds are not considered to be organometallic.
See *acetylides, ferrocenophanes, Grignard reagents, metallocenes*.
1995, *67*, 1353

organyl groups
Any organic substituent group, regardless of functional type, having one free valence at a carbon atom, e.g. CH_3CH_2-, $ClCH_2-$, $CH_3C(=O)-$, 4-pyridylmethyl. Organyl is also used in conjunction with other terms, as in organylthio- (e.g. MeS-) and organyloxy.
See *heterocyclyl groups, hydrocarbyl, organoheteryl*.
1995, *67*, 1354

origin of replication (ori)
A sequence of DNA at which replication is initiated on a *chromosome,* plasmid or *virus*.
1992, *64*, 162

ortho acids
Hypothetical compounds having the structure RC(OH)$_3$. Thus hydrated forms of *carboxylic acids*. Orthocarbonic acid, C(OH)$_4$, is generically included.
See also *ortho amides, ortho esters*.
1995, *67*, 1354

ortho amides
Hypothetical compounds having the structure RC(NH$_2$)$_3$, and *N*-substituted derivatives thereof. (R$_2$N)$_4$C are generically included, but such use is obsolescent.
1995, *67*, 1354

ortho- and *peri*-fused (polycyclic compounds)
Polycyclic compounds in which one ring contains two, and only two, atoms in common with each of two or more rings of a contiguous series of rings. Such compounds have *n* common faces and less than 2*n* common atoms.
Examples:

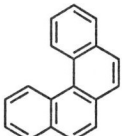

7 common faces
8 common atoms

5 common faces
6 common atoms

"*ortho*– and *peri*–fused" systems

B.B. 22

ortho esters
Compounds having the structure RC(OR')$_3$ (R' ≠ H), or the structure C(OR')$_4$ (R' ≠ H), e.g. HC(OCH$_3$)$_3$ trimethyl orthoformate, C(OCH$_3$)$_4$ tetramethyl orthocarbonate.
1995, *67*, 1354

ortho-fused (polycyclic compounds)
Polycyclic compounds in which two rings have two, and only two, atoms in common. Such compounds have *n* common faces and 2*n* common atoms.
Examples:

3 common faces
6 common atoms

"*ortho*–fused" system

B.B. 22

orthokinetic aggregation (in colloids)
The rate of *aggregation* is in general determined by the frequency of collisions and the probability of cohesion during collision. If the collisions are caused by hydrodynamic motions (e.g convection or sedimentation) this is then referred to as orthokinetic aggregation.
1972, *31*, 610

osazones
1,2-Bis(arylhydrazones) of *ketoaldoses* (aldoketoses) formed from *aldoses* and 2-*ketoses* by reaction with excess arylhydrazine.

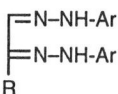

R = rest of monosaccharide chain

See *glycosides*.
1995, *67*, 1354

oscillating reaction
A reaction is said to be oscillating if the concentration of one or more *intermediates* does not vary monotonically, but passes through maximum and minimum values.
1996, *68*, 176

oscillator strength, f_{ij}
Measure for integrated intensity of electronic transitions and related to the Einstein transition probability coefficient A_{ij} by:
$$f_{ij} = 1.4992 \times 10^{-14} \, (A_{ij}/s^{-1}) \, (\lambda/nm)^2,$$
where λ is the transition wavelength. There are, however, differing uses of f.
G.B. 33; see also 1996, *68*, 2257

osmolality, *m*
Quotient of the negative natural logarithm of the rational activity of water and the molar mass of water.
1996, *68*, 985

osmolarity
See *osmotic concentration*.
1996, *68*, 985

osmotic coefficient, φ
Quantity characterizing the deviation of the *solvent* from ideal behaviour referenced to Raoult's law. The osmotic coefficient on a molality basis is defined by:
$$\varphi = \frac{\mu_A^* - \mu_A}{RTM_A \sum_i m_i}$$
and on an amount fraction basis by:
$$\varphi = \frac{\mu_A^* - \mu_A}{RT \ln x_A}$$
where μ_A^* and μ_A are the chemical potentials of the solvent as a pure substance and in solution, respectively, M_A is its molar mass, x_A its amount fraction, R the gas constant and T the temperaure. The latter osmotic coefficient is sometimes called the rational osmotic coefficient.
G.B. 51; 1994, *66*, 546

osmotic concentration, *c*
Product of the *osmolality* and the mass density of water. Formerly called osmolarity.
1996, *68*, 985

osmotic pressure, Π
Excess pressure required to maintain osmotic equilibrium between a solution and the pure solvent separated by a membrane permeable only to the solvent; $\Pi = c_B RT$ where R is the gas constant, T the thermo-

osmotic pressure, reduced
See *reduced osmotic pressure*.
1972, *31*, 615

+ osones
An obsolescent and non-recommended term for 1,2-ketoaldoses, usually derived by hydrolysis of *osazones*.
1995, *67*, 1354; W.B. 132

osotriazoles
The 1,2,3-triazoles formed on oxidizing *osazones*.

R = rest of monosaccharide chain
1995, *67*, 1354

Ostwald ripening
The growth of larger crystals from those of smaller size which have a higher *solubility* than the larger ones.
O.B. 86

outer electric potential, ψ
Electric potential outside the phase concerned. For a conducting sphere of excess charge Q and radius r in vacuum $\psi = Q/4\pi\varepsilon_0 r$.
G.B. 59

outer Helmholtz plane (OHP)
At an electrified interface, the locus of the electrical centres of non-specifically adsorbed ions in their position of closest approach.
1991, *63*, 900

outer-sphere electron transfer
An outer-sphere electron transfer is a reaction in which the electron transfer takes place with no or very weak (4–16 kJ mol^{-1}) electronic interaction between the reactants in the transition state. If instead the donor and the acceptor exhibit a strong electronic coupling, the reaction is described as inner-sphere electron transfer. The two terms derive from studies concerning metal complexes and it has been suggested that for organic reactions the term 'nonbonded' and 'bonded' electron transfer should be used.
See also *inner-sphere electron transfer*.
1994, *66*, 1148; 1996, *68*, 2257

outgassing (of a catalyst)
A form of pretreatment in which a *catalyst* is heated in vacuo to remove adsorbed or dissolved gas.
1976, *46*, 80

out-isomer
See *in-out isomerism*.
1996, *68*, 2212

out-of-plane bending coordinate (in molecular geometry)
Given by

$$\Delta\theta_{i-jkl} = \frac{\Delta z_j}{r_{eij}}\sin\Phi_{kil}$$

where the numbering of the atoms is given in the diagram. Φ_{kil} denotes the angle between the bonds ik and il, Δz_j the perpendicular distance of the atom j from the instantaneous plane ikl and r_{eij} the equilibrium length of the bond ij.
1978, *50*, 1710

output rate
The number of results that is produced by an instrument divided by time of operation.
1989, *61*, 1664

overall activation energy
See *energy of activation*.
1996, *68*, 151

overlap integral, S_{rs}
Integral over space of the type $\int \psi_r^* \psi_s \, d\tau$ where ψ_r and ψ_s are different wavefunctions.
G.B. 17

overpotential, η
Deviation of the potential of an electrode from its equilibrium value required to cause a given current to flow through the electrode.
G.B. 60

oxa-di-π-methane rearrangement
A photochemical reaction of a β,γ-unsaturated ketone to form a saturated α-cyclopropyl ketone. The rearrangement formally amounts to a 1,2-acyl shift and 'bond formation' between the former α and γ carbon atoms.

See also *di-π-methane rearrangement*.
1996, *68*, 2258

+ oxenium ions
An unnecessary and erroneous term for oxylium ions, RO$^+$ (:Ö: is monooxygen, not 'oxene').
1995, *67*, 1354

oxidant (in atmospheric chemistry)
A very qualitative term which includes any and all trace gases which have a greater oxidation potential than oxygen (for example, ozone, peroxyacetyl nitrate, hydrogen peroxide, organic peroxides, NO$_3$, etc.). It is recommended that alternative, more definitive terms be used which define the specific oxidant of interest whenever possible.
1990, *62*, 2204

oxidation
1. The complete, net removal of one or more electrons from a *molecular entity* (also called 'de-electronation').
2. An increase in the *oxidation number* of any atom within any *substrate*.
3. Gain of oxygen and/or loss of hydrogen of an organic substrate.

All oxidations meet criteria 1 and 2, and many meet criterion 3, but this is not always easy to demonstrate. Alternatively, an oxidation can be described as a *transformation* of an organic substrate that can be rationally dissected into steps or *primitive changes*. The latter consist in removal of one or several electrons from the substrate followed or preceded by gain or loss of water and/or *hydrons* or hydroxide ions, or by *nucleophilic* substitution by water or its reverse and/or by an *intramolecular molecular rearrangement*.

This formal definition allows the original idea of oxidation (combination with oxygen), together with its extension to removal of hydrogen, as well as processes closely akin to this type of transformation (and generally regarded in current usage of the term in organic chemistry to be oxidations and to be effected by 'oxidizing agents') to be descriptively related to definition 1. For example the oxidation of methane to chloromethane may be considered as follows:

$$CH_4 - 2e^- - H^+ + OH^- = CH_3OH$$

↘ reversal of hydrolysis

$$CH_3Cl$$

1994, *66*, 1148

oxidation number
Of a central atom in a coordination entity, the charge it would bear if all the ligands were removed along with the electron pairs that were shared with the central atom. It is represented by a Roman numeral.
Note:
The term Stock number is no longer recommended.
R.B. 148; see also R.B. 66

oxidation–reduction (redox) titration
See *titration*.
O.B. 47

oxidation state
A measure of the degree of oxidation of an atom in a substance. It is defined as the charge an atom might be imagined to have when electrons are counted according to an agreed-upon set of rules: (1) the oxidation state of a free element (uncombined element) is zero; (2) for a simple (monatomic) ion, the oxidation state is equal to the net charge on the ion; (3) hydrogen has an oxidation state of 1 and oxygen has an oxidation state of 2 when they are present in most compounds. (Exceptions to this are that hydrogen has an oxidation state of 1 in hydrides of active metals, e.g. LiH, and oxygen has an oxidation state of 1 in peroxides, e.g. H_2O_2; (4) the algebraic sum of oxidation states of all atoms in a neutral molecule must be zero, while in ions the algebraic sum of the oxidation states of the constituent atoms must be equal to the charge on the ion. For example, the oxidation states of sulfur in H_2S, S_8 (elementary sulfur), SO_2, SO_3, and H_2SO_4 are, respectively: 2, 0, +4, +6 and +6. The higher the oxidation state of a given atom, the greater is its degree of oxidation; the lower the oxidation state, the greater is its degree of reduction.
1990, *62*, 2204

oxidative acid-digestion (in spectrochemical analysis)
See *acid-digestion (in spectrochemical analysis)*.
1988, *60*, 1469

oxidative addition
The *insertion* of a metal of a metal complex into a covalent bond involving formally an overall two-electron loss on one metal or a one-electron loss on each of two metals, i.e.

$$L_nM^m + XY \rightarrow L_nM^{m+2}(X)(Y), \text{ or}$$
$$2L_nM^m + XY \rightarrow L_nM^{m+1}(X) + L_nM^{m+1}(Y)$$

In free-*radical* chemistry, the term is used to indicate a free radical addition to a carbon–carbon double bond, under oxidative conditions. For example:

$$RH + M^{n+}X \longrightarrow R\cdot + HX + M^{(n-1)}$$

$$R\cdot + \underset{/}{\overset{\backslash}{C}}=\underset{\backslash}{\overset{/}{C}} \longrightarrow -\underset{/}{\overset{R}{C}}-\underset{\backslash}{\overset{\cdot}{C}}-$$

$$-\underset{/}{\overset{R}{C}}-\underset{\backslash}{\overset{\cdot}{C}}- + M^{n+}X \longrightarrow -\underset{/}{\overset{R}{C}}-\underset{\backslash}{\overset{X}{C}}- + M^{(n-1)}$$

1994, *66*, 1149

oxidative coupling
The coupling of two *molecular entities* through an oxidative process, usually catalysed by a transition metal compound and involving dioxygen as the oxidant; e.g.

$$2CO + 2MeOH + \tfrac{1}{2}O_2$$
↓ catalyst
$$MeO_2CCO_2Me + H_2O$$

1994, *66*, 1149

oxidized species
A term used to characterize the degree of oxidation (or reduction) in atoms, molecules and ions. It can be applied to an atom in a molecule or an ion which has a high oxidation state. An element or atom in a compound can be oxidized by reaction with oxygen, while it can be reduced by reaction with hydrogen. An oxidized species may be formed also through the loss of electrons (either to the positive electrode in a cell, or through transfer to another atom or group of atoms). For example, the sulfur in H_2S is reduced sulfur

relative to elementary sulfur, while SO_2 and SO_3 are oxidized. Metallic iron (Fe) is a reduced state of iron, while the Fe^{2+} ion (ferrous ion) and Fe^{3+} ion (ferric ion) are oxidized states of iron. Fe^{3+} is in a higher oxidation state than Fe^{2+} which is in a higher oxidation state than Fe.
See *oxidation state*.
1990, *62*, 2204

oxidoreductases
Enzymes that catalyse electron transfer in oxidation-reduction reactions. Oxidoreductases are classified into several groups according to their respective donors or acceptors.
1992, *64*, 162

oxime *O*-ethers
O-Hydrocarbyl oximes $R_2C=NOR'$ (R' ≠ H).
1995, *67*, 1354

oximes
Compounds of structure $R_2C=NOH$ derived from condensation of *aldehydes* or *ketones* with hydroxylamine. Oximes from aldehydes may be called *aldoximes*; those from ketones may be called *ketoximes*.
1995, *67*, 1354

oxo acids
See *oxo carboxylic acids*.
1995, *67*, 1354

oxoacids
Oxoacids (and its variants oxyacids, oxo acids, oxyacids, oxiacids, oxacids) is a traditional name for any acid having oxygen in the acidic group. The term stands in contradistinction to 'hydracids' (e.g. HCl) lacking oxygen.
The term oxoacid now refers to a compound which contains oxygen, at least one other element, and at least one hydrogen bound to oxygen, and which produces a conjugate base by loss of positive hydrogen ion(s) (hydrons). E.g. $P(OH)_3$, $RC(=O)OH$, HOSOH, HOCl, HON=O, $(HO)_2SO_2$, $RP(=O)(OH)_2$.
See also *oxo carboxylic acids*.
1995, *67*, 1355; R.B. 123

oxocarbons
Compounds consisting wholly of carbon and oxygen, e.g. CO, O=C=O, O=C=C=C=O,

1995, *67*, 1355

oxo carboxylic acids
Compounds having a carboxy group as well as an aldehydic or ketonic group in the same molecule, e.g. $HC(=O)CH_2CH_2CH_2C(=O)OH$ 5-oxopentanoic acid. In an organic context the term is generally shortened to oxo acids. The full name should be used if confusion with *oxoacids* seems possible.
1995, *67*, 1354

oxo compounds
Compounds containing an oxygen atom, =O, doubly bonded to carbon or another element. The term thus embraces *aldehydes, carboxylic acids, ketones, sulfonic acids, amides* and *esters*. Oxo used as an adjective (and thus separated by a space) modifying another class of compound, as in *oxo carboxylic acids*, indicates the presence of an oxo substituent at any position. To indicate a double-bonded oxygen that is part of a ketonic structure, the term keto is sometimes used as a prefix, but such use has been abandoned by IUPAC for naming specific compounds. A traditional use of keto is for indicating oxidation of CHOH to C=O in a parent compound that contains OH groups, such as carbohydrates, e.g. 3-ketoglucose.
See *ketoaldonic acids, ketoaldoses*.
1995, *67*, 1355

oxonium ions
The parent ion H_3O^+ and substitution derivatives thereof.
See *onium compounds*.
1995, *67*, 1355

oxonium ylides
1. Compounds having the structure $R_2O^+-C^-R_2$.
See *ylides*.
2. A class of 1,3-*dipolar compounds* of general structure $R_2C=O^+-Y^-$, comprising *carbonyl imides, carbonyl oxides* and *carbonyl ylides*.
1995, *67*, 1355

oxyacid (oxy-acid)
See *oxoacids*.
1995, *67*, 1355

oxygen-flask combustion (in spectrochemical analysis)
In order to determine elements which are easily volatilized, or are present as volatile species, organic material may be ashed in oxygen in a closed oxygen combustion system. One such method is oxygen-flask combustion by which the *test sample* is burned in a closed flask containing oxygen and an absorbing solution in which the analytes are subsequently determined.
1988, *60*, 1468

oxylium ions
Species of the form RO^+, e.g. CH_3O^+ methoxylium, HO^+ hydroxylium.
1995, *67*, 1355

ozone hole
A region of the *stratosphere* over Antarctica in which a marked decrease in the concentration of ozone has been observed in the Antarctic spring in recent years. The origin of this phenomenon is not yet established, but several theories based on both physical (transport related) and chemical processes (involvement of the halocarbons and their products of oxidation) have been suggested. The latter explanation appears to be in better accord with recent findings.
1990, *62*, 2205

ozonides
The 1,2,4-trioxolanes formed by the reaction of ozone at a carbon–carbon double bond, or the analogous compounds derived from acetylenic compounds.

$$\underset{O-O}{R_2C\overset{O}{\diagup\diagdown}CR_2}$$

See *molozonides*.
1995, *67*, 1355

π–π* state
An excited state related to the ground state by a π → π* transition.
1996, 68, 2266

π → π* transition
An electronic transition described approximately as a promotion of an electron from a 'bonding' π orbital to an 'antibonding' π orbital designated as π*.
1996, 68, 2266

π → σ* transition
An electronic transition described approximately as a promotion of an electron from a 'bonding' π orbital to an 'antibonding' σ orbital designated as σ*. Such transitions generally involve high transition energies and appear close to or mixed with *Rydberg transitions*.
1996, 68, 2266

π-adduct (pi-adduct)
An *adduct* formed by electron-pair donation from a π-orbital into a σ-orbital, or from a σ-orbital into a π-orbital, or from a π-orbital into a π-orbital. For example:

Such an adduct has commonly been known as a 'π-complex', but, as the bonding is not necessarily weak, it is better to avoid the term *complex*.
See also *coordination*.
1994, 66, 1154; 1995, 67, 1312

π-bond (pi-bond)
See σ, π.
1994, 66, 1163

+ π-complex
See π-*adduct*.
1994, 66, 1154; 1995, 67, 1312

π-electron acceptor/donor group
A *substituent* capable of a +R (e.g. NO_2) or –R (e.g. OCH_3) effect, respectively.
See *electronic effect, polar effect, σ-constant*.
1994, 66, 1155

P, M
See *helicity*.
1996, 68, 2212

packed column (in chromatography)
A tube containing a solid packing.
1993, 65, 831

packing (in column chromatography)
The active solid, stationary liquid plus solid support, or swollen gel put in the column. The term packing refers to the conditions existing before the chromatographic run is started (i.e. to the material introduced into the column), whereas the stationary phase refers to the conditions during the run.

See also *totally porous packing, pellicular packing*.
O.B. 97; 1993, 65, 831

paddlanes
Tricyclic saturated *hydrocarbons* having two bridgehead carbon atoms joined through four bridges, systematically named tricyclo[$m.n.o.p^{1,(m+2)}$]alkanes, have been referred to as [$m.n.o.p$]paddlanes (when $p = 0$, the compounds are *propellanes*).

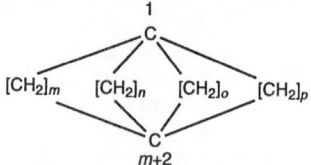

1995, 67, 1355

pair attenuation coefficient (in nuclear chemistry)
The *attenuation coefficient* when only the *pair production* process is taken into account.
1982, 54, 1547

pair correlation length (in thin films), ξ
A quantity which approximates to the *characteristic length* associated with certain number density profiles of diffuse layers. ξ is defined as the separation between two interacting entities (molecules, macromolecular segments, ions, spins), either in bulk or interfacial regions, beyond which an appropriate function, referred to as pair correlation function, either decays (e.g. to 1/e of an 'initial' value) or vanishes.
A characteristic length must be confused neither with the effective thickness of the *interfacial layer*, nor with the range (effective distance) of the intermolecular (pair) potential. This range is defined by the distance over which this potential and the corresponding correlation function both vanish.
1994, 66, 1674

pair production (in nuclear chemistry)
The simultaneous formation of an *electron* and a *positron* as a result of the interaction of a *photon* of sufficient energy (> 1.02 MeV) with the field of a *particle*.
1982, 54, 1547

Pallmann effect
Synonymous with *suspension effect (in an ion-selective electrode)*
1994, 66, 2533

PAN-based carbon fibres
Carbon fibres obtained from polyacrylonitrile (PAN) precursor fibres by *stabilization treatment, carbonization* and final heat treatment.
1995, 67, 499

+ paraffin
Obsolescent term for saturated *hydrocarbons*, commonly but not necessarily acyclic. Still widely used in the petrochemical industry, where the term desig-

nates acyclic saturated hydrocarbons, and stands in contradistinction to *naphthenes*.
1995, *67*, 1356

parallel-chain crystal (in polymers)
A type of crystal resulting from parallel packing of stems, irrespective of the stems' directional sense.
P.B. 84

parallel reactions
> See *composite reaction, simultaneous reactions*.
> 1993, *65*, 2293; 1993, *65*, 2297; 1994, *66*, 1098

paramagnetic
Substances having a magnetic susceptibility greater than 0 are paramagnetic. They are drawn into a magnetic field.
> See also *diamagnetic*.
> 1994, *66*, 1149

parent hydride
An unbranched acyclic or cyclic structure or an acyclic/cyclic structure having a semisystematic or trivial name to which only hydrogen atoms are attached.
B.B.(G) 13

parent ion (in mass spectrometry)
An electrically charged molecular moiety which may dissociate to form fragments, one or more of which may be electrically charged, and one or more neutral species. A parent ion may be a molecular ion or an electrically charged fragment of a molecular ion.
1991, *63*, 1549; O.B. 205

partial anodic (cathodic) current
When a single electrochemical reaction occurs at an electrode (i.e. all other electrochemical reactions may be neglected):
$$I = I_a + I_c$$
where I is the total current, I_a is the positive partial anodic current and I_c is the negative partial cathodic current. When more than one reaction is significant the reactions may be numbered and the numbers used as subscripts: $I_{1,a}$ $I_{1,c}$, $I_{2,a}$, etc. Then:
$$I = \Sigma I_{i,a} + \Sigma I_{i,a}$$
1974, *37*, 512

partial charge exchange reaction
> Synonymous with *partial charge transfer reaction*.
> 1991, *63*, 1556

partial charge transfer reaction
An ion/neutral species reaction wherein the charge on a multiply-charged reactant ion is reduced.
1991, *63*, 1556

partial decay constant (in nuclear chemistry)
For a radionuclide: the probability in unit time for the decay of one of its nuclei by one of several modes of decay.
1982, *54*, 1547

partial digestion (in spectrochemical analysis)
In partial digestion and/or *selective digestion* procedures only part or some of the analytes present are brought into solution. This may be preferred to total decomposition if relative concentrations of the analyte in the test samples provide sufficient information (e.g. materials for geochemical exploration).
1988, *60*, 1469

partial isotherm (or individual isotherm) (in surface chemistry)
The function relating, at constant temperature and pressure, the amount of a particular component in the *interfacial layer* per unit area (or per unit mass of *adsorbent*) with its mole fraction (or concentration) in the liquid phase. This function can be evaluated only when the location and the thickness of the interfacial layer have been defined.
1972, *31*, 594

partial kinetic current (in electrochemistry)
The current which would flow if mass transport were infinitely fast.
1974, *37*, 515

partially draining
An adjective describing a chain macromolecule that behaves in a hydrodynamic field as though the solvent within the domain of the macromolecule were progressively more immobilized with respect to the macromolecule in the direction from its outer fringes inward.
A freely draining macromolecule and a non-draining macromolecule are two extremes of the concept of a partially draining macromolecule.
P.B. 61

partial mass density, ρ_B
Change in mass due to addition of a small amount of component to a system divided by the change in volume of the system. The systematic name is partial volumic mass.
1996, *68*, 986

partial microscopic diffusion control (encounter control)
> See *microscopic diffusion control (encounter control)*.
> 1994, *66*, 2517; 1996, *68*, 173

partial molar Gibbs energy
> See *chemical potential*.
> G.B. 49

partial molar quantity
Change in the considered extensive quantity when an infinitesimal amount of substance is added to the system at constant temperature, pressure and amounts of all other constituents of the system divided by the added amount.
G.B. 49

partial order of reaction
> See *order of reaction*.
> 1981, *53*, 757

partial pressure
> See *pressure*.
> G.B. 12; 1994, *66*, 538

partial rate factor
The rate of substitution at one specific site in an *aromatic* compound relative to the rate of substitution at one position in benzene. For example, the partial rate factor f_p^Z for *para*-substitution in a monosubstituted benzene C_6H_5Z is related to the rate constants $k(C_6H_5Z)$ and $k(C_6H_6)$ for the total reaction (i.e. at all positions) of C_6H_5Z and benzene, respectively, and % *para* (the percentage *para*-substitution in the total product formed from C_6H_5Z) by the relation:

$$f_p^Z = \frac{6k\,(C_6H_5Z)}{k\,(C_6H_6)}\,\frac{\%\;para}{100}$$

Similarly for *meta*-substitution:

$$f_m^Z = \frac{6k\,(C_6H_5Z)}{2k\,(C_6H_6)}\,\frac{\%\;meta}{100}$$

(The symbols p_f^Z, m_f^Z, o_f^Z are also in use.) The term applies equally to the *ipso* position, and it can be extended to other substituted *substrates* undergoing parallel reactions at different sites with the same reagent according to the same *rate law*.
See also *selectivity*.
1994, *66*, 1149

partial specific volume, v_B
Change in volume of a system when a small amount of component is added divided by the mass of the added component. The systematic name is partial massic volume.
1996, *68*, 986

particle concentration (in atmospheric chemistry)
Commonly expressed in several ways: mass concentration (usually as $\mu g\,m^{-3}$) or number concentration (number of particles cm^{-3}); modern instrumentation allows measurement of the number of particles as a function of size as well as the total number present in a given air volume. For atmospheric aerosols, this is a complex distribution for which diameters range from below 0.01 to above 100 µm; the particles making the highest contribution to the total number density are in the size range below 0.1 µm, those contributing most to the total surface area are in the 0.1 to 1.0 µm range, while those with the highest contribution to the volume or mass of the aerosol come from both the 0.1 to 1.0 and 1.0 to 100 µm ranges.
1990, *62*, 2181

particle density (in nuclear chemistry)
The number of *particles (nuclear* or *elementary)* divided by the containing volume.
1982, *54*, 1547

particle, elementary (in nuclear chemistry)
See *elementary particle (in nuclear chemistry)*.
1982, *54*, 1547

particle induced X-ray emission analysis
A kind of analysis based on the measurement of the *energies* and intensities of *characteristic X-radiation* emitted by a test portion during *irradiation* with charged particles other than electrons.
1994, *66*, 2526

particle, nuclear
See *nuclear particle*.
1982, *54*, 1547

particle scattering factor
See *particle scattering function*.
P.B. 66

particle scattering function
The ratio of the intensity of radiation scattered at an angle of observation θ to the intensity of radiation scattered at an angle zero, i.e. $P(\theta) \equiv R(\theta)/R(0)$. An alternative recommended symbol is P_θ.
Synonymous with particle scattering factor.
P.B. 66

particle size (in atmospheric chemistry)
To describe the size of liquid or solid particles (*aerosol*) the average or equivalent diameter is used. For non-spherical particles collected in an impactor, for example, the aerodynamic diameter of a particle of arbitrary shape and density refers to the size of a spherical particle of unit density that would deposit on a given impactor surface.
1990, *62*, 2205

particle size distribution (in atmospheric chemistry)
The size of the liquid or solid particles in the atmosphere usually extends from > 0.01 to < 100 µm in diameter. In the earth's atmosphere the distribution function which describes the number of particles as a function of diameter, mass or surface area of the aerosol can be determined reasonably well with modern instrumentation.
1990, *62*, 2205

particular property
A property of a given object (phenomenon, body or substance).
'Particular property' includes the concept of particular quantity. The adjective 'particular' may be omitted, if no ambiguity is caused.
See also *property*.
1995, *67*, 1565

particulate carbon
A *carbon material* consisting of separated monolithic particles.
Notes:
Distinctions should be made between coarse particulate carbon or *granular carbon* (larger than about 100 µm, but smaller than about 1 cm in average size), fine particulate carbon or powder or flour (between 1 µm and 100 µm in average size) and *colloidal carbon* (below approximately 1 m in size in at least one direction), e.g. *carbon blacks* and colloidal carbon.
1995, *67*, 499

particulate matter (in atmospheric chemistry)
A general term used to describe airborne solid or liquid particles of all sizes. The term *aerosol* is rec-

ommended for general use in describing airborne particulate matter.
1990, *62*, 2205

partition
This term is often used as a synonym for distribution and extraction. However, an essential difference exists by definition between *distribution constant* or *partition ratio* and *partition constant*. The term partition should be, but is not invariably, applied to the distribution of a single definite chemical species between the two phases.
1993, *65*, 2378

partition chromatography
Chromatography in which separation is based mainly on differences between the solubility of the sample components in the stationary phase (*gas chromatography*), or on differences between the solubilities of the components in the mobile and stationary phases (*liquid chromatography*)
1993, *65*, 826; O.B. 94

+ partition coefficient
This term is not recommended and should not be used as a synonym for *partition constant*, *partition ratio* or *distribution ratio*.
See also *distribution constant*.
1993, *65*, 2385; 1996, *68*, 986

partition constant, K_D^o
The ratio of activity of a given species A in the extract to its activity in the other phase with which it is in equilibrium, thus:

$$(K_D^o)_A = a_{A,org}/a_{A,aq}$$

Its value should not vary with composition but depends on the choice of standard states and on the temperature (and eventually the pressure).
See *transfer activity coefficient, distribution constant*.
1993, *65*, 2385

partition function
For a molecule the ratio of the total number of molecules to their number in the non-degenerate ground state.
G.B. 39

partition ratio, K_D
The ratio of the concentration of a substance in a single definite form, A, in the extract to its concentration in the same form in the other phase at equilibrium, e.g. for an aqueous/organic system:

$$(K_D)_A = [A]_{org}/[A]_{aq}$$

Notes:
1. K_D is sometimes called the distribution constant; this is a good synonym. The terms *distribution coefficient*, *distribution ratio*, *partition constant* and extraction constant should not be used as synonyms for partition ratio.
2. The use of the inverse ratio (aqueous/organic) may be appropriate in certain cases, e.g. where the organic phase forms the feed but its use in such cases should be clearly specified. The ratio of the concentration in the denser phase to the less dense phase is not recommended as it can be ambiguous.
3. If the pure solvent and infinitely dilute feed are taken as the standard states, $K_D \rightarrow K_D^o$ as the total concentration of dissolved materials decreases.
1993, *65*, 2385

partly open film
See *open film*.
1994, *66*, 1671

pascal
SI derived unit of pressure, $Pa = N\ m^{-2} = kg\ m^{-1}\ s^{-2}$.
G.B. 72; 1996, *68*, 986

passivation (in electrochemical corrosion)
The process of transition from the active to the passive state by formation of the passivating film. Passivation is achieved by an anodic current which at the respective electrode potential must be larger than the maximum current, or by the presence of an oxidized substance in the neighbouring solution which passivates by being reduced (passivator).
See also *passive state, active state*.
1989, *61*, 21

passivation potential (in electrochemical corrosion)
The most negative electrode potential at which the passivating film is formed.
The passivation potential is equal to or more positive than the equilibrium potential of formation of the phase constituting the film. Usually, the corrosion current goes through a maximum at the passivation potential.
See also *corrosion, passive state*.
1989, *61*, 21

passive metal
A metal corroding in the *passive state*.
1989, *61*, 21

passive sampler
A device for preconcentration of trace substances from gaseous media based on molecular diffusion without controlled conveyance of the gas to be investigated (e.g. work place air).
1990, *62*, 2205

passive state (in electrochemical corrosion)
The passive state is characterized by a contiguous 'passivating film' of solid corrosion products separating the metallic phase from the adjacent electrolyte. Corrosion in the passive state involves growth of the passivating film and/or transfer of metal ions through the film into the electrolyte.
1989, *61*, 21

Paterno–Büchi reaction
The photocycloaddition of an electronically excited carbonyl group to a ground state olefin yielding an oxetane (see **I**).
1996, *68*, 2258

I (See **Paterno–Büchi reaction**)

paucidisperse system
A *colloidal* system in which only a few particle-sizes occur.
1972, *31*, 607

PDMR
See *ODMR*.
1996, *68*, 2258

peak (in chromatography)
The portion of a differential *chromatogram* recording the detector response when a single component is eluted from the column. If separation is incomplete, two or more components may be eluted as one unresolved peak.
1993, *65*, 835; O.B. 96, 101

peak analysis
The extraction of relevant peak parameters (i.e. position, area) from a measured spectrum.
1994, *66*, 2522; 1982, *54*, 547

peak area (in chromatography)
The area enclosed between the peak and the *base line*.
O.B. 101

peak area method
A kind of *peak analysis* in which a peak area is calculated by subtracting an estimate of the underlying continuum in a relevant part of a measured spectrum.
1994, *66*, 2523

peak base (in chromatography)
The interpolation, in a differential *chromatogram*, of the *base line* between the extremities of the peak.
O.B. 101

peak concentration (trace atmospheric component)
The highest concentration of a given trace component which was measured with a continuous analyser during a specified sampling period.
1990, *62*, 2205

peak current
In linear-sweep *voltammetry*, triangular-wave voltammetry, cyclic triangular-wave voltammetry, and similar techniques, the maximum value of the *faradaic current* due to the reduction or oxidation of a substance B during a single sweep. This maximum value is attained after an interval during which the concentration of B at the electrode-solution interface decreases monotonically, while the faradaic current due to the reduction or oxidation of B increases monotonically, with time. It is attained before an interval during which this current decreases monotonically with time because the rate of transport of B toward the electrode-solution interface is smaller than the rate at which it is removed from the interface by *electrolysis*.
The term has also been used to denote the maximum value of the faradaic current attributable to the reduction or oxidation of an electroactive substance in techniques such as ac *polarography*, differential pulse polarography, and derivative polarography. However, these techniques give curves that arise in ways different from that cited above, and the terms summit, *summit current*, and *summit potential* are therefore recommended for use in connection with such techniques.
See also *apex current*.
1985, *57*, 1498; O.B. 56

peak elution volume (time) (in column chromatography), $V_R, \bar{t}_R$
The volume of mobile phase entering the column between the start of the elution and the emergence of the peak maximum, or the corresponding time. In most of the cases, this is equal to the total retention volume (time). There are, however, cases when the elution process does not start immediately at sample introduction. For example, in liquid chromatography, sometimes the column is washed with a liquid after the application of the sample to displace certain components which are of no interest and during this treatment the sample does not move along the column. In gas chromatography, there are also cases when a liquid sample is applied to the top of the column but its elution starts only after a given period. This term is useful in such cases.
1993, *65*, 841; O.B. 104

+ peak enthalpimetry
See *flow injection enthalpimetry*.
1994, *66*, 2491

peak fitting
A kind of *peak analysis* in which a relevant part of a spectrum is fitted with a theoretical response function.
1994, *66*, 2523; O.B. 234

peak, full energy
See *full energy peak*.
1982, *54*, 1543

peak height (in chromatography)
The distance between the peak maximum and the peak base, measured in a direction parallel to the axis representing the detector response.
O.B. 102

peak maximum (in chromatography)
The point on the peak at which the distance to the *peak base*, measured in a direction parallel to the axis representing the detector response, is a maximum.
O.B. 102

peak potential
In linear-sweep *voltammetry*, triangular-wave voltammetry, cyclic triangular-wave voltammetry, and similar techniques, the potential of the *indicator electrode* at which the *peak current* is attained.
1985, *57*, 1502

peak resolution (in chromatography), R_s
The separation of two peaks in terms of their average peak width at base ($t_{R2} > t_{R1}$):

$$R_s = \frac{(t_{R2} - t_{R1})}{(w_{b1} + w_{b2})/2} = \frac{2(t_{R2} - t_{R1})}{w_{b1} + w_{b2}}$$

In the case of two adjacent peaks it may be assumed that $w_{b1} \approx w_{b2}$, and thus, the width of the second peak may be substituted for the average value:

$$R_s \approx (t_{R2} - t_{R1})/w_{b2}$$

1993, 65, 847; O.B. 108

peak widths (in chromatography)
Peak widths represent retention dimensions (time or volume) parallel to the baseline. If the baseline is not parallel to the axis representing time or volume, then the peak widths are to be drawn parallel to this axis. Three peak-width values are commonly used in chromatography. Peak width at base (w_b) is the segment of the peak base intercepted by the tangents drawn to the inflection points on either side of the peak. Peak width at half height (w_h) is the length of the line parallel to the peak base at 50% of the peak height that terminates at the intersection with the two limbs of the peak. Peak width at inflection points (w_i) is the length of the line drawn between the inflection points parallel to the peak base.

The peak width at base may be called the 'base width'. However, the peak width at half height must never be called the 'half width' because that has a completely different meaning. Also, the symbol $w_{1/2}$ should never be used instead of w_h.

1993, 65, 836; O.B. 101, 102

pectins
Polyuronic acids mostly derived from D-galacturonic acid (abundant in some fruits).
See also *uronic acids*.
1995, 67, 1356

Peierls transition
A *metal-insulator transition* in quasi one-dimensional solids that occurs as a result of a band gap opening up at the Fermi energy due to a displacive distortion of the regular array increasing the unit cell length, usually a dimerization; the decrease in electronic energy outweighs the increase in lattice energy.
Example: The transition in methylethylmorpholinium tetracyanoquinodimethanide at 335 K.
1994, 66, 588

pellicular packing (in chromatography)
In this case the stationary phase forms a porous outer shell on an impermeable particle.
1993, 65, 831

penams
Natural and synthetic antibiotics containing the 4-thia-1-azabicyclo[3.2.0]heptan-7-one structure, generally assumed to have the 5R configuration unless otherwise specified. The numbering of the penam skeleton differs from that of the von Baeyer named bicyclic system. Where they differ the von Baeyer numbering is shown in parentheses in the example.

See also *penems*.
1995, 67, 1356

pendant chain
See *branch*.
1996, 68, 2297

pendant group (side group)
An offshoot, neither *oligomeric* nor *polymeric* from a *chain*.
1996, 68, 2297

penems
2,3-Didehydropenams.

See *penams*.
1995, 67, 1356

penetration depth
See *depth of penetration*.
1996, 68, 2258

penicillins
Substituted *penams* having the basic structure shown, including the natural penicillins and synthetic analogues.

1995, 67, 1356

Penning gas mixture
A Penning gas mixture consists of a rare gas containing impurity atoms possibly at very low concentrations. The impurity atoms have an ionization potential V_{ion} which is lower than or equal to the metastable potential (V_{meta}) of the parent noble gas.

The Penning effect in a Penning gas mixture is the ionization by charge transfer (charge exchange) during collision between a metastable atom and a neutral atom which decreases the average energy to form an ion pair, e.g.

$$Cu + Ar_{meta} \rightarrow Cu^+ + Ar + e^{-1}$$

In a glow discharge, this results in an increase of the the ionization coefficient (Townsend first coefficient), a decrease in breakdown potential and a lowering of the cathode fall potential.

The magnetic Penning effect describes the increase of the ionization probability of gas in a low pressure electrical discharge resulting from the helical (spiral) movement of electrons in a magnetic field placed normal to the anode-cathode electrical field.
O.B. 148

pentads (in polymers)
See *constitutional sequence*.
1996, *68*, 2299

pentaprismo-
An affix used in names to denote ten atoms bound into a pentagonal prism.
R.B. 245; B.B. 465

peptides
Amides derived from two or more amino carboxylic acid molecules (the same or different) by formation of a covalent bond from the carbonyl carbon of one to the nitrogen atom of another with formal loss of water. The term is usually applied to structures formed from α-amino acids, but it includes those derived from any amino carboxylic acid.

$$H_2N-CH(R)-C(=O)-[NH-CH(R)-C(=O)]_n-OH$$

(R may be any organyl group, commonly but not necessarily one found in natural amino acids)
See also *proteins, retro, carboxylic acids*
1995, *67*, 1356; W.B. 48

peptidoglycan
A *glycosaminoglycan* formed by alternating residues of D-glucosamine and either muramic acid {2-amino-3-*O*-[(*S*)-1-carboxyethyl]-2-deoxy-D-glucose} or L-talosaminuronic acid (2-amino-2-deoxy-L-taluronic acid), which are usually *N*-acetylated or *N*-glycoloylated. The carboxyl group of the muramic acid is commonly substituted by a peptide containing residues of both L- and D-amino acids, whereas that of L-talosaminuronic acid is substituted by a peptide consisting of L-amino acids only.
1988, *60*, 1391; W.B. 85

peptization (in colloid chemistry)
See *deflocculation*.
1972, *31*, 610

+ per acids
An ambiguous term, which may imply either a higher oxidation state of a central atom, as in perchloric acid, or a derivative of hydrogen peroxide, as in $CH_3C(=O)OOH$. Accordingly, it is not recommended as a class name.
See *peroxy acids*.
1995, *67*, 1356

percent
One part in a hundred, % = 0.01.
G.B. 77

percentage error
See *percentage relative error*.
1994, *66*, 601

percentage exposed (in metallic catalysts)
The accessibility of the atoms of metal in metallic catalysts, supported or unsupported, depends upon the percentage of the total atoms of metal which are surface atoms. It is recommended that the term percentage exposed be employed for this quantity rather than the term *dispersion* which has been frequently employed.
1976, *46*, 80

percentage relative error, e_r (%)
The relative error expressed in percent. It can be calculated from the relative error by multiplying by 100.
Comment: The term 'percentage relative error' should always be quoted in full, rather than 'error' or 'percentage error' to avoid confusion.
1994, *66*, 601

percentage standard deviation, s_r (%), σ_r (%)
The value of the *relative standard deviation*, expressed in percent. It can be calculated from the relative standard deviation by multiplying by 100.
Comment: It is recommended that the 'relative standard deviation' be reported, rather than the 'percentage standard deviation', in order to avoid confusion where results themselves are expressed as percentages. The term 'coefficient of variation' in place of 'relative standard deviation' is not recommended.
1994, *66*, 601

percent diastereoisomer excess
See *diastereoisomer excess*.
1996, *68*, 2212

percent enantiomer excess
See *enantiomer excess*.
1996, *68*, 2212

perfectly polarized interphase
Differences from the behaviour of *interphases* containing only neutral species arise when the interphase may be described as perfectly polarized. This term is used when no charged component is common to both phases adjoining the interphase. This may arise as a result of the equilibrium conditions or from the kinetics of charge transfer and leads to an interphase impermeable to electric charge.
1983, *55*, 1260

pericyclic reaction
A *chemical reaction* in which *concerted* reorganization of bonding takes place throughout a cyclic array of continuously bonded atoms. It may be viewed as a reaction proceeding through a fully *conjugated* cyclic transition state. The number of atoms in the cyclic array is usually six, but other numbers are also possible. The term embraces a variety of processes, including *cycloadditions, cheletropic reactions, electrocyclic reactions* and *sigmatropic rearrangements*, etc. (provided they are concerted).
See also *multi-centre reaction*.
1994, *66*, 1150

perikinetic aggregation (in colloids)
The rate of *aggregation* is in general determined by the frequency of collisions and the probability of cohesion during collision. If the collisions are caused by *Brownian motion*, the process is called perikinetic aggregation.
1972, *31*, 610

period, T
Time for one cycle of a periodic phenomenon.
G.B. 11; 1996, *68*, 986

periodic copolymer
A *copolymer* consisting of *macromolecules* comprising more than two species of *monomeric units* in regular sequence.
1996, *68*, 2301

periodic copolymerization
A *copolymerization* in which a *periodic copolymer* is formed.
1996, *68*, 2307

periodic voltage
This general term is applicable to square, triangular, and other wave forms.
1985, *57*, 1505

peripheral atom (in organic reaction mechanisms)
A secondary *reference* (non-core) *atom* which at some point in the mechanistic sequence is situated in the molecule containing the *core atom(s)*.
1989, *61*, 27

periplanar
See *torsion angle*.
1996, *68*, 2213

periselectivity
The differentiation between two symmetry-allowed processes, for example the [2+4] vs. [4+6] *cycloaddition* of cyclopentadiene to tropone.
1994, *66*, 1150

peritectic reaction
An isothermal, reversible reaction between two phases, a liquid and a solid, that results, on cooling of a binary, ternary, ... , n system in one, two, ... $(n-1)$ new solid phases.
Synonymous with incongruent reaction.
1994, *66*, 588

peritectoid reaction
An isothermal, reversible reaction in the solid state, that, on cooling of a binary, ternary, ... , n system, results in one, two, ... $(n-1)$ new solid phases. For example, in a binary system containing two solids α' and α''

$$\alpha' + \alpha'' \to \beta$$

Synonymous with metatectoid reaction.
1994, *66*, 588

peritectoid temperature
The maximum temperature at which a *peritectoid reaction* can occur.
1994, *66*, 588

permeability, μ
Permeability of vacuum multiplied by the *relative permeability*.
G.B. 15

permeability of vacuum, μ_0
Fundamental physical constant by definition equal to $4\pi \times 10^{-7}$ H m^{-1}, exactly.
G.B. 15

permeation chromatography
Separation based mainly upon exclusion effects, such as differences in molecular size and/or shape (e.g. molecular-sieve chromatography) or in charge (e.g. ion-exclusion chromatography). The term gel-permeation chromatography is widely used for the process when the stationary phase is a gel. The term gel-filtration is not recommended.
O.B. 94

permeation tube
A device used for dynamic preparation of test gas mixtures by means of controlled permeation of a gaseous analyte out of a container through polymer material into a carrier gas stream. These devices containing certain condensable gases (e.g. NO_2, SO_2, etc.) when operated at closely controlled temperatures can be used as primary standards calibrated in terms of the weight loss per unit time.
1990, *62*, 2205

permittivity, ε
Permittivity of vacuum multiplied by the *relative permittivity*.
G.B. 14

permittivity of vacuum
Fundamental physical constant $\varepsilon_0 = 1/\mu_0 c_0^2$ exactly, where μ_0 is the *permeability of vacuum* and c_0 the speed of light in vacuum. It is equal to 8 854 187 816 ... $\times 10^{-12}$ F m^{-1}.
G.B. 14

permselectivity
A term used to define the preferential permeation of certain ionic species through *ion-exchange membranes*.
1993, *65*, 856

peroxides
Compounds of structure ROOR in which R may be any *organyl group*. (The term is also used in an inorganic sense to denote salts of the anion O_2^{2-}).
See also *hydroperoxides*
1995, *67*, 1356

peroxisome
Organelle, similar to a lysosome, characterized by its content of catalase (EC 1.11.1.6), peroxidase (EC 1.11.1.7) and other oxidative enzymes.
1993, *65*, 2080

peroxo compounds
See *carbonyl oxides*
1995, *67*, 1356

peroxy acids
Acids in which an acidic –OH group has been replaced by an –OOH group, e.g. $CH_3C(=O)OOH$ peroxyacetic acid, $PhS(=O)_2OOH$ benzeneperoxysulfonic acid.
1995, *67*, 1356

perpendicular effect
See *More O'Ferrall–Jencks diagram*.
1994, *66*, 1150

persistence length (in polymers)
The average projection of the end-to-end vector on the tangent to the chain contour at a chain end in the limit of infinite chain length.
The persistence length (symbol a) is the basic characteristic of the *worm-like chain*.
P.B. 51

persistent
A term used to characterize *radicals* which have lifetimes of several minutes or greater in dilute solution in inert solvents. Persistence is a kinetic or reactivity property. In contrast, radical *stability*, which is a thermodynamic property, is expressed in terms of the C–H bond strength of the appropriate hydrocarbon. The lifetime of a radical is profoundly influenced by steric shielding of the radical centre by bulky substituents.
See also *transient*.
1994, *66*, 1150

perspective formula
A geometric representation of stereochemical features of a molecule or model which appears as a view from an appropriate direction.
See *Fischer projection, Newman projection, projection formula, sawhorse projection, wedge projection, zig-zag projection*.
See also *stereochemical formula*.
1996, *68*, 2213

perturbed dimensions (in polymers)
The dimensions of an actual polymer *random coil* not in a theta state.
P.B. 48

pesticide
Strictly, a substance intended to kill pests: in common usage, any substance used for controlling, preventing, or destroying animal, microbiological or plant pests.
1993, *65*, 2081

pesticide residue
Any substance or mixture of substances in food for man or animals resulting from the use of a *pesticide*, including any specified derivatives, such as degradation and conversion products, metabolites, reaction products and impurities considered to be of toxicological significance.
1993, *65*, 2081

peta
SI prefix for 10^{15} (symbol: P).
G.B. 74

petroleum coke
A *carbonization* product of high-boiling hydrocarbon fractions obtained in petroleum processing (heavy residues). It is the general term for all special petroleum coke products such as *green*, *calcined* and *needle* petroleum *coke*.

Notes:
High-boiling hydrocarbon fractions (heavy residues) used as feedstock for petroleum coke are residues from distillation (atmospheric pressure, vacuum) or cracking (e.g. thermal, catalytic, steam-based) processes. The nature of feedstock has a decisive influence on the graphitizability of the calcined coke.
1995, *67*, 499

petroleum pitch
A residue from heat treatment and distillation of petroleum fractions. It is solid at room temperature, consists of a complex mixture of numerous predominantly aromatic and alkyl-substituted aromatic hydrocarbons, and exhibits a broad softening range instead of a defined melting temperature.
Notes:
The hydrogen aromaticity (ratio of aromatic to total hydrogen atoms) varies between 0.3 and 0.6. The aliphatic hydrogen atoms are typically present in alkyl groups substituted on aromatic rings or as naphthenic hydrogen.
1995, *67*, 499

pH
In the restricted range of dilute aqueous solutions having *amount concentrations* less than 0.1 mol dm^{-3} and being neither strongly acidic nor strongly alkaline (2 <pH <12).

$$pH = -\lg\{\gamma_\pm c(H^+)/\text{mol dm}^{-3}\} \pm 0.02$$

where $\gamma_\pm$ is the mean ionic *activity coefficient* of a typical univalent electrolyte on a concentration basis. The operational definition is based on the *electromotive force* measurement of the galvanic cell:

reference electrode |KCl(aq) ∥ solution X|H(g) |Pt
where the KCl solution has a molality greater than 3.5 mol kg^{-1} relative to a standard solution S, replacing X in the scheme above [usually KH$_2$PO$_4$ (aq) of molality 0.05 mol kg^{-1} which has a pH = 4.005].

$$pH(X) = pH(S) + (E_S - E_X)F/(RT \ln 10).$$

G.B. 62; 1996, *68*, 986; see also 1990, *62*, 2205; 1984, *56*, 569

pH$_{0.5}$ or pH$_{1/2}$ (in solvent extraction)
That value of pH in an aqueous phase at which the distribution ratio is unity at equilibrium. 50% of the solute is extracted ($E = 0.5$) only when the phase ratio is unity.
1993, *65*, 2385

pharmacodynamics
Study of pharmacological actions on living systems, including the reactions with and binding to cell constituents, and the biochemical and physiological consequences of these actions.
1993, *65*, 2081

pharmacokinetics
Process of the uptake of *drugs* by the body, the biotransformation they undergo, the distribution of the drugs and their metabolites in the tissues, and the elimination of the drugs and their metabolites from

the body. Both the amounts and the concentrations of the drugs and their metabolites are studied. The term has essentially the same meaning as toxicokinetics, but the latter term should be restricted to the study of substances other than drugs.
1993, 65, 2082

phase
An entity of a material system which is uniform in chemical composition and physical state.
1994, 66, 588

phase fluorimetry
See *lifetime of luminescence*.
O.B. 197

phase inversion
A term used in two senses which should be specified: *density inversion* or *continuity inversion*.
1993, 65, 2387

phase ratio (in chromatography), β
The ratio of the volume of the mobile phase to that of the stationary phase in a column:
$$\beta = V_0/V_s$$
In the case of open-tubular columns the geometric internal volume of the tube (V_c) is to be substituted for V_0.
1993, 65, 833; O.B. 100

phase ratio (in liquid-liquid distribution), r
The ratio of the quantity of the solvent to that of the other phase.
Notes:
1. Unless otherwise specified the phase ratio refers to the phase volume ratio.
2. If other aspects of the phase ratio are employed viz. phase mass ratio, phase flow ratio, these should be specified.
1993, 65, 2385

phase rule
The number of degrees of freedom, F, that a system containing C components can have when P phases are in equilibrium, is given as:
$$F = C - P + 2$$
1994, 66, 588

phase separation
The process by which a single solid (liquid) phase separates into two or more new phases.
1994, 66, 588

phase-space theory
This is a theory applied to unimolecular or bimolecular reactions proceeding through long-lived complexes. The probability of reaction is assumed to be proportional to the number of states available to a particular product *channel* divided by the number of states corresponding to all product channels. The theory is used to predict rates, product energy distributions, product velocity distributions, and product angular momentum distributions.
1996, 68, 176

phase-transfer catalysis
The phenomenon of rate enhancement of a reaction between *chemical species* located in different phases (immiscible liquids or solid and liquid) by addition of a small quantity of an agent (called the 'phase-transfer *catalyst*') that extracts one of the reactants, most commonly an anion, across the interface into the other phase so that reaction can proceed. These catalysts are salts of '*onium ions*' (e.g. tetraalkylammonium salts) or agents that complex inorganic cations (e.g. *crown ethers*). The catalyst cation is not consumed in the reaction although an anion exchange does occur.
1994, 66, 1150

phase transition
A change in the nature of a phase or in the number of phases as a result of some variation in externally imposed conditions, such as temperature, pressure, activity of a component or a magnetic, electric or stress field.
1994, 66, 588

phenolates
Synonymous with *phenoxides*. The term phenolate should not be used for solvates derived from a *phenol*, for the ending -ate often occurs in names for anions.
See *phenoxides*.
1995, 67, 1357

phenols
Compounds having one or more hydroxy groups attached to a benzene or other arene ring, e.g. 2-naphthol:

1995, 67, 1357

phenomenological equation
In the following only media which are isotropic with respect to mass transport (i.e. the transport coefficients are independent of direction) are being considered. In the linear range (not too far from equilibrium), for uniform temperature and neglecting external fields such as the earth's gravitational field, the flux density of species B is related to the gradients of the electrochemical potentials of all species by the phenomenological equation:
$$N_B - c_B v_A = -\sum_{i \neq A} L_{Bi}^A \nabla \mu_i \text{ with } i = B, C, \ldots$$
where $\nabla \mu_i$ is the gradient of the electrochemical potential of species i. The proportionality factors L_{Bi}^A are called phenomenological coefficients. Their values depend on the frame of reference. The latter is taken here to move with the velocity v_A of species A, and hence:
$$L_{Ai}^A = 0.$$
1981, 53, 1830

phenonium ions
The traditional generic name for those cyclohexadienyl cations that are spiro-annulated with a cyclopropane unit. Phenonium ions constitute a subclass of arenium ions.

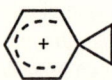

See also *bridged carbocation*.
1995, *67*, 1357

phenotype
The observable structural and functional characteristics of an organism determined by its *genotype* and modulated by its environment.
1992, *64*, 162

phenoxides
Salts or analogous metal derivatives of *phenols*; synonymous with *phenolates*. A preferable general term, however, is aryloxides.
1995, *67*, 1357

pheromone
Substance used in olfactory communication between organisms of the same species eliciting a change in sexual or social behaviour.
Synonymous with ectohormone, feromone.
1993, *65*, 2082

pH glass electrode
A hydrogen-ion responsive electrode usually consisting of a bulb, or other suitable form, of special glass attached to a stem of high resistance glass complete with *internal reference electrode* and *internal filling solution* system. Other geometrical forms may be appropriate for special applications, e.g. capillary electrode for measurement of blood *pH*.
1985, *57*, 540

pH gradient (in electrophoresis)
The differential change of *pH* with distance (dpH/dl).
1994, *66*, 895

phonon
Elementary excitation in the quantum mechanical treatment of vibrations in a crystal lattice.
1996, *68*, 2258

phosphanes
The saturated hydrides of tervalent phosphorus having the general formula P_nH_{n+2}. Individual members having an unbranched phosphorus chain are named phosphane, diphosphane, triphosphane, etc. The name of a saturated hydride of phosphorus wherein one or more phosphorus atoms have a bonding number of 5 is formed by prefixing locants and λ^5 symbols to the name of the corresponding phosphane. Hydrocarbyl derivatives of PH_3 belong to the class phosphines.
1995, *67*, 1357

phosphanylidenes
Recommended name for carbene analogues having the structure RP: (former IUPAC name is phosphinediyls). A common non-IUPAC synonym is phosphinidenes.
1995, *67*, 1357

phosphatidic acids
Derivatives of glycerol in which one hydroxy group, commonly but not necessarily primary, is esterified with phosphoric acid and the other two are esterified with *fatty acids*.

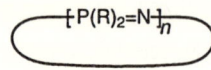

See *lecithins, phosphoglycerides, phospholipids*.
1995, *67*, 1357; W.B. 184

phosphazenes
Compounds containing a phosphorus–nitrogen double bond, i.e. derivatives of $H_3P=NH$ and $HP=NH$. A multiplicity of such bonds is present in various well-established chain, ring and cage compounds. E.g.

$$-[P(R)_2=N]_n-$$

$$\begin{bmatrix} OEt \\ | \\ P=N \\ | \\ OEt \end{bmatrix}_n$$

poly(diethoxyphosphazene)

1995, *67*, 1357

phosphine oxides
Compounds having the structure $R_3P=O \leftrightarrow R_3P^+-O^-$ (analogously, phosphine *imides* and phosphine *sulfides*).
See *imides (2)*.
1995, *67*, 1537

phosphines
PH_3 and compounds derived from it by substituting one, two or three hydrogen atoms by hydrocarbyl groups R_3P. RPH_2, R_2PH and R_3P ($R \neq H$) are called primary, secondary and tertiary phosphines, respectively. A specific phosphine is preferably named as a substituted phosphane, e.g. CH_3PH_2 methylphosphane.
See *phosphanes*.
1995, *67*, 1357

phosphinic acids
$H_2P(=O)OH$ (phosphinic acid) and its P-*hydrocarbyl* derivatives.
1995, *67*, 1357

phosphinidenes
See *phosphanylidenes*.
1995, *67*, 1358

phosphinous acids
H_2POH (phosphinous acid) and its P-*hydrocarbyl* derivatives.
1995, *67*, 1358

phospho
1. A prefix used in biochemical nomenclature in place of *phosphono* to denote the –P(=O)(OH)$_2$ group linked to a heteroatom, e.g. Me$_3$N$^+$–CH$_2$CH$_2$–OP(=O)(OH)O$^-$ phosphocholine.
2. An infix used in biochemical nomenclature to name phosphoric diesters, e.g. glycerophosphocholine.
See *phosphoglycerides*.
W.B. 256; 1995, *67*, 1358

phosphoglycerides
Phosphoric diesters, esters of phosphatidic acids, generally having a polar head group (OH or NH$_2$) on the esterified alcohol which typically is 2-aminoethanol, choline, glycerol, inositol, serine. The term includes *lecithins, cephalins*. E.g.

2-aminoethyl phosphatidates

1995, *67*, 1358

phospholipids
Lipids containing phosphoric acid as mono- or diesters, including *phosphatidic acids* and *phosphoglycerides*.
1995, *67*, 1358; see also W.B. 184

phosphonic acids
HP(=O)(OH)$_2$ (phosphonic acid) and its P-*hydrocarbyl* derivatives.
1995, *67*, 1358

phosphonitriles
Compounds of stoichiometric composition [X$_2$PN]$_n$, in which X is alkoxy, halogen, or other electronegative group, and *n* is a variable integer, the value of which may not be known.
See *phosphazenes*.
1995, *67*, 1358

phosphonium compounds
Salts (and hydroxides) [R$_4$P]$^+$X$^-$ containing tetracoordinate phosphonium ion and the associated anion.
See *onium compounds*.
1995, *67*, 1358

phosphonium ylides
Compounds having the structure R$_3$P$^+$–C$^-$R$_2$ ↔ R$_3$P=CR$_2$. Also known as Wittig reagents.
See *ylides*.
1995, *67*, 1358

phosphono
A prefix indicating the presence of the group –P(=O)(OH)$_2$.
See *phospho*.
1995, *67*, 1358

phosphonous acids
HP(OH)$_2$ (phosphonous acid) and its P-*hydrocarbyl* derivatives.
1995, *67*, 1358

phosphoramides
Compounds in which one or more of the OH groups of phosphoric acid have been replaced with an amino or substituted amino group; commonly confined to the phosphoric triamides, P(=O)(NR$_2$)$_3$, since replacement of one or two OH groups produces phosphoramidic acids P(=O)(OH)(NR$_2$)$_2$, P(=O)(OH)$_2$(NR$_2$).
1995, *67*, 1358

phosphoranes
The mononuclear hydride PH$_5$, systematically named λ^5-phosphane, and its hydrocarbyl derivatives. By extension, the literature also applies the term to *phosphonium ylides*.
See also *phosphanes*.
1995, *67*, 1358

phosphoranyl radicals
A tetracoordinate phosphorus species which has nine valence-shell electrons R$_4$P·.
1995, *67*, 1359

phosphorescence
From a phenomenological point of view, the term has been used to describe long-lived *luminescence*. In mechanistic photochemistry, the term designates luminescence involving change in spin multiplicity, typically from triplet to singlet or vice versa. The luminescence from a quartet state to a doublet state is also phosphorescence.
1996, *68*, 2258; 1984, *56*, 233; see also O.B. 184

phosphorimetry
See *flash fluorimetry*.
1984, *56*, 242

phosphoroscope
A *mechanical* device used to separate *phosphorescence* from *fluorescence*.
1984, *56*, 238

phosphorus ylides
See *phosphonium ylides*.
1995, *67*, 1359

+ phosphylenes
An obsolescent synonym for *phosphanylidenes*.
1995, *67*, 1359

photoacoustic detector
A *pressure-sensitive detector* used to detect intermittent radiation absorbed in a black body or in the sample concerned. The resulting rapid temperature change produces a transient pressure oscillation that is observed with the help of a microphone, or a piezoelectric device.
See also *radiation detector*.
1995, *67*, 1752

photoacoustic effect
Generation of heat after absorption of radiation, due to radiationless deactivation or chemical reaction.
See also *photoacoustic spectroscopy*.
1996, *68*, 2258

photoacoustic spectroscopy
A spectroscopic technique based on the *photoacoustic effect*. A photoacoustic spectrum consists of a plot of the intensity of the acoustic signal detected by a microphone or a 'piezoelectric' detector, against the excitation wavelength or another quantity related to the photon energy of the modulated excitation.
See also *isooptoacoustic point*.
1996, *68*, 2259

photoadsorption
Irradiation by light (usually visible or ultraviolet) may affect *adsorption*. In a system containing *adsorptive* and *adsorbent* exposure to light may lead to increased adsorption (photoadsorption) or it may lead to *desorption* of an *adsorbate* (photodesorption).
1976, *46*, 77

photoaffinity labelling
A technique in which a photochemically reactive *molecular entity*, specifically associated with a biomolecule, is photoexcited in order to covalently attach a label to the biomolecule, usually via intermediates.
1996, *68*, 2259

photoassisted catalysis
Catalytic reaction involving production of a catalyst by absorption of light.
See *photocatalysis*.
1996, *68*, 2259

photocatalysis
Catalytic reaction involving light absorption by a catalyst or by a substrate.
1996, *68*, 2259

photocatalytic cell
See *photoelectrolytic cell*.
1991, *63*, 593

photochemical detector
See *radiation detector*.
1995, *67*, 1748

photochemical equivalence
According to the principle of photochemical equivalence, there is a one-to-one relationship between the number of photons absorbed by a system and the number of excited species produced. The principle often fails for light of high intensity (e.g. laser beams), when molecules may absorb more than one photon.
1996, *68*, 176

photochemical hole burning
See *hole burning*.
1996, *68*, 2259

photochemical reaction
This term is generally used to describe a chemical reaction caused by absorption of ultraviolet, visible, or infrared radiation. There are many ground state reactions which have photochemical counterparts. Among these are photoadditions, photocycloadditions, photoeliminations, photoenolizations, photo-Fries rearrangements, photoisomerizations, photooxidations, photoreductions, photosubstitutions, etc.
1996, *68*, 2259; see also 1990, *62*, 2206

photochemical smog
The product of photochemical reactions caused by solar radiation and occurring in polluted air.
1996, *68*, 2259; see also 1990, *62*, 2206

photochemical yield
See *quantum yield*.
G.B. 57

photochemistry
The branch of chemistry concerned with the chemical effects of light (far UV to IR).
See *photochemical reaction*.
1996, *68*, 2259; see also 1990, *62*, 2206

photochromism
A photoinduced transformation of a molecular structure (e.g. of a solution), photochemically or thermally reversible, that produces a spectral change, typically, but not necessarily, of visible colour.
1996, *68*, 2259

photoconductive detector
In a photoconductive *detector* an electric potential is applied across the absorbing region and causes a current to flow in proportion to the irradiance if the photon energy exceeds the energy gap between the valence and the conduction band.
Depending on their *spectral responsivity function*, photoconductive detectors are divided into photoconductive detectors for the visible wavelength range e.g. cadmium sulfide or CdS photoconductive detectors, photoconductive detectors for the near infrared wavelength range e.g. lead sulfide or PbS photoconductive detectors, photoconductive detectors for the infrared wavelength range e.g. silicon doped with arsenide or Si:As photoconductive detectors, and the mercury-cadmium-telluride or HgCdTe photoconductive detector.
1995, *67*, 1754

photoconductivity
Electrical conductivity resulting from photoproduction of charge carriers.
1996, *68*, 2259

photocrosslinking
The formation of a covalent linkage between two macromolecules or between two different parts of one macromolecule.
1996, *68*, 2260

photocuring
Technical expressions for the photoinduced hardening of a monomeric, oligomeric or polymeric substrate normally in the form of a film.
1996, *68*, 2260

photocurrent yield
The quantum efficiency of electron transport between the two electrodes of a photovoltaic cell or photoelectrochemical cell.
1996, *68*, 2260

photodegradation
The photochemical transformation of a molecule into lower molecular weight fragments, usually in an oxi-

dation process. This term is widely used in the destruction (oxidation) of pollutants by UV-based processes.
1996, *68*, 2260

photodesorption
See *photoadsorption*.
1976, *46*, 77

photodetachment (of electrons)
Ejection of an electron from a negative ion upon photoexcitation.
1996, *68*, 2260

photodiode
A two-electrode, radiation-sensitive junction formed in a semiconductive material. A junction is formed by two successive regions of a semiconductive material having, respectively, an excess of electrons (n-type) or holes (p-type). A bias potential applied to the *detector* creates a region at the interface that is depleted of majority carriers. Each incident photon produces electron-hole pairs in the depletion region resulting in a measurable signal current. The photodiode can be operated either with zero bias in the photovoltaic mode where the photodiode is actually generating the electric potential supplied to the load. In a biased mode, the photoconductive mode, the reverse current is proportional to the irradiation.
A Schottky-barrier photodiode is constructed by deposition of a metal film on a semiconductor surface in such a way that no interface layer is present. The barrier thickness depends on the impurity dopant concentration in the semiconductor layer. The incident radiation generates electon-hole pairs within the depletion region of the barrier where they are collected efficiently and rapidly by the built-in field.
A PIN (p-intrinsic-n) diode is a planar diffused diode consisting of a single crystal having an intrinsic (undoped or compensated) region sandwiched between p- and n-type regions. A bias potential applied across the detector depletes the intrinsic region of charge carriers, constituting the radiation sensitive detector volume. The number of electron-hole pairs produced is dependent on the energy of the incident photons.
An avalanche photodiode is a photodiode in which the photogenerated electron-hole pairs are accelerated by a bias potential near to breakdown potential so that further electron-hole pairs are formed leading to saturation of the photocurrent. This operational mode for photon counting is the so-called Geiger mode, similar to that of the gas filled *Geiger counter*. Avalanche photodiodes can also be operated in the proportional mode.
1995, *67*, 1755

photodiode array
An arrangement of a number of *photodiodes* on a single chip.
1995, *67*, 1756

photodynamic effect
A term used in photobiology to refer to photoinduced damage requiring the simultaneous presence of light, photosensitizer and molecular oxygen. A sensitized photooxidation which involves molecular oxygen.
1996, *68*, 2260

photoelectrical effect
The ejection of an electron from a solid or a liquid by a photon.
1996, *68*, 2260; O.B. 234

photoelectric attenuation coefficient
The *attenuation coefficient* when only the photoelectric process is taken into account.
1982, *54*, 1548

photoelectric detector
See *radiation detector*.
1995, *67*, 1748

photoelectric peak
Of a spectrum of *γ-radiation*, the part of the spectral response curve corresponding to the *absorption* in the *radiation detector* by the *photoelectric effect* of the detected γ-energy. In most cases the peak also contains events caused by multiple processes and use of the expression *total absorption peak* or *full energy peak* is to be preferred.
Synonymous with photopeak.
1982, *54*, 1548

photoelectrochemical cell
An electrochemical cell in which current and a voltage are simultaneously produced upon absorption of light by one or more of the electrodes. Usually at least one of the electrodes is a semiconductor.
See also *photovoltaic cell, photoelectrolytic cell, photoelectrolysis cell, photogalvanic cell*
1996, *68*, 2260; 1991, *63*, 593

photoelectrochemical etching
The dissolution of a *semiconductor* in an electrolytic solution upon exposure to light. Used in the photopatterning of semiconductor surfaces.
1996, *68*, 2260

photoelectrochemistry
A term applied to a hybrid field of chemistry employing techniques which combine photochemical and electrochemical methods for the study of the oxidation-reduction chemistry of the ground or excited states of molecules or ions. In general, it is the chemistry resulting from the interaction of light with electrochemical systems.
See also *photoelectrochemical cell, photogalvanic cell, photovoltaic cell*.
1996, *68*, 2260

photoelectrolytic cell
A cell in which radiant energy causes a net chemical conversion in the cell, e.g. so as to produce hydrogen as a useful fuel. These cells can be classified as photosynthetic or photocatalytic. In the former case, radiant energy provides a Gibbs energy to drive a reaction such as $H_2O \rightarrow H_2 + \frac{1}{2}O_2$, and electrical or

thermal energy may be later recovered by allowing the reverse, spontaneous reaction to proceed.
In a photocatalytic cell the photon absorption promotes a reaction with $\Delta G < 0$ so there is no net storage of chemical energy, but the radiant energy speeds up a slow reaction.
 1991, *63*, 593; see also 1991, *63*, 596

photoelectron spectroscopy (PES)
A spectroscopic technique which measures the kinetic energy of electrons emitted upon the ionization of a substance by high energy monochromatic photons. A photoelectron spectrum is a plot of the number of electrons emitted versus their kinetic energy. The spectrum consists of bands due to transitions from the ground state of an atom or molecular entity to the ground and excited states of the corresponding radical cation. Approximate interpretations are usually based on 'Koopmans theorem' and yield orbital energies. PES and UPS (UV photoelectron spectroscopy) refer to the spectroscopy using vacuum ultraviolet sources, while ESCA (electron spectroscopy for chemical analysis) and XPS use X-ray sources.
 1996, *68*, 2261; 1976, *45*, 223; O.B. 246

photoelectron spectroscopy, X-ray
See *X-ray photoelectron spectroscopy*.
 O.B. 250

photoelectron yield
The number of photoelectrons emitted by the sample per incident photon.
 1979, *51*, 2247

photoemissive detector
A *detector* in which a photon interacts with a solid surface, which is called the photocathode, or a gas, releasing a photoelectron. This process is called the *external photoelectric effect*. The photoelectrons are collected by an electrode at positive electric potential, i.e. the anode.
 1995, *67*, 1752

photoexcitation
The production of an excited state by the absorption of ultraviolet, visible, or infrared radiation.
 1996, *68*, 2261

photo-Fries rearrangement
A photorearrangement of aryl or acyl esters to give the [1, 3]-rearranged product (as well as the [1,5]-rearranged product):

 1996, *68*, 2261

photogalvanic cell
An electrochemical cell in which current or voltage changes result from photochemically generated changes in the relative concentrations of reactants in a solution phase oxidation-reduction couple.

See also *photovoltaic cell*.
 1996, *68*, 2261; see also 1991, *63*, 596

photoimaging
See *imaging*.
 1996, *68*, 2261

photoinduced electron transfer
An *electron transfer* resulting from an electronic state produced by the resonant interaction of electromagnetic radiation with matter.
 1996, *68*, 2261

photoinduced polymerization
The *polymerization* of a monomer by a free radical or ionic chain reaction initiated by photoexcitation.
See *photoinitiation*.
 1996, *68*, 2261

photoinitiation
The photoproduction of a free radical or ion capable of initiating a chain reaction such as a polymerization.
See *photoinduced polymerization*.
 1996, *68*, 2261

photoionization
The term generally used to describe ionization of any species by photons. The process may, for example, be written

$$M + h\nu \rightarrow M^{+\bullet} + e$$

Electrons and photons do not 'impact' molecules or atoms; they interact with them in ways that result in various electronic excitations including ionization. For this reason it is recommended that in mass spectrometry the terms 'electron impact' and 'photon impact' not be used.
 1991, *63*, 1548; 1996, *68*, 2261; O.B. 203

photoionization detector (in gas chromatography)
Radiation from an ultraviolet lamp ionizes certain species in the carrier gas. A potential difference is applied and the resulting ionization current is detected. The detector is only useful for substances with ionization potentials below about 11 eV. This makes it quite useful for detecting one component of a combined eluent when the other component, for instance nitrogen, has a high ionization potential. The detector has a small linear dynamic range and is capable of detecting substances below 1 ppmv.
 1990, *62*, 2191

photoisomerization
A photochemical *isomerization*.
 1981, *53*, 762

photoluminescence
Luminescence arising from *photoexcitation*.
 1996, *68*, 2261

photolysis
The cleavage of one or more covalent *bonds* in a *molecular entity* resulting from absorption of light, or a photochemical process in which such cleavage is an essential part. For example:

$$Cl_2 \rightarrow 2Cl\cdot$$

The term is used incorrectly to describe irradiation of a sample, although in the combination flash photolysis this usage is accepted.

1994, *66*, 1150; see also 1990, *62*, 2206

photometry
The measurement of light over wavelengths which produce visual sensation (approximately 380–780 nm). Thus photometric quantities do not refer to a specific wavelength but to the light emitted by a standard source (formerly a 'standard international candle', now a blackbody radiator emitting at the temperature of solidifying platinum, 2042 K). *Luminance* and *illuminance* are the photometric analogues of the radiometric quantities *radiance* and *irradiance*, respectively, but conversion from photometric units (e.g. lm cm^{-2}) to radiometric units (e.g. J s^{-1} cm^{-2}) requires convolution over wavelength of the spectral radiation with the relative spectral response of the human eye. The standard response is called the 'spectral luminous efficiency of radiation', and is tabulated for daylight adapted vision in photopic response tables.

1990, *62*, 2206

photomultiplier tube
A *vacuum phototube* with additional amplification by electron multiplication. It consists of a photocathode, a series of dynodes, called a dynode chain on which a secondary-electron multiplication process occurs, and an anode. According to the desired *response time*, transit time, time spread, gain, or low *dark current*, different types of dynode structures have been developed, e.g. circular cage structure, linear focused structure, venetian blind structure, box and grid structure. Some special dynode structures permit combination with additional electric or magnetic fields.

A strip dynode photomultiplier tube consists of a photocathode followed by thin dynode material on an insulating substrate. In a continuous-strip photomultiplier, two strip dynodes are arranged in parallel. A potential applied to the ends of the two strips produces an electric field across the continuous strip dynodes, giving rise to electron multiplication along the dynodes. In a resistance-strip magnetic photomultiplier, a uniform magnetic field is applied to the planes of the strips, so that the electrons travel in the crossed electric and magnetic fields.

A channel photomultiplier tube photocathode consists of a channel electron multiplier (CEM) system for the photoelectrons, and an anode to collect the final electron current. The basic part of the CEM is a tube with a semiconducting inner surface. In general it is curved in order to inhibit the acceleration of positive ions towards the photocathode. A number of small channels called microchannels can be constructed in arrays for imaging applications.

1995, *67*, 1753; see also 1982, *54*, 1548

photon
Particle of zero charge, zero rest mass, spin quantum number 1, energy $h\nu$ and momentum $h\nu/c$ (h is the Planck constant, ν the frequency of radiation and c the speed of light), carrier of electromagnetic force.

G.B. 93; 1996, *68*, 2262

photon activation
See *activation (in nuclear chemistry)*.
1982, *54*, 1548

photon counting
Also called single photon counting. The recording of sequential single photon pulses counted by way of recording the electron emission events from a photosensitive layer (photocathode) and multiplied by means of a 'dynode' arrangement (photomultiplier). This technique is used for two purposes: 1. the sensitive measurement of low levels of radiation; 2. the recording of emission decays.

See *time-correlated single photon counting*.
1996, *68*, 2262

photon emittance
See *photon exitance*.
1996, *68*, 2262

photon exitance, M_p
The *photon flow*, Φ_p, emitted by an element of the surface containing the source point under consideration divided by the area (S) of that element. (dΦ_p/dS, simplified expression: $M_p = \Phi_p/S$ when the photon flow is constant over the surface area considered). The SI unit is s^{-1} m^{-2}. Alternatively, the term can be used with the amount of photons (mol or its equivalent einstein), the SI unit then being mol s^{-1} m^{-2}. Also called specific photon emission. Formerly called photon emittance.

See *spectral photon exitance*.
See also *radiant exitance*.
1996, *68*, 2262

photon exposure, H_p
The *photon irradiance*, E_p, integrated over the time of irradiation ($\int E_p \, dt$, simplified expression: $H_p = E_p t$ when the photon irradiance is constant over the time considered). The SI unit is m^{-2}. Alternatively, the term can be used with the amount of photons (mol or its equivalent einstein), the SI unit then being mol m^{-2}. For a parallel and perpendicularly incident beam, not scattered or reflected by the target or its surroundings, *photon fluence* H_p^0 is an equivalent term.

See also *fluence, radiant exposure*.
1996, *68*, 2262

photon flow, Φ_p
The number of photons (quanta, N) per unit time. (dN/dt, simplified expression: $\Phi_p = N/t$ when the number of photons is constant over the time considered). The SI unit is s^{-1}. Alternatively, the term can be used with the amount of photons (mol or its equivalent einstein), the SI unit then being mol s^{-1}.

See *spectral photon flow*.
See also *radiant power*.
1996, *68*, 2263

photon fluence, H_p^0

The integral of the amount of all photons (quanta) which traverse a small, transparent, imaginary spherical target, divided by the cross-sectional area of this target. The photon fluence rate, E_p^0, integrated over the duration of the irradiation ($\int E_p^0 \, dt$, simplified expression: $H_p^0 = E_p^0 t$ when E_p^0 is constant over the time considered). Photons per unit area (quanta m^{-2}). The SI unit is m^{-2}. Alternatively, the term can be used with the amount of photons (mol or its equivalent einstein), the SI unit then being mol m^{-2}.

See also *fluence*.
1996, *68*, 2263

photon fluence rate, E_p^0

The rate of *photon fluence*. Four times the ratio of the photon flow, Φ_p, incident on a small, transparent, imaginary spherical volume element containing the point under consideration divided by the surface of that sphere, S_K. ($\int_{4\pi} L_p \, d\omega$, simplified expression: $E_p^0 = 4\Phi_p/S_K$ when the photon flow is constant over the solid angle considered). The SI unit is m^{-2} s^{-1}. Alternatively, the term can be used with the amount of photons (mol or its equivalent einstein), the SI unit then being mol m^{-2} s^{-1}. It reduces to *photon irradiance* for a parallel and normally incident beam not scattered or reflected by the target or its suroundings.

See *photon radiance, fluence rate*.
1996, *68*, 2263

photon flux

Synonymous with *photon irradiance*.
1996, *68*, 2263

photon irradiance, E_p

The *photon flow*, Φ_p, incident on an infinitesimal element of surface containing the point under consideration divided by the area (S) of that element (dΦ_p/dS, simplified expression: $E_p = \Phi_p/S$ when the photon flow is constant over the surface considered). The SI unit is m^{-2} s^{-1}. Alternatively, the term can be used with the amount of photons (mol or its equivalent einstein), the SI unit then being mol^{-2} s^{-1}. For a parallel and perpendicularly incident beam not scattered or reflected by the target or its surroundings *photon fluence rate* (E_p^0) is an equivalent term.

See *spectral photon irradiance*.
See also *irradiance*.
1996, *68*, 2663; 1995, *67*, 1748

photon quantities

A group of quantities characterizing electromagnetic radiation in terms of the number of photons. The symbols of photon quantities may be distinguished by adding a subscript p.

See also *radiant* and *luminous quantities*.
G.B. 30

photon radiance, L_p

For a parallel beam, the *photon flow*, Φ_p, leaving or passing through an infinitesimal transparent element of surface in a given direction from the source divided by the orthogonally projected area of the element in a plane normal to the given direction of the beam, θ, [(dΦ_p/dS)/cos θ, simplified expression: $L_p = \Phi_p/(S \cos \theta)$ when the photon flow is constant over the surface area considered]. The SI unit is m^{-2} s^{-1}. For a divergent beam propagating in an elementary cone of the solid angle dω containing the direction θ, the photon radiance is d$^2\Phi_p$/ (dω dS cos θ) with SI unit m^{-2} s^{-1} sr^{-1}. Alternatively, the term can be used with the amount of photons (mol or its equivalent einstein), the SI units then being mol m^{-2} s^{-1} and mol m^{-2} s^{-1} sr^{-1}, respectively.

See *spectral photon radiance, radiance*.
1996, *68*, 2263

photooxidation

Oxidation reactions induced by light. Common processes are: 1. the loss of one or more electrons from a chemical species as a result of photoexcitation of that species; 2. the reaction of a substance with oxygen under the influence of light. When oxygen remains in the product this latter process is also called photooxygenation. Reactions in which neither the substrate nor the oxygen are electronically excited are sometimes called photoinitiated oxidations.

See also *photoreduction*.
1996, *68*, 2264

photooxygenation

The incorporation of molecular oxygen into a molecular entity. There are three common mechanisms:
Type I: the reaction of triplet molecular oxygen with radicals formed photochemically.
Type II: the reaction of photochemically produced singlet molecular oxygen with molecular entities to give rise to oxygen containing molecular entities.
The third mechanism proceeds by electron transfer producing superoxide anion as the reactive species.

See also *photooxidation*.
1996, *68*, 2264

photopeak

See *photoelectric peak*.
1982, *54*, 1548

photopeak efficiency, absolute

See *absolute photopeak efficiency*.
1982, *54*, 1548

photopeak efficiency, intrinsic

See *intrinsic photopeak efficiency*.
1982, *54*, 1548

photophoresis

In aerosol physics, the motion of particles due to the influence of light. In many cases, this amounts to a

special form of thermophoresis due to the heating of the particles by the light.
1990, *62*, 2206

photophosphorylation
The *transformation* of light (solar) energy for the reduction of CO_2 and the generation of adenosine 5′-triphosphate, ATP.
1992, *64*, 162

photophysical processes
Photoexcitation and subsequent events which lead from one to another state of a molecular entity through radiation and radiationless transitions. No chemical change results.
1996, *68*, 2264

photopolymerization
Polymerization processes requiring a photon for the propagation step.
See also *photoinduced polymerization*.
1996, *68*, 2264

photoreaction
See *photochemical reaction*.
1996, *68*, 2264

photoreduction
Reduction reactions induced by light. Common processes are: 1. addition of one or more electrons to a photoexcited species; 2. the photochemical hydrogenation of a substance. Reactions in which the substrate is not electronically excited are sometimes called photoinitiated reductions.
See also *photooxidation*.
1996, *68*, 2264

photoresist
A photoimaging material, generally applied as a thin film, whose local solubility properties can be altered photochemically. A subsequent development step produces an image which is useful for the fabrication of microelectronic devices (e.g. integrated circuits).
1996, *68*, 2265

photosensitization
The process by which a photochemical or photophysical alteration occurs in one molecular entity as a result of initial absorption of radiation by another molecular entity called a photosensitizer. In mechanistic photochemistry the term is limited to cases in which the photosensitizer is not consumed in the reaction.
See *energy transfer*.
1996, *68*, 2265; 1981, *53*, 762

photosensitizer
See *photosensitization*.
1996, *68*, 2265; 1981, *53*, 762

photostationary state
A steady state reached by a reacting chemical system when light has been absorbed by at least one of the components. At this state the rates of formation and disappearance are equal for each of the transient molecular entities formed.
1996, *68*, 2265; see also 1990, *62*, 2206

photosynthesis
A metabolic process involving plants and some types of bacteria (e.g. Chromataceae, Rhodospirillaceae, Chlorobiaceae) in which light energy absorbed by chlorophyll and other photosynthetic pigments results in the reduction of CO_2 followed by the formation of organic compounds. In plants the overall process involves the conversion of CO_2 and H_2O to carbohydrates (and other plant material) and the release of O_2.
1990, *62*, 2207; 1992, *64*, 162

photothermal effect
An effect produced by photoexcitation resulting partially or totally in the production of heat.
1996, *68*, 2265

photothermography
A process utilizing both light and heat, simultaneously or sequentially, for image recording.
1996, *68*, 2265

phototransistor
A bipolar transistor with its base-collector junction acting as a photodiode, which, if irradiated, controls the response of the device. Due to the inherent current gain (of the transistor) the responsivity of the phototransistor is greater than that of photodiodes.
A Darlington phototransistor consists of two separate transistors coupled in the high-impedance Darlington configuration with a phototransistor as the input transistor.
A field effect phototransistor or photo-FET is a field effect transistor (FET) that employs photogeneration of carriers in the channel region (the neutral region sandwiched between the insulator and the depletion region under the gate of the FET). It is characterized by high responsivity due to the high current gain of the FET.
1995, *67*, 1755

photovoltaic cell
A solid state device, usually a *semiconductor*, such as silicon, which absorbs photons with energies higher than or equal to the bandgap energy and simultaneously produces electric power.
See also *photogalvanic cell*.
1996, *68*, 2265; 1991, *63*, 593

pH-rate profile
A plot of observed rate coefficient, or more usually its decadic logarithm, against pH of solution, other variables being kept constant.
1994, *66*, 1150

pH standard
See *operational pH standard, primary pH standard, reference value pH standard*.
1985, *57*, 540

phthaleins
3,3-Bis(hydroxyaryl)-2-benzofuran-1(3*H*)-ones generally derived from the condensation of phthalic anhydride with phenols. E.g. phenolphthalein.

phenolphthalein

See *xanthenes*.
   1995, *67*, 1359

phthalides
3,3-Di(hydrocarbyl) or 3-hydrocarbylidene-2-benzofuran-1(3*H*)-ones.
   See *phthaleins*.
   1995, *67*, 1359

physical adsorption
   See *physisorption*.
   1972, *31*, 586

physical network
   See *network*.
   1996, *68*, 2298

physical quantity (measurable quantity)
An attribute of a phenomenon, body or substance that may be distinguished qualitatively and determined quantitatively.
   G.B. 3; VIM

physisorption (physical adsorption)
Adsorption in which the forces involved are *intermolecular* forces (*van der Waals forces*) of the same kind as those responsible for the imperfection of real gases and the condensation vapours, and which do not involve a significant change in the electronic orbital patterns of the species involved. The term van der Waals adsorption is synonymous with physical adsorption, but its use is not recommended.
   1972, *31*, 586; 1976, *46*, 75

phytotoxicant
An agent which produces a toxic effect in vegetation.
   1990, *62*, 2207

pi
For entries, see under π (beginning of 'p').

pico
SI prefix for 10^{-12} (symbol: p).
   G.B. 74; 1996, *68*, 987

picrates
Salts or *charge-transfer complexes* of picric acid (2,4,6-trinitrophenol).
   See *styphnates*.
   1995, *67*, 1359

piezoluminescence
Luminescence observed when certain solids are subjected to a change in pressure.
   See *triboluminescence*.
   1996, *68*, 2265

pile-up (in radioanalytical chemistry)
The processing by a *radiation* spectrometer of pulses resulting from the simultaneous adsorption of independent particles or photons in a *radiation detector*. As a result they are counted as one single particle or photon with *energy* between the individual energies and the sum of these energies.
   1994, *66*, 2523; O.B. 235

pinacols
Tetra(hydrocarbyl)ethane-1,2-diols, $R_2C(OH)-C(OH)R_2$, of which the tetramethyl example is the simplest one and is itself commonly known as pinacol (benzpinacol is the tetraphenyl analogue).
   1995, *67*, 1359

PIN (p-intrinsic-n) diode
   See *photodiode*
   1995, *67*, 1755

PIN semiconductor detector
A *semiconductor detector* consisting of a compensated region between a p and a n region. The compensated region is often referred to as intrinsic.
   1982, *54*, 1552

pitch
A residue from *pyrolysis* of organic material or tar distillation which is solid at room temperature, consisting of a complex mixture of numerous, essentially aromatic hydrocarbons and heterocyclic compounds. It exhibits a broad softening range instead of a defined melting temperature. When cooled from the melt, pitches solidify without crystallization.
Notes:
The ratio of aromatic to aliphatic hydrogen depends mainly on the source of the starting material. The hydrogen aromaticity (ratio of aromatic to total hydrogen atoms) varies between 0.3 and 0.9.
The aliphatic hydrogen in pitch is largely associated with alkyl side chains substituted on aromatic rings. The content of heterocyclic compounds in pitches varies depending on their origins. Also, the softening temperature can vary in a broad range between about 320 and 570 K depending on the molecular weight (relative molecular mass) and composition of the constituents.
   1995, *67*, 500

pitch-based carbon fibres
Carbon fibres obtained from *pitch* precursor fibres after *stabilization treatment*, *carbonization* and final heat treatment.
Notes:
The term pitch-based carbon fibres comprises the *isotropic pitch-based carbon fibres* as well as the anisotropic *mesophase pitch-based carbon fibres* (MPP-based carbon fibres). The isotropic type belongs to the *carbon fibres type LM* (low modulus) and is mainly used as filler in polymers and insulation materials and for similar applications. The anisotropic type (MPP-based carbon fibres) belongs to the *carbon fibres type HM* and is used mainly for reinforcement purposes due to its high Young's modulus value.
   1995, *67*, 500

pitting corrosion
A special type of non-uniform *corrosion* of passive metals, resulting in the formation of pits, which usually occurs in the presence of certain anions at electrode potentials positive to a critical pitting potential.
1989, *61*, 22

Pitzer strain
See *eclipsing strain*.
1996, *68*, 2213

planar chirality
Term used by some authorities to refer to *stereoisomerism* resulting from the arrangement of out-of-plane groups with respect to a plane (*chirality plane*). It is exemplified by the *atropisomerism* of (*E*)-cyclooctene (chiral plane = double bond and attached atoms) or monosubstituted paracyclophane (chiral plane = substituted ring). The configuration of molecular entities possessing planar chirality is specified by the stereodescriptors R_p and S_p (or P and M).

 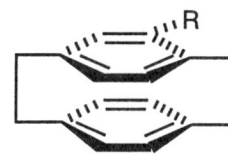

1996, *68*, 2213

planar chromatography
A separation technique in which the stationary phase is present as or on a plane. The plane can be a paper, serving as such or impregnated by a substrate as the stationary bed (paper chromatography, PC) or a layer of solid particles spread on a support e.g. a glass plate (thin layer chromatography, TLC). Sometimes planar chromatography is also termed open-bed chromatography.
1993, *65*, 825

planar film
A *film* whose boundaries are represented by ideal parallel planes. The planar film thickness is defined as the distance between these planes. The terms nanometre thick film, micrometre thick film and millimetre thick film refer specifically to film thickness in the ranges 0.1–100 nm, 0.1–100 μm and 0.1–100 mm, respectively.
1994, *66*, 1671

planar stereoisomerism
See *planar chirality*.
1996, *68*, 2213

Planck constant
Universal fundamental physical constant h = 6.626 0755 (40) × 10^{-34} J s, or $\hbar = h/2\pi$ = 1.054 572 66 (63) × 10^{-34} J s. The latter is used as *atomic unit of action*.
CODATA Bull., 1986, *63*, 1

plane angle
See *angle*
G.B. 11; 1996, *68*, 987

plasma (in spectrochemistry)
A gas which is at least partly ionized and contains particles of various types, viz. electrons, atoms, ions and molecules. The plasma as a whole is electrically neutral.
O.B. 119

plasma desorption ionization (in mass spectrometry)
The ionization of any species by interaction with heavy particles (which may be ions or neutral atoms) formed as a result of the fission of a suitable nuclide adjacent to a target supporting the sample.
1991, *63*, 1548

plasmid
An extrachromosomal genetic element consisting generally of a circular duplex of DNA which can replicate independently of chromosomal DNA. R-plasmids are responsible for the mutual transfer of antibiotic resistance among microbes. Plasmids are used as *vectors* for cloning DNA in *bacteria* or *yeast host* cells.
1992, *64*, 162

plastic flow
Steady flow occurring only above a certain finite stress.
1979, *51*, 1217

plastic transition
A *first-order transition* that occurs when a low-temperature phase transforms to an orientationally-disordered high-temperature phase with a strong variation of enthalpy.
1994, *66*, 588

plateau border (in surface chemistry)
Except for free-floating bubbles, films have to be supported by frames, bulk surfaces or other films. The transition zone separating these from the film proper, always containing some bulk liquid, is called a plateau border.
1972, *31*, 613

plate height (in chromatography), H_{eff}
The column length (L) divided by the plate number:
$$H = L/N$$
It is also called the height equivalent to one theoretical plate (HETP).
1993, *65*, 848

plate number (in chromatography), N
A number indicative of column performance, calculated from the following equations which depend on the selection of the peak width expression:
$$N = (V_R/\sigma)^2 = (t_R/\sigma)^2$$
$$N = 16(V_R/w_b)^2 = 16(t_R/w_b)^2$$
$$N = 5.545(V_R/w_h)^2 = 5.545(t_R/w_h)^2$$
The value of 5.545 stands for 8 ln 2. These expressions assume a Gaussian (symmetrical) peak. In these expressions the units for the quantities inside the brackets must be consistent so that their ratio is dimensionless: i.e. if the numerator is a volume, then

peak width must also be expressed in terms of volume. In former nomenclature the expressions 'number of theoretical plates' or 'theoretical plate number' were used for the same term. For simplification, the present name is suggested.
1993, *65*, 847

pleiotropic gene
A gene affecting more than one (apparently unrelated) characteristic of the *phenotype*.
1992, *64*, 163

ploidy
A term indicating the number of sets of *chromosomes* present in an organism, e.g. haploid (one) or diploid (two).
1992, *64*, 163

plug-flow (in catalysis)
A packed bed flow reactor is commonly called a fixed bed reactor and the term plug-flow is also used to indicate that no attempt is made to back-mix the reaction mixture as it passes through the catalyst bed.
1976, *46*, 80

+ plumbylenes
An older, no longer recommended name for *plumbylidenes*.
1995, *67*, 1359

plumbylidenes
Carbene analogues having the structure $R_2Pb:$.
1995, *67*, 1359

plume (in atmospheric chemistry)
The gaseous and aerosol effluents which are emitted from a chimney and the volume of space which they occupy (often visible). The shape of the plume and the concentrations of pollutants in it at various points along the path of the plume are sensitive functions of the meteorology, local topography and the chemistry which occurs in the plume. Urban plumes are observed over many urban areas and downwind of these areas in which a combination of man sources of pollution are concentrated.
1990, *62*, 2207

plus, minus
See:
1. *d*, *l*, *dl* for (+), (−) and (±)
2. *Helicity* for *P* and *M*.
1996, *68*, 2213

pneumatic detector
A *pressure-sensitive detector* based on the pressure increase of a gas. A special type is the Golay cell where the pressure change is detected by observing the deflection off one of the chamber walls.
1995, *67*, 1752

point group
The classification of the symmetry elements of an object. It is denoted in the Schoenflies notation by an italic symbol, such as C_3, D_2, T_d, etc.
1996, *68*, 2213

point of zero charge (p.z.c.)
A surface charge is at its point of zero charge when the surface charge density is zero. It is a value of the negative logarithm of the activity in the bulk of the charge-determining ions.
1991, 63, 902

poise
cgs unit of dynamic viscosity, P = 0.1 Pa s.
G.B. 112; 1996, *68*, 987

poison (in catalysis)
An inhibitory substance characterized by its propensity to attach very strongly, by a true chemical bond (e.g. covalent) to the surface atoms or ions constituting the catalytically active sites. Poisons act in minute quantities. Typical poisons are S, As, etc. In most cases, activity and/or selectivity cannot be recovered without a drastic change in operating conditions (most often a regeneration). Recovery, if at all, takes place very slowly and/or only partially.
1991, *63*, 1244

Poisson distribution
A discrete distribution with the differential mass-distribution function of the form:

$$f_w(x) = \frac{x}{a+1} \frac{e^{-a} a^{x-1}}{(x-1)!}$$

where x is a parameter characterizing the chain length, such as relative molecular mass or degree of polymerization and a is a positive adjustable parameter.
P.B. 56

+ polar aprotic solvent
See *dipolar aprotic*.
1994, *66*, 1150

polar effect
For a reactant molecule RY, the polar effect of the group R comprises all the processes whereby a substituent may modify the electrostatic forces operating at the reaction centre Y, relative to the standard R°Y. These forces may be governed by charge separations arising from differences in the *electronegativity* of atoms (leading to the presence of dipoles), the presence of unipoles, or electron *delocalization*. It is synonymous with *electronic effect* or 'electrical effect' of a substituent as distinguished from other substituent effects, e.g. *steric effects*.
Sometimes, however, the term 'polar effect' is taken to refer to the influence, other than steric, that non-conjugated substituents exert on reaction rates, i.e. effects connected with electron delocalization between a substituent and the molecular framework to which it is attached are excluded. Polar effect is then not synonymous with electronic effect
See also *field effect, inductive effect, mesomeric effect*.
1994, *66*, 1150

polarity
When applied to solvents, this rather ill-defined term covers their overall solvation capability (solvation power) for solutes (i.e. in chemical equilibria: reac-

tants and products; in reaction rates: reactants and *activated complex*; in light absorptions: ions or molecules in the ground and excited state), which in turn depends on the action of all possible, nonspecific and specific, intermolecular interactions between solute ions or molecules and solvent molecules, excluding such interactions leading to definite chemical alterations of the ions or molecules of the solute. Occasionally, the term solvent polarity is restricted to nonspecific solute/solvent interactions only (i.e. to van der Waals forces).

See also *Dimroth–Reichardt E_T parameter, Grunwald–Winstein equation, ionizing power, Kamlet–Taft solvent parameters, van der Waals forces, Z-value*.
1994, *66*, 1151

polarizability
The ease of distortion of the electron cloud of a *molecular entity* by an electric field (such as that due to the proximity of a charged reagent). It is experimentally measured as the ratio of induced dipole moment (μ_{ind}) to the field E which induces it:

$$\alpha = \mu_{ind}/E$$

The units of α are $C^2\,m^2\,V^{-1}$. In ordinary usage the term refers to the 'mean polarizability', i.e., the average over three rectilinear axes of the molecule. Polarizabilities in different directions (e.g. along the bond in Cl_2, called 'longitudinal polarizability', and in the direction perpendicular to the bond, called 'transverse polarizability') can be distinguished, at least in principle. Polarizability along the bond joining a substituent to the rest of the molecule is seen in certain modern theoretical approaches as a factor influencing chemical *reactivity*, etc., and parametrization thereof has been proposed.

See also *electric polarizability*.
1994, *66*, 1151

polarization
See *light polarization, transition polarization*.
1996, *68*, 2265

polarization (in electrochemistry), ζ
The difference of the electrode potential from the corrosion potential. The symbol η should not be used for polarization.
1989, *61*, 21

polarization error (in spectrochemical analysis)
An error which arises from the fact that the absorbance of a sample (especially a solid) could depend on the polarization of the incident radiation and from the polarization dependent response of the photodetector.
1988, *60*, 1456

polarized interphases
Ideally, interphases for which there are no common components between the phases or the exchange of these is hindered.
1983, *55*, 1253

polarography
A measure of current as a function of potential when the working electrode is a dropping mercury (or other liquid conductor) electrode and unstirred solutions are used.
O.B. 76

polar solvent
See *polarity*.
1994, *66*, 1151

pollution (pollutant)
See *air pollutant, air pollution*.
1990, *62*, 2172

pollution rose
See *wind rose*.
1990, *62*, 2219

polyaddition
A *polymerization* in which the growth of *polymer chains* proceeds by addition reactions between molecules of all degrees of *polymerization*.
Notes:
1. The growth steps are expressed by:

$$P_x + P_y \to P_{x+y}\ \{x\} \in \{1, 2, \ldots \infty\};$$
$$\{y\} \in \{1, 2, \ldots \infty\}$$

where P_x and P_y denote chains of degrees of polymerization x and y, respectively.
2. The earlier term 'addition polymerization' embraced both the current concepts of 'polyaddition' and 'chain polymerization', but did not include 'condensative chain polymerization'.
1996, *68*, 2307

polychromator
An extension of a *monochromator*, a number of exit slits being placed in the focal plane, so allowing a number of discrete bands to pass through (to fall, for example, upon a number of photomultipliers).
O.B. 156

polycondensation
A *polymerization* in which the growth of *polymer chains* proceeds by condensation reactions between molecules of all degrees of *polymerization*.
Notes:
1. The growth steps are expressed by:

$$P_x + P_y \to P_{x+y} + L\ \{x\} \in \{1, 2, \ldots \infty\};$$
$$\{y\} \in \{1, 2, \ldots \infty\}$$

where P_x and P_y denote chains of degree of polymerization x and y, respectively, and L a low-molar-mass by-product.
2. The earlier term 'polycondensation' was synonymous with 'condensation polymerization'. It should be noted that the current definitions of polycondensation and condensative chain polymerization were both embraced by the earlier term 'polycondensation'.
1996, *68*, 2307

polycrystalline graphite
A *graphite material* with coherent crystallographic domains of limited size regardless of the perfection and preferred orientation (texture) of their crystalline structure.

Notes:
The common use of the term polycrystalline graphite for *polygranular graphite* is in line with this definition but may be inexact because usually all grains of polygranular graphite are polycrystalline themselves. Polycrystalline graphite can exhibit a random orientation, more or less preferred orientation, or a highly oriented texture as in some *pyrolytic graphites*. There is no sharp transition, however, between the typical polycrystalline texture and the 'single crystal-like' texture of *highly oriented pyrolytic graphite* (HOPG).
 1995, *67*, 500

polycyclic system
A molecular system regarded as containing a number of rings equal to the number of scissions required to convert it into an open-chain compound.
 B.B. 33

polydent
 See *ambident*.
 1994, *66*, 1151

polydisperse medium
A *colloidal* system in which many particle sizes occur.
 1972, *31*, 607

polydisperse polymer
 See *non-uniform polymer*.
 1996, *68*, 2302

polyelectrolyte
A macromolecular substance which, on dissolving in water or another ionizing solvent, dissociates to give *polyions* (polycations or polyanions) — multiply charged ions — together with an equivalent amount of ions of small charge and opposite sign. Polyelectrolytes dissociating into polycations and polyanions, with no ions of small charge, are also conceivable. A polyelectrolyte can be a polyacid, a polybase, a polysalt or a polyampholyte.
 1972, *31*, 607; see also 1996, *68*, 2304

polyfunctional catalysis
 See *bifunctional catalysis*.
 1976, *46*, 78

polygranular carbon
A *carbon material* composed of grains, which can be clearly distinguished by means of optical microscopy.
Notes:
Industrial carbon materials (such as electrodes) are mostly polygranular, but special grades are agranular materials, such as *glass-like carbon*, *carbon fibres* or *pyrolytic carbon*. Such materials are covered by the term *agranular carbon*.
 1995, *67*, 501

polygranular graphite
A *graphite material* composed of grains which can be clearly distinguished by means of optical microscopy.
Notes:
From the viewpoint of crystallinity, a polygranular graphite is always a *polycrystalline graphite*, but not vice versa. Most industrial *graphite materials* are polygranular. Monogranular materials consist mostly of *non-graphitic carbon*, such materials are called monolithic or *agranular carbons*.
 1995, *67*, 501

polyhedranes
Polycyclic hydrocarbons of the $(CH)_n$ series having skeletons corresponding to the regular and semiregular geometrical solids, e.g. cubane,

 1995, *67*, 1359

polyions
Polycations or polyanions — multiply charged ions.
 1972, *31*, 607

polyketides
Natural compounds containing alternating carbonyl and methylene groups ('β-polyketones'), biogenetically derived from repeated condensation of acetyl coenzyme A (via malonyl coenzyme A), and usually the compounds derived from them by further condensations. Considered by many to be synonymous with the less frequently used terms acetogenins and ketides.

 1995, *67*, 1359

polymer
A substance composed of *macromolecules*.
 1996, *68*, 2289

polymer blend
A macroscopically homogeneous mixture of two or more different species of *polymer*.
Notes:
1. In most cases, blends are homogeneous on scales smaller than several times visual optical wavelengths.
2. For polymer blends, no account is taken of the miscibility or immiscibility of the constituent polymers, i.e. no assumption is made regarding the number of phases present.
3. The use of the term 'polymer alloy' for a polymer blend is discouraged.
 1996, *68*, 2305

polymer crystal
A crystalline domain usually limited by well-defined boundaries.
Notes:
1. Polymer crystals frequently do not display the perfection that is usual for low-molar-mass substances.

2. Twinned polymer crystals are, sometimes, erroneously referred to as 'crystals'.
3. Polymer crystals that can be manipulated individually are often called (polymer) single crystals. A single crystal may contain different fold domains.
P.B. 76

polymer crystallite
A small crystalline domain.
A crystallite may have irregular boundaries and parts of its constituent macromolecules may extend beyond its boundaries. This definition is not identical with that used in classical crystallography.
P.B. 76

polymeric
See *macromolecule*.
1996, *68*, 2289

polymerization
The process of converting a *monomer* or a mixture of monomers into a *polymer*.
1996, *68*, 2305

polymer molecule
See *macromolecule*.
1996, *68*, 2289

polymer network
See *network polymer*.
1996, *68*, 2305

polymer–polymer complex
A complex, at least two components of which are different *polymers*.
1996, *68*, 2305

polymer-poor phase
That phase of a two-phase equilibrium system, consisting of a polymer and low molecular weight material, in which the polymer concentration is lower.
The use of the name 'sol phase' is discouraged.
Synonymous with dilute phase.
P.B. 68

polymer-rich phase
That phase of a two-phase equilibrium system, consisting of a polymer and low molecular weight material, in which the polymer concentration is higher.
The use of the name 'gel phase' is discouraged.
Synonymous with concentrated phase.
P.B. 68

polymer–solvent interaction
The sum of the effects of all intermolecular interactions between polymer and solvent molecules in solution that are reflected in the Gibbs and Helmholtz energies of mixing.
P.B. 57

polymolecularity correction
A correction applied to relationships between a property and the molar mass or relative molecular mass, obtained from polymers non-uniform with respect to relative molecular mass, in order to obtain the corresponding relationship for polymers strictly uniform with respect to relative molecular mass.
P.B. 57

polymorphic transition
A *reversible transition* of a solid crystalline phase at a certain temperature and pressure (the *inversion point*) to another phase of the same chemical composition with a different crystal structure.
Note:
In a *liquid-crystal transition*, this term refers to phase changes in the *smectic state* or columnar discotic state.
Examples:
1. The transitions of SiO_2 (quartz-type) at 1143 K to SiO_2 (tridymite-type), and at 1743 K to SiO_2 (cristobalite-type).
2. The transition of β-AgI (wurtzite-type structure) to α-AgI (body-centred-cubic structure) at 418 K.
Synonymous with *enantiotropic transition*.
1994, *66*, 588

polypeptides
Peptides containing ten or more amino acid residues.
1995, *67*, 1359; W.B. 48

polyprenols
A subgroup of *prenols* in which *n* in the general formula H-$[CH_2$-$C(CH_3)$=CH-$CH_2]_n$-OH is greater than 4.
1987, *59*, 685; W.B. 252

polyquinanes (polyquinenes)
Saturated or unsaturated, respectively, polycyclic *hydrocarbons* consisting of fused five-membered rings, commonly but not necessarily incorporating the skeleton of quinacene (tricyclo[5.2.1.0^{4,10}]deca-2,5,8-triene):

E.g. bicyclo[3.3.0]octane, dodecahedrane.
1995, *67*, 1360

polysaccharides
Compounds consisting of a large number of *monosaccharides* linked glycosidically. This term is commonly used only for those containing more than ten monosaccharide residues. Also called *glycans*.
1995, *67*, 1360

polysulfanes
Compounds having an unbranched chain of sulfur atoms (S_2 or higher) terminating in H: HS_nH. Some chemists exclude disulfane, HS_2H, from the class polysulfanes.
1995, *67*, 1360

polysulfides
Compounds R–$[S]_n$–R, with a chain of sulfur atoms ($n \geq 2$) and R ≠ H. Some chemists exclude disulfides, RS_2R, from the class polysulfides.
1995, *67*, 1360

polytopal rearrangement
Stereoisomerization interconverting different or equivalent spatial arrangements of ligands about a central atom or of a cage of atoms, where the ligand

or cage defines the vertices of a polyhedron. For example *pyramidal inversion* of amines, *Berry pseudorotation* of PF_5, rearrangements of polyhedral boranes.
1996, *68*, 2213

polytypic transition
A transition of a crystalline structure into one or more forms which differ in the way identical layers of atoms are stacked.
Example: ZnS consists of two identical close packings, one of Zn atoms, the other S atoms, with the one displaced to the other along the *c*-axis through one-quarter of the layer spacing. In sphalerite-type ZnS the layers have the face-centred-cubic (ABC ABC) sequence, in wurtzite-type ZnS they have the hexagonal-close-packed (AB AB) sequence. The transition of sphalerite-type ZnS to wurtzite-type ZnS occurs at 1297 K.
1994, *66*, 589

pooled relative standard deviation
See *pooled standard deviation, relative standard deviation.*
1981, *53*, 1821

pooled standard deviation
A problem often arises when the combination of several series of measurements performed under similar conditions is desired to achieve an improved estimate of the imprecision of the process. If it can be assumed that all the series are of the same precision although their means may differ, the pooled standard deviations s_p from k series of measurements can be calculated as

$$s_p = \left(\frac{(n_1 - 1)s_1^2 + (n_2 - 1)s_2^2 + \ldots (n_k - 1)s_k^2}{n_1 + n_2 + \ldots n_k - k} \right)^{1/2}$$

The suffices 1, 2, ... k refer to the different series of measurements. In this case it is assumed that there exists a single underlying standard deviation σ of which the pooled standard deviation s_p is a better estimate than the individual calculated standard deviations $s_1, s_2, \ldots s_k$.
For the special case where k sets of duplicate measurements are available, the above equation reduces to

$$s_p = [\Sigma (x_{i1} - x_{i2})^2 / 2k]^{1/2}$$

Results from various series of measurements can be combined in the following way to give a pooled relative standard deviation $s_{r,p}$:

$$s_{r,p} = [(\Sigma (n_i - 1)s_{r,i}^2) / \Sigma (n_i - 1)]^{1/2}$$
$$= [(\Sigma (n_i - 1)s_i^2 x_i^{-2}) / \Sigma (n_i - 1)]^{1/2}$$

1981, *53*, 1821

population inversion
A situation in which a higher energy state is more populated than a lower energy state.
1996, *68*, 2266

pore size distribution
The distribution of pore volume with respect to pore size; alternatively, it may be defined by the related distribution of pore area with respect to pore size.
1976, *46*, 79

Porod–Kratky chain (in polymers)
See *worm-like chain (in polymers)*
P.B. 51

porosity
A concept related to texture, referring to the pore space in a material.
1976, *46*, 79

porous-layer open-tabular (PLOT) column (in chromatography)
A column in which there is a porous layer on the inner wall. Porosity can be achieved by either chemical means (e.g. etching) or by the deposition of porous particles on the wall from a suspension. The porous layer may serve as a support for a liquid stationary phase or as the stationary phase itself.
1993, *65*, 831

porphyrinogens
Hexahydroporphyrins in which the nitrogen atoms and four *meso* positions are saturated.
1987, *59*, 798; W.B. 294

porphyrins
Natural pigments containing a fundamental skeleton of four pyrrole nuclei united through the α-positions by four methine groups to form a macrocyclic structure (porphyrin is designated porphine in Chemical Abstracts indexes).

porphyrin

See also *corrinoids*.
1995, *67*, 1360; W.B. 281

position-sensitive photomultiplier tube
In position-sensitive *photomultiplier tubes* spatial resolution is obtained with the help of a partitioned photocathode.
1995, *67*, 1756

positive feedback
See *composite reaction*.
1996, *68*, 161

positive ion (in mass spectrometry)
This is an atom, radical, molecule or molecular moiety which has lost one of more electrons thereby attaining an electrically positive charge. The use of the term cation as an alternative is not recommended. The use of mass ion is not recommended.
See *negative ion*.
1991, *63*, 1549

positron
A positively charged *electron*.
1982, *54*, 1548

positronium
Atom-like pair of particles consisting of an anti-electron (positron) and an electron.
G.B. 93

post-column derivatization (in chromatography)
A version of reaction chromatography in which the separated components eluting from the column are derivatized prior to entering the detector. The derivatization process is generally carried out 'on-the-fly', i.e. during transfer of the sample components from the column to the detector. Derivatization may also be carried out before the sample enters the column or the planar medium; this is pre-column (preliminary) derivatization.
1993, *65*, 827

post-filter effect (in luminescence spectroscopy)
This effect arises when the exciting beam does not fill the cell completely and *luminescence* is absorbed by the analyte and interfering impurities in the non-illuminated region facing the detector.
1984, *56*, 244

postprecipitation
The subsequent *precipitation* of a chemically different species upon the surface of an initial *precipitate* usually, but not necessarily, including a common ion.
O.B. 86

potential, applied
See *applied potential*.
1985, *57*, 1502

potential at the point of zero charge (p.z.c.)
The value of the electric potential of an electrode at which one of the charges defined is zero. The *reference electrode* against which this is measured should always be clearly stated.
The potential difference with respect to the potential of zero charge is defined by

$$E_{pzc} = E - E_{\sigma=0}$$

where the potential at the point of zero charge is that for the given electrode in the absence of specific adsorption (other than that of the solvent).
1974, *37*, 509

potential-determining (p.d.) ions
Those species which by virtue of their electron distribution between the solid and liquid phase (or by their equilibrium with electrons in the solid) determine the difference in Galvani potential between these phases. This definition requires that adsorbed p.d. ions are part of the adsorbent and belong to the category of surface ions.
1991, *63*, 897

potential difference, electric
See *electric potential difference*.
1974, *37*, 504

potential energy, E_p, V
Energy of position or orientation in a field of force.
G.B. 12

potential-energy profile
A curve describing the variation of the potential energy of the system of atoms that make up the reactants and products of a reaction as a function of one geometric coordinate, and corresponding to the 'energetically easiest passage' from reactants to products (i.e. along the line produced by joining the paths of steepest descent from the *transition state* to the reactants and to the products). For an *elementary reaction* the relevant geometric coordinate is the *reaction coordinate*; for a *stepwise reaction* it is the succession of reaction coordinates for the successive individual reaction steps. (The reaction coordinate is sometimes approximated by a quasi-chemical index of reaction progress, such as 'degree of atom transfer' or *bond order* of some specified bond.)
See also *potential-energy (reaction) surface*, *Gibbs energy diagram*.
1994, *66*, 1151; 1996, *68*, 176

potential-energy (reaction) surface
A geometric hypersurface on which the potential energy of a set of reactants is plotted as a function of the coordinates representing the molecular geometries of the system.

For simple systems two such coordinates (characterizing two variables that change during the progress from reactants to products) can be selected, and the potential energy plotted as a contour map.

For simple *elementary reactions*, e.g. A–B + C → A + B–C, the surface can show the potential energy for all values of the A, B, C geometry, providing that the ABC angle is fixed.

For more complicated reactions a different choice of two coordinates is sometimes preferred, e.g. the *bond orders* of two different *bonds*. Such a diagram is often arranged so that reactants are located at the bottom left corner and products at the top right. If the trace of the representative point characterizing the route from reactants to products follows two adjacent edges of the diagram, the changes represented by the two coordinates take place in distinct succession; if the trace leaves the edges and crosses the interior of the diagram, the two changes are *concerted*. In many qualitative applications it is convenient (although not strictly equivalent) for the third coordinate to represent the standard Gibbs energy rather than potential energy.

Using bond orders is, however, an oversimplification, since these are not well-defined, even for the transition state. (Some reservations concerning the diagrammatic use of Gibbs energies are noted under *Gibbs energy diagram*.)

The energetically easiest route from reactants to products on the potential-energy contour map defines the *potential-energy profile*.

potential, half-peak
See *half-peak potential*.
1985, *57*, 1502

potential, half-wave
See *half-wave potential*.
1985, *57*, 1502

potential of a cell reaction
See *standard electrode potential*
G.B. 59

potential, peak
See *peak potential*.
1985, *57*, 1502

potential, quarter-transition-time
See *quarter-transition-time potential*.
1985, *57*, 1502

potential, summit
See *summit potential*.
1985, *57*, 1503

potential temperature
The temperature that a dry air parcel would have if lowered or raised adiabatically to a level of 1000 mbar pressure (or other arbitrary standard pressure).
1990, *62*, 2207

potentiation
Dependent action in which a substance or physical agent at a concentration or dose that does not itself have an adverse effect enhances the harm done by another substance or physical agent.
See also *antagonism, synergism*.
1993, *65*, 2085

potentiometer
1. In electronics, a sensitive voltage measuring device based on a null technique which provides infinite impedance at null.
2. In chemistry, potentiometric methods involve the measurement with a potentiometer of voltage generated in a cell. A high impedance digital voltmeter is often used today as a convenient alternative to a potentiometer.
1990, *62*, 2207

potentiometric detection method (in electrochemical analysis)
A detection method in which the potential of a cell is related to the concentration (activity) of a reactant which is a component of the cell fluid.
1990, *62*, 2186

potentiometric selectivity coefficient
A coefficient which defines the ability of an ion-selective electrode to distinguish a particular ion from others. The selectivity coefficient, $K^{pot}_{A,B}$, is evaluated by means of the emf response of the ion-selective electrode in mixed solutions of the primary ion, A, and interfering ion, B (fixed interference method) or less desirable, in separate solutions of A and B (separate solution method). The activities of the primary ion, A, and the interfering ion, B, at which $K^{pot}_{A,B}$ is determined should always be specified, as the value of $K^{pot}_{A,B}$ is defined by the modified Nikolsky–Eisenman equation. The smaller the value of $K^{pot}_{A,B}$ the greater the electrode's preference for the principal ion, A.
Comment: The terms selectivity constant and selectivity factor are frequently used instead of selectivity coefficient. However, in order to standardize the terminology associated with ion-selective electrodes, use of the term selectivity coefficient is recommended, as is the fixed interference method for its evaluation. This selectivity coefficient is not identical to the similar term used in separation science.
1994, *66*, 2532

power, *P*
Rate of energy transfer.
G.B. 13; 1996, *68*, 987

power level
Logarithm of the ratio of a power to a reference power. Levels are expressed in different ways: $L_P = \frac{1}{2}\ln(P/P_0)$Np, where Np is the symbol for the unit *neper* coherent with the SI, or $L_P = 10\lg(P/P_0)$ dB, where dB is the symbol for the unit *decibel*.
G.B. 79

pre-association
A step on the *reaction path* of some *stepwise reactions* in which the *molecular entity* C is already present in an *encounter pair* or *encounter complex* with A during the formation of B from A, e.g.

$$A + C \underset{}{\overset{\text{pre-association}}{\rightleftharpoons}} \underset{\text{encounter complex}}{(A \cdots C)} \overset{\text{rapid}}{\longrightarrow} B$$

In this mechanism the *chemical species* C may but does not necessarily assist the formation of B from A, which may itself be a *bimolecular* reaction with some other reagent. Pre-association is important when B is too short-lived to permit B and C to come together by diffusion.
See also *microscopic diffusion control, spectator mechanism*.
1994, *66*, 1152

precipitation
1. The sedimentation of a solid material (a precipitate) from a liquid solution in which the material is present in amounts greater than its solubility in the liquid.
1990, *62*, 2207; see also O.B. 84; 1994, *66*, 589
2. Electrostatic precipitation: Separation of particles or droplets suspended in a gas or air. A large potential difference (12 to 30 kV dc) is required between the spaced electrodes in the precipitator. The charged particles are attracted to an electrode of opposite charge and collected.
1990, *62*, 2207
3. Precipitation (in meteorology): Rain, snowfall, hail, etc.
1990, *62*, 2207

precipitation fractionation (of polymers)
A process in which a polymeric material, consisting of *macromolecules* differing in some characteristic affecting their solubility, is separated from solution into fractions by successively decreasing the solution power of the solvent, resulting in the repeated formation of a two-phase system in which the less soluble components concentrate in the *polymer-rich phase*.
P.B. 68

precipitation from homogeneous solution (pfhs) (in analysis)
The formation of a precipitate which is generated homogeneously and, generally, slowly by a precipitating agent within a solution.
O.B. 84

precipitation titration
See *titration*.
O.B. 47

precision
The closeness of agreement between independent test results obtained by applying the experimental procedure under stipulated conditions. The smaller the random part of the experimental errors which affect the results, the more precise the procedure. A measure of precision (or imprecision) is the *standard deviation*.
Comment:
As recognized in VIM, precision is sometimes misused for *accuracy*. This problem will be avoided if one recognizes that precision relates only to dispersion, not to deviation from the (conventional) true value. Imprecision has been defined as 'the standard error of the reported value.'
1994, *66*, 598; VIM; O.B. 6; 1990, *62*, 2207; see also 1990, *62*, 2174

precision (of a balance)
The *standard deviation* of the *instrument* for the stated load (e.g. l_{20} for a 20 g load). A statement of the procedure, conditions, and experience of the observer should be included.
O.B. 36

precision (of a weighing)
This depends upon the method of weighing and upon the *precision of indication*, l, for the *load* in question.
O.B. 37

precision of indication (of a balance)
The *standard deviation* of the *instrumental indication*, i, for a stated *load*.
O.B. 35

preconcentration (in trace analysis)
An operation (process) as a result of which the ratio of the concentration or the amount of microcomponents (*trace constituents*) and macrocomponents (matrix) increases. The term enrichment is not recommended.
1979, *51*, 1197

preconcentration coefficient (of a desired microcomponent in trace analysis)
This is defined as

$$K = (Q_T/Q_M)/(Q_T^0/Q_M^0)$$

where Q_T and Q_T^0 are the quantities of the microcomponent in the concentrate and in the sample, respectively (mass units or concentration units), and Q_M^0 and Q_M are the quantities of the matrix before and after *preconcentration*, respectively. If the *recovery* is 100%, $K = Q_M^0/Q_M$. The terms enrichment coefficient and enrichment factor are not recommended.
1979, *51*, 1198

precursor (in radioanalytical chemistry)
Of a *nuclide*, any *radioactive* nuclide which precedes that nuclide in a *decay chain*.
1994, *66*, 2523

precursor complex
May either indicate an *encounter complex* or a *collision complex*, but furthermore implies that this complex undergoes a reaction (e.g. *electron transfer*).
1996, *68*, 2266

precursor ion (in mass spectrometry)
Synonymous with *parent ion*.
1991, *63*, 1549; O.B. 205

predissociation
Dissociation occurring by tunnelling from a 'bound' to an 'unbound' vibronic state. In an absorption spectrum of a molecular entity, the appearance of a diffuse band region within a series of sharp bands is called predissociation, since irradiation with frequencies within the diffuse region leads to effective dissociation. The energy of the band is smaller than that of the dissociation continuum of the bound state.
1996, *68*, 2266

pre-equilibrium (in solvent extraction)
1. Preliminary treatment of a solvent in order to convert the extractants into a suitable chemical form.
2. Preliminary treatment of either phase with a suitable solution of the other phase (in the absence of main extractable solute(s)) so that when the subsequent equilibration is carried out changes in the (volume) phase ratio or in the concentrations of other components are minimized.
Notes:
1. The use of equilibration in this sense is confusing and should be avoided.
2. The term conditioning may be used as a synonym for pre-equilibration.
1993, *65*, 2378

pre-equilibrium (prior equilibrium)
A rapidly reversible step preceding the *rate-limiting step* in a *stepwise reaction*. For example:

$$H_2O + \,\diagdown\!\!\!C=N\!\!\diagup\, \underset{\substack{\downarrow \\ \text{rate-limiting} \\ \text{step}}}{\overset{\text{pre-equilibrium}}{\rightleftharpoons}}\, HO^- + \,\diagdown\!\!\!\overset{+}{C}=N\!\!\underset{H}{\diagup}$$

$$HO-\underset{|}{\overset{|}{C}}-N\!\!\underset{H}{\diagdown}\!\!\diagup$$

See also *kinetic equivalence*.
1994, *66*, 1152; see also 1996, *68*, 177

pre-exponential factor, A
Coefficient in front of the exponential factor expressing the empirical temperature dependence of the *rate coefficient*, k, on temperature, T, $k = A \exp(-E_a/RT)$, where E_a is the activation energy.
G.B. 56; 1996, *68*, 177

preferential sorption (in polymers)
An equilibrium phenomenon, operative in polymer solutions in multicomponent solvents and in polymer networks swollen by multicomponent solvents, that produces differences in solvent composition in the polymer-containing region and in the pure solvent which is in thermodynamic equilibrium with that region.
Synonymous with selective sorption.
P.B. 59

pre-filter effect (in luminescence spectroscopy)
This effect arises when the *luminescence* detector does not see a portion of the luminescent volume where the excitation beam enters the sample. Thus the exciting beam flux is reduced by absorption by the analyte and interfering impurities before it enters the volume observed by the detection system.
1984, *56*, 244

premium coke
An extremely well graphitizing carbon with a high degree of optical anisotropy (isochromatic areas of optical texture above about 100 µm) and which is characterized by a combination of the following properties which differ significantly from those of *regular coke*: high real density, low reversible thermal expansion and low ash content combined, in most cases, with low sulfur content.
Notes:
Premium coke is mainly produced from tars or residues from petrochemistry by the *delayed coking process*. Also, refined *coal tar pitches* are used as precursors for premium coke production.
1995, *67*, 501

premix burner (in flame spectroscopy)
A burner in which fuel and oxidant are thoroughly mixed inside the burner housing before they leave the burner ports and enter the primary-combustion or inner zone of the flame. This type of burner usually produces an approximately laminar flame, and is commonly combined with a separate unit for nebulizing the sample.
O.B. 165

prenols
Alcohols possessing the general formula H–[CH$_2$C(Me)=CHCH$_2$]$_n$OH in which the carbon skeleton is composed of one or more isoprene units (biogenetic precursors of the *isoprenoids*).
1995, *67*, 1360; W.B. 252

pre-polymer
A polymer or oligomer composed of *pre-polymer molecules*.
1996, *68*, 2304

pre-polymer molecule
A *macromolecule* or *oligomer molecule* capable of entering, through reactive groups, into further *polymerization*, thereby contributing more than one *monomeric unit* to at least one chain of the final *macromolecule*.
A pre-polymer molecule capable of entering into further polymerization through reactive *end groups*, often deliberately introduced, is known as a *telechelic molecule*.
1996, *68*, 2290

pressure, p
Normal force acting on a surface divided by the area of that surface. For a mixture of gases the contribution by each constituent is called the partial pressure $p_i = x_i p$, where x_i is the *amount fraction* of the ith constituent and p is the total pressure.
G.B. 12; 1996, *68*, 987

pressure flow
See *transport*.
1989, *61*, 1661

pressure gradient correction factor (in gas chromatography)
A factor that corrects for the compressibility of the *carrier gas*. The values of the measured quantities obtained after multiplication by the factor j are independent of the pressure drop in the column. If p_i, p_o are respectively the pressures of the carrier gas at the inlet and outlet of the column, then j is given by:
$$j = \frac{3[(p_i/p_o)^2 - 1]}{2[(p_i/p_o)^3 - 1]}$$
O.B. 101

pressure-induced transition
A transition induced by a change of pressure.
Example: The room-temperature transition of CdS with a wurtzite-type structure, at 1.74 GPa, to CdS with a NaCl-type structure.
1994, *66*, 589

pressure jump
See *relaxation*.
1993, *65*, 2297

pressure-sensitive detector
A *detector* which employs the pressure change which results from the absorption of radiation.
1995, *67*, 1752

pressure, static
See *static pressure*.
1990, *62*, 2208

pretreatment (of a catalyst)
Following the preparation of a *catalyst* or following its insertion into a catalytic reactor, it is often subjected to various treatments before the start of a catalytic run. The term pretreatment may, in general, be applied to this set of treatments. In some cases the word activation is used. It implies that the material is

converted into a catalyst or into a very much more effective one by the pretreatment.
1976, *46*, 80

primary crystallization
The first stage of *crystallization*, considered to be ended when most of the spherulite surfaces impinge on each other. In isothermal crystallization, primary crystallization is often described by the *Avrami equation*.
P.B. 85

primary current distribution
See *current distribution*.
1981, *53*, 1836

primary electrons (pe) (in *in situ* microanalysis)
Electrons generated by a thermal or field emission source used to bombard the specimen for generation of the analytical signals.
1983, *55*, 2026

primary excitation (of X-rays)
See *characteristic X-ray emission*.
1991, *63*, 737

primary isotope effect
A kinetic isotope effect attributable to isotopic substitution of an atom to which a bond is made or broken in the *rate-controlling step* or in a *pre-equilibrium* step of a specified reaction. The corresponding isotope effect on the equilibrium constant of a reaction in which one or more bonds to isotopic atoms are broken, is called a 'primary equilibrium isotope effect'.
See also *secondary isotope effect*.
1994, *66*, 1130

primary kinetic electrolyte effect
See *kinetic electrolyte effect*.
1994, *66*, 1152

primary kinetic isotope effect
See *isotope effect*.
1994, *66*, 1152

primary mixture
A mixture obtained directly from two or more components intended for the preparation of more dilute calibration mixtures (called secondary or tertiary mixtures).
1990, *62*, 2208

primary particle (in catalysis)
See *aggregate (in catalysis)*.
1991, *63*, 1231

primary photochemical process (primary photoreaction)
Any elementary chemical process undergone by an electronically excited *molecular entity* and yielding a primary photoproduct.
See *primary (photo)process*.
1996, *68*, 2266

primary pH standards
Certain substances which meet the criteria of:
(i) reproducible preparation in a highly pure state;
(ii) stability of solution over a reasonable period of time;
(iii) having a low value of the residual liquid junction potential;
(iv) having pH betwen 3 and 11.
O.B. 19

primary pollutant (in atmospheric chemistry)
A pollutant emitted directly into the air from identifiable sources (e.g. SO_2, NO, hydrocarbons, etc.). *Secondary pollutants,* such as ozone, are generated within the atmosphere through chemical changes which occur in primary pollutants.
1990, *62*, 2208

+ primary (photo)process
The term primary (photo)process for photophysical processes is apt to lead to inconsistencies, and its use is therefore discouraged.
See *primary photochemical process*.
1996, *68*, 2266

primary (photo)product
The first observable *chemical entity* which is produced in the primary photochemical process and which is chemically different from the reactant.
See *primary (photo)process*.
1996, *68*, 2266

primary sample
The collection of one or more increments or units initially taken from a population.
The portions may be either combined (composited or bulked sample) or kept separate (gross sample). If combined and mixed to homogeneity, it is a blended bulk sample. The term 'bulk sample' is commonly used in the sampling literature as the sample formed by combining increments. The term is ambiguous since it could also mean a sample from a bulk lot and it does not indicate whether the increments or units are kept separate or combined. Such use should be discouraged because less ambiguous alternative terms (composite sample, aggregate sample) are available. 'Lot sample' and 'batch sample' have also been used for this concept, but they are self limiting terms. The use of 'primary' in this sense is not meant to imply the necessity for multistage sampling.
1990, *62*, 1205

primary standard
See *standard solution*.
O.B. 47

primary structure
In the context of *macromolecules* such as *proteins*, the constitutional formula, usually abbreviated to a statement of the sequence and if appropriate crosslinking of chains.
See also *secondary structure, tertiary structure, quaternary structure*.
1996, *68*, 2213; see also W.B. 80

primary structure (of a segment of a polypeptide)
The amino acid sequence of the *polypeptide* chain(s), without regard to spatial arrangement (apart from configuration at the α-carbon atom).
Note: This definition does not include the positions of disulfide bonds and is therefore not identical with 'covalent structure'.
W.B. 80

primitive change
One of the conceptually simpler molecular changes into which an *elementary reaction* can be notionally dissected. Such changes include *bond* rupture, bond formation, internal rotation, change of bond length or bond angle, bond *migration*, redistribution of charge, etc.
The concept of primitive changes is helpful in the detailed verbal description of elementary reactions, but a primitive change does not represent a process that is by itself necessarily observable as a component of an elementary reaction.
1994, *66*, 1152

principal group
The characteristic group chosen for citation at the end of a name by a suffix or a class name.
B.B.(G) 13

principal ion (in mass spectrometry)
A molecular or fragment ion which is made up of the most abundant isotopes of each of its atomic constituents. In the case of compounds that have been artificially isotopically enriched in one or more positions such as $CH_3{}^{13}CH_3$ or CH_2D_2 the principal ion may be defined by treating the heavy isotopes as new atomic species. Thus, in the above two examples, the principal ions would be of masses 31 and 18, respectively.
1991, *63*, 1550

principal moments of inertia
Moments of inertia about the principal axes of the molecule chosen so as to make all the products of inertia equal to zero, $I_A \leq I_B \leq I_C$.
G.B. 23

principle of least nuclear motion
The hypothesis that, for given reactants, the reactions involving the smallest change in nuclear positions will have the lowest *energy of activation*. (It is also often simply referred to as principle of least motion.)
1994, *66*, 1134

principle of microscopic reversibility
In a reversible reaction, the mechanism in one direction is exactly the reverse of the mechanism in the other direction. This does not apply to reactions that begin with a photochemical excitation.
See also *chemical reaction, detailed balancing*.
1994, *66*, 1141

principle of minimum structural change
See *molecular rearrangement*.
1994, *66*, 1141

prior distribution, P_0
A product state distribution calculated on the basis of some physical model for the reaction.
The distribution usually referred to is that of the products at the instant of their formation.
1996, *68*, 178

prior equilibrium
See *pre-equilibrium*.
1994, *66*, 1152

priority
See *CIP priority*.
1996, *68*, 2213

probability, P
Number of favourable events divided by the total number of possible events. It is a number between 0 and 1.
G.B. 39

probability density, P
Product of the wave function and its complex conjugate.
G.B. 16

probe (in biotechnology)
A specific DNA or RNA sequence which has been labelled by radioactivity, fluorescence labels or chemiluminescence labels and which is used to detect complementary sequences by *hybridization* techniques such as *blotting* or colony hybridization.
1992, *64*, 163

process
A phenomenon by which change takes place in a system. In physiological systems, a process may be chemical, physical or both.
1992, *64*, 1572

prochirality
This term is used in different, sometimes contradictory ways; four are listed below.
1. The geometric property of an achiral object (or spatial arrangement of points or atoms) which is capable of becoming *chiral* in a single *desymmetrization* step. An achiral molecular entity, or a part of it considered on its own, is thus called prochiral if it can be made chiral by the replacement of an existing atom (or achiral group) by a different one.
An achiral object which is capable of becoming chiral in two desymmetrization steps is sometimes described as proprochiral. For example the proprochiral CH_3CO_2H becomes prochiral as CH_2DCO_2H and chiral as $CHDTCO_2H$.
2. The term prochirality also applies to an achiral molecule or entity which contains a trigonal system and which can be made chiral by the addition to the trigonal system of a new atom or achiral group. For example addition of hydrogen to one of the enantiotopic faces of the prochiral ketone $CH_3CH_2COCH_3$ gives one of the enantiomers of the chiral alcohol $CH_3CH_2CHOHCH_3$; the addition of CN^- to one of the *diastereotopic* faces of the chiral aldehyde shown below converts it into one of the diastereoisomers of

the cyanohydrin. The two faces of the trigonal system may be described as *Re* and *Si*.

(from attack of *Si* face)

(from attack of *Re* face)

3. The term prochiral also applies to a tetrahedral atom of an achiral or chiral molecule which is bonded to two *stereoheterotopic* groups. For example, the prochiral molecule CH_3CH_2OH can be converted into the chiral molecule CH_3CHDOH by the isotopic replacement of one of the two enantiotopic hydrogen atoms of the methylene group. The carbon atom of the methylene group is called prochiral. The prochiral molecule $HO_2CCH_2CHOHCH_2CO_2H$ can be converted into a chiral product by esterification of one of the two enantiotopic $-CH_2CO_2H$ groups. The carbon atom of the CHOH group is called prochiral. The chiral molecule $CH_3CHOHCH_2CH_3$ can be converted into one of the diastereoisomers of $CH_3CHOHCHDCH_3$ by the isotopic replacement of one of the two *diastereotopic* hydrogen atoms of the methylene group. The carbon atom of the methylene group is called prochiral. The stereoheterotopic groups in these cases may be described as *pro-R* or *pro-S*. Reference to the two stereoheterotopic groups themselves as prochiral, although common, is strongly discouraged.
See *chirality centre*.
4. The term prochirality is also applied to the enantiotopic faces of a trigonal system.
1996, *68*, 2213

prochirality centre
An atom of a molecule which becomes a *chirality centre* by replacing one of the two *stereoheterotopic* ligands attached to it by a different ligand, e.g. C-1 of ethanol; C-3 of butan-2-ol.
1996, *68*, 2214

product
A substance that is formed during a chemical reaction.
See *reactant*.
1996, *68*, 178

product-determining step
The step of a *stepwise reaction*, in which the product distribution is determined. The product-determining step may be identical to, or occur later than, the *rate-controlling step* on the reaction coordinate.
1994, *66*, 1153

product development control
The term is used for reactions under *kinetic control* where the *selectivity* parallels the relative (thermodynamic) stabilities of the products. Product development control is usually associated with a *transition state* occurring late on the *reaction coordinate*.
See also *steric-approach control, thermodynamic control*.
1994, *66*, 1153

product ion
Synonymous with *daughter ion*.
1991, *63*, 1550

productivity (in biotechnology), *r*
An economical figure denoting the mass of a product formed per unit reactor volume and unit time. Productivity is often referred to per unit of *enzyme* or *biomass*.
1992, *64*, 163

product state distribution
The distribution of available energy among the vibrational, rotational and translation degrees of freedom in the product molecules. The distribution usually referred to is that of the products at the instant of their formation.
1996, *68*, 178

pro-E, pro-Z
One of a pair of identical groups c attached to a double bond (as in $abC=Cc_2$) is described as *pro-E* if, when it is arbitrarily assigned *CIP priority* over the other group c, the *stereodescriptor* of the molecule becomes *E*. The other group c is then described as *pro-Z*.
1996, *68*, 2214

progenitor ion (in mass spectrometry)
Synonymous with *precursor ion (in mass spectrometry)*.
O.B. 205

program
Noun: A set of instructions enabling a device to perform an action.
Verb: To provide a set of instructions enabling a device to perform an action.
1989, *61*, 1659

programmed-flow chromatography (flow programming)
A procedure in which the rate of flow of the mobile phase is changed systematically during a part or whole of the separation.
1993, *65*, 827

programmed-pressure chromatography (pressure programming)
A procedure in which the inlet pressure of the mobile phase is changed systematically during a part or whole of the separation.
1993, *65*, 827

programmed-temperature chromatography (temperature programming)
A procedure in which the temperature of the column is changed systematically during a part or the whole of the separation.
1993, *65*, 827

projection formula
A formal two-dimensional representation of a three-dimensional molecular structure obtained by projection of bonds (symbolized as lines) onto a plane with or without the designation of the positions of relevant atoms by their chemical symbols.

A projection formula which indicates the spatial arrangement of bonds is called a *stereochemical formula* or stereoformula. Examples of stereoformulae are *Fischer projection, Newman projection, sawhorse projection, wedge projection* and *zig-zag projection*.
   See also *perspective formula*.
   1996, *68*, 2214

prolate trochoidal mass spectrometer
A *mass spectrometer* in which ions of different mass/charge are separated by means of crossed electric and magnetic fields in such a way that the selected ions follow a prolate trochoidal path. The usual term 'cycloidal' used sometimes is incorrect because the path used is not cycloidal. A cycloid is a special case of a trochoid.
   1991, *63*, 1545; O.B. 203

promoter (gene technology)
The DNA region, usually upsteam to the coding sequence of a *gene* or *operon*, which binds and directs RNA polymerase to the correct transcriptional start site and thus permits the initation of *transcription*.
   1992, *64*, 163

promoter (in catalysis)
A relatively small quantity of one or more substances, which when added to a *catalyst* improves the *activity*, the *selectivity*, or the useful lifetime of the catalyst. In general, a promoter may either augment a desired reaction or suppress an undesired one.
   1976, *46*, 79

promotion
   See *pseudo-catalysis*.
   1994, *66*, 1153

prompt coincidence
The occurrence of two or more events separated by a time interval which is less than a specified small value.
   1982, *54*, 1537

prompt neutrons
Neutrons accompanying the fission process without measurable delay.
   1982, *54*, 1547

propagating reaction
   See *chain-propagating reaction*.
   1996, *68*, 157

propagation
   See *chain reaction*.
   1994, *66*, 1153

propellanes
Tricyclic saturated hydrocarbons, systematically named tricyclo[$a.b.c.0^{1,(a+2)}$]alkanes; have been referred to as [$a.b.c$]propellanes.

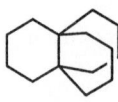 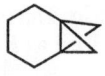

[4.4.4]propellane   [4.1.1]propellane

   See *paddlanes*.
   1995, *67*, 1360

property
A set of data elements (system, component, kind-of-property) common to a set of particular properties, e.g. substance concentration of glucose in blood plasma.
Information about identification, time and result is not considered.
   See also *particular property*.
   1995, *67*, 1565

prophage
The latent state of a phage *genome* in a lysogenic bacterium.
   1992, *64*, 163

proportional counter
A gas-filled X-ray detector in which the electric potential is high enough for the gain to reach a value in the range from 10^2 to 10^5. Each electron produced by the initial photo-ionization causes one avalanche. Since the number of avalanche events is proportional to the energy of the incident photons, the charge collected by the anode is proportional to the X-ray photon energy.
   1995, *67*, 1754

proportional counter tube
A *counter tube* operated under such conditions that the magnitude of each pulse is proportional to the amount of energy deposited in it.
   1982, *54*, 1538

proportional gas-scintillation counter
A *proportional counter* coupled to an ultraviolet sensitive photomultiplier tube. Initial electrons produced by the interaction of the high-energy photon with the counter fill-gas are accelerated by a high electric field where they acquire sufficient energy to excite the noble gas atoms. The resulting UV radiation is observed by a *photomultiplier tube*.
   1995, *67*, 1754

proprochirality
   See *prochirality*.
   1996, *68*, 2214

pro-R, pro-S
A *stereoheterotopic* group c (as in tetrahedral Xabc$_2$) is described as *pro-R* if, when it is arbitrarily assigned *CIP priority* over the other stereoheterotopic group c, the configuration of the thus generated chiral centre is assigned the *stereodescriptor R*. The other group c is then described as *pro-S*. This method for distinguishing between stereoheterotopic groups can be applied to other kinds of prochiral molecular entities or *prochiral* parts of molecular entities considered on their own.

See *prochirality centre*.
1996, *68*, 2214

pros (in histidine nomenclature)
The nitrogen atoms of the imidazole ring of histidine are denoted by *pros* ('near', abbreviated π) and *tele* ('far', abbreviated τ) to show their position relative to the side chain. This recommendation arose from the fact that two different systems of numbering the atoms in the imidazole ring of histidine had both been used for a considerable time (biochemists generally numbering as 1 the nitrogen atom adjacent to the side chain, and organic chemists designating it as 3). The carbon atom between the two ring nitrogen atoms is numbered 2 (as in imidazole), and the carbon atom next to the τ nitrogen is numbered 5. The carbon atoms of the aliphatic chain are designated α and β. This numbering should also be used for the decarboxylation product histamine and for substituted histidine.

histidine

W.B. 43

prostaglandins
Naturally occurring compounds derived from the parent C_{20} acid, prostanoic acid.

See *icosanoids, prostanoids*.
1995, *67*, 1360

prostanoids
The family of natural prostaglandins and prostaglandin-like compounds.
1995, *67*, 1360

prosthetic group
The non-amino acid portion of a conjugated protein. Examples include *cofactors* such as flavines or *cytochromes*, as well as lipids and polysaccharides, which are the prosthetic groups of lipoproteins and glycoproteins, respectively.
1992, *64*, 163

proteases
Enzymes that catalyse the hydrolysis of proteins. Usually several proteolytic enzymes are necessary for the complete breakdown of polypeptides to their amino acids.
1992, *64*, 163

protected lyophobic colloid
A *colloidally stable* mixture of a *lyophobic* and a *lyophilic* colloid.
1972, *31*, 610

protective action (in colloid chemistry)
Addition of small amounts of *hydrophilic* colloid to a hydrophobic sol may make the latter more sensitive to *flocculation* by electrolyte. Higher concentrations of the same hydrophilic colloid usually protect the hydrophobic sol from flocculation. This phenomenon is called protective action.
1972, *31*, 610

protein engineering
A technique used to produce proteins with altered or novel amino acid sequences. The methods used are:
1. *Transcription* and *translation* systems from synthesized lengths of DNA or RNA with novel sequences.
2. Chemical modification of 'normal' proteins.
3. Solid-state polypeptide synthesis to form proteins.
1992, *64*, 163

proteins
Naturally occurring and synthetic *polypeptides* having molecular weights greater than about 10000 (the limit is not precise).
See also *peptides*.
1995, *67*, 1361; see also W.B. 48

proteoglycan
A subclass of protein in which the carbohydrate units are polysaccharides that contain amino sugars. The protein is glycosylated by one or more (up to about 100) *glycosaminoglycans* [linear polymers of up to about 200 repeating disaccharide units that consist of a hexosamine (D-glucosamine or D-galactosamine) alternating with a uronic acid (D-glucuronic or L-iduronic) or a neutral sugar (D-galactose)].
1988, *60*, 1391; W.B. 85

protic
See *protogenic*.
1994, *66*, 1153

protide
See *protium*.
1988, *60*, 1116

protio
See *protium*.
1988, *60*, 1116

protium
A specific name for the atom ^1H. The cation $^1H^+$ is a *proton*, the species ^1H is a *protide* anion, and ^1H is the *protio* group. The word proton should not be used for H^+ in its natural abundance.
1988, *60*, 1116

protogenic (solvent)
Capable of acting as a proton donor (strongly or weakly acidic as a *Brønsted acid*). The term is preferred to the synonym protic or the more ambiguous expression acidic by itself.
1983, *55*, 1348

+ protolysis
This term has been used synonymously with proton (hydron)-transfer reaction. Because of its misleading similarity to *hydrolysis, photolysis*, etc., its use is discouraged.
See also *autoprotolysis*.
1994, *66*, 1153

proton
Nuclear particle of charge number +1, spin quantum number $\frac{1}{2}$ and rest mass of 1.007 276 470(12) u.
　See also *protium*.
　G.B. 93

proton affinity
The negative of the enthalpy change in the gas phase reaction (real or hypothetical) between a proton (more appropriately *hydron*) and the *chemical species* concerned, usually an electrically neutral species to give the *conjugate acid* of that species. Proton affinity is often, but unofficially, abbreviated as PA.
　See also *gas phase basicity*.
　1994, *66*, 1153

protonated molecule (in mass spectrometry)
An ion formed by interaction of a molecule with a proton abstracted from an ion, as often occurs in chemical ionization according to the reaction: M + XH$^+$ → MH$^+$ + X. The symbolism [M + H]$^+$ may also be used to represent the protonated molecule. The widely used term 'protonated molecular ion' to describe the MH$^+$ ion is not recommended, since it suggests an associated product of a proton with a molecular ion.
　1991, *63*, 1550

protonation constant
The equilibrium constant K_{Hn} for the addition of the *n*th proton to a charged or uncharged ligand. The cumulative protonation constant, β_{Hn}, is the equilibrium constant for the formation of H_nL from nH^+ and L.
The same equilibrium constant may be described in several ways, depending on how the ligand, L, or the acid, has been defined.
　O.B. 13

proton charge
　See *elementary charge*.
　CODATA Bull., 1986, *63*, 1

proton magnetic moment
Atomic fundamental physical constant μ_p = 1.410 607 61 (47) × 10^{-26} J T^{-1}.
　CODATA Bull., 1986, *63*, 1

proton magnetogyric ratio
Atomic fundamental physical constant γ_p = 2.675 221 28 (81) s^{-1} T^{-1}.
　CODATA Bull., 1986, *63*, 1

proton number
　See *atomic number*.
　G.B. 20

proton rest mass
Atomic fundamental physical constant m_p = 1.672 6231 (10) × 10^{-27} kg.
　CODATA Bull., 1986, *63*, 1

proton transfer reaction
A *chemical reaction*, the main feature of which is the *intermolecular* or *intramolecular* transfer of a proton (*hydron*) from one *binding site* to another. For example:

$$CH_3CO_2H + CH_3\overset{O}{\overset{\|}{C}}CH_3 \longrightarrow CH_3CO_2^- + CH_3\overset{O^+H}{\overset{\|}{C}}CH_3$$

In the detailed description of proton transfer reactions, especially of rapid proton transfers between electronegative atoms, it should always be specified whether the term is used to refer to the overall process (including the more-or-less *encounter-controlled* formation of a hydrogen bonded complex and the separation of the products; see *microscopic diffusion control*) or just to the proton transfer event (including solvent rearrangement) by itself.
　See also *autoprotolysis, tautomerism*.
　1994, *66*, 1153

protophilic (solvent)
Capable of acting as proton acceptor, strongly or weakly basic (as a *Brønsted base*). Also called HBA (hydrogen bond acceptor) solvent.
　See also *protogenic solvent*.
　1994, *66*, 1153

protoplast
A spherical, osmotically-sensitive *cell* without its cell wall but retaining an intact cell membrane. Protoplasts are used to create hybrid cells via protoplast *fusion*.
　1992, *64*, 164

prototrophs
Microoganisms capable of growing at or below a temperature optimum of 15 °C.
　1992, *64*, 164

prototropic rearrangement (or prototropy)
　See *tautomerism*.
　1994, *66*, 1153

pseudo acids
Potentially acidic compounds that require some structural reorganization, such as keto–enol tautomerization, having a non-negligible activation energy, in order to show normal acidic properties, e.g. nitroalkanes:

$$RCH_2NO_2 \xrightarrow{(slow)} RCH=NO_2H \xrightarrow{(fast)} RCHNO_2^- + H^+$$

　1995, *67*, 1361

pseudo-asymmetric carbon atom
The traditional name for a tetrahedrally coordinated carbon atom bonded to four different entities, two and only two of which have the same constitution but opposite chirality sense.
The *r/s* descriptors of pseudo-asymmetric carbon atoms are invariant on reflection in a mirror (i.e. *r* remains *r*, and *s* remains *s*), but are reversed by the exchange of any two entities (i.e. *r* becomes *s*, and *s* becomes *r*).

An example is C-3 of ribaric (C-3 is *r*) or xylaric acid (C-3 is *s*) or hyoscyamine (C-3 is *r*). The hyphen in pseudo-asymmetric may be omitted.

ribaric acid

xylaric acid

hyoscyamine

1996, *68*, 2214

pseudo-axial
See *axial, equatorial*.
1996, *68*, 2215

pseudo bases
Hydroxy compounds that give salts with acids by formation of water accompanied by a change of constitution.

See *anhydro bases*.
1995, *67*, 1361

pseudo-catalysis
If an acid or base is present in nearly constant concentration throughout a reaction in solution (owing to buffering or the use of a large excess), it may be found to increase the rate of that reaction and also to be consumed during the process. The acid or base is then not a *catalyst* and the phenomenon cannot be called *catalysis* according to the well-established meaning of these terms in chemical kinetics, although the *mechanism* of such a process is often intimately related to that of a catalysed reaction. It is recommended that the term pseudo-catalysis be used in these and analogous cases (not necessarily involving acids or bases). For example, if a *Brønsted acid* accelerates the hydrolysis of an ester to a carboxylic acid and an alcohol, this is properly called acid catalysis, whereas the acceleration, by the same acid, of hydrolysis of an amide should be described as pseudo-catalysis by the acid: the 'acid pseudo-catalyst' is consumed during the reaction through formation of an ammonium ion. The terms 'general acid pseudo-catalysis' and 'general base pseudo-catalysis' may be used as the analogues of *general acid catalysis* and *general base catalysis*.

The term 'base-promoted', 'base-accelerated' or 'base-induced' is sometimes used for reactions that are pseudo-catalysed by bases. However, the term 'promotion' also has a different meaning in other chemical contexts.
1994, *66*, 1153

pseudo-co-oligomer
An *irregular oligomer*, the molecules of which are derived from only one species of *monomer* but which display a variety of structural features more appropriate for description in *co-oligomer* terms.
1996, *68*, 2300

pseudo-copolymer
An *irregular polymer*, the molecules of which are derived from only one species of *monomer* but which display a variety of structural features more appropriate for description in *copolymer* terms.
Note:
Where appropriate, adjectives specifying the types of 'copolymer' may be applied to 'pseudo-copolymer'. The term *statistical pseudo-copolymer*, for instance, may be used to describe an irregular polymer in the molecules of which the sequential distribution of configurational units obeys known statistical laws.
1996, *68*, 2300

pseudo-equatorial
See *axial, equatorial*.
1996, *68*, 2215

pseudo-first-order rate coefficient
See *order of reaction*.
1994, *66*, 1154

pseudo-first-order reaction
See *order of reaction*.
1994, *66*, 1154

pseudohalogens
Compounds that resemble the halogen elements, X_2, in their chemistry, e.g. $(CN)_2$ cyanogen, $(SCN)_2$ thiocyanogen, ICN iodine cyanide. Certain ions that have sufficient resemblance to halide ions are sometimes referred to as pseudohalide ions, e.g. N_3^-, SCN^-, CN^-.
1995, *67*, 1361

+ pseudomolecular rearrangement
The use of this awkwardly formed term is discouraged. It is synonymous with 'intermolecular rearrangement'.
See *molecular rearrangement*.
1994, *66*, 1154

pseudopericyclic
A *concerted transformation* is pseudopericyclic if the primary changes in bonding occur within a cyclic array of atoms at one (or more) of which nonbonding and bonding atomic orbitals interchange roles. A formal example is the enol → enol prototropy of pentane-2,4-dione (acetylacetone).
Because the π- and σ-atomic orbitals that interchange roles are orthogonal, such a reaction does not proceed through a fully *conjugated transition state* and is thus

not a pericyclic reaction and therefore not governed by the rules that express *orbital symmetry* restrictions applicable to pericyclic reactions.
1994, *66*, 1154

pseudo rate constant
A term sometimes used for a rate coefficient. For example, if the rate equation is:
$$v = k[A][B]$$
the function $k[A]$ is the pseudo first-order rate constant (or first-order rate coefficient) with respect to B. Similarly, $k[B]$ is the pseudo first order rate constant (or first-order rate coefficient) with respect to A.
The term pseudo *rate constant* is particularly useful when the concentration of one reactant is much greater than that of another.
See also *rate coefficient*.
1996, *68*, 178

pseudorotation
Stereoisomerization resulting in a structure that appears to have been produced by rotation of the entire initial molecule and is *superposable* on the initial one, unless different positions are distinguished by substitution, including isotopic substitution.
One example of pseudorotation is a facile interconversion between the many envelope and twist conformers of a cyclopentane due to the out of plane motion of carbon atoms.
Another example of pseudorotation (Berry pseudorotation) is a polytopal rearrangement that provides an intramolecular mechanism for the isomerization of trigonal bipyramidal compounds (e.g. λ^5-phosphanes), the five bonds to the central atom E being represented as e^1, e^2, e^3, a^1 and a^2. Two *equatorial* bonds move apart and become *apical* bonds at the same time as the apical bonds move together to become equatorial.

A related conformational change of a trigonal bipyramidal structure is described as turnstile rotation. The process may be visualized as follows.

An apical and an equatorial bond rotate as a pair *ca.* 120° relative to the other three bonds. (Doubts have been expressed about the distinct physical reality of this mechanism.)
1996, *68*, 2215

+ pseudo-unimolecular
A term sometimes used as synonymous with pseudo-first-order, but is inherently meaningless.
See *molecularity, order of reaction*.
1994, *66*, 1154

+ pseudoureas
An obsolescent synonym for isoureas.
1995, *67*, 1361

pseudo-zero-order reaction
See *order of reaction*.
1994, *66*, 1154

psychosine
See *glycolipids*.
W.B. 129

psychrometric hygrometer
Instrument by which the relative humidity of the atmosphere may be determined. It is generally composed of two temperature sensors, one of which measures the temperature of the air; the other sensor is moistened with water and senses the cooling due to evaporation of water. The temperature difference between the two sensors is a function of relative humidity; sometimes referred to as a 'wet-and-dry bulb' thermometer.
1990, *62*, 2196

psychrometry
The use of a wet-and-dry bulb thermometer for measurement of atmospheric humidity.
1990, *62*, 2208

puffing
An irreversible expansion of some *carbon artifacts* during *graphitization heat treatment* between 1650 and 2700 K.
Notes:
Puffing is caused by the release of heteroatoms, for instance sulfur atoms, from the *coke* in association with specific microstructural rearrangements.
1995, *67*, 502

puffing inhibitor
Metals or metal compounds with a high chemical affinity for the heteroatoms in the carbons. They are distributed as fine particles within the *carbon materials* to be graphitized.
Notes:
Iron and iron compounds are most frequently used as puffing inhibitors when *puffing* is related to sulfur.
1995, *67*, 502

pulsatance
See *angular frequency*.
G.B. 11

pulse amplifier, linear
See *linear pulse amplifier*.
1982, *54*, 1545

pulse amplitude analyser
A sub-assembly for determining the distribution function of a set of pulses in terms of their amplitude. Synonymous with pulse height analyser.
1982, *54*, 1548

pulse amplitude selector
A circuit which gives an output pulse for each input pulse whose amplitude lies within a chosen interval. Synonymous with single channel pulse height analyser.
1982, *54*, 1548

pulse duration (in electroanalytical chemistry)
In pulse *polarography*, differential pulse polarography, Kalousek polarography, and related techniques, the duration of an interval during which the excitation signal deviates from the *base line*. This interval includes the *sampling interval*.
1985, *57*, 1503

pulse height analyser
See *pulse amplitude analyser*.
1982, *54*, 1548

pulse height analyser, multi-channel
See *multi-channel pulse height analyser*.
1982, *54*, 1548

pulse height analyser, single-channel
See *single-channel pulse height analyser*.
1982, *54*, 1548

pulse reactor (in catalysis)
In a pulse reactor, a *carrier gas*, which may be inert or possibly one of the reactants, flows over the *catalyst* and small amounts of the other reactant or reactants are injected into the carrier gas at intervals. A pulse reactor is useful for exploratory work but kinetic results apply to a transient rather than to the steady state conditions of the catalyst.
1976, *46*, 80

pump-probe technique
A flash *photolysis* technique in which the light beam (probe) used for spectral analysis is generated from a portion of the excitation (pump) beam. A time delay in the latter allows the obtention of kinetic data.
1996, *68*, 2266

purine bases
Purine and its substitution derivatives, especially naturally occurring examples. The customary numbering shown below is not systematic.

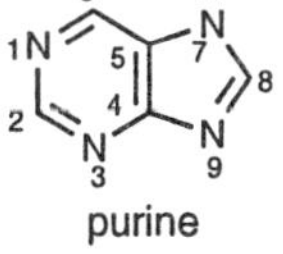

purine

1995, *67*, 1361

purity, radiochemical
See *radiochemical purity*.
1994, *66*, 2523

purity, radionuclidic
See *radionuclidic purity*.
1994, *66*, 2523

pyramidal inversion
A *polytopal rearrangement* in which the change in bond directions to a three-coordinate *central atom* having a pyramidal arrangement of bonds (tripodal arrangement) causes the central atom (apex of the pyramid) to appear to move to an equivalent position on the other side of the base of the pyramid. If the three ligands to the central atom are different pyramidal inversion interconverts *enantiomers*.

1996, *68*, 2215

pyranoses
Cyclic hemiacetal forms of *monosaccharides* in which the ring is six-membered (a tetrahydropyran).
See *furanoses*.
1995, *67*, 1361

pyrimidine bases
Pyrimidine and its substitution derivatives, especially naturally occurring examples.

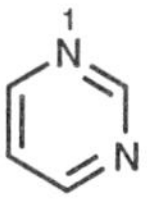

pyrimidine

1995, *67*, 1361

pyro
A prefix designating compounds formed by heating a compound, usually with the elimination of water, carbon dioxide, or other simple molecule, e.g. pyroglutamic acid from glutamic acid.
1995, *67*, 1361

pyroelectric detector
A detector (based on the temperature dependence of ferroelectricity in some crystals) which produces an electrical signal proportional to the energy flux on the collector surface.
O.B. 193; see also 1995, *67*, 1752

pyrolysis
A chemical degradation reaction that is caused by thermal energy. The term pyrolysis generally refers to an inert environment.
See also *analytical pyrolysis, flash vacuum pyrolysis, thermolysis*.
1993, *65*, 2407; 1994, *66*, 1154

pyrolysis-gas chromatography
A version of reaction chromatography in which a sample is thermally decomposed to simpler fragments before entering the column.
1993, *65*, 827

pyrolytic carbon
A *carbon material* deposited from gaseous hydrocarbon compounds on suitable underlying substrates (carbon materials, metals, ceramics) at temperatures ranging from 1000 to 2500 K (chemical vapour deposition).

Notes:
A wide range of microstructures, e.g. isotropic, lamellar, substrate-nucleated and a varied content of remaining hydrogen, can occur in pyrolytic carbons, depending on the deposition conditions (temperature, type, concentration and flow rate of the source gas, surface area of the underlying substrate, etc.).

'Pyrocarbon' which is synonymous with *pyrolytic carbon* was introduced as a trademark and should not be used as a term.

The term pyrolytic carbon does not describe the large range of carbon materials obtained by thermal degradation (thermolysis, pyrolysis) of organic compounds when they are not formed by chemical vapour deposition (CVD). Also, carbon materials, obtained by physical vapour deposition (PVD) are not covered by the term pyrolytic carbon.

1995, *67*, 502

pyrolytic graphite
A *graphite material* with a high degree of preferred crystallographic orientation of the *c*-axes perpendicular to the surface of the substrate, obtained by *graphitization heat treatment* of *pyrolytic carbon* or by chemical vapour deposition at temperatures above 2500 K.

Notes:
'Pyrographite', a synonym for pyrolytic graphite, was introduced as a trademark and should not be used as a term. Hot working of pyrolytic graphite (by heat treatment under compressive stress at temperatures above 3000 K) results in *highly oriented pyrolytic graphite* (HOPG).

1995, *67*, 502

pyrromethenes
A less preferred term for *dipyrrins*. (The term 'pyrromethane' has been used for the system with a $-CH_2-$ linkage).

See *dipyrrins*.
1995, *67*, 1361

QSAR
See *quantitative structure–activity relationships*.
1994, *66*, 1155

Q-switched laser
A *laser* in which the state of the device introducing important losses in the resonant cavity and preventing lasing operation is suddenly switched to a state where the device introduces very low losses. This increases rapidly the Quality factor of the cavity, allowing the build-up of a short and very intense laser pulse. Typical pulse durations are in the ns range. The Q-switching may be active (a rotating mirror or electro-optic device) or passive (a saturable absorber).
See also *free-running laser*.
1996, *68*, 2266

quadratic mean, $\bar{x}_q$
The square root of the expression in which the sum of squared observations is divided by the number n. It can be calculated by the formula:

$$\bar{x}_q = \left(\frac{\sum x_i^2}{n} \right)^{1/2}$$

Comment: This quantity is also sometimes directly (but inappropriately) calculated, for example, when an observable is proportional to the square of concentration. The quadratic mean, also known as the root mean square, is sometimes appropriate, however, as in certain of the formulae connected with linear calibration functions.
1994, *66*, 602

quadro-
An affix used in names to denote four atoms bound into a quadrangle (e.g. square).
R.B. 245; B.B. 465

quadrupole ion storage trap (Quistor)
An arrangement in which ions with a desired range of quotients mass/charge are trapped by making them describe stable paths under the effect of a static and a high frequency electric quadrupole field.
1991, *63*, 1546

quadrupole mass analyser
An arrangement in which ions with a desired quotient mass/charge are made to describe a stable path under the effect of a static and a high-frequency electric quadrupole field, and are then detected. Ions with a different mass/charge are separated from the detected ions because of their unstable paths.
1991, *63*, 1545; O.B. 202

quadrupole splitting (in Mössbauer spectroscopy)
The measured Doppler velocity difference between the two peaks seen in quadrupole split spectra from *nuclides* such as ^{57}Fe and ^{119}Sn.
1976, *45*, 214

qualitative analysis
Analysis in which substances are identified or classified on the basis of their chemical or physical properties, such as chemical reactivity, solubility, molecular weight, melting point, radiative properties (emission, absorption), mass spectra, nuclear half-life, etc.
See also *quantitative analysis*.
1995, *67*, 1701

qualitative elemental specificity (in analysis)
Ability of a method to detect one element in the presence of another element.
1979, *51*, 2247

quality assurance
The guarantee that the quality of a product (analytical data set, etc.) is actually what is claimed on the basis of the *quality control* applied in creating that product. Quality assurance is not synonymous with quality control. Quality assurance is meant to protect against failures of quality control.
1990, *62*, 2208

quality control
The maintenance and statement of the quality of a product (data set, etc.) specifically that it meets or exceeds some minimum standard based on known, testable criteria.
1990, *62*, 2208

quality factor (in nuclear analytical chemistry)
The *linear-energy-transfer*-dependent factor by which *absorbed dose* is multiplied to obtain *dose equivalent*.
1982, *54*, 1549

quality of solvent (in polymer chemistry)
A qualitative characterization of the *polymer–solvent interaction*. A solution of a polymer in a 'better' solvent is characterized by a higher value of the second *virial coefficient* than a solution of the same polymer in a 'poorer' solvent.
P.B. 57

quantized internal energy
The quantized internal energy of a molecule in its electronic ground or excited state can be approximated with sufficient accuracy for analytical purposes by:

$$E_{int} = E_{el} + E_{vib} + E_{rot}$$

where E_{el} is the electronic, E_{vib} the vibrational and E_{rot} the rotational energy, respectively.
1988, *60*, 1452

quantitative analysis
Analyses in which the amount or concentration of an analyte may be determined (estimated) and expressed as a numerical value in appropriate units. *Qualitative analysis* may take place without quantitative analysis, but quantitative analysis requires the identification (qualification) of the analytes for which numerical estimates are given.
1995, *67*, 1701

quantitative structure–activity relationships (QSAR)
The building of structure–biological activity models by using regression analysis with physicochemical constants, indicator variables or theoretical calculations. The term has been extended by some authors

to include chemical reactivity, i.e. activity is regarded as synonymous with reactivity. This extension is, however, discouraged.
See also *correlation analysis*.
1994, 66, 1155

quantity
Attribute of a phenomenon, body or substance that may be distinguished qualitatively and determined quantitatively.
Notes:
1. The term quantity may refer to a quantity in a general sense, for example length, mass, or to a particular quantity, for example length of a particular rod, mass of a specified object.
2. The term *kind-of-quantity* refers to the abstract concept of a quantity common to a group of related quantities.
1996, 68, 987

quantity calculus
Algebra with quantities where the symbols of quantities represent products of numerical values and their units.
G.B. 3

quantity of dimension one (dimensionless quantity)
Quantity in the dimensional expression of which all the exponents of the dimensions of the base quantities are zero.
1996, 68, 987

quantum, γ
See *γ-quantum*.
1982, 54, 1543

quantum (of radiation)
An elementary particle of electromagnetic energy in the sense of the wave-particle duality.
See *photon*.
1996, 68, 2267

quantum counter
A medium emitting with a *quantum yield* independent of the excitation energy over a defined spectral range (e.g, concentrated rhodamine 6G solutions between 300 and 600 nm). Also used for devices producing an electrical signal proportional to the *photon flux* absorbed in a medium.
1996, 68, 2267; O.B. 193; 1984, 56, 238

quantum efficiency
See *efficiency*.
For a primary *photochemical process*, quantum efficiency is identical to *quantum yield*.
1996, 68, 2267

quantum-mechanical tunnelling
See *tunnelling*.
1996, 68, 178

quantum well multilayer
See *superlattice*.
1994, 66, 1672

quantum yield, Φ
The number of defined events which occur per *photon* absorbed by the system. The integral quantum yield is:

Φ = (number of events)
    /(number of photons absorbed)
For a photochemical reaction:

Φ = (amount of reactant consumed or product formed)/(amount of photons absorbed)
The differential quantum yield is:

$$\Phi = \frac{d[x]/dt}{n}$$

where $d[x]/dt$ is the rate of change of a measurable quantity, and n the amount of photons (mol or its equivalent einstein) absorbed per unit time. Φ can be used for photophysical processes or photochemical reactions.
See also *efficiency*.
1996, 68, 2267; 1994, 66, 1155; 1996, 68, 178; G.B. 57; see also 1990, 62, 2208

quarter-transition-time potential
In *chronopotentiometry* (at constant *current density*), the potential of the *indicator electrode* at the instant when the time that has elapsed since the application of current is equal to one-fourth of the *transition time*. Appropriate correction for double-layer charging is needed in practice.
1985, 57, 1502

quartet state
A state having a total electron spin quantum number equal to $3/2$.
See *multiplicity*.
1996, 68, 2267

quartz–iodine lamp
A tungsten filament high-intensity incandescent lamp which contains iodine in a quartz envelope. Used primarily as a source of visible light.
1996, 68, 2268

quasi-axial
See *axial, equatorial*.
1996, 68, 2215

quasi-classical trajectory (QCT) method
A procedure for calculating trajectories in which the quantization of the reactants is taken into account, but in which the course of the reaction is treated classically.
1996, 68, 179

quasi-enantiomers
Constitutionally different yet closely related chemical species MX and MY having the opposite chirality sense of the large common chiral moiety M. For example (R)-2-bromobutane is a quasi-enantiomer of (S)-2-chlorobutane.
See also *quasi-racemic compound*.
1996, 68, 2215

quasi-equatorial
See *axial, equatorial*.
1996, *68*, 2216

quasi-equilibrium
In conventional *transition-state theory* it is assumed that *activated complexes* are formed in a state of equilibrium with the reactants. They are not in classical equilibrium with the reactants; if they were, addition of more activated complexes to the system would cause the equilibrium to shift in favour of the reactants. This would not occur for an activated complex, and the term quasi-equilibrium is used to denote this special type of equilibrium.
1996, *68*, 179

quasi-molecular ion (in mass spectrometry)
A protonated molecule or an ion formed from a molecular ion by loss of a hydrogen atom. The use of the term 'pseudo-molecular ion' is not recommended.
1991, *63*, 1550

quasi-racemic compound
The crystalline product of a 1:1 association between *quasi-enantiomers*.
1996, *68*, 2216

quasi-single-strand polymer
A regular linear polymer that can be described by a preferred *constitutional repeating unit* in which only one terminal constituent subunit is connected through a single atom to the other identical constitutional repeating units or to an end group.
P.B. 110

quaternary ammonium compounds
Derivatives of ammonium compounds, $NH_4^+Y^-$, in which all four of the hydrogens bonded to nitrogen have been replaced with *hydrocarbyl groups*. Compounds having a carbon–nitrogen double bond (i.e. $R_2C=N^+R_2Y^-$) are more accurately called *iminium compounds*. e.g. $[(CH_3)_4N]^+OH^-$, tetramethylammonium hydroxide.
See *onium compounds*.
1995, *67*, 1361

quaternary structure
The defined organization of two or more macromolecules with tertiary structure such as a protein that are held together by hydrogen bonds and van der Waals and coulombic forces.
See also *primary structure, secondary structure, tertiary structure*.
1996, *68*, 2216; see also W.B. 80

quaterpolymer
See *copolymer*.
1996, *68*, 2300

quencher
A molecular entity that deactivates (quenches) an excited state of another molecular entity, either by energy transfer, electron transfer, or by a chemical mechanism.
See *quenching, Stern–Volmer kinetic relationships*.
1996, *68*, 2268

quenching
1. Arresting the course of a chemical reaction by chemical or physical means.
1993, *65*, 2297
2. (in photochemistry) The deactivation of an excited molecular entity intermolecularly by an external environmental influence (such as a quencher) or intramolecularly by a substituent through a non-radiative process. When the external environmental influence (quencher) interferes with the behaviour of the excited state after its formation, the process is referred to as dynamic quenching. Common mechanisms include energy transfer, charge transfer, etc. When the environmental influence inhibits the excited state formation the process is referred to as static quenching.
See *Stern–Volmer kinetic relationships*.
1996, *68*, 2268; O.B. 120; 1994, *66*, 2523
3. (in radiation chemistry) The process of inhibiting continuous or multiple discharges following a single ionizing event in certain types of radiation detectors, particularly in Geiger–Müller counter tubes.
1994, *66*, 2523

quenching constant (in photochemistry)
See *quencher, quenching, Stern–Volmer kinetic relationships*.
1996, *68*, 2268

quenching correction (in photochemistry)
Correction for errors due to different *quenching* for standards and test portions. When using *liquid scintillation detectors*, these corrections can be based e.g on the standard addition or sample channels ratio method or the use of automated external standardization.
1994, *66*, 2523

quinarenes
Mancude assemblies of three carbocyclic rings, a six-membered quinonoid ring bonded at the 1,4-positions to odd-membered rings which differ in ring size by two, e.g.

[3.6.5]quinarene

[5.6.7]quinarene
1995, *67*, 1362

quinhydrones
Molecular complexes of one equivalent amount of a *quinone* with one equivalent amount of the corresponding hydroquinone.

1995, *67*, 1362

quinomethanes
Methylidenecyclohexadienones and dimethylidenecyclohexadienes, formally derived from *quinones* by replacement of one or both of the quinone oxygens by methylidene groups. (The diradical, triplet state of quinodimethanes can also be called *o*- or *p*-xylylenes).

p-quinomethane

p-quinodimethane
1995, *67*, 1362

+ quinomethides (quinone methides)
Undesirable names for *quinomethanes*.
1995, *67*, 1362

quinone diazides
A potentially confusing term for *diazooxides*; the presence of an azido group, $-N_3$, is falsely suggested.
1995, *67*, 1362

quinones
Compounds having a fully conjugated cyclic dione structure, such as that of benzoquinones, derived from *aromatic compounds* by conversion of an even number of –CH= groups into –C(=O)– groups with any necessary rearrangement of double bonds (polycyclic and heterocyclic analogues are included).

p-benzoquinone
1995, *67*, 1362

quinonimines (quinone imines)
Imines derived from quinones by replacement of one or both oxygens by =NR. The term may include both types unless the infix mono or di is included, e.g. *o*-benzoquinone diimine:

1995, *67*, 1362

quinonoximes
Mono- or di-oximes of *quinones*; quinone monooximes are tautomeric with nitrosophenols.
See *oximes*.
1995, *67*, 1362

Quistor
See *quadrupole ion storage trap*.
1991, *63*, 1546

ρ-value (rho-value)

A measure of the susceptibility to the influence of *substituent* groups on the rate constant or equilibrium constant of a particular organic reaction involving a family of related *substrates*. Defined by Hammett for the effect of ring substituents in *meta*- and *para*-positions of aromatic side-chain reactions by the empirical 'ρσ-equation' of the general form:

$$\log (k_X/k_H) = \rho\sigma_X$$

in which σ_X is a constant characteristic of the substituent X and of its position in the reactant molecule. More generally (and not only for aromatic series), ρ-values (modified with appropriate subscripts and superscripts) are used to designate the susceptibility of reaction series for families of various organic compounds to any substituent effects, as given by the modified set of σ-constants in an empirical ρσ-correlation.

Reactions with a positive ρ-value are accelerated (or the equilibrium constants of analogous equilibria are increased) by substituents with positive σ-constants. Since the sign of σ was defined so that substituents with a positive σ increase the acidity of benzoic acid, such substituents are generally described as attracting electrons away from the aromatic ring. It follows that reactions with a positive ρ-value are considered to involve a transition state (or reaction product) so that the difference in energy between this state and the reactants is decreased by a reduction in electron density at the reactive site of the substrate.

See also *Hammett equation*, *σ-constant*, *Taft equation*.
1994, *66*, 1161

ρσ-equation (rho-sigma equation)
See *Hammett equation*, *ρ-value*, *σ-constant*, *Taft equation*.
1994, *66*, 1162

R_F value (in chromatography)
See *retardation factor (in planar chromatography)*.
1993, *65*, 845; O.B. 105

R_M value (in planar chromatography)
A logarithmic function of the R_F value (*retardation factor*):

$$R_M = \log \frac{1 - R_F}{R_F} = \log \left[\frac{1}{R_F} - 1\right]$$

1993, *65*, 845

rabbit
A small container propelled pneumatically or hydraulically through a tube leading from the laboratory to a location in a *nuclear reactor* or other device where *irradiation* of a sample can take place.
1982, *54*, 1549

rac
See *racemate*.
1996, *68*, 2216

racemate
An equimolar mixture of a pair of *enantiomers*. It does not exhibit *optical activity*. The chemical name or formula of a racemate is distinguished from those of the enantiomers by the prefix (±)– or *rac*- (or *racem*-) or by the symbols *RS* and *SR*.
1996, *68*, 2216

racemic
Pertaining to a *racemate*.
1996, *68*, 2216

racemic compound
A crystalline *racemate* in which the two *enantiomers* are present in equal amounts in a well defined arrangement within the lattice of a homogeneous crystalline addition compound.
1996, *68*, 2216

racemic conglomerate
An equimolar mechanical mixture of crystals each one of which contains only one of the two *enantiomers* present in a *racemate*. The process of its formation on crystallization of a racemate is called spontaneous resolution, since pure or nearly pure enantiomers can often be obtained from the conglomerate by sorting.
1996, *68*, 2216

+ racemic mixture
The term is confusing since it has been used as a synonym for both *racemate* and *racemic conglomerate*. Usage strongly discouraged.
1996, *68*, 2216

racemization
The production of a *racemate* from a *chiral* starting material in which one *enantiomer* is present in excess.
1996, *68*, 2216

racemo structures (in polymers)
See *meso structures (in polymers)*.
P.B. 37

rad
Non-SI unit of absorbed dose of radiation, rad = 0.01 Gy.
G.B. 113; 1996, *68*, 987

radial development
See *radial elution*.
1993, *65*, 829

radial electrostatic field analyser (in mass spectrometry)
An arrangement of two conducting sheets forming a capacitor and giving a radial electrostatic field which is used to deflect and focus ion beams of different energies. The capacitor may be cylindrical, spherical, or toroidal.
O.B. 201

radial elution (radial development) or circular elution (circular development) (in planar chromatography)
A mode of operation in which the sample is spotted at a point source at or near the middle of the plane

and is carried outward in a circle by the mobile phase, also applied at that place.
1993, 65, 829

radian
SI derived unit of plane angle, rad = 1.
G.B. 72; 1996, 68, 988

radiance, L
At a point on a surface and in a given direction, the radiant intensity of an element of the surface, divided by the area of the orthogonal projection this element on a plane perpendicular to the given direction.
See also *photon radiance, spectral radiance, spherical radiance.*
G.B. 31; ISO 31-5: 1992; see also 1996, 68, 2268; O.B. 120; 1990, 62, 2208

radiant emittance
See *radiant exitance.*
1996, 68, 2268

radiant energy, Q
The total energy emitted, transferred or received as radiation in a defined period of time ($Q = \int Q_\lambda \, d\lambda$). It is the product of *radiant power*, P, and time, t: $Q = Pt$ when the radiant power is constant over the time considered.
See also *spectral radiant energy.*
1996, 68, 2268; 1996, 68, 988; G.B. 30; ISO 31-5: 1992

radiant energy density, ρ, w
Radiant energy divided by the volume.
G.B. 30

radiant exitance, M
At a point on a surface, the radiant energy flux leaving the element of the surface, divided by the area of that element.
See also *photon exitance, spectral radiant exitance.*
G.B. 31; 1996, 68, 988; see also 1996, 68, 2269

radiant exposure, H
The *irradiance*, E, integrated over the time of irradiation ($\int E \, dt$, simplified expression $H = Et$ when the irradiance is constant over the time considered). For a parallel and perpendicularly incident beam not scattered or reflected by the target or its surroundings, *fluence* (H_0) is an equivalent term.
1996, 68, 2269

radiant flux
See *radiant power.*
1996, 68, 988

radiant (energy) flux, P, Φ
Although flux is generally used in the sense of the 'rate of transfer of fluid, particles or energy across a given surface,' the radiant energy flux has been adopted by IUPAC as equivalent to *radiant power*, P. ($P = \Phi = dQ/dt$, simplified expression: $P = \Phi = Q/t$ when the radiant energy, Q, is constant over the time considered). In photochemistry, Φ is reserved for *quantum yield.*
See also *photon flow, photon radiance, radiance, radiant energy, spectral radiant flux, radiant power, quantum yield.*
1996, 68, 2268

radiant intensity
See *intensity.*
G.B. 31

radiant power, P, Φ
Same as *radiant (energy) flux*, Φ. Power emitted, transferred or received as radiation.
See *spectral radiant power.*
1996, 68, 2269; G.B. 31; 1996, 68, 988

radiant quantities
Group of quantities characterizing electromagnetic radiation in terms of energy. The symbols of radiant quantities may be distinguished by adding a subscript e.
G.B. 30

radiation
A term embracing electromagnetic waves as well as fast moving particles. In *radioanalytical chemistry* the term usually refers to radiation emitted during a nuclear process (*radioactive decay*, nuclear reaction, *nuclear fission*, accelerators).
1994, 66, 2523

radiation, background
See *background radiation.*
O.B. 213

radiation chemistry
The part of chemistry which deals with the chemical effects of *ionizing radiation*, as distinguished from photochemistry associated with visible and ultraviolet electromagnetic radiation.
1994, 66, 2523; O.B. 236

radiation constants
Fundamental physical constants characterizing black body radiation. The first radiation constant is $c_1 = 2\pi h c_0^2 = 3.741\ 7749\ (22) \times 10^{-16}$ W m^2, the second is $c_2 = hc_0/k = 1.438\ 769\ (12) \times 10^{-2}$ m K, where h is the *Planck constant*, c_0 the speed of light and k the *Boltzmann constant.*
CODATA Bull., 1986, 63, 1

radiation continuum (in spectrochemistry)
Continuous (in the wavelength, not time sense) *radiation* arising for example from non-quantized free-free transitions of electrons in the fields of the ions, free-bound transitions or radiative recombinations of electrons and ions, incandescent radiation emitted by hot solids (when the radiation distribution conforms to that described by Planck's law, it is considered black-body radiation) and unresolvable band spectra, i.e. where the spectral lines are wider than the spacings between them.
1985, 57, 1462

radiation counter
Radiation measuring assembly comprising a *radiation detector* in which individual ionizing events cause electrical pulses and the associated equipment

for processing and counting the pulses. Often an expression is added indicating the type of radiation *detector* (e.g. *scintillation, semiconductor*).
1994, *66*, 2516

radiation detector
A device in which incident radiation produces a measurable effect.
If this effect is a rise in temperature it is called a thermal detector. If it is a rise in pressure it is called a photoacoustic detector. In the case where an electrical signal is produced it is called a photoelectric detector. Photoelectric detectors can be classified as photo-emissive detectors and semiconductor detectors. Where the radiation produces a chemical reaction, it is termed a photochemical detector.
A detector yielding an output signal that is independent of the wavelength of the radiation over a specific region is called a nonselective detector. Where it is wavelength specific it is a selective detector. A detector having a quantum efficiency independent of the wavelength is a nonselective quantum counter.
Certain detectors are able to distinguish between different quantum energies. This property is described by the energy resolution ΔE and the energy resolving power $E/\Delta E$. These detectors are called energy dispersive detectors. In X-ray spectroscopy, the reciprocal $\Delta E/E$ is often used but this is discouraged.
1995, *67*, 1748; 1994, *66*, 2518

radiation hazard
Hazard that exists in a region where there is a *radiation* field, other than what is considered to be normal *background radiation*.
1982, *54*, 1549

radiationless deactivation (decay)
The loss of electronic excitation energy without photon emission or chemical change.
See *energy transfer, internal conversion, intersystem crossing*.
1996, *68*, 2269

radiationless transition
A transition between two states of a system without photon emission or absorption.
See also *radiative transition*.
1996, *68*, 2269; O.B. 185; see also 1984, *56*, 234

radiation, natural
See *natural radiation*.
1982, *54*, 1549

radiation spectrum
The components of *radiation* arranged in order of their wavelengths, frequencies or quantum energies. For particle radiation they are arranged in order of their kinetic energies.
1982, *54*, 1549

radiative absorption (in spectrochemistry)
A process by which a particle in the ground state or an excited state may undergo transition to a higher energy level by *absorption* of a photon. For a given particle in the lower state of probability per second of such a transition in a field with a continuous spectrum is proportional to the spectral radiant energy density of the absorption line. The proportionality constant is termed the transition probability for absorption.
The transition probability for stimulated emission is defined in a similar way for the reverse *radiative de-excitation* process that is induced by the same radiation field.
1985, *57*, 1463

radiative capture
Capture of a *particle* by a *nucleus* followed by immediate emission of *gamma radiation*.
1982, *54*, 1537

radiative de-excitation (in spectrochemistry)
The change in the internal energy of a particle may be due to radiative processes, i.e. the emission or absorption of a photon. A particle in an excited state may undergo a transition to a lower energy level by emission of a photon. This is known as radiative de-excitation. If such a transition occurs spontaneously its probability per second for a given excited particle is termed the transition probability for spontaneous emission.
1985, *57*, 1463

radiative energy transfer
Transfer of excitation energy by radiative deactivation of a donor molecular entity and reabsorption of the emitted light by an acceptor molecular entity. The probability of transfer is given approximately by:

$$P_{r,t} \propto [A]\chi J$$

where J is the spectral overlap integral, $[A]$ is the concentration of the acceptor, and χ is the specimen thickness. This type of energy transfer depends on the shape and size of the vessel utilized. Same as trivial energy transfer.
See also *Dexter excitation transfer, energy transfer, Förster excitation transfer*.
1996, *68*, 2269

radiative lifetime, τ_0
The lifetime of an excited molecular entity in the absence of radiationless transitions. It is the reciprocal of the first-order rate constant for the radiative step, or of the sum of these rate constants if there is more than one such step. The equivalent term, natural lifetime, is discouraged. Approximate expressions exist relating τ_0 to the oscillator strength of the emitting transition.
1996, *68*, 2270

radiative transition
A transition between two states of a *molecular entity*, the energy difference being emitted or absorbed as *photons*.
In principle, radiative and radiationless transitions can be distinguished in molecules. The first occur by absorption or emission of light quanta, and the latter is the result of the transformation of electronic excitation energy into vibrational/rotational energy.

See also *radiationless transition*
O.B. 185; 1996, *68*, 2270; 1984, *56*, 234

radical (free radical)
A *molecular entity* such as ·CH$_3$, ·SnH$_3$, Cl· possessing an unpaired electron. (In these formulae the dot, symbolizing the unpaired electron, should be placed so as to indicate the atom of highest spin density, if this is possible.) Paramagnetic metal ions are not normally regarded as radicals. However, in the 'isolobal analogy' the similarity between certain paramagnetic metal ions and radicals becomes apparent.

At least in the context of physical organic chemistry, it seems desirable to cease using the adjective 'free' in the general name of this type of *chemical species* and *molecular entity*, so that the term 'free radical' may in future be restricted to those radicals which do not form parts of radical pairs.

Depending upon the core atom that possesses the unpaired electron, the radicals can be described as carbon-, oxygen-, nitrogen-, metal-centred radicals. If the unpaired electron occupies an orbital having considerable s or more or less pure p character, the respective radicals are termed σ- or π-radicals.

In the past, the term 'radical' was used to designate a *substituent group* bound to a molecular entity, as opposed to 'free radical', which nowadays is simply called radical. The bound entities may be called *groups* or *substituents*, but should no longer be called radicals.

See also *biradical*.
1994, *66*, 1155; 1995, *67*, 1362; see also 1990, *62*, 2209

radical anion
See *radical ion*.
1994, *66*, 1155

radical cation
See *radical ion*.
1994, *66*, 1155

radical centre(s)
The atom (or group of atoms) in a polyatomic radical on which an unpaired electron is largely localized. Attachment of a monovalent atom to a radical centre gives a molecule for which it is possible to write a *Lewis formula* in which the normal stable valencies are assigned to all atoms.
1994, *66*, 1156

radical combination
See *colligation*.
1994, *66*, 1156

radical copolymerization
A *copolymerization* which is a *radical polymerization*.
1996, *68*, 2308

radical disproportionation
See *disproportionation*.
1983, *55*, 1350

radical ion
A radical that carries an electric charge. A positively charged radical is called a 'radical cation' (e.g. the benzene radical cation C$_6$H$_6$$^{·+}$); a negatively charged radical is called a 'radical anion' (e.g. the benzene radical anion C$_6$H$_6$$^{·-}$ or the benzophenone radical anion Ph$_2$C–O$^{·-}$). Commonly, but not necessarily, the odd electron and the charge are associated with the same atom.

Unless the positions of unpaired spin and charge can be associated with specific atoms, superscript dot and charge designations should be placed in the order ·+ or ·– suggested by the name 'radical ion'. (e.g. C$_3$H$_6$$^{·+}$).

Note: In the previous edition of this Compendium, it was recommended to place the charge designation directly above the centrally placed dot. However, this format is now discouraged because of the difficulty of extending it to ions bearing more than one charge, and/or more than one unpaired electron.

In mass spectroscopic usage the symbol for the charge precedes the dot representing the unpaired electron.
1994, *66*, 1156; 1995, *67*, 1363

radical pair (geminate pair)
The term is used to identify two *radicals* in close proximity in solution, within a solvent *cage*. They may be formed simultaneously by some *unimolecular* process, e.g. peroxide decomposition, or they may have come together by diffusion. While the radicals are together, correlation of the unpaired electron spins of the two species cannot be ignored: this correlation is responsible for the *CIDNP* phenomenon. A radical pair is called geminate provided that each radical partner is a descendant of the same parental pair.

See also *geminate recombination*.
1994, *66*, 1156; 1995, *67*, 1363; 1996, *68*, 2270

radical polymerization
A *chain polymerization* in which the kinetic-chain carriers are radicals.
Note:
Usually, the growing chain end bears an unpaired electron.
1996, *68*, 2308

radicofunctional name
See *functional class name*.
B.B.(G) 14

radioactive
The property of a *nuclide* of undergoing spontaneous nuclear transformations with the emission of *radiation*.
1994, *66*, 2523

radioactive age
The time, estimated from measurement of the isotopic composition, during which the content of a radioactive species within an object has remained unchanged except for *nuclear decay*.
1982, *54*, 1549

radioactive chain
See *decay chain*.
1994, *66*, 2518

radioactive contamination
A radioactive substance in a material or place where it is undesirable.
1982, *54*, 1549

radioactive cooling
Of a strongly *radioactive* material, the decrease of its *activity* by *nuclear decay*.
1982, *54*, 1538

radioactive dating
The determination of the *radioactive age* of an object from its content of *radioactive* substances and of their *daughter products*.
1982, *54*, 1539

radioactive decay
Nuclear decay in which particles or electromagnetic *radiation* are emitted or the nucleus undergoes spontaneous fission or electron capture.
1994, *66*, 2518

radioactive equilibrium
Among the members of a *decay chain*, the state which prevails when the ratios between the *activities* of successive members remain constant. (This is not an equilibrium in the strict sense since *radioactive decay* is an irreversible process).
1994, *66*, 2519

radioactive fallout
The deposition on the ground of radioactive substances from nuclear explosions and other injections of radioactive material into the atmosphere.
1982, *54*, 1541

radioactive half life
See *half life (of a radionuclide)*.
1994, *66*, 2520

radioactive series
See *decay chain*.
1994, *66*, 2518

radioactive source
Any quantity of *radioactive* material which is intended for use as a source of *ionizing radiation*.
1994, *66*, 2525

radioactive tracer
A *tracer* containing a radioactive *label*.
1982, *54*, 1550

radioactive tracer technique (in analysis)
A technique for investigating recovery and loss of a microcomponent, in which a radioactively-labelled element or compound chemically identical with the microcomponent is added to the sample before *preconcentration*, and its behaviour is followed by sensitive, rapid and selective *radioactivity* measurements.
1979, *51*, 1200

radioactive waste
Unwanted radioactive materials obtained in the processing or handling of radioactive materials.
1982, *54*, 1550

radioactivity
The property of certain *nuclides* of showing *radioactive decay*.
1994, *66*, 2523; 1996, *68*, 988; G.B. 22

radioanalytical chemistry
See *analytical radiochemistry*.
1982, *54*, 1550

radiochemical activation analysis
A kind of *activation* analysis in which, after the *irradiation*, chemical or physical separation is applied.
1994, *66*, 2515

radiochemical purification
Chemical separation applied to a radioactive preparation in order to improve the *radiochemical purity*.
1982, *54*, 1550

radiochemical purity
For a material, the fraction of the stated *isotope* present in the stated chemical form.
1994, *66*, 2523

radiochemical separation
Separation by a chemical means of the *radioactive isotopes* of a specific element from a mixture of *radionuclides*.
1994, *66*, 2525

radiochemical yield
The yield of a *radiochemical separation* expressed as a fraction of the *activity* originally present.
1994, *66*, 2526

radiochemistry
That part of chemistry which deals with *radioactive* materials. It includes the production of *radionuclides* and their compounds by processing irradiated materials or naturally occurring radioactive materials, the application of chemical techniques to nuclear studies, and the application of radioactivity to the investigation of chemical, biochemical or biomedical problems.
1994, *66*, 2524

radiochromatograph
A measuring assembly designed to measure the spatial or time distribution of the *activity* of a mixture of radioactive components after separation by a chromatographic method.
1982, *54*, 1550

radiocolloid
A *colloid* in which some atoms are radioactive.
1982, *54*, 1550

radioenzymatic assay
An *assay* of the *catalytic activity* of an *enzyme* based on the use of a *radioactive* substrate.
1994, *66*, 2524

radiograph
A visual representation of an object produced by placing the object between a source of *ionizing radiation* and a photographic plate or film.
1994, *66*, 2524

radiogravimetric analysis
A kind of quantitative analysis in which the *activity* of a precipitate is used as a measure of its mass.
1994, *66*, 2524

radioimmunoassay
An *assay* based on the reversible and non-covalent binding of an *antigen (hapten)* by a specific *antibody* employing radioactivity labelled antigen (hapten) to measure the fraction of the antigen (hapten) bound to a substoichiometric amount of antibody.
1994, *66*, 2524

radioimmunoassay, solid phase antibody
See *solid phase antibody radioimmunoassay*.
1994, *66*, 2524

radioiodination
The process of incorporating the *radionuclides* of iodine (usually ^{125}I, ^{131}I or ^{123}I) into, or of covalently linking a radioiodinated substance to, a substance.
1994, *66*, 2524

radioisotope
A *radioactive isotope* of a specified element.
1994, *66*, 2524

radioisotope dilution analysis
A kind of *isotope dilution analysis* making use of a *radionuclide*.
1994, *66*, 2524

radioisotope induced X-ray emission analysis
A kind of analysis based on the measurement of the *energies* and intensities of *characteristic X-radiation* emitted by a test portion during *irradiation* with a *radioactive source*.
1994, *66*, 2526

radioluminescence
Luminescence arising from excitation by high energy particles or radiation.
1996, *68*, 2270

radiolysis
The cleavage of one or several bonds resulting from exposure to high-energy radiation. The term is also often used loosely to specify the method of irradiation ('pulse radiolysis') used in any radiochemical reaction, not necessarily one involving bond cleavage.
1994, *66*, 1156; 1994, *66*, 2524

radiometric analysis
A kind of quantitative analysis in which measurement of the *activity* is an essential step.
1994, *66*, 2524

radiometric titration
A titration in which a *radioactive* indicator is used to monitor the end-point of the titration.
1994, *66*, 2524

radiometry
The measurement of quantities associated with *radiant energy*. The quantities may also describe the variation of the energy with respect to other variables such as wavelength, time, position, direction (solid angle), area normal to the light or projected area of emitting or receiving surfaces. If the light is monochromatic, it is sometimes convenient to replace the radiant energy by the corresponding number of photons (or quanta) which is obtained by dividing the energy by hc/λ where h is the Planck constant, c the velocity of light and λ the wavelength of the light.
See *intensity, radiance, irradiance*.
1990, *62*, 2209

radionuclide
A *nuclide* that is *radioactive*.
1994, *66*, 2524

radionuclidic purity
For a material, that fraction of the total *activity* which is present in the form of the stated *radionuclide*, including *daughter products*.
1994, *66*, 2523

radioreceptor assay
An *assay* employing a radioactively *labelled* receptor protein as a *tracer*.
1994, *66*, 2524

radiorelease analysis
A kind of quantitative analysis based on the release of *radioactivity* from the reagent by reaction with the analyte.
1994, *66*, 2524

radiosonde
A miniature radio transmitter with instruments in a package that is carried aloft (e.g. by an unmanned balloon) for broadcasting every few seconds by means of precise tone signals or other suitable method, the humidity, temperature, pressure or other parameter.
1990, *62*, 2209

radius of gyration, s
A parameter characterizing the size of a particle of any shape.
For a rigid particle consisting of mass elements of mass m_i, each located at a distance r_i from the centre of mass, the radius of gyration, s, is defined as the square root of the mass-average of r_i^2 for all the mass elements, i.e.

$$s = \left(\frac{\sum_i m_i r_i^2}{\sum_i m_i} \right)^{1/2}$$

For a non-rigid particle, an average over all conformations is considered, i.e.

$$\langle s^2 \rangle^{1/2} = \frac{\langle \sum_i m_i r_i^2 \rangle^{1/2}}{\left(\sum_i m_i \right)^{1/2}}$$

The subscript zero is used to indicate unperturbed dimensions, as in $\langle s^2 \rangle_0^{1/2}$.
P.B. 48

raffinate
The phase remaining after extraction of some specified solute(s). When necessary it should be further specified, e.g. scrub raffinate. The original meaning of raffinate as a 'refined product' has become extended and changed by common usage. The term should normally be applied only to waste streams but the latter may form the feed to a further extraction process for another solute.
1993, *65*, 2388

rain out (in atmospheric chemistry)
The mechanism by which small particles in the clouds are removed by the formation of raindrops; this is a different mechanism from wash out which is a process which occurs below cloud level. Both the terms rain out and *wash out* have not always been used in accordance with these definitions. For clarity they should be replaced by the terms, in-cloud scavenging and below-cloud scavenging, respectively.
1990, *62*, 2209

random coil (in polymers)
The complete set of spatial arrangements of a chain molecule with a large number of segments that randomly change mutual orientation with time, under conditions in which it is free from external constraints that would affect its conformation.
If the solution of the chain molecules is not in a theta state, the segments change mutual orientation only approximately randomly.
Synonymous with statistical coil.
P.B. 50

random coincidence (in nuclear chemistry)
A coincidence of events occurring in not physically connected *nuclei*.
1982, *54*, 1550

random copolymer
A *copolymer* consisting of *macromolecules* in which the probability of finding a given *monomeric unit* at any given site in the *chain* is independent of the nature of the adjacent units.
Note:
In a random copolymer, the sequence distribution of monomeric units follows Bernoullian statistics.
1996, *68*, 2301

random copolymerization
A *copolymerization* in which a *random copolymer* is formed.
1996, *68*, 2307

random error
Result of a measurement minus the mean that would result from an infinite number of measurements of the same measurand carried out under repeatability conditions.
See also *measurement result*.
VIM

random sample
The sample so selected that any portion of the population has an equal (or known) chance of being chosen.
Haphazard or arbitrary choice of units is generally insufficient to guarantee randomness.
1990, *62*, 1202

+ range (in analysis)
The difference between the highest and lowest members of a series. This term is not recommended.
O.B. 5; see also 1994, *66*, 599

range, extrapolated (in radiochemistry)
See *extrapolated range (in radiochemistry)*.
1982, *54*, 1550

range, mean linear (in nuclear chemistry)
See *mean linear range (in nuclear chemistry)*.
1982, *54*, 1550

range, mean mass (in nuclear chemistry)
See *mean mass range (in nuclear chemistry)*.
1982, *54*, 1550

range of measurement (of an analyser)
The range of concentration between the measurement threshold and the maximum usable indication.
1990, *62*, 2174

Raoult's law
See *Henry's law*.
1984, *56*, 571

rate
Derived quantity in which time is a denominator quantity. Rate of x is dx/dt.
1996, *68*, 988

rate coefficient
See *order of reaction, kinetic equivalence*.
1994, *66*, 1157

rate constant, k
See *order of reaction*.
1994, *66*, 1157

rate-controlling step
A rate-controlling (rate-determining or rate-limiting) step in a reaction occurring by a composite reaction sequence is an *elementary reaction* the rate constant for which exerts a strong effect — stronger than that of any other rate constant — on the overall rate. It is recommended that the expressions rate-controlling, *rate-determining* and *rate-limiting* be regarded as synonymous, but some special meanings sometimes given to the last two expressions are considered under a separate heading.
A rate-controlling step can be formally defined on the basis of a control function (or control factor) CF, identified for an elementary reaction having a rate constant k_i by:

$$CF = (\partial \ln v / \partial \ln k_i)_{K_j, k_j}$$

where v is the overall rate of reaction. In performing the partial differentiation all equilibrium constants K_j and all rate constants except k_i are held constant. The elementary reaction having the largest control factor exerts the strongest influence on the rate v, and a step having a CF much larger than any other step may be said to be rate-controlling.

A rate-controlling step defined in the way recommended here has the advantage that it is directly related to the interpretation of *kinetic isotope effects*.

As formulated this implies that all rate constants are of the same dimensionality. Consider however the reaction of A and B to give an intermediate C, which then reacts further with D to give products:

$$A + B \underset{k_{-1}}{\overset{k_1}{\rightleftharpoons}} C \quad (1)$$

$$C + D \xrightarrow{k_2} \text{Products} \quad (2)$$

Assuming that C reaches a *steady state*, then the observed rate is given by:

$$v = \frac{k_1 k_2 [A][B][D]}{k_{-1} + k_2 [D]}$$

Considering $k_2[D]$ a pseudo-first order rate constant, then $k_2[D] \gg k_{-1}$, and the observed rate $v = k_1[A][B]$ and $k_{obs} = k_1$. Step (1) is said to be the rate-controlling step.

If $k_2[D] \ll k_{-1}$, then the observed rate:

$$v = \frac{k_1 k_2}{k_{-1}} [A][B][D]$$

$$= K k_2 [A][B][D]$$

where K is the equilibrium constant for the pre-equilibrium (1) and is equal to k_1/k_{-1}, and $k_{obs} = K k_2$. Step (2) is said to be the rate-controlling step.

See also *Gibbs energy diagram, microscopic diffusion control, mixing control, rate-determining step, rate-limiting step.*

1994, *66*, 1156; see also 1996, *68*, 182

rate-determining step (rate-limiting step)
These terms are best regarded as synonymous with *rate-controlling step*. However, other meanings that have been given to them should be mentioned, as it is necessary to be aware of them in order to avoid confusion.

Sometimes the term rate-determining is used as a special case of rate-controlling, being assigned only to an initial slow step which is followed by rapid steps. Such a step imposes an upper limit on the rate, and has also been called rate-limiting.

In view of the considerable danger of confusion when special meanings are applied to rate-determining and rate-limiting, it is recommended that they be regarded as synonymous, with the meaning explained under the entry *rate-controlling step*.

See *Michaelis–Menten kinetics*.
1994, *66*, 1157; see also 1996, *68*, 183

rate law (empirical differential rate equation)
An expression for the *rate of reaction* of a particular reaction in terms of concentrations of *chemical species* and constant parameters (normally *rate coefficients* and partial *orders of reaction*) only. For examples of rate laws see equations (1)–(3) under *kinetic equivalence*, and (1) under *steady state*.
1994, *66*, 1157

ratemeter (in radiochemistry)
An electronic sub-assembly which gives a continuous indication proportional to the average counting rate over a predetermined time interval (time constant).
1982, *54*, 1550

rate of appearance
See *rate of reaction*.
1994, *66*, 1157

rate of change of a quantity
The time derivative dQ/dt of the value of the quantity. The differential quotient may also be called derivative (or instantaneous) rate of change.
Examples are: rate of change of mass, dm/dt; rate of change of amount of substance, dn/dt.
1992, *64*, 1572

rate of change ratio
The quotient of two rates where the quantities are of the same kind in the same system for different components:

$$(dQ_1/dt)/(dQ_2/dt)$$

For finite time intervals, mean rate of change ratio is:

$$(\Delta Q_1/\Delta t)/(\Delta Q_2/\Delta t) = (\Delta Q_1/\Delta Q_2)\Delta t$$

Rate of change ratio has the dimension one. The denominator is often called the reference quantity.
Examples are: mass rate ratio, $(dm_1/dt)/(dm_2/dt)$; amount of substance rate ratio, $(dn_1/dt)/(dn_2/dt)$.
1992, *64*, 1572

rate of consumption, $v_{n,B}$ or $v_{c,B}$
The rate of consumption of a specified *reactant* may be defined in two ways:
1. As the negative of the time derivative of the amount of reactant; thus for a reactant B, present at any time in amount n_B, the rate of its consumption may be defined as:

$$v(n_B) = -(dn_B/dt)$$

This definition is particularly appropriate for open systems.

Here and elsewhere, when a rate is defined in terms of a time derivative, it must be understood that the definition relates to the process occurring in isolation. In a flow system there may be no actual changes with time, and the time derivative must be inferred. Such an inference is also required for a reaction occurring by a composite mechanism.

2. For kinetics in closed systems it is more usual to define a rate of consumption per unit volume; thus for a reactant B the rate of consumption, $v(c_B)$ is given by:

$$v(c_B) = -(1/V)(dn_B/dt)$$

When the volume V is constant this reduces to:

$$v(c_B) = -(d[B]/dt)$$

When the volume is not constant the relationship $n_B = [B]V$ may be differentiated to give:

$$dc_B = Vd[B] + [B]dV$$
and therefore:
$$v(c_B) = -(d[B]/dt) - ([B]/V)(dV/dt)$$
In contrast to the rate of conversion and the rate of reaction, the rate of consumption of a reactant may be specified even for a reaction of time-dependent *stoichiometry* or of unknown stoichiometry.

The rate of consumption of a reactant is often called its rate of disappearance. However, the former expression is to be preferred since the word disappearance is not appropriately translatable into certain languages. When English is used the word disappearance might be reserved for cases where the reactant is almost completely removed.
1996, *68*, 179

rate of conversion, $\dot\xi$

The rate of conversion for a reaction occurring in a closed system is defined as the time derivative of the extent of reaction:

$$\dot\xi = d\xi/dt$$

In view of the definition of *extent of reaction* it follows that with reference to any species in a reaction showing time-independent *stoichiometry*:

$$\dot\xi = d\xi/dt = (1/v_i)(dn_i/dt)$$

where n_i is the amount of the species at any time and v_i is its *stoichiometric coefficient*.
1996, *68*, 180; 1992, *64*, 1573; G.B. 55; 1996, *68*, 989

rate of disappearance
See *rate of reaction*.
1994, *66*, 1157; see also 1996, *68*, 181

rate of flow, average (in polarography)
See *average rate of flow (in polarography)*.
1985, *57*, 1503

rate of flow, instantaneous (in polarography)
See *instantaneous rate of flow (in polarography)*.
1985, *57*, 1503

rate of fluid consumption (in flame emission and absorption spectrometry), q_V
The volume of fluid consumed by the nebulizer divided by time (in $\mu L\ s^{-1}$).
Volume rate has also been called volume flow, flux, aspiration rate, aspiration flow or aspiration flux. The fluid consumed is either aspirated or injected. The volume rate of fluid consumption by nebulizers is usually between 10 and 100 $\mu L\ s^{-1}$.
1986, *58*, 1739

rate of formation, $v_{n,y}$ or $v_{c,y}$
Like the *rate of consumption*, the rate of formation of a specified product may be defined in two ways:
1. As the time derivative of the amount of a product. Thus for a product Y, present at any time in amount n_Y, the rate of its formation may be given by:

$$v(n_Y) = (dn_Y/dt)$$

This definition is particularly appropriate for open systems.
2. For kinetics in closed systems it is more usual to define a rate of formation per unit volume, denoted $v(c_Y)$:

$$v(c_Y) = (1/V)(dn_Y/dt)$$

When the volume is constant this reduces to:

$$v(c_Y) = (1/V)(dn_Y/dt) = (d[Y]/dt)$$

When the volume is not constant the relationship $n_Y = [Y]V$ may be differentiated to give:

$$dn_Y = Vd[Y] + [Y]dV$$

and the rate of formation becomes:

$$v(c_Y) = (d[Y]/dt) + ([Y]/V)(dV/dt)$$

A rate of formation may be specified even for a reaction of time dependent *stoichiometry* or of unknown stoichiometry.
1996, *68*, 181

rate of liquid consumption (in flame spectroscopy)
The volume of liquid sample that is consumed by the nebulizing system per unit of time.
O.B. 166

rate of migration (in electrophoresis), v
The distance of migration divided by time, sometimes called velocity of migration. The symbol v is also used for velocity.
1994, *66*, 895

rate of nucleation
The number of *nuclei* formed in unit time per unit volume of the liquid phase.
O.B. 84; see also 1972, *31*, 608

rate of reaction, v
For the general *chemical reaction*:

$$aA + bB \to pP + qQ\ ...$$

occurring under constant-volume conditions, without an appreciable build-up of reaction *intermediates*, the rate of reaction v is defined as:

$$v = -\frac{1}{a}\frac{d[A]}{dt} = -\frac{1}{b}\frac{d[B]}{dt} = +\frac{1}{p}\frac{d[P]}{dt} = +\frac{1}{q}\frac{d[Q]}{dt}$$

where symbols placed inside square brackets denote amount (or amount of substance) concentrations (conventionally expressed in units of mol dm^{-3}). The symbols R and r are also commonly used in place of v. It is recommended that the unit of time should always be the second.

In such a case the rate of reaction differs from the rate of increase of concentration of a product P by a constant factor (the reciprocal of its coefficient in the stoichiometric equation, p) and from the rate of decrease of concentration of the reactant A by α^{-1}.
The quantity:

$$\dot\xi = \frac{d\xi}{dt}$$

defined by the equation:

$$\dot{\xi} = -\frac{1}{a}\frac{dn_A}{dt} = -\frac{1}{b}\frac{dn_B}{dt} = +\frac{1}{p}\frac{dn_P}{dt} = +\frac{1}{q}\frac{dn_Q}{dt}$$

(where n_A designates the amount of substance A, conventionally expressed in units of mole) may be called the 'rate of conversion' and is appropriate when the use of concentrations is inconvenient, e.g. under conditions of varying volume. In a system of constant volume, the rate of reaction is equal to the rate of conversion per unit volume throughout the reaction.

For a *stepwise reaction* this definition of 'rate of reaction' (and 'extent of reaction', ξ) will apply only if there is no accumulation of intermediate or formation of side products. It is therefore recommended that the term 'rate of reaction' be used only in cases where it is experimentally established that these conditions apply. More generally, it is recommended that, instead, the terms 'rate of disappearance' or 'rate of consumption' of A (i.e. $-d[A]/dt$, the rate of decrease of concentration of A) or 'rate of appearance' of P (i.e. $d[P]/dt$, the rate of increase of concentration of product P) be used, depending on the concentration change of the particular *chemical species* that is actually observed. In some cases reference to the *chemical flux* observed may be more appropriate.

The symbol v (without lettered subscript) should be used only for rate of reaction; v with a lettered subscript (e.g. v_A) refers to a rate of appearance or rate of disappearance (e.g. of the chemical species A).

See also *chemical relaxation, lifetime, order of reaction*.
1994, *66*, 1157; see also 1996, *68*, 181

ratio
Quotient of quantities of the same kind for different components within the same system.
1996, *68*, 989; 1985, *57*, 1309

raw coke
The term raw coke is equivalent to *green coke* although it is now used less frequently.
1995, *67*, 503

Rayleigh ratio
The quantity used to characterize the scattered intensity at the scattering angle θ, defined as $R(\theta) = i_\theta r^2/(I \cdot f \cdot V)$, where I is the intensity of the incident radiation, i_θ is the total intensity of scattered radiation observed at an angle θ and a distance r from the point of scattering and V is the scattering volume. The factor f takes account of polarization phenomena. It depends on the type of radiation employed.
1. For light scattering, dependent on the polarization of the incident beam, $f = 1$ for vertically polarized light, $f = \cos^2\theta$ for horizontally polarized light and $f = (1 + \cos^2\theta)/2$ for unpolarized light.
2. For small-angle neutron scattering, $f = 1$.
3. For small-angle X-ray scattering, $f \approx 1$, if $\theta <$ ca. 5°.

Notes:
1. The dimension of $R(\theta)$ is (length)$^{-1}$.
2. In small-angle neutron scattering the term cross-section is often used instead of $R(\theta)$; the two quantities are identical.
3. An alternative recommended symbol is R_θ.
P.B. 65

Rayleigh scattering
The scattering of light by particles and molecules which are much smaller than the wavelength of the light. In the ideal case, the process is one of pure dipole and induced dipole interactions with the electric field of the light wave. The scattering cross section for light of wavelength λ is proportional to λ^{-4}.
1990, *62*, 2209

rayon-based carbon fibres
Carbon fibres made from rayon (cellulose) precursor fibres.
Notes:
Rayon-based carbon fibres have a more isotropic structure than similarly heat-treated polyacrylonitrile (PAN)- or *mesophase pitch (MPP)-based carbon fibres*. Their Young's modulus values are therefore drastically lower (100 GPa; tensile strength 100 MPa). Rayon-based carbon fibres can be transformed into anisotropic carbon fibres with high strength and Young's modulus values by hot-stretching treatment at temperatures of approximately 2800 K.
1995, *67*, 503

reactance, X
Imaginary part of the *impedance*.
G.B. 15

reactant
A substance that is consumed in the course of a chemical reaction. It is sometimes known, especially in the older literature, as a reagent, but this term is better used in a more specialized sense as a test substance that is added to a system in order to bring about a reaction or to see whether a reaction occurs (e.g. an analytical reagent).
1996, *68*, 183

reacting bond rules
1. For an internal motion of a *molecular entity* corresponding to progress over a *transition state* (energy maximum), any change that makes the motion more difficult will lead to a new molecular geometry at the energy maximum, in which the motion has proceeded further. Changes that make the motion less difficult will have the opposite effect. (This rule corresponds to the *Hammond principle*).
2. For an internal motion of a molecular entity that corresponds to a vibration, any change that tends to modify the equilibrium point of the vibration in a particular direction will actually shift the equilibrium in that direction.
3. Effects on reacting bonds (bonds made or broken in the reaction) are the most significant. The bonds nearest the site of structural change are those most strongly affected.

See also *More O'Ferrall–Jencks diagram*.
1994, *66*, 1158

reaction
See *chemical reaction*.
1994, *66*, 1158

reaction barrier
The energy barrier to chemical reaction. In vibrationally *adiabatic transition-state theory* its height is the zero-point energy of the *activated complex* minus that of the reactants.
1996, *68*, 183

reaction chromatography
A technique in which the identities of the sample components are intentionally changed between sample introduction and detection. The reaction can take place upstream of the column when the chemical identity of the individual components passing through the column differs from that of the original sample, or between the column and the detector when the original sample components are separated in the column but their identity is changed prior to entering the detection device.
1993, *65*, 827

reaction coordinate
A geometric parameter that changes during the conversion of one (or more) reactant *molecular entities* into one (or more) product molecular entities and whose value can be taken for a measure of the progress of an *elementary reaction* (for example, a bond length or bond angle or a combination of bond lengths and/or bond angles; it is sometimes approximated by a non-geometric parameter, such as the *bond order* of some specified bond).
In the formalism of 'transition-state' theory', the reaction coordinate is that coordinate in a set of curvilinear coordinates obtained from the conventional ones for the reactants which, for each reaction step, leads smoothly from the configuration of the reactants through that of the transition state to the configuration of the products. The reaction coordinate is typically chosen to follow the path along the gradient (path of shallowest ascent/deepest descent) of potential energy from reactants to products.
The term has also been used interchangeably with the term transition coordinate, applicable to the coordinate in the immediate vicinity of the potential energy maximum. Being more specific, the name transition coordinate is to be preferred in that context.
See also *potential-energy profile, potential-energy reaction surface, minimum-energy reaction path*.
1994, *66*, 1158; 1996, *68*, 183

reaction cross-section, σ_r
A quantity used in collision theories of reactions in order to interpret calculated or experimental rates.
In some collision theories the reaction cross-section is considered to be related to b_{max}, the maximum value of the *impact parameter* that allows reaction to occur, by:

$$\sigma_r = P_r \pi b_{max}^2$$

where P_r is the reaction probability.
See *collision cross-section*.
1996, *68*, 183

reaction dynamics
A branch of chemical kinetics that is concerned with the intermolecular and intramolecular motions that occur in the elementary act of chemical change, and with the details of the relationships between the quantum states of the reactant molecules and those of the product molecules. It is also known as molecular dynamics.
1996, *68*, 184

reaction intermediate
See *intermediate*.
1994, *66*, 1126; see also 1996, *68*, 170

reaction layer, thickness of (in electrochemistry)
See *thickness of reaction layer (in electrochemistry)*.
1985, *57*, 1504

reaction mechanism
See *mechanism*.
1994, *66*, 1159

reaction path
1. A synonym for *mechanism*.
2. A trajectory on the *potential-energy surface*.
3. A sequence of synthetic steps.
See also *minimum-energy reaction path*.
1994, *66*, 1159

reaction path degeneracy
A factor that is introduced into rate theory to take account of the fact that the process may be able to occur in different but equivalent ways. Thus the process:

$$Cl + H_2 \rightarrow HCl + H$$

has a reaction path degeneracy of 2 since the chlorine atom can abstract either of the two hydrogen atoms in the hydrogen molecule. The reaction path degeneracy has also been referred to as the statistical factor for the reaction.
1996, *68*, 184

reaction probability, P_r
The probability that a reaction occurs when two particles undergo a collision. If N_c is the number of collisions occurring in unit time, and N_r the number of reactions occurring in unit time: $P_r = N_r/N_c$.
1996, *68*, 184

reaction stage
A set of one or more (possibly experimentally inseparable) *reaction steps* leading to and/or from a detectable or presumed reaction *intermediate*.
1994, *66*, 1159

reaction step
An *elementary reaction*, constituting one of the stages of a *stepwise reaction* in which a reaction intermediate (or, for the first step, the reactants) is converted into the next reaction intermediate (or, for the last

step, the products) in the sequence of intermediates between reactants and products.
See also *rate-limiting step, reaction stage*.
1994, *66*, 1159

reaction time
The period of time that elapses between the start of the reaction and the attainment of a given extent of a reaction.
1993, *65*, 2297

reactive (reactivity)
As applied to a *chemical species*, the term expresses a kinetic property. A species is said to be more reactive or to have a higher reactivity in some given context than some other (reference) species if it has a larger rate constant for a specified *elementary reaction*. The term has meaning only by reference to some explicitly stated or implicitly assumed set of conditions. It is not to be used for reactions or reaction patterns of compounds in general.
The term is also more loosely used as a phenomenological description not restricted to elementary reactions. When applied in this sense the property under consideration may reflect not only rate, but also equilibrium, constants.
See also *stable, unreactive, unstable*.
1994, *66*, 1159

reactive adsorption
Reactive adsorption (and its reverse, reactive desorption) resembles *dissociative adsorption* (and its reverse, associative desorption) but one fragment adds to an *adsorbate* rather than to a surface site.

$$H_2C=CH_2 + D-D(g) \longrightarrow \underset{**}{H_2C} \overset{CH_2D}{\underset{*}{|}} \quad \underset{*}{D}$$

1976, *46*, 84

reactive complex
A species of very short life that occurs as an intermediate in a chemical reaction. An *activated complex* (or transition state) is a special case.
1996, *68*, 184

reactive desorption
See *reactive adsorption*.
1976, *46*, 84

reactive scattering
If in a *molecular-beam* experiment a chemical reaction occurs there is said to be reactive scattering.
1996, *68*, 184

reactivity index
Any numerical index derived from quantum mechanical model calculations that permits the prediction of relative reactivities of different molecular sites. Many indices are in use, based on a variety of theories and relating to various types of reaction. The more successful applications have been to the *substitution reactions* of *conjugated systems* where relative reactivities are determined largely by changes of π-electron energy.
1994, *66*, 1159

reactivity–selectivity principle (RSP)
This idea may be expressed loosely as: the more *reactive* a reagent is, the less selective it is.
Consider two substrates S^1 and S^2 undergoing the same type of reaction with two reagents R^1 and R^2, S^2 being more reactive than S^1, and R^2 more reactive than R^1 in the given type of reaction. The relative reactivities (in log units, see *selectivity*) for the four possible reactions may notionally be represented as show in the diagram.

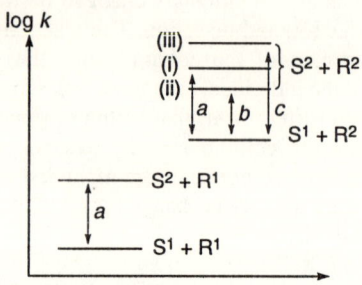

With the positions of ($S^1 + R^1$), ($S^2 + R^1$) and ($S^1 + R^2$) fixed, there are three types of positions for ($S^2 + R^2$).
In position (i) the selectivity of R^2 for the two substrates, measured by *a*, is the same as the selectivity of R^1 for the two substrates, also *a*.
In position (ii) the selectivity of R^2 for the two substrates, measured by *b*, is less than the selectivity of R^1 for the two substrates, i.e. $b < a$. It is this situation which is in accord with the RSP.
In position (iii) the selectivity of R^2 for the two substrates, measured by *c*, is greater than the selectivity of R^1 for the two substrates, i.e. $c > a$. This situation may be described as anti-RSP.
There are many examples in which the RSP is followed, but there are also many examples corresponding to situations (i) and (iii). The RSP is in accord with intuitive feeling and certainly holds in the limiting case when reactivity is controlled by diffusion. However, the validity of the RSP is a matter of great controversy.
1994, *66*, 1159

reactor, nuclear
See *nuclear reactor*.
1982, *54*, 1550

readability (of a balance)
The smallest fraction of a division to which the index scale can be read with ease either by estimation or by the use of a vernier. It should normally be expressed in divisions of the pointer scale.
O.B. 36

reading
Datum provided by an instrument.
1989, *61*, 1662

reagent
See *reactant*.
1996, *68*, 183

real (electrified) interphase
See *electrified interphase*.
1986, *58*, 439

real potential of a species in a phase
Defined for species B in phase β as

$$\alpha_B^\beta = \mu_B^\beta - z_B F \psi^\beta$$

where μ_B^β is the *electrochemical potential of species B in phase* β, z_B is the charge number of species B, F is the *Faraday constant*, and ψ^β is the *outer electric potential of phase* β. Since ψ^β is zero when the charge on the phase β is zero, the real potential may be regarded as the value of the electrochemical potential of the uncharged phase.
1974, *37*, 506

real surface (interface) area
See *extent of an interface (surface)*.
1986, *58*, 439

rearrangement
See *degenerate rearrangement, molecular rearrangement, sigmatropic rearrangement*.
1994, *66*, 1160

rearrangement ion (in mass spectrometry)
An electrically charged species, involving a molecular or parent ion, in which atoms or groups of atoms have transferred from one portion of a molecule or molecular moiety to another during the ionization fragmentation processes.
1991, *63*, 1550; O.B. 205

rearrangement stage
The *elementary reaction* or *reaction stage* (of a *molecular rearrangement*) in which there is both making and breaking of bonds between atoms common to a reactant and a reaction product or a reaction *intermediate*. If the rearrangement stage consists of a single elementary reaction, this is a 'rearrangement step'.
1994, *66*, 1160

rebound reaction
If in a chemical process, occurring in a molecular beam, the reaction products are scattered backwards with respect to the moving centre of mass of the system, the reaction is called a rebound reaction.
1996, *68*, 185

receptor
A molecule or a polymeric structure in or on a *cell* which specifically recognizes and binds to a compound acting as a molecular messenger (neurotransmitter, hormone, lymphokine, *lectin*, drug, etc.).
1992, *64*, 164

recognition site
1. A *nucleotide* sequence to which a protein binds specifically.
2. An amino acid sequence in an *antibody* molecule to which the specific *antigen* binds specifically.
1992, *64*, 164

recoil (in radioanalytical chemistry)
The motion acquired by a particle through a collision with, or the emission of, another particle or electromagnetic *radiation*.
1994, *66*, 2524

recoil-free fraction (in Mössbauer spectrometry)
The fraction of all Mössbauer gamma rays of the transition which are emitted or absorbed without significant recoil energy loss.
1976, *45*, 214

recoil labelling
Labelling by a chemical reaction initiated by *recoil*.
1994, *66*, 2521

recombinant DNA technology
See *gene manipulation*.
1992, *64*, 164

recombination fluorescence
See *delayed fluorescence*.
1984, *56*, 234

reconstructive transition
A transition which involves a major reorganization of the crystal structure and a change of local topology, during which primary bonds are broken and reformed so that there is no immediate relationship between the crystal structures of the parent and product phases. Examples: The transition of Mg_2SiO_4 (olivine) to Mg_2SiO_4 (spinel); the transition of diamond to graphite.
1994, *66*, 589

recovery
1. In toxicology, the process leading to partial or complete restoration of a cell, tissue, organ or organism following its damage from exposure to a harmful substance or agent.
2. Term used in analytical and preparative chemistry to denote the fraction of the total quantity of a substance recoverable following a chemical procedure.
1993, *65*, 2089

+ recovery factor (in an extraction process)
The fraction or percentage ($R_\%$) of the total quantity of a substance extracted (usually into the organic solvent phase) under specified conditions. Specifically, $R_A = Q_A/(Q_A)'$, where $(Q_A)'$ and Q_A are the original and final quantities of the substance A.
This term is not recommended. *Fraction extracted* should be used.
O.B. 90; 1993, *65*, 2385

recrystallization
1. (chemical): The repeated *dissolution* and *precipitation* of a solid from a liquid solvent.
2. (metallurgical): The process in which a crystalline solid with a high degree of disorder develops a new microstructure by the nucleation of relatively imperfection-free regions, and their propagation throughout the complete solid.
1994, *66*, 589

redox ion exchangers
Conventional ion exchangers in which reversible redox couples have been introduced as counter-ions either by sorption or complex formation. They closely resemble *redox polymers* in their behaviour.
1993, *65*, 855

redox polymers
Polymers containing functional groups which can be reversibly reduced or oxidized. Electron exchanger may be used as a synonym.
1993, *65*, 855

red shift
Informal term for *bathochromic shift*.
1996, *68*, 2270

reduced adsorption
Of component i, defined by the equation
$$\Gamma_i^{(n)} = \Gamma_i^\sigma - \Gamma^\sigma \left\{ \frac{c_i^\alpha - c_i^\beta}{c^\alpha - c^\beta} \right\}$$
where Γ^σ, c^α and c^β are, respectively, the total Gibbs surface concentration and the total concentrations in the bulk phases α and β:
$$\Gamma^\sigma = \sum_i \Gamma_i^\sigma$$
$$c^\alpha = \sum_i c_i^\alpha$$
$$c^\beta = \sum_i c_i^\beta$$
The reduced adsorption is invariant to the location of the *Gibbs surface*.
Alternatively, the reduced adsorption may be regarded as the Gibbs surface concentration of i when the Gibbs surface is chosen so that Γ^σ is zero, i.e. the Gibbs surface is chosen so that the reference system has not only the same volume, but also contains the same total amount of substance (n) as the real system.
1972, *31*, 591

reduced limiting sedimentation coefficient
See *sedimentation coefficient*.
1972, *31*, 617

reduced mass, μ
Effective mass in equations of motion for a many particle system.
G.B. 12

reduced mobile phase velocity (in chromatography), v
A term used mainly in liquid chromatography. It compares the mobile phase velocity with the velocity of diffusion into the pores of the particles (the so-called diffusion velocity, u_D):
$$v = \bar{u}/u_D = \bar{u} d_p / D_M$$
In open-tubular chromatography:
$$v = \bar{u} d_c / D_M$$
1993, *65*, 840

reduced osmotic pressure
The *osmotic pressure* divided by mass concentration.
1972, *31*, 615

reduced sample
A representative part of the primary (composited or gross) sample obtained by a division and reduction process.
Typically the mass approximates that of the final laboratory sample.
1990, *62*, 1205

reduced sedimentation coefficient
See *sedimentation coefficient*.
1972, *31*, 617

reduced species
A term used to characterize the degree of reduction (or oxidation) in atoms, molecules and ions. An atom in a molecule or an ion which has a low *oxidation state*. An element or atom in a compound can be reduced by the reaction of an element or compound with hydrogen, while it can be oxidized by reaction with oxygen. A reduced species can be formed also through the gain of electrons (either at the negative electrode in a cell or through transfer from another atom, ion or group of atoms in a chemical reaction). For example, the carbon atom in CH_4 and other hydrocarbons is in a reduced state, while the carbon in CO_2 is in an oxidized state. Similarly the sulfur atom in H_2S is in a reduced state while that in sulfuric acid (H_2SO_4) is in an oxidized state.
See *oxidation state*.
1990, *62*, 2210

reduced viscosity (of a polymer)
The ratio of the *relative viscosity* increment to the mass concentration of the polymer, c, i.e. η_i/c, where η_i is the relative viscosity increment.
Notes:
1. The unit must be specified; $cm^3 \, g^{-1}$ is recommended.
2. This quantity is neither a viscosity nor a pure number. The term is to be looked on as a traditional name. Any replacement by consistent terminology would produce unnecessary confusion in the polymer literature.
Synonymous with viscosity number.
P.B. 63

reducing (in analytical chemistry)
Decreasing the size of the laboratory sample or individual particles, or both.
Division of the size of the laboratory sample may be accomplished manually by coning and quartering or by riffling or mechanically by rotary dividers. Reduction of particle size may be accomplished by milling or grinding. Simultaneous division and reduction may also be achieved with mills having stream diverters.
1990, *62*, 1204

reduction
The complete transfer of one or more electrons to a *molecular entity* (also called 'electronation'), and,

more generally, the reverse of the processes described under *oxidation* (2) and (3).
1994, *66*, 1160

reductive elimination
The reverse of *oxidative addition*.
1994, *66*, 1160

reductones
Compounds containing an enediol structure stabilized by conjugation and hydrogen bonding with an adjacent carbonyl group, $RC(OH)=C(OH)C(=O)R$. They are strong reducing agents, fairly strong acids and commonly derived from *saccharides* by oxidation at the carbon atom alpha to the carbonyl function, e.g. ascorbic acid.
1995, *67*, 1363

+ re-extraction
Since the prefix 're' can signify 'back' as well as 'again' this term is ambiguous and should be avoided, except where the process of *extraction* (e.g. from aqueous solution to an organic phase) in a single direction is repeated (following *stripping*). It should not be used as a synonym for stripping or back-extraction.
1993, *65*, 2379

referee sample
See *umpire sample*.
1990, *62*, 1203

reference atom (in organic reaction mechanisms)
The atom to which a *primitive change* is nucleophilic, electrophilic, nucleofugal or homolytic.
1989, *61*, 55

reference cell (or blank cell)(in spectrochemical analysis)
See *matched cells*.
1988, *60*, 1453

reference electrode
1. An external electrode system which comprises an inner element, usually mercury–mercury(I) chloride (calomel), silver–silver chloride or thallium amalgam–thallium(I) chloride, a chamber containing the appropriate filling solution and a device for forming a liquid junction, e.g. capillary, ceramic plug, fritted disc or ground glass sleeve.
O.B. 27
2. An electrode that maintains a virtually invariant potential under the conditions prevailing in an electrochemical measurement, and that serves to permit the observation, measurement, or control of the potential of the indicator (or test) or working electrode.
O.B. 60; see also O.B. 65; 1994, *66*, 2533

reference material
A substance or mixture of substances, the composition of which is known within specified limits, and one or more of the properties of which is sufficiently well established to be used for the calibration of an apparatus, the assessment of a measuring method or for assigning values to materials. Reference materials are available from national laboratories in many countries [e.g. National Institute for Standards and Technology (NIST), USA.; Community Bureau of Reference, UK].
1990, *62*, 2210

reference method
A method having small, estimated inaccuracies relative to the end use requirement. The accuracy of a reference method must be demonstrated through direct comparison with a *definitive method* or with a primary reference material.
1995, *67*, 1702

reference procedure (in analysis of trace air constituents)
Agreed method for determining one or more air quality characteristics where it is not practical to produce a reference material; the result obtained is defined as the measure of the air quality characteristic. For example, a gas such as NO_2 may be somewhat unstable when stored in a tank at low mixing ratios, so that the use of a more stable NO standard mixture in N_2 may be oxidized to NO_2 by O_3 in a specified manner referred to as the NO_2 reference procedure.
1990, *62*, 2210

reference state (of an element)
The state in which the element is stable at a chosen standard-state pressure and for a given temperature.
1982, *54*, 1249

reference value pH standard
An aqueous solution of 0.05 mol potassium hydrogen phthalate per kg water which is the reference value for the definition of pH(RVS) values in the temperature range 0 – 95 °C.
See also *operational pH standard, primary pH standard*.
1985, *57*, 540

reflectance, ρ
Ratio of the reflected to the incident *radiant power*. Also called reflection factor.
G.B. 32; see also 1996, *68*, 989

reflection electron energy loss spectroscopy (REELS)
Any technique in which a specimen is bombarded with a focused low energy ($E_0 < 10$ eV) electron beam and the energy distribution of the reflected electrons is measured. This energy distribution contains features corresponding to discrete losses of energy of the reflected electrons due to excitation of vibrational and plasmon states and provides information on the type and geometric structure of compounds at the surface of the specimen.
1983, *55*, 2025

reflection factor
See *reflectance*.
G.B. 32

reflection high energy electron diffraction (RHEED)
Any technique which measures the angular intensity distribution of electrons 'reflected' from a crystalline surface under bombardment with high energy electrons near grazing incidence. The diffraction pattern provides very surface sensitive information (information depth 1 nm) on the atomic arrangement of the top layers of a solid.
1983, *55*, 2025

refraction effects
The *luminescence* flux emitted from the interior of a rectangular sample reaching a photodetector place at some distance from the sample is decreased by a factor of approximately n^2 (where n is the *refractive index* of the medium) compared with a medium whose refractive index is 1.0. Such effects are termed refraction effects.
1984, *56*, 244

refractive index, *n*
The ratio of the speed of light in vacuum to that in a given medium.
G.B. 30; 1996, *68*, 989

refractive index increment (in polymer chemistry)
The change of the solution refractive index, n, with solute concentration, C, i.e. $\partial n/\partial C$.
Notes:
1. The solute concentration is most frequently expressed in terms of mass concentration, molality or volume fraction. If expressed in terms of mass concentration or molality, the corresponding refractive index increments are referred to as specific or molal refractive index increments, respectively.
2. Following use of the full name, the abbreviated name refractive increment may be used.
P.B. 65

regeneration (of a catalyst)
Reversal of catalytic deactivation with restoration of the original catalytic activity.
1976, *46*, 84

regioselectivity (regioselective)
A regioselective reaction is one in which one direction of bond making or breaking occurs preferentially over all other possible directions. Reactions are termed completely (100%) regioselective if the discrimination is complete, or partially (*x*%), if the product of reaction at one site predominates over the product of reaction at other sites. The discrimination may also semi-quantitatively be referred to as high or low regioselectivity. (Originally the term was restricted to *addition* reactions of unsymmetrical reagents to unsymmetrical alkenes.)
In the past, the term 'regiospecificity' was proposed for 100% regioselectivity. This terminology is not recommended owing to inconsistency with the terms *stereoselectivity* and *stereospecificity*.

See also *chemoselectivity*.
1994, *66*, 1160

regular block (in a polymer)
A *block* that can be described by only one species of *constitutional repeating unit* in a single sequential arrangement.
P.B. 17

regular coke
A *petroleum coke* with good graphitizability, characterized by a combination of properties which differ significantly from those of *metallurgical coke* but do not reach the quality level of *premium coke*. These properties are: optical anisotropy, medium reversible thermal expansion and low ash content.
Regular coke is mainly used for the production of synthetic carbon and *graphite materials*.
1995, *67*, 503

regular comb macromolecule
See *comb macromolecule*.
1996, *68*, 2296

regular macromolecule
A *macromolecule* the structure of which essentially comprises the repetition of a single *constitutional unit* with all units connected identically with respect to directional sense.
1996, *68*, 2289

regular oligomer molecule
An *oligomer molecule* the structure of which essentially comprises the repetition of a single *constitutional unit* with all units connected identically with respect to directional sense.
1996, *68*, 2290

regular polymer
A polymer composed of *regular macromolecules*, *regular star macromolecules* or *regular comb macromolecules*.
A polymer consisting of star macromolecules with arms identical with respect to constitution and degree of polymerization is considered to be regular; see *star macromolecule*, note 2. Analogously, a polymer consisting of comb macromolecules with the sub-chains between the branch points of the main chain and the terminal sub-chains of the main chain identical with respect to constitution and degree of polymerization and the side-chains identical with respect to constitution and degree of polymerization is considered to be regular; see *comb macromolecule*, note 1.
1996, *68*, 2302

regular single-strand polymer
A regular linear polymer that can be described by a preferred *constitutional repeating unit* in which both terminal constituent subunits are connected through single atoms to the other identical constitutional repeating units or to an end group.
P.B. 110

regular star macromolecule
See *star macromolecule*.
1996, *68*, 2296

regulator gene
A gene which codes for a protein (an activator of repressor) having the ability to induce or repress the *transcription* of other genes.
1992, *64*, 164

Reissert compounds
Compounds formed by formal addition of an acyl group and a cyano group to a nitrogen atom and carbon atom, respectively, of a nitrogen–carbon bond in quinolines, isoquinolines, and related nitrogen heterocycles, e.g.

1995, *67*, 1363

rel
See *relative configuration*.
1996, *68*, 2216

relative
Quotient of quantities having the same kind-of-quantity in different systems.
Notes:
1. The denominator quantity is often called the reference quantity.
2. Preferentially the same component is found in the numerator and denominator quantity.
1996, *68*, 989

relative activity
See *activity*.
G.B. 49, 22

relative adsorption
If Γ_i^σ and Γ_1^σ are the *Gibbs surface* concentrations of components i and 1, respectively, with reference to the same, but arbitrarily chosen, Gibbs surface, then the relative adsorption of component i with respect to component 1, is defined as

$$\Gamma_i^{(1)} = \Gamma_i^\sigma - \Gamma_1^\sigma \left\{ \frac{c_i^\alpha - c_i^\beta}{c_1^\alpha - c_1^\beta} \right\}$$

and is invariant to the location of the Gibbs surface. Alternatively, $\Gamma_i^{(1)}$ may be regarded as the Gibbs surface concentration of i when the Gibbs surface is chosen so that Γ_1^σ is zero, i.e. the Gibbs surface is chosen so that the reference system contains the same amount of component 1 as the real system. Hence $\Gamma_1^{(1)} \equiv 0$.
1972, *31*, 591

relative atomic mass (atomic weight), A_r
The ratio of the average mass of the atom to the *unified atomic mass unit*.
See also *standard atomic weight*.
G.B. 41; 1996, *68*, 990

relative biological effectiveness (of radiation)
The ratio of the appropriate value of the biological effectiveness of the radiation in question to that of X-radiation with an average *specific ionization* of 100 ion pairs per micron of water, for the particular biological effect under consideration and for the condition under which the *radiation* is received.
1982, *54*, 1551

relative configuration
1. The *configuration* of any *stereogenic* (*asymmetric*) *centre* with respect to any other *stereogenic* centre contained within the same molecular entity. Unlike *absolute configuration*, relative configuration is reflection-invariant. Relative configuration, distinguishing *diastereoisomers*, may be denoted by the configurational descriptors R^*, R^* (or l) and R^*, S^* (or u) meaning, respectively, that the two centres have identical or opposite configurations. For molecules with more than two asymmetric centres the prefix *rel-* may be used in front of the name of one enantiomer where R and S have been used. If any centres have known absolute configuration then only R^* and S^* can be used for the relative configuration.
See also α (*alpha*), β (*beta*) (1 and 3).
2. Two different molecules Xabcd and Xabce, may be said to have the same relative configurations if e takes the position of d in the tetrahedral arrangement of ligands around X (i.e. the pyramidal fragments Xabc are superposable). By the same token the *enantiomer* of Xabce may be said to have the opposite relative configuration to Xabcd. The terms may be applied to *chiral* molecular entities with central atoms other than carbon but are limited to cases where the two related molecules differ in a single ligand.
Both definitions can be generalized to include *stereogenic units* other than asymmetric centres.
1996, *68*, 2217

relative counting (in nuclear chemistry)
A measurement in which the *activity* of a test portion is derived from the ratio between the count rates observed for the test portion and for a *radioactive source* of known activity.
1994, *66*, 2525

relative density, *d*
Ratio of density to a reference density, usually the density of water at 4 °C; in the older literature this is called *specific gravity*.
G.B. 12; 1990, *62*, 2210; see also 1996, *68*, 990

relative detection limit
(Often incorrectly referred to as *sensitivity*). Smallest amount of material detectable (3σ-criterion) in a matrix relative to the amount of material analysed — given in atomic, mole or weight fractions.
1979, *51*, 2247

relative electrode potential
The electrode potential of a single electrode is not amenable to direct experimental determination and can only be measured against a reference system. The measured quantity, E, is the relative electrode potential.
1986, *58*, 956

relative elongation
See *linear strain*.
G.B. 12

relative error
Error of measurement divided by the true value of the *measurand*. Since a true value cannot be determined, in practice the *conventional true value* is used.
VIM; see also 1994, *66*, 601

relative humidity
The ratio, often expressed as a percentage, of the partial pressure of water in the atmosphere at some observed temperature, to the saturation vapour pressure of pure water at this temperature.
1990, *62*, 2210

relative local current density
See *current distribution*.
1981, *53*, 1836

relative micellar mass
The relative molecular mass (M_r) of a micelle is called the relative micellar mass or micellar weight and is defined as the mass of a mole of micelles divided by the mass of $1/12$ mole of ^{12}C.
1972, *31*, 612

relative molar mass
Molar mass divided by 1 g mol^{-1} (the latter is sometimes called the standard molar mass).
G.B. 41

relative molecular mass, M_r
Ratio of the mass of a molecule to the *unified atomic mass unit*. Sometimes called the molecular weight or *relative molar mass*.
G.B. 41; 1996, *68*, 990

relative permeability, μ_r
Ratio of the *magnetic flux density* in a given medium to that in vacuum.
G.B. 15

relative permittivity, ε_r
Ratio of the *electric field strength* in vacuum to that in a given medium. It was formerly called the dielectric constant.
G.B. 14

relative preconcentration (in trace analysis)
An operation (process) as a result of which the ratio of concentration or amounts of microcomponents and main macrocomponents increases. This is a particular example of a separation, but at greatly differing concentrations of components. The ratio of the masses of the initial and final samples is not of great importance. Frequently, the main purpose of relative *preconcentration* is the replacement of a matrix unsuitable for analysis by a suitable medium. In some cases, it is difficult to trace a boundary between the absolute and the relative preconcentrations as, for example, in preconcentration by zone melting.
1979, *51*, 1197

relative responsivity
See *responsivity*.
1995, *67*, 1749

relative retardation (in planar chromatography), R_{rel}
A term which is equivalent to relative retention used in column chromatography: the ratio of the R_F value of a component to the R_F value of a standard (reference) substance. Since the mobile phase front is common for the two components, the R_F value can be expressed directly as the ratio of the distances travelled by the spot of the compound of interest (b_i) and the reference substance (b_{st}) respectively:
$$R_{rel} = R_{F(i)}/R_{F(st)} = b_i/b_{st}$$
In former nomenclatures the symbol R_s was used to express relative retardation in planar chromatography. Because of its identity with the symbol for peak resolution the symbol R_{rel} is suggested for relative retardation in planar chromatography.
1993, *65*, 845

relative retention (in column chromatography), r
The ratio of the adjusted or net retention volume (time) or retention factor of a component relative to that of a standard, obtained under identical conditions:
$$r = V'_{Ri}/V'_{R(st)} = V_{Ni}/V_{N(st)}$$
$$= t'_{Ri}/t'_{R(st)} = k_i/k_{st}$$
Depending on the relative position of the peak corresponding to the standard compound in the chromatogram, the value of r may be smaller, larger or identical to unity.
1993, *65*, 843; O.B. 105

relative selectivity (in catalysis)
See *fractional selectivity (in catalysis)*.
1976, *46*, 81

relative spectral responsivity
See *action spectrum, responsivity*.
1996, *68*, 2270; 1995, *67*, 1749

relative standard deviation, s_r, σ_r
The standard deviation divided by the mean of the series. It can be calculated with the formula:
$$s_r = s/\bar{x}$$
1994, *66*, 601; O.B. 5

relative uncertainty
Uncertainty divided by the mean value.
VIM

relative viscosity
The ratio of the viscosity of the solution, η, to the viscosity of the solvent, η_s, i.e. $\eta_r = \eta/\eta_s$.
Synonymous with viscosity ratio.
P.B. 62

relative viscosity increment
The ratio of the difference between the viscosities of solution and solvent to the viscosity of the solvent, i.e. $\eta_i = (\eta - \eta_s)/\eta_s$, where η is the viscosity of the solution and η_s is the viscosity of the solvent.
The use of the term 'specific viscosity' for this quantity is discouraged, since the relative viscosity incre-

ment does not have the attributes of a specific quantity.
P.B. 63

relative volumic mass
Synonymous with *relative density*.
1979, *51*, 2472

relaxation
If a system is disturbed from its state of equilibrium it relaxes to that state, and the process is referred to as relaxation. The branch of kinetics concerned with such processes is known as *relaxation kinetics*. Relaxation techniques include *temperature jump* and *pressure jump*.
See also *chemical relaxation*.
1996, *68*, 185; 1994, *66*, 1160; 1993, *65*, 2297

relaxation kinetics
See *relaxation*.
1996, *68*, 185; 1994, *66*, 1160; 1993, *65*, 2297

relaxation time
1. In magnetic resonance spectroscopy the longitudinal relaxation time, T_1, is associated with spin-lattice relaxation, and the transverse relaxation time, T_2, with spin-spin relaxation. The definitions are: $dM_z/dt = -(M_z - M_{z,e})/T_1$ and $dM_x/dt = -M_x/T_2$, where M_z and M_x are the components of magnetization parallel and perpendicular to the static field B and the subscript e denotes the equilibrium value.
G.B. 25
2. In a chemical reaction, the time, τ, in which a concentration perturbation falls to 1/e of its initial value.
G.B. 55; see also 1996, *68*, 185

releaser (in analytical flame spectroscopy)
A substance which reduces solute-volatilization interferences by forming a compound preferentially with the interferent, thus preventing the reaction of the analyte or interferent from entering a thermally stable compound.
O.B. 172

rem
Non-SI unit of dose equivalent (rem = 0.01 Sv).
G.B. 113; 1996, *68*, 990

Renner–Teller effect
Splittings in the vibrational levels of molecular entities due to even terms in the vibronic perturbation expansion. This is generally a minor effect for non-linear molecular entities compared to the *Jahn–Teller effect* which is due to the odd terms. For linear molecular entities it is the only possible vibronic effect characteristic of degenerate electronic states.
1996, *68*, 2271

reorganization (in polymers)
The molecular process by which (i) amorphous or poorly ordered regions of a polymer specimen become incorporated into crystals, or (ii) a change to a more stable crystal structure takes place, or (iii) defects within the crystal decrease.

Notes:
1. Secondary crystallization may be involved in the reorganization process.
2. Reorganization may result from annealing.
3. Points (i) and (ii) above may also be called crystal perfection.
P.B. 85

reorganization energy, λ
In a one-electron transfer reaction:

$$A + D \rightleftarrows A^{-} + D^{+}$$

the reorganization energy λ is the energy required for all structural adjustments (in the reactants and in the surrounding solvent molecules) which are needed in order that A and D assume the configuration required for the transfer of the electron.
See *intrinsic barrier, Marcus equation*.
1994, *66*, 1160; see also 1996, *68*, 2271

repeatability
The closeness of agreement between independent results obtained with the same method on identical test material, under the same conditions (same operator, same apparatus, same laboratory and after short intervals of time). The measure of repeatability is the standard deviation qualified with the term: 'repeatability' as *repeatability standard deviation*.
In some contexts repeatability may be defined as the value below which the absolute difference between two single test results obtained under the above conditions, may be expected to lie with a specified probability.
See also *reproducibility*.
1994, *66*, 598; see also 1995, *67*, 1701; 1990, *62*, 2210

(chain) repeating distance
See *(chain) identity period*.
P.B. 77

repeating distance
See *(chain) identity period*.
P.B. 77

repetency
See *wavenumber*.
1996, *68*, 990

replacement name
A name in which the replacement of an atom or a group of a parent structure by another atom or group is indicated by affixes attached to or inserted into the name of the parent structure. There are two main types of replacement names.
1. Skeletal replacement name. A name in which the replacement of skeletal atoms and their associated hydrogen atoms is indicated by non-detachable prefixes. When carbon atoms are replaced by heteroatoms this method has been called 'a' nomenclature since the prefixes end in 'a'.
Certain names in which the prefix 'thio-', 'seleno-' or 'telluro-' indicates replacement of a skeletal oxygen atom by a sulfur, selenium or tellurium atom, respectively, are also skeletal replacement names.

2. Functional replacement name. A name containing prefixes or infixes which indicate the replacement of an oxygen atom or hydroxy group of a characteristic group, functional parent or class name by other atoms or groups.
B.B.(G) 15

replacement operation (in organic nomenclature)
The exchange of one group of atoms or a single non-hydrogen atom for another.
B.B.(G) 23

replicate (duplicate) sample
Multiple (or two) samples taken under comparable conditions. This selection may be accomplished by taking units adjacent in time or space.
Although the replicate samples are expected to be identical, often the only thing replicated is the act of taking the physical sample. A duplicate sample is a replicate sample consisting of two portions. The umpire sample is usually used to settle a dispute; the replicate sample is usually used to estimate sample variability.
1990, *62*, 1203

replication
The duplication of *DNA* as part of the reproductive cycle of a *cell* or *virus*. During replication the two DNA strands in the double helix separate, and each strand then acts as a *template* specifying the base sequence of newly synthesized complementary strands.
1992, *64*, 164

report (in analysis)
A combination of specimen information and results. The report should contain information about unequivocal identification of the source and type of material analysed and the requesting agency. It may contain such other information that is pertinent to the correct interpretation of a result (e.g. confidence interval, reference data and interpretative information).
1989, *61*, 1659

reprecipitation
The planned repetition of a *precipitation* in order to remove chemically different species from a previous *precipitate* or to improve its stoichiometry.
O.B. 86

representative sample
A sample resulting from a sampling plan that can be expected to adequately reflect the properties of interest of the parent population.
A representative sample may be a random sample or, for example, a stratified sample, depending upon the objective of sampling and the characteristics of the population. The degree of representativeness of the sample may be limited by cost or convenience.
1990, *62*, 1202

repression
Binding of a protein (repressor) to the operator sequence in an *operon* thus preventing the *transcription* of the following structural *gene(s)* into mRNA and consequently sysnthesis of protein (cf. *enzyme repression*).
1992, *64*, 164

reproducibility
The closeness of agreement between independent results obtained with the same method on identical test material but under different conditions (different operators, different apparatus, different laboratories and/or after different intervals of time). The measure of reproducibility is the standard deviation qualified with the term 'reproducibility' as reproducibility standard deviation.
In some contexts reproducibility may be defined as the value below which the absolute difference between two single test results on identical material obtained under the above conditions, may be expected to lie with a specified probability.
Note that a complete statement of reproducibility requires specification of the experimental conditions which differ.
See also *repeatability*.
1994, *66*, 598; see also 1995, *67*, 1707; 1990, *62*, 2210

repulsive potential-energy surface
A surface for an *exergonic reaction* A + BC in which the col corresponds to considerable separation between the products A–B + C. The energy barrier in the *potential-energy profile* is in the later stages of the reaction path. On such a surface most of the energy is released after A–B is formed.
Repulsive surfaces are also called late-downhill surfaces, and the barrier in such a surface is called a Type-II barrier.
1996, *68*, 185

reserve sample
See *umpire sample*.
1990, *62*, 1203

Re, Si
A *stereoheterotopic* face of a trigonal atom is designated *Re* if the ligands of the trigonal atom appear in a clockwise sense in order of *CIP priority* when viewed from that side of the face. The opposite arrangement is termed *Si*.
See also *enantiotopic*, and illustrated under *prochirality*.
1996, *68*, 2216

residence time
1. (instrumentation): The time required for air or reagent parcel to pass from the entrance to the exit of an instrument. Often this is approximated as the ratio of the interior volume of the device to the flow rate.
2. (atmospheric): The average time a molecule or aerosol spends in the atmosphere after it is released or generated there. For compounds with well defined sources and emission rates, this is estimated by the ratio of the average global concentration of a substance to its production rate on a global scale. It is a

function of not only the emission rates but the loss rates by chemical and physical removal processes.
1990, *62*, 2211

residence time (hydraulic retention time) (in biotechnology), t_r
The average time a particle or volume element of the culture resides in a *bioreactor* (or other device) through which a liquid medium continously flows. Residence time of the liquid is the reciprocal of *dilution rate*.
1992, *64*, 164

residual current
The current that flows, at any particular value of the *applied potential*, in the absence of the substance whose behaviour is being investigated (i.e. in a *blank solution*).
1985, *57*, 1499

residual fuel/oil
The liquid or semi-liquid, high-boiling fraction of residue from the distillation of petroleum which is used as a fuel. After removal of the lower boiling fraction of crude oil, sold as petroleum gas, the somewhat higher boiling fraction becomes gasoline and diesel oil. A portion of the higher boiling fraction is 'cracked' to yield additional gasoline. Still heavier oils become lubricants. Paraffin, asphalt, etc., are also extracted from the crude oil. It is the remaining residual oil, not economically usable for other industrial purposes, which is then sold as a relatively low cost fuel for burning.
1990, *62*, 2211

residual liquid junction (potential) error (in pH measurement)
Error arising from breakdown in the assumption that the *liquid junction* potential remains constant when solution X is substituted for solution S in the operational cell.
1985, *57*, 541

residual spectrum/background spectrum (in mass spectrometry)
Set of peaks recorded in the absence of a sample and due either to small air leaks or to the presence of molecules desorbed from the walls of the introduction device or the source, or from the pump fluid.
1990, *62*, 2211

resistance, R
Electric potential difference divided by the electric current when there is no electromotive force in a conductor.
1996, *68*, 990

resistivity, ρ
Electric field strength divided by the *current density*. This quantity is a tensor in an anisotropic medium.
G.B. 15; 1996, *68*, 990

resolution (in gas chromatography)
A characteristic of the separation of two adjacent peaks. It may be expressed according to the equation:

$$R_{AB} = 2\frac{|d_R(B) - d_R(A)|}{|w(B) + w(A)|}$$

where R_{AB} is the resolution, $d_R(A)$ and $d_R(B)$ are the retention distances (time or volume) of each eluted component A and B, and $w(A)$ and $w(B)$ are the respective widths of each peak at its base.
1990, *62*, 2211

resolution (in mass spectroscopy)
1. (energy): By analogy with the peak width definition for mass resolution, a peak showing the number of ions as a function of their translational energy should be used to give a value for the energy resolution.
2. (10 per cent valley definition): Let two peaks of equal height in a mass spectrum at masses m and $m - \Delta m$ be separated by a valley which at its lowest point is just 10 per cent of the height of either peak. For similar peaks at a mass exceeding m, let the height of the valley at its lowest point be more (by any amount) than ten per cent of either peak height. Then the resolution (10 per cent valley definition) is $m/\Delta m$. It is usually a function of m. The ratio $m/\Delta m$ should be given for a number of values of m.
3. (peak width definition): For a single peak made up of singly charged ions at mass m in a mass spectrum, the resolution may be expressed as $m/\Delta m$, where Δm is the width of the peak at a height which is a specified fraction of the maximum peak height. It is recommended that one of three values 50%, 5% or 0.5% should always be used. For an isolated symmetrical peak recorded with a system which is linear in the range between 5% and 10% levels of the peak, the 5% peak width definition is technically equivalent to the 10% valley definition. A common standard is the definition of resolution based upon Δm being Full Width of the peak at Half its Maximum height, sometimes abbreviated 'FWHM'. This acronym should preferably be defined the first time it is used.
1991, *63*, 1554; O.B. 203

resolution (in optical spectroscopy)
Wavenumber, wavelength or frequency difference of two still distinguishable lines in a spectrum.
G.B. 31

resolution (in stereochemistry)
The separation of a *racemate* into the component *enantiomers*.
See also *kinetic resolution, racemic conglomerate*.
1996, *68*, 2217

resolving power (in mass spectrometry)
The ability to distinguish between ions differing in the quotient mass/charge by a small increment. It may be characterized by giving the peak width, measured in mass units, expressed as a function of mass, for at least two points on the peak, specifically at fifty percent and at five percent of the maximum peak height.
O.B. 203; 1978, *50*, 72; 1991, *63*, 1554

resolving power (in optical spectroscopy), R
Transition wavenumber (or wavelength or frequency) divided by the *resolution*.
G.B. 31

resolving time (in nuclear analytical chemistry)
The smallest time interval which must elapse between the occurrence of two consecutive ionizing events or signal pulses, in order that the measuring device be capable of fulfilling its function of each of them separately.
1982, *54*, 1551

resolving time, coincidence (in nuclear analytical chemistry)
See *coincidence resolving time (in nuclear analytical chemistry)*.
1982, *54*, 1551

resolving time correction (in nuclear analytical chemistry)
Correction to be applied to the observed number of pulses in order to take into account the number of pulses lost during the *resolving time*.
1982, *54*, 1551

resonance
In the context of chemistry, the term refers to the representation of the electronic structure of a *molecular entity* in terms of *contributing structures*. Resonance among contributing structures means that the wavefunction is represented by 'mixing' the wavefunctions of the contributing structures. The concept is the basis of the quantum mechanical valence bond methods. The resulting stabilization is linked to the quantum mechanical concept of 'resonance energy'. The term resonance is also used to refer to the delocalization phenomenon itself.
See also *mesomerism*.
1994, *66*, 1161

resonance absorption technique
The monitoring of atoms or radicals generated in the gas phase by observing the attenuation of the radiation from a lamp emitting the characteristic resonance radiation of the observed species.
1996, *68*, 2271

resonance broadening (of a spectral line)
See *collisional broadening*.
O.B. 122

resonance cross-section (in Mössbauer spectrometry)
The cross-section for resonant absorption of the Mössbauer transition γ ray.
1976, *45*, 214

resonance effect
See *mesomeric effect*.
1994, 66, 1139

resonance effect magnitude (in Mössbauer spectrometry)
Ratio of the difference in the transmitted or scattered intensity at resonance maximum and off-resonance, relative to the intensity off-resonance.
1976, *45*, 214

resonance emission
See *resonance line (in X-ray spectroscopy)*.
1991, *63*, 739

resonance energy
The difference in potential energy between the actual *molecular entity* and the *contributing structure* of lowest potential energy. The resonance energy cannot be measured, but only estimated, since contributing structures are not observable molecular entities.
See *resonance*.
1994, *66*, 1161

resonance energy (in radiochemistry)
The *energy* of a particle entering a nuclear reaction, this energy being just sufficiently high to lead to the formation of reaction products in one of their excited states.
1994, *66*, 2525

resonance fluorescence
Fluorescence from the primary excited atomic or molecular species at the wavelength of the exciting radiation (no relaxation within the excited manifold). This term is also used to designate the radiation emitted by an atom of the same wavelength as the longest one capable of exciting its fluorescence, e.g. 122.6 nm in the case of the hydrogen atom, and 253.7 nm in the case of the mercury atom.
See also *resonance line (in photochemistry)*.
1996, *68*, 2271

resonance fluorescence technique
The monitoring of atoms or radicals generated in the gas phase by observing the intensity of fluorescence (exitance) emitted by the species after excitation with radiation of the same wavelength.
1996, *68*, 2271

resonance integral, H_{rs}
In Hückel molecular orbital theory, the integral over space of the type $\int \psi_r^* \hat{H} \psi_s \, d\tau$ where $\hat{H}$ is the hamiltonian operator and ψ_r and ψ_s are different wavefunctions.
G.B. 17

resonance integral (in radiochemistry)
The integral, over all or some specified portion of the *resonance energy* range, of the *cross-section* divided by the *energy* of a *radiation*.
1994, *66*, 2525

resonance lamp
A lamp emitting resonance radiation of atoms and their ions. Depending on the requirements the lamp is filled either with pure vapour of the element or with a mixture of it and other gases e.g. Hg (253.7 nm), Cd (228.8 and 643.8 nm), Na (589.0 nm), Zn (213.8,

330.0, 334.5, and 636.2 nm), Kr (116.5 and 123.6 nm), Xe (129.6 and 147.0 nm).
1996, *68*, 2271

resonance line (in photochemistry)
The longest *wavelength* capable of exciting *fluorescence* in an atom.
See also *resonance fluorescence*.
1996, *68*, 2271

resonance line (in X-ray spectroscopy)
The radiative decay of an excitation level may proceed to the neutral ground state and would thus occur at the same energy as the corresponding line in the absorption spectrum. Such a line is called a resonance line and the process is called resonance emission.
1991, *63*, 739

resonance neutrons
Neutrons whose *energy* corresponds to the resonance energy of a specified nuclide or element. If the *nuclide* is not specified, the term refers to resonance neutrons of ^{239}U.
1994, *66*, 2522

resonance radiation
Synonymous with *resonance fluorescence*.
1996, *68*, 2271

response constant (in electroanalytical chemistry)
A quantity whose expression includes a current, whose value is characteristic of a charge-transfer process and the experimental conditions under which it is investigated, and whose nature depends on the technique that is employed. Some typical response constants are the *diffusion current constant* in *polarography*, the *voltammetric constant* in linear-sweep *voltammetry*, and the *chronopotentiometric constant* in *chronopotentiometry*.
1985, *57*, 1504

response time (of a detector), τ_R
The time required for the detector output to go from the initial value to a percentage (e.g. 99%) of the final value. In the case of an exponential behaviour of the detector τ_R can be related to the *time constant* τ_c. The *rise time* τ_r is the time required for the detector output to vary between given percentages (e.g. from 10% to 90%) of the final value. Similarly, the *fall time* τ_f is the time required for the detector output to vary between given percentages (e.g. from 90% to 10%) of the initial value.
1995, *67*, 1751

response time (of an analyser)
Time which elapses, when there is a stepwise change in the quantity to be measured, between the moment when this change is produced and the moment when the indication reaches a value conventionally fixed at 90% of the final change in indication.
1990, *62*, 2174

responsivity (in detection of radiation), R
Detector input can be e.g. radiant power, irradiation, radiant energy. It produces a measurable detector output which may be e.g. an electrical charge, an electrical current or potential or a change in pressure. The ratio of the detector output and the detector input is defined as the responsivity. It is given in e.g. ampere/watt, volt/watt. The responsivity is a special case of the general term *sensitivity*.
Dark current is the term for the electrical output of a detector in the absence of input. This is a special case of the general term *dark output*. For photoconductive detectors the term *dark resistance* is used.
If the responsivity is normalized with regard to that obtained from a reference radiation the resulting ratio is called *relative responsivity*. For measurements with monochromatic radiation at a given wavelength λ the term spectral responsivity $R(\lambda)$ is used. In some cases the relative spectral responsivity, where the spectral responsivity is normalized with respect to the responsivity at some given wavelength, is used. The dependence of the spectral responsivity on the wavelength is described by the *spectral responsivity function*. The useful spectral range of the detector should be given as the wavelength range where the relative responsivity does not fall below a specified value.
1995, *67*, 1749

rest point (of a balance)
The position of the pointer with respect to the pointer scale when the motion of the beam has ceased.
O.B. 36

restricted rotation
See *free rotation, hindered rotation, restricted rotation*.
1996, *68*, 2217

restriction enzymes
Endonucleases which recognize specific *base* sequences within a DNA helix, creating a double-strand break of DNA. Type I restriction enzymes bind to these *recognition sites* but subsequently cut the DNA at different sites. Type II restriction enzymes both bind and cut within their recognition or target sites.
1992, *64*, 164

result (in analysis)
The final value reported for a measured or computed quantity, after performing a measuring procedure including all subprocedures and evaluations.
1994, *66*, 597

retardation factor (in column chromatography), R
The fraction of the sample component in the mobile phase at equilibrium; it is related to the retention factor and other fundamental chromatography terms:
$$R = 1/(k + 1)$$
1993, *65*, 843

retardation factor (in planar chromatography), R_F
Ratio of the distance travelled by the centre of the spot to the distance simultaneously travelled by the mobile phase:
$$R_F = b/a$$
By definition the R_F values are always less than unity. They are usually given to two decimal places. In order

to simplify this presentation the hR_F values may be used: they correspond to the R_F values multiplied by 100. Ideally, R_F values are identical to the R values used in column chromatography.
   1993, *65*, 845

retardation, relative (in planar chromatography), R_{rel}
   See *relative retardation* (R_{rel}) (*in planar chromatography*).
   1993, *65*, 845

retarder
This term, used mainly in polymerization kinetics, refers to a substance that reduces the rate of reaction.
   Synonymous with *inhibitor*.
   1996, *68*, 185

retentate
   See *dialysate*.
   1972, *31*, 608

retention (in nuclear chemistry)
Of atoms undergoing a *nuclear transformation*, that fraction which remains in or reverts to its initial chemical form.
   1982, *54*, 1551

retention efficiency (in particle separation)
The ratio of the quantity of particles retained by a separator to the quantity entering it (generally expressed as a percentage).
   1990, *62*, 2211

retention factor (in column chromatography), k
A measure of the time the sample component resides in the stationary phase relative to the time it resides in the mobile phase; it expresses how much longer a sample component is retarded by the stationary phase than it would take to travel through the column with the velocity of the mobile phase. Mathematically, it is the ratio of the *adjusted retention volume* (time) and the *hold-up volume* (time):

$$k = V'_R / V_M = t'_R / t_M$$

If the *distribution constant* is independent of sample component concentration, then the retention factor is also equal to the ratio of the amounts of a sample component in the stationary and mobile phases respectively, at equilibrium:

$$k = \frac{\text{amount of component in stationary phase}}{\text{amount of component in mobile phase}}$$

If the fraction of the sample component in the mobile phase is R, then the fraction in the stationary phase is $(1 - R)$; thus

$$k = (1 - R)/R$$

In former nomenclatures and in the literature one may find the expressions partition ratio, capacity ratio, capacity factor or mass distribution ratio to describe this term.
In the literature the symbol k' is often used for the retention factor, particularly in liquid chromatography. The original reason for this was to clearly distinguish it from the partition coefficient (distribution constant) for which the symbol K had been utilized. Since, however, the distribution constants are all identified with a subscript, there is no reason to add the prime sign to this symbol. It should be emphasized that all the recognized nomenclatures (IUPAC, BS, ASTM) have always clearly identified the capacity factor with the symbol k and not k'.
The logarithm of the retention factor is equivalent to the R_M value used in planar chromatography. The symbol κ is suggested to express $\log k$:

$$\kappa = \log k = \log [(1 - R)/R]$$

   1993, *65*, 843

retention index (in column chromatography), I
The retention index of a sample component is a number, obtained by interpolation (usually logarithmic), relating the *adjusted retention volume* (time) or the retention factor of the sample component to the adjusted retention volumes (times) of two standards eluted before and after the peak of the sample component.
In the Kováts index or Kováts retention index used in gas chromatography, n-alkenes serve as the standards and logarithmic interpolation is utilized:

$$I = 100 \left[\frac{\log X_i - \log X_z}{\log X_{(z+1)} - \log X_z} + z \right]$$

where X refers to the adjusted retention volumes or times, z is the number of carbon atoms of the n-alkane eluting before and $(z + 1)$ is the number of carbon atoms of the n-alkene eluting after the peak of interest:

$$V'_{Rz} < V'_{Ri} < V'_{R(z+1)}$$

The Kováts (retention) index expresses the number of carbon atoms (multiplied by 100) of a hypothetical normal alkene which would have an adjusted retention volume (time) identical to that of the peak of interest when analysed under identical conditions.
The Kováts retention index is always measured under isothermal conditions. In the case of temperature-programmed gas chromatography a similar value can be calculated utilizing direct numbers instead of their logarithm. Since both the numerator and denominator contain the difference of two values, here we can use the total retention volumes (times). Sometimes this value is called the linear retention index.

$$I^T = 100 \left[\frac{t_{Ri}^T - t_{Rz}^T}{t_{R(z+1)}^T - t_{Rz}^T} + z \right]$$

where t_R^T refers to the total retention times (chart distances) measured under the conditions of temperature programming. The value of I^T will usually differ from the value of I measured for the same compound under isothermal conditions, using the same two phases.
   1993, *65*, 844; O.B. 108

retention of configuration
See *Walden inversion*.
1996, *68*, 2217

retention, relative (in column chromatography), *r*
See *relative retention (in column chromatography)*.
1993, *65*, 843

retention temperature (in chromatography)
The column temperature when the *peak maximum* for a component has been reached in *temperature-programmed chromatography*.
O.B. 105

retention time (in chromatography)
See *hold-up volume (time), relative retention, retention factor, retention index, retention volumes, total retention volume (time)*.
1993, *65*, 841

retention volume, corrected (in gas chromatography)
See *corrected retention volume (in gas chromatography)*.
O.B. 104

retention volumes (in chromatography)
Retention measurements (and measurements of hold-up volume and peak width) may be made in terms of times or chart distances as well as volumes. If flow and recorder speeds are constant, the volumes are directly proportional to the times and chart distances. The following definitions are drawn up in terms of volume, and it is recommended that theoretical discussion should be couched in the same terms wherever possible.
The total retention volume, V_R, is the volume of eluent carrier gas admitted to the column between the injection of the sample and the emergence of the peak maximum of the specified component. It includes the hold-up volume. In gas chromatography, the volume of carrier gas is specified at the outlet pressure and temperature of the column.
Note: the word 'total' in this definition allows retention time to be used as a general term when specification of a particular quantity is not required.
The adjusted retention volume, V'_R, is the total retention volume less the hold-up volume, V_M, i.e.

$$V'_R = V_R - V_M = \bar{V} - V_I$$

where $\bar{V}$ is the peak elution volume and V_I the interstitial volume.
The net retention volume, V_N, is the adjusted retention volume multiplied by the pressure-gradient correction factor:

$$V_N = jV'_R$$

The specific retention volume, V_g, is the net retention volume per gram of stationary liquid, active solid or solvent-free gel. In liquid chromatography, except when conducted at very high pressures, the compression of the mobile phase is negligible, and the adjusted and net retention volumes are identical. The specific retention volume is then the adjusted retention volume per gram of stationary liquid, active solid, or solvent-free gel. It is recommended that, when appropriate, authors specify the drying conditions. At 0 °C,

$$V_g = 273\, V_N/w_L T$$

where w_L is the mass of the stationary liquid phase.
O.B. 103, 104; see also 1993, *65*, 841

retention volume, total (time) (in column chromatography), V_R, t_R
See *total retention volume (time) (in column chromatography)*.
1993, *65*, 841

reticulation (in catalysis)
See *encapsulation (in catalysis)*.
1991, *63*, 1230

retinoids
Oxygenated derivatives of 3,7-dimethyl-1-(2,6,6-trimethylcyclohex-1-enyl)nona-1,3,5,7-tetraene as shown and derivatives thereof.

R = -CH₂OR', -C(=O)H, -CH₂OH, -C(=O)OH

See also *carotenoids, retro*.
1995, *67*, 1363; see also W.B. 247

retro
1. A prefix which indicates a shift by one position of all single and double bonds in a conjugated polyene system. It has been applied only to *carotenoids* and *retinoids*, e.g. *retro*-carotenoids, *retro*-retinoids.

β,ε-Carotene

6',7-*retro*-β,ε-Carotene

2. A prefix which indicates the reverse sequence of a *peptide*.
1995, *67*, 1363; B.B.(G) 32; see also W.B. 229, 249, 64

+ retroaddition
See *cycloelimination*.
1994, *66*, 1161

+ retrocycloaddition
See *cycloelimination*.
1994, *66*, 1161

retro-ene reaction
See *ene reaction*.
1994, *66*, 1161

reversed direct-injection burner (in analytical flame spectroscopy)
A direct-injection burner combines the function of nebulizer and burner. Most commonly the oxidant is also used for aspirating and nebulizing the sample. However, when the fuel is used for this purpose, the term reversed direct-injection burner is applied.
O.B. 166

reversed (radiochemical) isotope dilution analysis
Isotope dilution analysis used for the determination of the *isotopic carrier* in a solution of a *radionuclide* with the aid of one of its stable isotopes.
1994, *66*, 2521

reversed-phase chromatography
An elution procedure used in liquid chromatography in which the mobile phase is significantly more polar then the stationary phase, e.g. a microporous silica-based material with chemically bonded alkyl chains. The term 'reverse phase' is an incorrect expression to be avoided.
1993, *65*, 826; O.B. 93

reverse micelle (or reversed micelle)
See *inverted micelle*.
1994, *66*, 1161

reverse transcriptases
Enzymes found in retroviruses that can synthesize complementary single-strands of DNA from an mRNA sequence as *template*. They are used in *genetic engineering* to produce specific cDNA molecules from purified preparations of mRNA.
1992, *64*, 165

reversible transition
A transition that returns a system directly to its original state by reversing the process that changed it.
1994, *66*, 589

revolutions per minute (rpm)
Non-SI unit for rotational frequency.
  See *rotational frequency*.
  1996, *68*, 991

rheology
The study of the flow and deformation of matter under the influence of a mechanical force. It is concerned particularly with material behaviour which cannot be described by the simple linear models of hydrodynamics and elasticity. Some of these departures are due to the presence of colloidal particles or to the influence of surfaces.
1979, *51*, 1215

rheopexy
A phenomenon in which the time of solidification, after discontinuation of a relatively high *shear rate*, is shortened by applying a small shear rate.
1979, *51*, 1217

rho
See entries under ρ (beginning of 'r').

rhodamine dyes
Dyes derived from condensation of phthalic anhydride with *m*-dialkylaminophenols.

See *xanthenes*.
1995, *67*, 1363

rhombohedral graphite
A thermodynamically unstable allotropic form of *graphite* with an ABCABC stacking sequence of the layers. The exact crystallographic description of this allotropic form is given by the space group D_{3d}^5–$R\bar{3}m$, (unit cell constants: $a = 256.6$ pm, $c = 1006.2$ pm).
Notes:
The structure of rhombohedral graphite can be best considered as an extended stacking fault in hexagonal graphite. Rhombohedral graphite can not be isolated in pure form (*natural graphite* and laboratory preparations contain less than 40% of rhombohedral graphite in combination with *hexagonal graphite*). It is produced by shear deformation of hexagonal graphite and transforms progressively to the hexagonal (ABAB) modification on heating above 1600 K.
1995, *67*, 504

ribonucleic acids (RNA)
Naturally occurring polyribonucleotides.

See also *nucleic acids, nucleosides, nucleotides, ribonucleotides, messenger RNA, transfer RNA, ribosomal RNA*.
1995, *67*, 1364; see also 1992, *64*, 165; W.B. 110

ribonucleotides
Nucleotides in which the *glycosyl group* is a ribosyl group.

See also *nucleotides*.
1995, *67*, 1364; see also W.B. 109

ribosomal RNA (rRNA)
RNA molecules which are essential structural and functional components of *ribosomes*, the subcellular units responsible for protein synthesis.
1992, *64*, 165

ribosomes
Subcellular units composed of specific RNA molecules and a large number of proteins that are responsible for protein synthesis.
1992, *64*, 165

Rice–Ramsperger–Kassel (RRK) theory
A theory of unimolecular gas reactions in which the rate with which the energized reactant molecule breaks down is treated as a function of the energy ε that it contains. The theory assumes that the rate is proportional to the number of ways of distributing ε among the internal degrees of freedom of the reactant molecule, in such a manner that the critical energy ε_c is localized in one particular degree of freedom.
1996, *68*, 185

Rice–Ramsperger–Kassel–Marcus (RRKM) theory
An improved form of *Rice–Ramsperger–Kassel (RRK) theory* in which account is taken of the way in which the various normal-mode vibrations and rotations contribute to reaction, and allowance is made for the zero-point energies. In this theory the energy ε^* in an energized molecule is classified as either active, ε^*_{active}, or inactive, $\varepsilon^*_{inactive}$. The rate depends upon $P(\varepsilon^*_{active})/N(\varepsilon^*)$, where $N(\varepsilon^*)$ is the density of states having energy between ε^* and $\varepsilon^* + d\varepsilon^*$, and $P(\varepsilon^*_{active})$ is the sum of the active quantum states of the activated complex. This extension of RRK theory brings it in line with *transition-state theory*.
1996, *68*, 186

riffling (in analytical chemistry)
The separation of a free-flowing sample into (usually) equal parts by means of a mechanical device composed of diverter chutes.
1990, *62*, 1204

ring assembly
Two or more cyclic systems (single rings or fused systems) which are directly joined to each other by double or single bonds are named ring assemblies when the number of such direct ring junctions is one less than the number of cyclic systems involved.
Examples:

Ring assemblies

Fused polycyclic system

B.B. 42

Ringelmann chart (in atmospheric chemistry)
A chart which has been used in *air pollution* evaluation for assigning the degree of blackness of smoke emanating from a source. The observer compares the shades of grey (white to black) with a series of shade diagrams formed by horizontal and vertical black grid lines on a white background. A corresponding number, the Ringelmann number, is then assigned to the describe the best match; numbers range from 0 (white) to 5 (black). This method of pollution evaluation, although relevant to the enforcement of legislation still in force in many countries, is not recommended for use today since more quantitative indicators of the efficiency of the combustion are now available.
1990, *62*, 2211

ring-opening copolymerization
A *copolymerization* which is a *ring-opening polymerization* with respect to at least one *monomer*.
1996, *68*, 2308

ring-opening polymerization
A *polymerization* in which a *cyclic monomer* yields a *monomeric unit* which is acyclic or contains fewer cycles than the monomer.
Note:
If the monomer is polycyclic, the opening of a single ring is sufficient to classify the reaction as ring-opening polymerization.
1996, *68*, 2308

ring reversal (ring inversion)
The interconversion by rotation about single bonds (coupled with angle deformation in the transition state) of cyclic conformers having equivalent ring shapes (e.g. *chair* → *chair*) though not necessarily equivalent spatial positions of substituents (e.g. *equatorial* → *axial*).
1996, *68*, 2217

ring-sector
A sequence of ring atoms, linked together as an unbranched chain and connecting two *bridgeheads* or junctions, but not itself containing any bridgehead or junction, is termed an atomic ring-sector.
A bond linking together directly two bridgeheads or junctions is termed a bond ring-sector.
B.B. 499

rise time (of a radiation detector)
See *response time*.
1995, *67*, 1751

rise time (of an analyser)
The difference between the *response time* and the *dead time*.
1990, *62*, 2174

(vertical) rise velocity (in flame emission and absorption spectrometry), v_f
The vertical component of the velocity of the volatilized component in the observation space (in mm s^{-1}). It depends on the flame temperature, the solute nebulized, the observation height and the gas flow rate.
1986, *58*, 1742

Ritchie equation
The *linear free energy relation*:

$$\log k_N = \log k_0 + N_+$$

applied to the reactions between nucleophiles and certain large and relatively stable organic cations, e.g. arenediazonium, triarylmethyl and aryltropylium cations in various solvents. k_N is the rate constant for reaction of a given cation with a given nucleophilic system (i.e. given nucleophile in a given solvent). k_0 is the rate constant for the same cation with water in water, and N_+ is a parameter which is characteristic of the nucleophilic system and independent of the cation. A surprising feature of the equation is the absence of a coefficient of N_+, characteristic of the substrate (*cf.* the s in the *Swain–Scott equation*), even though values of N_+ vary over 13 log units. The equation thus involves a gigantic breakdown of the *reactivity–selectivity principle*. The equation has been extended both in form and in range of application.

1994, *66*, 1161

RNA
See *ribonucleic acids*.
1992, *64*, 165

röntgen
Non-SI unit of exposure to X- or γ-radiation: R = 2.58 × 10^{-4} C kg.
G.B. 113; 1996, *68*, 991

root-mean-square end-to-end distance (in polymers), $<r^2>^{1/2}$
The square root of the mean-square end-to-end distance of a linear polymer chain averaged over all conformations of the chain. For a freely jointed chain consisting of N segments each of length L, it is given by:

$$<r^2>^{1/2} = N^{1/2} L$$

The subscript zero is used to indicate unperturbed dimensions, as in $<r^2>_0^{1/2}$. If this term is used repeatedly, and if it is not confusing, the abbreviated name 'end-to-end distance' may be used.
P.B. 49

Rosanoff convention
See *Fischer–Rosanoff convention*.
1996, *68*, 2217

rotamer
One of a set of *conformers* arising from restricted rotation about one single bond.
1996, *68*, 2217

rotational barrier
In a rotation of groups about a bond, the potential energy barrier between two adjacent minima of the molecular entity as a function of the *torsion angle*.
1996, *68*, 2217

rotational constants
Coefficients of quantum numbers in the rotational *term* expression and inversely proportional to the *principal moments of inertia*. Symbols: A, B, C. $\tilde{A} = h/8\pi^2 c I_A$ (dimension wavenumber), $A = h/8\pi^2 I_A$ (dimension frequency) where h is the Planck constant and c the speed of light in vacuum.
G.B. 23

rotational diffusion
A process by which the equilibrium statistical distribution of the overall orientation of molecules or particles is maintained or restored.
Rotational diffusion may be compared to translational diffusion through which the equilibrium statistical distribution of position in space is maintained or restored.
P.B. 61

rotational diffusion coefficient
Defined by the equation:

$$D_\theta = \frac{t_\theta}{(\partial f(\theta,\Phi)/\partial \theta) \sin\theta}$$

where $f(\theta,\Phi)\sin\theta\, d\theta\, d\Phi$ is the fraction of particles whose axes make an angle between θ and $\theta + d\theta$ with the direction $\theta = 0$, and have an azimuth between Φ and $\Phi + d\Phi$; $t_\theta d\Phi$ is the fraction of particles having an azimuth between Φ and $\Phi + d\Phi$ whose axis passes from values $< \theta$ to values $> \theta$ in unit time. The axis whose rotational diffusion is considered has to be clearly indicated.
1972, *31*, 617

rotational frequency (in centrifugation), f_{rot}
The number of rotations divided by time:

$$f_{rot} = dN/dt$$

The synonyms rate of rotation, rate of revolution, centrifugal speed, centrifugation speed, and the traditional units of rotational frequency such as revolutions per minute, r.p.m., ppm, rev. min^{-1}, r min^{-1}, are not recommended.
1994, *66*, 904; 1996, *68*, *991*

rotational term, F
Rotational energy divided by the product of the Planck constant and the speed of light.
See *term*.
G.B. 23

rotator phase transition
A transition from a fully-ordered crystal of globular or quasi-spherical molecules to a crystal that retains translational order but exhibits dynamic orientational disorder and is usually mechanically soft (plastic phase).
Example: The transition of CBr$_4$ to a plastic phase which exists from 319 to 365 K.
1994, *66*, 589

rotatory power
The quantitative measure of *optical activity*.
1996, *68*, 2217

rotaxanes
Molecules in which a ring encloses another, rod-like molecule having end groups too large to pass through the ring opening, and thus holds the rod-like molecule in position without covalent bonding.

G—|—G

G = end groups

See *catenanes*.
1995, *67*, 1364

rotenoids
Naturally occurring substances containing a *cis*-fused tetrahydrochromeno[3,4-*b*]chromene nucleus. Many rotenoids contain an additional ring, e.g rotenone:

1995, *67*, 1364

rotometer (in atmospheric chemistry)
A device, based on *Stokes law*, for measuring rate of fluid flow. It is a tapered vertical tube having a circular cross section in which a float moves in a vertical path to a height dependent upon the rate of fluid flow upward through the tube.
1990, *62*, 2211

roughness factor (rugosity) (of a surface)
The ratio:
$$f_r = A_r/A_g$$
where A_r is the *real (true, actual) surface (interface) area* and A_g is the *geometric surface (interface) area*.
1986, *58*, 439

rovibronic state
A state corresponding to a particular rotational sublevel of a particular vibrational level of a particular electronic state.
1996, *68*, 2271

rRNA
See *ribosomal RNA*.
1992, *64*, 165

RS
See *racemate*.
1996, *68*, 2216

R, S
The approved designations (devised by Cahn, Ingold and Prelog) of absolute configuration at four-coordinate (quadriligant) and six-coordinate (sexiligant) *stereogenic centres*.
1996, *68*, 2216

r, s
Stereodescriptors of *pseudo-asymmetric atoms*.
1996, *68*, 2216

R*, S*
See *relative configuration*.
1996, *68*, 2216

R_p, S_p
See *planar chirality*.
1996, *68*, 2217

ruby laser
A pulsed source of coherent radiation emitting mainly at 694.3 nm from chromium ions (Cr^{3+}) in aluminium oxide.
See *laser, solid state lasers*.
1996, *68*, 2272

rupture (of a thin film)
The formation of a hole which permits *coalescence* or direct contact of the two phases which it separates.
1972, *31*, 613

Rutherford backscattering (RBS)
Any technique using high energy particles directed toward a sample, in which the bombarding particles are detected and recorded as function of energy and/or angle. The technique is mostly used for determining depth distributions of elements based on the energy of the backscattered particle. In general, He^+ or H^+ particles are used at energies in the order of 100 keV to some MeV.
Also referred to as backscattering spectrometry (BSS).
1979, *51*, 2246

Rydberg constant
Atomic fundamental physical constant $R_\infty = E_h/2hc_0$ = 1.097 373 1534 (13) × 10^7 m^{-1}, where E_h is the Hartree energy, h the Planck constant, and c_0 the speed of light.
CODATA Bull., 1986, *63*, 1

Rydberg orbital
For an atom, an orbital with principal quantum number greater than that of any occupied orbital of the ground state. For a molecular entity, a molecular orbital which correlates with a Rydberg atomic orbital in an atomic fragment produced by dissociation. Typically, the extension of the Rydberg orbital is large compared to the size of the atom or molecular entity.
1996, *68*, 2272

Rydberg transition
An electronic transition described approximately as promotion of an electron from a 'bonding' orbital to a *Rydberg orbital*. Spectral bands corresponding to Rydberg transitions approximately fit the Rydberg formula:
$$\sigma = I - R/(n - \Delta)^2$$
where σ is the wavenumber, I the ionization potential of the atom or molecular entity, n a principal quantum number, R the Rydberg constant, and Δ the quantum defect which differentiates between s, p, d, etc., orbitals. The notation used is, e.g. $\pi \rightarrow$ ns.
1996, *68*, 2272

RYDMR
See *ODMR*.
1996, *68*, 2272

σ, π (sigma, pi)

The terms are symmetry designations, π molecular orbitals being antisymmetric with respect to a defining plane containing at least one atom (e.g. the molecular plane of ethene) and σ molecular orbitals symmetric with respect to the same plane. In practice the terms are used both in this rigorous sense (for orbitals encompassing the entire molecule) and also for localized two-centre orbitals or bonds, and it is necessary to make a clear distinction between the two usages.

In the case of two-centre bonds, a π-bond has a nodal plane that includes the internuclear bond axis, whereas a σ-bond has no such nodal plane. (A δ-bond in organometallic or inorganic molecular species has two nodes.) *Radicals* are classified by analogy into σ- and π-radicals.

Such two-centre orbitals may take part in molecular orbitals of σ- or π-symmetry. For example, the methyl group in propene contains three C–H bonds, each of which is of local σ-symmetry (i.e. without a nodal plane including the internuclear axis), but these three 'σ-bonds' can in turn be combined to form a set of group orbitals one of which has π-symmetry with respect to the principal molecular plane and can accordingly interact with the two-centre orbital of π-symmetry (π-bond) of the double-bonded carbon atoms, to form a molecular orbital of π-symmetry.

Such an interaction between the CH_3 group and the double bond is an example of what is called *hyperconjugation*. This cannot rigorously be described as 'σ–π conjugation' since σ and π here refer to different defining planes, and interaction between orbitals of different symmetries (with respect to the same defining plane) is forbidden.

1994, *66*, 1163

σ → σ* transition

An electronic transition described approximately as promotion of an electron from a 'bonding' σ orbital to an 'antibonding' σ orbital designated as σ*. Such transitions generally involve high transition energies, and appear close to or mixed with *Rydberg transitions*.

1996, *68*, 2279

σ-adduct (sigma adduct)

The product formed by the *attachment* of an *electrophilic* or *nucleophilic entering group* or of a *radical* to a ring carbon of an aromatic species so that a new σ-bond is formed and the original conjugation is disrupted. (This has generally been called a 'σ-complex', but adduct is more appropriate than complex according to the definitions given.) The term may also be used for analogous adducts to unsaturated (and conjugated) systems in general.

See also *Meisenheimer adduct*.
1994, *66*, 1170

σ-bond (sigma bond)

See σ, π.
1994, *66*, 1171

σ-constant (sigma constant)

Specifically the substituent constant for *meta*- and for *para*-substituents in benzene derivatives as defined by Hammett on the basis of the ionization constant of a substituted benzoic acid in water at 25 °C, i.e. log (K_a/K_a^o), where K_a is the ionization constant of a *m*- or *p*-substituted benzoic acid and K_a^o that of benzoic acid itself.

The term is also used as a collective description for related electronic substituent constants based on other standard reaction series, of which, σ^+, σ^- and σ^o are typical; also constants which represent dissected electronic effects such as σ_I and σ_R. For this purpose it might be better always to spell out the term in full, i.e. as 'Hammett sigma constant', and restrict σ-constants to the scale of substituent constants which is based on benzoic acid. A large positive σ-value implies high electron-withdrawing power by inductive and/or resonance effect, relative to H; a large negative σ-value implies high electron-releasing power relative to H.

See also *Hammett equation, ρ-value, Taft equation*.
1994, *66*, 1171

σ-orbital (sigma orbital)

See σ, π.
1994, *66*, 1171

S

See *R, S*.
1996, *68*, 2217

saccharides

The *monosaccharides* and di-, *oligo*- and *polysaccharides*, which are made up of *n* monosaccharide units linked to each other by a glycosidic bond. Considered by some to be synonymous with *carbohydrates*.
1995, *67*, 1364

Sackur–Tetrode constant

Fundamental physical constant representing the translational contribution to molar entropy; $S_0/R = -1.151\ 693(21)$ at 1 K and standard pressure of 100 kPa, where R is the gas constant.
CODATA Bull., 1986, *63*, 1

sacrificial acceptor

Molecular entity that acts as the electron acceptor in a *photoinduced electron transfer* process and is not restored in a subsequent oxidation process but is destroyed by irreversible chemical conversion.
1996, *68*, 2272

sacrificial donor

Molecular entity that acts as the electron donor in a *photoinduced electron transfer* process and is not restored in a subsequent reduction process but is destroyed by irreversible chemical conversion.
1996, *68*, 2272

sacrificial hyperconjugation

See *hyperconjugation*.
1994, *66*, 1162

saddle point
See *col*.
1996, *68*, 186

salt
A chemical compound consisting of an assembly of cations and anions.
R.B. 118

+ salt effect
See *kinetic electrolyte effect*.
1994, *66*, 1162

salt form of an ion exchanger
The ionic form of an ion exchanger in which the counter-ions are neither hydrogen nor hydroxide ions. When only one valence is possible for the counter-ion, or its exact form or charge is not known, the symbol or the name of the counter-ion without charge is used, e.g. sodium-form or Na-form, tetramethylammonium-form, orthophosphate-form. When one of two or more possible forms is exclusively present, the oxidation state may be indicated by a Roman numeral, e.g. Fe^{II} form, Fe^{III} form.
1993, *65*, 855

salting out
The addition of particular electrolytes to an aqueous phase in order to increase the distribution ratio of a particular solute.
Notes:
1. The addition of electrolytes to improve phase separation behaviour should not be referred to as salting out.
2. The term is also used for the addition of electrolytes to reduce the mutual partial miscibility of two liquids.
3. It has no connection with synergism.
1993, *65*, 2379

sample (in analytical chemistry)
A portion of material selected from a larger quantity of material. The term needs to be qualified, e.g. *bulk sample, representative sample, primary sample,* bulked sample, *test sample.*
The term 'sample' implies the existence of a sampling error, i.e. the results obtained on the portions taken are only estimates of the concentration of a constituent or the quantity of a property present in the parent material. If there is no or negligible sampling error, the portion removed is a test portion, aliquot or specimen. The term 'specimen' is used to denote a portion taken under conditions such that the sampling variability cannot be assessed (usually because the population is changing), and is assumed, for convenience, to be zero. The manner of selection of the sample should be prescribed in a sampling plan.
1990, *62*, 1200; 1989, *61*, 1660; see also 1988, *60*, 1465

sample, bulk
See *bulk sample*.
1988, *60*, 1463

sample, composite
See *composite sample*.
1988, *60*, 1465

sample, laboratory
See *laboratory sample*.
1988, *60*, 1465

sample, test
See *test sample (in spectrochemical analysis)*.
1988, *60*, 1465; 1990, *62*, 1206

sample error (in spectrochemical analysis)
An error which arises when the absorbance of some samples changes with time, e.g. as a result of photochemical reaction, formation of aggregates or adsorption on the cell wall.
1988, *60*, 1457

sample handling (in analysis)
Any action applied to the sample before the analytical procedure. Such actions include the addition of preservatives, separation procedures, storage at low temperature, protection against light and irradiation, loading, etc.
1989, *61*, 1661

sample injector (in chromatography)
A device by which a liquid or gaseous sample is introduced into the apparatus. The sample can be introduced directly into the carrier-gas stream, or into a chamber temporarily isolated from the system by values which can be changed so as to make an instantaneous switch of the gas stream through the chamber. The latter is a *by-pass injector*.
O.B. 98

sampler
A device used to withdraw and deliver a volume or an amount of a sample.
1989, *61*, 1661

sampler, dichotomous
See *dichotomous sampler*.
1990, *62*, 2212

sample unit
The discrete identifiable portion suitable for taking as a sample or as a portion of a sample. These units may be different at different stages of sampling.
1988, *60*, 1463

sampling, cryogenic
See *cryogenic sampling*.
1990, *62*, 2212

sampling error
That part of the total error (the estimate from a sample minus the population value) associated with using only a fraction of the population and extrapolating to the whole, as distinct from analytical or test error. It arises from a lack of homogeneity in the parent population.
In chemical analysis, the final test result reflects the value only as it exists in the test portion. It is usually assumed that no sampling error is introduced in preparing the test sample from the laboratory sample. Therefore, the sampling error is usually associated exclusively with the variability of the laboratory sample.
Sampling error is determined by replication of the laboratory samples and their multiple analyses. Since

sampling error is always associated with analytical error, it must be isolated by the statistical procedure of analysis of variance.
1990, *62*, 1201

sampling, grab
See *grab sampling*.
1990, *62*, 2212

sampling interval (in electroanalysis)
In Tast *polarography*, square-wave polarography, and similar techniques, the interval during which the current is measured or recorded.
1985, *57*, 1504

sampling plan (in analytical chemistry)
A predetermined procedure for the selection, withdrawal, preservation, transportation and preparation of the portions to be removed from a population as samples.
Summarizing the test values or observations from the selected portions yields an estimate for the concentration of an analyte or a value for a property determined with a calculable degree of uncertainty at a specified confidence level. A sampling plan includes the designation of the number, location and size of the portions, and instructions for the extent of compositing and for the reduction (in amount and fineness) of the portions to a laboratory sample and to test portions. It may also contain acceptance criteria. Some sampling plans do not include more than instructions for the statistical selection of portions to be removed. Such plans should properly be designated as 'statistical sampling plans'.
1990, *62*, 1201

sampling time (in electroanalysis)
In Tast *polarography*, square-wave polarography, and similar techniques, the duration of the *sampling interval*.
1985, *57*, 1504

sandwich compounds
Compounds in which a metal atom is located between the faces of two parallel and planar (or nearly so) ring structures, e.g. bis(η^5-cyclopentadienyl)iron (ferrocene), dibenzenechromium.
See also *metallocenes*.
1995, *67*, 1364

sanitary land fill
An engineered burial of refuse. The refuse is dumped into trenches and compacted by bulldozer, where, it is hoped, aerobic metabolism by microorganisms decomposes the organic matter to stable compounds (H_2O, CO_2, etc.). Moisture is essential for the biological degradation and groundwater assists the process except when it fills air voids and prevents the transport of oxygen to the refuse. Land fills of unsatisfactory design can be major sources of air, water and soil pollution.
1990, *62*, 2212

saprophyte
An microorganism that feeds on dead and decaying organic matter. Saprophytes excrete *enzymes* that digest organic residues externally, the low molecular weight compounds formed then being absorbed.
1992, *64*, 165

saturated solution
A solution which has the same concentration of a *solute* as one that is in equilibrium with undissolved solute at specified values of the temperature and pressure.
O.B. 83

saturation
The state of a *saturated solution*.
O.B. 18

saturation (in radioanalytical chemistry)
Of an irradiated element for a specified isotope, the steady state reached when the *disintegration rate* of the *nuclide* formed is equal to its production rate.
1994, *66*, 2525

saturation activity
For a specified *isotope*, the value of the *activity* of an irradiated element, whence state of *saturation* is reached.
1994, *66*, 2525

saturation capacity
See *loading capacity*.
1993, *65*, 2385

saturation fraction, s_B
Amount-of-substance of a component (solute) in a solution divided by the amount-of-substance of the component when it is saturating the system at constant temperature and pressure. Also referred to simply as saturation, for example oxygen saturation.
1996, *68*, 991

saturation loading
See *loading capacity*.
1993, *65*, 2385

saturation transfer
A term used in nuclear magnetic resonance. When a nucleus is strongly irradiated, its spin population may partly be transferred to another nucleus by an exchange process.
See *magnetization transfer*.
1994, *66*, 1162

saturation vapour pressure
The pressure exerted by a pure substance (at a given temperature) in a system containing only the vapour and condensed phase (liquid or solid) of the substance.
1990, *62*, 2212

sawhorse projection
A *perspective formula* indicating the spatial arrangement of bonds on two adjacent carbon atoms. The bond between the two atoms is represented by a diagonal line, the left-hand bottom end of which locates the atom nearer the observer and the right-hand top

end the atom that is further away. In general a *Newman* or *zig-zag projection* is preferred.

staggered   eclipsed

1996, *68*, 2217

sc

See *torsion angle*.
1996, *68*, 2218

Saytzeff rule
Dehydrohalogenation of secondary- and tertiary-alkyl halides proceeds by the preferential removal of the β-hydrogen from the carbon that has the smallest number of hydrogens. Originally formulated by A. Saytzeff (Zaitsev) to generalize the orientation in β-*elimination reactions* of alkyl halides, this rule has been extended and modified, as follows: When two or more olefins can be produced in an elimination reaction, the thermodynamically most *stable* alkene will predominate. Exceptions to the Saytzeff rule are exemplified by the *Hofmann rule*.
See also *Markownikoff rule*.
1994, *66*, 1162

scaler
A sub-assembly for counting electrical pulses and containing one or more *scaling circuits*.
1982, *54*, 1551

scaling circuit
An electronic circuit which produces an output pulse for each time a specified number of pulses has been received at its input.
1982, *54*, 1551

scanning electron microscopy (SEM)
Any analytical technique which involves the generation and evaluation of *secondary electrons* (and to a lesser extent *back scattered electrons)* by a finely focused electron beam (typically 10 nm or less) for high resolution and high depth of field imaging.
1983, *55*, 2024

scanning method (in mass spectrometry)
This term refers to the sequence of control over operating parameters of a mass spectrometer that results in a spectrum of masses, velocities, momenta or energies.
1991, *63*, 1546; 1990, *62*, 2212

scanning transmission electron microscopy (STEM)
A special TEM-technique in which an electron transparent sample is bombarded with a finely focused electron beam (typically of a diameter of less than 10 nm) which can be scanned across the specimen or rocked across the optical axis and transmitted, secondary, back scattered and diffracted electrons as well as the characteristic X-ray spectrum can be observed. STEM essentially provides high resolution imaging of the inner microstructure and the surface of a thin sample (or small particles), as well as the possibility of chemical and structural characterization of micrometer and sub-micrometer domains through evaluation of the X-ray spectra and the electron diffraction pattern.
1983, *55*, 2025

scattering
A process in which a change in direction or energy of an incident *radiation* is caused by interaction with a *particle*, a system of particles, or a *photon*.
1982, *54*, 1551; 1990, *62*, 2212

scattering angle, θ
The angle between the forward direction of the incident beam and a straight line connecting the scattering point and the detector.
Synonymous with angle of observation.
P.B. 65; 1990, *62*, 2212

scattering, Compton
See *Compton effect*.
1982, *54*, 1551

scattering cross-section, σ_{scat}
The scattering coefficient per particle (cm^2/particle); $b_{scat} = n\sigma_{scat}$ where n is the number concentration of particles (particles cm^{-3}) and σ_{scat} is the scattering cross-section. b_{scat}, the scattering component of extinction due to gas and particles, is measured in the atmosphere using a nephelometer. For a homogeneous atmosphere it is related in theory to the meteorological range (L_v): $L_v = 3.9/b_{scat}$; b_{scat} and b_{abs} represent the scattering and absorption coefficients per unit length for a light beam (of path length L) which has a spectral radiance (intensity) I_0 incident on a sample of air and I is the transmitted spectral radiance (intensity), $\ln (I_0/I) = L(b_{scat} + b_{abs})$.
1990, *62*, 2212

scattering error (in spectrochemical analysis)
The presence of particulate matter, emulsions, micelles, etc. may cause radiation scattering which will result in further *attenuation* of the transmitted beam and the measured absorbance will be too high. This error is called the *scattering error*.
1988, *60*, 1457

scattering matrix
The *Stokes parameters* of scattered light are given by the matrix equation

$$(s_0, s_1, s_2, s_3) = F (s_0^0, s_1^0, s_2^0, s_3^0)$$

where the 4, 4 scattering matrix F is comprised of 16 scattering matrix elements. These matrix elements, which may originate in theory or experiment, provide a complete description of the scattered *radiation* in terms of the incident radiation.
1983, *55*, 934

scattering plane
The plane containing the incident light beam and the line from the centre of the scattering system to the observer.
1983, *55*, 932

scattering vector
The vector difference between the wave propagation vectors of the incident and the scattered beam, both of length $2\pi/\lambda$, where λ is the wavelength of the scattered radiation in the medium.
P.B. 65

scavenger
A substance that reacts with (or otherwise removes) a trace component (as in the scavenging of trace metal ions) or traps a reactive reaction *intermediate*.
See also *inhibition*.
1994, *66*, 1162

scavenging
1. (in radiation chemistry): Binding radicals or free electrons with a receptive (or reactive) material.
2. (in radiochemistry): The use of a precipitate to remove from solution by absorption or coprecipitation, a large fraction of one or more *radionuclides*.
1994, *66*, 2525
3. (in atmospheric chemistry): The removal of pollutants from the atmosphere by natural processes, including scavenging by cloud water, rainout and washout. This type of removal process is termed *precipitation scavenging*. Scavenging of airborne pollutants at the surfaces of plant, soil, etc., is termed *dry deposition*
1990, *62*, 2213

Schenck sensitization mechanism
The mechanism of chemical transformation of one molecular entity caused by photoexcitation of a sensitizer which undergoes temporary covalent bond formation with the molecular entity.
1996, *68*, 2272

Schiff bases (Schiff's bases)
Imines bearing a *hydrocarbyl* group on the nitrogen atom $R_2C=NR'$ ($R' \neq H$). Considered by many to be synonymous with *azomethines*.
1995, *67*, 1364

Schiller layers
In some systems, sedimenting particles form layers separated by approximately equal distances of the order of the wavelength of light. This gives rise to strong colours when observed in reflected light and the system is said to form iridescent layers or Schiller layers.
1972, *31*, 611

Schottky-barrier photodiode
See *photodiode*.
1995, *67*, 1755

Schulze–Hardy rule
The generalization that the critical *coagulation* concentration for a typical *lyophobic sol* is extremely sensitive to the valence of the *counter-ions* (high valence gives a low critical coagulation concentration).
1972, *31*, 610

Schulz–Zimm distribution
In an assembly of macromolecules, a continuous distribution with the differential mass-distribution function of the form:
$$f_w(x)\,dx = \frac{a^{b+1}}{\Gamma(b+1)} x^b \exp(-ax)\,dx$$
where x is a parameter characterizing the chain length, such as relative molecular mass or degree of polymerization, a and b are positive adjustable parameters, and $\Gamma(b+1)$ is the gamma function of $(b+1)$.
P.B. 56

scintillation
Burst of *luminescence* of short duration caused by an individual energetic particle.
1994, *66*, 2525

scintillation counter
A scintillator coupled to a *photomultiplier tube*. Incident X-ray photons are converted in the scintillator into bursts of visible light photons, some of which fall on the photocathode and can be measured. For incident photons having energies higher than the absorption edge of the elements contained in the scintillator, an escape peak can be observed.
1995, *67*, 1753

scintillation detector
A *radiation detector* using a medium in which a burst of *luminescence radiation*, produced along the path of an ionizing particle, is quantified.
1994, *66*, 2525

scintillation spectrometer
A measuring assembly incorporating a *scintillation detector* and a *pulse amplitude analyser*, used for determining the energy *spectrum* of certain types of *radiation*.
1982, *54*, 1551

scintillators
Materials used for the measurement of *radioactivity*, by recording the *radioluminescence*. They contain compounds (*chromophores*) which combine a high fluorescence quantum efficiency, a short fluorescence lifetime, and a high solubility. These compounds are employed as solutes in aromatic liquids and polymers to form organic liquid and plastic scintillators, respectively.
1996, *68*, 2272; 1994, *66*, 2525

s-*cis*, s-*trans*
The spatial arrangement of two conjugated double bonds about the intervening single bond is described as s-*cis* if *synperiplanar* and s-*trans* if *antiperiplanar*. This term should not be applied to other systems such as *N*-alkyl amides (use *E/Z* or *sp/ap*).

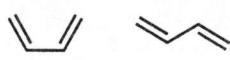

1996, *68*, 2217

scrambling
See *isotopic scrambling*.
1994, *66*, 1162

scrubber (in atmospheric chemistry)
An apparatus used in sampling and in flue gas cleaning. The gas is passed through a space containing wetted 'packing' or spray. In general, particles are collected in scrubbers by one or a combination of the following: impingement of particles on a liquid medium; diffusion of the particles onto a liquid medium; condensation of liquid medium vapours on the particles; partitioning of the gas into extremely small elements to allow collection of the particles by Brownian diffusion and gravitation settling on the gas-liquid interface. The devices include spray towers, jet scrubbers, Venturi scrubbers, cyclonic scrubbers, inertial scrubbers, mechanical scrubbers and packed scrubbers. Normally the gas flow in the scrubber is counter to the liquid flow. Efficient scrubbers will collect particles as small as 1 to 2 µm in diameter.
1990, *62*, 2213

scrubbing
1. (in solvent extraction): The process of selectively removing contaminating solutes (impurities) from an extract that contains these as well as the main extractable solute by treatment with a new immiscible liquid phase. The term stripping has a different meaning and should not be used in this sense although this usage has been customary in certain industries.
See also *crowding, selective stripping*.
1993, *65*, 2388
2. (in atmospheric chemistry): A process used in gas sampling or gas cleaning in which components in the gas stream are removed by contact with a liquid surface or a wetted packing, on spray drops, droplets, or in a bubbler, etc.
1990, *62*, 2213

seco-
Cleavage of a ring with addition of one or more hydrogen atoms at each terminal group thus created is indicated by the prefix 'seco'.
Examples:

5β-Androstane
↓
2,3-Seco-5β-androstane

B.B.(G) 31

second
SI base unit of time (symbol: s). The second is the duration of 9 192 631 770 periods of the radiation corresponding to the transition between the two hyperfine levels of the ground state of the caesium-133 atom.
G.B. 70; 1996, *68*, 991

second (of arc)
Non-SI unit of plane angle: $1" = (1/3600)° = 4.848\ 137\ ... \times 10^{-6}$ rad.
G.B. 113; 1996, *68*, 991

secondary crystallization
Crystallization occurring after *primary crystallization*, usually proceeding at a lower rate.
P.B. 85

secondary current distribution
See *current distribution*.
1981, *53*, 1836

secondary electron multiplier (in mass spectrometry)
A device to multiply current in an electron beam (or in a photon or particle beam by first conversion to electrons) by incidence of accelerated electrons upon the surface of an electrode which yields a number of secondary electrons greater than the number of incident electrons. These electrons are then accelerated to another electrode (or another part of the same electrode), which in turn emits further secondary electrons so that the process can be repeated. It is recommended that one should refer to the abundance of an ion, to the intensity of an ion beam, and to the height or area of a peak.
1991, *63*, 1554; O.B. 204

secondary electrons (se) (in *in situ* microanalysis)
All electrons emitted from the surface of a solid except back scattered primary electrons.
In practice, electrons emitted from the surface of a solid under particle bombardment which have a kinetic energy of less than 50 eV.
1983, *55*, 2026

secondary electron yield (in *in situ* microanalysis)
The number of *secondary electrons* generated per *primary electron* for a given specimen and experimental conditions. It depends on the (mean) atomic number of the excited area of the sample, the angle between electron beam and sample surface, the primary electron energy, thickness of the sample and sample potentials.
1983, *55*, 2026

secondary excitation (of X-rays)
See *characteristic X-ray emission*.
1991, *63*, 737

secondary fluorescence (in X-ray emission spectroscopy)
Ionization of the analyte element in a sample by characteristic *radiation* from other elements in the sample resulting in an enhancement of the signal measured.
1980, *52*, 2547

secondary ionization (in mass spectrometry)
The process in which ions are ejected from a sample surface (which may be a solid or substrate dissolved in a solvent matrix) as a result of bombardment by a primary beam of atoms or ions.
1991, *63*, 1548

secondary isotope effect
A kinetic isotope effect that is attributable to isotopic substitution of an atom to which bonds are neither made nor broken in the *rate-controlling step* or in a *pre-equilibrium step* of a specified reaction, and is therefore not a primary isotope effect. One speaks of α, β (etc.) secondary isotope effects, where α, β (etc.) denote the position of isotopic substitution relative to the reaction centre. The corresponding isotope effect on the equilibrium constant of such a reaction is called a 'secondary equilibrium isotope effect'.
Secondary isotope effects have been discussed in terms of the conventional electronic effects of physical organic chemistry, e.g. induction, *hyperconjugation, hybridization*, etc., since these properties are determined by the electron distribution, that depends on vibrationally averaged bond lengths and angles which vary slightly with isotopic substitution. While this usage is legitimate, the term 'electronic isotope effect' should be avoided, because of the misleading implication that such an effect is electronic rather than vibrational in origin.
See also *steric isotope effect*.
1994, *66*, 1131

secondary kinetic electrolyte effect
See *kinetic electrolyte effect*.
1994, *66*, 1162

secondary kinetic isotope effect
See *isotope effect*.
1994, *66*, 1162

secondary metabolites
Metabolites which are produced by routes other than the normal metabolic pathways, mostly after the phase of active growth and under conditions of deficiency. The biological significance of many secondary metabolites is not exactly known.
1992, *64*, 165

secondary pollution (emissions)
The products of the primary pollutants which form through photochemical and thermal reactions in the atmosphere (O_3, peroxyacetyl nitrate, etc.).
1990, *62*, 2213

secondary radiation
Radiation emitted by any matter irradiated with electromagnetic or *ionizing radiation*.
1982, *54*, 1552

secondary standard
See *standard solution*.
O.B. 47

secondary structure
The conformational arrangement (α-helix, β-pleated sheet, etc.) of the backbone segments of a macromolecule such as a polypeptide chain of a protein without regard to the conformation of the side chains or the relationship to other segments.
See also *primary structure, tertiary structure, quaternary structure*.
1996, *68*, 2218

second-order transition
A transition in which a crystal structure undergoes a continuous change and in which the first derivatives of the Gibbs energies (or chemical potentials) are continuous but the second derivatives with respect to temperature and pressure (i.e. heat capacity, thermal expansion, compressibility) are discontinuous.
Example:
The *order-disorder transition* in metal alloys, for example, CuZn.
Synonymous with continuous transition.
1994, *66*, 589

secular equilibrium
Radioactive equilibrium where the *half life* of the *precursor isotope* is so long that the change of its *activity* can be ignored during the period of interest and all activities remain constant.
1982, *54*, 1541

sediment
The highly concentrated *suspension* which may be formed by the *sedimentation* of a dilute suspension.
1972, *31*, 610

sedimentation
In the atmospheric sciences, the process of removal of an air borne particle from the atmosphere due to the effect of gravity.
1990, *62*, 2213

sedimentation coefficient, s
Velocity of sedimentation divided by the acceleration of the force field (gravitation or centrifugation).
G.B. 63; 1996, *68*, 992; see also P.B. 62; 1994, *66*, 904

sedimentation equilibrium
The equilibrium established in a centrifugal field when there is no net flux of any component across any plane perpendicular to the centrifugal force.
P.B. 62

sedimentation field strength
The potential difference E_{sed} per unit length in a *sedimentation* or centrifugation cell. As the contributions of the interfacial potential differences at the electrodes are not included in E_{sed} this quantity, although theoretically important, is not accessible to measurement.
1972, *31*, 620

sedimentation potential difference (sedimentation potential)
The potential difference E_{sed} or E at zero current caused by the *sedimentation* of particles in the field of gravity or in a centrifuge, between two identical electrodes at different levels (or at different distances from the centre of rotation). E is positive if the lower (peripheral) electrode is negative. This is also called the Dorn effect.
1972, *31*, 620

sedimentation velocity, v_B
Velocity of a component in a fluid relative to the fluid in the direction of gravitational or centrifugal acceleration. Also called sedimentation rate.
1996, *68*, 992; 1994, *66*, 904

sedimentation velocity method
A method by which the velocity of motion of solute component(s) or dispersed particles is measured and the result is expressed in terms of its (their) sedimentation coefficient(s).
P.B. 62

sedimentation volume
The volume V_{sed} of *sediment* formed in a *suspension*. If the sediment is formed in a centrifugal field, the strength of this field should be explicitly indicated, otherwise normal gravity is understood.
1972, *31*, 616

segment (in analytical chemistry)
1. (for bulk materials): Each of the single, large portions of material pre-existing either in space (e.g. bags, bales, drums) or accumulated during a fixed time (e.g. discharge from a conveyor belt) or formable as increments by a sampling device. Segments may be actual or conceptual.
1990, *62*, 1202
2. The set of samples which can be analysed between two successive calibrations. A segment includes samples, control materials and blank samples. This term is of particular importance in clinical chemistry.
1989, *61*, 1661

segregated star macromolecule
See *star macromolecule*.
1996, *68*, 2296

segregation
The process that differentiates the composition at an interface or surface from the average or bulk composition.
Note:
The composition of the segregated phases need not be uniform, for example, when concentration gradients are established in the phases.
1994, *66*, 590

segregation (in polymers)
The rejection of a fraction of macromolecules, or of impurities, or both, from growing crystals.
The rejected macromolecules are usually those of insufficient relative molecular mass, or differing in constitution or configuration (e.g. branching, tacticity, etc.).
P.B. 85

selected area electron diffraction (SAED)
See *transmission high energy electron diffraction (THEED)*.
O.B. 248

selected ion monitoring (in mass spectrometry)
This term is used to describe the operation of the mass spectrometer in which the intensities of several specific ion beams are recorded rather than the entire mass spectrum. An alternative recommended term is multiple ion (peak) monitoring. The use of the terms multiple ion detection and mass fragmentography are not recommended, because in the case of the former, it is the monitoring of several peaks which should be emphasized rather than the detection of different ions, and in the case of the latter it is often not fragments that are being monitored but peaks from molecular ions.
1991, *63*, 1551

selection (in biotechnology)
A laboratory method applying a mixture of *microorganisms* to particular growth conditons under which only the *cells* with particular characteristics can survive and may be isolated.
1992, *64*, 165

selection rule
A rule that states whether a given transition is allowed or forbidden, on the basis of the symmetry or spin of the wavefunctions of the initial and final states.
1996, *68*, 2273

selective (in analysis)
A term which expresses qualitatively the extent to which other substances interfere with the determination of a substance according to a given procedure.
1983, *55*, 555

selective corrosion
Corrosion of a single phase having more than one component, when the ratio of the corrosion rates of the components differs from the ratio of bulk mole fractions.
Selective corrosion changes the composition of the material in the interfacial region.
1989, *61*, 21

selective detector
See *radiation detector*.
1995, *67*, 1748

selective digestion (in spectrochemical analysis)
See *partial digestion (in spectrochemical analysis)*.
1988, *60*, 1469

selective elution (in chromatography)
An elution procedure in which a specific eluent is used, e.g. a complexing agent that forms stable nonsorbable complexes with one or a group of the com-

pounds to be separated, but affects the other components only to a negligible extent.
O.B. 92

selective inhibition (in catalysis)
See *selective poisoning (in catalysis)*.
1976, *46*, 83

selectively labelled
An *isotopically labelled* compound is designated as selectively labelled when a mixture of *isotopically substituted* compounds is formally added to the analogous *isotopically unmodified* compound in such a way that the position(s) but not necessarily the number of each labelling *nuclide* is defined. A selectively labelled compound may be considered as a mixture of *specifically labelled* compounds.
B.B. 515; 1981, *53*, 1893

selective micro-sample (in spectrochemical analysis)
A sample which results where a small portion has been separated from the *lot* or *laboratory sample* by selective means such as magnetic-, density-, or manual separation, by micro-drilling, or by centrifugation, e.g. the separation of magnetic materials from a geological material, or the separation of metal particles from a lubricating oil.
If individual particles are analysed the term *individual particle analysis* is applied.
1988, *60*, 1467

selective poisoning (in catalysis)
In selective poisoning or selective inhibition, a poison retards the rate of one *catalysed reaction* more than that of another or it may retard only one of the reactions.
1976, *46*, 83

selective preconcentration (in trace analysis)
An operation (process) as a result of which microcomponents are selectively isolated from a sample. It is used when the simultaneous presence of several components in the concentrate may distort the results of analysis. Selective preconcentration usually is achieved by isolation of the microcomponent to be determined.
1979, *51*, 1197

selective sample
A sample that is deliberately chosen by using a sampling plan that screens out materials with certain characteristics and/or selects only material with other relevant characteristics.
1990, *62*, 1203

selective solvent (in polymer chemistry)
A medium that is a solvent for at least one component of a mixture of polymers, or for at least one block of a *block* or *graft polymer*, but a non-solvent for the other component(s) or block(s).
P.B. 59

selective sorption (in polymer chemistry)
See *preferential sorption*.
P.B. 59

selectivity
This term is used in two different ways:
1. It sometimes refers to the discrimination shown by a given reactant A when it reacts with two alternative reactants B and C, or in two different ways (e.g. at two different sites) with a reactant B.
2. The term also sometimes refers to the ratio of products obtained from given reactants. This meaning is of importance for catalysts, which can have a wide range of selectivities.
Selectivity is quantitatively expressed by ratios of rate constants for the alternative reactions, or by the decadic logarithms of such ratios.
See also *isoselective relationship, partial rate factor, regioselectivity, selectivity factor, stereoselectivity*.
1996, *68*, 186; 1994, *66*, 1162

selectivity (in analysis)
1. (qualitative): The extent to which other substances interfere with the determination of a substance according to a given procedure.
2. (quantitative): A term used in conjunction with another substantive (e.g. constant, coefficient, index, factor, number) for the quantitative characterization of *interferences*.
1983, *55*, 555; see also 1990, *62*, 2174

selectivity (of a reagent)
See *selectivity factor*.
1994, *66*, 1162

selectivity coefficient (in ion exchange chromatography), $k_{A/B}$
The equilibrium coefficient obtained by application of the law of mass action to ion exchange and characterizing quantitatively the ability of an ion exchanger to select one of two ions present in the same solution. The ions involved in the exchange should be specified as subscripts.
Examples:
Exchange: $Mg^{2+} - Ca^{2+}$

$$k_{Mg/Ca} = \frac{[Mg]_S/[Ca]_S}{[Mg]_M/[Ca]_M}$$

Exchange: $SO_4^{2+} - Cl^-$

$$k_{SO_4/Cl} = \frac{[SO_4]_S/[Cl]_S^2}{[SO_4]_M/[Cl]_M^2}$$

In the above equations subscript S refers to the ion exchanger ('stationary phase') and M to the external solution ('mobile phase'). For exchanges involving counter-ions differing in their charges, the numerical value of $k_{A/B}$ depends on the choice of the concentration scales in the ion exchanger and the external solution (molal scale, molar scale, mole fraction scale, etc.). Concentration units must be clearly stated for an exchange of ions of differing charges. The corrected selectivity coefficient ($k_{A/B}{}^a$) is calculated in a way identical to the selectivity coefficient except that the concentrations in the external solutions are replaced by activities.

This term should not be used as a synonym for *separation factor*.

1993, *65*, 857; 1993, *65*, 2385

selectivity factor, S_f

A quantitative representation of *selectivity* in *aromatic substitution reactions* (usually *electrophilic*, for monosubstituted benzene derivatives). If the *partial rate factor*, f, expresses the reactivity of a specified position in the aromatic compound PhX relative to that of a single position in benzene, then the selectivity factor S_f (expressing discrimination between p- and m-positions in PhX) is defined as:

$$S_f = \log(f_p^X/f_m^X)$$

1994, *66*, 1162

selectivity factor (in ion exchange chromatography)

See *selectivity coefficient*.

1993, *65*, 857

selectivity ratio

Synonymous with *selectivity coefficient*.

1993, *65*, 2386

selenenic acids

Compounds having the structure RSeOH (R ≠ H), e.g. ArSeOH, areneselenenic acids.

1995, *67*, 1364

selenides

1. Compounds having the structure RSeR (R ≠ H). They are thus selenium analogues of ethers.
2. Metal salts of selane, H_2Se.

1995, *67*, 1364

seleninic acids

Compounds having the structure RSe(=O)OH.

1995, *67*, 1365

selenocyanates

Salts and *esters* of selenocyanic acid, HSeCN, e.g. CH_3CH_2SeCN ethyl selenocyanate.

1995, *67*, 1365

selenols

Compounds having the structure RSeH (R ≠ H).

1995, *67*, 1365

selenones

Compounds having the structure $R_2Se(=O)_2$.

1995, *67*, 1365

selenonic acids

Compounds having the structure $RSe(=O)_2OH$.

1995, *67*, 1365

selenoxides

Compounds having the structure $R_2Se=O$ (R ≠ H).

1995, *67*, 1365

self-absorption

Occurs in emission sources of finite thickness when radiant energy quanta emitted by atoms (or molecules) are absorbed by atoms of the same kind present in the same source. The absorbed energy is usually dissipated by collisional transfer of energy, or through emission of radiant energy of the same or other frequencies. In consequence, the observed *radiant intensity* of a *spectral line* (or band component) emitted by a source may be less than the radiant intensity would be from an optically thin source having the same number of emitting atoms. Self-absorption may occur in all emitting sources to some degree, whether they are homogeneous or not.

O.B. 242; 1996, *68*, 2273; 1994, *66*, 2525; see also 1982, *54*, 1552; 1985, *57*, 1464

self-absorption broadening (of a spectral line)

Photons emitted in one region of a source are partly absorbed in their passage through the plasma. Because of the fact that the absorption profile is of the same shape as the emission profile, energy is selectively absorbed from the emission line, i.e. the absorption coefficient is a maximum at the centre of the line or central wavelength. The actual line profile is changed as a result of the lowering of the maximum intensity accompanied by a corresponding increase in apparent halfwidth.

O.B. 122

self-absorption effect (in luminescence spectroscopy)

The reabsorption of *luminescence* by the analyte and interfering impurities within the excitation volume.

1984, *56*, 244

self-absorption factor (of a radiation source)

The ratio between the quantity of the radiation emitted by the source and the quantity of the radiation as produced by the *radioactive* nuclei present in the source.

Synonymous with source efficiency.

1994, *66*, 2525

self-diffusion coefficient

The *diffusion coefficient* D_i^* of species i in the absence of a *chemical potential* gradient. It is related to the diffusion coefficient D_i by

$$D_i^* = D_i \frac{\partial \ln c_i}{\partial \ln a_i}$$

where a_i is the activity of i in the solution, and c_i is the concentration of i. If an *isotopically labelled* species (i^*) is used to study diffusion, the tracer diffusion coefficient, D_i^*, is practically identical to the self-diffusion coefficient provided that the *isotope effect* is sufficiently small.

1972, *31*, 617

self inductance, L

For a thin conducting loop, the *magnetic flux* through the loop, caused by an electric current in the loop, divided by that current.

G.B. 15; ISO 31-4: 1992

self-poisoning (in catalysis)

A product of a reaction may cause *poisoning* or *inhibition*. The phenomenon is called self-poisoning or autopoisoning.

1976, *46*, 83

self-quenching
Quenching of an excited atom or molecular entity by interaction with another atom or molecular entity of the same species in the ground state.
See also *Stern–Volmer kinetic relationships*.
1996, *68*, 2273

self-reversal
A case of *self-absorption*, when a line is self-absorbed to such an extent that the peak or central wavelength intensity is less than at the wings or non-central wavelengths.
O.B. 122

self-shielding
The lowering of the *flux density* in the inner part of an object due to absorption in its outer layers.
1994, *66*, 2525

selones
Compounds having the structure $R_2C=Se$ (R ≠ H). They are thus selenium analogues of *ketones*.
1995, *67*, 1365

semicarbazones
Compounds having the structure $R_2C=N-NHC(=O)NH_2$, formally derived by condensation of aldehydes or ketones with semicarbazide $[NH_2NHC(=O)NH_2]$.
1996, *68*, 1365

semicoke
A carbonaceous material intermediate between a fusible mesophase pitch and a non-deformable *green coke* produced by incomplete *carbonization* at temperatures between the onset of fusion (of coal, *ca.* 620 K), and complete devolatilization. Semicoke still contains volatile matter, therefore.
Notes:
Semicoke may be conceived as covering a continuous range from coal that has not yet been fused to coke breeze. Semicoke can also be used as a *filler* in carbon mixtures.
See also *coal tar pitch*.
1995, *67*, 504

semiconductor
Material whose *conductivity*, due to charges of both signs, is normally in the range between that of metals and insulators and in which the electric charge carrier density can be changed by external means.
1982, *54*, 1552

semiconductor detector
A *radiation detector* using a semiconductor, in which free electric charges are produced along the path of an ionizing particle, in combination with a high voltage and electrodes to collect the induced electric charges.
1994, *66*, 2525; see also 1995, *67*, 1754

semiconductor detector, diffused junction
See *diffused junction semiconductor detector*.
1982, *54*, 1540

semiconductor detector, PIN
See *PIN semiconductor detector*.
1982, *54*, 1552

semiconductor detector, surface barrier
See *surface barrier semiconductor detector*.
1982, *54*, 1552

+ semiconductor laser
Usage not recommended.
See *diode laser*.
1995, *67*, 1920

semiconductor-metal transition
Any transition from a semiconductor to a metallic state under the influence of a temperature or pressure change or both.
Examples:
1. Ti_2O_3: band-edge crossing in a semiconductor to semimetal transition.
2. SmS: localized level crossing a band edge. In SmS hydrostatic pressure above the critical pressure $P_c \approx$ 0.65 GPa broadens the Sm-5d band to make the band edge cross the Sm-4f^6 level.
3. A *localized-itinerant transition*.
4. A *switching transition*.
1994, *66*, 590

semi-interpenetrating polymer network (SIPN)
A *polymer* comprising one or more *networks* and one or more *linear* or *branched* polymer(s) characterized by the penetration on a molecular scale of at least one of the networks by at least some of the linear or branched macromolecules.
Note:
Semi-interpenetrating polymer networks are distinguished from *interpenetrating polymer networks* because the constituent linear or branched polymers can, in principle, be separated from the constituent polymer network(s) without breaking chemical bonds; they are polymer blends.
1996, *68*, 2305

semioxamazones
Compounds having the structure $R_2C=NNHC(=O)C(=O)NH_2$, formally derived from the condensation of *aldehydes* or *ketones* with semi-oxamazide (the hydrazide of oxamic acid), $H_2NNHC(=O)C(=O)NH_2$.
1995, *67*, 1365

semiquinones
Radical anions having the structure $^-O-Z-O\cdot$ where Z is an *ortho-* or *para-*arylene group or analogous heteroarylene group; they are formally generated by the addition of an electron to a *quinone*.
1995, *67*, 1365

semisystematic name (semitrivial name)
A name in which at least one part is used in a systematic sense.
B.B.(G) 14

seniority (senior)(in organic nomenclature)
Terms used in reference to priority in a prescribed hierarchical order, a senior feature being preferred.
  B.B.(G) 17

sensitive area (of a radiation detector)
That area of the *detector* where an incident *radiant power* results in a measurable output.
  1995, *67*, 1751

sensitive volume (of a radiation detector)
That volume of the *detector* where an incident *radiant power* produces a measurable output.
  1995, *67*, 1751

sensitivity (in mass spectrometry)
Two different measures of sensitivity are recommended. The first, which is suitable for relatively involatile materials as well as gases, depends upon the observed change in ion current for a particular amount or change of flow rate of sample though the ion source. A second method of stating sensitivity, that is most suitable for gases, depends upon the change of ion current related to the change of partial pressure of the sample in the ion source.

It is important that the relevant experimental conditions corresponding to sensitivity measurement should always be stated. These include in a typical case details of the instrument type, bombarding electron current, slit dimensions, angular collimation, gain of the multiplier detector, scan speed and whether the measured signal corresponds to a single mass peak or to the ion beam integrated over all masses. Some indication of the time involved in the determination should be given, e.g. counting time or band width. The sensitivity should be differentiated from the detection limit.
  1991, *63*, 1553; O.B. 206

sensitivity (in metrology and analytical chemistry), A
The slope of the *calibration curve*. If the curve is in fact a 'curve', rather than a straight line, then of course sensitivity will be a function of analyte concentration or amount. If *sensitivity* is to be a unique performance characteristic, it must depend only on the *chemical measurement process*, not upon scale factors.
  1995, *67*, 1703; 1990, *62*, 2174; see also O.B. 5, 36

sensitization
See *photosensitization*.
  1996, *68*, 2273

sensitization (in colloid chemistry)
Addition of small amounts of a *hydrophilic* colloid to a hydrophobic *sol* to possibly make the latter more sensitive to *flocculation* by electrolyte.
  1972, *31*, 610

sensitized luminescence
The deactivation of the primarily excited emitter which can lead to the activation of the quencher followed by radiative deactivation.
  1984, *56*, 235

sensitizer
See *photosensitizer*.
  1996, *68*, 2273

separability assumption
This expression refers to the assumption, essential to conventional *transition-state theory*, that the energy of the system may be expressed as the sum of components associated with different degrees of freedom. In transition-state theory it is assumed that the energy of the motion of the system through the *dividing surface* of the *potential-energy surface* is separable into various components. In many practical calculations it is assumed that the energy of the system is separable into electronic, vibrational, rotational and translational energy.
  1996, *68*, 186

separated flame (in flame spectroscopy)
In a *premix burner*, sometimes provision is made to screen the observed portion of the flame gases from direct contact with the surrounding air. This may be done either mechanically by placing a tube on top of the burner around the flame, which produces a zonal separation (separated flame), or aerodynamically by surrounding the flame with a sheath of inert gas that emerges from openings at the rim of the burner top (shielded flame).
  O.B. 166

+ separation coefficient
Synonymous with *separation factor*. Use of this term is not recommended.
  1993, *65*, 2386

separation factor (in column chromatography), α
The relative retention value calculated for two adjacent peaks ($V'_{R2} > V'_{R1}$):
$$\alpha = V'_{R2}/V'_{R1} = V_{N2}/V_{N1} = t'_{R2}/t'_{R1} = k_2/k_1$$
By definition, the value of the separation factor is always greater than unity. The separation factor is also identical to the ratio of the corresponding distribution constants.

The separation factor is sometimes also called the '*selectivity*'. The use of this expression is discouraged.
  1993, *65*, 844

separation factor (in liquid-liquid distribution), $\alpha_{A,B}$
The ratio of the respective distribution ratios of two extractable solutes measured under the same conditions.
$$\alpha_{A,B} = D_A/D_B$$
Notes:
1. By convention the solutes designated as A and B in the above are chosen so as to make $\alpha > 1$.

2. The term *separation coefficient* is not recommended.
3. The terms *selectivity coefficient* and selectivity ratio are not synonymous and should not be used.
1993, *65*, 2386; O.B. 107; see also O.B. 85

separation number (in chromatography), SN
This expresses the number of peaks which can be resolved in a given part of the chromatogram between the peaks of two consecutive *n*-alkanes with z and $(z + 1)$ carbon atoms in their molecules:

$$SN = \frac{t_{R(z+1)} - t_{Rz}}{w_{hz} + w_{h(z+1)}} - 1$$

In the German literature the symbol TZ (trennzahl) is commonly used to express the separation number. As the separation number depends on the *n*-alkanes used for the calculation, they always must be specified with any given SN value.
1993, *65*, 847

separation, radiochemical
See *radiochemical separation*.
1994, *66*, 2525

separation temperature (in chromatography)
The temperature of the chromatographic bed under isothermal operation. In column chromatography it is called the column temperature.
1993, *65*, 838

sequence rules
See *priority*.
1996, *68*, 2218

sequencing (proteins, nucleic acids)
Analytical procedures for the determination of the order of amino acids in a polypeptide chain or of nucleotides in a DNA or RNA molecule.
1992, *64*, 165

sequential analyser
A discontinuous analyser in which at least one sub-assembly operates sequentially.
1990, *62*, 2174

sequential indication (of an analyser)
An indication obtained following sequential sampling or received from a sequential cell or from data processing comprising a succession of predetermined repetitive operations (or a combination of the three).
1990, *62*, 2174

sequential measuring cell
A measuring cell which operates according to a succession of operations on the sample or on the sensitive elements (or on both), these operations being carried out according to one or more repetitive programs.
1990, *62*, 2178

sequential sample
Units, increments or samples taken one at a time or in successive predetermined groups, until the cumulative result of their measurements (typically applied to attributes), as assessed against predetermined limits, permits a decision to accept or reject the population or to continue sampling. The number of observations required is not determined in advance, but the decision to terminate the operation depends, at each stage, on the results of the previous observations. The plan may have a practical, automatic termination after a certain number of units have been examined.
1990, *62*, 1203

sequential spectrometer
A *spectrometer* which enables the intensity of several spectral bands of radiation to be measured one after the other in time, i.e. sequentially.
1995, *67*, 1729

series (of analytical results)
A number of measured values $(x_1, x_2, ... x_i, ... x_n)$ equivalent to each other with respect to statistical considerations, e.g. the results of repeated analyses using only one analytical method on a substance that is presumed to be homogeneous.
1994, *66*, 597; O.B. 4

sesquiterpenoids
Terpenoids having a C_{15} skeleton.
1995, *67*, 1365

sesterterpenoids
Terpenoids having a C_{25} skeleton. Sometimes erroneously referred to as sesterpenoids.
1995, *67*, 1365

settling chamber (in atmospheric chemistry)
Chamber designed to reduce the velocity of gases in order to permit the settling out of fly ash. It may be either part of, adjacent to, or external to an incinerator.
1990, *62*, 2213

settling error (in spectrochemical analysis)
An error caused by allowing insufficient time for the reading to settle or if an absorption peak is scanned too rapidly.
1988, *60*, 1456

settling velocity
The terminal rate of fall of a particle through a fluid as induced by gravity or other external force.
1990, *62*, 2213

shape selectivity (in catalysis)
Possibly observed in *catalysts* with very small pores. The selectivity is largely determined by the bulk or size of one or more reactants.
1976, *46*, 81

shear breakdown
See *work softening*.
1979, *51*, 1217

shear dependent viscosity
For systems showing non-Newtonian behaviour when measured in steady *simple shear*, a coefficient η equal to σ/D at a given value of the *shear rate D*, where σ is the stress; η_0 is the limiting *viscosity* at zero shear rate, and η_∞ the limiting viscosity at infinite shear rate; $[\eta_0]$ is the limit of *intrinsic viscosity* at zero shear.
1979, *51*, 1217

shear modulus, G
Shear stress divided by *shear strain*.
G.B. 12

shear rate
The velocity gradient in a flowing fluid.
1992, *64*, 165

shear strain, γ
Displacement of one surface with respect to another divided by the distance between them.
G.B. 12

shear stress, τ
Force acting tangentially to a surface divided by the area of the surface.
G.B. 12

shear thickening
See *shear thinning*.
1979, *51*, 1217

shear thinning
If viscosity is a univalued function of the rate of shear, a decrease of the viscosity with increasing rate of shear is called shear thinning, and an increase of the viscosity shear thickening.
1979, *51*, 1217

shear transition
A *diffusionless transition* that involves a change of the shape of the unit cell by a process that can be described as shear.
1994, *66*, 590

shear viscosity
For a *Newtonian fluid*, the shear viscosity η is often termed simply *viscosity* since in most situations it is the only one considered. It relates the shear components of stress and those of rate of *strain* at a point in the fluid by:

$$\sigma_{xy} = \sigma_{yx} = \eta(\partial v_x/\partial y + \partial v_y/\partial x) = 2\eta\dot{\gamma}_{xy}$$

where $\dot{\gamma}_{xy}$, the shear component of rate of strain is defined as follows:

$$\dot{\gamma}_{xy} = \tfrac{1}{2}(\partial v_x/\partial y + \partial v_y/\partial x)$$

Corresponding relations hold for σ_{xz} and σ_{yz}; σ_{xy} is the component of stress acting in the y-direction on a plate normal to the x-axis; v_x, v_y, v_z are the components of velocity.
See also *shear dependent viscosity*.
1979, *51*, 1216

shielded flame (in flame spectroscopy)
See *separated flame (in flame spectroscopy)*.
O.B. 166

shielding
In the context of NMR spectroscopy shielding is the effect of the electron shells of the observed and the neighbouring nuclei on the external magnetic field. The external field induces circulations in the electron cloud. The resulting magnetic moment is oriented in the opposite direction to the external field, so that the local field at the central nucleus is weakened, although it may be strengthened at other nuclei (deshielding). The phenomenon is the origin of the structural dependence of the resonance frequencies of the nuclei.
See also *chemical shift*.
1994, *66*, 1163

shielding constant, σ
In NMR the difference between the external magnetic flux density and the local magnetic flux density at a resonating nucleus affected by the neighbouring electrons divided by the external flux density, $\sigma = (B_0 - B)/B_0$.
G.B. 25

shish-kebab structure
A polycrystalline morphology of double habit consisting of fibrous crystals overgrown epitaxially by lamellar crystals, the stems of which are parallel to the fibre axis.
P.B. 83

short chain
A *chain* of low relative molecular mass.
See *oligomer molecule (1)*.
1996, *68*, 2294

short-chain branch
See *branch*.
1996, *68*, 2297

short-range intramolecular interactions (in polymers)
A steric or other interaction involving atoms or groups or both situated nearby in sequence along the chain. The interacting atoms or groups are typically separated by fewer than ten consecutive bonds in a chain. If no confusion can occur, the word 'intramolecular' may be omitted.
P.B. 48

shut-down state (in analysis)
The condition of an instrument when it is switched off to conserve energy or reagents or to protect working parts. This term is of particular importance in clinical chemistry.
1989, *61*, 1664

shut-down time (in analysis)
The time interval between production of the last result of an instrument and *shut-down state*.
1989, *61*, 1664

shuttle vector
A DNA molecule (e.g. *plasmid*) that is able to replicate in two different *host* organisms and can therefore be used to 'shuttle' or convey *genes* from one to the other.
1992, *64*, 165

SI
Système International d'Unités, the international system of units established in 1960 and based on seven base units for the quantities: length, time, mass, electric current, thermodynamic temperature, amount of substance and luminous intensity.
G.B. 69; see also 1996, *68*, 995

Si
See *Re, Si*.
1996, *68*, 2218

side chain
See *branch*.
1996, *68*, 2297

side group
See *pendant group*.
1996, *68*, 2297

siemens
SI derived unit of electric conductance, $S = \Omega^{-1} = m^{-2}$ kg^{-1} s^3 A^2.
G.B. 72; 1996, *68*, 992

sievert
SI derived unit of dose equivalent (the mean specific energy imparted to an element of matter corrected by a quality factor and a modifying factor to take into account the properties of irradiated matter) equal to one joule per kilogram and admitted for reasons of safeguarding human health.
G.B. 72; 1996, *68*, 992

sigma
For entries, see under σ (beginning of 's').

sigmatropic rearrangement
A *molecular rearrangement* that involves both the creation of a new σ-bond between atoms previously not directly linked and the breaking of an existing σ-bond. There is normally a concurrent relocation of π-bonds in the molecule concerned, but the total number of π- and σ- bonds does not change. The term was originally restricted to intramolecular pericyclic reactions, and many authors use it with this connotation. It is, however, also applied in a more general, purely structural, sense.

If such reactions are *intramolecular*, their *transition state* may be visualized as an *association* of two fragments connected at their termini by two partial σ-bonds, one being broken and the other being formed as, for example, the two allyl fragments in (a'). Considering only atoms within the (real or hypothetical) cyclic array undergoing reorganization, if the numbers of these in the two fragments are designated *i* and *j*, then the rearrangement is said to be a sigmatropic change of order [*i,j*] (conventionally [*i*] ≤ [*j*]). Thus the rearrangement (a) is of order [3,3], whilst reaction (b) is a [1,5]sigmatropic shift of hydrogen. (N.B. By convention square brackets [...] here refer to numbers of atoms, in contrast with current usage in the context of cycloaddition.)

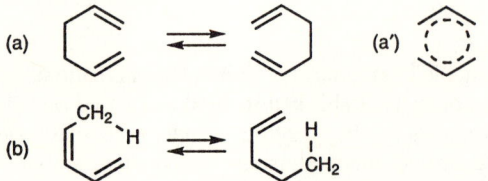

The descriptors a and s (*antarafacial* and *suprafacial*) may also be annexed to the numbers *i* and *j*; (b) is then described as a [1s,5s] sigmatropic rearrangement, since it is suprafacial with respect both to the hydrogen atom and to the pentadienyl system:

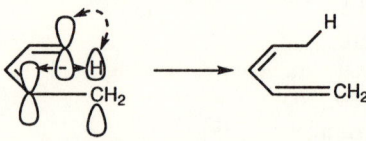

The prefix 'homo' (meaning one extra atom, interrupting *conjugation* — *cf.* 'homoaromaticity') has frequently been applied to sigmatropic rearrangements, but is misleading.
See also *cycloaddition, tautomerism*.
1994, *66*, 1163

signal (in analysis)
A representation of a quantity within an analytical instrument.
1989, *61*, 1662

signal sequence
See *leader sequence (2)*.
1992, *64*, 165

silanes
Saturated silicon hydrides, analogues of the alkanes; i.e. compounds of the general formula Si_nH_{2n+2}.
Silanes may be subdivided into silane, oligosilanes and polysilanes. Note *hydrocarbyl* derivatives and other derivatives are often referred to loosely as silanes.
1995, *67*, 1365

silanols
1. In a strict sense, hydroxy derivatives of silanes $Si_nH_{2n+1}OH$.
2. A name commonly applied to Si-*hydrocarbyl* derivatives, R_3SiOH, of silanol, H_3SiOH.
1995, *67*, 1365

silasesquiazanes
Compounds in which every silicon atom is linked to three nitrogen atoms and every nitrogen atom is linked to two silicon atoms, thus consisting of SiH and NH units, and having the general formula $(SiH)_{2n}(NH)_{3n}$. By extension *hydrocarbyl* derivatives are commonly included.
1995, *67*, 1366

silasesquioxanes
Compounds in which every silicon atom is linked to three oxygen atoms and every oxygen atom is linked to two silicon atoms, and having the general formula $(SiH)_{2n}O_{3n}$. By extension *hydrocarbyl* derivatives are commonly included.
1995, *67*, 1366

silasesquithianes
Compounds in which every silicon atom is linked to three sulfur atoms and every sulfur atom is linked to two silicon atoms, and having the general formula $(SiH)_{2n}S_{3n}$. By extension *hydrocarbyl* derivatives are commonly included.
1995, *67*, 1366

silathianes
Compounds having the structure $H_3Si[SSiH_2]_nSSiH_3$ and branched-chain analogues. They are analogous in structure to *siloxanes* with –S– replacing –O–. By extension *hydrocarbyl* derivatives are commonly included.
1995, *67*, 1366

silazanes
Saturated silicon-nitrogen hydrides, having straight or branched chains. They are analogous in structure to *siloxanes* with –NH– replacing –O–, e.g. $H_3SiNH\text{-}SiH_2NHSiH_3$ trisilazane. By extension hydrocarbyl derivatives are commonly included.
1995, *67*, 1366

silicones
Polymeric or oligomeric *siloxanes*, usually considered unbranched, of general formula $[-OSiR_2-]_n$ (R ≠ H).
1995, *67*, 1366

siloxanes
Saturated silicon–oxygen hydrides with unbranched or branched chains of alternating silicon and oxygen atoms (each silicon atom is separated from its nearest silicon neighbours by single oxygen atoms). The general structure of unbranched siloxanes is $H_3Si[OSiH_2]_nOSiH_3$. $H_3Si[OSiH_2]_nOSiH[OSiH_2OSiH_3]_2$ is an example of a branched siloxane. By extension *hydrocarbyl* derivatives are commonly included.
See also *silicones*.
1995, *67*, 1366

silver film
When viewed in reflected white light against a black background, transparent films show the classical interference colours of thin plates which permit an estimate of their thickness to be made. When of the order of 100 nm (1000 Å) in thickness they appear white (silver film) and when thinner they appear gradually less intensely white, then grey and finally black.
1972, *31*, 614

silylene
1. Generic name for $H_2Si\text{:}$ and substitution derivatives thereof, containing an electrically neutral bivalent silicon atom with two non-bonding electrons. (The definition is analogous to that given for *carbene*.)
2. The silanediyl group (H_2Si), analogous to the methylene group (H_2C).
1994, *66*, 1164

silyl groups
1. In a strict sense, the silyl group is H_3Si–.
2. A name commonly applied to *hydrocarbyl* derivatives of the silyl group R_3Si–.
1995, *67*, 1366

silyl radicals
1. In a strict sense, the silyl radical is $H_3Si\cdot$.
2. A name commonly applied to the silicon-centered *radicals* $R_3Si\cdot$.
1995, *67*, 1366

simple shear
An idealized treatment of a fluid between two large parallel plates (to permit ignoring edge effects) of area *A*, separated by a distance *h*. If one plate moves relative to the other with a constant velocity *V*, requiring a force *F* acting in the direction of movement, and the density, pressure, and *viscosity* throughout the fluid are constant, the Newtonian equation can be coupled with the equations of motion and of continuity to show that the velocity gradient in the fluid is constant (= V/h), and that $F/A = \eta V/h$ This idealized case (simple shear) is sometimes used to define shear viscosity.
1979, *51*, 1216

simulation technique (in analysis)
A technique for reducing or eliminating analytical errors resulting from *interferences*, using a *reference solution* sufficiently similar in quantitative composition to the sample solutions to be analysed that the interferences in the reference and sample solution are equivalent.
O.B. 172

simultaneous nucleation
See *nucleation*.
O.B. 84

simultaneous pair transitions
Simultaneous electronic transitions in two coupled absorbers or emitters. Because of the coupling, transitions which are spin-forbidden in one of the centres might become spin-allowed (spin flip).
1996, *68*, 2273

simultaneous reactions
Composite reactions, in which processes like

$$A \xrightarrow{1} Y \text{ and } B \xrightarrow{2} Z$$

occur in parallel, can be called simultaneous or *parallel reactions*. Sometimes there is competition involved, like in the scheme:

$$A + B \xrightarrow{1} Y \quad \text{and} \quad A + C \xrightarrow{2} Z$$

where B and C compete with one another for A.
1993, *65*, 2297

simultaneous spectrometer
A *spectrometer* which has more than one detector and enables the intensities of several spectral bands to be measured at the same time.
1995, *67*, 1729

single-beam (luminescence) spectrometer
A luminescence spectrometer in which a single beam of radiation is used for excitation and a single beam of luminescence radiation is taken from the sample.
1984, *56*, 238

single cell protein (SCP)
Microbial *biomass* or proteins extracted therefrom obtained from processes in which bacteria, yeasts,

other fungi or algae are cultivated in large quantities as human or animal protein supplement in animal feed or in human nutrition.
1992, *64*, 166

single channel pulse height analyser
See *pulse amplitude selector*.
1982, *54*, 1548

single-electron transfer mechanism (SET)
A reaction *mechanism* characterized by the transfer of a single electron between the species occurring on the *reaction coordinate* of one of the *elementary steps*.
1994, *66*, 1164

single escape peak
In a γ-ray spectrum, the peak due to *pair production* in the detector and escape, from the sensitive part of the detector, of one of the *photons* of 511 keV resulting from *annihilation*.
1982, *54*, 1541

single-focusing mass spectrometer
An instrument in which an ion beam with a given value of mass/charge is brought to a focus although the initial directions of the ions diverge.
1991, *63*, 1546; O.B. 201

single photon counting
See *photon counting*.
1996, *68*, 2273

single photon timing
See *time-correlated single photon counting*.
1996, *68*, 2273

single scattering
Radiation *scattering* in which there is only one scattering encounter.
See also *light scattering*.
1983, *55*, 932

single-step reaction
A reaction that proceeds through a single *transition state*.
1994, *66*, 1164

single-strand chain (in a polymer)
A *chain* that comprises *constitutional units* connected in such a way that adjacent constitutional units are joined to each other through two atoms, one on each constitutional unit.
1996, *68*, 2294

single-strand macromolecule
A *macromolecule* that comprises *constitutional units* connected in such a way that adjacent constitutional units are joined to each other through two atoms, one on each constitutional unit.
1996, *68*, 2294

single-strand polymer
A *polymer* the *macromolecules* of which are *single-strand macromolecules*.
1996, *68*, 2303

singlet molecular oxygen
The oxygen molecule (dioxygen), O_2, in an excited singlet state. The ground state of O_2 is a triplet $^3\Sigma_g^-$. The two metastable singlet states derived from the ground state configuration are $^1\Delta_g$ and $^1\Sigma_g^+$.
The term singlet oxygen alone, without mention of the chemical species is discouraged since it can also refer to an oxygen atom in a ^1S or ^1D excited state. While the oxygen atom ground state is a triplet ^3P state, the ^1S and ^1D states are also derived from the ground state configuration.
1996, *68*, 2273

singlet-singlet absorption
Absorption which results in the transition from the singlet ground state of the molecule into singlet excited states ($S_0 \to S_n$) and leads to the UV/visible absorption spectrum.
1984, *56*, 234

singlet-singlet annihilation
See *annihilation, spin conservation rule*.
1996, *68*, 2273

singlet-singlet energy transfer
Transfer of excitation from an electronically excited donor in a singlet state to produce an electronically excited acceptor in a singlet state.
See *electron exchange excitation transfer, Förster excitation transfer, radiative energy transfer*.
1996, *68*, 2273

singlet state
A state having a total electron spin quantum number equal to 0.
See *multiplicity*.
1996, *68*, 2273

singlet-triplet absorption
Absorption which takes place with the transition from the singlet ground state of the molecule to triplet states ($S_0 \to T_n$) and results in the singlet-triplet absorption spectrum.
1984, *56*, 234

singlet-triplet energy transfer
Transfer of excitation from an electronically excited donor in a singlet state to produce an electronically excited acceptor in a triplet state.
See *energy transfer, spin conservation rule*.
1996, *68*, 2273

singly labelled
A *specifically labelled* compound in which the *isotopically substituted* molecule has only one isotopically modified atom e.g. $CH_3-CH[^2H]-OH$.
B.B. 515; 1981, *53*, 1892

sink (in atmospheric chemistry)
The receptor for material which is removed from the atmosphere. Because of long range transport of many pollutants such as SO_2, sulfuric acid and its salts, the sink region can be many hundreds of kilometres from the source region of the pollutants.
1990, *62*, 2213

sintering
Coalescence of solid particles. The process by which fly ash produced in combustion of fuels such as coal

is baked (sintered) at a very high temperature. The sintered material is used in the manufacture of cinder blocks and other ceramic products. This term is also relevant to the enrichment of low grade ores and preparation of the charge for, e.g. a blast furnace. The sinter plant may be a significant source of pollution while serving to aid in the abatement of pollution for the blast furnace usually called sintering.

Catalysts often suffer during use from a gradual increase in the average size of the crystallites or growth of the *primary particles*. This is also usually called sintering.

1990, *62*, 2213; 1972, *31*, 611; 1976, *46*, 84

size-exclusion chromatography (SEC)
A separation technique in which separation mainly according to the hydrodynamic volume of the molecules or particles takes place in a porous non-adsorbing material with pores of approximately the same size as the effective dimensions in solution of the molecules to be separated.

P.B. 69

skeletal atom
An atom of a *skeletal structure*.

1996, *68*, 2296

skeletal bond
A bond connecting two *skeletal atoms*.

1996, *68*, 2296

skeletal structure
The sequence of atoms in the *constitutional unit(s)* of a *macromolecule*, an *oligomer molecule*, a *block* or a *chain* which defines the essential topological representation.

1996, *68*, 2296

skew
See *torsion angle* and *chair, boat, twist*.

1996, *68*, 2218

Slater-type orbital
An approximate *atomic orbital* that attempts to allow for electron-electron repulsion by scaling the nuclear charge for each orbital.

1994, *66*, 1164

slot burner
See *Bunsen burner*.

O.B. 166

slow neutrons
Neutrons of kinetic energy less than some specified value. This value may vary over a wide range and depends on the application. In reactor physics, the value is frequently chosen to be 1 eV; in dosimetry, the *effective cadmium cut-off* is used.

1982, *54*, 1547

small particle (in radiation scattering)
A particle much smaller than the wavelength of the radiation in the medium. In practice, all dimensions of a particle considered small must be less than about one-twentieth of the wavelength employed.

P.B. 65

smectic state
See *liquid-crystal transitions, mesomorphic phase*.

1994, *66*, 590; 1972, *31*, 613

SM-interference
Interference by a substance that produces a signal by a similar mechanism to the analyte, which in the given procedure (including the apparatus used) can not be distinguished from the signal given by the analyte. For example, in a colour reaction with an organic *ligand* by also reacting with the ligand and producing a coloured compound absorbing at the same wavelength as the compound produced by the analyte or in an atomic spectrochemical analysis by absorbing or emitting at the same wavelength as the analyte.

1983, *55*, 554

smog (in atmospheric chemistry)
The term originated in Great Britain as a popular derivation of 'smoke-fog' and appears to have been in common use before World War 1. It originally referred to the heavy pollution derived largely from coal burning (largely smoke filled air, rich in sulfur dioxide), and it probably was largely a reducing atmosphere. More common today in cities is an oxidizing atmosphere which contains ozone and other oxidants.

See *photochemical smog*.

1990, *62*, 2214

smog chamber (in atmospheric chemistry)
A large confined volume in which sunlight or simulated sunlight is allowed to irradiate air mixtures of atmospheric trace gases (hydrocarbons, nitrogen oxides, sulfur dioxide, etc.) which undergo oxidation. In theory these chambers allow the controlled study of complex reactions which occur in the atmosphere. However, ill-defined wall reactions which generate some molecular and radical species (e.g. HONO, CH_2O, HO-radicals, etc.) and remove certain products (H_2O_2, HNO_3, etc.), the use of reactant concentrations well above those in the atmosphere, ill-defined light intensities and wavelength distribution within the chamber, and other factors peculiar to chamber experiments require that caution be exercised in the extrapolation of results obtained from them to atmospheric systems.

1990, *62*, 2214

smog index (in atmospheric chemistry)
A mathematical correlation between *smog* and meteorological and/or pollutant concentrations associated with it. These are qualitative indices which are sometimes used in some urban communities to predict the degree of the air pollution which is expected for the coming day.

1990, *62*, 2214

smoke
An *aerosol* originating from combustion, thermal decomposition or thermal evaporation. Its particles may

be solid (magnesium oxide smoke) or liquid (tobacco smoke).
1972, *31*, 606; see also 1990, *62*, 2214

smooth transition
See *higher-order transition*.
1994, *66*, 590

soap
A salt of a fatty acid, saturated or unsaturated, containing at least eight carbon atoms or a mixture of such salts.
A neat soap is a lamellar structure containing much (e.g. 75%) soap and little (e.g. 25%) water.
1972, *31*, 612

soap curd
A gel-like mixture of fibrous soap-crystals ('curd-fibres') and their *saturated solution* (not a *mesomorphic phase*).
1972, *31*, 613

soap film
A term established by usage for A/W/A films stabilized by *surfactants* although it is not a film of *soap*, nor is the stabilizing surfactant necessarily a soap. (A/W/A means water film in air).
1972, *31*, 613

soft acid
See *hard acid*.
1994, *66*, 1164

soft base
See *hard base*.
1994, *66*, 1164

soiling (in atmospheric chemistry)
Visible damage to materials by deposition of air pollutants.
1990, *62*, 2214

sol
A fluid *colloidal* system of two or more components, e.g. a protein sol, a gold sol, an *emulsion*, a *surfactant* solution above the *critical micelle concentration*.
1972, *31*, 606

solar conversion efficiency
The ratio of the Gibbs energy gain per unit time per m^2 of surface exposed to the sun and the solar irradiance, E, integrated between $\lambda = 0$ and $\lambda = \infty$.
See also *photoelectrolysis cell*.
1996, *68*, 2274

solar flare
A bright eruption of energy from the sun's chromosphere.
1990, *62*, 2188

solar radiation (in atmospheric chemistry)
The electromagnetic radiation emitted by the sun. The total range of wavelengths of light emitted by the sun (99.9% in the range from 150 to 4000 nm) is filtered on entering the earth's atmosphere, largely through the absorption by oxygen, ozone, water vapour and carbon dioxide. Near sea level only light of wavelengths longer than about 290 nm is present. The light from 290–400 nm is effective in inducing important photochemical processes since absorption by the important trace gases, ozone, nitrogen dioxide, aldehydes, ketones, etc., is significant in this region.
1990, *62*, 2214

sol-gel transition
The transition of a suspension of solid particles in a liquid (sol) to an apparent solid, jelly-like material (gel).
1994, *66*, 590

solid angle, Ω, ω
Of a cone, the ratio of the area cut out on a spherical surface (with its centre at the apex of that cone) to the square of the radius of the sphere. It is a quantity of dimension one (dimensionless quantity) with unit steradian (sr = 1).
G.B. 11; 1996, *68*, 992

solidification
The transition of a liquid or gas into a solid.
1994, *66*, 590

solid phase antibody radioimmunoassay
A kind of *radioimmunoassay* employing an *antibody* bound to a solid phase.
1994, *66*, 2524

solid solution
See *mixed crystal*.
O.B. 85

solid state lasers
CW or pulsed lasers in which the active medium is a solid matrix (crystal or glass) doped with an ion (e.g. Nd^{3+}, Cr^{3+}, Er^{3+}). The emitted wavelength depends on the active ion, the selected optical transition, and the matrix. Some of these lasers are tunable within a very broad range (e.g. from 700 to 1000 nm for Ti^{3+} doped sapphire).
Pulsed lasers may be free-running, Q-switched, or mode-locked. Some CW lasers may be mode-locked.
1996, *68*, 2274

solid support (in column chromatography)
Normally an inert porous solid, which sorbs the *liquid phase*. The particle-size range of the *support* affects column efficiency and the pressure differential necessary to achieve a given flow rate. Modifications have been introduced for the achievement of special separations, in which the solid support is not inert but is an *active solid*. In capillary columns the inner wall of the column serves as the solid support and obviates the use of additional porous solids for this purpose.
O.B. 98; 1993, *65*, 830

solidus
A line on a binary phase diagram (or a surface on a ternary phase diagram) that indicates the temperature at which a system becomes completely solid on cooling or at which melting begins on heating under equilibrium conditions.
1994, *66*, 590

solid volume (in column chromatography)
The volume occupied by the *solid support* or the *active solid* in the *column*.
  O.B. 98

solubility
The analytical composition of a saturated solution, expressed in terms of the proportion of a designated solute in a designated solvent, is the solubility of that solute. The solubility may be expressed as a concentration, molality, mole fraction, mole ratio, etc.
  O.B. 84; see also 1990, *62*, 2214

solubility parameter (of a polymer), δ
A parameter which is a characteristic of a polymer used in predicting the solubility of that polymer in a given solvent.
Notes:
1. For a substance of low molecular weight, the value of the solubility parameter is often estimated from the enthalpy of vaporization; for a polymer, it is usually taken to be the value of the solubility parameter of the solvent producing the solution with maximum intrinsic viscosity or maximum swelling of a network of the polymer.
2. The solubility parameter is usually expressed in $(\text{cal cm}^{-3})^{1/2}$ or, preferably, $(\text{J cm}^{-3})^{1/2}$ units. The units must always be given; 1 $(\text{cal cm}^{-3})^{1/2} \approx 2.05$ $(\text{J cm}^{-3})^{1/2}$.
  P.B. 60

solubility product
The product of the ion activities raised to appropriate powers of an ionic solute in its saturated solution expressed with due reference to the dissociation equilibria involved and the ions present.
  O.B. 83

solubilization
  See *micellar solubilization*.
  1972, *31*, 613

solute
The minor component of a *solution* which is regarded as having been dissolved by the *solvent*.
  O.B. 83

solute-volatilization interference (in flame spectroscopy)
Interference due to changes in the volatilization rate of the dry *aerosol* particles in the case when volatilization of the analyte is incomplete in the presence and/or absence of the *concomitant*. This interference can either be specific, if the analyte and *interferent* form a new phase of different thermostability, as when Mg and Al form $MgAl_2O_4$ in an air–acetylene flame, or non-specific, if the analyte is simply dispersed in a large excess of the interferent, as when Ag is dispersed in ThO_2. If the interferent has a high boiling point, this latter is sometimes referred to as a blocking interference. It is often difficult to make sharp distinctions between the specific and non-specific solute-volatilization interferences.
  O.B. 136

solution
A liquid or solid phase containing more than one substance, when for convenience one (or more) substance, which is called the solvent, is treated differently from the other substances, which are called solutes. When, as is often but not necessarily the case, the sum of the mole fractions of solutes is small compared with unity, the solution is called a dilute solution. A superscript ∞ attached to the symbol for a property of a solution denotes the property in the limit of infinite dilution.
  1994, *66*, 535; O.B. 83; 1990, *62*, 2214

solvation
Any stabilizing interaction of a solute (or solute *moiety*) and the solvent or a similar interaction of solvent with groups of an insoluble material (i.e. the ionic groups of an ion-exchange resin). Such interactions generally involve electrostatic forces and *van der Waals forces*, as well as chemically more specific effects such as *hydrogen bond* formation.
  See also *cybotactic region*.
  1994, *66*, 1164

solvatochromic relationship
A *linear free-energy* relationship based on *solvatochromism*.
  See also *Kamlet–Taft solvent parameters*.
  1994, *66*, 1164

solvatochromism
The (pronounced) change in position and sometimes intensity of an electronic absorption or emission band, accompanying a change in the polarity of the medium. Negative (positive) solvatochromism corresponds to a *hypsochromic* (*bathochromic*) shift with increasing solvent *polarity*.
  See also *Dimroth–Reichardt E_T parameter, Z-value*.
  1994, *66*, 1164; 1996, *68*, 2274

solvent
  See *solution*.
  1994, *66*, 535

solvent (in liquid-liquid distribution)
The term applied to the whole initial liquid phase containing the extractant.
Notes:
1. The solvent may contain only one extractant or it may be a composite homogeneous mixture of extractant(s) with diluent(s) and also sometimes modifiers and accelerators.
2. The term solvent must not be used as a synonym for any of the individual components of a composite liquid phase even where, in the case of a single component (e.g. 3-methylbutan-2-one or tributyl phosphate), it becomes identical with the extractant.
3. The term may be qualified to denote the extract from a given processing step e.g. loaded solvent.
  1993, *65*, 2381

solvent extraction
The process of transferring a substance from any matrix to an appropriate liquid phase. If the substance is

initially present as a solute in an immiscible liquid phase the process is synonymous with *liquid-liquid extraction*.
Notes:
1. If the extractable material is present in a solid (such as a crushed mineral or an ore) the term leaching may be more appropriate. The extractable material may also be a liquid entrapped within or adsorbed on a solid phase.
2. Common usage has established this term as a synonym for liquid-liquid distribution. This is acceptable provided that no danger of confusion with extraction from solid phases exists in a given context.
1993, *65*, 2379

solvent front (in chromatography)
The front line of the *eluent*.
O.B. 99

+ solvent ion exchange (SIX)
This term is not recommended.
See *liquid ion exchange*.
1993, *65*, 2379

solvent isotope effect
A kinetic or equilibrium isotope effect resulting from change in the isotopic composition of the solvent.
1994, *66*, 1130

solvent migration-distance (in chromatography)
The distance travelled by the *solvent front*.
O.B. 99

solvent parameter
Quantitative measures of the capability of solvents for interaction with solutes. Such parameters have been based on numerous different physico-chemical quantities, e.g. *rate constants*, solvatochromic shifts in ultraviolet/visible spectra, solvent-induced shifts in infrared frequencies, etc. Some solvent parameters are purely empirical in nature, i.e. they are based directly on some experimental measurement. It may be possible to interpret such a parameter as measuring some particular aspect of solvent–solute interaction or it may be regarded simply as a measure of solvent *polarity*. Other solvent parameters are based on analysing experimental results. Such a parameter is considered to quantify some particular aspect of solvent capability for interaction with solutes.
See also *Dimroth–Reichardt E_T parameter, Grunwald–Winstein equation, Kamlet–Taft solvent parameters, Koppel–Palm solvent parameters, solvophobicity parameter, Z-value*.
1994, *66*, 1164

solvent regeneration (in extraction processes)
Treatment of the solvent for re-cycling, e.g. by removal of degradation products or non-strippable solutes. The term solvent purification is synonymous, but the terms scrubbing, stripping and washing should not be used in this context.
1993, *65*, 2389

solvent-separated ion pair
See *ion pair*.
1994, *66*, 1126

solvent-shared ion pair
See *ion pair*.
1994, *66*, 1126

solvent shift
A shift in the frequency of a spectral band of a chemical species arising from interaction with its solvent environment.
See *bathochromic shift, hypsochromic shift*.
1996, *68*, 2274

solvolysis
Generally, reaction with a solvent, or with a *lyonium ion* or *lyate ion*, involving the rupture of one or more bonds in the reacting solute. More specifically the term is used for *substitution, elimination* and *fragmentation reactions* in which a solvent species is the *nucleophile* ('alcoholysis' if the solvent is an alcohol, etc.).
1994, *66*, 1165

solvophobicity parameter, Sp
A *solvent parameter* defined by:
$$Sp = 1 - M/M(\text{hexadecane})$$
derived from the Gibbs energy of transfer ($\Delta_t G^o$) of a series of solutes from water to numerous aqueous-organic mixtures and to pure solvents:
$$\Delta_t G^o \text{ (to solvent)} = MR_T + D$$
where R_T is a solute parameter, and M and D characterize the solvent. The M values are used to define a solvent solvophobic effect so that Sp values are scaled from unity (water) to zero (hexadecane).
1994, *66*, 1165

solvus
A line on a binary phase diagram (or a surface on a ternary phase diagram) that defines the limit of solid solubility under equilibrium conditions.
1994, *66*, 590

SOMO
A Singly Occupied Molecular Orbital (such as the half-filled *HOMO* of a *radical*).
See also *frontier orbitals*.
1994, *66*, 1165

sonication
Irradiation with (often ultra)sound waves, e.g. to increase the rate of a reaction or to prepare vesicles in mixtures of surfactants and water.
1994, *66*, 1165

sonoluminescence
Luminescence induced by sound waves.
See *triboluminescence*.
1996, *68*, 2274

soot
A randomly formed *particulate carbon* material and may be coarse, fine and/or colloidal in proportions depending on its origin. Soot consists of variable quantities of carbonaceous and inorganic solids together with absorbed and occluded tars and resins.

Notes:
An unwanted by-product of incomplete combustion or pyrolysis. Soot generated within flames consists essentially of aggregates of spheres of carbon. Soot found in domestic fireplace chimneys contains few aggregates but may contain substantial amounts of particulate fragments of coke or *char*. Soot from diesel engines consists essentially of aggregates together with tars and resins. For historical reasons, the term soot is sometimes incorrectly used for *carbon black*. This misleading use should be avoided.
1995, *67*, 504; 1990, *62*, 2215

sorption (in colloid chemistry)
The process by which a substance (sorbate) is sorbed (adsorbed or absorbed) on or in another substance (sorbent).
See also *absorption, adsorption, desorption*.
1990, *62*, 2215; 1972, *31*, 584; see also 1993, *65*, 854

sorption isotherm (in ion exchange)
The concentration of a sorbed species in the ion exchanger, expressed as a function of its concentration in the external solution under specified conditions and at constant temperature.
1993, *65*, 854

sorption techniques (in trace analysis)
Techniques based on the distribution of components being separated between two phases, one of which is stationary and the other mobile. The advantage of some chromatographic methods is the possibility of combining the *preconcentration* and determination steps, as well as improving the speed of determination, and the possibility of separation of components with similar properties, and of achieving high values of the *preconcentration coefficient*. The methods enable analyses of microamounts of substances. *Ion-exchange chromatography* is not widely used owing to the great volumes of solutions being treated and, consequently, to a great degree to the variation in the blank, and to some diffusional limitations. Static ion-exchange, which is much simpler and more readily carried out, is practised quite widely.
1979, *51*, 1199

sorptive insertion (in surface catalysis)
This is analogous to the process of *ligand* insertion in coordination chemistry.

$$H^* + C_2H_4(g) \rightarrow {^*C_2H_5}.$$

This reaction might also be imagined to proceed by *adsorption* of C_2H_4 followed by ligand migration (an *associative surface reaction*).
1976, *46*, 84

source efficiency (of a radiation source)
See *self-absorption factor (of a radiation source)*.
1982, *54*, 1552

source, radioactive
See *radioactive source*.
1994, *66*, 2525

sp
See *torsion angle*.
1996, *68*, 2218

space charge (in a semiconductor)
At the surface region of a *semiconductor* the surface excess or deficiency of electrons constitutes a space charge.
1986, *58*, 443

space formula
Synonymous with *stereochemical formula*.
1996, *68*, 2218

space time (in catalysis)
The reciprocal of *space velocity*.
1976, *46*, 82

space velocity (in catalysis)
Defined as v_m, v_v, and v_a where the v_is represent the rate of feed of the given reactant fed per unit mass, volume or surface area of the catalyst.
1976, *46*, 82

spark (source) ionization (in mass spectrometry)
Occurs when a solid sample is vaporized and partially ionized by an intermittent discharge. It is recommended that the word 'source' be dropped from this term.
1991, *63*, 1548; O.B. 204

spatial-distribution interference (in flame spectroscopy)
Interference which may occur when changes in concentration of *concomitants* affect the mass flow rates or mass flow patterns of the analyte species in the flame. If they are caused by changes in volume and rise velocity of the gases formed by combustion, in extreme cases manifesting themselves by changes in the size and/or shape of the flame, they are non-specific and are called flame-geometry interferences.
O.B. 171

spatially resolving detector (of radiation)
A detector for the measurement of the *spatial distribution* of the radiation. Spatially resolving detectors can be divided into two groups:
1. the photosensitive area consists of a matrix of discrete photosensitive elements, the *pixels* (picture elements) forming an *array* with the facility to separately read out the information, simultaneously or sequentially.
2. the photosensitive area consists of a single photosensitive element that must be scanned (e.g. image dissection tube.)
1995, *67*, 1755

special salt effect
The initial steep rate increase observed in the *kinetic electrolyte effect* on certain solvolysis reactions, upon addition of some non-common ion salts, especially $LiClO_4$.
1994, *66*, 1165

speciation
Determination of the exact chemical form or compound in which an element occurs in a sample, for instance determination of whether arsenic occurs in the form of trivalent or pentavalent ions or as part of an organic molecule, and the quantitative distribution of the different chemical forms that may coexist.
1993, *65*, 2099

species (chemical)
See *chemical species*.
1994, *66*, 1165

species (taxonomic)
A taxonomic subdivision of a *genus*; a group of closely related, morphologically and physiologically similar individuals.
1992, *64*, 166

specific
Attribute to a physical quantity obtained by division by mass. Specific volume is the volume of a sample divided by its mass.
G.B. 7; see also 1996, *68*, 992

specific (in analysis)
A term which expresses qualitatively the extent to which other substances interfere with the determination of a substance according to a given procedure. Specific is considered to be the ultimate of *selective*, meaning that no *interferences* are supposed to occur.
1983, *55*, 555

specific acid–base catalysis
Catalysis by acids or bases in solution is said to be specific when the only observable catalytic effects are those due to the ions formed from the solvent itself (e.g. if when water is the solvent the only observable *catalysis* is that due to the H^+ and OH^- ions).
See *catalysis, general acid catalysis, general base catalysis*.
1996, *68*, 150

specific activity (in radiochemistry), *a*
For a specified *isotope*, or mixture of isotopes, the *activity* of a material divided by the mass of the material.
1994, *66*, 2515; 1996, *68*, 993

specific adsorption
Ions become specifically adsorbed when short-range interactions between them and the *interphase* become important. They are believed then to penetrate into the inner layer and may (but not necessarily) come into contact with the surface. They are usually assumed to form a partial or complete monolayer.
See *non-specific adsorption*.
1986, *58*, 448

specifically absorbing ion
Ions which possess a chemical affinity for the surface in addition to the Coulomb interaction, where chemical is a collective adjective, embracing all interactions other than purely Coulombic. Examples are van der Waals or hydrophobic bonding, pi-electron exchange and complex formation. Specifically adsorbing ions can adsorb on an initially uncharged surface and hence provide it with a charge.
The term specifically adsorbed applies to the sorption of all other ions having an affinity to the surface in addition to the purely Coulombic contribution.
See also *indifferent adsorbing ion*.
1991, *63*, 899

specifically labelled
An *isotopically labelled* compound is designated as specifically labelled when a unique *isotopically substituted* compound is formally added to the analogous *isotopically unmodified* compound. In such a case, both position(s) and number of each labelling *nuclide* are defined.
B.B. 514; 1981, *53*, 1891

specifically labelled tracer
A *tracer* in which the *label* is present in a specified position.
1994, *66*, 2526

specific burn-up
The total energy released through induced *nuclear transformations* divided by the mass of a nuclear fuel.
1982, *54*, 1537

specific catalysis
The acceleration of a reaction by a unique *catalyst*, rather than by a family of related substances. The term is most commonly used in connection with specific hydrogen-ion or hydroxide-ion (*lyonium ion* or *lyate ion*) catalysis.
See also *general acid catalysis, general base catalysis, pseudo-catalysis*.
1994, *66*, 1165

specific conductance
See *conductivity*.
1996, *68*, 993; G.B. 15

specific detector (in chromatography)
A detector which responds to a single sample component or to a limited number of components having similar chemical characteristics.
1993, *65*, 849

specific gravity
See *relative density*.
1990, *62*, 2215; 1979, *51*, 2472

specific heat capacity, *c*
Heat capacity divided by mass.
See *heat capacity*.
1996, *68*, 993

specific ionization (in nuclear chemistry)
The number of *ion pairs* formed per unit distance along the track of an ionizing *particle* passing through matter.
1982, *54*, 1552

specific permeability (in chromatography)
A term expressing the resistance of an empty tube or packed column to the flow of a fluid (the mobile phase).
1993, *65*, 833

specific photon emission
Synonymous with *photon exitance*.
1996, *68*, 2274

specific pore volume (of a catalyst)
The total internal void volume per unit mass of *adsorbent*.
1976, *46*, 80

specific retention volume (in chromatography)
See *retention volumes (in chromatography)*.
O.B. 103, 104

specific surface area (in surface chemistry)
When the area of the *interface* between two phases is proportional to the mass of one of the phases (e.g. for a solid *adsorbent*, for an *emulsion* or for an *aerosol*), the specific surface area (a, s or preferably a_s) is defined as the surface area divided by the mass of the relevant phase.
1972, *31*, 583; G.B. 63

specific volume, v
Volume of a substance divided by its mass. It is the reciprocal of *mass density*.
See *volume content*.
1996, *68*, 993

specific weight
See *mass density*.
1979, *51*, 2471

specimen (in analytical chemistry)
A specifically selected portion of a material taken from a dynamic system and assumed to be representative of the parent material at the time it is taken. Although the specimen may not be reproducible in time, e.g. it may be taken from a flowing stream or a portion of blood, no separable sampling error exists since this error is unavoidably included with the corresponding error of the estimate of the property, function or analyte being studied. A specimen may be considered as a special type of sample, taken primarily in time rather than in space.

The term 'specimen' has been used both as a representative unit and as a nonrepresentative (often better than most) unit of a population, usually in clinical, biological and mineralogical collections. 'Collections' in this case is used as either a noun or verb. This usage is almost always self-evident, and thus would not be confused with a time-type sample.
1990, *62*, 1202

spectator mechanism
A *pre-association* mechanism in which one of the *molecular entities*, C, is already present in an *encounter pair* with A during formation of B from A, but does not assist the formation of B, e.g.

$$\mathrm{A + C} \underset{}{\overset{\text{pre-association}}{\rightleftharpoons}} \underset{\text{encounter complex}}{\mathrm{(A \cdots C)}} \longrightarrow \underset{\text{encounter complex}}{\mathrm{(B \cdots C)}} \longrightarrow \text{product}$$

The formation of B from A may itself be a bimolecular reaction with some other reagent. Since C does not assist the formation of A, it is described as being present as a spectator, and hence such a mechanism is sometimes referred to as a spectator mechanism.
See also *microscopic diffusion control*.
1994, *66*, 1165

spectator-stripping reaction
An extreme type of *stripping reaction* in which one reaction product has almost the same direction and momentum as one of the reactant molecules had before the reactive collision occurred.
1996, *68*, 186

spectral bandwidth error (in spectrochemical analysis)
To measure the true shape, particularly the true maximum of an absorption band, the spectral bandwidth $\Delta\lambda$ of the instrument must be much less than the width of the absorption band. A spectral bandwidth error results from using too large a bandwidth relative to the absorption band being measured.
1988, *60*, 1456

spectral distribution
The variation of the *spectral radiance* with wavelength.
1988, *60*, 1453

spectral (photon) effectiveness
The reciprocal of the *photon fluence rate*, E_p^0, at wavelength λ, causing identical photoresponse, Δy, per unit time ($\Delta y/\Delta t$). The effectiveness spectrum is directly proportional to the conversion spectrum of the sensory pigment, if spectral attenuance is negligible.
1996, *68*, 2275

spectral intensity
See *spectral quantities*.
G.B. 31

spectral interference (in flame spectroscopy)
Interference due to the incomplete isolation of the *radiation* emitted or absorbed by the analyte from other radiation detected by the *instrument*. Its occurrence may be established by comparing the measures of the analyte-free *blank solution* and the *solvent blank*.
O.B. 169

spectral irradiance, E_λ
Irradiance, E, at wavelength λ per unit wavelength interval. The SI unit is W m^{-3}, but a commonly used unit is W m^{-2} nm^{-1}.
1996, *68*, 2275; see also 1990, *62*, 2198

spectral overlap
In the context of radiative energy transfer, the integral, $J = \int_0^\infty f_D'(\sigma) \varepsilon_A(\sigma) \, d\sigma$, which measures the overlap of the emission spectrum of the excited donor, D, and the absorption spectrum of the ground state acceptor, A; f_D' is the measured normalized emission

of D, $f_D' = f_D(\sigma)/\int_0^\infty f_D(\sigma)\,d\sigma$, $f_D(\sigma)$ is the photon exitance of the donor at wavenumber σ, and $\varepsilon_A(\sigma)$ is the decadic molar absorption coefficient of A at wavenumber σ.

In the context of *Förster excitation transfer*, J is given by:

$$J = \int_0^\infty [f_D'(\sigma)\,\varepsilon_A(\sigma)/\sigma^4]\,d\sigma$$

In the context of *Dexter excitation transfer*, J is given by:

$$J = \int_0^\infty f_D(\sigma)\,\varepsilon_A(\sigma)\,d\sigma$$

In this case f_D and ε_A, the emission spectrum of donor and absorption spectrum of acceptor, respectively, are both normalized to unity, so that the rate constant for energy transfer, k_{ET}, is independent of the oscillator strength of both transitions (contrast to Förster mechanism).

See *energy transfer*.
1996, 68, 2275

spectral photon exitance, $M_{p\lambda}$
The *photon exitance*, M_p, at wavelength λ per unit wavelength interval. The SI unit is $s^{-1}\,m^{-3}$, but a commonly used unit is $s^{-1}\,m^{-2}\,nm^{-1}$. Alternatively, the term can be used with the amount of photons (mol or its equivalent einstein) the SI unit then being mol $s^{-1}\,m^{-3}$ and the common unit mol $s^{-1}\,m^{-2}\,nm^{-1}$.
1996, 68, 2275

spectral photon flow, $\Phi_{p\lambda}$
The *photon flow*, Φ_p, at wavelength λ per unit wavelength interval. The SI unit is $s^{-1}\,m^{-1}$, but a commonly used unit is $s^{-1}\,nm^{-1}$. Alternatively, the term can be used with the amount of photons (mol or its equivalent einstein), the SI unit then being mol $s^{-1}\,m^{-1}$ and the common unit mol $s^{-1}\,nm^{-1}$.
1996, 68, 2275

spectral photon flux (photon irradiance), $E_{p\lambda}$
The *photon irradiance*, E_p, at wavelength λ per unit wavelength interval. The SI unit is $s^{-1}\,m^{-3}$, but a commonly used unit is $s^{-1}\,m^{-2}\,nm^{-1}$. Alternatively, the term can be used with the amount of photons (mol or its equivalent einstein), the SI unit then being mol $s^{-1}\,m^{-3}$ and the common unit mol $s^{-1}\,m^{-2}\,nm^{-1}$.
1996, 68, 2275

spectral photon radiance, $L_{p\lambda}$
The *photon radiance*, L_p, at wavelength λ per unit wavelength interval. The SI unit is $s^{-1}\,m^{-3}\,sr^{-1}$, but a commonly used unit is $s^{-1}\,m^{-2}\,sr^{-1}\,nm^{-1}$. Alternatively, the term can be used with the amount of photons (mol or its equivalent einstein) the SI unit then being mol $s^{-1}\,m^{-3}\,sr^{-1}$ and the common unit mol $s^{-1}\,m^{-2}\,sr^{-1}\,nm^{-1}$.
1996, 68, 2276

spectral quantities
Quantities characterizing electromagnetic radiation (radiant power, energy, energy density, intensity, exitance, radiance, irradiance, etc.) derived by differentiation with respect to wavelength, frequency or wavenumber, e.g. spectral (concentration of) irradiance is the derivative of the irradiance with respect to wavelength (E_λ), frequency (E_ν) or wavenumber ($E_{\tilde{\nu}}$).
G.B. 31

spectral radiance, L_λ
The *radiance*, L, at wavelength λ per unit wavelength interval. The SI unit is $W\,m^{-3}\,sr^{-1}$, but a commonly used unit is $W\,m^{-2}\,sr^{-1}\,nm^{-1}$.
1996, 68, 2276

spectral radiant excitance, M_λ
The *radiant exitance*, M, at wavelength λ per unit wavelength interval. The SI unit is $W\,m^{-3}$, but a commonly used unit is $W\,m^{-2}\,nm^{-1}$.
1996, 68, 2276

spectral radiant flux
Synonymous with *spectral radiant power*.
1996, 68, 2276

spectral radiant intensity, I_λ
The radiation intensity, I, at wavelength λ per unit wavelength interval. The SI unit is $W\,m^{-1}\,sr^{-1}$, but a commonly used unit is $W\,nm^{-1}\,sr^{-1}$.
1996, 68, 2276

spectral radiant power, P_λ
The *radiant power* at wavelength λ per unit wavelength interval. The SI unit is $W\,m^{-1}$, but a commonly used unit is $W\,nm^{-1}$.
1996, 68, 2276; see also 1985, 57, 112

spectral responsivity
The spectral output quantity of a system such as a photomultiplier, diode array, photoimaging device, or biological unit divided by the spectral irradiance $s(\lambda) = dy(\lambda)/dE(\lambda)$, simplified expression: $s(\lambda) = Y_\lambda/E_\lambda$, where Y_λ is the magnitude of the output signal for irradiation at wavelength λ and E_λ is the spectral irradiance of the parallel and perpendicular incident beam at the same wavelength.

See also *responsivity*.
1996, 68, 2276; 1995, 67, 1749

spectral responsivity function
See *responsivity*.
1995, 67, 1749

spectral sensitization
The process of increasing the spectral responsivity of a (photoimaging) system in a given wavelength region.
1996, 68, 2276

spectral spheradiance
Alternative term suggested for *actinic flux*.
1990, 62, 2215

spectrochemical buffer (in atomic spectroscopy)
Buffer added to samples and reference samples with the intention of making the measure of the analytical element less sensitive to changes in concentration of an *interferent*.
O.B. 159

spectrochemical carrier (in atomic spectroscopy)
An *additive* which gives rise to a gas which can help to transport the vapour of the sample material into the excitation region of the source, e.g. carbon in an air atmosphere when carbon dioxide is formed.
O.B. 160

spectrogram
A *spectrum* as recorded by a *spectrometer*.
1982, *54*, 1552

spectrograph
A combination of a spectral apparatus and a camera, which enables an image of a spectrum to be recorded. Spectra are recorded by a photographic emulsion or other means, e.g. two-dimensional electronic image sensors.
1995, *67*, 1729

spectrometer
A general term for describing a combination of spectral apparatus with one or more detectors to measure the intensity of one or more spectral bands.
See also *sequential spectrometer, simultaneous spectrometer, multiplex spectrometer, filter spectrometer*.
1995, *67*, 1729

spectrometer, anti-Compton γ-ray
See *anti-Compton γ-ray spectrometer*.
1982, *54*, 1552

spectrometer, crystal diffraction
See *crystal diffraction spectrometer*.
1982, *54*, 1552

spectrometer, filter
See *filter spectrometer*
1995, *67*, 1729

spectrometer, multiplex
See *multiplex spectrometer*
1995, *67*, 1729

spectrometer, α, β, γ-ray
See *α, β, γ-ray spectrometer*.
1982, *54*, 1552

spectrometer, scintillation
See *scintillation spectrometer*.
1982, *54*, 1552

spectrometer, sequential
See *sequential spectrometer*.
1995, *67*, 1729

spectrometer, simultaneous
See *simultaneous spectrometer*.
1995, *67*, 1729

spectrometry
See *spectroscopy*.
1986, *58*, 1738

spectroscope
A device which enables visual observation and evaluation of optical spectra (usually confined to the visible spectral region).
1995, *67*, 1729

spectroscopy
The study of physical systems by the electromagnetic radiation with which they interact or that thay produce.
Spectrometry is the measurement of such radiations as a means of obtaining information about the systems and their components. In certain types of optical spectroscopy, the radiation originates from an external source and is modified by the system, whereas in other types, the radiation originates within the system itself.
1986, *58*, 1738

spectrum analysis
The interpretation of the information present in an energy *spectrum* in terms of *radiation energy* and *intensity*.
1982, *54*, 1552

spectrum, measured
Synonymous with *spectrogram*.
1982, *54*, 1552

specular reflectance (reflection factor) (in optical spectroscopy), ρ
Radiant power specularly reflected from the surface of a system, P_{refl}, divided by the incident power, P_0:
$$\rho = P_{\text{refl}}/P_0$$
1985, *57*, 115

speed, v, c
Scalar quantity equal to the absolute value of the *velocity* vector.
G.B. 11; 1996, *68*, 993

speed distribution function, $F(c)$
Probability density of finding a particle with speed within an interval between c and $c + dc$.
G.B. 39

speed of light in a vacuum
Universal fundamental physical constant representing the speed of electromagnetic waves in vacuum; by definition c_0 is exactly = 299 792 458 m s^{-1}.
CODATA Bull., 1986, *63*, 1

spheradiance, spectral
See *spectral spheradiance*.
1990, *62*, 2215

spherical carbonaceous mesophase
A term which describes the morphology of *carbonaceous mesophase* which is formed in the isotropic *pitch* matrix. The spherical carbonaceous mesophase usually has a lamellar structure consisting of flat aromatic molecules arranged in parallel layers which are perpendicular to the sphere/isotropic phase interface. On coalescence, this spherical mesophase loses its

characteristic morphology and is converted to the *bulk mesophase*.
1995, *67*, 505

spherical radiance
Same as *radiant exitance*, M. It is the integration of the *radiant power*, P, leaving a source over the solid angle and over the whole wavelength range.
1996, *68*, 2276

spherical radiant exposure
Synonymous with *fluence*.
1996, *68*, 2276

spherulite
A polycrystalline, roughly spherical morphology consisting of lath, fibrous or lamellar crystals emanating from a common centre.
Space filling is achieved by branching, bending or both, of the constituent fibres or lamellae.
P.B. 83

spin adduct
See *spin trapping*.
1994, *66*, 1166

spin-allowed electronic transition
An electronic transition which does not involve a change in the spin part of the wavefunction.
1996, *68*, 2276

spin conservation rule (Wigner rule)
For both radiative and radiationless transitions, the principle that transitions between terms of the same multiplicity are spin-allowed, while transitions between terms of different multiplicity are spin-forbidden.
O.B. 185; 1996, *68*, 2277; 1984, *56*, 234

spin counting
See *spin trapping*.
1994, *66*, 1166

spin density
The unpaired *electron density* at a position of interest, usually at carbon, in a *radical*. It is often measured experimentally by electron paramagnetic resonance [EPR, ESR (electron spin resonance)] spectroscopy through hyperfine coupling constants of the atom or an attached hydrogen.
See also *radical centre*.
1994, *66*, 1166

spin-flip transition
A rotation of electron spins, above a critical magnetic field, H_c, in an antiferromagnet from parallel to largely perpendicular alignment, relative to an applied magnetic field, H_a, for $H_a > H_c$.
See also *simultaneous pair transitions*.
1994, *66*, 590

spin-flop transition
See *Morin transition*.
1994, *66*, 590

spin-glass transition
A *second-order transition* from a paramagnetic or ferromagnetic state to a spin-glass state in which spins from moment-bearing solute atoms become ordered randomly in a non-magnetic host such that the net magnetization of any region is zero.
Examples: Au-Fe, Cu-Mn and Mo-Fe
1994, *66*, 591

spin label
A *stable* paramagnetic *group* (typically a nitryl radical) that is attached to a part of a *molecular entity* whose microscopic environment is of interest and may be revealed by the electron spin resonance (ESR) spectrum of the spin label. When a simple paramagnetic molecular entity is used in this way <u>without</u> covalent attachment to the molecular entity of interest it is frequently referred to as a 'spin probe'.
1994, *66*, 1166

spinodal decomposition
A clustering reaction in a homogeneous, supersaturated solution (solid or liquid) which is unstable against infinitesimal fluctuations in density or composition. The solution therefore separates spontaneously into two phases, starting with small fluctuations and proceeding with a decrease in the Gibbs energy without a nucleation barrier.
1994, *66*, 591

spin-orbit coupling
The interaction of the electron spin magnetic moment with the magnetic moment due to the orbital motion of the electron. One consequence of spin-orbit coupling is the mixing of zero-order states of different multiplicity. This effect may result in fine structure called spin-orbit splitting.
1996, *68*, 2277

spin-orbit coupling constant, *A*
Coefficient in the spin-orbit coupling term in the *hamiltonian*.
G.B. 23

spin-orbit splitting
Removal of state degeneracy by spin-orbit coupling.
1996, *68*, 2277

spin-Peierls transition
A magneto-elastic transition that occurs in quasi one-dimensional antiferromagnetic materials when the magnetic free energy decrease due to the formation of singlet spin pairs outweighs the increase in lattice free energy occurring as a result of the dimerization of the regular array.
Example: The transition in tetrathiafulvalenium bis(dithiolene)cuprate at 12 K.
1994, *66*, 591

spin probe
See *spin label*.
1994, *66*, 1166

spin-spin coupling
The interaction between the spin magnetic moments of different electrons and/or nuclei. It causes, e.g. the multiplet pattern in nuclear magnetic resonance spectra.
1996, *68*, 2277

spin–spin coupling constant, J_{AB}
Coefficient of the indirect spin-spin coupling between two nuclei (A and B) in a magnetic resonance *hamiltonian*.
G.B. 15

spin-state transition
An electronic transition from a high-spin state to a low-spin state, or vice versa.
Example: With an increase in temperature Co^{3+} ions in $LaCoO_3$ transform from a low-spin state ($t_{2g}^6 e_g^0$) to the high-spin state ($t_{2g}^4 e_g^2$).
1994, *66*, 591

spin trapping
In certain reactions in solution a *transient radical* will interact with a diamagnetic reagent to form a more *persistent* radical. The product radical accumulates to a concentration where detection and, frequently, identification are possible by EPR/ESR spectroscopy. The key reaction is usually one of *attachment*; the diamagnetic reagent is said to be a 'spin trap' and the persistent product radical is then the 'spin *adduct*'. The procedure is referred to as spin trapping, and is used for monitoring reactions involving the intermediacy of *reactive* radicals at concentrations too low for direct observation. Typical spin traps are *C*-nitroso compounds and nitrones, to which reactive radicals will rapidly add to form nitryl radicals. A quantitative development, in which essentially all reactive radicals generated in a particular system are intercepted, has been referred to as 'spin counting'. Spin trapping has also been adapted to the interception of radicals generated in both gaseous and solid phases. In these cases the spin adduct is in practice transferred to a liquid solution for observation in order to facilitate interpretation of the EPR/ESR spectra of the radicals obtained.
1994, *66*, 1166

spiro atom
See *spiro union*.
B.B. 37

spiro chain (in a polymer)
A *double-strand chain* consisting of an uninterrupted sequence of rings, with adjacent rings having only one atom in common.
Alternatively, a spiro chain is a double-strand chain with adjacent *constitutional units* joined to each other through three atoms, two on one side and one on the other side of each constitutional unit.
1996, *68*, 2295

spiro compounds
Compounds having one atom (usually a quaternary carbon) as the only common member of two rings, e.g.

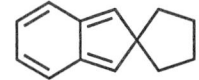

spiro[cyclopentane-1,2'-indene]

spiro[4.4]nonane

1995, *67*, 1366

spiro macromolecule
A *double-strand macromolecule* consisting of an uninterrupted sequence of rings, with adjacent rings having only one atom in common.
Alternatively, a spiro macromolecule is a double-strand macromolecule with *adjacent constitutional units* joined to each other through three atoms, two on one side and one on the other side of each constitutional unit.
1996, *68*, 2295

spiro polymer
See *double-strand polymer*.
1996, *68*, 2304

spiro union
A union formed by a single atom which is the only common member of two rings. The common atom is designated as the spiro atom.
A free spiro union is one which constitutes the only union direct or indirect between the two rings.
B.B. 37

splicing
1. (of RNA): The procedure by which *introns* are removed from eukaryotic precursor mRNA molecules and adjacent *exon* sequences are joined together (spliced).
2. (of DNA): Manipulation for joining together double-stranded DNA fragments with protruding single-stranded '*sticky ends*' by means of *ligases*.
1992, *64*, 166

spontaneous emission
That mode of *emission* which occurs even in the absence of a perturbing external electromagnetic field. The transition between states, n and m, is governed by the Einstein coefficient of spontaneous emission, A_{nm}.
See also *stimulated emission*.
1996, *68*, 2277

spontaneous fission
Nuclear *fission* which occurs without the addition of *particles* or energy to the *nucleus*.
1982, *54*, 1542

spontaneous resolution
See *racemic conglomerate*.
1996, *68*, 2218

spot (in chromatography)
A zone in paper and thin-layer chromatography of approximately circular appearance.
1993, *65*, 834; O.B. 96

spot sampling
See *instantaneous (spot) sampling*.
1990, *62*, 2197

sprayer (in flame spectroscopy)
The particular part of a nebulizer where the aspirated liquid is disrupted by the gas-jet into a spray.
O.B. 165

spreading function (in chromatography)
The normalized signal produced, as a function of elution volume at the outlet of a size-exclusion chromatography set-up, by an instantaneous injection of a uniform sample.
P.B. 69

spreading wetting
A process in which a drop of liquid spreads over a solid or liquid substrate.
1972, *31*, 597

spread monolayer
It may happen that a component is virtually insoluble in both of two adjoining phases, but is present as a monolayer between them. Such a layer can be produced by spreading.
1972, *31*, 591

sputtering
Removal of surface material (atoms, clusters and molecules) by particle bombardment.
1979, *51*, 2247

sputter yield
The number of particles sputtered from the surface of a target per primary ion.
1979, *51*, 2247

square-wave current
In square-wave *polarography*, the component of the current that is associated with the presence of a substance B. This component may be either faradaic (if B is electroactive) or non-faradaic (if B is surface-active).
1985, *57*, 1499

SR
See *racemate*.
1996, *68*, 2218

stability, static
See *static stability*.
1990, *62*, 2215

stable
As applied to *chemical species*, the term expresses a thermodynamic property, which is quantitatively measured by relative molar standard Gibbs energies. A chemical species A is more stable than its isomer B if $\Delta_r G^o > 0$ for the (real or hypothetical) reaction A → B, under standard conditions. If for the two reactions:

$$P \rightarrow X + Y \quad (\Delta_r G_1^o)$$
$$Q \rightarrow X + Z \quad (\Delta_r G_2^o)$$

$\Delta_r G_1^o > \Delta_r G_2^o$, P is more stable relative to the product Y than is Q relative to Z. Both in qualitative and quantitative usage the term stable is therefore always used in reference to some explicitly stated or implicitly assumed standard.

The term should not be used as a synonym for *unreactive* or 'less reactive' since this confuses thermodynamics and kinetics. A relatively more stable chemical species may be more *reactive* than some reference species towards a given reaction partner.
See also *inert, unstable*.
1994, *66*, 1166

stable film, metastable film
A film often thins gradually to a thickness at which it either ruptures or converts abruptly to an *equilibrium film*. Rupture under these conditions characterizes unstable films, whereas transition to an equilibrium film characterizes (meta)stable films.
1972, *31*, 614

stable ion (in mass spectrometry)
An ion which is not sufficiently excited to dissociate spontaneously into a daughter ion and associated neutral fragment(s) or to react further in any other way within the time scale of the experiment, e.g. until hitting the detector.
1991, *63*, 1550

staggered conformation
The *conformation* of groups attached to two adjacent atoms if the *torsion angles* are such that the groups are as far away as possible from an *eclipsed* arrangement.
1996, *68*, 2218

stagnant inversion (in atmospheric chemistry)
See *temperature inversion*.
1990, *62*, 2215

standard acceleration of free fall
Acceleration defined as $g_n = 9.806\ 65$ m s^{-2}.
CODATA Bull., 1986, *63*, 1

standard atmosphere
Pressure defined as 101 325 Pa and used as unit of pressure with the symbol atm.
CODATA Bull., 1986, *63*, 1; 1996, *68*, 993

standard atomic weights
Recommended values of *relative atomic masses* of the elements revised biennially by the IUPAC Commission on Atomic Weights and Isotopic Abundances and applicable to elements in any normal sample with a high level of confidence. A normal sample is any reasonably possible source of the element or its compounds in commerce for industry and science and has not been subject to significant modification of isotopic composition within a geologically brief period.
G.B. 94

standard chemical potential
Of substance B at temperature T, $\mu_B^o(T)$, is the value of the chemical potential under specified standard conditions (in the standard state).
G.B. 53; see also 1994, *66*, 536

standard concentration
Chosen value of amount concentration denoted by $c^{\ominus}$ or c^o and usually equal to 1 mol dm^{-3}.
See also *standard molality*.
G.B. 54

standard conditions for gases
Temperature, 273.15 K (°C) and pressure of 10^5 pascals. IUPAC recommends that the former use of the pressure of 1 atm as standard pressure (equivalent to 1.01325×10^5 Pa) should be discontinued.
 See *STP*.
 1990, *62*, 2216

standard deviation, *s*
The positive square root of the sum of the squares of the deviations between the observations and the mean of the series, divided by one less than the total number in the series. The standard deviation is the positive square root of the variance, a more fundamental statistical quantity.
 O.B. 5; see also 1994, *66*, 599; 1990, *62*, 2216

standard electrode potential, E°
The value of the standard emf of a cell in which molecular hydrogen under *standard pressure* is oxidized to solvated protons at the left-hand electrode.
 G.B. 59

standard electromotive force
Quantity defined by $E^\circ = -\Delta_r G^\circ / nF = (RT/nF) \ln K^\circ$, where $\Delta_r G^\circ$ is the standard Gibbs energy of the cell reaction in the direction in which reduction occurs at the right-hand electrode in the diagram representing the cell ('reduction at right'), K° is the standard equilibrium constant for this reaction, n its charge number, F the Faraday constant, R the gas constant and T the thermodynamic temperature.
 G.B. 58

standard entropy of activation, $\Delta^\ddagger S^\circ, \Delta^\ddagger S$
Entropy change associated with the *activation reaction*.
 See *standard reaction quantities*.
 G.B. 57

standard equilibrium constant, K°, K
Quantity defined by $K^\circ = \exp(-\Delta_r G^\circ / RT)$, where $\Delta_r G^\circ$ is the standard reaction Gibbs energy, R the gas constant and T the thermodynamic temperature. Some chemists prefer the name thermodynamic equilibrium constant and the symbol K.
 G.B. 50; see also 1994, *66*, 548

standard Gibbs energy of activation, $\Delta^\ddagger G^\circ, \Delta^\ddagger G$
Standard Gibbs energy change associated with the *activation reaction*.
 See *standard reaction quantities*.
 G.B. 57

standard hydrogen electrode
For solutions in protic solvents, the universal *reference electrode* for which, under standard conditions, the standard electrode potential (H$^+$/H$_2$) is zero at all temperatures.
The *absolute electrode potential* of the hydrogen electrode under standard conditions can be expressed in terms of thermodynamic quantities by applying a suitable Born–Haber cycle, thus:

$$E^\circ(\text{H}^+/\text{H}_2)(\text{abs}) = (\Delta_{at}G^\circ + \Delta_{ion}G^\circ + \alpha_{\text{H}^+}^{\circ, S}/F)$$

where $\Delta_{at}G^\circ$ and $\Delta_{ion}G^\circ$ are the atomization and ionization Gibbs energies of H$_2$, $\alpha_{\text{H}^+}^{\circ, S}$ is the real potential of H$_2$ in solvent S and F is the Faraday constant.
The recommended *absolute electrode potential* of the hydrogen electrode is:

$$E^\circ (\text{H}^+/\text{H}_2)^{\text{H}_2\text{O}} (\text{abs}) = (4.44 \pm 0.02) \text{ V}$$
$$\text{at } 298.15 \text{ K}$$

 1986, *58*, 957

standard molality, $m^\ominus, m^\circ$
Chosen value of molality and usually equal to 1 mol kg^{-1}.
 G.B. 54; see also 1994, *66*, 538

standard potential of an electrode reaction
Synonymous with *standard electrode potential*.
 G.B. 59

standard potential of the reaction in a chemical cell
Synonymous with *standard electromotive force*.
 G.B. 58

standard pressure
Chosen value of pressure denoted by $p^\ominus$ or p°. In 1982 IUPAC recommended the value 10^5 Pa, but prior to 1982 the value 101 325 Pa (= 1 atm) was usually used.
 G.B. 54; 1994, *66*, 536

standard rate constant of an electrode reaction
 See *electrode reaction rate constants*.
 1974, *37*, 516

standard reaction quantities
Infinitesimal changes in thermodynamic functions with extent of reaction divided by the infinitesimal increase in the extent when all the reactants and products are in their standard states. For the quantity X they should be denoted by $\Delta_r X^\circ$, but usually only ΔX° is used. For specific types of reactions the subscript r is replaced by: f for formation, c for combustion, a for atomization and superscript $\ddagger$ for activation.
 G.B. 52

standard solution
A solution of accurately known concentration, prepared using standard substances in one of several ways. A primary standard is a substance of known high purity which may be dissolved in a known volume of solvent to give a primary standard solution. If stoichiometry is used to establish the strength of a titrant, it is called a secondary standard solution. The term secondary standard can also be applied to a substance whose active agent contents have been found by comparison against a primary standard. Concentrations of standard solutions may be expressed in kmol m^{-3}, mol dm^{-3}, or in terms more closely related to those used in specific titrations (as titres).
 O.B. 48

standard state
State of a system chosen as standard for reference by convention. Three standard states are recognized:

For a gas phase it is the (hypothetical) state of the pure substance in the gaseous phase at the *standard pressure* $p = p°$, assuming ideal behaviour.

For a pure phase, or a mixture, or a solvent in the liquid or solid state it is the state of the pure substance in the liquid or solid phase at the standard pressure $p = p°$.

For a solute in solution it is the (hypothetical) state of solute at the standard molality $m°$, standard pressure $p°$ or standard concentration $c°$ and exhibiting infinitely dilute solution behaviour.

For a pure substance the concept of standard state applies to the substance in a well defined state of aggregation at a well defined but arbitrarily chosen *standard pressure*.

G.B. 53

standard subtraction method (in electroanalytical chemistry)
A variation of the *standard addition method*. In this procedure changes in the potential resulting from the addition of a known amount of a species which reacts stoichiometrically with the ion of interest (e.g. a complexing agent) are employed to determine the original activity or concentration of the ion.

O.B. 171

standard thermodynamic quantities
Values of thermodynamic functions in the *standard state* characterized by a standard pressure, molality or amount concentration, but not by temperature. Standard quantities are denoted by adding the superscript ⊖ or ° to the symbol of the quantity.

G.B. 53; see also 1994, *66*, 536

standard uncertainty
Uncertainty of measurement expressed as one standard deviation.

VIM

stand-by state (in analysis)
The condition of an instrument where the analytical procedure can begin immediately.
See *shut-down state, start-up time*.
1989, *61*, 1664

stannoxanes
Compounds having the structure $H_3S_n[OSnH_2]_n$-$OSnH_3$. Thus tin analogues of *siloxanes*.
1995, *67*, 1366

stannylenes
See *stannylidenes*.
1995, *67*, 1366

stannylidenes
Carbene analogues having the structure R_2Sn:. The older synonymous term stannylenes is no longer recommended.
1995, *67*, 1366

star copolymer
A *copolymer* where chains having different constitutional or configurational features are linked through a central moiety.
1985, *57*, 1435

Stark effect
The splitting or shift of spectral lines in an electric field. Also called electrochromic effect.
1996, *68*, 2277

star macromolecule
A *macromolecule* containing a single *branch point* from which *linear chains* (arms) emanate.
Notes:
1. A star macromolecule with n linear chains (arms) attached to the branch point is termed an n-*star macromolecule*, e.g. five-star macromolecule.
2. If the arms of a star macromolecule are identical with respect to constitution and degree of polymerization, the macromolecule is termed a *regular star macromolecule*.
3. If different arms of a star macromolecule are composed of different monomeric units, the macromolecule is termed a *variegated star macromolecule*.
1996, *68*, 2296

star polymer
A polymer composed of *star macromolecules*.
1996, *68*, 2304

starting line (in chromatography)
The point or line on a chromatographic layer where the substance to be chromatographed is applied.
O.B. 77

starting point (in chromatography)
See *starting line (in chromatography)*.
O.B. 77

start-up time (in analysis)
The time interval between turn-on of the instrument and its stand-by state.
See *shut-down state, stand-by state*.
1989, *61*, 1664

state crossing
See *avoided crossing, surface crossing*.
1996, *68*, 2277

state diagram
See *Jablonski diagram*.
1996, *68*, 2277

state function
See *wavefunction*.
G.B. 16

state-to-state kinetics
A branch of chemical kinetics concerned with the dynamics of reactions in which the reactant species are in known quantum states, and in which the quantum states of the products are determined.
1996, *68*, 186

static fields mass spectrometer
An instrument which can separate selected ion beams with fields which do not vary with time. The fields are generally both electric and magnetic.
1991, *63*, 1546; O.B. 202

static pressure
The pressure of a fluid at rest or in motion exerted perpendicularly to the direction of flow.
1990, *62*, 2208

static quenching
See *quenching*.
1996, *68*, 2277

static stability
In meteorology, the stability of the atmosphere in the vertical direction to vertical displacements. If Γ is the dry adiabatic lapse rate, dry air is stable if the lapse rate $(-dT/dz) < \Gamma$, unstable if $(-dT/dz) > \Gamma$ and neutral if $(-dT/dz) = \Gamma$.
See *lapse rate, inversion*.
1990, *62*, 2215

stationary phase (fermentation)
The phase of a culture of *microorganisms* or animal and plant cells cultured *in vitro* that follows the *exponential growth phase* and in which there is little or no growth. In some cases it is a phase of product formation, e.g. formation of *secondary metabolites*.
1992, *64*, 166

stationary phase (in chromatography)
One of the two phases forming a chromatographic system. It may be a solid, a gel or a liquid. If a liquid, it may be distributed on a solid. This solid may or may not contribute to the separation process. The liquid may also be chemically bonded to the solid (bonded phase) or immobilized onto it (immobilized phase).
The expression chromatographic bed or sorbent may be used as a general term to denote any of the different forms in which the stationary phase is used.
Particularly in gas chromatography where the stationary phase is most often a liquid, the term liquid phase is used for it as compared to the gas phase, i.e. the mobile phase. However, particularly in the early development of liquid chromatography, the term 'liquid phase' had also been used to characterize the mobile phase as compared to the 'solid phase' i.e. the stationary phase. Due to this ambiguity, the use of the term 'liquid phase' is discouraged. If the physical state of the stationary phase is to be expressed, the use of the adjective forms such as liquid stationary phase and solid stationary phase, bonded phase or immobilized phase is proposed.
1993, *65*, 823; see also O.B. 98

stationary-phase fraction
The volume of the stationary phase per unit volume of a packed column.
O.B. 100

stationary phase, mass (weight) of the, W_s
See *mass (weight) of the stationary phase (in chromatography)*.
1993, *65*, 833

stationary phase volume (in chromatography), V_s
The volume of the liquid stationary phase or the active solid in the column. The volume of any solid support is not included. In the case of partition chromatography with a liquid stationary phase, it is identical to the liquid phase volume.
1993, *65*, 833

stationary state
1. (in quantum mechanics): A state that does not evolve with time.
2. (in kinetics): See *steady state*.
1994, *66*, 1166

statistical coil (in polymers)
See *random coil*.
P.B. 50

statistical copolymer
A *copolymer* consisting of *macromolecules* in which the sequential distribution of the *monomeric units* obeys known statistical laws.
Note:
An example of a statistical copolymer is one consisting of macromolecules in which the sequential distribution of monomeric units follows Markovian statistics.
1996, *68*, 2301

statistical copolymerization
A *copolymerization* in which a *statistical copolymer* is formed.
1996, *68*, 2307

statistical factor
See *reaction path degeneracy*.
1996, *68*, 187

statistical pseudo-copolymer
See *pseudo-copolymer*.
1996, *68*, 2300

statistical segment (in polymers)
A segment of an actual polymer chain which behaves, with respect to some property, virtually as a segment of a freely jointed chain.
P.B. 50

statistical weight
See *degeneracy*.
G.B. 39

steady state (in liquid-liquid distribution)
The state of a continuous process when it is operating in such a way that the concentration of solutes in exit streams remains constant with respect to time for constant feed concentrations, even though the two phases are not necessarily in thermodynamic equilibrium in any part of the process. The term equilibrium should not be used to describe this situation.
1993, *65*, 2390

steady state (stationary state)
1. In a kinetic analysis of a complex reaction involving *unstable* intermediates in low concentration, the rate of change of each such *intermediate* is set equal to zero, so that the *rate equation* can be expressed as a function of the concentrations of *chemical species* present in macroscopic amounts. For example, assume that X is an unstable intermediate in the reaction sequence:

$$A \underset{k_{-1}}{\overset{k_1}{\rightleftharpoons}} X$$

$$X + C \xrightarrow{k_2} D$$

Conservation of mass requires that:
$$[A] + [X] + [D] = [A]_0$$
which, since $[A]_0$ is constant, implies:
$$-d[X]/dt = d[A]/dt + d[D]/dt.$$
Since [X] is negligibly small, the rate of formation of D is essentially equal to the rate of disappearance of A, and the rate of change of [X] can be set equal to zero. Applying the steady state approximation ($d[X]/dt = 0$) allows the elimination of [X] from the kinetic equations, whereupon the rate of reaction is expressed:

$$d[D]/dt = -d[A]/dt = \frac{k_1 k_2 [A][C]}{k_{-1} + k_2 [C]} \quad (1)$$

Note:
1. The steady-state approximation does not imply that [X] is even approximately constant, only that its absolute rate of change is very much smaller than that of [A] and [D]. Since according to the reaction scheme $d[D]/dt = k_2[X][C]$, the assumption that [X] is constant would lead, for the case in which C is in large excess, to the absurd conclusion that formation of the product D will continue at a constant rate even after the reactant A has been consumed.
2. In a stirred flow reactor a steady state implies a regime so that all concentrations are independent of time.
1994, *66*, 1166; 1993, *65*, 2298; see also 1996, *68*, 187; 1990, *62*, 2216

steady state approximation (treatment)
See *steady state (stationary state)*.
1994, *66*, 1166

Stefan–Boltzmann constant
Fundamental physical constant $\sigma = 2\pi^5 k^4/15h^3 c_0^2 = 5.670\,51\,(19) \times 10^{-8}$ W m^{-2} K^{-4}.
CODATA Bull., 1986, *63*, 1

stem (in polymer crystals)
A crystallized, rodlike portion of a polymer chain connected to non-rodlike portions, or chain ends, or both.
P.B. 83

step (in chromatography)
The portion of an integral chromatogram recording the amount of a component, or the corresponding change in the signal from the detector as the component emerges from the column.
O.B. 96

step height (in chromatography)
The distance perpendicular to the time or volume axis, through which the *base line* moves as a result of a *step* on an integral *chromatogram*.
O.B. 96

stepped surface
See *extent of an interface (surface)*.
1986, *58*, 439

stepwise elution (in chromatography)
A elution process in which the composition of the mobile phase is changed in steps during a single chromatographic run.
1993, *65*, 827; O.B. 92

stepwise reaction
A *chemical reaction* with at least one reaction *intermediate* and involving at least two consecutive *elementary reactions*.
See also *composite reaction*.
1994, *66*, 1167; 1996, *68*, 187

steradian
SI derived unit of solid angle, sr = 1.
G.B. 72; see also 1996, *68*, 993

stereoblock macromolecule
A *block macromolecule* composed of *stereoregular*, and possibly *non-stereoregular*, blocks.
1996, *68*, 2293

stereoblock polymer
A polymer composed of *stereoblock macromolecules*.
1996, *68*, 2303

stereochemical formula (stereoformula)
A three-dimensional view of a molecule either as such or in a projection.
See *projection formula*.
1996, *68*, 2218

stereoconvergence
The predominant formation of the same *stereoisomer* or stereoisomer mixture of a reaction product when two different stereoisomers of the reactant are used in the same reaction. When that product involved in the reaction is one *enantiomer* the result has been called enantioconvergence.
1996, *68*, 2218

stereodescriptor
A prefix to specify *configuration* (absolute or relative) or *conformation*. For example *R*, *S*; *r*, *s*; *P*, *M*; *Re*, *Si*; *E*, *Z*; *ap*, *sp*; etc.
1996, *68*, 2218

stereoelectronic
Pertaining to the dependence of the properties (especially the energy) of a *molecular entity* in a particular electronic state (or of a *transition state*) on relative nuclear geometry. The electronic ground state is usually considered, but the term can apply to excited states as well. Stereoelectronic effects arise from the

different alignment of electronic orbitals in different arrangements of nuclear geometry.
1996, *68*, 2218; see also 1994, *66*, 1167

stereoelectronic control
Control of the nature of the products of a chemical reaction (or of its rate) by *stereoelectronic* factors. The term is usually applied in the framework of an orbital approximation. The variations of *molecular orbital* energies with relative nuclear geometry (along a *reaction coordinate*) are then seen as consequences of variations in basis-orbital overlaps.
1996, *68*, 2218; see also 1994, *66*, 1167

stereoformula
See *projection formula*.
1996, *68*, 2218

stereogenic unit (stereogen/stereoelement)
A grouping within a molecular entity that may be considered a focus of stereoisomerism. At least one of these must be present in every enantiomer (though the presence of stereogenic units does not conversely require the corresponding chemical species to be chiral). Three basic types are recognized for molecular entities involving atoms having not more than four substituents:
(a) A grouping of atoms consisting of a central atom and distinguishable ligands, such that the interchange of any two of the substituents leads to a stereoisomer. An *asymmetric atom* (*chirality centre*) is the traditional example of this stereogenic unit.
(b) A chain of four non-coplanar atoms (or rigid groups) in a stable conformation, such that an imaginary or real (restricted) rotation (with a change of sign of the torsion angle) about the central bond leads to a stereoisomer.
(c) A grouping of atoms consisting of a double bond with substituents which give rise to *cis-trans isomerism*.
1996, *68*, 2219

stereoheterotopic
Either *enantiotopic* or *diastereotopic*. (In contrast the term constitutionally heterotopic has been used to describe identical groups located in constitutionally non-equivalent positions).
See also *pro-R, pro-S*.
1996, *68*, 2219

stereohomosequence (in a polymer)
A *configurational homosequence* in which the *relative* or *absolute configuration* is defined at all sites of stereoisomerism in the main chain of a *polymer* molecule.
P.B. 36

stereoisomerism
Isomerism due to differences in the spatial arrangement of atoms without any differences in connectivity or bond multiplicity between the isomers.
1996, *68*, 2219

stereoisomers
Isomers that possess identical *constitution*, but which differ in the arrangement of their atoms in space.
See *enantiomer, diastereoisomer, cis-trans isomers*.
1996, *68*, 2219

stereomutation
A change of *configuration* at a *stereogenic unit* brought about by physical or chemical means.
See *epimerization*.
1996, *68*, 2219

stereoregular macromolecule
A *regular macromolecule* essentially comprising only one species of *stereorepeating unit*.
1996, *68*, 2292

stereoregular polymer
A polymer composed of *stereoregular macromolecules*.
1996, *68*, 2302

stereorepeating unit (in a polymer)
A *configurational repeating unit* having defined configuration at all sites of stereoisomerism in the *main chain* of a *regular macromolecule*, a *regular oligomer molecule*, a *regular block* or a *regular chain*.
1996, *68*, 2291

stereoselective polymerization
Polymerization in which a polymer molecule is formed from a mixture of stereoisomeric monomer molecules by incorporation of only one stereoisomeric species.
P.B. 19

stereoselective synthesis
A chemical reaction (or reaction sequence) in which one or more new elements of chirality are formed in a substrate molecule and which produces the stereoisomeric (enantiomeric or diastereoisomeric) products in unequal amounts. Traditionally called *asymmetric synthesis*.
See also *stereoselectivity*.
1996, *68*, 2219

stereoselectivity
The preferential formation in a chemical reaction of one stereoisomer over another. When the stereoisomers are enantiomers, the phenomenon is called enantioselectivity and is quantitatively expressed by the *enantiomer excess;* when they are diastereoisomers, it is called diastereoselectivity and is quantitatively expressed by the *diastereoisomer excess*.
1996, *68*, 2219; see also 1994, *66*, 1167

stereosequence (in a polymer)
A *configurational sequence* in which the *relative* or *absolute configuration* is defined at all sites of stereoisomerism in the main chain of a *polymer* molecule.
P.B. 36

stereospecifically labelled tracer
A *tracer* in which the *label* is present in a *stereospecific* position.
1994, *66*, 2526

stereospecificity (stereospecific)
1. A reaction is termed stereospecific if starting materials differing only in their *configuration* are converted into stereoisomeric products. According to this definition, a stereospecific process is necessarily *stereoselective* but not all stereoselective processes are stereospecific. Stereospecificity may be total (100%) or partial. The term is also applied to situations where reaction can be performed with only one stereoisomer. For example, the exclusive formation of *trans*-1,2-dibromocyclohexane upon bromination of cyclohexene is a stereospecific process, although the analogous reaction with (*E*)-cyclohexene has not been performed.
2. The term has also been applied to describe a reaction of very high stereoselectivity, but this usage is unnecessary and is discouraged.
  1994, *66*, 1167; 1996, *68*, 2219

stereospecific polymerization
Polymerization in which a *tactic polymer* is formed. However, polymerization in which stereoisomerism present in the *monomer* is merely retained in the *polymer* is not to be regarded as *stereospecific*. For example, the polymerization of a chiral monomer, e.g. D-propylene oxide (D-methyloxirane), with retention of *configuration* is not considered to be a stereospecific reaction; however, selective polymerization, with retention, of one of the *enantiomers* present in a mixture of D- and L-propylene oxide molecules is so classified.
  P.B. 19, 32

steric-approach control
Control of *stereoselectivity* of a reaction by steric hindrance towards attack of the reagent, which is directed to the less hindered face of the molecule. Partial bond making is strong enough at the *transition state* for steric control to take place. This suggests that the transition state should not be close to products.
  See also *product development control*.
  1994, *66*, 1168

steric effect
The effect on a chemical or physical property (structure, rate or equilibrium constant) upon introduction of *substituents* having different steric requirements. The steric effect in a reaction is ascribed to the difference in steric energy between, on the one hand, reactants and, on the other hand, a *transition state* (or products). A steric effect on a rate process may result in a rate increase ('steric acceleration') or a decrease ('steric retardation'). (The adjective 'steric' is not to be confused with stereochemical.)
Steric effects arise from contributions ascribed to *strain* as the sum of (1) non-bonded repulsions, (2) bond angle strain and (3) bond stretches or compressions.
For the purpose of *correlation analysis* or *linear free-energy relations* various scales of steric parameters have been proposed, notably *A* values, Taft's E_s and Charton's *v* scales.

In a reactant molecule RY and an appropriate reference molecule R°Y, the 'primary steric effect' of R is the direct result of differences in compressions which occur because R differs from R° in the vicinity of the reaction centre Y. A 'secondary steric effect' involves the differential moderation of electron delocalization by non-bonded compressions.
Some authors make a distinction between 'steric' effects attributed to van der Waals repulsions alone, and 'strain' effects, attributed to deviations of bond angles from 'ideal' values.
  See *Taft equation, van der Waals forces*.
  1994, *66*, 1168

steric factor
A factor introduced into simple versions of the *collision theory* of reactions to take care of the fact that the reaction probability depends on the certain mutual orientations of the reactant molecules.
  See also *collision frequency*.
  1996, *68*, 187; see also 1990, *62*, 2216

steric factor (in polymers)
The ratio, σ, of the *root-mean-square end-to-end distance* of a polymer chain with unperturbed dimensions, $\langle r^2 \rangle_0^{1/2}$, to that of a *freely rotating chain* with the same structure, $\langle r^2 \rangle_{0,f}^{1/2}$, i.e. $(\langle r^2 \rangle_0 / \langle r^2 \rangle_{0,f})^{1/2}$, in the limit of infinite chain length.
The steric factor reflects the effect of hindrance to free rotation.
  P.B. 50

steric hindrance
The original term for a *steric effect* arising from crowding of *substituents*.
  1994, *66*, 1168

steric isotope effect
A secondary isotope effect attributed to the different vibrational amplitudes of *isotopologues*. For example, both the mean and mean-square amplitudes of vibrations associated with C–H bonds are greater than those of C–D bonds. The greater effective bulk of molecules containing the former may be manifested by a *steric effect* on a rate or equilibrium constant.
  1994, *66*, 1131

steric strain
  1. See *strain*.
  1994, *66*, 1077
  2. See *angle strain*.
  3. See *eclipsing strain*.
  4. See *transannular strain*.
  1996, *68*, 2219

Stern layer
Counter and *co-ions* in immediate contact with a surface are said to be located in the Stern layer, and form with the fixed charge a molecular capacitor.
  1972, *31*, 618

Stern–Volmer kinetic relationships
This term applies broadly to variations of *quantum yields* of photophysical processes (e.g. *fluorescence* or *phosphorescence*) or *photochemical reactions* (usually reaction *quantum yield*) with the concentra-

tion of a given reagent which may be a substrate or a *quencher*. In the simplest case, a plot of Φ^0/Φ (or M^0/M for emission) vs. concentration of quencher, [Q], is linear obeying the equation:

$$\Phi^0/\Phi \text{ or } M^0/M = 1 + K_{sv}[Q] \quad (1)$$

In equation (1) K_{sv} is referred to as the Stern–Volmer constant. Equation (1) applies when a quencher inhibits either a photochemical reaction or a photophysical process by a single reaction. Φ^0 and M^0 are the quantum yield and emission intensity (*radiant exitance*), respectively, in the absence of the quencher Q, while Φ and M are the same quantities in the presence of the different concentrations of Q. In the case of dynamic quenching the constant K_{sv} is the product of the true *quenching* constant k_q and the *excited state lifetime*, τ^0, in the absence of quencher. k_q is the bimolecular reaction rate constant for the elementary reaction of the excited state with the particular quencher Q. Equation (1) can therefore be replaced by the expression (2):

$$\Phi^0/\Phi \text{ or } M^0/M = 1 + k_q\tau^0[Q] \quad (2)$$

When an excited state undergoes a bimolecular reaction with rate constant k_r to form a product, a double-reciprocal relationship is observed according to the equation:

$$1/\Phi_p = (1 + 1/k_r \tau^0[S]) [1/(A \cdot B)]$$

where Φ_p is the quantum efficiency of product formation, A the efficiency of forming the reactive excited state, B the fraction of reactions of the excited state with substrate S which leads to product, and [S] is the concentration of reactive ground-state substrate. The intercept/slope ratio gives $k_r\tau^0$. If [S] = [Q], and if a photophysical process is monitored, plots of equations (2) and (3) should provide independent determinations of the product-forming rate constant k_r. When the lifetime of an excited state is observed as a function of the concentration of S or Q, a linear relationship should be observed according to the equation:

$$\tau^0/\tau = 1 + k_q \tau^0 [Q]$$

where τ^0 is the lifetime of the excited state in the absence of the quencher Q.
See also *self-quenching*.
1996, 68, 2277

steroids
Naturally occurring compounds and synthetic analogues, based on the cyclopenta[*a*]phenanthrene carbon skeleton, partially or completely hydrogenated; there are usually methyl groups at C-10 and C-13, and often an *alkyl group* at C-17 (see diagram I). By extension, one or more bond scissions, ring expansions and/or ring contractions of the skeleton may have occurred. Natural steroids are derived biogenetically from *triterpenoids*.
1995, 67, 1367; see also W.B. 193

I (See **steroids**)

sterols
Natural products derived from the *steroid* skeleton and containing a hydroxy group in the 3 position, closely related to cholestan-3-ol.
1995, 67, 1367; see also W.B. 193

stibanes
The saturated hydrides of tervalent antimony, having the general formula Sb_nH_{n+2}. *Hydrocarbyl* derivatives of SbH_3 belong to the class *stibines*.
1995, 67, 1367

stibanylidenes
Recommended name for *carbene analogues* having the structure RSb: (former IUPAC name is stibinediyls). A common non-IUPAC synonym is stibinidenes.
1995, 67, 1367

stibines
SbH_3 and compounds derived from it by substituting one, two or three hydrogen atoms by *hydrocarbyl* groups R_3Sb. $RSbH_2$, R_2SbH and R_3Sb (R ≠ H) are called primary, secondary and tertiary stibines, respectively. A specific stibine is preferably named as a substituted stibane. E.g. $(CH_2=CH)_3Sb$ trivinylstibane.
See *stibanes*.
1995, 67, 1367

stibinidenes
See *stibanylidenes*.
1995, 67, 1367

stibonium compounds
Salts (including hydroxides) containing an atom of tetracoordinate antimony, of the form $[R_4Sb]^+X^-$.
See *onium compounds*.
1995, 67, 1367

sticking coefficient (in surface chemistry)
The ratio of the rate of *adsorption* to the rate at which the *adsorptive* strikes the total surface, i.e. covered and uncovered. It is usually a function of surface coverage, of temperature and of the details of the surface structure of the *adsorbent*.
1976, 46, 78

sticking probability (in surface chemistry)
Often used with the same meaning as *sticking coefficient*, but in principle it is a microscopic quantity concerned with the individual collision process.
1976, 46, 78

sticky ends (in biotechnology)
The staggered ends of complementary sequences of DNA which result from cleavage by *restriction enzymes*.
1992, 64, 166

stimulated emission
That part of the emission which is induced by a resonant perturbing electromagnetic field. The transition between states, n and m, is governed by the Einstein coefficient of stimulated emission, B_{nm}. CIDNP emission and lasing action are examples of processes which require stimulated emission.
See also *spontaneous emission*.
1996, *68*, 2279

stirred flow reactor (in catalysis)
A reactor within which effective mixing is achieved, often by placing the *catalyst* in a rapidly-rotating basket.
1976, *46*, 80

stochastic sampling
Synonymous with *random sampling*.
1990, *62*, 2216

stochastic theories
Theories that treat reaction rates in terms of the probabilities of transitions between the various energy levels in the reactant molecules.
1981, *53*, 767

Stockholm convention
IUPAC convention on signs of electromotive forces and electrode potentials.
Physical Chemistry Division, unpublished

+ Stock number
Usage not recommended.
See *oxidation number*.
R.B. 66, 148

stoichiometric
Involving chemical combination in simple integral ratios. Characterized by having no excess of reactants or products over that required to satisfy the balanced chemical equation representing the given chemical reaction.
1990, *62*, 2216

stoichiometric capacity
See *ultimate capacity*.
1993, *65*, 2386

stoichiometric concentration, c
The stoichiometric concentration, c_B, of a component B in a system is given by the expression:
$$c_B = n_B/V$$
where n_B is the stoichiometric amount of substance of component B in the system and V is the volume of the system.
The component added to the system may dissociate or react with other components to form a series of derived components and only a fraction of the original component may actually exist in a free form in the system. It is therefore essential to distinguish between the stoichiometric concentration and the amount-of-substance concentration of the free form of the component in the system. Sometimes stoichiometric quantities are indicated by a subscript ($_o$), e.g. $n_{o,B}$, $c_{o,B}$.

In clinical chemistry, the term stoichiometric concentration is rarely employed. Instead, the name of the component is modified to indicate inclusion of the various derived forms, e.g. mixtures of a defined chemical component and its derivatives may be denoted by the plural form of the name of the pure unchanged substance, or to indicate the sum of components specified in individual quantities the specification 'total' may be employed.
1984, *56*, 568

stoichiometric mean molal activity coefficient (practical activity coefficient) (in electrochemistry)
The practical activity coefficient of electrolyte B is given by
$$\gamma_\pm = a_\pm/(v_+^v v_-^v)^{1/2}(m_B/m^+)$$
where $a_\pm$ is the mean activity of B in solution, m_B is the *molality* of B, $m^+ = 1$ mol kg^{-1}, v_+ is the number of cations and v_- the number of anions in the chosen group B which is taken as the electrolyte.
$$v = v_+ + v_-$$
1974, *37*, 510

stoichiometric number, v
A chemical reaction of known *stoichiometry* can be written in general as:
$$aA + bB + ... \rightarrow ... yY + zZ$$
For the reaction products Y and Z the numbers y and z are known as the stoichiometric numbers, v_Y and v_Z, for Y and Z respectively. For the reactants the stoichiometric numbers are the negatives of the numbers appearing in the equation; for example the stoichiometric number v_A for the reactant A is $-a$. In other words, the stoichiometric numbers are positive for products and negative for reactants.
1996, *68*, 187; G.B. 42; 1996, *68*, 993; see also 1996, *68*, 1168

stoichiometry
The term refers to the relationship between the amounts of substances that react together in a particular chemical reaction, and the amounts of products that are formed. The general stoichiometric equation:
$$aA + bB + ... \rightarrow ... yY + zZ$$
provides the information that a moles of A reacts with b moles of B to produce y moles of Y and z moles of Z.
The stoichiometry of a reaction may be unknown, or may be very complex. For example, the thermal decomposition of acetaldehyde yields mainly methane and carbon monoxide, but also a variety of minor products such as ethane, acetone and diacetyl. The stoichiometric equation:
$$CH_3CHO \rightarrow CH_4 + CO$$
is therefore only an approximate one.
Even when the overall stoichiometry of a reaction is well defined, it may be time-dependent in that it varies during the course of a reaction. Thus if a reaction

occurs by the mechanism A → X → Y, and X is formed in substantial amounts during the course of the process, the relationship between the amounts of A, X and Y will vary with time, and no one stoichiometric equation can represent the reaction at all times.
   1996, *68*, 187

stokes
cgs unit of kinematic viscosity, St = 10^{-4} m^2 s^{-1}.
   G.B. 112; 1996, *68*, 993

Stokes law
$F = 6\pi\eta rv$, where F is the force exerted on a sphere of radius r which is moving through a fluid of viscosity η with a relative velocity v; this equation holds at low velocities which are free from turbulence (called the Stokes region).
   1990, *62*, 2216

Stokes number, *St*
Sometimes referred to as the inertial parameter; it is an index of the impactability of an aerosol particle. It is defined by the equation: $St = 2\tau(V_t - v_t)D_p$ where D_p is the diameter of a small drop, $V_t - v_t$ is the difference in fall velocities of the drop and aerosol particles and τ is the characteristic relaxation time of a particle.
   1990, *62*, 2216

Stokes parameters
Of elliptically polarized incident *radiation*, these are given by
$$s_0^0 = E_1^0 + E_2^0$$
$$s_1^0 = E_1^0 - E_2^0$$
$$s_2^0 = 2(E_1^0 E_2^0)^{1/2} \cos \delta^0$$
$$s_3^0 = 2(E_1^0 E_2^0)^{1/2} \sin \delta^0$$
where E_1^0 and E_2^0 specify the *irradiances* of the incident light polarized with their electric vectors vibrating perpendicular and parallel to the *scattering plane*, respectively and δ^0 is the phase difference between these electric vectors.
   See also *scattering matrix*.
   1983, *55*, 933

Stokes shift
The difference (usually in frequency units) between the spectral positions of the band maxima (or the band origin) of the absorption and luminescence arising from the same electronic transition. Generally, the luminescence occurring at a longer wavelength than the absorption is stronger than the opposite. The latter may be called an anti-Stokes shift.
   See *coherent radiation*.
   1996, *68*, 2278

Stokes type radiation
Fluorescence radiation occurring at wavelengths longer than absorption, i.e. the normal case, is said to be of the Stokes type.
   1984, *56*, 236

stopped flow
A technique for following the kinetics of reactions in solution (usually in the millisecond time range) in which two reactant solutions are rapidly mixed by being forced through a mixing chamber. The flow of the mixed solution along a uniform tube is then suddenly arrested. At a fixed position along the tube the solution is monitored (as a function of time following the stoppage of the flow) by some method with a rapid response (e.g. photometry).
   See *mixing control*.
   1994, *66*, 1168; 1993, *65*, 2298

stopped-flow cell (in spectrochemical analysis)
A small-volume absorption cell connected to a rapid mixing chamber.
   1988, *60*, 1454

stopping power
Of a substance, for charged *particles* of specified energy, the average energy loss in passing through a thin layer of that substance, divided by the thickness of that layer.
   1982, *54*, 1553

STP
Abbreviation for standard temperature (273.15 K or 0 °C) and pressure (10^5 Pa); usually employed in reporting gas volumes. Note that flow meters calibrated in standard gas volumes per unit time often refer to volumes at 25 °C, not 0 °C.
   See also *standard conditions for gases*.
   1990, *62*, 2217

strain
Strain is present in a *molecular entity* or *transition structure* if the energy is enhanced because of unfavourable bond lengths, bond angles or dihedral angles ('torsional strain') relative to a standard.
It is quantitatively defined as the standard enthalpy of a structure relative to a strainless structure (real or hypothetical) made up from the same atoms with the same types of bonding. (The enthalpy of formation of cyclopropane is 53.6 kJ mol^{-1}, whereas the enthalpy of formation based on three 'normal' methylene groups, from acyclic models, is -62 kJ mol^{-1}. On this basis cyclopropane is destabilized by *ca.* 115 kJ mol^{-1} of strain energy.)
   See *molecular mechanics calculation*.
   1994, *66*, 1168

strain-induced transition
A transition in a solid that is initiated by plastic strain.
   1994, *66*, 591

stratified film
A film in which more than two thicknesses coexist in a fixed configuration over significant periods of time.
   1972, *31*, 614

stratified sample
A sample consisting of portions obtained from identified subparts (strata) of the parent population. Within each stratum, the samples are taken randomly. The objective of taking stratified samples is to obtain a more representative sample than that which might otherwise be obtained by random sampling.
1990, *62*, 1203

stratocumulus cloud
A layer of patches composed of laminae or globular masses arranged in groups, lines or waves and having a soft, grey appearance; very often the rolls are so close together that their edges join and give the undersurface a wavy character; the process of formation (cumulogenesis) involves the spreading out of the tops of cumulus clouds, the latter having disappeared; < 2 000 m; usually warmer than 5 °C; vertical velocities usually < 0.1 m s^{-1}.
1990, *62*, 2180

stratopause
That region of the atmosphere which lies between the *stratosphere* and the *mesosphere* and in which a maximum in the temperature occurs.
1990, *62*, 2216

stratosphere
The atmospheric shell lying just above the *troposphere* which is characterized by an increasing temperature with altitude. The stratosphere begins at the *tropopause* (about 10–15 km height) and extends to a height of about 50 km, where the lapse rate changes sign at the stratopause and the beginning of the *mesosphere*.
1990, *62*, 2217

stratus cloud
A uniform, featureless layer of cloud resembling fog but not resting on the ground; when this very low layer is broken up into irregular shreds, it is designated as fractostratus; 300–600 m; widespread irregular stirring and lifting of the shallow layer of cool, damp air near the ground.
1990, *62*, 2180

stray radiation error (in spectrochemical analysis)
A spectrometer set to pass radiation of a particular wavelength band always has a small amount of stray radiation of other wavelengths. Since the sample may absorb more (or less) of this stray radiation than of the radiation at the selected wavelength an error can occur. This error is called a stray radiation error.
1988, *60*, 1456

streak camera
See *streak tube*.
1995, *67*, 1758

streak tube
An *image converter* adapted to provide scanning or time-resolved images. If the image is recorded the whole device is an example of a streak camera.
1995, *67*, 1758

streaming birefringence
The birefringence induced by flow in liquids, solutions and dispersions of optically anisotropic, anisometric or deformable flow molecules or particles due to a non-random orientation of the molecules or particles.
Synonymous with *flow birefringence*.
P.B. 61

streaming current
The electric current I flowing in a streaming cell if the electrodes, which are supposed to be ideally depolarized, are short-circuited. I is positive if the current in the membrane, plug, etc., is from high to low pressure side (and in the outside lead from low to high pressure side).
1972, *31*, 620

streaming potential difference (streaming potential)
The potential difference E_{st} or E at zero current caused by the flow of liquid under a pressure gradient through a membrane, plug or capillary. Identical electrodes must be used on both sides of the membrane, plug, etc. E is positive if the higher potential is on the high pressure side.
1972, *31*, 620

stress-assisted transition
A transition that takes place when an applied stress assists the transition to the new phase.
1994, *66*, 591

stress graphitization
The solid-state transformation of *non-graphitic carbon* into *graphite* by heat treatment combined with application of mechanical stress, resulting in a defined degree of *graphitization* being obtained at a lower temperature and/or after a shorter time of heat treatment than in the absence of applied stress.
Notes:
Stress graphitization may also occur in volume elements of a carbon body in the process of heat treatment as a result of the action of internal residual or thermal stresses.
1995, *67*, 505

strip dynode photomultiplier tube
See *photomultiplier tube*.
1995, *67*, 1753

stripping
The process of removing solute(s) from a loaded solvent or extract. Generally this refers to the main solute(s) present.
Notes:
1. Where appropriate, e.g. when liquid-liquid distribution is used for stripping, the term back-extraction can be used. The terms back-washing and re-extraction are not recommended.
2. The recent application of selective stripping of solutes as a separation method leads to some confusion between the terms stripping and scrubbing. It is recommended that the term scrubbing be reserved for

the operation of removing contaminants (impurities) from an extract (where the scrub raffinate is often recycled to the loading step) and the term selective stripping be used where two or more main solutes are stripped successively from an extract, usually with different stripping agents, with a view to their subsequent separate recovery from solution for analysis.
1993, *65*, 2390; O.B. 91

stripping isotherm
See *distribution isotherm*.
Note:
In the graphical representation of stripping isotherms, the axes are often interchanged from those used to represent the phases for extraction isotherms. It is essential that the axes be clearly labelled.
1993, *65*, 2391

stripping ratio (in solvent extraction)
See *distribution ratio*.
Notes:
1. This term is usually defined as the inverse ratio to the distribution ratio, i.e. in aqueous-organic systems the aqueous phase concentration of solute is the numerator and the organic phase concentration is the denominator. Their usage should be clearly defined.
2. The term stripping coefficient is not recommended.
1993, *65*, 2391

stripping reaction
A chemical process, studied in a *molecular beam*, in which the reaction products are scattered forward with respect to the moving centre of mass of the system.
1996, *68*, 188

stripping solution (in liquid-liquid distribution)
A solution (usually aqueous, sometimes water alone) used for back-extracting the distribuend from the extract (usually organic).
O.B. 91

strong collision
A collision between two molecules in which the amount of energy transferred from one to the other is large compared with $k_B T$, where k_B is the Boltzmann constant and T the absolute temperature.
1996, *68*, 188

structural disorder
Any deviation from the ideal three-dimensional regularity of the crystal structure.
P.B. 79

structural formula
A formula which gives information about the way the atoms in a molecule are connected and arranged in space.
R.B. 46

structural transition
A *reversible* or *irreversible transition* that involves a change of the crystal structure.
Example: The transition of NH_4Cl at 469 K from a CsCl-type structure to a NaCl-type structure.
1994, *66*, 591

structure (of a catalyst)
The distribution in space of the atoms or ions in the material part of the *catalyst* and, in particular, the distribution at the surface.
1976, *46*, 79

styphnates
Salts or charge-transfer complexes of styphnic acid (2,4,6-trinitrobenzene-1,3-diol).
See *picrates*.
1995, *67*, 1367

subchain (of a polymer)
An arbitrarily chosen contiguous sequence of *constitutional units* in a *chain*.
Note:
The term 'subchain' may be used to define designated subsets of the constitutional units in a chain.
1996, *68*, 2293

subgroup-supergroup transition
A transition in which the space-group symmetry of the lower symmetry phase is a subgroup of that of the higher symmetry phase.
Example: The transition of the low-temperature polymorph of quartz characterized by space-group symmetry $P3_12$ (trigonal), to the high temperature polymorph of quartz with space-group symmetry, $P6_222$ (hexagonal).
Synonymous with symmetry-breaking transition (note: this term, strictly speaking, is no longer in use).
1994, *66*, 591

subjacent orbital
The *N*ext-to-*H*ighest *O*ccupied *M*olecular *O*rbital ('NHOMO', also called 'HOMO') and the *S*econd *L*owest *U*noccupied *M*olecular *O*rbital (SLUMO). Subjacent orbitals are sometimes found to play an important role in the interpretation of molecular interactions according to the *frontier orbital approach*.
1994, *66*, 1169

sublation (in solvent extraction)
A flotation process in which the material of interest, adsorbed on the surface of gas bubbles in a liquid, is collected on an upper layer of immiscible liquid. There is no liquid-phase mixing in the bulk of the system; as a result recoveries can approach 100%.
1993, *65*, 2379

sublimation
The direct transition of a solid to a vapour without passing through a liquid phase.
Example: The transition of solid CO_2 to CO_2 vapour.
1994, *66*, 591

subphase (in thin films)
See *substrate (in thin films)*.
1994, *66*, 1671

subsample
A subsample may be: (a) a portion of the sample obtained by selection or division; (b) an individual unit of the lot taken as part of the sample; (c) the final unit of multistage sampling.

The term 'subsample' is used either in the sense of a 'sample of a sample' or as a synonym for 'unit'. In practice, the meaning is usually apparent from the context or is defined.
1990, *62*, 1205

substance concentration
See *amount (of substance) concentration*.
G.B. 42; see also 1996, *68*, 994

substance content, *n/m*
Amount-of-substance of a component divided by the mass of the system.
Notes:
1. The term *component* is recommended for clinical chemistry, ISO uses *substance*.
2. The term *system* is recommended for clinical chemistry, ISO uses *mixture*.
3. Use of the term level as a synonym for content is deprecated.
4. In describing a quantity, *content* must be clearly differentiated from *concentration*.
1996, *68*, 994

substance flow rate
Amount-of-substance of a component crossing a surface divided by the time. For clinical chemistry the term *component* is recommended. ISO uses *substance*.
1996, *68*, 994

substance fraction, *x*
Ratio of the amount-of-substance of the component to the total amount-of-substance in the system containing the component.
Notes:
1. The term *substance fraction* is recommended for clinical chemistry, IUPAC also uses mole fraction and *amount fraction*.
2. For clinical chemistry the term *component* is recommended, ISO uses *substance*.
3. For clinical chemistry the term *system* is recommended, ISO uses *mixture*.
1996, *68*, 994

substituent atom (group)
An atom (group) that replaces one or more hydrogen atoms attached to a parent structure or characteristic group except for hydrogen atoms attached to a chalcogen atom.
B.B.(G) 13; 1994, *66*, 1169

substituent electronegativity
See *electronegativity*.
1994, *66*, 1169

substitution reaction
A *reaction, elementary* or *stepwise*, in which one atom or group in a *molecular entity* is replaced by another atom or group. For example,

$CH_3Cl + OH^- \rightarrow CH_3OH + Cl^-$

1994, *66*, 1169

substitutive name
A name which indicates the exchange of one or more hydrogen atoms attached to a skeletal atom of a parent structure or to an atom of a characteristic group for another atom or group, which may be expressed by a suffix or by prefixes.
B.B.(G) 15

substoichiometric extraction
Solvent extraction in which the amount of reagent used is lower than that dictated by stoichiometry. If the constants of formation and extraction of the complexes are high, the amount of extracted metal is dictated by the amount of extractant introduced.
1993, *65*, 2379

substoichiometric isotope dilution analysis
A kind of *isotope dilution analysis* in which the final isotopic abundance is estimated from the amount of the *nuclide* present in a known quantity of the relevant element separated from the test portion, where this quantity is smaller than the total amount of that element present in the test portion.
1994, *66*, 2521

substrate
A *chemical species*, the reaction of which with some other chemical reagent is under observation (e.g. a compound that is transformed under the influence of a *catalyst*). The term should be used with care. Either the context or a specific statement should always make it clear which chemical species in a reaction is regarded as the substrate.
See also *transformation*.
1994, *66*, 1169; 1993, *65*, 2298

substrate (in biocatalysis)
1. The chemical entity whose conversion to a product or products is catalysed by one or several *enzymes*.
2. A solution or dry mixture containing all ingredients which are necessary for the growth of a microbial culture or for product formation.
3. Component in the *nutrient medium*, supplying the organisms with carbon (C-substrate), nitrogen (N-substrate), etc.
1992, *64*, 166

substrate (in thin films)
The terms *substrate* and *subphase* refer to a condensed phase that physically supports a film or layer. The term substrate should be used with care in the light of the existing double meaning of the term substrate in film science and technology, on the one hand, and in biochemistry, on the other. The term subphase applies, mainly, to a liquid phase underlying an adsorption layer or a floating spread monolayer.
1994, *66*, 1671

subtractive name
A name for a modified parent structure in which prefixes and/or suffixes indicate the removal of atoms or groups and, where required, replacement by an appropriate number of hydrogen atoms.
B.B.(G) 16

successor complex
The radical ion pair which forms by the transfer of an electron from the donor D to the acceptor A after

these species have diffused together to form the precursor or encounter complex:

A + D → (A D) → (A·⁻ D·⁺)
1994, 66, 1169

sudden potential-energy surface
See *gradual (sudden) potential-energy surface*.
1996, 68, 188

sugars
A loose term applied to *monosaccharides* and lower *oligosaccharides*.
1995, 67, 1367

suicide inhibition
See *mechanism-based inhibition*.
1994, 66, 1169

suicide metabolism
See *lethal synthesis*.
1993, 65, 2063

sulfamic acids
$H_2NS(=O)_2OH$ (sulfamic acid) and its N-*hydrocarbyl* derivatives. Sulfamic acid is called sulfamidic acid in inorganic chemistry.
1995, 67, 1367

+ sulfanes
A term including *hydropolysulfides, polysulfanes, polysulfides*. Use is discouraged because of confusion with the newer systematic name sulfane, H_2S, and the numerous names derived therefrom.
1995, 67, 1367

sulfatides
Hydrogen sulfate *esters* of glycosphingolipids. Specific compounds should be named as glycerosphingolipid derivatives, e.g.

R = fatty acid acyl group or H
R' = CH₃[CH₂]₁₄CH(OH)-

1995, 67, 1367; see also W.B. 187

sulfenamides
Compounds derived from *sulfenic acids*, RSOH (R ≠ H), by replacement of –OH by –NR₂. May alternatively be considered as alkylsulfanylamines, e.g. $C_2H_5SNH_2$ ethanesulfenamide or ethylsulfanylamine.
1995, 67, 1368

sulfenes
S,S-Dioxides of *thioaldehydes* and *thioketones*, $R_2C=SO_2$.
1995, 67, 1368

sulfenic acids
Compounds having the structure RSOH (R ≠ H), e.g. PhSOH benzenesulfenic acid.
1995, 67, 1368

sulfenium ions
A term sometimes erroneously used for *sulfenylium ions*.
1995, 67, 1368

sulfenyl groups
Groups having the structure RS– (R ≠ H). This term is derived from sulfenic acids. The synonymous term sulfanyl groups is derived from sulfane, H_2S. E.g. CH_3–S– methanesulfenyl, methylthio or methylsulfanyl.
1995, 67, 1368

sulfenylium ions
Cations having the structure RS⁺ (R ≠ H). This term is derived from sulfenic acids. The synonymous term sulfanylium ions is derived from sulfane, H_2S. E.g. CH_3S^+ methylsulfanylium or methanesulfenylium ion.
1995, 67, 1368

sulfenyl nitrenes
See *hydrocarbylsulfanyl nitrenes*.
1995, 67, 1368

sulfenyl radicals
Sulfur-centered radicals having the structure RS· (R ≠ H). This term is derived from *sulfenic acids*.
The synonymous term sulfanyl radicals is derived from sulfane, H_2S. E.g. $CH_3S·$ methylsulfanyl or methanesulfenyl radical. An older synonymous term is alkylthio radical.
1995, 67, 1368

sulfides
1. Compounds having the structure RSR (R ≠ H). Such compounds were once called thioethers. See also *thioacetals*.
2. In an inorganic sense, salts or other derivatives of hydrogen sulfide.
3. A term used in additive nomenclature, see *imides (2)*.
1995, 67, 1368

sulfilimines
See *sulfimides (1)*.
1995, 67, 1368

sulfimides
1. H_2S=NH, sulfimide, and its *hydrocarbyl* derivatives. They are thus related to *sulfoxides* in the same way that imines are related to *aldehydes* or *ketones*, e.g. $(C_2H_5)_2S$=NPh *S,S*-diethyl-*N*-phenylsulfimide.
2. A term used in Chemical Abstracts Service Index Nomenclature for *sulfonylamines*, RN=S(=O)₂.
1995, 67, 1368

+ sulfimines
A non-recommended synonym for *sulfimides (1)*.
1995, 67, 1368

sulfinamides
Amides of *sulfinic acids*, RS(=O)OH; thus RS(=O)NR₂, e.g. PhS(=O)NHCH₃ *N*-methylbenzenesulfinamide.
See also *amides*.
1995, 67, 1368

sulfinamidines
Amidines of *sulfinic acids*, RS(=O)OH; thus RS(=NR)NR$_2$, e.g. PhS(=NH)NH$_2$ benzenesulfinamidine.
  1995, *67*, 1368

+ sulfines
S-Oxides of *thioaldehydes* and *thioketones*. Not recommended because in English the -ine termination should be reserved for *amines*, *imines*, etc., e.g. PhC(=S=O)H thiobenzaldehyde *S*-oxide.
  1995, *67*, 1369

sulfinic acids
HS(=O)OH, sulfinic acid, and its *S*-hydrocarbyl derivatives, e.g. (CH$_3$)$_2$CHS(=O)OH propane-2-sulfinic acid.
  1995, *67*, 1369

sulfinic anhydrides
Compounds having the structure RS(=O)OS(=O)R.
  See also *acid anhydrides*.
  1995, *67*, 1369

+ sulfinimines
An obsolescent term for *sulfimides (1)*.
  1995, *67*, 1369

sulfinylamines
Compounds having the structure RN=S=O. Formerly called thionylamines.
  1995, *67*, 1369

sulfolipids
Sulfate esters of *glycolipids*.
  See also *sulfatides*.
  1995, *67*, 1369

sulfonamides
Amides of *sulfonic acids* RS(=O)$_2$NR'$_2$, e.g. PhS(=O)$_2$NHCH$_3$ *N*-methylbenzenesulfonamide.
  1995, *67*, 1369

+ sulfonamidines
A term abandoned because of its ambiguous use to mean either RS(=O)(=NH)NH$_2$ (sulfonimidamide) or RS(=NH)$_2$NH$_2$ (sulfonodiimidamide).
  1995, *67*, 1369

sulfonediimines
Compounds having the structure, RS(=NR)$_2$R, formally derived from sulfones by replacing (=O)$_2$ by (=NR)$_2$, e.g. Ph$_2$S(=NH)$_2$ diphenyl sulfonediimine.
  1995, *67*, 1369

sulfones
Compounds having the structure, RS(=O)$_2$R (R ≠ H), e.g. C$_2$H$_5$S(=O)$_2$CH$_3$ ethyl methyl sulfone.
  1995, *67*, 1369

sulfonic acids
HS(=O)$_2$OH, sulfonic acid, and its *S*-hydrocarbyl derivatives.
  1995, *67*, 1369

sulfonic anhydrides
Compounds having the structure RS(=O)$_2$O-S(=O)$_2$R', e.g. PhS(=O)$_2$OS(=O)$_2$Ph benzenesulfonic anhydride.
  See also *acid anhydrides*.
  1995, *67*, 1369

+ sulfonimides
A seldom used synonym of *sulfoximides*; its use is discouraged.
  1995, *67*, 1369

sulfonium compounds
Compounds having the structure R$_3$S$^+$ and associated anion (generally, but not necessarily, all three R groups are *hydrocarbyl*). E.g. [(CH$_3$)$_3$S]$^+$Cl$^-$, trimethylsulfonium chloride, or

1-thioniabicyclo[2.2.1]heptane bromide
  See *onium compounds*.
  1995, *67*, 1369

sulfonphthaleins
3,3-Bis(hydroxyaryl)-3*H*-2,1-benzoxathiole *S*,*S*-dioxides, formed by condensation of *o*-sulfobenzoic acid with *phenols* or related compounds, e.g.

  See *phthaleins*.
  1995, *67*, 1370

sulfonylamines
Compounds having the structure RN=S(=O)$_2$. (Indexed by Chemical Abstracts Service at the heading sulfimide). E.g. CH$_3$N=S(=O)$_2$ *N*-sulfonylmethylamine.
  1995, *67*, 1370

sulfoxides
Compounds having the structure R$_2$S=O (R ≠ H), e.g. Ph$_2$S=O diphenyl sulfoxide.
  1995, *67*, 1370

sulfoximides
Compounds having the structure R$_2$S(=O)=NR. (Indexed by Chemical Abstracts Service at the heading sulfoximines). E.g. (CH$_3$)$_2$S(=O)=NPh *S*,*S*-dimethyl-*N*-phenylsulfoximide.
  1995, *67*, 1370

sulfoximines
  See *sulfoximides*.
  1995, *67*, 1370

sulfur diimides
The parent compound HN=S=NH and *hydrocarbyl* derivatives.
  1995, *67*, 1370

sultams

Sulfonamides in which the S–N bond is part of a ring.

1995, *67*, 1370

sultims

Tautomeric forms of *sultams*, having a sulfur–nitrogen double bond as part of a ring.

1995, *67*, 1370

sultines

Intramolecular cyclic esters of hydroxy sulfinic acids. (Named by supposed analogy with *lactones* and *sultones*; but the -ine ending is unfortunate).

See *sulfinic acids*.
1995, *67*, 1370

sultones

Intramolecular cyclic esters of hydroxy sulfonic acids, analogous to *lactones*.

See *sulfonic acids*.
1995, *67*, 1370

summit current (in polarography)

In ac *polarography*, differential pulse polarography, derivative polarography, square-wave polarography, and similar techniques, the maximum value of the component of the current that is associated with the presence of a substance B. Normally this component of the current is faradaic, and the maximum arises because the rate of variation (with *applied potential*) of the rate of the charge-transfer process passes through a maximum. Similar maxima arise when this component is non-faradaic (and when B is surface-active rather than electroactive). In a case known to be of the latter type, the term *apex current* is recommended as being more specific.
1985, *57*, 1499

summit potential

In ac *polarography*, differential pulse polarography, derivative polarography, and similar techniques, the potential of the *indicator electrode* at which the *summit current* is attained.
1985, *57*, 1503

sum of states, $P(\varepsilon)$

The total number of states for a system corresponding to an energy ε less than or equal to a specified value.
1996, *68*, 188

superacid

A medium having a high *acidity*, generally greater than that of 100 wt.% sulfuric acid. The common superacids are made by dissolving a powerful *Lewis acid* (e.g. SbF_5) in a suitable *Brønsted acid*, such as HF or HSO_3F. (An equimolar mixture of HSO_3F and SbF_5 is known by the trade name 'magic acid'.) In a biochemical context 'superacid catalysis' is sometimes used to denote catalysis by metal ions analogous to catalysis by hydrogen ions. By analogy, a compound having a very high *basicity*, such as lithium diisopropylamide, is called a 'superbase'.
1994, *66*, 1169

superbase

See *superacid*.
1994, *66*, 1169

superconducting transition

A transition at the critical temperature, T_c, below which the resistance of electrical conductors becomes zero and magnetic flux is excluded.
Examples: The transitions of Nb_3Sn, Nb_3Al and V_3Si to superconductors (15 K < T_c < 20 K) and $YBa_2Cu_3O_{7-x}$ ($x \approx 0.2$) at 90 K.
1994, *66*, 592

supercritical fluid

The defined state of a compound, mixture or element above its critical pressure (p_c) and critical temperature (T_c).
1993, *65*, 2399

supercritical fluid chromatography (SFC)

A separation technique in which the mobile phase is a fluid above and relatively close to its critical temperature and pressure.
In general the terms and definitions used in gas or liquid chromatography are equally applicable to supercritical fluid chromatography.
1993, *65*, 825

superequivalent adsorption

This occurs when the specifically adsorbed amount of charge in the *inner Helmholtz plane* is higher than the charge on the metal phase, taken with the reverse sign.
1986, *58*, 448

superexchange interaction

Electronic interaction between two molecular entities mediated by one or more different molecules or ions.
1996, *68*, 2279

superficial work (in surface chemistry)

The reversible work of formation of unit area of new surface by cleavage. It has previously been called specific surface work, surface energy, surface tension or surface free energy. The symbol recommended is γ_π in preference to σ because of the confusion with

surface charge density. The superficial work γ_π is a scalar quantity and consequently is isotropic.
1983, *55,* 1265

superlattice
A periodic *multilayer* which is synthetic and where a unit cell, consisting of successive layers that are chemically different from their adjacent neighbours, is repeated. The term quantum well multilayer applies to the superlattices with artificially created electronic band structures.
1994, *66,* 1672

superposability
The ability to bring two particular *stereochemical formulae* (or models) into coincidence (or to be exactly superposable in space, and for the corresponding molecular entities or objects to become exact replicas of each other) by no more than translation and rigid rotation.
1996, *68,* 2220

superradiance
Spontaneous emission amplified by a single pass through a population inverted medium. It is distinguished from true laser action by its lack of coherence. The term superradiance is frequently used in laser technology.
See *coherent radiation.*
1996, *68,* 2279

supersaturation
1. In chemistry, an unstable system which has a greater concentration of a material in solution than would exist at equilibrium is said to be supersaturated.
2. In meteorology, supersaturation of an air mass with respect to H_2O vapour is of special interest. It is the saturation ratio minus one, or the percent supersaturation is the percent relative humidity minus 100.
1990, *62,* 2217; see also O.B. 84

support (of a catalyst)
In multiphase catalysts, the active catalytic material is often present as the minor component dispersed upon a support sometimes called a *carrier*. The support may be catalytically inert but it may contribute to the overall catalytic activity.
1976, *46,* 79

support-coated open-tubular (SCOT) column (in chromatography)
A version of a *porous-layer open-tabular* (PLOT) column in which the porous layer consists of support particles and was deposited from a suspension.
1993, *65,* 831

supporting electrolyte
An electrolyte solution, whose constituents are not electroactive in the range of *applied potentials* being studied, and whose *ionic strength* (and, therefore, contribution to the conductivity) is usually much larger than the concentration of an *electroactive substance* to be dissolved in it.
1985, *57,* 1501

support plate (in chromatography)
The plate that supports the thin layer in *thin-layer chromatography*.
O.B. 98

suppressor
A substance which reduces emission, absorption, or *light scattering* by an *interferent*, thus removing or lowering *spectral interference*.
O.B. 172

suprafacial
See *antarafacial.*
1994, *66,* 1084

supramolecule
A system of two or more *molecular entities* held together and organized by means of intermolecular (noncovalent) binding interactions.
1994, *66,* 1169

surface
The boundary between two phases. It is recommended that for the purpose of surface analysis a distinction be made between 'surface' in general, 'physical surface' and 'experimental surface':
Surface — The 'outer portion' of a sample of undefined depth; to be used in general discussions of the outside regions of the sample.
Physical Surface — That atomic layer of a sample which, if the sample were placed in a vacuum, is the layer 'in contact with' the vacuum; the outermost atomic layer of a sample.
Experimental Surface — That portion of the sample with which there is significant interaction with the particles or radiation used for excitation. It is the volume of sample required for analysis or the volume corresponding to the escape for the emitted radiation or particle, whichever is larger.
See also *interface*.
O.B. 251; 1979, *51,* 2246

surface active agent
See *surfactant.*
1972, *31,* 611

surface amount, n^s, n^a
Amount of substance adsorbed on a surface.
G.B. 63

surface barrier semiconductor detector
A *semiconductor detector* utilizing a junction due to a surface inversion layer.
1982, *54,* 1552

surface catalysis
See *homogeneous catalysis.*
1981, *53,* 763

surface charge density, σ
Electric charge on a surface divided by the surface area.
G.B. 14

surface charge layer
See *interfacial double-layer.*
1994, *66,* 1674

surface chemical potential
Defined by:

$$\mu_i^\sigma = \left(\frac{\partial A^\sigma}{\partial n_i^\sigma}\right)_{T, A_s, n_j^\sigma} = \left(\frac{\partial G^\sigma}{\partial n_i^\sigma}\right)_{T, p, \gamma, n_j^\sigma}$$

$$\mu_i^s = \left(\frac{\partial A^s}{\partial n_i^s}\right)_{T, V^s, A_s, n_j^s} = \left(\frac{\partial G^s}{\partial n_j^s}\right)_{T, p, \gamma, n_j^s}$$

where A^σ is the *surface excess Helmholtz energy*, G^σ is the *surface excess Gibbs energy*, A^s is the *interfacial Helmholtz energy*, G^s is the *interfacial Gibbs energy*, and A_s is the surface area. The quantities thus defined can be shown to be identical, and the conditions of equilibrium of component i in the system to be

$$\mu_i^\alpha = \mu_i^\sigma = \mu_i^s = \mu_i^\beta$$

where μ_i^α and μ_i^β are the *chemical potentials* of i in the bulk phases α and β. (μ_i^α or μ_i^β have to be omitted from this equlibrium condition if component i is not present in the respective bulk phase.)

The surface chemical potentials are related to the *Gibbs energy* functions by the equations

$$G^\sigma = \sum_i n_i^\sigma \mu_i^\sigma$$

$$G^s = \sum_i n_i^s \mu_i^s$$

1972, *31*, 602

surface concentration, Γ
Amount of substance adsorbed on a surface divided by the surface area.
See *concentration in experimental surface*.
G.B. 42

surface (excess) conductivity
The excess *conductivity* in a surface per unit length and width, symbol κ^σ.
1972, *31*, 621

surface contamination (in surface analysis)
Material in the experimental surface which is either not characteristic of the sample or which would not be present if the sample had been prepared in an absolute vacuum by methods not contacting other substances to the sample.
1979, *51*, 2247

surface coverage, θ
Number of adsorbed molecules on a surface divided by the number of molecules in a filled monolayer on that surface.
G.B. 63; 1979, *51*, 2247

surface crossing
In a diagram of electronic energy versus molecular geometry, the electronic energies of two states of different symmetry may be equal at certain geometrical parameters. At this point (unidimensional representation), line or surface (more than one dimension), the two potential-energy surfaces are said to cross one another.
See *avoided crossing*.
1996, *68*, 2279

surface density, ρ_A, ρ_S
Mass divided by area.
G.B. 12; 1996, *68*, 994

surface dilatational viscosity
See *surface shear viscosity*.
1979, *51*, 1218

surface dipole layer
Particles in the surface region of a phase are subjected to orienting forces as a result of the anisotropic force field. Polar molecules (e.g. permanent dipoles) may thus be preferentially oriented in the surface region, while polarizable molecules may be polarized (induced dipoles). The array of oriented polar and/or polarized molecules is called the surface dipole layer with which an electric potential drop is associated, called the surface potential of the phase.
1983, *55*, 1254

surface electric potential
Given by

$$\chi^\beta = \Phi^\beta - \psi^\beta$$

where ψ^β and Φ^β are the outer and inner electric potentials of phase β, respectively.
1974, *37*, 506; G.B. 59

surface excess, n^σ
For an *interface*, the adsorption or surface excess of a given component is defined as the difference between the amount of component actually present in the system, and that which would be present (in a reference system) if the bulk concentration in the adjoining phases were maintained up to a chosen geometrical dividing surface (*Gibbs dividing surface*).

For a solid/liquid interface in which no component of the liquid phase penetrates into the solid, the surface excess (or adsorption) of component i is defined as:

$$n_i^\sigma = n_i - V^l c_i^l$$

where n_i is the total amount of i in the system, V^l is the volume of an arbitrarily chosen amount of bulk liquid (in the framework of the so-called algebraic method) and c_i^l is its bulk concentration in the liquid.
See *Gibbs adsorption*.
1986, *58*, 969; G.B. 63

surface excess concentration (at an interface), Γ
A quantity defined by

$$\Gamma_i = n_i^\sigma / A_s$$

where n_i^σ is the *surface excess* (or *adsorption*) of component i and A_s is the area of the dividing surface or *interface*.
1986, *58*, 970; G.B. 63

surface excess energy
Defined by:

$$U^\sigma = U - U^\alpha - U^\beta$$

$$= U - V^\alpha \left(\frac{U_m^\alpha}{V_m^\alpha}\right) - V^\beta \left(\frac{U_m^\beta}{U_m^\beta}\right)$$

where V^α and V^β satisfy the condition $V^\alpha + V^\beta = V$, the total volume of the system. (U_m^α/V_m^α) and (U_m^β/V_m^β) are the energy densities in the two bulk phases where U_m^α and U_m^β are the mean molar energies and V_m^α and V_m^β are the mean molar volumes of the two phases.
1972, *31*, 599

surface excess enthalpy
Defined by:
$$H^\sigma = U^\sigma - \gamma A_s$$
where U^σ is the *surface excess energy*, γ is the *surface tension*, and A_s is the area relative to a *Gibbs surface*.
1972, *31*, 599

surface excess entropy
Defined by:
$$S^\sigma = S - S^\alpha - S^\beta$$
$$= S - V^\alpha \left(\frac{S_m^\alpha}{V_m^\alpha}\right) - V^\beta \left(\frac{S_m^\beta}{V_m^\beta}\right)$$

(S_m^α/V_m^α) and (S_m^β/V_m^β) are the entropy densities in the two bulk phases, where S_m^α and S_m^β are the mean molar entropies and V_m^α and V_m^β are the mean molar volumes of the two phases.
1972, *31*, 599

surface excess Gibbs energy
Defined by
$$G^\sigma = H^\sigma - TS^\sigma = A^\sigma - \gamma A_s$$
where H^σ is the *surface excess enthalpy*, T is the thermodynamic temperature, S^σ is the *surface excess entropy*, A^σ is the *surface excess Helmholtz energy*, γ is the *surface tension*, and A_s is the area relative to a *Gibbs surface*.
1972, *31*, 599

surface excess Helmholtz energy
Defined by:
$$A^\sigma = U^\sigma - TS^\sigma$$
where U^σ is the *surface excess energy*, T is the thermodynamic temperature, and S^σ is the *surface excess entropy*.
1972, *31*, 599

surface excess isotherm
The function relating, at constant temperature and pressure, $\Gamma_i^{(n)}$, $\Gamma_i^{(1)}$ or $\Gamma_i^{(v)}$, or the respective specific quantities $n^1\Delta x_i^1/m$, $A_s\Gamma_i^{(1)}/m$ or $V^1\Delta c_i^1/m$ to the mole fraction (or concentration) of component i in the equilibrium liquid phase. With solutions of more than two components such isotherms are unequivocal functions only when the ratios of the mole fractions (or concentrations) of all other components except i are kept constant. (A_s is the area of interface, $\Gamma_i^{(n)}$, $\Gamma_i^{(1)}$, $\Gamma_i^{(v)}$ are relative adsorptions, Δx_i^1 is the change in mole fraction of i resulting from bringing a specified mass m of solid into contact with a specified amount of solution n^1, Δc_i^1 is the change in concentration resulting from bringing a specified mass of solid into contact with a specified amount of solution of volume V^1.)
1972, *31*, 593

surface ionization (in mass spectrometry)
Takes place when an atom or molecule is ionized when it interacts with a solid surface. Ionization occurs only when the work function of the surface, the temperature of the surface, and the ionization energy of the atom or molecule have an appropriate relationship.
1991, *63*, 1548

surface ions
Ions that are constituents of the surface or which have a particularly high affinity for the surface or surface sites.
1991, *63*, 899

surface layer (or interfacial layer)
The region of space comprising and adjoining the phase boundary within which the properties of matter are significantly different from the values in the adjoining bulk phases.
1972, *31*, 583

surface of a phase (free surface)
The plane ideally marking the boundary between a phase and its environment.
1983, *55*, 1252

surface of tension
The mechanical properties of an *interfacial layer* between two fluid phases may be expressed in terms of those of a geometrical surface of uniform tension called the surface of tension, whose location is dependent on the distribution of the stress tensor within the interfacial layer.
1972, *31*, 596

surface potential
See *surface dipole layer*.
1983, *55*, 1254

surface pressure
The change of interfacial tension caused by addition of a given species to a base solution. When an area of liquid covered with a spread substance is separated from a clean area of surface by a mechanical barrier, the force acting on unit length of the barrier is called the surface pressure, π or π_s, and is equal to $\gamma^0 - \gamma$ where γ^0 is the *surface tension* of the clean surface and γ that of the covered surface.
1972, *31*, 598; 1983, *55*, 1264; G.B. 63

surface reaction (in electrochemistry)
See *kinetic current*.
1985, *57*, 1504

surface region
The tridimensional region, extending from the free surface of a condensed phase towards the interior, where the properties differ from the bulk. The term *surface* may also be used in this sense.
1983, *55*, 1253

surface rheology
See *bulk rheology*.
1979, *51*, 1215

surface shear viscosity
For steady state deformations a surface shear viscosity η^s, and an area viscosity or surface dilatational viscosity ζ^s can be defined. In a Cartesian system with the x-axis normal to the surface, they are defined by the equations:

$$\eta^s = \frac{\sigma_{xy}}{\partial v_y/\partial v_x}$$

and

$$\zeta^s = \Delta\gamma \frac{d(\ln A)}{dt}$$

where σ_{xy} is the shear component of the surface stress tensor, v_x and v_y are the x and y components of the surface velocity vector, respectively, A is the surface area, t is the time, and $\Delta\gamma$ is the difference between the (steady state) dynamic surface tension and the equilibrium surface tension.
1979, *51*, 1218

surface states
Energy levels localized in the surface region of semiconductors, which do not bear any direct relation to the bulk energy distribution, but which can exchange electrons with the bulk. In such a case a *space charge* may arise even when σ (the free charge density) is zero, i.e. $\sigma = \sigma_{sc} + \sigma_{ss}$, where σ_{sc} is the space charge density and σ_{ss} is the charge density associated with the surface states.
1986, *58*, 443

surface stress, T_{ij}
The work required to form unit area of new surface by stretching under equilibrium conditions. It is numerically equal to the force acting in the jth direction per unit length of exposed edge, the edge being normal to the ith direction, that must be applied to a terminating surface to keep it in equilibrium, the ith and jth directions lying in the plane of the surface.
1986, *58*, 451

surface tension, γ, σ
Work required to increase a surface area divided by that area. When two phases are studied it is often called interfacial tension.
G.B. 12; 1996, *68*, 995

surface work
Surface tension is the intensive factor in the differential expression for the work required to increase the area of the *surface of tension*. Measured under reversible conditions at constant temperature (and normally constant pressure) and referred to unit area, this work, the so-called (differential) surface work, is equal to the static surface tension.
1972, *31*, 597

surfactant (surface active agent)
A substance which lowers the *surface tension* of the medium in which it is dissolved, and/or the *interfacial tension* with other phases, and, accordingly, is positively adsorbed at the liquid/vapour and/or at other interfaces.
1972, *31*, 611

surprisal, s
A function relating the prior distribution P_0 of product states to an observed or calculated product distribution P.
It is defined by:

$$s = -\ln(P/P_0).$$

1996, *68*, 188

surprisal analysis
The study of product state distributions in terms of *surprisals* has been referred to as surprisal analysis.
1996, *68*, 188

suspended matter (in atmospheric chemistry)
All particulate material which persists in the atmosphere or in a flue gas stream for lengthy periods because the particles are too small in size to have an appreciable falling velocity.
1990, *62*, 2217

suspension
A liquid in which solid particles are dispersed.
1972, *31*, 606

suspension effect (in an ion-selective electrode)
An effect which occurs when *ion-selective electrodes* are used in concentrated, space-filled suspensions while the external reference electrode remains in the supernatant (suspension-free) solution. The suspensions are specifically solvent-swollen ion exchangers or other materials, like soils and clays, that concentrate ions by adsorption and absorption. Space-filled, gravity-packed suspensions act like a second phase and form apparently an interfacial potential difference (PD) with respect to the supernatant. The measured ion activity in the suspension differs from the value in the supernatant by the interfacial PD, and corresponds to a higher value approximating the activity inside the ion exchanger gel. The effect nearly disappears when the outer reference electrode is placed in the same region of the suspension as the sensor electrode. There are some changes in the junction potential differences of the reference electrode, between suspension and supernatant.
1994, *66*, 2533

Swain–Lupton equation
A dual parameter approach to the correlation analysis of substituent effects, which involves a field constant (F) and a resonance constant (R). The original treatment was modified later.
The procedure has been considerably applied, but also much criticized.
1994, *66*, 1170

Swain–Scott equation
The *linear free-energy relation* of the form:

$$\lg(k/k_0) = sn$$

applied to the variation of *reactivity* of a given *electrophilic substrate* towards a series of nucleophilic

reagents. n is characteristic of the reagent (i.e. a measure of its *nucleophilicity*) and s is characteristic of the substrate (i.e. a measure of its sensitivity to the nucleophilicity of the reagent). A scale of n values is based on the rate coefficients k for the reaction of methyl bromide with nucleophiles in water at 25 °C, s being defined as 1.00 for these reactions and n being defined as 0.00 for the hydrolysis of methyl bromide. (Other scales have been devised.)
1994, *66*, 1169

swelling (in colloid and surface chemistry)
The increase in volume of a *gel* or solid associated with the uptake of a liquid or gas.
1972, *31*, 615

swelling pressure (in colloid and surface chemistry)
That pressure difference Π_{sw} or Π which has to be established between a *gel* and its equilibrium liquid, to prevent further *swelling* of the gel.
1972, *31*, 615

switchboard model (in polymer crystals)
A model of crystallinity in which the crystallized segments of a macromolecule belong to the same crystal, although the stems are connected randomly.
P.B. 84

switching transition
A transition in semiconductive glasses in which, beyond a critical applied voltage, there is an avalanche breakdown of conducting electrons that causes local melting and hence local crystallization that gives metallic conductivity.
1994, *66*, 592

sydnone imines
Imines of *sydnones* (i.e. having an imino group in place of the exocyclic oxygen atom).

(These structures represent only two canonical forms.)
1995, *67*, 1371

sydnones
Mesoionic compounds having the 1,2,3-oxadiazole skeleton bearing an oxygen atom attached to the 5 position.

(These structures represent only two canonical forms).
1995, *67*, 1370

sym-
An affix used in names to denote symmetrical.
R.B. 245; B.B. 465

symbiosis
The term was originally applied to describe the maximum flocking of either hard or soft *ligands* in the same complexes. For hydrocarbon molecules, symbiosis implies that those containing a maximum number of C–H bonds (e.g. CH_4) or C–C bonds (e.g. Me_4C) are the most stable.
1994, *66*, 1170

symbol (in quantities and units)
Conventional sign designating a quantity or a unit of measurement.
Notes:
1. Symbols for quantities are denoted by letters of the Latin or Greek alphabet printed in italic (sloping) type; pH is an exception.
2. Symbols for units are denoted by letters of the Latin or Greek alphabet printed in roman (upright) type.
1996, *68*, 995

symmetrical films
The properties of fluid films depend on the nature of the film phase and that of each of the two neighbouring bulk phases. These films should be described, where appropriate, by three capital letters such as A for air, W for water, O for oil, and S for solid, separated by solidi, the middle letter indicating the film phase. For symmetrical films the first and last symbols are the same, e.g. A/W/A: water film in air, or W/O/W: oil film in water, whereas for unsymmetrical films these are different, e.g. W/O/A: oil film between water and air.
See also *asymmetric film*.
1972, *31*, 613

symmetry-breaking transition
See *subgroup-supergroup transition*.
1994, *66*, 592

symmetry-conserving transition
A transition in which the cell dimensions and/or angles of the one phase differ from those in the other phase, but where the space-group symmetry is conserved.
Example: The transition of face-centred cubic Ce, upon cooling, to a face-centred cubic phase that is 10% denser. Upon cooling, enough contraction takes place to allow an overlap of the fsp^2 configuration and the change from an isolated non-bonding magnetic f electron to a bonding non-magnetic electron pair.
1994, *66*, 592

symmetry number, s
The symmetry number of a molecule is obtained by imagining all identical atoms to be labelled, and then counting the number of different but equivalent arrangements that can be obtained by rotating (but not reflecting) the molecule.
In the statistical-mechanical treatment of chemical equilibrium, the partition function for each molecular species must be divided by its symmetry number.
1996, *68*, 189; G.B. 40

symproportionation
Synonymous with *comproportionation*.
1994, *66*, 1170

syn
1. See *torsion angle*.
2. See *endo, exo, syn, anti*.
3. See *anti*.
4. Formerly used to describe the stereochemistry of oximes and related systems. See *E,Z*.
1996, *68*, 2220

synartetic acceleration
See *neighbouring group participation*.
1994, *66*, 1170

synchronization (principle of nonperfect synchronization)
This principle applies to reactions in which there is a lack of synchronization between bond formation or bond rupture and other *primitive changes* that affect the stability of products and reactants, such as *resonance, solvation*, electrostatic, *hydrogen bonding* and *polarizability* effects. The principle states that a product-stabilizing factor whose development lags behind bond changes at the *transition state*, or a reactant-stabilizing factor whose loss is ahead of bond changes at the transition state, increases the *intrinsic barrier* and decreases the 'intrinsic rate constant' of a reaction. For a product-stabilizing factor whose development is ahead of bond changes, or reactant factors whose loss lags behind bond changes, the opposite relations hold. The reverse effects are observable for factors that destabilize a reactant or product.
See also *imbalance, synchronous*.
1994, *66*, 1170

synchronous
A *concerted* process in which the *primitive changes* concerned (generally bond rupture and bond formation) have progressed to the same extent at the *transition state* is said to be synchronous. The term figuratively implies a more or less synchronized progress of the changes. However, the progress of the bonding change (or other primitive change) has not been defined quantitatively in terms of a single parameter applicable to different bonds or different bonding changes. The concept is therefore in general only qualitatively descriptive and does not admit an exact definition except in the case of concerted processes involving changes in two identical bonds.
See also *imbalance*.
1994, *66*, 1170

synchronously excited (fluorescence, phosphorescence) spectrum
A two-dimensional spectrum obtained by varying both the excitation and emission wavelengths simultaneously and which corresponds to the curve where a plane, parallel to the z-axis, intersects the *excitation-emission spectrum* (EES).
1984, *56*, 242

synchrotron radiation
X-Radiation which results from the acceleration of charged particles in circular orbits by strong electric and magnetic fields.
1991, *63*, 737

synclinal
See *torsion angle*.
1996, *68*, 2220

syndet
A synthetic detergent; a detergent other than soap.
1972, *31*, 612

syndiotactic macromolecule
A *tactic macromolecule*, essentially comprising alternating enantiomeric *configurational base units*, which have chiral or prochiral atoms in the *main chain* in a unique arrangement with respect to their adjacent *constitutional units*.
Notes:
1. In a syndiotactic macromolecule, the configurational repeating unit consists of two configurational base units that are enantiomeric.
2. A syndiotactic macromolecule consists of *racemo diads*.
1996, *68*, 2292

syndiotactic polymer
A polymer composed of *syndiotactic macromolecules*.
1996, *68*, 2302

syndiotactic triads (in polymers)
See *triads (in polymers)*.
P.B. 40

syneresis
The spontaneous shrinking of a *gel* with exudation of liquid.
1972, *31*, 615

synergism (in solvent extraction)
A term describing the co-operative effect of two (or more) extractants where the distribution ratio for the combination is greater than the largest individual distribution ratio (measured under comparable conditions).
Notes:
1. The corresponding adjective is synergic and the term synergistic should not be used.
2. No standard method for quantification of the phenomenon has been agreed and any approach should be clearly defined in a given situation.
1993, *65*, 2379

synergism (in toxicology)
Pharmacological or toxicological interaction in which the combined biological effect of two or more substances is greater than expected on the basis of the simple summation of the toxicity of each of the individual substances.
1993, *65*, 2103

synoptic scale
In meteorology, the size or scale of ordinary weather systems or cyclones; typically 1000 km horizontally.
1990, *62*, 2217

synperiplanar
See *torsion angle*.
1996, *68*, 2220

syntectic reaction
A reversible reaction that involves the conversion of two liquid phases, l' and l", into a solid phase α on cooling:

$$l' + l'' \rightleftharpoons \alpha$$

The maximum temperature at which this reaction can occur is the congruent melting point of the solid phase.
Example: The conversion of co-existing K-rich and Zn-rich phases in the K–Zn system to form an intermediate solid phase KZn_{13}.
1994, *66*, 592

synthetic graphite
A material consisting of *graphitic carbon* which has been obtained by graphitizing of *non-graphitic carbon*, by chemical vapour deposition (CVD) from hydrobcarbons at temperatures above 2500 K, by decomposition of thermally unstable carbides or by crystallizing from metal melts supersaturated with carbon.
Notes:
The term *artificial graphite* is often used as a synonym for synthetic graphite. The term synthetic graphite is preferred, however, since graphite crystals can be considered to consist of carbon macromolecules. Although the term synthetic graphite also covers the CVD product *pyrolytic graphite* as well as the residues of carbide decomposition, it is predominantly used for graphitized carbon. Such common use is in line with the above definition. Synonyms for this most important type of synthetic graphite are Acheson graphite and electrographite.
1995, *67*, 505

system
Arbitrarily defined part of the universe, regardless of form or size, e.g. for clinical chemistry, patient, patient plasma, patient urine.
1996, *68*, 995

systematic error
Mean that would result from an infinite number of measurements of the same measurand carried out under repeatability conditions minus a true value of the measurand.
VIM; see also 1995, *67*, 1705; 1990, *62*, 2188

systematic name
A name composed wholly of specially coined or selected syllables, with or without numerical prefixes; e.g. pentane, oxazole.
B.B. xviii

system of units (of measurement)
Set of base units, together with derived units, defined in accordance with given rules, for a given system of quantities, e.g. Système International d'Unités (*SI*).
1996, *68*, 995

Système International d'Unités
See *SI*.
G.B. 69; 1996, *68*, 995

Szilard–Chalmers effect
The rupture of the chemical bond between an atom and the molecule of which the atom is part, as a result of a nuclear reaction of that atom.
1994, *66*, 2525

θ state (in polymers)
The state of a polymer solution for which the second *virial coefficient* is zero.

Notes:
1. In some respects, a polymer solution in the θ state resembles an ideal solution and the θ state may be referred to as a pseudo-ideal state. However, a solution in the θ state must not be identified with an ideal solution.
2. The solvent involved is often referred to as 'θ solvent'.
3. It is assumed that the molar mass of the polymer is high.

P.B. 57

θ temperature (in polymers)
The temperature at which a solution is in the *θ state*.

P.B. 58

tactic block (in a polymer)
A regular block that can be described by only one species of *configurational repeating unit* in a single sequential arrangement.

P.B. 17, 34

tactic block polymer
A *polymer*, the molecules of which consist of *tactic blocks* connected linearly.

P.B. 17, 34

tacticity
The orderliness of the succession of *configurational repeating units* in the *main chain* of a *regular macromolecule*, a *regular oligomer molecule*, a *regular block* or a *regular chain*.

1996, 68, 2292

tactic macromolecule
A *regular macromolecule* in which essentially all the *configurational (repeating) units* are identical.

1996, 68, 2292

tactic polymer
A polymer composed of tactic *macromolecules*.

1996, 68, 2302

Taft equation
Various equations are associated with R.W. Taft, but the term is most often used to designate the family of equations that emerged from Taft's analysis of the reactivities of aliphatic esters, and which involved the polar substituent constant σ^* and the steric substituent constant E_s:

$$\log k = \log k_0 + \rho^* \sigma^* + \delta E_s$$

or the one-parameter forms applicable when the role of either the polar term or the steric term may be neglected. Nowadays σ^* is usually replaced by the related constant σ_I.

See also *Hammett equation*, *ρ-value*, *σ-constant*.

1994, 66, 1171

tagged
Made identifiable by a *label*.

1982, 54, 1553

tailing (in chromatography)
Asymmetry of a *peak* such that, relative to the *base line*, the front is steeper than the rear. In paper chromatography and in *thin-layer chromatography*, the distortion of a *zone* showing a diffuse region behind the zone in the direction of travel.

O.B. 96; 1993, 65, 837

tandem mass spectrometer
An arrangement in which ions are subjected to two or more sequential stages of analysis (which may be separated spatially or temporally) according to the quotient mass/charge. A hybrid mass spectometer is an instrument which combines analysers of different types, e.g. magnetic plus electric sector combined with quadrupole. The study of ions involving two stages of mass analysis has been termed mass spectrometry/mass spectrometry.

1991, 63, 1546

+ tautomeric effect
See *electromeric effect*.

1994, 66, 1171

tautomerism
Isomerism of the general form:

$$G-X-Y=Z \rightleftarrows X=Y-Z-G$$

where the isomers (called tautomers) are readily interconvertible; the atoms connecting the groups X,Y,Z are typically any of C, H, O or S, and G is a group which becomes an *electrofuge* or *nucleofuge* during isomerization. The commonest case, when the electrofuge is H^+, is also known as 'prototropy'. Examples, written so as to illustrate the general pattern given above, include:

Keto–enol tautomerism, such as:

$$\underset{\text{enol}}{H-O-\underset{|}{\overset{H_3C}{C}}=\underset{|}{\overset{CO_2Et}{C}}-H} \rightleftarrows \underset{\text{keto}}{O=\underset{|}{\overset{H_3C}{C}}-\underset{|}{\overset{CO_2Et}{C}}-H}$$

($G = H$, $X = O$, $Y = CCH_3$, $Z = CHCO_2Et$)

$$H-\underset{|}{\overset{Ar}{C}}-N=\underset{|}{\overset{Ar'}{C}}-H \rightleftarrows \underset{|}{\overset{Ar}{C}}=N-\underset{|}{\overset{Ar'}{C}}-H$$

($G = H$, $X = CHAr$, $Y = N$, $Z = CHAr'$)

The grouping Y may itself be a three-atom (or five-atom) chain extending the conjugation, as in:

$$HO-\langle\text{pyridine}\rangle-N \rightleftarrows O=\langle\text{ring}\rangle-N-H$$

($G = H$, $X = O$, $Y = C-CH=CH$, $Z = N$)

The double bond between Y and Z may be replaced by a ring, when the phenomenon is called ring-chain tautomerism, as in:

$$\text{H-O-C(CO}_2\text{H)(C(CH}_3)_2\text{CCH(CH}_3)\text{C(=O)-O)} \rightleftharpoons \text{O=C(CO}_2\text{H)(C(CH}_3)_2\text{CCH(CH}_3)\text{C(=O)-O-H)}$$

(G = H, X = O, (Y = Z) = –C(CO$_2$H)–O)–(CH$_3$)$_2$CCH(CH$_3$)C(=O))

See also *ambident, sigmatropic rearrangement, tautomerization, valence tautomerization*.
1994, *66*, 1171

tautomerization
The *isomerization* by which tautomers are interconverted. It is a *heterolytic molecular rearrangement* and is frequently very rapid.
   See *tautomerism*.
   1994, *66*, 1172

tele (in histidine nomenclature)
   See *pros (in histidine nomenclature)*.
   W.B. 43

telechelic molecule
   See *pre-polymer molecule*.
   1996, *68*, 2290

tele-substitution
A *substitution reaction* in which the *entering group* takes up a position more than one atom away from the atom to which the *leaving group* was attached:

   See also *cine-substitution*.
   1994, *66*, 1172

tellurides
1. Compounds having the structure RTeR (R ≠ H). Thus tellurium analogues of ethers.
2. Metal salts of tellurane, H_2Te.
   1995, *67*, 1371

tellurones
Compounds having the structure $R_2Te(=O)_2$. Thus tellurium analogues of sulfones.
   1995, *67*, 1371

telomer
   See *oligomer*.
   1996, *68*, 2300

telomerization
The formation of an *addition* oligomer, having uniform end groups X′ ... X″, by a *chain reaction* in which a *chain transfer* limits the length of the polymer ('telomer') produced. An example is the polymerization of styrene in bromotrichloromethane solution (X′ = CCl$_3$, X″ = Br), where Cl$_3$C· radicals are formed in the initiation step to produce Cl$_3$C[CH$_2$CHPh]$_n$Br, with *n* greater than 1 and often less than *ca.* 10:

$$\text{CCl}_3 + \text{CH}_2=\text{CHPh} \rightarrow \text{Cl}_3\text{CCH}_2\dot{\text{C}}\text{HPh} \quad \text{chain propagation}$$
$$\text{Cl}_3\text{CCH}_2\dot{\text{C}}\text{HPh} + \text{CH}_2=\text{CHPh} \rightarrow \text{Cl}_3\text{CCH}_2\text{CHPhCH}_2\dot{\text{C}}\text{HPh}$$
$$\text{Cl}_3\text{C(CH}_2\text{CHPh)}_{n-1}\text{CH}_2\dot{\text{C}}\text{HPh} + \text{CH}_2=\text{CHPh} \rightarrow \text{Cl}_3\text{C(CH}_2\text{CHPh)}_n\text{CH}_2\dot{\text{C}}\text{HPh}$$
$$\text{Cl}_3\text{C(CH}_2\text{CHPh)}_n\text{CH}_2\dot{\text{C}}\text{HPh} + \text{BrCCl}_3 \rightarrow \text{Cl}_3\text{C(CH}_2\text{CHPh)}_{n+1}\text{Br} + \text{Cl}_3\text{C·} \quad \text{chain transfer}$$

See also *oligomerization*.
1994, *66*, 1173

temperature
Short form for *thermodynamic, Celsius* or *Fahrenheit temperature*.
G.B. 48

temperature coefficient of responsivity
The dependence of a detector on temperature can be described by the temperature coefficient of responsivity and is expressed as percentage change in output per K. In the case of a nonlinear dependence the temperature and the temperature range should also be stated for which the stated temperature coefficient of responsivity is applicable.
   1995, *67*, 1751

temperature effect (in luminescence spectroscopy)
A term used for changes in the *luminescence* parameters caused by changes in temperature, since the type of *solvent* and its temperature can effect the luminescence yield from an analyte as a result of *quenching*, *exciplex* formation, *aggregation*, etc.
   1984, *56*, 244

temperature inversion (in atmospheric chemistry)
A departure from the normal decrease of temperature with increasing altitude. A temperature inversion may be produced, for example, by the movement of a warm air mass over a cool one. Intense surface inversions may form over the land during nights with clear skies and low winds due to the radiative loss of heat from the surface of the earth. The temperature increases as a function of height in this case. Poor mixing of the pollutants generally occurs below the inversion, since the normal convective process which drives the warmer and lighter air at ground level to higher altitudes is interrupted as the rising air parcels encounter the warmer air above. Temperature inversions near the surface are particularly effective in trapping ground level emissions.
   1990, *62*, 2197

temperature jump
A relaxation technique in which the temperature of a chemical system is suddenly raised. The system then relaxes to a new state of equilibrium, and analysis of the relaxation processes provides rate constants.
   See also *relaxation*.
   1996, *68*, 189; 1993, *65*, 2297

temperature lapse rate (in atmospheric chemistry)
The rate of change of temperature with altitude (dT/dz). The rate of temperature decrease with increase in altitude which is expected to occur in an unperturbed dry air mass is 9.8×10^3 °C min^{-1}. This is called the *dry adiabatic lapse rate*. The lapse rate is taken as positive when temperature decreases with increasing height. For air saturated with H$_2$O, the lapse rate is less because of the release of the latent heat of water as it condenses. The average tropospheric lapse rate is about 6.5×10^3 °C min^{-1}. The lapse rate has a negative value within an inversion layer.
    1990, *62*, 2199

temperature-programmed chromatography
A procedure in which the temperature of the *column* is changed systematically during a part or the whole of the separation.
    O.B. 92

temperature, thermodynamic
    See *thermodynamic temperature*.
    1990, *62*, 2217

template (in biotechnology)
The nucleic acid single strand that is copied during *replication* or *transcription*.
    1992, *64*, 166

temporary poisoning (in catalysis)
Weak and reversible *adsorption* of a *poison*, such that removal of the poison from the fluid phase results in restoration of the original catalytic activity.
    1976, *46*, 83

tera
SI prefix for 10^{12} (symbol: T).
    G.B. 74

term, T
Energy divided by the product of the Planck constant and the speed of light, when of wavenumber dimension, or energy divided by the Planck constant, when of frequency dimension.
    G.B. 23

term (in X-ray spectroscopy)
A set of levels which have the same electron configuration and the same value of the quantum numbers for total spin S and total orbital angular momentum L.
    1991, *63*, 737

termination
The steps in a *chain reaction* in which reactive *intermediates* are destroyed or rendered inactive, thus ending the chain.
    1994, *66*, 1173

terminator (in biotechnology)
A sequence of DNA lying beyond the 3' end of the coding segment of a *gene* which is recognized by *RNA* polymerase as a signal to stop synthesizing mRNA.
    1992, *64*, 166

termolecular
    See *molecularity*.
    1993, *65*, 2296

term symbols
Symbols characterizing the states of atoms and molecules in terms of multiplicity, symmetry of the total electronic wavefunction, and sometimes total (orbital + spin) angular momentum. Examples: $^2P_{1/2}$ denotes an atomic state of multiplicity 2 (doublet), electron orbital angular momentum quantum number $L = 1$, and orbital plus spin angular momentum quantum number $J = 1/2$.
    G.B. 28

terpenes
Hydrocarbons of biological origin having carbon skeletons formally derived from isoprene [CH$_2$=C(CH$_3$)CH=CH$_2$]. This class is subdivided into the C$_5$ hemiterpenes, C$_{10}$ monoterpenes, C$_{15}$ sesquiterpenes, C$_{20}$ diterpenes, C$_{25}$ sesterterpenes, C$_{30}$ triterpenes, C$_{40}$ tetraterpenes (carotenoids) and C$_{5n}$ polyterpenes.
    See also *carotenes, carotenoids, isoprenoids, prenols, retinoids, steroids, terpenoids*.
    1995, *67*, 1371

terpenoids
Natural products and related compounds formally derived from isoprene units (see *isoprenoids*). They contain oxygen in various functional groups. This class is subdivided according to the number of carbon atoms in the same manner as are terpenes. The skeleton of terpenoids may differ from strict additivity of isoprene units by the loss or shift of a fragment, generally a methyl group.
    1995, *67*, 1371; see also W.B. 255

terpolymer
    See *copolymer*.
    1996, *68*, 2300

tertiary current distribution
    See *current distribution*.
    1981, *53*, 1836

tertiary structure
The spatial organization (including conformation) of an entire protein molecule or other macromolecule consisting of a single chain.
    See also *primary structure, secondary structure, quaternary structure*.
    1996, *68*, 2220; W.B. 80

tesla
SI derived unit of magnetic flux density, T = Wb m^{-2} = m^2 kg s^{-2} A^{-1}.
    G.B. 72

test portion
The amount or volume of the *test sample* taken for analysis, usually of known weight or volume.
    1989, *61*, 1660; see also 1990, *62*, 1206

test sample
The sample, prepared from the *laboratory sample*, from which *test portions* are removed for testing or for analysis.
1990, *62*, 1206; see also 1988, *60*, 1465

test solution (in analysis)
The solution prepared from the *test portion* for the analytical procedure. The proportions of test portion and solvent are normally known.
1989, *61*, 1660; see also 1990, *62*, 1206

tetracyclines
A subclass of polyketides having an octahydro-tetracene-2-carboxamide skeleton, substituted with many hydroxy and other groups, e.g. chlortetracycline:

1995, *67*, 1371

tetrads (in polymers)
See *triads (in polymers)*.
P.B. 40

tetrahedral intermediate
A reaction *intermediate* in which the bond arrangement around an initially double-bonded carbon atom (typically a carbonyl carbon atom) has been transformed from trigonal to tetrahedral. For example, aldol in the *condensation reaction* of acetaldehyde (but most tetrahedral intermediates have a more fleeting existence).
1994, *66*, 1173

tetrahedro-
An affix used in names to denote four atoms bound into a tetrahedron.
R.B. 245; B.B. 465

tetrapyrroles
Natural pigments containing four pyrrole rings joined by one-carbon units linking position 2 of one pyrrole ring to position 5 of the next. *Porphyrins* are macrocyclic tetrapyrroles. E.g. bilin (a linear tetrapyrrole).

1995, *67*, 1371; see also W.B. 279

tetraterpenoids
Terpenoids having a C_{40} skeleton.
See also *carotenes, carotenoids, xanthophylls*.
1995, *67*, 1371

texture (of a catalyst)
The detailed geometry of the void space in the particles of a catalyst.
1976, *46*, 79

theoretical plate number
See *plate number (in chromatography)*.
1993, *65*, 847

thermal analysis
A group of techniques in which a physical property of a substance and/or its reaction product is measured as a function of temperature while the substance is subjected to a controlled temperature program.
See also *enthalpimetric analysis*
1994, *66*, 2488; O.B. 38

thermal black
A special type of *carbon black* produced by pyrolysis of gaseous hydrocarbons in a preheated chamber in the absence of air. Thermal black consists of relatively large individual spheres (100–500 nm diameter) and aggregates of a small number of pseudo-spherical particles. The preferred alignment of the layer planes is parallel to the surface of the spheres.
1995, *67*, 505

thermal column (in nuclear chemistry)
A large body of *moderator*, adjacent to or inside a *reactor* to provide *thermal neutrons* for experiments.
1982, *54*, 1553

thermal conductance, G
Heat *flow rate* divided by the temperature difference.
G.B. 65

thermal conductivity, λ
Tensor quantity relating the heat flux, J_q, to the temperature gradient, $J_q = -\lambda \, \text{grad} \, T$.
G.B. 37

thermal conductivity detector (in gas chromatography)
In general, two cells arranged in a bridge configuration detect the change in thermal conductivity of the gas at the output of the column. This detector is sensitive to any substance with thermal conductivity different from that of the carrier gas. The lowest detectable limit is between 0.5 and 100 ppmv. The linear dynamic range is of the order of 10^3. This type of detector is often used for measuring components at relatively high concentrations.
1990, *62*, 2192

thermal detector
See *radiation detector*.
1995, *67*, 1748

thermal fission
Fission caused by *thermal neutrons*.
1982, *54*, 1542

thermal ionization (in mass spectrometry)
Takes place when an atom or molecule interacts with a heated surface or is in a gaseous environment at high temperature. Examples of the latter may be a capillary arc plasma, a microwave plasma or an inductively coupled plasma.
1991, *63*, 1548; O.B. 203

thermal lensing
A technique that determines the alteration in the refractive index of a medium as a result of the temperature rise in the path of a laser beam absorbed by the medium. The lens produced (usually divergent)

causes a change (usually a decrease) in the irradiance measured along the laser beam axis.
See also *photothermal effects*.
1996, *68*, 2279

thermally activated delayed fluorescence
See *delayed fluorescence*.
1996, *68*, 2279

thermally-induced transition
A transition that is induced by a change in temperature.
Example: RbNO$_3$ undergoes three *structural transitions* below 573 K:
(i) trigonal → CsCl-type structure at 439 K,
(ii) CsCl-type → hexagonal structure at 501 K, and
(iii) hexagonal structure → NaCl-type structure at 551 K.
1994, *66*, 592

thermal neutrons
Neutrons in thermal equilibrium with the medium in which they exist, in general at room temperature.
1994, *66*, 2522

thermal resistance, *R*
Reciprocal of the *thermal conductance*.
G.B. 65

thermoacoustimetry
A technique in which the characteristics of imposed acoustic waves are measured as a function of temperature after passing through a substance (and/or its reaction product(s)) whilst the substance is subjected to a controlled temperature program.
1985, *57*, 1740

+ thermochemical analysis
See *enthalpimetric analysis*.
1994, *66*, 2489

thermochemical calorie
See *calorie*.
G.B. 112; 1996, *68*, 996

thermochromism
A thermally induced transformation of a molecular structure or of a system (e.g. of a solution), thermally reversible, that produces a spectral change, typically, but not necessarily, of visible colour.
1996, *68*, 2279

thermocouple
A device based on the thermoelectric effect, by which two junctions between dissimilar conductors (metallic or heavily doped semiconductors) kept at different temperatures generate an electric potential.
1995, *67*, 1751

thermodilatometry
A technique in which a dimension of a substance under negligible load is measured as a function of temperature while the substance is subjected to a controlled temperature program.
Linear thermodilatometry and volume thermodilatometry are distinguished on the basis of the dimensions measured.
O.B. 45

thermodynamically equivalent sphere (in polymer chemistry)
A sphere, impenetrable to other spheres, displaying the same excluded volume as an actual polymer molecule.
P.B. 48

thermodynamic control (of product composition)
The term characterizes conditions that lead to reaction products in a proportion governed by the equilibrium constant for their interconversion and/or for the interconversion of reaction *intermediates* formed in or after the *rate-limiting step*. (Some workers prefer to describe this phenomenon as 'equilibrium control'.)
See also *kinetic control*.
1994, *66*, 1173

thermodynamic energy
See *internal energy*.
G.B. 48

thermodynamic equilibrium constant
See *standard equilibrium constant*.
G.B. 50

thermodynamic isotope effect
The effect of isotopic substitution on an equilibrium constant is referred to as a thermodynamic (or equilibrium) isotope effect.
For example, the effect of isotopic substitution in reactant A that participates in the equilibrium:

$$A + B \rightleftharpoons C$$

is the ratio K^l/K^h of the equilibrium constant for the reaction in which A contains the light isotope to that in which it contains the heavy isotope. The ratio can be expressed as the equilibrium constant for the isotopic exchange reaction:

$$A^l + C^h \rightleftharpoons A^h + C^l$$

in which reactants such as B that are not isotopically substituted do not appear.
The potential energy surfaces of isotopic molecules are identical to a high degree of approximation, so thermodynamic isotope effects can only arise from the effect of isotopic mass on the nuclear motions of the reactants and products, and can be expressed quantitatively in terms of partition function ratios for nuclear motion:

$$\frac{K^l}{K^h} = \frac{(Q^l_{nuc}/Q^h_{nuc})_C}{(Q^l_{nuc}/Q^h_{nuc})_A}$$

Although the nuclear partition function is a product of the translational, rotational and vibrational partition functions, the isotope effect is determined almost entirely by the last named, specifically by vibrational modes involving motion of isotopically different atoms. In the case of light atoms (i.e. protium vs. deuterium or tritium) at moderate temperatures, the isotope effect is dominated by zero-point energy differences.
See also *fractionation factor*.
1994, *66*, 1131

thermodynamic quality of solvent (in polymer chemistry)
See *quality of solvent*.
P.B. 57

thermodynamic temperature, T
Base quantity in the system of quantities upon which SI is based.
G.B. 48; 1996, *68*, 996; see also 1990, *62*, 2217

thermoelectrometry
A technique in which an electrical characteristic of a substance (and/or its reaction product(s)) is measured as a function of temperature whilst the substance is subjected to a controlled temperature program.
The most common measurements are of resistance, conductance or capacitance.
1985, *57*, 1740

+ thermogram
See *enthalpogram*.
1994, *66*, 2490

thermogravimetry (TG)
A technique in which the mass of a substance (and/or its reaction product(s)) is measured as a function of temperature whilst the substance is subjected to a controlled temperature program.
1985, *57*, 1738; O.B. 39

thermoluminescence
Luminescence arising from a reaction between species trapped in a rigid matrix and released as a result of an increase in temperature.
1996, *68*, 2279

thermolysis
The uncatalysed cleavage of one or more covalent *bonds* resulting from exposure of a compound to a raised temperature, or a process in which such cleavage is an essential part.
See also *pyrolysis*.
1994, *66*, 1173

thermomagnetometry
A technique in which a magnetic characteristic of a substance (and/or its reaction product(s)) is measured as a function of temperature whilst the substance is subjected to a controlled temperature program.
1985, *57*, 1740

thermomechanical measurement
A technique in which the deformation of a substance (and/or its reaction product(s)) under non-oscillatory load is measured as a function of temperature whilst the substance is subjected to a controlled temperature program.
1985, *57*, 1740

+ thermometric
See *enthalpimetric analysis, thermometric titration*.
1994, *66*, 2489

+ thermometric enthalpy titration
See *thermometric titration*.
1994, *66*, 2490

thermometric titration
An analytical method in which one reactant (the titrant) is added continuously or stepwise to an adiabatic or isoperibol vessel containing another reactant. The enthalpy change(s) of the ensuing reaction(s) causes a temperature change which, when plotted versus volume of titrant, may be used to find the titration endpoint(s).
This is the preferred term for experiments producing plots of temperature versus volume of titrant in which the main goal is a quantitative determination. Nonetheless, when a calorimetric vessel is used, such that the heat capacity is known, thermodynamic parameters may also be estimated from such experiments. An acceptable synonym in that case is enthalpimetric titration. The use of the adjective thermometric is justified because of widespread historical and current usage, and because a titration of necessity implies a chemical reaction. The term thermometric enthalpy titration has been used, but is not recommended. A method in which the titrant is a catalyst for an indicator reaction that occurs after the endpoint for the analyte reaction should be called a thermometric titration with catalytic endpoint detection, not a catalytic thermometric titration.
1994, *66*, 2490

thermoparticulate analysis
A thermoanalytical technique in which the release of particulate matter from a substance (and/or its reaction product(s)) is measured as a function of temperature whilst the substance is subjected to a controlled temperature programme.
1985, *57*, 1739

thermophile
An organism that can tolerate high temperatures and that grows optimally at temperatures above 45 °C.
1992, *64*, 166

thermophotometry
See *thermoptometry*.
O.B. 46

thermopile
Several *thermocouples* connected in series to increase the magnitude of the electric potential.
1995, *67*, 1752

thermoptometry
A technique in which an optical characteristic of a substance (and/or its reaction product(s)) is measured as a function of temperature whilst the substance is subjected to a controlled temperature program.
Measurements of total light, light of specific wavelength(s), refractive index and *luminescence* lead to thermophotometry, thermospectrometry, thermorefractometry and thermoluminescence, respectively; observation under the microscope leads to thermomicroscopy.
1985, *57*, 1740

thermosonimetry
A technique in which the sound emitted by a substance (and/or its reaction product(s)) is measured as a function of temperature whilst the substance is subjected to a controlled temperature program.
 1985, *57*, 1740

thermosphere
Atmospheric shell extending from the top of the *mesosphere* to outer space. It is a region of more or less steadily increasing temperature with height, starting at 70 or 80 km. It includes the exosphere and most or all of the ionosphere (not the D region).
 1990, *62*, 2217

theta
 For entries, see under θ (beginning of 't').

thiazynes
 See *hydrocarbylsulfanyl nitrenes*.
 1995, *67*, 1371

thick film
 See *thin film*.
 1994, *66*, 1672

thickness of diffusion layer (in electrochemistry)
The distance from the electrode where the ratio $(c - c_e)/(c_0 - c_e)$ reaches a given value. (c_e is the *interfacial concentration*, c is the concentration, c_0 is the value in the bulk solution.) If this ratio is selected as, say, 0.99 the corresponding thickness of the diffusion layer is denoted by $\delta_{0.99}$.
 See also *diffusion layer*.
 1981, *53*, 1837

thickness of electrical double layer
The length characterizing the decrease with distance of the potential in the double layer = characteristic Debye length in the corresponding electrolyte solution = κ^{-1}:

$$\frac{1}{\kappa} = [\varepsilon_r \varepsilon_0 RT/(F^2 \sum_i c_i z_i^2)]^{1/2}$$

(rationalized four-quantity system)

$$\frac{1}{\kappa} = [\varepsilon_r RT/(4\pi F^2 \sum_i c_i z_i^2)]^{1/2}$$

(three-quantity electrostatic system)
where ε = static permittivity = $\varepsilon_r \varepsilon_0$, ε_r = relative static permittivity of solution; ε_0 = permittivity of vacuum, R = gas constant, T = *thermodynamic temperature*, F = *Faraday constant*, c_i = concentration of species i, z_i = ionic charge on species i.
 1972, *31*, 619

thickness of reaction layer (in electrochemistry)
When a *kinetic current* flows, the concentrations of the electroactive substance B and its precursor C at very small distances from the electrode surface are influenced both by mass transfer and by the finite rate of establishment of the chemical equilibrium. As the distance from the electrode surface increases, the chemical equilibrium is more and more nearly attained. The thickness of the reaction layer is the distance from the electrode surface beyond which deviations from the chemical equilibrium between C and B are taken to be negligibly small.
 1985, *57*, 1504

thin film
A *film* whose thickness is of the order of a *characteristic scale* or smaller. Since a film may 'look' operationally thin or thick, according to the procedure applied, it is also recommended that the measurement procedure employed be specified (e.g. ellipsometrically thin film, optically thin film, etc.). It is recommended that the physical specification of the film thickness be used, whenever possible (e.g. thick compared to the electron mean free path, thin compared to the optical wavelength, etc.)
 1994, *66*, 1672

thin-layer chromatography
Chromatography carried out in a layer of *adsorbent* spread on a support e.g. a glass plate.
 O.B. 94

thio
This prefix properly denotes replacement of an oxygen by a sulfur, e.g. PhC(=S)NH$_2$, thiobenzamide.
 See also entries prefixed thio-.
 1995, *67*, 1371

thioacetals
A term including monothioacetals having the structure R$_2$C(OR')(SR') (subclass monothioketals, R ≠ H); and dithioacetals having the structure R$_2$C(SR')$_2$ (subclass dithioketals, R ≠ H, R' ≠ H).
 See also *thiohemiacetals*.
 1995, *67*, 1371

thioaldehydes
Compounds in which the oxygen of an *aldehyde* has been replaced by divalent sulfur, RC(=S)H, e.g. CH$_3$CH$_2$C(=S)H propanethial.
 1995, *67*, 1372

thioaldehyde S-oxides
Compounds having the structure RC(=S=O)H. (Also known by the disapproved term *sulfines* which includes thioaldehyde S-oxides and thioketone S-oxides).
 1995, *67*, 1372

thioanhydrides
Compounds having the structure acyl-S-acyl. Also called diacylsulfanes, e.g. CH$_3$C(=O)SC(=S)-CH$_2$CH$_3$ acetic thiopropanoic thioanhydride.
 See *acyl groups*.
 1995, *67*, 1372

thiocarboxylic acids
Compounds in which one or both oxygens of a carboxy group have been replaced by divalent sulfur RC(=O)SH or RC(=S)OH monothiocarboxylic acids, RC(=S)SH dithiocarboxylic acids.
 1995, *67*, 1372

thiocyanates
Salts and esters of thiocyanic acid HSC≡N, e.g. $CH_3SC{\equiv}N$ methyl thiocyanate.
　See *isothiocyanates*.
　1995, *67*, 1372

thioethers
Former name for *sulfides* RSR (R ≠ H).
　1995, *67*, 1372

thiohemiacetals
Compounds of structure $R_2C(SR')OH$ or $R_2C(OR')SH$ (monothiohemiacetals), or $R_2C(SR')SH$ (dithiohemiacetals), R' ≠ H.
　1995, *67*, 1372

thioketones
Compounds in which the oxygen of a *ketone* has been replaced by divalent sulfur $R_2C{=}S$ (R ≠ H), e.g. $CH_3C(=S)CH_2CH_3$ butane-2-thione.
　1995, *67*, 1372

thioketone *S*-oxides
Compounds having the structure $R_2C{=}S{=}O$ (R ≠ H). (Also known by the disapproved term *sulfines* which comprises thioaldehyde *S*-oxides and thioketone *S*-oxides.)
　1995, *67*, 1372

thiolates
Derivatives of *thiols*, in which a metal (or other cation) replaces the hydrogen attached to sulfur, e.g. CH_3S^- Na^+ sodium methanethiolate.
　1995, *67*, 1372

thiols
Compounds having the structure RSH (R ≠ H), e.g. $MeCH_2SH$ ethanethiol. Also known by the term mercaptans (abandoned by IUPAC).
　1995, *67*, 1372

thionylamines
　See *sulfinylamines*.
　1995, *67*, 1372

third body
A species, other than the reactant itself, which brings about the energization of a molecule that can undergo a unimolecular process, or brings about a combination reaction between atoms or radicals. In the latter case a third body is also called a *chaperon*.
　1996, *68*, 189

thixotropy
　See *work softening*.
　1979, *51*, 1217

+ thiyl radicals
Synonymous with *sulfenyl radicals*. Due to inconsistencies in use the term is not recommended.
　1995, *67*, 1372

threo
　See *erythro*.
　1996, *68*, 2220

threo **structures (in a polymer)**
　See *erythro structures (in a polymer)*.
　P.B. 36

threshold energy, E_0
Potential energy gap between *reactants* and the *transition state*, sometimes involving the zero point energies, but usually not.
　See also *critical energy*.
　G.B. 56; 1996, *68*, 189

through-bond electron transfer
Intramolecular *electron transfer* for which the relevant electronic interaction between the donor and acceptor sites is mediated by through-bond interaction, i.e. via the covalent bonds interconnecting these sites, as opposed to through-space interaction.
　See also *through-space electron transfer*.
　1996, *68*, 2280

+ throughput rate
Usage not recommended.
　See *input rate, output rate*.
　1989, *61*, 1664

through-space electron transfer
Electron transfer for which the relevant electronic interaction between the donor and acceptor sites is mediated either by direct orbital overlap or by *superexchange interaction* via intervening molecular entities not covalently bound to the donor or acceptor sites.
　See also *through-bond electron transfer*.
　1996, *68*, 2280

throwing power
　See *current distribution*.
　1981, *53*, 1836

TICT emission
Electronic emission emanating from a *TICT* state.
　See also *TICT state, twisted internal charge transfer*.
　1996, *68*, 2280

TICT state
The acronym derives from Twisted Internal Charge Transfer state, proposed to be responsible for strongly Stokes-shifted fluorescence from certain aromatics, particularly in a polar medium.
　See *twisted internal charge transfer*.
　1996, *68*, 2280

tie molecule (in polymers)
A molecule that connects at least two different crystals
　P.B. 83

tight ion pair
　See *ion pair*.
　1994, *66*, 1173

time, *t*
Base quantity in the system of quantities upon which SI is based.
　G.B. 11; 1996, *68*, 996

time constant (of a detector), τ_c
If the output of a *detector* changes exponentially with time, the time required for it to change from its initial value by the fraction $[1 - \exp(-t/\tau_c)]$ (for $t = \tau_c$) of the final value, is called the time constant.
　1995, *67*, 1751

time-correlated single photon counting
A technique for the measurement of the time histogram of a sequence of photons with respect to a periodic event, e.g. a flash from a repetitive nanosecond lamp or a CW operated laser (mode locked laser). The essential part is a time-to-amplitude-converter (TAC) which transforms the arrival time between a start and a stop pulse into a voltage. Sometimes called single photon timing.
    1996, *68*, 2280

time-dependent stoichiometry
A state in which if intermediates are formed in significant amounts during the course of a reaction, the overall stoichiometric equation does not apply throughout.
    1981, *53*, 754

time-independent stoichiometry
A state in which the stoichiometric equation applies throughout the course of the reaction.
    1981, *53*, 754

time of centrifugation, *t*
The time difference from switching on until switching off. The time for deceleration is not included.
    1994, *66*, 905

time of deactivation (in heterogeneous catalysis)
    See *decay time (in heterogeneous catalysis)*.
    1976, *46*, 83

time-of-flight mass spectrometer
An arrangement using the fact that ions of different mass/charge need different times to travel through a certain distance in a field-free region after they have all been initially given the same translational energy.
    1991, *63*, 1546; O.B. 202

time of solidification (or time of thixotropic recovery)
The time in which a certain *viscosity* or *yield stress* is reached after discontinuation of a shear. These times depend on the values of viscosity or yield stress chosen by the experimenter.
    1979, *51*, 1217

time-resolved microwave conductivity (TRMC)
Technique allowing the quantitative and qualitative detection of radiation-induced *charge separation* by time-resolved measurement of the changes in microwave absorption resulting from the production and decay of charged and dipolar molecular entities.
    1996, *68*, 2280

time-resolved spectroscopy
The recording of spectra at a series of time intervals after the excitation of the system with a light pulse (or other perturbation) of appropriately short duration.
    1996, *68*, 2280

titrant
The solution containing the active agent with which a *titration* is made.
    O.B. 47

titration
The process of determining the quantity of a substance A by adding measured increments of substance B, with which it reacts (almost always as a standardized solution called the titrant, but also by electrolytic generation, as in coulometric titration) with provision for some means of recognizing (indicating) the endpoint at which essentially all of A has reacted. If the endpoint coincides with the addition of the exact chemical equivalence, it is called the equivalence point or stoichiometric or theoretical endpoint, thus allowing the amount of A to be found from known amounts of B added up to this point, the reacting weight ratio of A to B being known from stoichiometry or otherwise.
Terms for varieties of titration can reflect the nature of the reaction between A and B. Thus, there are acid–base, complexometric, chelatometric, oxidation–reduction, and precipitation titrations.
Additionally, the term can reflect the nature of the titrant, such as acidimetric, alkalimetric, and iodometric titrations as well as coulometric titrations, in which the titrant is generated electrolytically rather than being added as a standard solution.
    O.B. 47; see also 1990, *62*, 2217

titration curve
A plot of a variable related to a relevant concentration (activity) as the ordinate vs. some measure of the amount of titrant, usually titration volume (titre) as the abscissa. If the variable is linearly related to concentrations, such as the electrical conductance or the photometric absorbance, the term linear titration curve is used. When a logarithmic expression of the concentration or activity is used, such as the pH, pM, or the electrical potential in mV, the curve is referred to as a logarithmic titration curve.
    O.B. 49

titration error
The difference in the amount of *titrant*, or the corresponding difference in the amount of substance being titrated, represented by the expression:

    (*end-point* value − *equivalence-point* value).
    O.B. 48

titration, radiometric
    See *radiometric titration*.
    1982, *54*, 1553

titre (titer)
The reacting strength of a standard solution, usually expressed as the weight (mass) of the substance equivalent to 1 cm^3 of the solution.
    O.B. 48

T-jump
    See *chemical relaxation*.
    1994, *66*, 1171

tNRA
    See *transfer RNA*.
    1992, *64*, 167

tonne
Non-SI unit of mass, t = 10^3 kg.
  G.B. 111

topochemical reaction
A reversible or irreversible reaction that involves the introduction of a guest species into a host structure and that results in significant structural modifications to the host, for example, the breakage of bonds.
Example:
The insertion of lithium at 50 °C into the spinel Li[Mn_2]O_4 with symmetry $F\bar{d}3m$ to yield a layered structure with symmetry $P\bar{3}m1$.

3 Li + Li[Mn_2]O_4 → 2Li_2MnO_2

Alternative terms: topotactic reaction, *insertion reaction*.
  1994, *66*, 592

topomerization
The identity reaction leading to exchange of the positions of identical ligands. The indistinguishable molecular entities involved are called topomers. For example in the reaction below the two identical ligating atoms N_a and N_b are interchanged by rotation about the C-aryl bond but may be identified by NMR spectroscopy. This is a degenerate isomerization.

  1996, *68*, 2220

topomers
  See *topomerization*.
  1996, *68*, 2220

topotactic reaction
  See *topochemical reaction*.
  1994, *66*, 592

topotactic transition
A transition in which the crystal lattice of the product phase shows one or more crystallographically equivalent, orientational relationships to the crystal lattice of the parent phase.
Example: Transitions in which the anionic array is unchanged during the transition but cation reorganization occurs, as in:

β-Li_2ZnSiO_4 → γ-Li_2ZnSiO_4
  1994, *66*, 593

torque, *T*
Sum of *moments of forces* not acting along the same line.
  G.B. 12; ISO 31-2: 1992

torquoselectivity
The preference for 'inward' or 'outward' rotation of substituents in conrotatory or disrotatory electrocyclic ring opening reactions.
  1994, *66*, 1173

torr
Non-SI unit of pressure, torr ≈ 133.322 Pa.
  G.B. 112; 1996, *68*, 996

torsional braid analysis
A particular case of dynamic thermomechanometry in which the material is supported on a braid.
  O.B. 45

torsional stereoisomers
Stereoisomers that can be interconverted (actually or conceptually) by torsion about a bond axis. This includes *E,Z*-isomers of alkenes, *atropisomers* and *rotamers*.
  1996, *68*, 2221

torsional strain
  See *eclipsing strain*.
  1996, *68*, 2221

torsion angle
In a chain of atoms A–B–C–D, the *dihedral angle* between the plane containing the atoms A,B,C and that containing B,C,D. In a *Newman projection* the torsion angle is the angle (having an absolute value between 0° and 180°) between bonds to two specified (*fiducial*) groups, one from the atom nearer (proximal) to the observer and the other from the further (distal) atom. The torsion angle between groups A and D is then considered to be positive if the bond A–B is rotated in a clockwise direction through less than 180° in order that it may *eclipse* the bond C–D: a negative torsion angle requires rotation in the opposite sense. Stereochemical arrangements corresponding to torsion angles between 0° and ± 90° are called syn (*s*), those corresponding to torsion angles between ± 90° and 180° anti (*a*). Similarly, arrangements corresponding to torsion angles between 30° and 150° or between –30° and –150° are called clinal (*c*) and those between 0° and 30° or 150° and 180° are called periplanar (*p*). The two types of terms can be combined so as to define four ranges of torsion angle; 0° to 30° synperiplanar (*sp*); 30° to 90° and –30° to –90° synclinal (*sc*); 90° to 150°, and –90° to –150° anticlinal (*ac*); ± 150° to 180° antiperiplanar (*ap*).

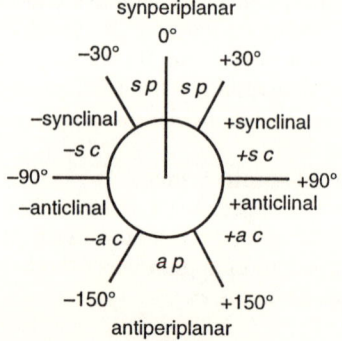

The synperiplanar conformation is also known as the *syn*- or *cis*-conformation; antiperiplanar as *anti* or *trans* and synclinal as gauche or skew. For macromolecular usage the symbols T, C, G^+, G^-, A^+ and

A⁻ are recommended (*ap*, *sp*, *+sc*, *−sc*, *+ac* and *−ac* respectively).
1996, *68*, 2220

total absorption peak
See *full energy peak*.
1982, *54*, 1553

total chemiflux
See *chemical flux*.
1994, *66*, 1095

total consumption time (in flame emission and absorption spectrometry), t_{tot}
The time necessary to consume the prepared sample completely (in s). It is equal to the volume of the prepared sample divided by the rate of fluid consumption.
1986, *58*, 1740

total ion current (in mass spectrometry)
1. (after mass analysis): The sum of the separate ion currents carried by the different ions contributing to the spectrum (this is sometimes called the reconstructed ion current).
2. (before mass analysis): The sum of all the separate ion currents for ions of the same sign before mass analysis.
1991, *63*, 1554; O.B. 206

totally porous packing (in chromatography)
Packing in which the stationary phase permeates each porous particle.
1993, *65*, 831

total radiant power
See *spectral radiant power*.
1985, *57*, 112

total retention volume (time) (in column chromatography), V_R, t_R
The volume of mobile phase entering the column between sample injection and the emergence of the peak maximum of the sample component of interest, or the corresponding time. It includes the hold-up volume (time):

$$t_R = V_R/F_c$$

where F_c is the mobile phase flow rate at column temperature.
1993, *65*, 841

toxicity
1. Capacity to cause injury to a living organism defined with reference to the quantity of substance administered or absorbed, the way in which the substance is administered (inhalation, ingestion, topical application, injection) and distributed in time (single or repeated doses), the type and severity of injury, the time needed to produce the injury, the nature of the organism(s) affected and other relevant conditions.
2. Adverse effects of a substance on a living organism defined with reference to the quantity of substance administered or absorbed, the way in which the substance is administered (inhalation, ingestion, topical application, injection) and distributed in time (single or repeated doses), the type and severity of injury, the time needed to produce the injury, the nature of the organism(s) affected, and other relevant conditions
3. Measure of incompatibility of a substance with life: this quantity may be expressed as the reciprocal of the absolute value of median lethal dose (1/LD$_{50}$) or concentration (1/LC$_{50}$).
1993, *65*, 2107

toxicodynamics
Study of toxic actions on living systems, including the reactions with and binding to cell constituents, and the biochemical and physiological consequences of these actions.
1993, *65*, 2108

toxicokinetics
Process of the uptake of potentially toxic substances by the body, the biotransformation they undergo, the distribution of the substances and their metabolites in the tissues, and the elimination of the substances and their metabolites from the body. Both the amounts and the concentrations of the substances and their metabolites are studied. The term has essentially the same meaning as pharmacokinetics, but the latter term should be restricted to the study of pharmaceutical substances.
See also *pharmacokinetics*.
1993, *65*, 2108

toxicology
Scientific discipline involving the study of the actual or potential danger presented by the harmful effects of substances (poisons) on living organisms and ecosystems, of the relationship of such harmful effects to exposure, and of the mechanisms of action, diagnosis, prevention and treatment of intoxications.
1993, *65*, 2108

toxin
Poisonous substance produced by a biological organism such as a microbe, animal or plant.
1993, *65*, 2109

toxinology
Scientific discipline involving the study of the chemistry, biochemistry, pharmacology and toxicology of toxins.
See also *toxicology*.
1993, *65*, 2109

traceability
The property of a result or measurement whereby it can be related to appropriate standards, generally international or national standards, through an unbroken chain of comparisons.
1995, *67*, 1701

trace element
Any element having an average concentration of less than about 100 parts per million atoms (ppma) or less than 100 µg per g.
1979, *51*, 2246

tracer
A foreign substance mixed with or attached to a given substance to enable the distribution or location of the

latter to be determined subsequently. There are several types of tracers which are used: (i) A physical tracer is one that is attached by physical means to the object being traced; (ii) A chemical tracer is a chemical with properties similar to those of the substance being traced with which it is mixed homogeneously; (iii) An isotopic tracer is a unique isotope, either radioactive or an enriched, uncommon stable isotope, of the element to be traced; (iv) A radioactive tracer is a physical or chemical tracer having radioactivity as its distinctive property which allows detection at small concentrations and hence after large transport distances. The composition of aerosols in the troposphere has been used as a qualitative tracer of air masses. The elemental analyses (determined by neutron activation, X-ray fluorescence, etc.) of the aerosols transported from various sources or source regions sometimes have characteristic patterns which are used to define qualitatively the origin of tropospheric aerosols collected in other geographical regions.
   1990, *62*, 2217; 1994, *66*, 2525

tracer, generally labelled
   See *generally labelled tracer*.
   1994, *66*, 2526

tracer, isotopic
   See *isotopic tracer*.
   1994, *66*, 2526

tracer, nominally labelled
   See *nominally labelled tracer*.
   1994, *66*, 2526

tracer, specifically labelled
   See *specifically labelled tracer*.
   1994, *66*, 2526

tracer, stereospecifically labelled
   See *stereospecifically labelled tracer*.
   1994, *66*, 2526

tracer, uniformly labelled
   See *uniformly labelled tracer*.
   1994, *66*, 2526

track (nuclear)
The path of an ionizing particle as revealed by a *track detector*.
   1982, *54*, 1553

track detector (nuclear)
A detector which makes the paths of ionizing particles visible, either directly (e.g. cloud chamber) or after suitable treatment (photographic emulsion, polymers).
   1982, *54*, 1553

trajectory (in reaction dynamics)
A path taken by a reaction system over a *potential-energy surface*, or a diagram or mathematical description that represents that path. A trajectory can also be called a reaction path.
   1996, *68*, 189

trans
   See *cis, trans*.
   1996, *68*, 2221

***trans-* (in inorganic nomenclature)**
A prefix designating two groups directly across a central atom from each other, i.e. in the polar positions on a sphere, not now generally recommended for precise nomenclature purposes.
   R.B. 245

transannular strain
In medium-sized ring compounds, the strain due to repulsive *non-bonded interactions* between substituents or hydrogen atoms attached to non-adjacent ring atoms.
   1996, *68*, 2221

***trans* conformation (in polymers)**
   See *cis conformation (in polymers)*.
   P.B. 41

transcription (in biotechnology)
The process by which the genetic information encoded in a linear sequence of *nucleotides* in one strand of DNA is copied into an exactly complementary sequence of RNA.
   1992, *64*, 167

transducer
An analytical instrument which provides an output quantity having a given relationship to the input quantity.
   1989, *61*, 1662

transduction (in biotechnology)
The transfer of genetic information from one *bacterium* to another by means of a transducing *bacteriophage*. When the *phage* is grown on the first *host*, a fragment of the host DNA can be incorporated into the phage particles. This foreign DNA can be transferred to the second host upon infection with progeny phage from the first experiment.
   1992, *64*, 167

transfer
Movement of a component within a system or across its boundary. It may be expressed using different kinds of quantities, e.g rates of change dQ/dt or $\Delta Q/\Delta t$.
Examples are: mass rate, dm_B/dt or $\Delta m_B/\Delta t$; substance rate, dn_B/dt or $\Delta n_B/\Delta t$.
   1992, *64*, 1573

transfer (in analysis)
The action of moving material between containers, or containers between transport devices. This term is of particular importance in clinical chemistry.
   1989, *61*, 1662

transferability
Transferability assumes invariance of properties, associated conceptually with an atom or a fragment present in a variety of molecules. The property, such as *electronegativity*, nucleophilicity, NMR *chemical shift*, etc. is held as retaining a similar value in all these occurrences.
   1994, *66*, 1173

transfer activity coefficient, γ_t
A term used to quantify the difference in the free energy of a solute ion in two different standard states often in two different liquid phases. The relationship is $\Delta_t G^o = \nu RT \ln \gamma_t$ where $\Delta_t G^o$ is the transfer Gibbs energy and ν is the number of ions in the solute.
Notes:
1. It should not be confused with the mass transfer coefficient which represents the specific rate of transfer of a species from one phase to another.
2. It does not necessarily imply the physical transfer of a solute between two liquid phases.
See also *partition constant*.
1993, *65*, 2386

transferases
Enzymes that catalyse reactions in which a group is transferred from one compound to another. Groups that are transferred are C_1, aldehydic or ketonic residues, acyl, glycosyl, alkyl, nitrogenous, phosphorus and sulfur-containing groups.
1992, *64*, 167

transfer line
Line provided to carry the sample to be analysed from the sample point to the analytical unit without altering the composition of the sample.
1990, *62*, 2218

transfer RNA (tRNA)
A single-stranded RNA molecule containing about 70–90 *nucleotides*, folded by intrastrand *base pairing* into a characteristic secondary ('cloverleaf') structure that carries a specific amino acid and matches it to its corresponding *codon* on an mRNA during protein synthesis.
1992, *64*, 167

transformation
The conversion of a *substrate* into a particular product, irrespective of reagents or *mechanisms* involved. For example, the transformation of aniline ($C_6H_5NH_2$) into *N*-phenylacetamide (C_6H_5NH-$COCH_3$) may be effected by use of acetyl chloride or acetic anhydride or ketene. A transformation is distinct from a reaction, the full description of which would state or imply all the reactants and all the products.
1994, *66*, 1173

transformation (in gene technology)
A process for genetic alteration of a *cell* following incorporation of foreign DNA.
1992, *64*, 167

transformation, nuclear
See *nuclear transformation*.
1982, *54*, 1553

***trans*-fused**
See *cis-fused*.
B.B. 479

transient phase (induction period)
This expression relates to the period of time that elapdes prior to the establishment of a *steady state*. Initially the concentration of a reactive intermediate is zero, and it rises to the steady-state concentration during the transient phase.
1996, *68*, 189

transient (chemical) species
Relating to a short-lived reaction *intermediate*. It can be defined only in relation to a time scale fixed by the experimental conditions and the limitations of the technique employed in the detection of the intermediate. The term is a relative one.
Transient species are sometimes also said to be 'metastable'. However, this latter term should be avoided, because it relates a thermodynamic term to a kinetic property, although most transients are also thermodynamically *unstable* with respect to reactants and products.
See also *persistent*.
1994, *66*, 1174

transient species
A species occurring as a short-lived intermediate in a reaction occurring by a composite mechanism.
1996, *68*, 189

transient spectroscopy
A technique for the spectroscopic observation of transient species (excited-state molecular entities or reactive intermediates) generated by a pulse of short duration.
See also *flash photolysis, time-resolved spectroscopy*.
1996, *68*, 2280

transition
See $n \rightarrow \sigma^*$ *transition*, $\pi \rightarrow \pi^*$ *transition*, $\pi \rightarrow \sigma^*$ *transition*, $\sigma \rightarrow \sigma^*$ *transition*.
1988, *60*, 1079, 1089, 1099

transition coordinate
The *reaction coordinate* at the *transition state* corresponding to a vibration with an imaginary frequency. Motion along it in the two opposite senses leads towards the reactants or towards the products.
See also *reaction coordinate, transition state*.
1994, *66*, 1174

transition element
An element whose atom has an incomplete d sub-shell, or which can give rise to cations with an incomplete d sub-shell.
R.B. 43

transition interval (in titrimetric analysis)
The range of concentration over which the eye is able to perceive change in hue, colour intensity, fluorescence or other property of a visual indicator arising from the varying ratio of the two relevant forms of the indicator.
O.B. 48

transition, isomeric (in nuclear chemistry)
See *isomeric transition (in nuclear chemistry)*.
1982, *54*, 1553

transition layer
See *interfacial double-layer*.
1994, *66*, 1674

transition (dipole) moment, M_{nm}
An oscillating electric or magnetic moment can be induced in an atom or molecular entity by an electromagnetic wave. Its interaction with the electromagnetic field is resonant if the frequency of the latter corresponds to the energy difference between the initial and final states of a transition ($\Delta E = h\nu$). The amplitude of this moment is referred to as the transition moment. It can be calculated from an integral taken over the product of the wavefunctions of the initial (m) and final (n) states of a spectral transition and the appropriate dipole moment operator ($\vec{D}$) of the electromagnetic radiation:

$$M_{nm} = e \int \Psi_n^* \sum_i Z_i \vec{r}_i \ \Psi_m \, d\tau$$

where the summation is over the coordinates of all charged particles (electrons and nuclei). Its sign is arbitrary, its direction in the molecular framework defines the direction of transition polarization, and its square determines the strength of the transition. If e is omitted one obtains R_{nm} in the sense used in oscillator strength. The SI unit of the transition dipole moment is C m. The common unit is debye (D).
1996, *68*, 2281; G.B. 24

transition, nuclear
See *nuclear transition*.
1982, *54*, 1553

transition polarization
The direction of the *transition moment* in the molecular framework.
1996, *68*, 2281

transition probability for absorption (in spectrochemistry)
See *radiative absorption (in spectrochemistry)*.
1985, *57*, 1463

transition probability for spontaneous emission (in spectrochemistry)
See *radiative de-excitation (in spectrochemistry)*.
1985, *57*, 1463

transition probability for stimulated emission (in spectrochemistry)
See *radiative absorption (in spectrochemistry)*.
1985, *57*, 1463

transition, radiationless
See *radiative transition*.
O.B. 45

transition species
Any intermediate species in an *elementary reaction*. An *activated complex* is an example of a transition species, but the latter term also includes other intermediate species.
Transition species have also been called transition states, but there is then danger of confusion with a true *activated complex*, which is also commonly known as a transition state.
1996, *68*, 190

transition state
In theories describing *elementary reactions* it is usually assumed that there is a transition state of more positive molar Gibbs energy between the reactants and the products through which an assembly of atoms (initially composing the *molecular entities* of the reactants) must pass on going from reactants to products in either direction. In the formalism of 'transition state theory' the transition state of an elementary reaction is that set of states (each characterized by its own geometry and energy) in which an assembly of atoms, when randomly placed there, would have an equal probability of forming the reactants or of forming the products of that elementary reaction. The transition state is characterized by one and only one imaginary frequency. The assembly of atoms at the transition state has been called an activated complex. (It is not a *complex* according to the definition in this Compendium.)
It may be noted that the calculations of reaction rates by the transition state method and based on calculated *potential-energy surfaces* refer to the potential energy maximum at the saddle point, as this is the only point for which the requisite separability of transition state coordinates may be assumed. The ratio of the number of assemblies of atoms that pass through to the products to the number of those that reach the saddle point from the reactants can be less than unity, and this fraction is the 'transmission coefficient' κ. (There are also reactions, such as the gas-phase *colligation* of simple *radicals*, that do not require 'activation' and which therefore do not involve a transition state.)
See also *Gibbs energy of activation, Hammond principle, potential energy profile, transition structure, activated complex*.
1994, *66*, 1174

transition state analogue
A substrate designed to mimic the properties or the geometry of the *transition state* of reaction.
1994, *66*, 1174

transition state theory
A theory of the rates of *elementary reactions* which assumes a special type of equilibrium, having an equilibrium constant $K^{\ddagger}$, to exist between reactants and activated complexes. According to this theory the rate constant is given by:

$$k = (k_B T/h) K^{\ddagger}$$

where k_A is the Boltzmann constant and h is the Planck constant. The rate constant can also be expressed as:

$$k = (k_B T/h) \exp(\Delta^{\ddagger} S^o/R) \exp(-\Delta^{\ddagger} H^o/RT)$$

where $\Delta^{\ddagger} S^o$, the entropy of activation, is the standard molar change of entropy when the activated complex is formed from reactants and $\Delta^{\ddagger} H^o$, the enthalpy of activation, is the corresponding standard molar change of enthalpy. The quantities E_a (the *energy of activation*) and $\Delta^{\ddagger}H^o$ are not quite the same, the relationship between them depending on the type of reaction. Also:

$$k = (k_B T/h) \exp(-\Delta^{\ddagger}G^o/RT)$$

where $\Delta^{\ddagger}G^o$, known as the *Gibbs energy of activation*, is the standard molar Gibbs energy change for the conversion of reactants into activated complex. A plot of standard molar Gibbs energy against a reaction coordinate is known as a Gibbs-energy profile; such plots, unlike *potential-energy profiles*, are temperature-dependent.

In principle the equations above must be multiplied by a transmission coefficient, κ, which is the probability that an activated complex forms a particular set of products rather than reverting to reactants or forming alternative products.

It is to be emphasized that $\Delta^{\ddagger}S^o$, $\Delta^{\ddagger}H^o$ and $\Delta^{\ddagger}G^o$ occurring in the former three equations are not ordinary thermodynamic quantities, since one degree of freedom in the activated complex is ignored.

Transition-state theory has also been known as absolute rate theory, and as activated-complex theory, but these terms are no longer recommended.
1996, *68*, 190

transition structure
A saddle point on a *potential-energy surface*. It has one negative force constant in the harmonic force constant matrix.
See also *activated complex, transition state*.
1994, *66*, 1174

transition time (in electroanalytical chemistry)
In *chronopotentiometry* and related techniques, the time that elapses between the instant at which current is applied and the instant at which the concentration of an electroactive substance B at the electrode-solution interface becomes indistinguishable from zero. In experimental practice the latter time is often taken to be the instant at which the rate of variation of the potential of the *indicator electrode* attains a maximum value.
1985, *57*, 1504

transition wavenumber, $\tilde{\nu}$
Difference of *term* values for the corresponding states (upper minus lower).
G.B. 23

transit time (in flame emission and absorption spectrometry), t_{ts}
The time needed for the component to pass throught the observation space (in s). It depends on the geometry of the observation space, the flame temperature and the solute nebulized.
1986, *58*, 1742

translation (in biotechnology)
The unidirectional process that takes place on the *ribosomes* whereby the genetic information present in an *mRNA* is converted into a corresponding sequence of amino acids in a protein.
1992, *64*, 167

translational spectroscopy
A technique to investigate the distribution of the velocity of product ions from reactions of ions and neutral species.
1991, *63*, 1556

transmission
See *transmittance*.
1996, *68*, 996

transmission (in mass spectrometry)
The ratio of the number of ions leaving a region of a mass spectrometer to the number entering that region.
1991, *63*, 1546

transmission coefficient
See *transition state, transition state theory*.
1994, *66*, 1174; 1996, *68*, 190

transmission electron energy loss spectroscopy (TEELS)
Any technique in which an electron transparent specimen is bombarded with a finely focused electron beam and the energy distribution of the transmitted electrons is measured. This energy spectrum contains features corresponding to discrete losses of energy of the transmitted electrons due to excitation of electronic or plasmon states and provides information on the identity (and in some cases) chemical bonding of the elements in the sample.
1983, *55*, 2025

transmission electron microscopy (TEM)
Any technique in which an electron transparent sample is bombarded with an electron beam and the intensity of the transmitted electrons which is determined by scattering phenomena (electron absorption phenomena) in the interior of the sample is recorded. TEM essentially provides a high resolution image of the microstructure of a thin sample.
This technique is often just called electron microscopy. The term transmission electron microscopy is however recommended for the sake of a clear distinction from other electron microscopic techniques.
1983, *55*, 2024

transmission factor
See *transmittance*.
G.B. 32

transmission high energy electron diffraction (THEED)
Any technique which is based on the diffraction of high energy electrons (E_0 = 10–200 keV) in crystalline materials and evaluation of the angular distribution of the transmitted electrons. The diffraction pattern represents an image of the reciprocal lattice and therefore contains information about crystal structure. This technique is also often called selected area electron diffraction (SAED). For consistency with other

electron diffraction techniques, however, the term THEED is recommended.
O.B. 248

transmittance, T, τ
The ratio of the transmitted radiant power (P_λ) to that incident on the sample (P_λ^0):
$$T = P_\lambda/P_\lambda^0$$
Internal transmittance refers to energy loss by absorption, whereas the total transmittance is that due to absorption, reflection, scatter, etc.
See *absorbance, attenuance, Beer–Lambert law*.
1996, *68*, 2281; G.B. 32; 1996, *68*, 996

+ transoid conformation
Usage strongly discouraged.
See s-*cis*, s-*trans*; see *cisoid* for use of *transoid* as a stereodescriptor.
1996, *68*, 2221

transport (in analysis)
The action of moving materials within the analytical instrument. Transport can involve any of several means including pressure flow, where the materials are moved by fluid pressure, either continuously or discontinuously; centrifugal flow, where the materials are moved by centrifugal force; spontaneous motion, where the materials are moved due to their intrinsic properties, e.g. diffusion, capillarity; and discrete transport, where materials are enclosed within a moving container.
1989, *61*, 1662

transport control
See *microscopic diffusion control*.
1994, *66*, 1174

transport interference (in flame spectroscopy)
Transport interferences affect the amount of desolvated sample passing through the horizontal flame cross-section per unit time at the observation height. They include factors affecting the rate of liquid consumption, F_1, the efficiency of nebulization, ε_n, and the fraction desolvated β_s. They may be classified as non-specific (and physical).
O.B. 170

transport number, t
Of ions B, the current density due to ions B divided by the sum of current densities of all the ions in the electrolyte. ISO recommends the name current fraction of ions B.
G.B. 60

transposon
A movable DNA element that can be inserted at new sites into *plasmids* or *chromosomes* independently of the *host* cell recombination system. Prokaryotic transposons can carry *genes* that confer new phenotypic properties, such as resistance to antibiotics on the host.
1992, *64*, 167

transtactic polymer
A *tactic polymer* in which the main-chain double bonds of the *configurational base units* are entirely in the *trans*-arrangement.
P.B. 33

trapping
The interception of a *reactive* molecule or reaction *intermediate* so that it is removed from the system or converted into a more *stable* form for study or identification.
See also *scavenger*.
1994, *66*, 1174

travel time (in flame emission and absorption spectrometry), t_{tv}
The time needed for the component to be carried from the burner tip to the observation space (in s).
Travel time depends on the observation height and the flame rise velocity.
1986, *58*, 1742

treated solution (in analytical chemistry)
The *test solution* that has been subjected to reaction or separation procedures prior to measurement of some property.
1990, *62*, 1206

triads
See *constitutional sequence*.
1996, *68*, 2299

triangulo-
An affix used in names to denote three atoms bound into a triangle.
R.B. 245; B.B. 465

triazanes
Triazane, NH_2NHNH_2, and its *hydrocarbyl* derivatives.
1995, *67*, 1372

triazenes
Triazene, $NH_2N=NH$, and its *hydrocarbyl* derivatives.
See also *diazoamino compounds*.
1995, *67*, 1372

triboluminesence
Luminescence resulting from the rubbing together of the surface of certain solids. It can be produced, for example, when solids are crushed.
See *sonoluminescence*.
1996, *68*, 2281

trimethylenemethanes
The *diradical* $CH_2=C(\dot{C}H_2)_2$, 2-methylenepropane-1,3-diyl, for which no Kekulé structure can be written, and its *hydrocarbyl* derivatives.
1995, *67*, 1373

trioxides
Organic derivatives of trioxidane, HOOOH, e.g. ROOOR'. When R' = H, the compound is a hydrotrioxide.
1995, *67*, 1373

triple point
The point in a one-component system at which the temperature and pressure of three phases are in equilibrium. If there are p possible phases, there are $p!/(p-3)!3!$ triple points.
Example: In the sulfur system four possible *triple points* (one metastable) exist for the four phases comprising rhombic S (solid), monoclinic S (solid), S (liquid) and S (vapour).
1994, *66*, 593

triplet state
A state having a total electron spin quantum number of 1.
See *multiplicity*.
1996, *68*, 2281

triplet-triplet absorption
Absorption which takes place with the transition from the lowest triplet state of the molecule to higher triplet states (T → T_n) thus leading to the triplet-triplet absorption spectrum.
1984, *56*, 234

triplet-triplet annihilation
Two atoms or molecular entities both in a triplet state often interact (usually upon collision) to produce one atom or molecular entity in an excited singlet state and another in its ground singlet state. This is often, but not always, followed by delayed fluorescence.
See also *annihilation, spin conservation rule*.
1996, *68*, 2281

triplet-triplet energy transfer
Energy transfer from an electronically excited triplet donor to produce an electronically excited acceptor in its triplet state.
See *spin conservation rule*.
1996, *68*, 2282

triplet-triplet transitions
Electronic transitions in which both the initial and final states are triplet states.
1996, *68*, 2282

triprismo-
An affix used in names to denote six atoms bound into a triangular prism.
R.B. 245; B.B. 465

tritactic polymer
A *tactic polymer* that contains three sites of defined stereoisomerism in the main chain of the *configurational base unit*.
Example:

poly[3-(methoxycarbonyl)-4-methyl-*cis*-but-1-enylene]

P.B. 32

triterpenoids
Terpenoids having a C_{30} skeleton.
1995, *67*, 1373

tritide
See *tritium*.
1988, *60*, 1116

tritio
See *tritium*.
1988, *60*, 1116

tritium
A specific name for the atom ^{3}H. The cation ^{3}H$^+$ is a triton, the species ^{3}H$^-$ is a tritide anion and ^{3}H is the tritio group.
1988, *60*, 1116

triton
See *tritium*.
1988, *60*, 1116

trivial energy transfer
Synonymous with *radiative energy transfer*.
1996, *68*, 2282

trivial name
A name having no part used in a systematic sense.
B.B.(G) 14

Troe expression
A semiempirical description of the rate coefficient for a specific three body reaction [e.g. HO + NO$_2$ (+ M) → HONO$_2$ (+ M)] which represents well its pressure and temperature dependence in the region of transition between second and third order kinetics.
1990, *62*, 2218

+ tropilidenes
This term has inconsistently been used to mean cyclohepta-1,3,5-trienes or to mean cyclohepta-1,3,5-trienes in dynamic equilibrium with bicyclo-[4.1.0]hepta-2,4-dienes, so that the structure of the bulk substance is indeterminate in the time scale of the method of observation. Use of this term is therefore discouraged.

1995, *67*, 1373

tropolones
2-Hydroxycyclohepta-2,4,6-trienones and derivatives formed by substitution.

See *tropones*.
1995, *67*, 1373

tropones
Compounds that contain the cyclohepta-2,4,6-trienone ring system.

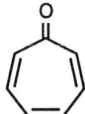

1995, *67*, 1373

tropopause
The region of the atmosphere which joins the *troposphere* and *stratosphere*, and where the decreasing temperature with altitude, characteristic of the troposphere ceases, and the temperature increase with height which is characteristic of the stratosphere begins.
1990, *62*, 2218

troposphere
The lowest layer of the atmosphere, ranging from the ground to the base of the *stratosphere* (*tropopause*) at 10–15 km of altitude depending on the latitude and meteorological conditions. About 70% of the mass of the atmosphere is in the troposphere. This is where most of the weather features occur and where the chemistry of the reactive anthropogenic species released into the atmosphere takes place.
1990, *62*, 2218

tropylium ions
The delocalized carbenium ion, cycloheptatrienylium, $C_7H_7^+$, derived formally by detachment of one hydride ion from the CH_2 group of cyclohepta-1,3,5-triene and substitution derivatives thereof.
1995, *67*, 1373

tropyl radicals
The delocalized radicals derived formally by abstraction of one hydrogen from the CH_2 group of cyclohepta-1,3,5-triene and substitution derivatives thereof.
1995, *67*, 1373

true coincidence (in radiochemistry)
A coincidence of events occurring in the same atom or in physically connected atoms.
O.B. 216

'true' rate constant
See *Frumkin effect*.
1980, *52*, 239

true value (in analysis), τ
The value that characterizes a quantity perfectly in the conditions that exist when that quantity is considered. It is an ideal value which could be arrived at only if all causes of measurement error were eliminated, and the entire population was sampled.
See also *measurement result*.
1994, *66*, 598; see also 1996, *68*, 996; VIM

tub conformation
A *conformation* (of symmetry group D_{2d}) of an eight-membered ring in which the four atoms forming one pair of diametrically opposite bonds in the ring lie in one plane and all other ring atoms lie to one side of that plane. It is analogous to the boat conformation of cyclohexane.

See also *crown conformation*.
1996, *68*, 2221

tunable laser (in spectrochemical analysis)
A high intensity source with narrow spectral bandwidth, with and without frequency doubling and/or Raman shifting. Its use may enable the spectral apparatus to be omitted. Tunable lasers may be either continuous (CW) or pulsed in nature.
1988, *60*, 1453

Tung distribution (of a macromolecular assembly)
A continuous distribution with the differential mass-distribution function of the form:

$$f_w(x)\,dx = abx^{b-1} \exp(-ax^b)\,dx$$

where x is a parameter characterizing the chain length, such as *relative molecular mass* or *degree of polymerization* and a and b are positive adjustable parameters.
P.B. 56

tunnelling
The process by which a particle or a set of particles crosses a barrier on its *potential-energy surface* without having the energy required to surmount this barrier. Since the rate of tunnelling decreases with increasing reduced mass, it is significant in the context of *isotope effects* of hydrogen isotopes.
1994, *66*, 1174; see also 1996, *68*, 2282

turbidimetric titration
The process in which a precipitant is added incrementally to a highly dilute polymer solution and the intensity of light scattered by, or the turbidity due to, the finely dispersed particles of the *polymer-rich phase* is measured as a function of the amount of precipitant added.
P.B. 67

turbidity (in light scattering), τ
The apparent absorbance of the incident radiation due to scattering. For small particles, direct proportionality exists between turbidity and the *Rayleigh ratio*.
P.B. 66; see also 1990, *62*, 2218

turnover frequency (in catalysis)
Commonly called the turnover number, N, and defined, as in enzyme catalysis, as molecules reacting per *active site* in unit time.
1976, *46*, 81

turnstile rotation
See *pseudorotation*.
1996, *68*, 2215

turntable reactor
See *merry-go-round reactor*.
1996, *68*, 2282

twisted internal charge transfer (TICT)
Intramolecular, photoinduced *charge transfer* between *chromophores* interconnected by a single bond leading to an *excited state* (a *TICT state*) in which the chromophores interact only weakly because of a considerable twist about the interconnecting bond.
1996, *68*, 2282

twist form
See *chair, boat, twist*.
1996, *68*, 2221

two-colour indicator
See *colour indicator*.
O.B. 48

two-dimensional chromatography
A procedure in which parts or all of the separated sample components are subjected to additional separation steps. This can be done e.g. by conducting a particular fraction eluting from the column into another column (system) having different separation characteristics.

When combined with additional steps, this may be described as multi-dimensional chromatography.

In planar chromatography two-dimensional chromatography refers to the chromatographic process in which the components are caused to migrate first in one direction and subsequently in a direction at right angles to the first one; the two elutions are carried out with different eluents.
1993, *65*, 827

two-photon excitation
Excitation resulting from successive or simultaneous absorption of two photons by an atom or molecular entity. This term is used for successive absorption only if some of the excitation energy of the first photon remains in the atom or molecular entity before absorption of the second photon. The simultaneous two-photon absorption can also be called biphotonic excitation.
See *two-photon process*.
1996, *68*, 2282

two-photon process
A photophysical or photochemical event triggered by a *two-photon excitation*.
1996, *68*, 2282

two-site immunoradiometric assay
Immunoradiometric assay involving two sets of *antibodies*, one of which is *labelled*, that combine with different immunoreactive sites of an *antigen* molecule.
1994, *66*, 2520

u

See *l, u*.
1996, *68*, 2221

ulosonic acids
See *ketoaldonic acids*.
1995, *67*, 1373

ultimate capacity (in solvent extraction)
The theoretical maximum capacity of a solvent containing a given concentration of extractant for a solute under any conditions. Where appropriate the term stoichiometric capacity can be used.
1993, *65*, 2386

ultrafiltrate
The solution obtained by *ultrafiltration*, in general not of the same composition as the *equilibrium solution*.
1972, *31*, 608

ultrafiltration
A separation process whereby a solution containing a solute of molecular size significantly greater than that of the solvent molecule is removed from the solvent by the application of a hydraulic pressure which forces only the solvent to flow through a suitable membrane, usually having a pore size in the range 0.001–0.1 µm.
1992, *64*, 167

ultrasonic detector (in gas chromatography)
Sound is generated in a reference cell and a measuring cell. An eluted component passing through the measuring cell changes the velocity of sound in the cell. This change is detected by a phase shift of acoustic signals between the two cells. This detector is usually employed for inorganic gases in the region where thermal conductivity detectors are not sufficiently sensitive. The detection threshold is of the order of 0.1 ppmv. The linear dynamic range is of the order of 10^3.
1990, *62*, 2192

umpire sample
A *sample* taken, prepared and stored in an agreed upon manner for the purpose of settling a dispute.
The agreement usually extends beyond the sample to the basis for reaching a decision (e.g. quantity of material from which taken, use of a third party and criteria serving as the basis for acceptance, rejection or economic adjustment).
The term 'reference sample' has been used in this context but this term more properly should be used in conjunction with a '*reference material*' or 'reference standard' which has a true or assigned value for a constituent or property. One of the characteristics of a reference material or reference standard is that it must have a negligible sampling error between test portions.
1990, *62*, 1203

umpolung
Any process by which the normal alternating donor and acceptor reactivity pattern of a chain, which is due to the presence of O or N heteroatoms, is interchanged. Reactivity umpolung is most often achieved by temporary exchange of heteroatoms (N, O) by others, such as P, S and Se.
The original meaning of the term has been extended to the reversal of any commonly accepted reactivity pattern. For example, reaction of R–C≡CX (X = halide) as a synthon for 'R–C≡C$^+$ (i.e. *electrophilic* acetylene) is an umpolung of the normal more common acetylide, R–C≡C$^-$ (i.e. *nucleophilic*) reactivity.
1994, *66*, 1174

unactivated adsorption process
If the temperature coefficient of the rate of *adsorption* is very small, the adsorption process is said to be unactivated (i.e. to have a negligible *activation energy*). In this case the *sticking coefficient* at low coverages may be near unity particularly for smaller molecules.
1976, *46*, 79

uncertainty of measurement
Parameter, associated with the result of a measurement, that characterizes the dispersion of the values that could reasonably be attributed to the *measurand* (the quantity being measured).
VIM

unified atomic mass unit
Non-SI unit of mass (equal to the atomic mass constant), defined as one twelfth of the mass of a carbon-12 atom in its ground state and used to express masses of atomic particles, u ≈ 1.660 5402(10) × 10^{-27} kg.
G.B. 75; 1996, *68*, 997

uniform corrosion
Corrosion is uniform if the time average of the corrosion current through a unit area of any macroscopic dimension is independent of the position on the surface.
1989, *61*, 21

uniformly labelled
A *selectively labelled* compound where all atoms of a particular element are labelled in the same isotopic ratio.
1981, *53*, 1895

uniformly labelled tracer
A *tracer* in which the *label* is uniformly distributed over its possible positions.
1994, *66*, 2526

uniform polymer
A *polymer* composed of molecules uniform with respect to *relative molecular mass* and constitution.
Notes:
1. A polymer comprising a mixture of *linear* and *branched chains*, all of uniform relative molecular mass, is not uniform.
2. A *copolymer* comprising linear molecules of uniform relative molecular mass and uniform elemental composition but different sequential arrangements of the various types of *monomeric units,* is not uniform (e.g. a copolymer comprising molecules with a ran-

dom arrangement as well as a block arrangement of monomeric units).

3. A polymer uniform with respect only to either relative molecular mass or constitution may be termed 'uniform', provided a suitable qualifier is used (e.g. 'a polymer uniform with respect to relative molecular mass').

4. The adjectives 'monodisperse' and 'polydisperse' are deeply rooted in the literature, despite the former being non-descriptive and self-contradictory. They are in common usage and it is recognized that they will continue to be used for some time; nevertheless, more satisfactory terms are clearly desirable. After an extensive search for possible replacements, the terms 'uniform' and 'non-uniform' have been selected and they are now the preferred adjectives.
 1996, *68*, 2301

unimolecular
 See *molecularity*.
 1994, *66*, 1175

unit (item/portion/individual) (in analytical chemistry)
Each of the discrete, identifiable portions of material suitable for removal from a population as a *sample* or as a portion of a sample, and which can be individually considered, examined or tested, or combined.

In the case of sampling bulk materials (or large packages), the units are increments, created by a sampling device. In the case of packaged materials, the unit may vary with the level of commercial distribution. For example, an individual piece of candy is the sampling unit at the consumer level; a package of individual pieces is the sampling unit at the retail level; a carton of packages is the sampling unit at the wholesale level; a pallet of cartons is the shipping unit at the distribution centre level; and a truckload of pallets is the consignment unit at the manufacture level. Before packaging, the bin containing the individual pieces would be the bulk lot (or batch) for sampling.
 1990, *62*, 1202

unit (of measurement)
Particular quantity, defined and adopted by convention, with which other quanitities of the same kind are compared in order to express their magnitudes relative to that quantity. Units have conventionally assigned names and symbols.
 VIM; 1996, *68*, 997

unit cell
The smallest, regularly repeating material portion contained in a parallelepiped from which a crystal is formed by parallel displacements in three dimensions.
Notes:
1. Unlike in the case of low molar mass substances, the unit cell of polymer crystals usually comprises only parts of the polymer molecules and the regularity of the periodic repetition may be imperfect.
2. In the case of parallel-chain crystals, the chain axis is usually denoted by *c* or, sometimes, *b*.

3. This definition applies to the so-called primitive unit cell. In practice, the effective unit cell may consist of more than one primitive unit cell.
 P.B. 76

unit system
Set of base units, together with derived units, defined in accordance with given rules, for a given system of quantities.
 VIM

universal calibration (in chromatography)
A calibration of a *size-exclusion chromatography* set-up based on the finding that the retention volume of a molecular or particulate species is a single-valued function of an appropriate size parameter of this molecule or particle, irrespective of its chemical nature and structure.

The product of the *intrinsic viscosity* and molar mass, $[\eta]M$, has been widely used as the size parameter.
 P.B. 69

universal detector (in chromatography)
A *detector* which responds to every component in the column effluent except the mobile phase.
 1993, *65*, 849

unperturbed dimensions (in polymers)
The dimensions of an actual polymer random coil in a θ state.
 P.B. 48

unreactive
Failing to react with a specified *chemical species* under specified conditions. The term should not be used in place of *stable*, since a relatively more stable species may nevertheless be more *reactive* than some reference species towards a given reaction partner.
 1994, *66*, 1175

unresolved peak (in chromatography)
 See *peak (in chromatography)*.
 O.B. 96

unstable
The opposite of *stable*, i.e. the *chemical species* concerned has a higher molar Gibbs energy than some assumed standard. The term should not be used in place of *reactive* or *transient*, although more reactive or transient species are frequently also more unstable. (Very unstable chemical species tend to undergo exothermic *unimolecular* decompositions. Variations in the structure of the related chemical species of this kind generally affect the energy of the *transition states* for these decompositions less than they affect the stability of the decomposing chemical species. Low stability may therefore parallel a relatively high rate of unimolecular decomposition.)
 1994, *66*, 1175

unstable film
 See *stable film, metastable film*.
 1972, *31*, 614

unstable ion (in mass spectrometry)
An ion which is sufficiently excited to dissociate within the ion source.
1991, *63*, 1550

unsymmetrical films
See *symmetrical films*.
1972, *31*, 613

unzipping
See *depolymerization*.
1996, *68*, 2309

upconversion
A *non-linear optical effect* in which light frequency is increased.
1996, *68*, 2282

upfield
See *chemical shift*.
1994, *66*, 1175

upper limit of measurement (in atmospheric trace component analysis)
Highest value of the air quality characteristic which can be measured by an instrument; its variations, caused for example by instability, are expected to lie within specified limits. The difference between the lower detection limit and the upper limit of measurement constitutes the dynamic range of the instrument.
1990, *62*, 2201

ureides
N-Acyl or *N,N'*-diacyl ureas.

H$_2$N–C(=O)–NH–C(=O)–R or R–C(=O)–NH–C(=O)–NH–C(=O)–R'

1995, *67*, 1373

urethanes (urethans)
An alternative term for the compounds R$_2$N-C(=O)OR' (R' ≠ H), *esters* of carbamic acids, R$_2$N-C(=O)OH, in strict use limited to the ethyl esters, but widely used in the general sense, e.g. 'polyurethane resins'.
See *carbamates*.
1995, *67*, 1373

uronic acids
Monocarboxylic acids formally derived by oxidation to a carboxy group of the terminal –CH$_2$OH group of aldoses, e.g. D-glucuronic acid.

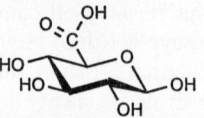

1995, *67*, 1373; see also W.B. 141

uronium salts
Salts of *O*-hydronated urea, having the structure [HOC(=NH$_2$)NH$_2$]$^+$ X$^-$ (and *O*- and *N*-hydrocarbyl derivatives).
1995, *67*, 1374

useful spectral range
See *responsivity*.
1995, *67*, 1749

UV dose
A dose of ultraviolet (UV) radiation.
1996, *68*, 2283

UV photoelectron spectroscopy (UPS)
See *photoelectron spectroscopy*.
1996, *68*, 2282; O.B. 250

UV stabilizer
A substance added to a sample to prevent photodeterioration by ultraviolet (UV) light.
See *photochemical reaction*.
1996, *68*, 2283

vaccine
An agent containing *antigens* produced from killed, attenuated or live pathogenic *microorganisms*, synthetic peptides or by recombinant organisms, used for stimulating the immune system of the recipient to produce specific *antibodies* providing active immunity and/or passive immunity in the progeny.
1992, *64*, 167

vacuum phototube
A *photo-emissive detector* inside an evacuated envelope with a transparent window, the photocathode, and the anode. The photocathode can be opaque or semitransparent.
1995, *67*, 1752

vacuum system (in mass spectrometry)
Those components used to lower the pressure within a *mass spectrometer* are all parts of the vacuum system. This includes not only the various pumping components but also valves, gauges and associated electronic or other control devices, the chamber in which ions are formed and detected, and the vacuum envelope.
1991, *63*, 1546

valence
The maximum number of univalent atoms (originally hydrogen or chlorine atoms) that may combine with an atom of the element under consideration, or with a fragment, or for which an atom of this element can be substituted.
1994, *66*, 1175

valence band
The highest energy continuum of energy levels in a semiconductor that is fully occupied by electrons at 0 K.
See *bandgap, conduction band, Fermi level*.
1996, *68*, 2283

valence isomer
A constitutional isomer interrelated with another by *pericyclic reactions*. For example, Dewar benzene, prismane and benzvalene are valence isomers of benzene.
See *tautomerism*.
1994, *66*, 1175

valence tautomerization
The term describes simple reversible and generally rapid *isomerizations* or *degenerate rearrangements* involving the formation and rupture of single and/or double *bonds*, without *migration* of atoms or *groups*; e.g.

See also *fluxional, tautomerism*.
1994, *66*, 1175

valence transition
A transition observed in certain rare-earth and actinide materials in which the electronic occupation of the 4f or 5f orbital changes with external conditions for example, temperature and pressure.
Example: The transition at approximately 0.65 GPa where black, semiconducting $Sm^{2+}S^{2-}$ changes to golden metallic $(Sm^{3+} + e^-)S^{2-}$.
1994, *66*, 593

value (of a quantity)
Magnitude of a particular quantity generally expressed as a unit of measurement multiplied by a number.
1996, *68*, 997

value of a division (of a precision balance scale)
The reciprocal of the *sensitivity*; like the latter, it usually varies with load. It is determined by empirical calibration.
O.B. 35

van der Waals adsorption
See *physisorption*.
1972, *31*, 586; 1976, *46*, 75

van der Waals broadening (of a spectral line)
See *collisional broadening (of a spectral line)*.
1985, *57*, 1463

van der Waals forces
The attractive or repulsive forces between *molecular entities* (or between groups within the same molecular entity) other than those due to *bond* formation or to the electrostatic inteaction of ions or of ionic groups with one another or with neutral molecules. The term includes: *dipole–dipole, dipole–induced dipole* and *London* (instantaneous induced dipole–induced dipole) forces.
The term is sometimes used loosely for the totality of nonspecific attractive or repulsive intermolecular forces.
1994, *66*, 1175

vaporization temperature (in electrothermal atomization), T_{vap}
The temperature of the atomization surface at which analyte loss becomes statistically significant.
1992, *64*, 257

vapour phase interference (in analysis)
Interference caused by a change in the fraction of analyte dissociated, ionized or excited in the gaseous phase. (Note: here 'dissociation' means the formation of free neutral atoms from free molecules in the gaseous phase. The term atomization is here not appropriate because the latter also covers the formation of free atomic ions.) These interferences may be called dissociation, ionization and excitation interference, respectively.
O.B. 171

variable, x
The quantity or characteristic measured or computed. The corresponding numerical value may be taken for statistical handling; it may, for example, be a measured value or result.

Comment:
Note that statistical usage employs capital letters for random variables, and lower case for particular or observed values. In circumstances where the choice of 'x' as the symbol may cause confusion, another symbol may be used.

See also *dependent* and *independent* variables
1994, *66*, 597

variable pathlength cell (in spectrochemical analysis)
A cell whose pathlength can be varied either continuously or in steps by means of spacers.
1988, *60*, 1454

variance, V, σ^2
The sum of the squares of the *standard deviations*, expressing the contributions to the overall precision from various sources of uncertainty.
O.B. 5; 1994, *66*, 601; 1990, *62*, 2218

variational transition state theory
This expression is applied to modifications of conventional *transition state theory* in which the position of the *dividing surface* in the potential-energy surface is varied. The rate is calculated with the surface at various positions, and the lowest rate calculated is taken to be closest to the truth.
1996, *68*, 190

variegated star macromolecule
See *star macromolecule*.
1996, *68*,, 2296

Vavilov rule
See *Kasha–Vavilov rule*.
1996, *68*, 2283

vector (in biotechnology)
1. A DNA molecule (*plasmid*, virus, bacteriophage, artificial or cut DNA molecule) capable of being replicated and bearing cloning sites for the introduction of foreign DNA, used to introduce this DNA into *host cells*.
2. Any organism that transmits a disease between two hosts.
1992, *64*, 168

velocity, v, c
Vector quantity equal to the derivative of the position vector with respect to time (symbols u, v, w for components of c).
G.B. 11; ISO 31-1: 1992; 1996, *68*, 997

velocity (in mass transport)
The *flux density* N_B is related to the velocity by the equation

$$N_B = c_B v_B$$

The vector v_B is the macroscopic average velocity at which the species B moves (which is to be distinguished from the random molecular velocity); c_B is the concentration of species B (mol m^{-3}). Note that the velocity is defined with respect to a frame of reference.
1981, *53*, 1830

verdazyl radicals
Relatively stable delocalized *radicals*, derived from hydrazyl, H_2NNH, of the types shown, in particular the six-membered ring:

R most commonly *aryl* groups

1995, *67*, 1374

vertical ionization
A process whereby an electron is removed from a molecule in its ground or an excited state so rapidly that a positive ion is produced without change in the positions or momenta of the atoms. The resultant ion is often in an excited state.
1991, *63*, 1548

vertical transition
See *Franck–Condon principle*.
1996, *68*, 2283

Verwey transition
An electron-ordering transition occurring in a mixed-valent system that results in an ordering of formal valence states in the low-temperature phase.
Example: The prototype system, first identified by Verwey, is the ferrospinel magnetite, Fe^{3+}-$[Fe^{3+}Fe^{2+}]O_4$ in which an ordering of Fe^{3+} and Fe^{2+} ions within octahedral sites is thought to occur below $T_v \approx 120$ K.
1994, *66*, 593

vibrationally adiabatic transition-state theory
See *adiabatic transition-state theory*.
1996, *68*, 190

vibrational redistribution
Intramolecular redistribution of energy among the vibrational modes usually giving a statistical distribution of their populations, characterized by the 'vibrational temperature'. For large molecules, this process does not require collisions.
1996, *68*, 2283

vibrational relaxation
The loss of vibrational excitation energy by a molecular entity through energy transfer to the environment caused by collisions. The molecular entity relaxes into vibrational equilibrium with its environment.
See *relaxation*.
1996, *68*, 2283

vibrational term, G
Vibrational energy divided by the product of the Planck constant and the speed of light.
See *term*.
G.B. 23

vibronic coupling
Interaction between electronic and vibrational motions in a molecular entity.
See *Jahn–Teller effect, Renner–Teller effect.*
1996, *68*, 2283

vibronic transition
A transition which involves a change in both the electronic and vibrational quantum numbers of a molecular entity, as opposed to purely electronic or purely vibrational transitions. The transition occurs between two states, just as in a purely electronic transition, but involves a change in both electronic and vibrational energy.
1996, *68*, 2283

vidicon
A vacuum tube containing a photosensitive area, or target, and an electron gun to read the signal from the target. The silicon target consists of a two-dimensional array of Si-photodiodes having a common cathode and isolated anodes. Irradiation of the target causes the production of electron-hole pairs which, by recombination, leads to a depletion of the surface charge.
1995, *67*, 1757

vinyl carbenes
Carbenes having a *vinylic group* on a carbenic carbon atom $R_2C=CR\dot{C}R \leftrightarrow R_2C^+-CR=C^-R$.
E.g. $H_2C=CH\ddot{C}H$ prop-2-en-1-ylidene.
See 1,3-*dipolar compounds.*
1995, *67*, 1374

vinylic cations
Carbocations having the structure $R_2C=C^+-R$.
1995, *67*, 1374

vinylic groups
The vinyl group ($CH_2=CH-$) and derivatives formed by substitution. Informally, a group, such as –OH, attached to the free valence of a (substituted) vinyl group is sometimes referred to as 'vinylic'.
1995, *67*, 1374

vinylidenes
Carbenes in which the carbenic carbon atom has a double bond to another carbon atom $R_2C=C:$, e.g. $H_2C=C:$ ethenylidene.
1995, *67*, 1374

viologens
1,1′-Di(hydrocarbyl)-4,4′-bipyridinium salts.

$$R-\overset{+}{N}\text{=}\!\!\!\!\bigcirc\!\!\!\!\text{=}\!\!\!\!\bigcirc\!\!\!\!\text{=}\overset{+}{N}-R\quad 2X^-$$

1995, *67*, 1374

virial coefficients
Coefficients (B, C, ...) in the virial equation of state for a real gas $pV_m = RT(1 + B/V_m + C/V_m^2 + ...)$ where p is the pressure, V_m the molar volume, R the gas constant and T the thermodynamic temperature.
G.B. 49; see also P.B. 58

virtual transition
A non-observable *liquid-crystal transition* that occurs below the crystallization temperature and is determined by extrapolation in binary phase diagrams.
1994, *66*, 593

viscosity
See *dynamic viscosity.*
G.B. 13; 1996, *68*, 997

viscosity function, Φ
A coefficient connecting the *intrinsic viscosity*, the *radius of gyration* and the molar mass of a chain macromolecule, according to the equation:

$$[\eta] = \Phi 6^{3/2} <s^2>^{3/2}/M$$

where [η] is the intrinsic viscosity, s is the radius of gyration and M is the molar mass. The viscosity function is often referred to as the Flory constant.
P.B. 64

viscosity number (of a polymer)
See *reduced viscosity.*
P.B. 63

viscosity ratio
See *relative viscosity.*
P.B. 62

visibility (in atmospheric chemistry)
Defined as the greatest distance at which a black object of suitable dimensions can be seen and recognized against the horizon sky, or, in the case of night observations, could be seen and recognized if the general illumination were raised to the normal daylight level. The criterion of recognizing the object, not just seeing the object without recognition, is used. Transmissometers, telephotometers and sun photometers are devices used to measure the degree of transmission of light. Nephelometers (integrating) are used to measure visibility by way of the light scattering from aerosols in the air mass.
1990, *62*, 2218

volatilization
The conversion of a solid or liquid to a gas or vapour by application of heat, by reducing pressure, by chemical reaction or by a combination of these processes.
Example: $W(s) + 3/2 O_2(g) \rightarrow WO_3(g)$
1994, *66*, 593

volatilizer
Material added to a sample to increase its volatilization or that of some component of it. Volatilizers increase the fraction volatilized either by forming more volatile compounds or by increasing the total surface area of all analyte particles (e.g. by explosive disintegration or by dispersal of the analyte in a highly volatile matrix).
O.B. 159 and 172

volt
SI derived unit of electric potential, $V = J\ C^{-1} = m^2\ kg\ s^{-3}\ A^{-1}$.
G.B. 72; 1996, *68*, 997

voltage (in electroanalysis)
The use of this term is discouraged, and the term *applied potential* should be used instead, for non-periodic signals. However, it is retained here for sinusoidal and other periodic signals because no suitable substitute for it has been proposed.
1985, *57*, 1505

voltage, alternating
See *alternating voltage*.
1985, *57*, 1505

voltage, alternating amplitude
See *alternating amplitude voltage*.
1985, *57*, 1505

voltage, periodic
See *periodic voltage*.
1985, *57*, 1505

voltammetric constant
In linear-sweep *voltammetry* and related techniques, the empirical quantity defined by the equation
$$i_p/Av^{1/2}c_B \ (= j_p/v^{1/2}c_B)$$
where i_p is the *peak current*, A is the area of the electrode-solution interface, v is the rate of change of applied potential, and c_B is the bulk concentration of the substance B whose reduction or oxidation is responsible for the peak in question.
1985, *57*, 1505

Volta potential difference
See *contact potential difference*.
1986, *58*, 442

volume content, V_c/m_s
The volume of an (isolated) specified component (C) divided by the mass of the system (S), e.g. patient–blood volume content. Volume content should not be confused with specific volume (or massic volume), which is the volume of the system divided by the mass of the system.
1979, *51*, 2474

volume flow rate, q_v
Volume of a component crossing a surface divided by the time.
1996, *68*, 997

volume fraction, φ
Volume of a constituent of a mixture divided by the sum of volumes of all constituents prior to mixing.
See *fractions*.
G.B. 41; 1996, *68*, 998

volume of activation, $\Delta^{\ddagger}V$
A quantity derived from the pressure dependence of the *rate constant* of a reaction (mainly used for reactions in solution), defined by the equation:
$$\Delta^{\ddagger}V = -RT(\partial \ln k/\partial p)_T$$
providing that the rate constants of all reactions (except first-order reactions) are expressed in pressure-independent concentration units, such as mol dm^{-3} at a fixed temperature and pressure.
The volume of activation is interpreted, according to *transition state* theory, as the difference between the partial molar volumes of the transition state (V) and the sums of the partial volumes of the reactants at the same temperature and pressure, i.e.
$$\Delta^{\ddagger}V = {}^{\ddagger}V - \Sigma(rV_R)$$
where r is the order in the reactant R and V_R its partial molar volume.
1994, *66*, 1175; see also 1996, *68*, 191; G.B. 56

volume of the stationary phase (in chromatography)
The volume of the stationary liquid phase or of the *active solid* or of the *gel* in the *column*. The volume of any solid support is not included.
O.B. 100

volume rate
See *clearance*.
1996, *68*, 998

volume reaction (in electrochemistry)
See *kinetic current*.
1985, *57*, 1504

volume strain (bulk strain), θ
The change of volume divided by the original volume.
G.B. 12

volumetric flowrate (in gas chromatography)
The flowrate of the mobile phase (cm^3 min^{-1}), normally specified at the column temperature and outlet pressure, although the measurement may be made at ambient temperature and must be corrected accordingly (and possibly also for water vapour present in the flowmeter).
O.B. 102

volume viscosity (or dilatational viscosity)
A quantity ζ which enters into equations at any point where the flow involves a change in volume, i.e. is dilatational. If the deformation is purely dilatational, the average of three normal stress components is:
$$\sigma = (1/3)(\sigma_{xx} + \sigma_{yy} + \sigma_{zz})$$
$$= -p + \zeta(\partial v_x/\partial x + \partial v_y/\partial y + \partial v_z/\partial z)$$
where p is the hydrostatic pressure at the point considered in the absence of motion.
1979, *51*, 1216

volumic
Attribute to a *physical quantity* obtained by division by volume. Charge density may thus be called volumic charge which is the *electric charge* in a section of space divided by the volume of that section.
ISO 31-0: 1992; 1985, *57*, 1310; 1996, *68*, 999

Wagner number
See *current distribution*.
1981, *53*, 1836

Walden inversion
Retention of configuration is the preservation of integrity of the spatial arrangement of bonds to an chiral centre during a chemical reaction or transformation. It is also the configurational correlation when a chemical species Xabcd is converted into the chemical species Xabce having the same *relative configuration*. The configurational change when a chemical species Xabcd (where X is typically carbon), having a tetrahedral arrangement of bonds to X, is converted into the chemical species Xabce having the opposite relative configuration (or when it undergoes an identity reaction in which Xabcd of opposite configuration is produced) is called a Walden inversion or inversion of configuration. The occurrence of a Walden inversion during a chemical transformation is sometimes indicated in the chemical equation by the symbol shown below in place of a simple arrow pointing from reactants to products.

$$e^- \quad {}_c^{\text{\tiny III}}X{-}d \overset{a}{\underset{b}{\longrightarrow}} e{-}X{}_{\text{\tiny III}}{}_c \quad d^-$$

1996, *68*, 2221

wall-coated open-tubular (WCOT) column (in chromatography)
A column in which the liquid stationary phase is coated on the essentially unmodified smooth inner wall of the tube.
1993, *65*, 831

wash out (in atmospheric chemistry)
The removal from the *atmosphere* of gases and sometimes particles by their solution in or attachment to raindrops as they fall.
See also *rain out (in atmospheric chemistry)*.
1990, *62*, 2219

watt
SI derived unit of power, $W = J\ s^{-1} = m^2\ kg\ s^{-3}$.
G.B. 72; 1996, *68*, 998

wavefunction (state function), Ψ, ψ, φ
The solution of the Schrödinger equation, eigenfunction of the hamiltonian operator. Complex conjugate functions are denoted by an asterisk read as 'star'.
G.B. 16

wave height (electrochemical)
The *limiting current* of an individual wave, frequently expressed in arbitrary units for convenience.
1985, *57*, 1505

wavelength, λ
Distance in the direction of propagation of a periodic wave between two successive points where at a given time the phase is the same.
G.B. 30; ISO 31-5: 1992; 1996, *68*, 998; see also 1996, *68*, 2283; 1985, *57*, 109

wavelength converter
A wavelength converter converts *radiation* at one wavelength to radiation at another detectable wavelength or at a wavelength of improved *responsivity* of the detector. The classical wavelength converter consists of a screen of luminescent material that absorbs radiation and radiates at a longer wavelength. Such materials are often used to convert ultraviolet to visible radiation for detection by conventional phototubes. In X-ray spectroscopy a converter that emits optical radiation is called a *scintillator*. In most cases wavelength conversion is from short to long wavelength, but in the case of conversion of long to short wavelength the process is sometimes called upconversion. Wavelengths of coherent sources can be converted using nonlinear optical techniques. A typical example is frequency doubling.
1995, *67*, 1758

wavelength dispersion (in X-ray emission spectroscopy)
Spatial separation of characteristic X-rays according to their wavelengths.
1980, *52*, 2547

wavelength-dispersive X-ray fluorescence analysis
A kind of *X-ray fluorescence analysis* involving the measurement of the wavelength spectrum of the emitted *radiation* e.g. by using a diffraction grating or crystal.
1994, *66*, 2526

wavelength error (in spectrochemical analysis)
The error in *absorbance* which may occur if there is a difference between the (mean) wavelength of the radiation entering the *sample cell* and the indicated wavelength on the spectrometer scale.
1988, *60*, 1456

wavenumber, $\sigma, \tilde{\nu}$
The reciprocal of the *wavelength*, λ, or the number of waves per unit length along the direction of propagation. The SI unit is m^{-1}, but a commonly used unit is cm^{-1}. Symbols $\tilde{\nu}$ in a vacuum, σ in a medium.
1996, *68*, 2283; 1985, *57*, 109; G.B. 30

weak collision
A collision between two molecules in which the amount of energy transferred from one to the other is not large compared to $k_B T$ (k_B is the *Boltzmann constant* and T the absolute temperature).
See *strong collision*.
1996, *68*, 191

weber
SI derived unit of *magnetic flux*, $Wb = V\ s = m^2\ kg\ s^{-2}\ A^{-1}$.
G.B. 72; 1996, *68*, 998

wedge projection
A stereochemical projection, roughly in the mean plane of the molecule, in which bonds are represented by open wedges, tapering off from the nearer atom to the farther atom. It is mainly used to illustrate the

conformation of larger cycloalkanes e.g. cyclotetradecane:

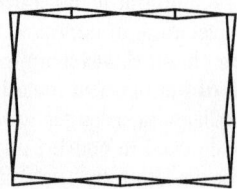

equivalent to

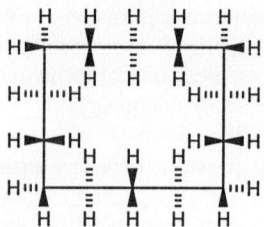

1996, *68*, 2222

weight, *G*
Force of gravity acting on a body, $G = mg$, where m is its mass and g the acceleration of free fall.
G.B. 12; 1996, *68*, 999

weight-distribution function
See *mass-distribution function*.
P.B. 56

weighted mean
If in a series of observations a statistical weight (w_i) is assigned to each value, a weighted mean $\bar{x}_w$ can be calculated by the formula:

$$\bar{x}_w = \frac{\sum w_i x_i}{\sum w_i}$$

Comment: Unless the weights can be assigned objectively, the use of the weighted mean is not normally recommended.
1994, *66*, 599

Weller correlation
Empirical correlation for the energy of full *charge-transfer-exciplexes* relative to the ground-state in *n*-hexane as a function of the electrochemical one electron oxidation and reduction potentials measured in a polar solvent for the donor (D) and the acceptor (A) involved:

$\Delta H(D^+A^-)_{hexane}$
$= E^0(D/D^+) - E^0(A/A^-) + (0.15 \pm 0.10)$ eV
1996, *68*, 2284

Westcott cross-section
See *effective thermal cross-section*.
1994, *66*, 2517; O.B. 218

wet bulb temperature
In *psychrometry*, the temperature of the sensor or the bulb of a thermometer in which a constantly renewed film of water is evaporating. The temperature of the water used to renew the film must be at the temperature of the gas.
See *psychrometry*.
1990, *62*, 2219

wetting tension (or work of immersional wetting per unit area)
The work done on a system when the process of *immersional wetting* involving unit area of phase β is carried out reversibly:

$$w_W{}^{\alpha\beta\delta} = \gamma^{\beta\delta} - \gamma^{\alpha\beta}$$

where $\gamma^{\alpha\beta}$ and $\gamma^{\beta\delta}$ are the *surface tensions* between two bulk phases α, β and β, δ, respectively.
1972, *31*, 598

Wheland intermediate
See *Meisenheimer adduct*, σ-*adduct*, *arenium ions*.
1994, *66*, 1176; 1995, *67*, 1374

Wiegner effect
Synonymous with *suspension effect (in an ion-selective electrode)*.
1994, *66*, 2533

Wigner rule
See *spin conservation rule*.
1996, *68*, 2284

Wilzbach labelling
Labelling of a substance by exposing it to tritium gas.
1994, *66*, 2521

wind rose
A diagram designed to show the distribution of wind direction experienced at a given location over a considerable period of time. Usually shown in polar coordinates (distance from the origin being proportional to the probability of the wind direction being at the given angle usually measured from the north). Similar diagrams are sometimes used to summarize the average concentrations of a given pollutant seen over a considerable period of time as a function of direction from a given site (sometimes called a pollution rose).
1990, *62*, 2219

Wittig reagents
See *phosphonium ylides*.
1995, *67*, 1374

Wood horn
A mechanical device that acts by absorption as a perfect photon trap.
1996, *68*, 2284

Wood lamp
A term used to describe a low-pressure mercury arc.
See *lamp*.
1996, *68*, 2284

Woodward–Hoffmann rules
See *orbital symmetry*.
1994, *66*, 1176

work, *w*, *W*
Scalar product of force, F, and position change, dr, $w = \int F \times dr$.
G.B. 48; 1996, *68*, 999

work hardening
Opposite of *work softening*, in which shear results in a permanent increase of *viscosity* or consistency with time.
1979, *51*, 1217

working electrode
An electrode that serves as a transducer responding to the excitation signal and the concentration of the substance of interest in the solution being investigated, and that permits the flow of current sufficiently large to effect appreciable changes of bulk composition within the ordinary duration of a measurement.
O.B. 60

work of adhesion
The work of adhesion per unit area, $w_A^{\alpha\beta\delta}$, is the work done on the system when two condensed phases α and β, forming an interface of unit area are separated reversibly to form unit areas of each of the $\alpha\delta$- and $\beta\delta$- interfaces.

$$w_A^{\alpha\beta\delta} = \gamma^{\alpha\delta} + \gamma^{\beta\delta} - \gamma^{\alpha\beta}$$

where $\gamma^{\alpha\beta}$, $\gamma^{\alpha\delta}$ and $\gamma^{\beta\delta}$ are the *surface tensions* between two bulk phases α, β; α, δ and β, δ respectively.
The work of adhesion as defined above, and traditionally used, may be called the work of separation.
1972, *31*, 597

work of cohesion per unit area
Of a single pure liquid or solid phase α, w_C^α is the work done on the system when a column α of of unit area is split, reversibly, normal to the axis of the column to form two new surfaces each of unit area in contact with the equilibrium gas phase.

$$w_C^\alpha = 2\gamma^\alpha$$

where γ^α is the surface tension between phase α and its equilibrium vapour or a dilute gas phase.
1972, *31*, 597

work of immersional wetting per unit area
See *wetting tension*.
1972, *31*, 598

work of separation
See *work of adhesion*.
1972, *31*, 598

work softening
The application of a finite shear to a system after a long rest may result in a decrease of the *viscosity* or the consistency. If the decrease persists when the shear is discontinued, this behaviour is called work softening (or shear breakdown), whereas if the original viscosity or consistency is recovered this behaviour is called thixotropy.
1979, *51*, 1217

worm-like chain (in polymers)
A hypothetical *linear macromolecule* consisting of an infinitely thin chain of continuous curvature; the direction of curvature at any point is random.
The model describes the whole spectrum of chains with different degrees of chain stiffness from rigid rods to random coils, and is particularly useful for representing stiff chains. In the literature this chain is sometimes referred to as a Porod–Kratky chain.
Synonymous with continuously curved chain.
P.B. 51

ξ- (xi-)
A symbol used to denote unknown configuration at a chiral centre.
1984, *56*, 601

xanthates
See *xanthic acids*.
1995, *67*, 1374

xanthene dyes
Dyes derived by condensation of phthalic anhydride with resorcinol (and derivatives) or *m*-aminophenol (and derivatives), of which fluorescein is the prototype (all such dyes have the xanthene nucleus).

fluorescein

See also *phthaleins, rhodamine dyes*.
1995, *67*, 1374

+ xanthic acids
Compounds having the structure ROC(=S)SH. Thus *O*-esters of dithiocarbonic acid. Salts and *esters* of xanthic acid are xanthates. Use of this term is not recommended.
1995, *67*, 1374

xanthophylls
A subclass of *carotenoids* consisting of the oxygenated carotenes.
W.B. 226; 1995, *67*, 1374

xenobiotics
Man-made compounds with chemical structures foreign to a given organism.
1992, *64*, 168

xenon lamp
An intense source of ultraviolet, visible and near-infrared light produced by electrical discharge in xenon under high pressure.
See also *antimony–xenon lamp (arc), lamp, mercury–xenon lamp (arc)*.
1996, *68*, 2284

xerogel
A term used for the dried out open structures which have passed a *gel* stage during preparation (e.g. silica gel); and also for dried out compact macromolecular gels such as gelatin or rubber.
1972, *31*, 606

xi
See ξ (beginning of 'x').

XPS
See *photoelectron spectroscopy*.
1996, *68*, 2284

X-radiation
Radiation resulting from the interaction of high energy particles or photons with matter.
1991, *63*, 737

X-radiation, characteristic
See *characteristic X-ray emission*.
1991, *63*, 737

X-ray emission analysis, particle induced
See *particle induced X-ray emission analysis*.
1994, *66*, 2526

X-ray emission analysis, radioisotope induced
See *radioisotope induced X-ray emission*.
1994, *66*, 2526

X-ray emission, characteristic
See *characteristic X-ray emission*.
1991, *63*, 737

X-ray emission, particle induced (PIXE)
See *characteristic X-ray emission*.
1991, *63*, 737

X-ray escape peak
In a gamma or X-ray spectrum, the peak due to the *photoelectric effect* in the detector and escape, from the sensitive part of the detector, of the X-ray photon emitted as a result of the photoelectric effect.
1982, *54*, 1541

X-ray fluorescence
The emission of characteristic X-radiation by an atom as a result of the interaction of electromagnetic *radiation* with its orbital electrons.
1994, *66*, 2526

X-ray fluorescence analysis
A kind of analysis based on the measurement of the energies and intensities of characteristic X-radiation emitted by a test portion during *irradiation* with electromagnetic radiation.
1994, *66*, 2526

X-ray fluorescence analysis, energy-dispersive
See *energy-dispersive X-ray fluorescence analysis*.
1994, *66*, 2526

X-ray fluorescence analysis, wavelength-dispersive
See *wavelength-dispersive X-ray fluorescence analysis*.
1994, *66*, 2526

X-ray intensity
Essentially all X-ray measurements are made by photon counting techniques but the results are seldom converted to radiant flux or *irradiance* or *radiant exposure*. The term *photon flux* would be appropriate if the measurements were corrected for detector efficiency but this is seldom done for X-ray chemical analysis. Therefore the term X-ray intensity, I, is commonly used and expressed as photons/unit time detected. Likewise the term relative X-ray intensity, I_r, is used to mean the intensity for the analyte in an unknown specimen divided by the intensity for a known concentration of the analyte element.
1980, *52*, 2544

X-ray level
An electronic state occuring as the initial or final state of a process involving the absorption or emission of X-ray radiation. It represents a many-electron state

which, in the purely atomic case, has total *angular momentum* ($J = L + S$) as a well defined quantum number.
1991, *63*, 737

X-ray photoelectron spectroscopy (XPS)
Any technique in which the sample is bombarded with X-rays and photoelectrons produced by by the sample are detected as a function of energy.
ESCA (Electron Spectroscopy for Chemical Analysis) refers to the use of this technique to identify elements, their concentrations, and their chemical state within the sample.
O.B. 250

X-ray satellite
A weak line in the same energy region as a normal X-ray line. Another name used for weak features is non-diagram line. Recommendations as to the use of these two terms have conflicted. Using the term *diagram line* as defined here, *non-diagram line* may well be used for all lines with a different origin. The majority of these lines originate form the dipole-allowed de-excitation of multiply ionized or excited states, and are called multiple-ionization satellites. A line where the initial state has two vacencies in the same shell, notably the K-shell, is called a hypersatellite. Other mechanisms leading to weak spectral features in X-ray emission are, e.g. resonance emission, the radiative Auger effect, magnetic dipole and electric quadrupole transitions and, in metals, plasmon excitation. Atoms with open electron shells, i.e. transition metals, lanthanides and actinides, show a splitting of certain X-ray lines due to the electron interaction involving this open shell. Structures originating in all these ways as well as structures in the valence band of molecules and solid chemical compounds have in the past been given satellite designations.
1991, *63*, 739

X-ray spectroscopy
X-ray spectroscopy consists of three steps: (a) excitation to produce emission lines characteristic of the elements in the material, (b) measurement of their intensity, and (c) conversion of *X-ray intensity* to concentration by a calibration procedure which may include correction for *matrix effects*.
1980, *52*, 2544

x unit
Non-SI unit of length: $X \approx 1.002 \times 10^{-13}$ m.
G.B. 110

xylylenes
See *quinomethanes*.
1995, *67*, 1374

YAG
See *neodymium laser*.
1996, *68*, 2284

yard
Non-SI unit of length, yd = 3 ft = 0.9144 m.
G.B. 110

year
Non-SI unit of time, a ≈ 31 556 952 s. The year is not commensurable with the day and not a constant. The value given corresponds to the Gregorian calendar year (a = 365.2425 d).
G.B. 111

yield (in biotechnology), Y
Ratio expressing the efficiency of a mass conversion process. The yield coefficient is defined as the amount of *cell* mass (kg) or product formed (kg, mol) related to the consumed *substrate* (carbon or nitrogen source or oxygen in kg or moles) or to the intracellular ATP production (moles).
1992, *64*, 168

yield, fluorescence
See *fluorescence yield*.
1994, *66*, 2526

yield, radiochemical
See *radiochemical yield*.
1994, *66*, 2526

yield stress
The *shear stress* σ_0 or τ_0 at which yielding starts abruptly. Its value depends on the criterion used to determine when yielding occurs.
1979, *51*, 1217

ylides
Compounds in which an anionic site Y^- (originally on carbon, but now including other atoms) is attached directly to a heteroatom X^+ (usually nitrogen, phosphorus or sulfur) carrying a formal positive charge. They are thus 1,2-dipolar species of the type $R_mX^+-Y^-R_n$. If X is a saturated atom of an element from the first row of the periodic system, the ylide is commonly represented by a charge-separated form; if X is a second, third, etc. row element uncharged canonical forms are available $R_mX=YR_n$. If X is an unsaturated atom, doubly bonded to another first row element Z, the negative charge on Y may be stabilized by π-conjugation, $Z=X^+-Y^-R_n \leftrightarrow Z^--X^+=YR_n$. Such ylides belong to the class 1,3-*dipolar compounds*. However, 1,3-dipolar compounds with only sextet-containing canonical forms (e.g. *vinylcarbenes*) are not ylides. E.g. $Ph_3P^+-C^-H_2 \leftrightarrow Ph_3P=CH_2$ (often called a Wittig reagent), $(CH_3)_3N^+-C^-H_2$, $RC\equiv N^+N^--R$, $(CH_3)_2S=CHPh \leftrightarrow (CH_3)_2S^+-C^-HPh$. Note that ylide is a complete word, not to be confused with the suffix -ylide, used for some radical anions.

Subclasses of ylides:
Ylides $R_mX^+-C^-R_2$ having the negative charge on carbon are classified by citing the name of the element X before the word ylide. E.g. *nitrogen ylide, phosphorus ylide, oxygen ylide, sulfur ylide*. A further specification may be achieved by citing the class name of R_mX before the word ylide. Thus nitrogen ylides include *amine ylides*, $R_3N^+-C^-R_2$, *azomethine ylides*, $R_2C=N^+R-C^-R_2$, *nitrile ylides*, $RC\equiv N^+-C^-R_2$. Some authors, who wish to express the positive charge on X, prefer e.g. *ammonium ylides* over amine ylides; such usage varies according to the heteroatom X and to national custom. The ylides $R_mX^+-Y^- \leftrightarrow R_mX=Y$ (Y = O, S, Se, Te, NR) are usually named by citing the name of R_mX followed by the additive nomenclature term for Y (oxide, sulfide, selenide, telluride, imide, respectively). E.g. *amine imides*; use of the less systematic synonyms amine imines and aminimines is discouraged. Some classes of ylides are known by trivial names e.g. *nitrones*, Wittig reagents (synonymous with *phosphonium ylides*).
See also *betaines, dipolar compounds*.
1995, *67*, 1375; 1994, *66*, 1176

ynamines
N,N-Disubstituted alk-1-yn-1-amines, $RC\equiv CNR_2$. By usage, restricted to this type of acetylenic amine.
See *enamines*.
1995, *67*, 1375

ynols
Alk-1-yn-1-ols, $RC\equiv COH$; tautomeric with *ketenes*, $RCH=C=O$.
1995, *67*, 1375

yocto
SI prefix for 10^{-24} (symbol: y).
G.B. 74

yotta
SI prefix for 10^{24} (symbol: Y).
G.B. 74

Young's modulus
See *modulus of elasticity*.
G.B. 12

Yukawa–Tsuno equation
A multiparameter extension of the *Hammett equation* to quantify the role of *enhanced resonance* effects on the reactivity of *meta-* and *para-*substituted benzene derivatives, e.g.

$$\lg k = \lg k_0 + \rho[\sigma + r(\sigma^+ - \sigma)]$$

The parameter r gives the enhanced resonance effect on the scale $(\sigma^+-\sigma)$ or $(\sigma^--\sigma)$, respectively.
See also ρ-*value*, σ-*constant*.
1994, *66*, 1176

ζ-potential
See *electrokinetic potential*.
1996, *68*, 999

Z
See *E, Z*.
1996, *68*, 2222

Z-value
An index of the *ionizing power* of a solvent based on the frequency of the longest wavelength electronic absorption maximum of 1-ethyl-4-methoxycarbonyl-pyridinium iodide in the solvent. The Z-value is defined by:

$$Z = 2.859 \times 10^4 / \lambda$$

where Z is in kcal mol^{-1} and λ is in nm.
See also *Dimroth–Reichardt E_T parameter, Grunwald–Winstein equation*.
1994, *66*, 1176

Zaitsev rule
See *Saytzeff rule*.
1994, *66*, 1176

Zeeman effect
The splitting or shift of spectral lines due to the presence of an external magnetic field.
1996, *68*, 2284; O.B. 122

zepto
SI prefix for 10^{-21} (symbol: z).
G.B. 74

zero field splitting
The separation of multiplet sublevels in the absence of an external magnetic field.
1996, *68*, 2284

zero point (of a glass electrode)
Value of the *pH* of a solution, which in combination with a stated outer *reference electrode*, gives zero emf from the operational cell.
1985, *57*, 541

zero point of scale (of a balance)
The *rest point* of the properly adjusted balance with no load on the pans and the rider (or chain) in the zero position.
O.B. 36

zero–zero (0–0) absorption or emission
A purely electronic transition occurring between the lowest vibrational levels of two electronic states.
1996, *68*, 2284

zeta
See ζ (beginning of 'z').

zetta
SI prefix for 10^{21} (symbol: Z).
G.B. 74

zig-zag projection
A stereochemical projection (see **I**) for an acyclic molecule (or portion of a molecule) where the main chain is represented by a zig-zag line in the plane and the substituents are shown above or below the plane.
1996, *68*, 2222

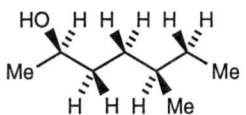

I (See **zig-zag projection**)

Zimm plot
A diagrammatic representation of data on scattering from large particles, corresponding to the equation:

$$\frac{Kc}{\Delta R(\theta)} = \frac{1}{\overline{M}_w \, P(\theta)} + 2A_2 c + \ldots$$

and used for the simultaneous evaluation of the mass average molar mass, $\overline{M}_w$, the second *virial coefficient* of the chemical potential, A_2, and (usually) the z-average *radius of gyration*, $<s^2>_z^{1/2}$; c is the mass concentration of the solute, $\Delta R(\theta)$ the excess *Rayleigh ratio*, and $P(\theta)$ the particle scattering function that comprises (usually) the z-average radius of gyration. K depends on the solute, the temperature and the type of radiation employed.
Several modifications of the Zimm plot are in frequent use; the most common one uses the excess scattering instead of the excess Rayleigh ratio.
P.B. 66

zone (in chromatography)
A region in the chromatographic bed where one or more components of the sample are located. The term band may also be used for it.
1993, *65*, 824; O.B. 96

zone melting method of preconcentration
A technique based on the different solubility of microcomponents in the liquid and solid matrix. It has a limited field of application for the analysis of fusible (and stable) substances.
1979, *51*, 1200

Zucker–Hammett hypothesis
This hypothesis states that, if in an acid catalysed reaction, lg k_1 (first-order rate constant of the reaction) is linear in H_0 (Hammett *acidity function*), water is not involved in the *transition state* of the *rate-controlling step*. However, if lg k_1 is linear in lg [H$^+$] then water is involved. This has been shown to be incorrect by Hammett himself.
1994, *66*, 1176

zwitterionic compounds/zwitterions
Neutral compounds having formal unit electrical charges of opposite sign. Some chemists restrict the term to compounds with the charges on non-adjacent atoms. Sometimes referred to as inner salts, dipolar ions (a misnomer). E.g. H$_3$N$^+$CH$_2$C(=O)O$^-$ ammonioacetate (glycine), (CH$_3$)$_3$N$^+$–O$^-$ trimethylamine oxide.
See *betaines, dipolar compounds, ylides*.
1995, *67*, 1375; 1994, *66*, 1176

zygote
The *cell* obtained as a result of complete or partial *fusion* of meiotically (cf. *meiosis*) produced cells.
1992, *64*, 168

Source Documents

The first seven are compilations of IUPAC or IUBMB (International Union of Biochemistry and Molecular Biology) recommendations in book form. Several IUPAC-approved sources then follow. The remainder are articles in *Pure and Applied Chemistry* (PAC) in chronological order.

1. B.B. (Blue Book)
IUPAC Nomenclature of Organic Chemistry.
Sections A, B, C, D, E, F and H.
Pergamon Press, Oxford, 1979.

2. B.B.(G) [Blue Book (Guide)]
A Guide to IUPAC Nomenclature of Organic Compounds.
Blackwell Scientific Publications, Oxford, 1993.

3. G.B. (Green Book)
IUPAC Quantities, Units and Symbols in Physical Chemistry.
Second Edition, Blackwell Scientific Publications, Oxford, 1993.

4. O.B. (Orange Book)
IUPAC Compendium of Analytical Nomenclature.
Second Edition, Blackwell Scientific Publications, Oxford, 1987.

5. P.B. (Purple Book)
IUPAC Compendium of Macromolecular Nomenclature.
Blackwell Scientific Publications, Oxford, 1991.

6. R.B. (Red Book)
IUPAC Nomenclature of Inorganic Chemistry.
Third Edition, Blackwell Scientific Publications, Oxford, 1990.

7. W.B. (White Book)
IUBMB Biochemical Nomenclature and Related Documents.
Second Edition, Portland Press, London, 1992.

VIM: International Vocabulary of Basic and General Terms in Metrology (VIM), Second Edition, ISO, 1993.

E.R. Cohen and B.N. Taylor, The 1986 Adjustment of the Fundamental Physical Constants, *CODATA Bull.*, 1986, *63*, 1

Physical Chemistry Division, unpublished: this denotes a few physicochemical terms not mentioned in the Green Book which were defined specially for this Compendium by the Physical Chemistry Division of IUPAC.

International Organization for Standardization, Geneva.
ISO 31-0:1992, Quantities and Units - Part 0: General Principles Units and Symbols.
ISO 31-1:1992, Quantities and Units - Part 1: Space and Time.
ISO 31-2:1992, Quantities and Units - Part 2: Periodic and Related Phenomena.
ISO 31-4:1992, Quantities and Units - Part 4: Heat.
ISO 31-5:1992, Quantities and Units - Part 5: Electricity and Magnetism.
ISO 31-10:1992, Quantities and Units - Part 10: Nuclear Reactions and Ionizing Radiations.

PAC documents

PAC 1972, *30*, 681
Nomenclature of inorganic boron compounds.

PAC 1972, *31*, 577
Manual of symbols and terminology for physicochemical quantities and units. Appendix II, Part I: definitions, terminology and symbols in colloid and surface chemistry.

PAC 1974, *37*, 499
Manual of symbols and terminology for physicochemical quantities and units. Appendix III: electrochemical nomenclature.

PAC 1976, *45*, 211
Nomenclature and conventions for reporting Mössbauer spectroscopic data.

PAC 1976, *45*, 221
Nomenclature and spectral presentation in electron spectroscopy resulting from excitation by photons.

PAC 1976, *46*, 71
Manual of symbols and terminology for physicochemical quantities and units. Appendix II, Part II: heterogeneous catalysis.

PAC 1978, *50*, 587
Proposed terminology and symbols for the quantity representing the transfer of solutes from one solute to another.

PAC 1978, *50*, 1707
Definition and symbolism of molecular force constants.

PAC 1979, *51*, 1195
Separation and preconcentration of trace substances. I–Preconcentration for inorganic trace analysis.

PAC 1979, *51*, 1213
Terminology and symbols in colloid and surface chemistry. Part 1.13: definitions, terminology and symbols for rheological properties.

PAC 1979, *51*, 2451
Quantities and units in clinical chemistry.

PAC 1980, *52*, 233
Electrode reaction orders, transfer coefficients and rate constants. Amplification of definitions and recommendations for publication of parameters.

PAC 1980, *52*, 2541
Nomenclature, symbols, units and their usage in spectrochemical analysis – IV: X–ray emission spectroscopy.

PAC 1981, *53*, 1805
Assignment and presentation of uncertainties of the numerical results of thermodynamic measurements.

PAC 1981, *53*, 1827
Nomenclature for transport phenomena in electrolytic systems.

PAC 1981, *53*, 1887
Nomenclature of inorganic chemistry: II. 1– Isotopically modified compounds.

PAC 1982, *54*, 1239
Notation for states and processes, significance of the word standard in chemical thermodynamics, and remarks on commonly tabulated forms of thermodynamic functions.

PAC 1982, *54*, 1533
Glossary of terms used in nuclear analytical chemistry.

PAC 1982, *54*, 2553
Recommendations on use of the term amplification reactions.

PAC 1983, *55*, 409
Revision of the extended Hantzsch–Widman system of nomenclature for heteromonocycles.

PAC 1983, *55*, 553
Recommendations for the usage of selective, selectivity and related terms in analytical chemistry.

PAC 1983, *55*, 931
Manual of symbols and terminology for physicochemical quantities and units. Appendix II, Part 1.14: light scattering.

PAC 1983, *55*, 1251
Interphases in systems of conducting phases.

PAC 1983, *55*, 2023
Nomenclature, symbols and units recommended for *in situ* microanalysis.

PAC 1984, *56*, 232
Nomenclature, symbols, units and their usage in spectrochemical analysis – VI: Molecular luminescence spectroscopy.

PAC 1984, *56*, 567
Physiochemical quantities and units in clinical chemistry with special emphasis on activities and activity coefficients.

PAC 1984, *56*, 769
Treatment of variable valence in organic nomenclature (lambda convention).

PAC 1985, *57*, 105
Names, symbols, definitions and units of quantities in optical spectroscopy.

PAC 1985, *57*, 531
Definition of pH scales, standard reference values, measurement of pH and related terminology.

PAC 1985, *57*, 1453
Nomenclature, symbols, units and their usage in spectrochemical analysis – V: radiation sources.

PAC 1985, *57*, 1491
Recommended terms, symbols and definitions for electroanalytical chemistry.

PAC 1985, *57*, 1737
Nomenclature for thermal analysis – IV.

PAC 1986, *58*, 437
Interphases in systems of conducting phases.

PAC 1986, *58*, 955
The absolute electrode potential: an explanatory note.

PAC 1986, *58*, 967
Reporting data on adsorption from solution at the solid/solution interface.

PAC 1986, *58*, 1405
Recommendations for the presentation of thermodynamic and related data in biology

PAC 1986, *58*, 1737
Quantities and units in clinical chemistry: nebulizer and flame properties in flame emission and adsorption spectrometry.

PAC 1987, *59*, 683
Nomenclature of prenols.

PAC 1987, *59*, 779
Nomenclature of tetrapyrroles.

PAC 1987, *59*, 833
Nomenclature and symbols for folic acid and related compounds.

PAC 1988, *60*, 1115
Names for hydrogen atoms, ions and groups, and for reactions involving them.

PAC 1988, *60*, 1389
Nomenclature of glycoproteins, glycopeptides and peptidoglycans.

PAC 1988, *60*, 1395
Nomenclature for cyclic organic compounds with contiguous formal double bonds.

PAC 1988, *60*, 1449
  Nomenclature, symbols, units and their usage in spectrochemical analysis – VII: molecular absorption spectroscopy, ultraviolet and visible (UV/VIS).

PAC 1988, *60*, 1461
  Nomenclature, symbols, units and their usage in spectrochemical analysis – X: preparation of materials for analytical atomic spectroscopy and other related techniques.

PAC 1989, *61*, 19
  Electrochemical corrosion nomenclature.

PAC 1989, *61*, 23
  System for symbolic representation of reaction mechanisms.

PAC 1989, *61*, 1657
  Nomenclature for automated and mechanised analysis.

PAC 1989, *61*, 1783
  Nomenclature of steroids.

PAC 1989, *61*, 2195
  Recommendations for EPR/ESR nomenclature and conventions for presenting experimental data in publications.

PAC 1990, *62*, 1193
  Nomenclature for sampling in analytical chemistry.

PAC 1990, *62*, 2167
  Glossary of atmospheric chemistry terms.

PAC 1991, *63*, 301
  Proposals for the description and measurement of carry-over effects in clinical chemistry.

PAC 1991, *63*, 569
  Terminology in semiconductor electrochemistry and photoelectrochemical energy conversion.

PAC 1991, *63*, 735
  Nomenclature, symbols, units and their usage in spectrochemical analysis – VIII. Nomenclature system for X-ray spectroscopy.

PAC 1991, *63*, 887
  English-derived abbreviations for experimental techniques in surface science and chemical spectroscopy.

PAC 1991, *63*, 895
  Nomenclature, symbols, definitions and measurements for electrified interfaces in aqueous dispersions of solids.

PAC 1991, *63*, 1227
  Manual on catalyst characterisation.

PAC 1991, *57*, 1307
  Nomenclature of derived quantities.

PAC 1991, *63*, 1541
  Recommendations for nomenclature and symbolism for mass spectroscopy (including an appendix of terms used in vacuum technology).

PAC 1992, *64*, 143
  Glossary for chemists of terms used in biotechnology.

PAC 1992, *64*, 253
  Nomenclature, symbols, units and their usage in spectrochemical analysis – XII. Terms related to electrothermal atomisation.

PAC 1992, *64*, 261
  Nomenclature, symbols, units and their usage in spectrochemical analysis – XIII. Terms related to chemical vapour generation.

PAC 1992, *64*, 1569
  Quantities and units for metabolic processes as a function of time.

PAC 1993, *65*, 819
  Nomenclature for chromatography.

PAC 1993, *65*, 2003
  Glossary for chemists of terms used in toxicology.

PAC 1993, *65*, 2291
  Nomenclature of kinetic methods of analysis.

PAC 1993, *65*, 2373
  Nomenclature for liquid-liquid distribution (solvent extraction).

PAC 1993, *65*, 2397
  Special terminology used in supercritical-fluid chromatography and extraction.

PAC 1993, *65*, 2405
  Nomenclature and terminology for analytical pyrolysis.

PAC 1994, *66*, 533
  Standard quantities in chemical thermodynamics.

PAC 1994, *66*, 577
  Definition of terms relating to phase transitions of the solid state.

PAC 1994, *66*, 595
  Nomenclature for the presentation of results of chemical analysis.

PAC 1994, *66*, 891
  Quantities and units for electrophoresis in the clinical laboratory.

PAC 1994, *66*, 897
  Quantities and units for centrifugation in the clinical laboratory.

PAC 1994, *66*, 1077
  Glossary of terms used in physical organic chemistry.

PAC 1994, *66*, 1667
  Thin films including layers – terminology in relation to their preparation and characterisation.

PAC 1994, *66*, 2487
  Nomenclature of thermometric and enthalpimetric methods in chemical analysis.

PAC 1994, *66*, 2493
   Classification and definition of analytical methods based on flowing media.

PAC 1994, *66*, 2513
   Nomenclature for radioanalytical chemistry.

PAC 1994, *66*, 2528
   Nomenclature of ion-selective electrodes.

PAC 1994, *66*, 2587
   Glossary of terms in bioanalytical nomenclature.

PAC 1995, *67*, 473
   Recommended terminology for the description of carbon as a solid.

PAC 1995, *67*, 1307
   Glossary of class names of organic compounds and reactive intermediates based on structure.

PAC 1995, *67*, 1563
   Properties and units in the clinical laboratory sciences. I. Syntax and semantic rules.

PAC 1995, *67*, 1699
   Nomenclature in evaluation of analytical methods including detection and quantification capabilities.

PAC 1995, *67*, 1725
   Nomenclature, symbols, units and their usage in spectrochemical analysis – IX. Instrumentation for the spectral dispersion and isolation of optical radiation.

PAC 1995, *67*, 1745
   Nomenclature, symbols, units and their usage in spectrochemical analysis – XI. Detection of radiation.

PAC 1995, *67*, 1913
   Nomenclature, symbols, units and their usage in spectrochemical analysis – XV. Laser-based molecular spectroscopy for chemical analysis: laser fundamentals.

PAC 1996, *68*, 149
   Glossary of terms used in chemical kinetics, including reaction dynamics.

PAC 1996, 68, 957
   Glossary of terms in quantities and units in clinical chemistry.

PAC 1996, *68*, 2193
   Basic terminology of stereochemistry.

PAC 1996, *68*, 2223
   Glossary of terms used in photochemistry.

PAC 1996, *68*, 2287
   Glossary of basic terms in polymer science.